Summary of Project Analysis Methods

Analysis Method	Description	Single Project Evaluation	Mutually Exclusive Projects	
			Revenue Projects	Service Projects
Payback period PP	A method for determining when in a project's history it breaks even. Management sets the benchmark PP°.	PP	shortest PP	
Discounted payback period PP(i)	A variation of payback period when factors in the time value of money. Management sets the benchmark PP*.	PP(i) < PP*	Select the one with shortest PP(i)	
Present worth PW(i)	An equivalent method which translates a project's cash flows into a net present value	PW(i) > 0	Select the one with the largest PW	Select the one with the least negative PW
Future worth FW(i)	An equivalence method variation of the PW: a project's cash flows are translated into a net future value	FW(i) > 0	Select the one with the largest FW	Select the one with the least negative FW
Capitalized equivalent CE(i)	An equivalence method variation of the PW of perpetual or very long-lived project that generates a constant annual net cash flow	CE(i) > 0	Select the one with the largest CE	Select the one with the least negative CE
Annual equivalence AE(i)	An equivalence method and variation of the PW: a project's cash flows are translated into an annual equivalent sum	AE(i) > 0	Select the one with the largest AE	Select the one with the least negative AE
Internal rate of return IRR	A relative percentage method which measures the yield as a percentage of investment over the life of a project: The IRR must exceed the minimum required rate of return (MARR).	IRR > MARR	Incremental analysis: If IRR$_{A2\text{-}A1}$ > MARR, select the higher cost investment project, A2.	
Benefit–cost ratio BC(i)	An equivalence method to evaluate public projects by finding the ratio of the equivalent benefit over the equivalent cost	BC(i) > 1	Incremental analysis: If BC(i)$_{A2\text{-}A1}$ > 1, select the higher cost investment project, A2.	

Contemporary
Engineering
Economics

Contemporary

Engineering
Economics

Fourth Edition

Chan S. Park
*Department of Industrial
and Systems Engineering
Auburn University*

PEARSON

Prentice
Hall

Upper Saddle River, NJ 07458

Library of Congress Cataloging-in-Publication Data on File

Vice President and Editorial Director, ECS: *Marcia J. Horton*
Senior Editor: *Holly Stark*
Editorial Assistant: *Nicole Kunzmann*
Executive Managing Editor: *Vince O'Brien*
Managing Editor: *David A. George*
Production Editor: *Scott Disanno*
Director of Creative Services: *Paul Belfanti*
Creative Director: *Juan López*
Art Director: *Heather Scott*
Interior and Cover Designer: *Tamara Newnam*
Art Editor: *Xiaohong Zhu*
Manufacturing Manager: *Alexis Heydt-Long*
Manufacturing Buyer: *Lisa McDowell*
Marketing Manager: *Tim Galligan*

The author and publisher of this book have used their best efforts in preparing this book. These
efforts include the development, research, and testing of the theories and programs to determine
their effectiveness. The author and publisher make no warranty of any kind, expressed or
implied, with regard to these programs or the documentation contained in this book. The author
and publisher shall not be liable in any event for incidental or consequential damages in connection with, or arising out of, the furnishing, performance, or use of these programs.

Printed in the United States of America

10   9   8   7   6   5   4   3   2   1

ISBN 0-13-187628-7

Pearson Education Ltd., London
Pearson Education Australia Pte. Ltd., Sydney
Pearson Education Singapore, Pte. Ltd.
Pearson Education North Asia Ltd., Hong Kong
Pearson Education Canada, Inc., Toronto
Pearson Educación de Mexico, S.A. de C.V.
Pearson Education–Japan, Tokyo
Pearson Education Malaysia, Ptd. Ltd.
Pearson Education, Inc., Upper Saddle River, New Jersey

To my wife, Kim (Inkyung); and my children, Michael and Edward

CONTENTS

PART 2 EVALUATION OF BUSINESS AND ENGINEERING ASSETS   203

Chapter 5   Present-Worth Analysis   204

PART 3 ANALYSIS OF PROJECT CASH FLOWS 385

Chapter 8 Cost Concepts Relevant to Decision Making 386

Chapter 9   Depreciation and Corporate Taxes   428

Chapter 10   Developing Project Cash Flows   490

PART 4   HANDLING RISK AND UNCERTAINTY   541

Chapter 11   Inflation and Its Impact on Project Cash Flows   542

Chapter 12   Project Risk and Uncertainty   584

Chapter 13 Real-Options Analysis 664

PART 5 SPECIAL TOPICS IN ENGINEERING ECONOMICS 715

Chapter 14 Replacement Decisions 716

Chapter 15 Capital-Budgeting Decisions 776

Chapter 16 Economic Analysis in the Service Sector 822

PREFACE

What is "Contemporary" About Engineering Economics?

Decisions made during the engineering design phase of product development determine the majority of the costs associated with the manufacturing of that product (some say that this value may be as high as 85%). As design and manufacturing processes become more complex, engineers are making decisions that involve money more than ever before. With more than 80% of the total GDP (Gross Domestic Product) in the United States provided by the service sector, engineers work on various economic decision problems in the service sector as well. Thus, the competent and successful engineer in the twenty-first century must have an improved understanding of the principles of science, engineering, and economics, coupled with relevant design experience. Increasingly, in the new world economy, successful businesses will rely on engineers with such expertise.

Economic and design issues are inextricably linked in the product/service life cycle. Therefore, one of my strongest motivations for writing this text was to bring the realities of economics and engineering design into the classroom and to help students integrate these issues when contemplating many engineering decisions. Of course my underlying motivation for writing this book was not simply to address contemporary needs, but to address as well the ageless goal of all educators: to help students to learn. Thus, thoroughness, clarity, and accuracy of presentation of essential engineering economics were my aim at every stage in the development of the text.

Changes in the Fourth Edition

Much of the content has been streamlined to provide materials in depth and to reflect the challenges in contemporary engineering economics. Some of the highlighted changes are as follows:

- Chapter 13 "Real Options Analysis" is new and provides a new perspective on how engineers should manage risk in their strategic economic decision problems. Traditionally, risk is avoided in project analysis, which is a passive way of handling the matter. The goal of the real options approach is to provide a contemporary tool that will assist engineers so that they can actively manage the risk involved in long-term projects.
- Chapter 12 has been significantly revised to provide more probabilistic materials for the analytical treatment of risk and uncertainty. Risk simulation has been introduced by way of using @RISK.
- Three chapters have been merged with various materials from other chapters. Chapter 3 on cost concepts and behaviors has been moved to Part III and now appears as Chapter 8 "Cost Concepts Relevant to Decision Making"; it is now part of project cash flow analysis. Chapter 6 on principles of investing is now part of Chapter 4 "Understanding Money and Its Management." Materials from various chapters have been merged into a single chapter and now appear as Chapter 9 "Depreciation and Corporate Income Taxes".
- The chapter on the economic analysis in the public sector has been expanded and now appears as Chapter 16 "Economic Analysis in the Service Sector"; this revised chapter now provides economic analysis unique to service sectors beyond the government

sector. Increasingly, engineers seek their career in the service sector, such as health-care, financial institutions, transportation, and logistics. In this chapter, we present some unique features that must be considered when evaluating investment projects in the service sector.

- All the end-of-chapter problems are revised to reflect the materials changes in the main text.
- All the chapter opening vignettes—a trademark of *Contemporary Engineering Economics*—have been completely replaced with more current and thought-provoking case studies.
- Self-study problems and FE practice questions are available as interactive quizzes with instant feedback as part of the book's new OneKey CourseCompass site. OneKey is an online resource for instructors and students; more detailed information about OneKey can be found in the *OneKey* section of this Preface. OneKey can be accessed via www.prenhall.com/onekey.
- Various Excel spreadsheet modeling techniques are introduced throughout the chapters and the original Excel files are provided online at the OneKey site .

Overview of the Text

Although it contains little advanced math and few truly difficult concepts, the introductory engineering economics course is often a curiously challenging one for the sophomores, juniors, and seniors who take it. There are several likely explanations for this difficulty.

1. The course is the student's first analytical consideration of money (a resource with which he or she may have had little direct contact beyond paying for tuition, housing, food, and textbooks).
2. The emphasis on theory may obscure for the student the fact that the course aims, among other things, to develop a very practical set of analytical tools for measuring project worth. This is unfortunate since, at one time or another, virtually every engineer—not to mention every individual—is responsible for the wise allocation of limited financial resources.
3. The mixture of industrial, civil, mechanical, electrical, and manufacturing engineering, and other undergraduates who take the course often fail to "see themselves" in the skills the course and text are intended to foster. This is perhaps less true for industrial engineering students, whom many texts take as their primary audience, but other disciplines are often motivationally shortchanged by a text's lack of applications that appeal directly to them.

Goal of the Text

This text aims not only to provide sound and comprehensive coverage of the concepts of engineering economics but also to address the difficulties of students outlined above, all of which have their basis in inattentiveness to the practical concerns of engineering economics. More specifically, this text has the following chief goals:

1. To build a thorough understanding of the theoretical and conceptual basis upon which the practice of financial project analysis is built.

2. To satisfy the very practical needs of the engineer toward making informed financial decisions when acting as a team member or project manager for an engineering project.

3. To incorporate all critical decision-making tools—including the most contemporary, computer-oriented ones that engineers bring to the task of making informed financial decisions.

4. To appeal to the full range of engineering disciplines for which this course is often required: industrial, civil, mechanical, electrical, computer, aerospace, chemical, and manufacturing engineering, as well as engineering technology.

Prerequisites

The text is intended for undergraduate engineering students at the sophomore level or above. The only mathematical background required is elementary calculus. For Chapters 12 and 13, a first course in probability or statistics is helpful but not necessary, since the treatment of basic topics there is essentially self-contained.

Taking Advantage of the Internet

The integration of computer use is another important feature of *Contemporary Engineering Economics*. Students have greater access to and familiarity with the various spreadsheet tools, and instructors have greater inclination either to treat these topics explicitly in the course or to encourage students to experiment independently.

A remaining concern is that the use of computers will undermine true understanding of course concepts. This text does not promote the use of trivial spreadsheet applications as a replacement for genuine understanding of and skill in applying traditional solution methods. Rather, it focuses on the computer's productivity-enhancing benefits for complex project cash flow development and analysis. Specifically, *Contemporary Engineering Economics* includes a robust introduction to computer automation in the form of Computer Notes, which are included in the optional OneKey course (www.prenhall.com/onekey).

Additionally, spreadsheets are introduced via Microsoft Excel examples. For spreadsheet coverage, the emphasis is on demonstrating a chapter concept that embodies some complexity that can be much more efficiently resolved on a computer than by traditional longhand solutions.

OneKey

Available as a special package, OneKey is Prentice Hall's exclusive new resource for instructors and students. Instructors have access online to all available course supplements and can create and assign tests, quizzes, or graded homework assignments. Students have access to interactive exercises, quizzes, and more. The following resources are available when an instructor adopts the text plus OneKey package:

- Interactive self-study quizzes organized by chapter with instant feedback, plus interactive FE Exam practice questions
- Computer notes with Excel files of selected example problems from the text.
- Case Studies: A collection of actual cases, two personal-finance and six industry-based, is now available. The investment projects detailed in the cases relate to a

variety of engineering disciplines. Each case is based on multiple text concepts, thus encouraging students to synthesize their understanding in the context of complex, real-world investments. Each case begins with a list of engineering economic concepts utilized in the case and concludes with discussion questions to test students' conceptual understanding.

- **Analysis Tools**: A collection of various financial calculators is available. Cash Flow Analyzer is an integrated online Java program that is menu driven for convenience and flexibility; it provides (1) a flexible and easy-to-use cash flow editor for data input and modifications, and (2) an extensive array of computational modules and user-selected graphic outputs.
- **Instructor Resources**: Instructors Solutions Manual, PowerPoint Lecture Notes, Case Studies and more.

Please contact your Prentice Hall representative for details and ordering information for OneKey packages. Detailed instructions about how to access and use this content can be found at the site, which can be accessed at: www.prenhall.com/onekey.

The Financial Times

We are please to announce a special partnership with *The Financial Times*. For a small additional charge, Prentice Hall offers students a 15-week subscription to *The Financial Times*. Upon adoption of a special package containing the book and the subscription booklet, professors will receive a free one-year subscription. Please contact your Prentice Hall representative for details and ordering information.

Acknowledgments

This book reflects the efforts of a great many individuals over a number of years. In particular, I would like to recognize the following individuals, whose reviews and comments on prior editions have contributed to this edition. Once again, I would like to thank each of them:

Kamran Abedini, *California Polytechnic–Pomona*
James Alloway, *Syracuse University*
Mehar Arora, *U. Wisconsin–Stout*
Joel Arthur, *California State University–Chico*
Robert Baker, *University of Arizona*
Robert Barrett, *Cooper Union and Pratt Institute*
Tom Barta, *Iowa State University*
Charles Bartholomew, *Widener University*
Richard Bernhard, *North Carolina State University*
Bopaya Bidanda, *University of Pittsburgh*
James Buck, *University of Iowa*
Philip Cady, *The Pennsylvania State University*
Tom Carmichal, *Southern College of Technology*
Jeya Chandra, *The Pennsylvania State University*

Max C. Deibert, *Montana State University*
Stuart E. Dreyfus, *University of California, Berkeley*
Philip A. Farrington, *University of Alabama at Huntsville*
W. J. Foley, *RPI*
Jane Fraser, *University of Southern Colorado*
Terry L Friesz, Penn State University
Anil K. Goyal, *RPI*
Bruce Hartsough, *University of California, Davis*
Carl Hass, *University of Texas, Austin*
John Held, *Kansas State University*
T. Allen Henry, *University of Alabama*
R.C. Hodgson, *University of Notre Dame*
Scott Iverson, *University of Washington*
Peter Jackson, *Cornell University*
Philip Johnson, *University of Minnesota*
Harold Josephs, *Lawrence Tech*
Henry Kallsen, *University of Alabama*
W. J. Kennedy, *Clemson University*
Oh Keytack, *University of Toledo*
Wayne Knabach, *South Dakota State University*
Stephen Kreta, *California Maritime Academy*
John Krogman, *University of Wisconsin–Platteville*
Dennis Kroll, *Bradley University*
Michael Kyte, *University of Idaho*
Gene Lee, *University of Central Florida*
William Lesso, *University of Texas–Austin*
Martin Lipinski, *Memphis State University*
Robert Lundquist, *Ohio State University*
Richard Lyles, *Michigan State University*
Gerald T. Mackulak, *Arizona State University*
Abu S. Masud, *The Wichita State University*
Sue McNeil, *Carnegie-Mellon University*
James Milligan, *University of Idaho*
Richard Minesinger, *University of Massachusetts, Lowell*
Gary Moynihan, *The University of Alabama*
James S. Noble, *University of Missouri, Columbia*
Michael L. Nobs, *Washington University, St. Louis*
Wayne Parker, *Mississippi State University*
Elizabeth Pate-Cornell, *Stanford University*
Cecil Peterson, *GMI*
George Prueitt, *U.S. Naval Postgraduate School*
J.K. Rao, *California State University-Long Beach*
Susan Richards, *GMI*
Bruce A. Reichert, *Kansas State University*
Mark Roberts, *Michigan Tech*
John Roth, *Vanderbilt University*
Paul L. Schillings, *Montana State University*

Bill Shaner, *Colorado State University*
Fred Sheets, *California Polytechnic, Pomona*
Dean Shup, *University of Cincinnati*
Milton Smith, *Texas Tech*
David C. Slaughter, *University of California, Davis*
Charles Stavridge, *FAMU/FSU*
Junius Storry, *South Dakota State University*
Frank E. Stratton, *San Diego State University*
George Stukhart, *Texas A&M University*
Donna Summers, *University of Dayton*
Joe Tanchoco, *Purdue University*
Deborah Thurston, *University of Illinois at Urbana-Champaign*
Lt. Col. James Treharne, *U.S. Army*
L. Jackson Turaville, *Tennessee Technological University*
Thomas Ward, *University of Louisville*
Theo De Winter, *Boston University*
Yoo Yang, *Cal Poly State University*

Special Acknowledgement

Personally, I wish to thank the following individuals for their additional inputs to the fourth edition: Michael L. Nobs, *Washington University, St. Louis*, Terry L. Friesz, *Penn State University*, Gene Lee, *University of Central Florida*, Gerald T. Mackulak, *Arizona State University*, and Phillip A. Farrington, *University of Alabama, Huntsville*. Major Hyun Jin Han who helped me in developing the Instructor Manual; Holly Stark, my editor at Prentice Hall, who assumed responsibility for the overall project; Scott Disanno, my production editor at Prentice Hall, who oversaw the entire book production. I also acknowledge that many of the financial terminologies found in the marginal notes are based on the online glossary defined by *Investopedia* and *Investorwords.com*. Finally, I would like to thank Dr. Alice E. Smith, Chair of Industrial & Systems Engineering at Auburn University, who provided me with the resources.

CHAN S. PARK
AUBURN, ALABAMA

1

Basics of Financial Decisions

ONE

Engineering Economic Decisions

Google Cofounder Sergey Brin Comes to Class at Berkeley[1] Sergey Brin, cofounder of Google, showed up for class as a guest speaker at Berkeley on October 3, 2005. Casual and relaxed, Brin talked about how Google came to be, answered students' questions, and showed that someone worth $11 billion (give or take a billion) still can be comfortable in an old pair of blue jeans. Indistinguishable in dress, age, and demeanor from many of the students in the class, Brin covered a lot of ground in his remarks, but ultimately it was his unspoken message that was most powerful: To those with focus and passion, all things are possible. In his remarks to the class, Brin stressed simplicity. Simple ideas sometimes can change the world, he said. Likewise, Google started out with the simplest of ideas, with a global audience in mind. In the mid-1990s, Brin and Larry Page were Stanford students pursuing doctorates in computer science. Brin recalled that at that time there were some five major Internet search engines, the importance of searching was being de-emphasized, and the owners of these major search sites were focusing on creating portals with increased content offerings. "We believed we could build a better search. We had a simple idea—that not all pages are created equal. Some are more important," related Brin. Eventually, they developed a unique approach to solving one of computing's biggest challenges: retrieving relevant information from a massive set of data.

According to Google lore,[1] by January of 1996 Larry and Sergey had begun collaboration on a search engine called BackRub, named for its unique ability to analyze the "back links" pointing to a given website.

[1]*UC Berkeley News*, Oct. 4, 2005, UC Regents and Google Corporate Information: http://www. google.com/corporate/history.html.

How Google Works

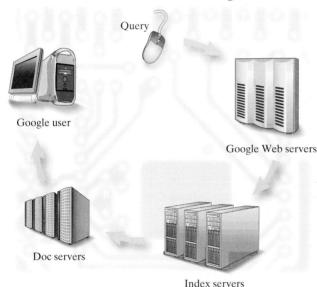

Query

Google user

Google Web servers

Doc servers

Index servers

1. The Web server sends the query to the index servers—it tells which pages contain the words that match the query.

2. The query travels to the Doc servers (which retrieve the stored documents) and snippets are generated to describe each search result.

3. The search results are returned to the user in a fraction of a second.

Larry, who had always enjoyed tinkering with machinery and had gained some "notoriety" for building a working printer out of Lego® bricks, took on the task of creating a new kind of server environment that used low-end PCs instead of big expensive machines. Afflicted by the perennial shortage of cash common to graduate students everywhere, the pair took to haunting the department's loading docks in hopes of tracking down newly arrived computers that they could borrow for their network. A year later, their unique approach to link analysis was earning BackRub a growing reputation among those who had seen it. Buzz about the new search technology began to build as word spread around campus. Eventually, in 1998 they decided to create a company named "Google" by raising $25 million from venture capital firms Kleiner Perkins Caufield & Byers and Sequoia Capital. Since taking their Internet search engine public in August 2004, the dynamic duo behind Google has seen their combined fortune soar to $22 billion. At a recent $400, Google trades at 90 times trailing earnings, after starting out at $85. The success has vaulted both Larry and Sergey into *Forbes* magazine's list of the 400 wealthiest Americans. The net worth of the pair is estimated at $11 billion each.

A Little Google History

- 1995
  - Developed in dorm room of Larry Page and Sergey Brin, graduate students at Stanford University
  - Nicknamed BackRub
- 1998
  - Raised $25 million to set up Google, Inc.
  - Ran 100,000 queries a day out of a garage in Menlo Park
- 2005
  - Over 4,000 employees worldwide
  - Over 8 billion pages indexed

The story of how the Google founders got motivated to invent a search engine and eventually transformed their invention to a multibillion-dollar business is a typical one. Companies such as Dell, Microsoft, and Yahoo all produce computer-related products and have market values of several billion dollars. These companies were all started by highly motivated young college students just like Brin. One thing that is also common to all these successful businesses is that they have capable and imaginative engineers who constantly generate good ideas for capital investment, execute them well, and obtain good results. You might wonder about what kind of role these engineers play in making such business decisions. In other words, what specific tasks are assigned to these engineers, and what tools and techniques are available to them for making such capital investment decisions? We answer these questions and explore related issues throughout this book.

CHAPTER LEARNING OBJECTIVES

After completing this chapter, you should understand the following concepts:

- The role of engineers in business.
- Types of business organization.
- The nature and types of engineering economic decisions.
- What makes the engineering economic decisions difficult.
- How a typical engineering project idea evolves in business.
- Fundamental principles of engineering economics.

1.1 Role of Engineers in Business

Yahoo, Apple Computer, Microsoft Corporation, and Sun Microsystems produce computer products and have a market value of several billion dollars each. These companies were all started by young college students with technical backgrounds. When they went into the computer business, these students initially organized their companies as proprietorships. As the businesses grew, they became partnerships and were eventually converted to corporations. This chapter begins by introducing the three primary forms of business organization and briefly discusses the role of engineers in business.

1.1.1 Types of Business Organization

As an engineer, you should understand the nature of the business organization with which you are associated. This section will present some basic information about the type of organization you should choose should you decide to go into business for yourself.

The three legal forms of business, each having certain advantages and disadvantages, are proprietorships, partnerships, and corporations.

Proprietorships

A **proprietorship** is a business owned by one individual. This person is responsible for the firm's policies, owns all its assets, and is personally liable for its debts. A proprietorship has two major advantages. First, it can be formed easily and inexpensively. No legal and organizational requirements are associated with setting up a proprietorship, and organizational costs are therefore virtually nil. Second, the earnings of a proprietorship are taxed at the owner's personal tax rate, which may be lower than the rate at which corporate income is taxed. Apart from personal liability considerations, the major disadvantage of a proprietorship is that it cannot issue stocks and bonds, making it difficult to raise capital for any business expansion.

Partnerships

A **partnership** is similar to a proprietorship, except that it has more than one owner. Most partnerships are established by a written contract between the partners. The contract normally specifies salaries, contributions to capital, and the distribution of profits and losses. A partnership has many advantages, among which are its low cost and ease of formation. Because more than one person makes contributions, a partnership typically has a larger amount of capital available for business use. Since the personal assets of all the partners stand behind the business, a partnership can borrow money more easily from a bank. Each partner pays only personal income tax on his or her share of a partnership's taxable income.

On the negative side, under partnership law each partner is liable for a business's debts. This means that the partners must risk all their personal assets—even those not invested in the business. And while each partner is responsible for his or her portion of the debts in the event of bankruptcy, if any partners cannot meet their pro rata claims, the remaining partners must take over the unresolved claims. Finally, a partnership has a limited life, insofar as it must be dissolved and reorganized if one of the partners quits.

Corporations

A **corporation** is a legal entity created under provincial or federal law. It is separate from its owners and managers. This separation gives the corporation four major advantages: (1) It can raise capital from a large number of investors by issuing stocks and bonds; (2) it permits easy transfer of ownership interest by trading shares of stock; (3) it allows limited liability—personal liability is limited to the amount of the individual's investment in the business; and (4) it is taxed differently than proprietorships and partnerships, and under certain conditions, the tax laws favor corporations. On the negative side, it is expensive to establish a corporation. Furthermore, a corporation is subject to numerous governmental requirements and regulations.

As a firm grows, it may need to change its legal form because the form of a business affects the extent to which it has control of its own operations and its ability to acquire funds. The legal form of an organization also affects the risk borne by its owners in case of bankruptcy and the manner in which the firm is taxed. Apple Computer, for example, started out as a two-man garage operation. As the business grew, the owners felt constricted by this form of organization: It was difficult to raise capital for business expansion; they felt that the risk of bankruptcy was too high to bear; and as their business income grew, their tax burden grew as well. Eventually, they found it necessary to convert the partnership into a corporation.

In the United States, the overwhelming majority of business firms are proprietorships, followed by corporations and partnerships. However, in terms of total business volume (dollars of sales), the quantity of business transacted by proprietorships and partnerships is several times less than that of corporations. Since most business is conducted by corporations, this text will generally address economic decisions encountered in that form of ownership.

1.1.2 Engineering Economic Decisions

What role do engineers play within a firm? What specific tasks are assigned to the engineering staff, and what tools and techniques are available to it to improve a firm's profits? Engineers are called upon to participate in a variety of decisions, ranging from manufacturing, through marketing, to financing decisions. We will restrict our focus, however, to various economic decisions related to engineering projects. We refer to these decisions as **engineering economic decisions**.

In manufacturing, engineering is involved in every detail of a product's production, from conceptual design to shipping. In fact, engineering decisions account for the majority (some say 85%) of product costs. Engineers must consider the effective use of capital assets such as buildings and machinery. One of the engineer's primary tasks is to plan for the acquisition of equipment (**capital expenditure**) that will enable the firm to design and produce products economically.

With the purchase of any fixed asset—equipment, for instance—we need to estimate the profits (more precisely, cash flows) that the asset will generate during its period of service. In other words, we have to make capital expenditure decisions based on predictions about the future. Suppose, for example, you are considering the purchase of a deburring machine to meet the anticipated demand for hubs and sleeves used in the production of gear couplings. You expect the machine to last 10 years. This decision thus involves an implicit 10-year sales forecast for the gear couplings, which means that a long waiting period will be required before you will know whether the purchase was justified.

An inaccurate estimate of the need for assets can have serious consequences. If you invest too much in assets, you incur unnecessarily heavy expenses. Spending too little on fixed assets, however, is also harmful, for then the firm's equipment may be too obsolete to produce products competitively, and without an adequate capacity, you may lose a portion of your market share to rival firms. Regaining lost customers involves heavy marketing expenses and may even require price reductions or product improvements, both of which are costly.

1.1.3 Personal Economic Decisions

In the same way that an engineer can play a role in the effective utilization of corporate financial assets, each of us is responsible for managing our personal financial affairs. After we have paid for nondiscretionary or essential needs, such as housing, food, clothing, and transportation, any remaining money is available for discretionary expenditures on items such as entertainment, travel, and investment. For money we choose to invest, we want to maximize the economic benefit at some acceptable risk. The investment choices are un-limited and include savings accounts, guaranteed investment certificates, stocks, bonds, mutual funds, registered retirement savings plans, rental properties, land, business ownership, and more.

How do you choose? The analysis of one's personal investment opportunities utilizes the same techniques that are used for engineering economic decisions. Again, the challenge is predicting the performance of an investment into the future. Choosing wisely can be very rewarding, while choosing poorly can be disastrous. Some investors in the energy stock Enron who sold prior to the fraud investigation became millionaires. Others, who did not sell, lost everything.

A wise investment strategy is a strategy that manages risk by diversifying invest-ments. With such an approach, you have a number of different investments ranging from very low to very high risk and are in a number of business sectors. Since you do not have all your money in one place, the risk of losing everything is significantly reduced. (We discuss some of these important issues in Chapters 12 and 13.)

1.2 What Makes the Engineering Economic Decision Difficult?

The economic decisions that engineers make in business differ very little from the finan-cial decisions made by individuals, except for the scale of the concern. Suppose, for ex-ample, that a firm is using a lathe that was purchased 12 years ago to produce pump shafts. As the production engineer in charge of this product, you expect demand to con-tinue into the foreseeable future. However, the lathe has begun to show its age: It has bro-ken frequently during the last 2 years and has finally stopped operating altogether. Now you have to decide whether to replace or repair it. If you expect a more efficient lathe to be available in the next 1 or 2 years, you might repair the old lathe instead of replacing it. The major issue is whether you should make the considerable investment in a new lathe now or later. As an added complication, if demand for your product begins to decline, you may have to conduct an economic analysis to determine whether declining profits from the project offset the cost of a new lathe.

Let us consider a real-world engineering decision problem on a much larger scale, as taken from an article from *The Washington Post*.[2] Ever since Hurricane Katrina hit the city of New Orleans in August 2005, the U.S. federal government has been under pres-sure to show strong support for rebuilding levees in order to encourage homeowners and businesses to return to neighborhoods that were flooded when the city's levees crumbled under Katrina's storm surge. Many evacuees have expressed reluctance to rebuild without assurances that New Orleans will be made safe from future hurricanes, including Category 5 storms, the most severe. Some U.S. Army Corps of Engineers officers estimated that it would cost more than $1.6 billion to restore the levee system merely to its design strength—that is, to withstand a Category 3 storm. New design features would include floodgates on several key canals, as well as stone-and-concrete fortification of some of

---

[2] Joby Warrick and Peter Baker, "Bush Pledges $1.5 Billion for New Orleans—Proposal Would Double Aid From U.S. for Flood Protection," *The Washington Post*, Dec. 16, 2005, p. A03.

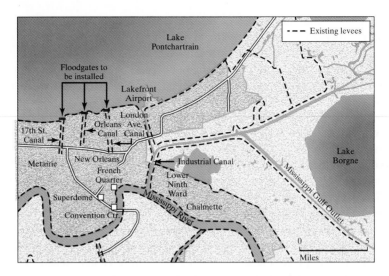

Figure I.I The White House pledged $1.5 billion to armor existing New Orleans levees with concrete and stone, build floodgates on three canals, and upgrade the city's pumping system.

the city's earthen levees, as illustrated in Figure 1.1. Donald E. Powell, the administration's coordinator of post-Katrina recovery, insisted that the improvements would make the levee system much stronger than it had been in the past. But he declined to say whether the administration would support further upgrades of the system to Category 5 protection, which would require substantial reengineering of existing levees at a cost that could, by many estimates, exceed $30 billion.

Obviously, this level of engineering decision is far more complex and more significant than a business decision about when to introduce a new product. Projects of this nature involve large sums of money over long periods of time, and it is difficult to justify the cost of the system purely on the basis of economic reasoning, since we do not know when another Katrina-strength storm will be on the horizon. Even if we decide to rebuild the levee systems, should we build just enough to withstand a Category 3 storm, or should we build to withstand a Category 5 storm? Any engineering economic decision pertaining to this type of extreme event will be extremely difficult to make.

In this book, we will consider many types of investments—personal investments as well as business investments. The focus, however, will be on evaluating engineering projects on the basis of their economic desirability and on dealing with investment situations that a typical firm faces.

1.3 Economic Decisions versus Design Decisions

Economic decisions differ in a fundamental way from the types of decisions typically encountered in engineering design. In a design situation, the engineer utilizes known physical properties, the principles of chemistry and physics, engineering design correlations, and engineering judgment to arrive at a workable and optimal design. If the judgment is sound, the calculations are done correctly, and we ignore technological advances, the design is time invariant. In other words, if the engineering design to meet a particular need is done today, next year, or in five years' time, the final design would not change significantly.

In considering economic decisions, the measurement of investment attractiveness, which is the subject of this book, is relatively straightforward. However, the information required in such evaluations always involves predicting or forecasting product sales, product selling prices, and various costs over some future time frame—5 years, 10 years, 25 years, etc.

All such forecasts have two things in common. First, they are never completely accurate compared with the actual values realized at future times. Second, a prediction or forecast made today is likely to be different from one made at some point in the future. It is this ever-changing view of the future that can make it necessary to revisit and even change previous economic decisions. Thus, unlike engineering design, the conclusions reached through economic evaluation are not necessarily time invariant. Economic decisions have to be based on the best information available at the time of the decision and a thorough understanding of the uncertainties in the forecasted data.

1.4 Large-Scale Engineering Projects

In the development of any product, a company's engineers are called upon to translate an idea into reality. A firm's growth and development depend largely upon a constant flow of ideas for new products, and for the firm to remain competitive, it has to make existing products better or produce them at a lower cost. In the next section, we present an example of how a design engineer's idea eventually turned into an innovative automotive product.

1.4.1 How a Typical Project Idea Evolves

The Toyota Motor Corporation introduced the world's first mass-produced car powered by a combination of gasoline and electricity. Known as the Prius, this vehicle is the first of a new generation of Toyota cars whose engines cut air pollution dramatically and boost fuel efficiency to significant levels. Toyota, in short, wants to launch and dominate a new "green" era for automobiles—and is spending $1.5 billion to do it. Developed for the Japanese market initially, the Prius uses both a gasoline engine and an electric motor as its motive power source. The Prius emits less pollution than ordinary cars, and it gets more mileage, which means less output of carbon dioxide. Moreover, the Prius gives a highly responsive performance and smooth acceleration. The following information from *BusinessWeek*[3] illustrates how a typical strategic business decision is made by the engineering staff of a larger company. Additional information has been provided by Toyota Motor Corporation.

Why Go for a Greener Car?

Toyota first started to develop the Prius in 1995. Four engineers were assigned to figure out what types of cars people would drive in the 21st century. After a year of research, a

---

[3] Emily Thornton (Tokyo), Keith Naughton (Detroit), and David Woodruff, "Japan's Hybrid Cars—Toyota and rivals are betting on pollution fighters—Will they succeed?" *BusinessWeek*, Dec. 4, 1997.

chief engineer concluded that Toyota would have to sell cars better suited to a world with scarce natural resources. He considered electric motors. But an electric car can travel only 215 km before it must be recharged. Another option was fuel-cell cars that run on hydrogen. But he suspected that mass production of this type of car might not be possible for another 15 years. So the engineer finally settled on a hybrid system powered by an electric motor, a nickel–metal hydride battery, and a gasoline engine. From Toyota's perspective, it is a big bet, as oil and gasoline prices are bumping along at record lows at that time. Many green cars remain expensive and require trade-offs in terms of performance. No carmaker has ever succeeded in selling consumers en masse something they have never wanted to buy: *cleaner air*. Even in Japan, where a liter of regular gas can cost as much as $1, carmakers have trouble pushing higher fuel economy and lower carbon emissions. Toyota has several reasons for going green. In the next century, as millions of new car owners in China, India, and elsewhere take to the road, Toyota predicts that gasoline prices will rise worldwide. At the same time, Japan's carmakers expect pollution and global warming to become such threats that governments will enact tough measures to clean the air.

What Is So Unique in Design?

It took Toyota engineers two years to develop the current power system in the Prius. The car comes with a dual engine powered by both an electric motor and a newly developed 1.5-liter gasoline engine. When the engine is in use, a special "power split device" sends some of the power to the driveshaft to move the car's wheels. The device also sends some of the power to a generator, which in turn creates electricity, to either drive the motor or recharge the battery. Thanks to this variable transmission, the Prius can switch back and forth between motor and engine, or employ both, without creating any jerking motion. The battery and electric motor propel the car at slow speeds. At normal speeds, the electric motor and gasoline engine work together to power the wheels. At higher speeds, the battery kicks in extra power if the driver must pass another car or zoom up a hill.

When the car decelerates and brakes, the motor converts the vehicle's kinetic energy into electricity and sends it through an inverter to be stored in the battery. (See Figure 1.2.)

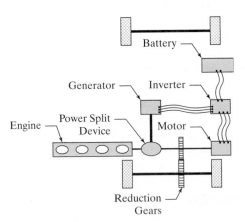

Figure 1.2 Arrangement of components of the Toyota Hybrid System II.
(*Source: Evaluation of 2004 Toyota Prius Hybrid Electric Drive System*, Oak Ridge National Laboratory, ONR/TM2004/247, U.S. Department of Energy.)

- When engine efficiency is low, such as during start-up and with midrange speeds, motive power is provided by the motor alone, using energy stored in the battery.
- Under normal driving conditions, overall efficiency is optimized by controlling the power allocation so that some of the engine power is used for turning the generator to supply electricity for the motor while the remaining power is used for turning the wheels.
- During periods of acceleration, when extra power is needed, the generator supplements the electricity being drawn from the battery, so the motor is supplied with the required level of electrical energy.
- While decelerating and braking, the motor acts as a generator that is driven by the wheels, thus allowing the recovery of kinetic energy. The recovered kinetic energy is converted to electrical energy that is stored in the battery.
- When necessary, the generator recharges the battery to maintain sufficient reserves.
- When the vehicle is not moving and when the engine moves outside of certain speed and load conditions, the engine stops automatically.

So the car's own movement, as well as the power from the gasoline engine, provides the electricity needed. The energy created and stored by deceleration boosts the car's efficiency. So does the automatic shutdown of the engine when the car stops at a light. At higher speeds and during acceleration, the companion electric motor works harder, allowing the gas engine to run at peak efficiency. Toyota claims that, in tests, its hybrid car has boosted fuel economy by 100% and engine efficiency by 80%. The Prius emits about half as much carbon dioxide, and about one-tenth as much carbon monoxide, hydrocarbons, and nitrous oxide, as conventional cars.

Is It Safe to Drive on a Rainy Day?

Yet, major hurdles remain to be overcome in creating a mass market for green vehicles. Car buyers are not anxious enough about global warming to justify a "sea-level change" in automakers' marketing. Many of Japan's innovations run the risk of becoming impressive technologies meant for the masses, but bought only by the elite. The unfamiliarity of green technology can also frighten consumers. The Japanese government sponsors festivals at which people can test drive alternative-fuel vehicles. But some turned down that chance on a rainy weekend in May because they feared that riding in an electric car might electrocute them. An engineer points out, "It took 20 years for the automatic transmission to become popular in Japan." It certainly would take that long for many of these technologies to catch on.

How Much Would It Cost?

The biggest question remaining about the mass production of the vehicle concerns its production cost. Costs will need to come down for Toyota's hybrid to be competitive around the world, where gasoline prices are lower than in Japan. Economies of scale will help as production volumes increase, but further advances in engineering also will be essential. With its current design, Prius's monthly operating cost would be roughly twice that of a conventional automobile. Still, Toyota believes that it will sell more than 12,000 Prius models to Japanese drivers during the first year. To do that, it is charging just $17,000 a car. The company insists that it will not lose any money on the Prius, but rivals and analysts estimate that, at current production levels, the cost of building a Prius could be as much as $32,000. If so, Toyota's low-price strategy will generate annual losses of more than $100 million on the new compact.

Will There Be Enough Demand?

Why buy a $17,000 Prius when a $13,000 Corolla is more affordable? It does not help Toyota that it is launching the Prius in the middle of a sagging domestic car market. Nobody really knows how big the final market for hybrids will be. Toyota forecasts that hybrids will account for a third of the world's auto market as early as 2005, but Japan's Ministry of International Trade and Industry expects 2.4 million alternative-fuel vehicles, including hybrids, to roam Japan's backstreets by 2010. Nonetheless, the Prius has set a new standard for Toyota's competitors. There seems to be no turning back for the Japanese. The government may soon tell carmakers to slash carbon dioxide emissions by 20% by 2010. And it wants them to cut nitrous oxide, hydrocarbon, and carbon monoxide emissions by 80%. The government may also soon give tax breaks to consumers who buy green cars.

Prospects for the Prius started looking good: Total sales for Prius reached over 7,700 units as of June 1998. With this encouraging sales trend, Toyota finally announced that the Prius would be introduced in the North American and European markets by the year 2000. The total sales volume will be approximately 20,000 units per year in the North American and European market combined. As with the 2000 North American and European introduction, Toyota is planning to use the next two years to develop a hybrid vehicle optimized to the usage conditions of each market.

What Is the Business Risk in Marketing the Prius?

Engineers at Toyota Motors have stated that California would be the primary market for the Prius outside Japan, but they added that an annual demand of 50,000 cars would be necessary to justify production. Despite Toyota management's decision to introduce the hybrid electric car into the U.S. market, the Toyota engineers were still uncertain whether there would be enough demand. Furthermore, competitors, including U.S. automakers, just do not see how Toyota can achieve the economies of scale needed to produce green cars at a profit. The primary advantage of the design, however, is that the Prius can cut auto pollution in half. This is a feature that could be very appealing at a time when government air-quality standards are becoming more rigorous and consumer interest in the environment is strong. However, in the case of the Prius, if a significant reduction in production cost never materializes, demand may remain insufficient to justify the investment in the green car.

1.4.2 Impact of Engineering Projects on Financial Statements

Engineers must understand the business environment in which a company's major business decisions are made. It is important for an engineering project to generate profits, but it also must strengthen the firm's overall financial position. How do we measure Toyota's success in the Prius project? Will enough Prius models be sold, for example, to keep the green-engineering business as Toyota's major source of profits? While the Prius project will provide comfortable, reliable, low-cost driving for the company's customers, the bottom line is its financial performance over the long run.

Regardless of a business's form, each company has to produce basic financial statements at the end of each operating cycle (typically a year). These financial statements provide the basis for future investment analysis. In practice, we seldom make investment decisions solely on the basis of an estimate of a project's profitability, because we must also consider the overall impact of the investment on the financial strength and position of the company.

Suppose that you were the president of the Toyota Corporation. Suppose further that you even hold some shares in the company, making you one of the company's many owners. What objectives would you set for the company? While all firms are in business in hopes of making a profit, what determines the market value of a company are not profits per se, but cash flow. It is, after all, available cash that determines the future investments and growth of the firm. Therefore, one of your objectives should be to increase the company's value to its owners (including yourself) as much as possible. To some extent, the market price of your company's stock represents the value of your company. Many factors affect your company's market value: present and expected future earnings, the timing and duration of those earnings, and the risks associated with them. Certainly, any successful investment decision will increase a company's market value. Stock price can be a good indicator of your company's financial health and may also reflect the market's attitude about how well your company is managed for the benefit of its owners.

1.4.3 A Look Back in 2005: Did Toyota Make the Right Decision?

Clearly, there were many doubts and uncertainties about the market for hybrids in 1998. Even Toyota engineers were not sure that there would be enough demand in the U.S. market to justify the production of the vehicle. Seven years after the Prius was introduced, it turns out that Toyota's decision to go ahead was the right decision. The continued success of the vehicle led to the launching of a second-generation Prius at the New York Motor Show in 2003. This car delivered higher power and better fuel economy than its predecessor. Indeed, the new Prius proved that Toyota has achieved its goal: to create an eco-car with high-level environmental performance, but with the conventional draw of a modern car. These features, combined with its efficient handling and attractive design, are making the Prius a popular choice of individuals and companies alike. In fact, Toyota has already announced that it will double Prius production for the U.S. market, from 50,000 to 100,000 units annually, but even that may fall short of demand if oil prices continue to increase in the future.

   Toyota made its investors happy, as the public liked the new hybrid car, resulting in an increased demand for the product. This, in turn, caused stock prices, and hence shareholder wealth, to increase. In fact, the new, heavily promoted, green car turned out to be a market leader in its class and contributed to sending Toyota's stock to an all-time high in late 2005.[4] Toyota's market value continued to increase well into 2006. Any successful investment decision on Prius's scale will tend to increase a firm's stock prices in the marketplace and promote long-term success. Thus, in making a large-scale engineering project decision, we must consider its possible effect on the firm's market value. (In Chapter 2, we discuss the financial statements in detail and show how to use them in our investment decision making.)

1.5 Common Types of Strategic Engineering Economic Decisions

The story of how the Toyota Corporation successfully introduced a new product and became the market leader in the hybrid electric car market is typical: Someone had a good idea, executed it well, and obtained good results. Project ideas such as the Prius can originate from many different levels in an organization. Since some ideas will be good, while others

---

[4] Toyota Motor Corporation, *Annual Report, 2005.*

will not, we need to establish procedures for screening projects. Many large companies have a specialized project analysis division that actively searches for new ideas, projects, and ventures. Once project ideas are identified, they are typically classified as (1) equipment or process selection, (2) equipment replacement, (3) new product or product expansion, (4) cost reduction, or (5) improvement in service or quality. This classification scheme allows management to address key questions: Can the existing plant, for example, be used to attain the new production levels? Does the firm have the knowledge and skill to undertake the new investment? Does the new proposal warrant the recruitment of new technical personnel? The answers to these questions help firms screen out proposals that are not feasible, given a company's resources.

The Prius project represents a fairly complex engineering decision that required the approval of top executives and the board of directors. Virtually all big businesses face investment decisions of this magnitude at some time. In general, the larger the investment, the more detailed is the analysis required to support the expenditure. For example, expenditures aimed at increasing the output of existing products or at manufacturing a new product would invariably require a very detailed economic justification. Final decisions on new products, as well as marketing decisions, are generally made at a high level within the company. By contrast, a decision to repair damaged equipment can be made at a lower level. The five classifications of project ideas are as follows:

- **Equipment or Process Selection.** This class of engineering decision problems involves selecting the best course of action out of several that meet a project's requirements. For example, which of several proposed items of equipment shall we purchase for a given purpose? The choice often hinges on which item is expected to generate the largest savings (or the largest return on the investment). For example, the choice of material will dictate the manufacturing process for the body panels in the automobile. Many factors will affect the ultimate choice of the material, and engineers should consider all major cost elements, such as the cost of machinery and equipment, tooling, labor, and material. Other factors may include press and assembly, production and engineered scrap, the number of dies and tools, and the cycle times for various processes.

- **Equipment Replacement.** This category of investment decisions involves considering the expenditure necessary to replace worn-out or obsolete equipment. For example, a company may purchase 10 large presses, expecting them to produce stamped metal parts for 10 years. After 5 years, however, it may become necessary to produce the parts in plastic, which would require retiring the presses early and purchasing plastic molding machines. Similarly, a company may find that, for competitive reasons, larger and more accurate parts are required, making the purchased machines become obsolete earlier than expected.

- **New Product or Product Expansion.** Investments in this category increase company revenues if output is increased. One common type of expansion decision includes decisions about expenditures aimed at increasing the output of existing production or distribution facilities. In these situations, we are basically asking, "Shall we build or otherwise acquire a new facility?" The expected future cash inflows in this investment category are the profits from the goods and services produced in the new facility. A second type of expenditure decision includes considering expenditures necessary to produce a new product or to expand into a new geographic area. These projects normally require large sums of money over long periods.

- **Cost Reduction.** A cost-reduction project is a project that attempts to lower a firm's operating costs. Typically, we need to consider whether a company should buy equipment to perform an operation currently done manually or spend money now in order to save more money later. The expected future cash inflows on this investment are savings resulting from lower operating costs.
- **Improvement in Service or Quality.** Most of the examples in the previous sections were related to economic decisions in the manufacturing sector. The decision techniques we develop in this book are also applicable to various economic decisions related to improving services or quality of product.

1.6 Fundamental Principles of Engineering Economics

This book is focused on the principles and procedures engineers use to make sound economic decisions. To the first-time student of engineering economics, anything related to money matters may seem quite strange when compared to other engineering subjects. However, the decision logic involved in solving problems in this domain is quite similar to that employed in any other engineering subject. There are fundamental principles to follow in engineering economics that unite the concepts and techniques presented in this text, thereby allowing us to focus on the logic underlying the practice of engineering economics.

- **Principle 1: A nearby penny is worth a distant dollar.** A fundamental concept in engineering economics is that money has a time value associated with it. Because we can earn interest on money received today, it is better to receive money earlier than later. This concept will be the basic foundation for all engineering project evaluation.

Time Value of Money

If you receive $100 now, you can invest it and have more money available six months from now

$100   Earning opportunity $100

Today   Six months later

- **Principle 2: All that counts are the differences among alternatives.** An economic decision should be based on the *differences* among the alternatives considered. All that is common is irrelevant to the decision. Certainly, any economic decision is no better than the alternatives being considered. Thus, an economic decision should be based on the objective of making the best use of limited resources. Whenever a choice is made, something is given up. The opportunity cost of a choice is the value of the best alternative given up.

Comparing Buy versus Lease

Whatever you decide, you need to spend the same amount of money on
fuel and maintenance

Option	Monthly Fuel Cost	Monthly Maintenance	Cash Outlay at Signing	Monthly Payment	Salvage Value at End of Year 3
Buy	$960	$550	$6,500	$350	$9,000
Lease	$960	$550	$2,400	$550	0

Irrelevant items in decision making

- **Principle 3: Marginal revenue must exceed marginal cost.** Effective decision
making requires comparing the additional costs of alternatives with the additional
benefits. Each decision alternative must be justified on its own economic merits
before being compared with other alternatives. Any increased economic activity
must be justified on the basis of the fundamental economic principle that marginal
revenue must exceed marginal cost. Here, *marginal revenue* means the additional revenue made possible by increasing the activity by one unit (or small unit). *Marginal
cost* has an analogous definition. Productive resources—the natural resources, human
resources, and capital goods available to make goods and services—are limited.
Therefore, people cannot have all the goods and services they want; as a result, they
must choose some things and give up others.

Marginal cost:
The cost
associated
with one
additional unit
of production,
also called the
incremental
cost.

Marginal Analysis

To justify your action, marginal revenue must exceed marginal cost

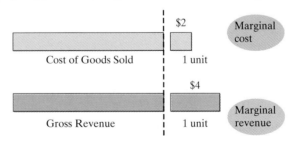

- **Principle 4: Additional risk is not taken without the expected additional return.**
For delaying consumption, investors demand a minimum return that must be greater
than the anticipated rate of inflation or any perceived risk. If they didn't receive
enough to compensate for anticipated inflation and the perceived investment risk, investors would purchase whatever goods they desired ahead of time or invest in assets
that would provide a sufficient return to compensate for any loss from inflation or
potential risk.

Risk and Return Trade-Off

Expected returns from bonds and stocks are normally higher than the expected return from a savings account

Investment Class	Potential Risk	Expected Return
Savings account (cash)	Low/None	1.5%
Bond (debt)	Moderate	4.8%
Stock (equity)	High	11.5%

Risk-return tradeoff: Invested money can render higher profits only if it is subject to the possibility of being lost.

The preceding four principles are as much statements of common sense as they are theoretical precepts. These principles provide the logic behind what is to follow. We build on them and attempt to draw out their implications for decision making. As we continue, keep in mind that, while the topics being treated may change from chapter to chapter, the logic driving our treatment of them is constant and rooted in the four fundamental principles.

SUMMARY

■ This chapter has given us an overview of a variety of engineering economic problems that commonly are found in the business world. We examined the role, and the increasing importance, of engineers in the firm, as evidenced by the development at Toyota of the Prius, a hybrid vehicle. Commonly, engineers are called upon to participate in a variety of strategic business decisions ranging from product design to marketing.

■ The term **engineering economic decision** refers to any investment decision related to an engineering project. The facet of an economic decision that is of most interest from an engineer's point of view is the evaluation of costs and benefits associated with making a capital investment.

■ The five main types of engineering economic decisions are (1) equipment or process selection, (2) equipment replacement, (3) new product or product expansion, (4) cost reduction, and (5) improvement in service or quality.

■ The factors of **time** and **uncertainty** are the defining aspects of any investment project.

■ The four fundamental principles that must be applied in all engineering economic decisions are (1) the time value of money, (2) differential (incremental) cost and revenue, (3) marginal cost and revenue, and (4) the trade-off between risk and reward.

TWO

Understanding Financial Statements

Dell Manages Profitability, Not Inventory[1] In 1994, Dell was a struggling second-tier PC maker. Like other PC makers, Dell ordered its components in advance and carried a large amount of component inventory. If its forecasts were wrong, Dell had major write-downs. Then Dell began to implement a new business model. Its operations had always featured a build-to-order process with direct sales to customers, but Dell took a series of ingenious steps to eliminate its inventories. The results were spectacular.

Over a four-year period, Dell's revenues grew from $2 billion to $16 billion, a 50 % annual growth rate. Earnings per share increased by 62 % per year. Dell's stock price increased by more than 17,000 % in a little over eight years. In 1998, Dell's return on invested capital was 217 %, and the company had $1.8 billion in cash. (The rapid growth

1983–	Michael Dell starts business of preformatting IBM PC HDs on weekends
1985–	**$6 million** sales, upgrading IBM compatibles for local businesses
1986–	**$70 million** sales; focus on assembling own line of PCs
1990–	**$500 million** sales with an extensive line of products
1996–	Dell goes online; $1 million per day in online sales; **$5.3B** in annual sales
1997–	Dell online sales at $3 million per day; 50% growth rate for third consecutive year, **$7.8B** in total annual sales.
2005–	**$49.2B** in sales

[1] Jonathan Byrnes, "Dell Manages Profitability, Not Inventory," Harvard Business School, Working Knowledge, June 2, 2003.

continued, and the company's sales revenues finally reached $50 billion in 2005.)

Profitability management—coordinating a company's day-to-day activities through careful forethought and attentive oversight—was at the core of Dell's transformation in this critical period. Dell created a tightly aligned business model that enabled it to dispense with the need for its component inventories. Not only was capital not needed, but the change generated enormous amounts of cash that Dell used to fuel its growth. How did Dell do it?

Account selection. Dell purposely selected customers with relatively predictable purchasing patterns and low service costs. The company developed a core competence in targeting customers and kept a massive database for this purpose.

The remainder of Dell's business involved individual consumers. To obtain stable demand in this segment, Dell used higher-end products and those with the latest technology to target second-time buyers who had regular upgrade purchase patterns, required little technical support, and paid by credit card.

Demand management. Dell's core philosophy of actively managing demand in real time, or "selling what you have," rather than making what you want to sell, was a critical driver of Dell's successful profitability management. Without this critical element, Dell's business model simply would not have been effective.

Product life-cycle management. Because Dell's customers were largely high-end repeat buyers who rapidly adopted new technology, Dell's marketing could focus on managing product life-cycle transitions.

Supplier management. Although Dell's manufacturing system featured a combination of build-product-to-order and buy-component-to-plan processes, the company worked closely with its suppliers to introduce more flexibility into its system.

Forecasting. Dell's forecast accuracy was about 70 to 75%, due to its careful account selection. Demand management, in turn, closed the forecast gap. When in doubt, Dell managers overforecast on high-end products because it was easier to sell up and high-end products had a longer shelf life.

Liquidity management. Direct sales were explicitly targeted toward high-end customers who paid with a credit card. These sales had a 4-day cash conversion cycle, while Dell took 45 days to pay its vendors. This approach generated a huge amount of liquidity that helped finance Dell's rapid growth and limited its external financing needs. Dell's cash engine was a key underlying factor that enabled it to earn such extraordinarily high returns.

If you want to explore investing in Dell stock, what information would you go by? You would certainly prefer that Dell have a record of accomplishment of profitable operations, earning a profit (net income) year after year. The company would need a steady stream of cash coming in and a manageable level of debt. How would you determine whether the company met these criteria? Investors commonly use the financial statements contained in the annual report as a starting point in forming expectations about future levels of earnings and about the firm's riskiness.

Before making any financial decision, it is good to understand an elementary aspect of your financial situation—one that you'll also need for retirement planning, estate planning, and, more generally, to get an answer to the question, "How am I doing?" It is called your **net worth**. If you decided to invest $10,000 in Dell stocks, how would that affect your net worth? You may need this information for your own financial planning, but it is routinely required whenever you have to borrow a large sum of money from a financial institution. For example, when you are buying a home, you need to apply for a mortgage. Invariably, the bank will ask you to submit your net-worth statement as a part of loan processing. Your net-worth statement is a snapshot of where you stand financially at a given point in time. The bank will determine how creditworthy you are by examining your net worth. In a similar way, a corporation prepares the same kind of information for its financial planning or to report its financial health to stockholders or investors. The reporting document is known as the financial statements. We will first review the basics of figuring out the personal net worth and then illustrate how any investment decision will affect this net-worth statement. Understanding the relationship between net worth and investing decisions will enhance one's overall understanding of how a company manages its assets in business operations.

> **Net worth** is the amount by which a *company's* or individual's assets exceed the company's or individual's liabilities.

CHAPTER LEARNING OBJECTIVES

After completing this chapter, you should understand the following concepts:

- The role of accounting in economic decisions.
- Four types of financial statements prepared for investors and regulators.
- How to read the balance sheet statement.
- How to use the income statement to manage a business.
- The sources and uses of cash in business operation.
- How to conduct the ratio analysis and what the numbers really mean.

2.1 Accounting: The Basis of Decision Making

We need financial information when we are making business decisions. Virtually all businesses and most individuals keep accounting records to aid in making decisions. As illustrated in Figure 2.1, accounting is the information system that measures business activities, processes the resulting information into reports, and communicates the results to decision makers. For this reason, we call accounting "the language of business." The better you understand this language, the better you can manage your financial well-being, and the better your financial decisions will be.

Personal financial planning, education expenses, loans, car payments, income taxes, and investments are all based on the information system we call accounting. The uses of accounting information are many and varied:

- **Individual people** use accounting information in their day-to-day affairs to manage bank accounts, to evaluate job prospects, to make investments, and to decide whether to rent an apartment or buy a house.
- **Business managers** use accounting information to set goals for their organizations, to evaluate progress toward those goals, and to take corrective actions if necessary. Decisions based on accounting information may include which building or equipment to purchase, how much merchandise to keep on hand as inventory, and how much cash to borrow.
- **Investors and creditors** provide the money a business needs to begin operations. To decide whether to help start a new venture, potential investors evaluate what income they can expect on their investment. Such an evaluation involves analyzing the financial statements of the business. Before making a loan, banks determine the borrower's ability to meet scheduled payments. This kind of evaluation includes a projection of future operations and revenue, based on accounting information.

Figure 2.1 The accounting system, which illustrates the flow of information.

An essential product of an accounting information system is a series of financial statements that allows people to make informed decisions. For personal use, the net-worth statement is a snapshot of where you stand financially at a given point in time. You do that by adding your assets—such as cash, investments, and pension plans—in one column and your liabilities—or debts—in the other. Then subtract your liabilities from your assets to find your net worth. In other words, your net worth is what you would be left with if you sold everything and paid off all you owe. For business use, financial statements are the documents that report financial information about a business entity to decision makers. They tell us how a business is performing and where it stands financially. Our purpose is not to present the bookkeeping aspects of accounting, but to acquaint you with financial statements and to give you the basic information you need to make sound engineering economic decisions through the remainder of the book.

2.2 Financial Status for Businesses

Just like figuring out your personal wealth, all businesses must prepare their financial status. Of the various reports corporations issue to their stockholders, the annual report is by far the most important, containing basic financial statements as well as management's opinion of the past year's operations and the firm's future prospects. What would managers and investors want to know about a company at the end of the fiscal year? Following are four basic questions that managers or investors are likely to ask:

- What is the company's financial position at the end of the fiscal period?
- How well did the company operate during the fiscal period?
- On what did the company decide to use its profits?
- How much cash did the company generate and spend during the fiscal period?

As illustrated in Figure 2.2, the answer to each of these questions is provided by one of the following financial statements: the balance sheet statement, the income statement, the statement of retained earnings, and the cash flow statement. The fiscal year (or operating

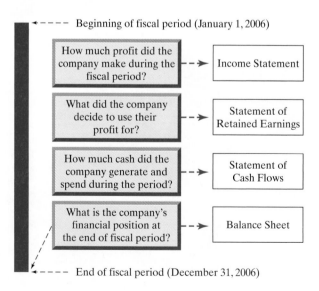

Figure 2.2 Information reported on the financial statements.

cycle) can be any 12-month term, but is usually January 1 through December 31 of a calendar year.

As mentioned in Section 1.1.2, one of the primary responsibilities of engineers in business is to plan for the acquisition of equipment (capital expenditure) that will enable the firm to design and produce products economically. This type of planning will require an estimation of the savings and costs associated with the acquisition of equipment and the degree of risk associated with execution of the project. Such an estimation will affect the business' **bottom line** (profitability), which will eventually affect the firm's stock price in the marketplace. Therefore, engineers should understand the various financial statements in order to communicate with upper management regarding the status of a project's profitability. The situation is summarized in Figure 2.3.

For illustration purposes, we use data taken from Dell Corporation, manufacturer of a wide range of computer systems, including desktops, notebooks, and workstations, to discuss the basic financial statements. In 1984, Michael Dell began his computer business at the University of Texas in Austin, often hiding his IBM PC in his roommate's bathtub when his family came to visit. His dorm-room business officially became Dell Computer Corporation on May 3, 1984. Since 2001, Dell has become the number-one and fastest growing among all major computer system companies worldwide, with 55,200 employees around the globe. Dell's pioneering "direct model" is a simple concept involving selling personal computer systems directly to customers. It offers (1) in-person relationships with corporate and institutional customers; (2) telephone and Internet purchasing (the latter averaging $50 million a day in 2001); (3) built-to-order computer systems; (4) phone and on-line technical support; and (5) next-day on-site product service.

The company's revenue in 2005 totaled $49.205 billion. During fiscal 2005, Dell maintained its position as the world's number-one supplier of personal computer systems, with a performance that continued to outpace the industry. Over the same period, Dell's global market share of personal computer sales reached 17.8%. In the company's 2005 annual report, management painted an even more optimistic picture for the future,

Bottom line is slang for net income or accounting profit.

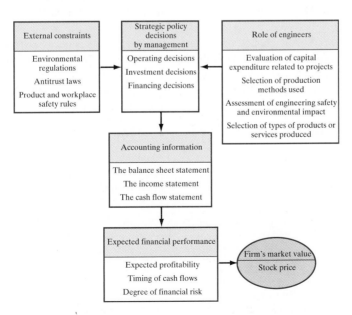

Figure 2.3   Summary of major factors affecting stock prices.

stating that Dell will continue to invest in information systems, research, development, and engineering activities to support its growth and to provide for new competitive products. Of course, there is no assurance that Dell's revenue will continue to grow at the annual rate of 50% in the future.

What can individual investors make of all this? Actually, we can make quite a bit. As you will see, investors use the information contained in an annual report to form expectations about future earnings and dividends. Therefore, the annual report is certainly of great interest to investors.

2.2.1 The Balance Sheet

What is the company's financial position at the end of the reporting period? We find the answer to this question in the company's **balance sheet statement**. A company's balance sheet, sometimes called its **statement of financial position**, reports three main categories of items: assets, liabilities, and stockholders' equity. Assets are arranged in order of liquidity. The most liquid assets appear at the top of the page, the least liquid assets at the bottom of the page. (See Figure 2.4.) Because cash is the most liquid of all assets, it is always listed first. Current assets are so critical that they are separately broken out and totaled. They are what will hold the business afloat for the next year.

Liabilities are arranged in order of payment, the most pressing at the top of the list, the least pressing at the bottom. Like current assets, current liabilities are so critical that they are separately broken out and totaled. They are what will be paid out during the next year.

A company's financial statements are based on the most fundamental tool of accounting: the accounting equation. The **accounting equation** shows the relationship among assets, liabilities, and owners' equity:

$$\text{Assets} = \text{Liabilities} + \text{Owners' Equity}$$

Every business transaction, no matter how simple or complex, can be expressed in terms of its effect on the accounting equation. Regardless of whether a business grows or contracts, the equality between the assets and the claims against the assets is always maintained. In other words, any change in the amount of total assets is necessarily accompanied by an equal change on the other side of the equation—that is, by an increase or decrease in either the liabilities or the owners' equity.

As shown in Table 2.1, the first half of Dell's year-end 2005 and 2004 balance sheets lists the firm's assets, while the remainder shows the liabilities and equity, or claims against those assets.

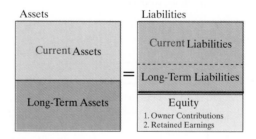

Figure 2.4 Using the four quadrants of the balance sheet.

TABLE 2.1 Consolidated Statements of Financial Position (in millions) for Dell, Inc.

	January 28, 2005	January 30, 2004
Assets		
Current assets:		
Cash and cash equivalents	$ 4,747	$ 4,317
Short-term investments	5,060	835
Accounts receivable, net	4,414	3,635
Inventories	459	327
Other	2,217	1,519
Total current assets	16,897	10,633
Property, plant, and equipment, net	1,691	1,517
Investments	4,319	6,770
Other noncurrent assets	308	391
Total assets	$ 23,215	$ 19,311
Liabilities and Stockholders' Equity		
Current liabilities:		
Accounts payable	$ 8,895	$ 7,316
Accrued and other	5,241	3,580
Total current liabilities	14,136	10,896
Long-term debt	505	505
Other noncurrent liabilities	2,089	1,630
Total liabilities	16,730	13,031
Commitments and contingent liabilities (Note 8)	—	—
Stockholders' equity:		
Preferred stock and capital in excess of $.01 par value; shares issued and outstanding: none	—	—
Common stock and capital in excess of $.01 par value; shares authorized: 7,000; shares issued: 2,769 and 2,721, respectively	8,195	6,823
Treasury stock, at cost; 284 and 165 shares, respectively	(10,758)	(6,539)
Retained earnings	9,174	6,131
Other comprehensive loss	(82)	(83)
Other	(44)	(52)
Total stockholders' equity	6,485	6,280
Total liabilities and stockholders' equity	$ 23,215	$ 19,311

Source: *Annual Report*, Dell Corporation, 2005.

Assets

The dollar amount shown in the assets portion of the balance sheet represents how much the company owns at the time it issues the report. We list the asset items in the order of their "liquidity," or the length of time it takes to convert them to cash.

Current assets account
represents the value of all assets that are reasonably expected to be converted into cash within one year.

- **Current assets** can be converted to cash or its equivalent in less than one year. Current assets generally include three major accounts:

  1. The first is *cash*. A firm typically has a cash account at a bank to provide for the funds needed to conduct day-to-day business. Although assets are always stated in terms of dollars, only cash represents actual money. Cash-equivalent items are also listed and include marketable securities and short-term investments.

  2. The second account is *accounts receivable*—money that is owed the firm, but that has not yet been received. For example, when Dell receives an order from a retail store, the company will send an invoice along with the shipment to the retailer. Then the unpaid bill immediately falls into the accounts receivable category. When the bill is paid, it will be deducted from the accounts receivable account and placed into the cash category. A typical firm will have a 30- to 45-day accounts receivable, depending on the frequency of its bills and the payment terms for customers.

  3. The third account is *inventories*, which show the dollar amount that Dell has invested in raw materials, work in process, and finished goods available for sale.

- **Fixed assets** are relatively permanent and take time to convert into cash. Fixed assets reflect the amount of money Dell paid for its plant and equipment when it acquired those assets. The most common fixed asset is the physical investment in the business, such as land, buildings,[2] factory machinery, office equipment, and automobiles. With the exception of land, most fixed assets have a limited useful life. For example, buildings and equipment are used up over a period of years. Each year, a portion of the usefulness of these assets expires, and a portion of their total cost should be recognized as a depreciation expense. The term *depreciation* denotes the accounting process for this gradual conversion of fixed assets into expenses. *Property, plant and equipment, net* thus represents the current book value of these assets after deducting depreciation expenses.

- Finally, **other assets** include investments made in other companies and intangible assets such as goodwill, copyrights, franchises, and so forth. Goodwill appears on the balance sheet only when an operating business is purchased in its entirety. Goodwill indicates any additional amount paid for the business above the fair market value of the business. (Here, the fair market value is defined as the price that a buyer is willing to pay when the business is offered for sale.)

Liabilities and Stockholders' Equity (Owners' Net Worth)

The claims against assets are of two types: liabilities and stockholders' equity. The liabilities of a company indicate where the company obtained the funds to acquire its assets and to operate the business. Liability is money the company owes. Stockholders' equity is that portion of the assets of a company which is provided by the investors (owners). Therefore, stockholders' equity is the liability of a company to its owners.

---

[2] Land and buildings are commonly called **real assets** to distinguish them from equipment and machinery.

- **Current liabilities** are those debts which must be paid in the near future (normally, within one year). The major current liabilities include accounts and notes payable within a year. Also included are accrued expenses (wages, salaries, interest, rent, taxes, etc., owed, but not yet due for payment), and advance payments and deposits from customers.

- **Other liabilities** include *long-term liabilities*, such as bonds, mortgages, and long-term notes, that are due and payable more than one year in the future.

- **Stockholders' equity** represents the amount that is available to the owners after all other debts have been paid. Generally, stockholders' equity consists of preferred and common stock, treasury stock, capital surplus, and retained earnings. Preferred stock is a hybrid between common stock and debt. In case the company goes bankrupt, it must pay its preferred stockholders after its debtors, but before its common stockholders. Preferred dividend is fixed, so preferred stockholders do not benefit if the company's earnings grow. In fact, many firms do not use preferred stock. The common stockholders' equity, or **net worth**, is a residual:

$$\text{Assets} - \text{Liabilities} - \text{Preferred stock} = \text{Common stockholders' equity}$$
$$\$11,471 - \$6,163 - \$0 = \$5,308.$$

- **Common stock** is the aggregate par value of the company's stock issued. Companies rarely issue stocks at a discount (i.e., at an amount below the stated par). Normally, corporations set the par value low enough so that, in practice, stock is usually sold at a premium.

- **Paid-in capital** (capital surplus) is the amount of money received from the sale of stock that is over and above the par value of the stock. Outstanding stock is the number of shares issued that actually are held by the public. If the corporation buys back part of its own issued stock, that stock is listed as *treasury stock* on the balance sheet.

- **Retained earnings** represent the cumulative net income of the firm since its inception, less the total dividends that have been paid to stockholders. In other words, retained earnings indicate the amount of assets that have been financed by plowing profits back into the business. Therefore, retained earnings belong to the stockholders.

Current liabilities are bills that are due to creditors and suppliers within a short period of time.

Treasury stock is not taken into consideration when calculating earnings per share or dividends.

Retained earnings refers to earnings not paid out as dividends but instead are reinvested in the core business or used to pay off debt.

2.2.2 The Income Statement

The second financial report is the **income statement**, which indicates whether the company is making or losing money during a stated *period*, usually a year. Most businesses prepare quarterly and monthly income statements as well. The company's accounting period refers to the period covered by an income statement.

Basic Income Statement Equation

For Dell, the accounting period begins on February 1 and ends on January 31 of the following year. Table 2.2 gives the 2005 and 2004 income statements for Dell.

TABLE 2.2 Consolidated Statements of Income (in millions, except per share amounts) Dell, Inc.

	Fiscal Year Ended		
	January 28, 2005	January 30, 2004	January 31, 2003
Net revenue	$ 49,205	$ 41,444	$ 35,404
Cost of revenue	40,190	33,892	29,055
Gross margin	9,015	7,552	6,349
Operating expenses:			
Selling, general, and administrative	4,298	3,544	3,050
Research, development, and engineering	463	464	455
Total operating expenses	4,761	4,008	3,505
Operating income	4,254	3,544	2,844
Investment and other income, net	191	180	183
Income before income taxes	4,445	3,724	3,027
Income tax provision	1,402	1,079	905
Net income	$ 3,043	$ 2,645	$ 2,122
Earnings per common share:			
Basic	$ 1.21	$ 1.03	$ 0.82
Diluted	$ 1.18	$ 1.01	$ 0.80
Weighted average shares outstanding:			
Basic	2,509	2,565	2,584
Diluted	2,568	2,619	2,644

Source: *Annual Report*, Dell Corporation, 2005.

Reporting Format

Typical items that are itemized in the income statement are as follows:

- **Revenue** is the income from goods sold and services rendered during a given accounting period.
- **Net revenue** represents gross sales, less any sales return and allowances.
- Shown on the next several lines are the expenses and costs of doing business, as deductions from revenue. The largest expense for a typical manufacturing firm is the expense it incurs in making a product (such as labor, materials, and overhead), called the **cost of revenue** (or cost of goods sold).
- Net revenue less the cost of revenue gives the **gross margin**.
- Next, we subtract any other operating expenses from the operating income. These other operating expenses are expenses associated with paying interest, leasing machinery or equipment, selling, and administration. This results in the operating income.
- Finally, we determine the **net income** (or net profit) by subtracting the income taxes from the taxable income. Net income is also commonly known as *accounting income*.

Gross margin represents the amount of money the company generated over the cost of producing its goods or services.

Earnings per Share

Another important piece of financial information provided in the income statement is the **earnings per share** (EPS).[3] In simple situations, we compute the EPS by dividing the available earnings to common stockholders by the number of shares of common stock outstanding. Stockholders and potential investors want to know what their share of profits is, not just the total dollar amount. The presentation of profits on a per share basis allows the stockholders to relate earnings to what they paid for a share of stock. Naturally, companies want to report a higher EPS to their investors as a means of summarizing how well they managed their businesses for the benefits of the owners. Interestingly, Dell earned $1.21 per share in 2005, up from $1.03 in 2004, but it paid no dividends.

> **EPS** is generally considered to be the single most important variable in determining a share's price.

Retained Earnings

As a supplement to the income statement, many corporations also report their retained earnings during the accounting period. When a corporation makes some profits, it has to decide what to do with those profits. The corporation may decide to pay out some of the profits as dividends to its stockholders. Alternatively, it may retain the remaining profits in the business in order to finance expansion or support other business activities.

When the corporation declares dividends, preferred stock has priority over common stock. Preferred stock pays a stated dividend, much like the interest payment on bonds. The dividend is not a legal liability until the board of directors has declared it. However, many corporations view the dividend payments to preferred stockholders as a liability. Therefore, "available for common stockholders" reflects the net earnings of the corporation, less the preferred stock dividends. When preferred and common stock dividends are subtracted from net income, the remainder is retained earnings (profits) for the year. As mentioned previously, these retained earnings are reinvested into the business.

EXAMPLE 2.1 Understanding Dell's Balance Sheet and Income Statement

With revenue of $49,205 million for fiscal year 2005, Dell is the world's leading direct computer systems company. Tables 2.2 and 2.3 show how Dell generated its net income during the fiscal year.

The Balance Sheet. Dell's $23,215 million of total assets shown in Table 2.1 were necessary to support its sales of $49,205 million.

- Dell obtained the bulk of the funds it used to buy assets
  1. By buying on credit from its suppliers (accounts payable).
  2. By borrowing from financial institutions (notes payable and long-term bonds).
  3. By issuing common stock to investors.
  4. By plowing earnings into the business, as reflected in the retained earnings account.

---

[3] In reporting EPS, the firm is required to distinguish between "basic EPS" and "diluted EPS." Basic EPS is the net income of the firm, divided by the number of shares of common stock outstanding. By contrast, the diluted EPS includes all common stock equivalents (convertible bonds, preferred stock, warrants, and rights), along with common stock. Therefore, diluted EPS will usually be less than basic EPS.

- The net increase in fixed assets was $174 million ($1,691 − $1,517; Table 2.1). However, this amount is after a deduction for the year's depreciation expenses. We should add depreciation expense back to show the increase in gross fixed assets. From the company's cash flow statement in Table 2.3, we see that the 2005 depreciation expense is $334 million; thus, the acquisition of fixed assets equals $508 million.

- Dell had a total long-term debt of $505 million (Table 2.1), consisting of several bonds issued in previous years. The interest Dell paid on these long-term debts was about $16 million.

- Dell had 2,509 million shares of common stock outstanding. Investors actually provided the company with a total capital of $8,195 million (Table 2.1). However, Dell has retained the current as well as previous earnings of $9,174 million since it was incorporated. Dell also held $10,758 million worth of treasury stock, which was financed through the current as well as previous retained earnings. The combined net stockholders' equity was $6,485 million, and these earnings belong to Dell's common stockholders (Table 2.1).

- On the average, stockholders have a total investment of $2.58 per share ($6,485 million/2,509 million shares) in the company. The $2.58 figure is known as the stock's *book value*. In the fall of 2005, the stock traded in the general range from $32 to $40 per share. Note that this market price is quite different from the stock's book value. Many factors affect the market price, the most important one being how investors expect the company to perform in the future. Certainly, the company's direct made-to-order business practices have had a major influence on the market value of its stock.

The Income Statement. Dell's net revenue was $49,205 million in 2005, compared with $41,444 million in 2004, a gain of 18.73% (Table 2.2). Profits from operations (operating income) rose 20.03% to $4,254 million, and net income was up 15.05% to $3,043 million.

- Dell issued no preferred stock, so there is no required cash dividend. Therefore, the entire net income of $3,043 million belongs to the common stockholders.

- Earnings per common share climbed at a faster pace than in 2004, to $1.21, an increase of 17.48% (Table 2.2). Dell could retain this income fully for reinvestment in the firm, or it could pay it out as dividends to the common stockholders. Instead of either of these alternatives, Dell repurchased and retired 56 million common stocks for $1,012 million. We can see that Dell had earnings available to common stockholders of $3,043 million. As shown in Table 2.1, the beginning balance of the retained earnings was $6,131 million. Therefore, the total retained earnings grew to $9,174 million.

2.2.3 The Cash Flow Statement

The income statement explained in the previous section indicates only whether the company was making or losing money during the reporting period. Therefore, the emphasis was on determining the net income (profits) of the firm for supporting its operating activities. However, the income statement ignores two other important business activities for the period: financing and investing activities. Therefore, we need another financial statement—the cash flow statement, which details how the company generated the cash it received and how the company used that cash during the reporting period.

Sources and Uses of Cash

The difference between the sources (inflows) and uses (outflows) of cash represents the net cash flow during the reporting period. This is a very important piece of information, because investors determine the value of an asset (or, indeed, of a whole firm) by the cash flows it generates. Certainly, a firm's net income is important, but cash flows are even more important, particularly because the company needs cash to pay dividends and to purchase the assets required to continue its operations. As mentioned in the previous section, the goal of the firm should be to maximize the price of its stock. Since the value of any asset depends on the cash flows produced by the asset, managers want to maximize the cash flows available to investors over the long run. Therefore, we should make investment decisions on the basis of cash flows rather than profits. For such investment decisions, it is necessary to convert profits (as determined in the income statement) to cash flows. Table 2.3 is Dell's statement of cash flows, as it would appear in the company's annual report.

Reporting Format

In preparing the cash flow statement such as that in Table 2.3, many companies identify the sources and uses of cash according to the types of business activities. There are three types of activities:

- **Operating activities.** We start with the net change in operating cash flows from the income statement. Here, operating cash flows represent those cash flows related to production and the sales of goods or services. All noncash expenses are simply added back to net income (or after-tax profits). For example, an expense such as depreciation is only an accounting expense (a bookkeeping entry). Although we may charge depreciation against current income as an expense, it does not involve an actual cash outflow. The actual cash flow may have occurred when the asset was purchased. (Any adjustments in **working capital**[4] will also be listed here.)
- **Investing activities.** Once we determine the operating cash flows, we consider any cash flow transactions related to investment activities, which include purchasing new fixed assets (cash outflow), reselling old equipment (cash inflow), and buying and selling financial assets.
- **Financing activities.** Finally, we detail cash transactions related to financing any capital used in business. For example, the company could borrow or sell more stock, resulting in cash inflows. Paying off existing debt will result in cash outflows.

By summarizing cash inflows and outflows from three activities for a given accounting period, we obtain the net change in the cash flow position of the company.

---

[4] The difference between the increase in current assets and the spontaneous increase in current liabilities is the **net change in net working capital**. If this change is positive, then additional financing, over and above the cost of the fixed assets, is needed to fund the increase in current assets. This will further reduce the cash flow from the operating activities.

TABLE 2.3 Consolidated Statements of Cash Flows (in millions) Dell, Inc.

	Fiscal Year Ended		
	January 28, 2005	January 30, 2004	January 31, 2003
Cash flows from operating activities:			
Net income	$ 3,043	$ 2,645	$ 2,122
Adjustments to reconcile net income to net cash provided by operating activities:			
Depreciation and amortization	334	263	211
Tax benefits of employee stock plans	249	181	260
Effects of exchange rate changes on monetary assets and liabilities denominated in foreign currencies	(602)	(677)	(537)
Other	78	113	60
Changes in:			
Operating working capital	1,755	872	1,210
Noncurrent assets and liabilities	453	273	212
Net cash provided by operating activities	5,310	3,670	3,538
Cash flows from investing activities:			
Investments:			
Purchases	(12,261)	(12,099)	(8,736)
Maturities and sales	10,469	10,078	7,660
Capital expenditures	(525)	(329)	(305)
Purchase of assets held in master lease facilities	—	(636)	—
Cash assumed in consolidation of Dell Financial Services, L.P.	—	172	—
Net cash used in investing activities	(2,317)	(2,814)	(1,381)
Cash flows from financing activities:			
Repurchase of common stock	(4,219)	(2,000)	(2,290)
Issuance of common stock under employee plans and other	1,091	617	265
Net cash used in financing activities	(3,128)	(1,383)	(2,025)
Effect of exchange rate changes on cash and cash equivalents	565	612	459
Net increase in cash and cash equivalents	430	85	591
Cash and cash equivalents at beginning of period	4,317	4,232	3,641
Cash and cash equivalents at end of period	$ 4,747	$ 4,317	$ 4,232

Source: *Annual Report,* Dell Corporation, 2005.

EXAMPLE 2.2 Understanding Dell's Cash Flow Statement

As shown in Table 2.3, Dell's cash flow from operations amounted to $5,310 million. Note that this is significantly more than the $3,043 million earned during the reporting period. Where did the extra money come from?

- The main reason for the difference lies in the accrual-basis accounting principle used by the Dell Corporation. In **accrual-basis accounting**, an accountant recognizes the impact of a business event as it occurs. When the business performs a service, makes a sale, or incurs an expense, the accountant enters the transaction into the books, regardless of whether cash has or has not been received or paid. For example, an increase in accounts receivable of $4,414 million − $3,635 million = $779 million during 2005 represents the amount of total sales on credit (Table 2.1). Since the $779 million figure was included in the total sales in determining the net income, we need to subtract it to determine the company's true cash position. After adjustments, the net cash provided from operating activities is $5,310 million.

> **Accrual-basis accounting** measures the performance and position of a company by recognizing economic events regardless of when cash transactions occur.

- As regards investment activities, there was an investment outflow of $525 million in new plant and equipment. Dell sold $10,469 million worth of stocks and bonds during the period, and reinvested $12,261 million in various financial securities. The net cash flow provided from these investing activities amounted to −$2,317 million, which means an outflow of money.

- Financing activities produced a net outflow of $4,219 million, including the repurchase of the company's own stocks. (Repurchasing its own stock is equivalent to investing the firm's idle cash from operation in the stock market. Dell could have bought another company's stock, such as IBM or Microsoft stock, with the money, but Dell liked its own stock better than any other stocks on the market.) Dell also raised $1,091 million by issuing shares of common stock. The net cash used in financing activities amounted to $3,128 million (outflow).

- Finally, there was the effect of exchange rate changes on cash for foreign sales. This amounted to a net increase of $565 million. Together, the three types of activities generated a total cash flow of $430 million. With the initial cash balance of $4,317 million, the ending cash balance thus increased to $4,747 million. This same amount denotes the change in Dell's cash position, as shown in the cash accounts in the balance sheet.

2.3 Using Ratios to Make Business Decisions

As we have seen in Dell's financial statements, the purpose of accounting is to provide information for decision making. Accounting information tells what happened at a particular point in time. In that sense, financial statements are essentially historical documents. However, most users of financial statements are concerned about what will happen in the future. For example,

- Stockholders are concerned with future earnings and dividends.
- Creditors are concerned with the company's ability to repay its debts.
- Managers are concerned with the company's ability to finance future expansion.
- Engineers are concerned with planning actions that will influence the future course of business events.

Although financial statements are historical documents, they can still provide valuable information bearing on all of these concerns. An important part of financial analysis is the calculation and interpretation of various financial ratios. In this section, we consider some of the ratios that analysts typically use in attempting to predict the future course of events in business organizations. We may group these ratios into five categories (debt management, liquidity, asset management, profitability, and market trend) as outlined in Figure 2.5. In all financial ratio calculations, we will use the 2005 financial statements for Dell Computer Corporation, as summarized in Table 2.4.

2.3.1 Debt Management Analysis

All businesses need assets to operate. To acquire assets, the firm must raise capital. When the firm finances its long-term needs externally, it may obtain funds from the capital markets. Capital comes in two forms: **debt** and **equity**. Debt capital is capital borrowed from financial institutions. Equity capital is capital obtained from the owners of the company.

The basic methods of financing a company's debt are through bank loans and the sale of bonds. For example, suppose a firm needs $10,000 to purchase a computer. In this situation, the firm would borrow money from a bank and repay the loan, together with the interest specified, in a few years. This kind of financing is known as *short-term debt*

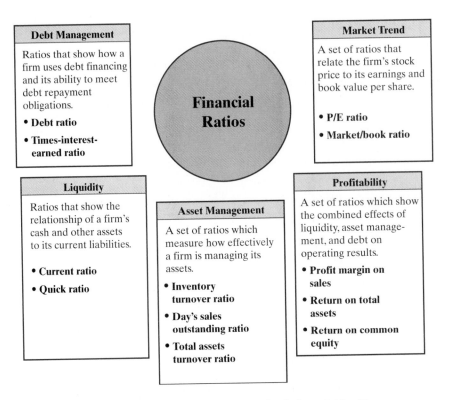

Figure 2.5   Types of ratios used in evaluating a firm's financial health.

TABLE 2.4 Summary of Dell's Key Financial Statements

Balance Sheet	January 28, 2005	January 30, 2004
Cash and cash equivalent	4,747	4,317
Accounts receivables, net	4,414	3,635
Inventories	459	327
Total current assets	16,897	10,633
Total assets	23,215	19,311
Total current liabilities	14,136	10,896
Long-term debt	505	505
Total liabilities	16,730	13,031
Common stock	8,195	6,823
Retained earnings	9,174	6,131
Total stockholders' equity	6,485	6,280
Income Statement		
Net revenue	49,205	41,444
Gross income (margin)	9,015	7,522
Operating income (margin)	4,254	3,544
Net income (margin)	3,043	2,645
Statements of Retained Earnings		
Beginning retained earnings	6,131	3,486
Net income	3,043	2,645
Ending retained earnings	9,174	6,131
Statement of Cash Flows		
Net cash from operating activities	5,310	3,670
Net cash used in investing activities	(2,317)	(2,814)
Net cash used in financing activities	(3,128)	(1,383)
Effect of exchange rate changes	565	612
Beginning cash position	4,317	4,232
Ending cash position	4,747	4,317

financing. Now suppose that the firm needs $100 million for a construction project. Normally, it would be very expensive (or require a substantial mortgage) to borrow the money directly from a bank. In this situation, the company would go public to borrow money on a long-term basis. When investors lend capital to a company and the company consents to repay the loan at an agreed-upon interest rate, the investor is the creditor of the corporation. The document that records the nature of the arrangement between the issuing company and the investor is called a **bond**. Raising capital by issuing a bond is called *long-term debt financing*.

Similarly, there are different types of equity capital. For example, the equity of a proprietorship represents the money provided by the owner. For a corporation, equity capital comes in two forms: *preferred stock* and *common stock*. Investors provide capital to a corporation and the company agrees to endow the investor with fractional ownership in the corporation. Preferred stock pays a stated *dividend*, much like the interest payment on bonds. However, the dividend is not a legal liability until the company declares it. Preferred stockholders have preference over common stockholders as regards the receipt of dividends if the company has to liquidate its assets. We can examine the extent to which a company uses debt financing (or financial leverage) in the operation of its business if we

- Check the balance sheet to determine the extent to which borrowed funds have been used to finance assets, and
- Review the income statement to see the extent to which fixed charges (interests) are covered by operating profits.

Two essential indicators of a business's ability to pay its long-term liabilities are the *debt ratio* and the *times-interest-earned ratio*.

Debt Ratio

Debt ratio: A ratio that indicates what proportion of debt a *company* has relative to its assets.

The relationship between total liabilities and total assets, generally called the **debt ratio**, tells us the proportion of the company's assets that it has financed with debt:

$$\text{Debt ratio} = \frac{\text{Total debt}}{\text{Total assets}}$$

$$= \frac{\$16,730}{\$23,215} = 72.07\%.$$

Total debt includes both current liabilities and long-term debt. If the debt ratio is unity, then the company has used debt to finance all of its assets. As of January 28, 2005, Dell's debt ratio was 72.07%; this means that its creditors have supplied close to 72% of the firm's total financing. Certainly, most creditors prefer low debt ratios, because the lower the ratio, the greater is the cushion against creditors' losses in case of liquidation. If a company seeking financing already has large liabilities, then additional debt payments may be too much for the business to handle. For such a highly leveraged company, creditors generally charge higher interest rates on new borrowing to help protect themselves.

Times-Interest-Earned Ratio

The most common measure of the ability of a company's operations to provide protection to the long-term creditor is the times-interest-earned ratio. We find this ratio by dividing earnings before interest and income taxes (EBIT) by the yearly interest charges that must be met. Dell issued $500 million worth of senior notes and long-term bonds with a combined interest rate of 2.259%. This results in $11.29 million in interest expenses[5] in 2005:

$$\text{Times-interest-earned ratio} = \frac{\text{EBIT}}{\text{Interest expense}}$$

$$= \frac{\$4,445 + \$11.29}{\$11.29} = 394.72 \text{ times.}$$

The times-interest-earned ratio measures the extent to which operating income can decline before the firm is unable to meet its annual interest costs. Failure to meet this obligation can bring legal action by the firm's creditors, possibly resulting in the company's bankruptcy. Note that we use the earnings before interest and income taxes, rather than net income, in the numerator. Because Dell must pay interest with pretax dollars, Dell's ability to pay current interest is not affected by income taxes. Only those earnings remaining after all interest charges are subject to income taxes. For Dell, the times-interest-earned ratio for 2005 would be 395 times. This ratio is exceptionally high compared with the rest of the industry's 65.5 times during the same operating period.

2.3.2 Liquidity Analysis

If you were one of the many suppliers to Dell, your primary concern would be whether Dell will be able to pay off its debts as they come due over the next year or so. Short-term creditors want to be repaid on time. Therefore, they focus on Dell's cash flows and on its working capital, as these are the company's primary sources of cash in the near future. The excess of current assets over current liabilities is known as **working capital**, a figure that indicates the extent to which current assets can be converted to cash to meet current obligations. Therefore, we view a firm's net working capital as a measure of its *liquidity* position. In general, the larger the working capital, the better able the business is to pay its debt.

Current Ratio

We calculate the **current ratio** by dividing current assets by current liabilities:

$$\text{Current ratio} = \frac{\text{Current assets}}{\text{Current liabilities}}$$

$$= \frac{\$16,897}{\$14,136} = 1.1953 \text{ times.}$$

The **current ratio** measures a company's ability to pay its short-term obligations.

If a company is getting into financial difficulty, it begins paying its bills (accounts payable) more slowly, borrowing from its bank, and so on. If current liabilities are rising faster than current assets, the current ratio will fall, and that could spell trouble. What is an acceptable current ratio? The answer depends on the nature of the industry. The general rule of thumb calls for a current ratio of 2 to 1. This rule, of course, is subject to many exceptions, depending heavily on the composition of the assets involved.

---

[5] Unless the interest expenses are itemized in the income statement, you will find them in the firm's annual report.

Quick (Acid-Test) Ratio

The quick ratio tells us whether a company could pay all of its current liabilities if they came due immediately. We calculate the quick ratio by deducting inventories from current assets and then dividing the remainder by current liabilities:

$$\text{Quick ratio} = \frac{\text{Current assets} - \text{Inventories}}{\text{Current liabilities}}$$

$$= \frac{\$16,897 - \$459}{\$14,136} = 1.1628 \text{ times}.$$

The quick ratio measures how well a company can meet its obligations without having to liquidate or depend too heavily on its inventory. Inventories are typically the least liquid of a firm's current assets; hence, they are the assets on which losses are most likely to occur in case of liquidation. Although Dell's current ratio may appear to be below the average for its industry, 1.4, its liquidity position is relatively strong, as it has carried very little inventory in its current assets (only $459 million out of $16,897 million of current assets, or 2.7%). We often compare against industry average figures and should note at this point that an industry average is not an absolute number that all firms should strive to maintain. In fact, some very well managed firms will be above the average, while other good firms will be below it. However, if we find that a firm's ratios are quite different from the average for its industry, we should examine the reason for the difference.

2.3.3 Asset Management Analysis

The ability to sell inventory and collect accounts receivables is fundamental to business success. Therefore, the third group of ratios measures how effectively the firm is managing its assets. We will review three ratios related to a firm's asset management: (1) the inventory turnover ratio, (2) the day's sales outstanding ratio, and (3) the total asset turnover ratio. The purpose of these ratios is to answer this question: Does the total amount of each type of asset, as reported on the balance sheet, seem reasonable in view of current and projected sales levels? The acquisition of any asset requires the use of funds. On the one hand, if a firm has too many assets, its cost of capital will be too high; hence, its profits will be depressed. On the other hand, if assets are too low, the firm is likely lose profitable sales.

Inventory Turnover

Inventory turnover: A ratio that shows how many times the inventory of a firm is sold and replaced over a specific period.

The inventory turnover ratio measures how many times the company sold and replaced its inventory over a specific period—for example, during the year. We compute the ratio by dividing sales by the average level of inventories on hand. We compute the average inventory figure by taking the average of the beginning and ending inventory figures. Since Dell has a beginning inventory figure of $327 million and an ending inventory figure of $459 million, its average inventory for the year would be $393 million, or ($327 + $459)/2. Then we compute Dell's inventory turnover for 2005 as follows:

$$\text{Inventory turnover ratio} = \frac{\text{Sales}}{\text{Average inventory balance}}$$

$$= \frac{\$49,205}{\$393} = 125.20 \text{ times}.$$

As a rough approximation, Dell was able to sell and restock its inventory 125.20 times per year. Dell's turnover of 125.20 times is much faster than its competitor HPQ (Hewlett Packard), 9.5 times. This suggests that HPQ is holding excessive stocks of inventory; excess stocks are, of course, unproductive, and they represent an investment with a low or zero rate of return.

Day's Sales Outstanding (Accounts Receivable Turnover)

The day's sales outstanding (DSO) is a rough measure of how many times a company's accounts receivable have been turned into cash during the year. We determine this ratio, also called the **average collection period**, by dividing accounts receivable by average sales per day. In other words, the DSO indicates the average length of time the firm must wait after making a sale before receiving cash. For Dell,

Average collection period is often used to help determine if a company is trying to disguise weak sales.

$$DSO = \frac{\text{Receivables}}{\text{Average sales per day}} = \frac{\text{Receivables}}{\text{Annual sales}/365}$$

$$= \frac{\$4,414}{\$49,205/365} = \frac{\$4,414}{\$134.81}$$

$$= 32.74 \text{ days}.$$

Thus, on average, it takes Dell 32.74 days to collect on a credit sale. During the same period, HPQ's average collection period was 43–52 days. Whether the average of 32.74 days taken to collect an account is good or bad depends on the credit terms Dell is offering its customers. If the credit terms are 30 days, we can say that Dell's customers, on the average, are not paying their bills on time. In order to improve their working-capital position, most customers tend to withhold payment for as long as the credit terms will allow and may even go over a few days. The long collection period may signal either that customers are in financial trouble or that the company manages its credit poorly.

Total Assets Turnover

The total assets turnover ratio measures how effectively the firm uses its total assets in generating its revenues. It is the ratio of sales to all the firm's assets:

Asset turnover is a measure of how well assets are being used to produce revenue.

$$\text{Total assets turnover ratio} = \frac{\text{Sales}}{\text{Total assets}}$$

$$= \frac{\$49,205}{\$23,215} = 2.12 \text{ times}.$$

Dell's ratio of 2.12 times, compared with HPQ's 1.1, is almost 93% faster, indicating that Dell is using its total assets about 93% more intensively than HPQ is. In fact, Dell's total investment in plant and equipment is about one-fourth of HPQ's. If we view Dell's ratio as the industry average, we can say that HPQ has too much investment in inventory, plant, and equipment compared to the size of sale.

2.3.4 Profitability Analysis

One of the most important goals for any business is to earn a profit. The ratios examined thus far provide useful clues about the effectiveness of a firm's operations, but the profitability ratios show the combined effects of liquidity, asset management, and debt on operating results. Therefore, ratios that measure profitability play a large role in decision making.

Profit Margin on Sales

The **profit margin** measures how much out of every dollar of sales a company actually keeps in *earnings*.

We calculate the profit margin on sales by dividing net income by sales. This ratio indicates the profit per dollar of sales:

$$\text{Profit margin on sales} = \frac{\text{Net income available to common stockholders}}{\text{Sales}}$$

$$= \frac{\$3,043}{\$49,205} = 6.18\%.$$

Thus, Dell's profit margin is equivalent to 6.18 cents for each sales dollar generated. Dell's profit margin is greater than HPQ's profit margin of 3.6%, indicating that, although HPQ's sales are about 76% more than Dell's revenue during the same operating period, HPQ's operation is less efficient than Dell's. HPQ's low profit margin is also a result of its heavy use of debt and its carrying a very high volume of inventory. Recall that net income is income after taxes. Therefore, if two firms have identical operations in the sense that their sales, operating costs, and earnings before income tax are the same, but if one company uses more debt than the other, it will have higher interest charges. Those interest charges will pull net income down, and since sales are constant, the result will be a relatively low profit margin.

Return on Total Assets

The return on total assets—or simply, return on assets (ROA)—measures a company's success in using its assets to earn a profit. The ratio of net income to total assets measures the return on total assets after interest and taxes:

$$\text{Return on total assets} = \frac{\text{Net income} + \text{interest expense}(1 - \text{tax rate})}{\text{Average total assets}}$$

$$= \frac{\$3,043 + \$11.29(1 - 0.315)}{(\$23,215 + \$19,311)/2} = 14.35\%.$$

Adding interest expenses back to net income results in an adjusted earnings figure that shows what earnings would have been if the assets had been acquired solely by selling shares of stock. (Note that Dell's effective tax rate was 31.5% in 2005.) With this adjustment, we may be able to compare the return on total assets for companies with differing amounts of debt. Again, Dell's 14.35% return on assets is well above the 4.1% for HPQ. This high return results from (1) the company's high basic earning power and (2) its low use of debt, both of which cause its net income to be relatively high.

The **return on equity** reveals how much profit a company generates with the money its shareholders have invested in it.

Return on Common Equity

Another popular measure of profitability is rate of return on common equity. This ratio shows the relationship between net income and common stockholders' investment in the company—that is, how much income is earned for every $1 invested by the common stockholders. To compute the return on common equity, we first subtract preferred dividends from net income, yielding the net income available to common stockholders. We then divide this net income available to common stockholders by the average common stockholders' equity during the year. We compute average common equity by using the beginning and ending balances. At the beginning of fiscal-year 2005, Dell's common equity balance was $6,280 million; at the end of fiscal-year 2005, the balance was $6,485 million. The average balance is then simply $6,382.50 million, and we have

$$\text{Return on common equity} = \frac{\text{Net income available to common stockholders}}{\text{Average common equity}}$$

$$= \frac{\$3,043}{(\$6,485 + \$6,280)/2}$$

$$= \frac{\$3,043}{\$6,382.50} = 47.68\%.$$

The rate of return on common equity for Dell was 47.68% during 2005. Over the same period, HPQ's return on common equity amounted to 8.2%, a poor performance relative to the computer industry (12.6% in 2005) in general.

To learn more about what management can do to increase the return on common equity, or ROE, we may rewrite the ROE in terms of the following three components:

$$\text{ROE} = \frac{\text{Net income}}{\text{Stockholders' equity}}$$

$$= \frac{\text{Net income}}{\text{Sales}} \times \frac{\text{Sales}}{\text{Assets}} \times \frac{\text{Assets}}{\text{Stockholders' equity}}.$$

The three principal components can be described as the profit margin, asset turnover, and financial leverage, respectively, so that

$$\text{ROE} = (\text{Profit margin}) \times (\text{Asset turnover}) \times (\text{Financial leverage})$$

$$= (6.18\%) \times (2.12) \times \left(\frac{23,215}{6,382.5}\right)$$

$$= 47.68\%.$$

Financial leverage: The degree to which an *investor* or business is utilizing borrowed money.

This expression tells us that management has only three key ratios for controlling a company's ROE: (1) the earnings from sales (the profit margin); (2) the revenue generated from each dollar of assets employed (asset turnover); and (3) the amount of equity used to finance the assets in the operation of the business (financial leverage).

2.3.5 Market Value Analysis

When you purchase a company's stock, what are your primary factors in valuing the stock? In general, investors purchase stock to earn a return on their investment. This return consists of two parts: (1) gains (or losses) from selling the stock at a price that differs from the investors' purchase price and (2) dividends—the periodic distributions of profits to stockholders. The market value ratios, such as the price-to-earnings ratio and the market-to-book ratio, relate the firm's stock price to its earnings and book value per share, respectively. These ratios give management an indication of what investors think of the company's past performance and future prospects. If the firm's asset and debt management is sound and its profit is rising, then its market value ratios will be high, and its stock price will probably be as high as can be expected.

Price-to-Earnings Ratio

The price-to-earnings (P/E) ratio shows how much investors are willing to pay per dollar of reported profits. Dell's stock sold for $41.50 in early February of 2005, so with an EPS of $1.21, its P/E ratio was 34.29:

The ***P/E*** **ratio shows how much investors are willing to pay per dollar of earnings.**

$$P/E \text{ ratio} = \frac{\text{Price per share}}{\text{Earnings per share}}$$

$$= \frac{\$41.5}{\$1.21} = 34.29.$$

That is, the stock was selling for about 34.29 times its current earnings per share. In general, P/E ratios are higher for firms with high growth prospects, other things held constant, but they are lower for firms with lower expected earnings. Dell's expected annual increase in operating earnings is 30% over the next 3 to 5 years. Since Dell's ratio is greater than 25%, the average for other computer industry firms, this suggests that investors value Dell's stock more highly than most as having excellent growth prospects. However, all stocks with high P/E ratios carry high risk whenever the expected growths fail to materialize. Any slight earnings disappointment tends to punish the market price significantly.

Book Value per Share

Another ratio frequently used in assessing the well-being of the common stockholders is the book value per share, which measures the amount that would be distributed to holders of each share of common stock if all assets were sold at their balance-sheet carrying amounts and if all creditors were paid off. We compute the book value per share for Dell's common stock as follows:

$$\text{Book value per share} = \frac{\text{Total stockholders' equity} - \text{preferred stock}}{\text{Shares outstanding}}$$

$$= \frac{\$6,485 - \$0}{2,509} = \$2.58.$$

If we compare this book value with the current market price of $41.50, then we may say that the stock appears to be overpriced. Once again, though, market prices reflect expectations about future earnings and dividends, whereas book value largely reflects the results of events that occurred in the past. Therefore, the market value of a stock tends to exceed its book value. Table 2.5 summarizes the financial ratios for Dell Computer Corporation in comparison to its direct competitor Hewlett Packard (HPQ) and the industry average.

2.3.6 Limitations of Financial Ratios in Business Decisions

Business decisions are made in a world of uncertainty. As useful as ratios are, they have limitations. We can draw an analogy between their use in decision making and a physician's use of a thermometer. A reading of 102°F indicates that something is wrong with the

TABLE 2.5 Comparisons of Dell Computer Corporation's Key Financial Ratios with Those of Hewlett Packard (HPQ) and the Industry Average (2005)

Category	Financial Ratios	Dell	HPQ	Industry
Debt	Debt ratio	72.07%	9%	29%
Management	Time-interest earned	394.72	18.37	65.5
Liquidity	Current ratio	1.1953	1.4	1.4
	Quick ratio	1.16	0.90	1.0
	Inventory turnover	125.20	9.5	11.1
Asset	Day's sales			
Management	outstanding	32.74	52.43	34
	Total asset turnover	2.12	1.1	1.0
	Profit margin	6.18%	3.6%	5.0%
Profitability	Return on total asset	14.35%	4.1%	4.9%
	Return on common equity	47.68%	8.2%	12.6%
	P/E ratio	34.29	27.5	25.3
Market Trend	Book value-to-share			
	ratio	2.58	13.05	9.26

patient, but the temperature alone does not indicate what the problem is or how to cure it. In other words, ratio analysis is useful, but analysts should be aware of ever-changing market conditions and make adjustments as necessary. It is also difficult to generalize about whether a particular ratio is "good" or "bad." For example, a high current ratio may indicate a strong liquidity position, which is good, but holding too much cash in a bank account (which will increase the current ratio) may not be the best utilization of funds. Ratio analysis based on any one year may not represent the true business condition. It is important to analyze trends in various financial ratios, as well as their absolute levels, for trends give clues as to whether the financial situation is likely to improve or deteriorate. To do a **trend analysis**, one simply plots a ratio over time. As a typical engineering student, your judgment in interpreting a set of financial ratios is understandably weak at this point, but it will improve as you encounter many facets of business decisions in the real world. Again, accounting is a language of business, and as you speak it more often, it can provide useful insights into a firm's operations.

Trend analysis is based on the idea that what has happened in the past gives traders an idea of what will happen in the future.

SUMMARY

The primary purposes of this chapter were (1) to describe the basic financial statements, (2) to present some background information on cash flows and corporate profitability, and (3) to discuss techniques used by investors and managers to analyze financial statements. Following are some concepts we covered:

■ Before making any major financial decisions, it is important to understand their impact on your net worth. Your net-worth statement is a snapshot of where you stand financially at a given point in time.

■ The three basic financial statements contained in the annual report are the balance sheet, the income statement, and the statement of cash flows. Investors use the information provided in these statements to form expectations about future levels of earnings and dividends and about the firm's risk-taking behavior.

■ A firm's balance sheet shows a snapshot of a firm's financial position at a particular point in time through three categories: (1) assets the firm owns, (2) liabilities the firm owes, and (3) owners' equity, or assets less liabilities.

■ A firm's income statement reports the results of operations over a period of time and shows earnings per share as its "bottom line." The main items are (1) revenues and gains, (2) expenses and losses, and (3) net income or net loss (revenue less expenses).

■ A firm's statement of cash flows reports the impact of operating, investing, and financing activities on cash flows over an accounting period.

■ The purpose of calculating a set of financial ratios is twofold: (1) to examine the relative strengths and weaknesses of a company compared with those of other companies in the same industry and (2) to learn whether the company's position has been improving or deteriorating over time.

■ Liquidity ratios show the relationship of a firm's current assets to its current liabilities and thus its ability to meet maturing debts. Two commonly used liquidity ratios are the current ratio and the quick (acid-test) ratio.

■ Asset management ratios measure how effectively a firm is managing its assets. Some of the major ratios are inventory turnover, fixed assets turnover, and total assets turnover.

■ Debt management ratios reveal (1) the extent to which a firm is financed with debt and (2) the firm's likelihood of defaulting on its debt obligations. In this category are the debt ratio and the times-interest-earned ratio.

■ Profitability ratios show the combined effects of liquidity, asset management, and debt management policies on operating results. Profitability ratios include the profit margin on sales, the return on total assets, and the return on common equity.

■ Market value ratios relate the firm's stock price to its earnings and book value per share, and they give management an indication of what investors think of the company's past performance and future prospects. Market value ratios include the price-to-earnings ratio and the book value per share.

■ Trend analysis, in which one plots a ratio over time, is important, because it reveals whether the firm's ratios are improving or deteriorating over time.

PROBLEMS

Financial Statements

2.1 Consider the balance-sheet entries for War Eagle Corporation in Table P2.1.
    (a) Compute the firm's
        Current assets: $_____
        Current liabilities: $_____
        Working capital: $_____
        Shareholders' equity: $_____

TABLE P2.1

Balance Sheet Statement as of December 31, 2000		
Assets:		
Cash		$ 150,000
Marketable securities		200,000
Accounts receivables		150,000
Inventories		50,000
Prepaid taxes and insurance		30,000
Manufacturing plant at cost	$ 600,000	
Less accumulated depreciation	100,000	
Net fixed assets		500,000
Goodwill		20,000
Liabilities and shareholders' equity:		
Notes payable		50,000
Accounts payable		100,000
Income taxes payable		80,000
Long-term mortgage bonds		400,000
Preferred stock, 6%, $100 par value (1,000 shares)		100,000
Common stock, $15 par value (10,000 shares)		150,000
Capital surplus		150,000
Retained earnings		70,000

(b) If the firm had a net income of $500,000 after taxes, what is the earnings per share?

(c) When the firm issued its common stock, what was the market price of the stock per share?

2.2 A chemical processing firm is planning on adding a duplicate polyethylene plant at another location. The financial information for the first project year is shown in Table P2.2.

(a) Compute the working-capital requirement during the project period.

(b) What is the taxable income during the project period?

(c) What is the net income during the project period?

(d) Compute the net cash flow from the project during the first year.

TABLE P2.2 Financial Information for First Project Year

Sales		$1,500,000
Manufacturing costs		
Direct materials	$ 150,000	
Direct labor	200,000	
Overhead	100,000	
Depreciation	200,000	
Operating expenses		150,000
Equipment purchase		400,000
Borrowing to finance equipment		200,000
Increase in inventories		100,000
Decrease in accounts receivable		20,000
Increase in wages payable		30,000
Decrease in notes payable		40,000
Income taxes		272,000
Interest payment on financing		20,000

Financial Ratio Analysis

2.3   Table P2.3 shows financial statements for Nano Networks, Inc. The closing stock price for Nano Network was $56.67 (split adjusted) on December 31, 2005. On the basis of the financial data presented, compute the various financial ratios and make an informed analysis of Nano's financial health.

TABLE P2.3 Balance Sheet for Nano Networks, Inc.

	Dec. 2005 U.S. $ (000) (Year)	Dec. 2004 U.S. $ (000) (Year)
Balance Sheet Summary		
Cash	158,043	20,098
Securities	285,116	0
Receivables	24,582	8,056
Allowances	632	0
Inventory	0	0
Current assets	377,833	28,834
Property and equipment, net	20,588	10,569

(*Continued*)

	Dec. 2005 U.S. $ (000) (Year)	Dec. 2004 U.S. $ (000) (Year)
Depreciation	8,172	2,867
Total assets	513,378	36,671
Current liabilities	55,663	14,402
Bonds	0	0
Preferred mandatory	0	0
Preferred stock	0	0
Common stock	2	1
Other stockholders' equity	457,713	17,064
Total liabilities and equity	513,378	36,671
Income Statement Summary		
Total revenues	102,606	3,807
Cost of sales	45,272	4,416
Other expenses	71,954	31,661
Loss provision	0	0
Interest income	8,011	1,301
Income pretax	−6,609	−69
Income tax	2,425	2
Income continuing	−9,034	−30,971
Net income	**− 9,034**	**− 30,971**
EPS primary	**− $0.1**	**− $0.80**
EPS diluted	−$0.10	−$0.80
	−$0.05	−$0.40
	(split adjusted)	(split adjusted)

(a) Debt ratio
(b) Times-interest-earned ratio
(c) Current ratio
(d) Quick (acid-test) ratio
(e) Inventory turnover ratio
(f) Day's sales outstanding
(g) Total assets turnover
(h) Profit margin on sales
(i) Return on total assets

(j) Return on common equity

(k) Price-to-earnings ratio

(l) Book value per share

2.4 The balance sheet that follows summarizes the financial conditions for Flex, Inc., an electronic outsourcing contractor, for fiscal-year 2005. Unlike Nano Network Corporation in Problem 2.3, Flex has reported a profit for several years running. Compute the various financial ratios and interpret the firm's financial health during fiscal-year 2005.

	Aug. 2005 U.S. $ (000) (12 mos.)	Aug. 2004 U.S. $ (000) (Year)
Balance Sheet Summary		
Cash	1,325,637	225,228
Securities	362,769	83,576
Receivables	1,123,901	674,193
Allowances	5,580	−3,999
Inventory	1,080,083	788,519
Current assets	3,994,084	1,887,558
Property and equipment, net	1,186,885	859,831
Depreciation	533,311	−411,792
Total assets	4,834,696	2,410,568
Current liabilities	1,113,186	840,834
Bonds	922,653	385,519
Preferred mandatory	0	0
Preferred stock	0	0
Common stock	271	117
Other stockholders' equity	2,792,820	1,181,209
Total liabilities and equity	4,834,696	2,410,568
Income Statement Summary		
Total revenues	8,391,409	5,288,294
Cost of sales	7,614,589	4,749,988
Other expenses	335,808	237,063
Loss provision	2,143	2,254
Interest expense	36,479	24,759

(*Continued*)

	Aug. 2005 U.S. $ (000) (12 mos.)	Aug. 2004 U.S. $ (000) (Year)
Income pretax	432,342	298,983
Income tax	138,407	100,159
Income continuing	293,935	198,159
Discontinued	0	0
Extraordinary	0	0
Changes	0	0
Net income	**293,935**	**198,159**
EPS primary	**$1.19**	**$1.72**
EPS diluted	$1.13	$1.65

(a) Debt ratio

(b) Times-interest-earned ratio

(c) Current ratio

(d) Quick (acid-test) ratio

(e) Inventory turnover ratio

(f) Day's sales outstanding

(g) Total assets turnover

(h) Profit margin on sales

(i) Return on total assets

(j) Return on common equity

(k) Price-to-earnings ratio

(l) Book value per share

2.5 J. C. Olson & Co. had earnings per share of $8 in year 2006, and it paid a $4 dividend. Book value per share at year's end was $80. During the same period, the total retained earnings increased by $24 million. Olson has no preferred stock, and no new common stock was issued during the year. If Olson's year-end debt (which equals its total liabilities) was $240 million, what was the company's year-end debt-to-asset ratio?

2.6 If Company A uses more debt than Company B and both companies have identical operations in terms of sales, operating costs, etc., which of the following statements is *true*?

(a) Company B will definitely have a higher current ratio.

(b) Company B has a higher profit margin on sales than Company A.

(c) Both companies have identical profit margins on sales.

(d) Company B's return on total assets would be higher.

2.7 You are looking to buy stock in a high-growth company. Which of the following ratios best indicates the company's growth potential?

(a) Debt ratio

(b) Price-to-earnings ratio

(c) Profit margin

(d) Total asset turnover

2.8 Which of the following statements is *incorrect*?

(a) The quickest way to determine whether the firm has too much debt is to calculate the debt-to-equity ratio.

(b) The best rule of thumb for determining the firm's liquidity is to calculate the current ratio.

(c) From an investor's point of view, the rate of return on common equity is a good indicator of whether or a firm is generating an acceptable return to the investor.

(d) The operating margin is determined by expressing net income as a percentage of total sales.

2.9 Consider the following financial data for Northgate Corporation:

- Cash and marketable securities, $100
- Total fixed assets, $280
- Annual sales, $1,200
- Net income, $358
- Inventory, $180
- Current liabilities, $134
- Current ratio, 3.2
- Average correction period, 45 days
- Average common equity, $500

On the basis of these financial data, determine the firm's *return on (common) equity*.

(a) 141.60%

(b) 71.6%

(c) 76.0%

(d) 30%

Short Case Studies

ST2.1 Consider the two companies Cisco Systems and Lucent Technologies, which compete with each other in the network equipment sector. Lucent enjoys strong relationships among Baby Bells in the telephone equipment area and Cisco has a dominant role in the network router and switching equipment area. Get these companies' annual reports from their websites, and answer the following questions (*Note*: To download their annual reports, visit http://www.cisco.com and http://www.lucent.com) and look for "Investors' Relations":

(a) On the basis of the most recent financial statements, comment on each company's financial performance in the following areas:

- Asset management
- Liquidity

- Debt management
- Profitability
- Market trend

(b) Check the current stock prices for both companies. The stock ticker symbol is CSCO for Cisco and LU for Lucent. Based on your analysis in Problem 2.6(a), which company would you bet your money on and why? (Lucent and Alcatel were engaged in discussions about a potential merger of equals in late March of 2006.)

ST2.2 Compare XM Satellite Radio, Inc., and Sirius Satellite Radio, Inc., using a thorough financial ratios analysis.

(a) For each company, compute all the ratios listed in Figure 2.5 (i.e., debt management, liquidity, asset management, market trend, and profitability) for the year 2005.

(b) Compare and contrast the two companies, using the ratios you calculated from part (a).

(c) Carefully read and summarize the "risk management" or "hedging" practices described in the financial statements of each company.

(d) If you were a mutual-fund manager and could invest in only one of these companies, which one would you select and why? Be sure to justify your answer by using your results from parts (a), (b), and (c).

THREE

Interest Rate and Economic Equivalence

No Lump Sum for Lottery-Winner Grandma, 94[1] A judge denied a 94-year-old woman's attempt to force the Massachusetts Lottery Commission to pay her entire $5.6 million winnings up front on the grounds that she otherwise won't live long enough to collect it all. The ruling means that the commission can pay Louise Outing, a retired waitress, in installments over 20 years. After an initial gross payment of $283,770, Outing would be paid 19 annual gross checks of $280,000. That's about $197,000 after taxes. Lottery Executive Director Joseph Sullivan said all players are held to the same rules, which are printed on the back of Megabucks tickets. Lottery winners are allowed to "assign" their winnings to a state-approved financial company that makes the full payment—but only in return for a percentage of the total winnings. Outing, who won a Megabucks drawing in September, has seven grandchildren, nine great-grandchildren, and six great-great-grandchildren. "I'd like to get it and do what I want with it," she said. "I'm not going to live 20 years. I'll be 95 in March."

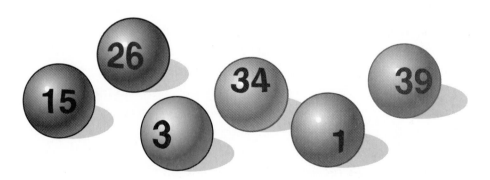

[1] "No Lump Sum for Lottery-Winner Grandma, 94," The Associated Press, December 30, 2004.

The next time you play a lottery, look at the top section of the play slip. You will see two boxes: "Cash Value" and "Annual Payments." You need to mark one of the boxes before you fill out the rest of the slip. If you don't, and you win the jackpot, you will automatically receive the jackpot as annual payments. That is what happened to Ms. Louise Outing. If you mark the "Cash Value box" and you win, you will receive the present cash value of the announced jackpot in one lump sum. This amount will be less than the announced jackpot. With the announced jackpot of $5.6 million, Ms. Outing could receive about 52.008%, or $2.912 million, in one lump sum (less withholding tax). This example is based on average market costs as of January 2005 of 20 annual payments funded by the U.S. Treasury Zero Coupon Bonds (or a 7.2% coupon rate). With this option, you can look forward to a large cash payment up front.

First, most people familiar with investments would tell Ms. Outing that receiving a lump amount of $2.912 million today is likely to prove a far better deal than receiving $280,000 a year for 20 years, even if the grandma lives long enough to collect the entire annual payments. After losing the court appeal, Ms. Outing was able to find a buyer for her lottery in a lump-sum amount of $2.467 million. To arrive at that price, the buyer calculated the return he wanted to earn—at that time about 9.5% interest, compounded annually—and applied that rate in reverse to the $5.6 million he stood to collect over 20 years. The buyer says the deals he strikes with winners applies a basic tenet of all financial transactions, the **time value of money**: A dollar in hand today is worth more than one that will be paid to you in the future.

In engineering economics, the principles discussed in this chapter are regarded as the underpinning for nearly all project investment analysis. This is because we always need to account for the effect of interest operating on sums of cash over time. Interest formulas allow us to place different cash flows received at different times in the same time frame and to compare them. As will become apparent, almost our entire study of engineering economics is built on the principles introduced in this chapter.

CHAPTER LEARNING OBJECTIVES

After completing this chapter, you should understand the following concepts:

■ The time value of money.

■ The difference between simple interest and the compound interest.

■ The meaning of economic equivalence and why we need it in economic analysis.

■ How to compare two different money series by means of the concept of economic equivalence.

■ The interest operation and the types of interest formulas used to facilitate the calculation of economic equivalence.

3.1 Interest: The Cost of Money

Most of us are familiar in a general way with the concept of interest. We know that money left in a savings account earns interest, so that the balance over time is greater than the sum of the deposits. We also know that borrowing to buy a car means repaying an amount over time, that that amount includes interest, and that it is therefore greater than the amount borrowed. What may be unfamiliar to us is the idea that, in the financial world, money itself is a commodity and, like other goods that are bought and sold, money costs money.

Market interest rate: Interest rate quoted by financial institutions.

The cost of money is established and measured by a **market interest rate**, a percentage that is periodically applied and added to an amount (or varying amounts) of money over a specified length of time. When money is borrowed, the interest paid is the charge to the borrower for the use of the lender's property; when money is lent or invested, the interest earned is the lender's gain from providing a good to another (Figure 3.1). **Interest**, then, may be defined as the cost of having money available for use. In this section, we examine how interest operates in a free-market economy and we establish a basis for understanding the more complex interest relationships that follow later on in the chapter.

Charge or Cost to Borrower

Interest Rate
8%

Profit or Earning to Lender

Figure 3.1   The meaning of *interest rate* to the lender (bank) and to the borrower.

	Account Value	Cost of Refrigerator
Case 1: Inflation exceeds earning power	$N = 0$ $100 $N = 1$ $106 (earning rate $= 6\%$)	$N = 0$ $100 $N = 1$ $108 (inflation rate $= 8\%$)
Case 2: Earning power exceeds inflation	$N = 0$ $100 $N = 1$ $106 (earning rate $= 6\%$)	$N = 0$ $100 $N = 1$ $104 (inflation rate $= 4\%$)

Figure 3.2 Gains achieved or losses incurred by delaying consumption.

3.1.1 The Time Value of Money

The "time value of money" seems like a sophisticated concept, yet it is a concept that you grapple with every day. Should you buy something today or save your money and buy it later? Here is a simple example of how your buying behavior can have varying results: Pretend you have $100, and you want to buy a $100 refrigerator for your dorm room. If you buy it now, you are broke. Suppose that you can invest money at 6% interest, but the price of the refrigerator increases only at an annual rate of 4% due to inflation. In a year you can still buy the refrigerator, and you will have $2 left over. Well, if the price of the refrigerator increases at an annual rate of 8% instead, you will not have enough money (you will be $2 short) to buy the refrigerator a year from now. In that case, you probably are better off buying the refrigerator now. The situation is summarized in Figure 3.2.

The time value of money: The idea that a dollar today is worth more than a dollar in the future because the dollar received today can earn interest.

Clearly, the rate at which you earn interest should be higher than the inflation rate to make any economic sense of the delayed purchase. In other words, in an inflationary economy, your purchasing power will continue to decrease as you further delay the purchase of the refrigerator. In order to make up this future loss in purchasing power, your earning interest rate should be sufficiently larger than the anticipated inflation rate. After all, time, like money, is a finite resource. There are only 24 hours in a day, so time has to be budgeted, too. What this example illustrates is that we must connect the "earning power" and the "purchasing power" to the concept of time.

When we deal with large amounts of money, long periods of time, or high interest rates, the change in the value of a sum of money over time becomes extremely significant. For example, at a current annual interest rate of 10%, $1 million will earn $100,000 in interest in a year; thus, to wait a year to receive $1 million clearly involves a significant sacrifice. When deciding among alternative proposals, we must take into account the operation of interest and the time value of money in order to make valid comparisons of different amounts at various times.

Purchasing power: The value of a currency expressed in terms of the amount of goods or services that one unit of money can buy.

The way interest operates reflects the fact that money has a time value. This is why amounts of interest depend on lengths of time; interest rates, for example, are typically given in terms of a percentage per year. We may define the principle of the time value of money as follows: The economic value of a sum depends on when it is received. Because money has both **earning** as well as **purchasing power** over time, as shown in Figure 3.3 (it can be put to work, earning more money for its owner), a dollar received today has a greater value than a dollar received at some future time.

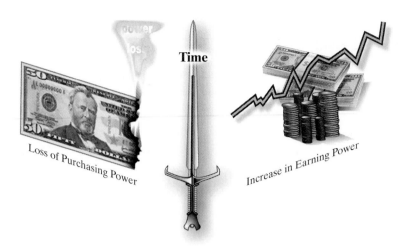

Figure 3.3 The time value of money. This is a two-edged sword whereby earning grows, but purchasing power decreases, as time goes by.

When lending or borrowing interest rates are quoted by financial institutions on the marketplace, those interest rates reflect the desired amounts to be earned, as well as any protection from loss in the future purchasing power of money because of inflation. (If we want to know the true desired earnings in isolation from inflation, we can determine the real interest rate. We consider this issue in Chapter 11. The earning power of money and its loss of value because of inflation are calculated by different analytical techniques.) In the meantime, we will assume that, unless otherwise mentioned, *the interest rate used in this book reflects the market interest rate,* which takes into account the earning power, as well as the effect of inflation perceived in the marketplace. We will also assume that all cash flow transactions are given in terms of **actual dollars**, with the effect of inflation, if any, reflected in the amount.

Actual dollars: The cash flow measured in terms of the dollars at the time of the transaction.

3.1.2 Elements of Transactions Involving Interest

Many types of transactions (e.g., borrowing or investing money or purchasing machinery on credit) involve interest, but certain elements are common to all of these types of transactions:

- An initial amount of money in transactions involving debt or investments is called the **principal**.
- The **interest rate** measures the cost or price of money and is expressed as a percentage per period of time.
- A period of time, called the **interest period**, determines how frequently interest is calculated. (Note that even though the length of time of an interest period can vary, interest rates are frequently quoted in terms of an annual percentage rate. We will discuss this potentially confusing aspect of interest in Chapter 4.)
- A specified length of time marks the duration of the transaction and thereby establishes a certain **number of interest periods**.
- A **plan for receipts or disbursements** yields a particular cash flow pattern over a specified length of time. (For example, we might have a series of equal monthly payments that repay a loan.)
- A **future amount of money** results from the cumulative effects of the interest rate over a number of interest periods.

For the purposes of calculation, these elements are represented by the following variables:

A_n = A discrete payment or receipt occurring at the end of some interest period.
  i = The interest rate per interest period.
  N = The total number of interest periods.
  P = A sum of money at a time chosen as time zero for purposes of analysis; sometimes referred to as the **present value** or **present worth**.
  F = A future sum of money at the end of the analysis period. This sum may be specified as F_N.
  A = An end-of-period payment or receipt in a uniform series that continues for N periods. This is a special situation where $A_1 = A_2 = \cdots = A_N$.
V_n = An equivalent sum of money at the end of a specified period n that considers the effect of the time value of money. Note that $V_0 = P$ and $V_N = F$.

> **Present value:** The amount that a future sum of money is worth today, given a specified rate of return.

Because frequent use of these symbols will be made in this text, it is important that you become familiar with them. Note, for example, the distinction between A, A_n, and A_N. The symbol A_n refers to a specific payment or receipt, at the end of period n, in any series of payments. A_N is the final payment in such a series, because N refers to the total number of interest periods. A refers to any series of cash flows in which all payments or receipts are equal.

Example of an Interest Transaction

As an example of how the elements we have just defined are used in a particular situation, let us suppose that an electronics manufacturing company buys a machine for $25,000 and borrows $20,000 from a bank at a 9% annual interest rate. In addition, the company pays a $200 loan origination fee when the loan commences. The bank offers two repayment plans, one with equal payments made at the end of every year for the next five years, the other with a single payment made after the loan period of five years. These two payment plans are summarized in Table 3.1.

- In Plan 1, the principal amount P is $20,000, and the interest rate i is 9%. The interest period is one year, and the duration of the transaction is five years, which means there are five interest periods ($N = 5$). It bears repeating that whereas one year is a common interest period, interest is frequently calculated at other intervals: monthly, quarterly, or

TABLE 3.1 Repayment Plans for Example Given in Text (for $N = 5$ years and $i = 9\%$)

End of Year	Receipts	Payments Plan 1	Payments Plan 2
Year 0	$20,000.00	$ 200.00	$ 200.00
Year 1		5,141.85	0
Year 2		5,141.85	0
Year 3		5,141.85	0
Year 4		5,141.85	0
Year 5		5,141.85	30,772.48

P = $20,000, A = $5,141.85, F = $30,772.48

Note: You actually borrow $19,800 with the origination fee of $200, but you pay back on the basis of $20,000.

semiannually, for instance. For this reason, we used the term **period** rather than **year** when we defined the preceding list of variables. The receipts and disbursements planned over the duration of this transaction yield a cash flow pattern of five equal payments *A* of $5,141.85 each, paid at year's end during years 1 through 5. (You'll have to accept these amounts on faith for now—the next section presents the formula used to arrive at the amount of these equal payments, given the other elements of the problem.)

- Plan 2 has most of the elements of Plan 1, except that instead of five equal repayments, we have a grace period followed by a single future repayment *F* of $30,772.78.

Cash Flow Diagrams

Problems involving the time value of money can be conveniently represented in graphic form with a cash flow diagram (Figure 3.4). **Cash flow diagrams** represent time by a horizontal line marked off with the number of interest periods specified. The cash flows over time are represented by arrows at relevant periods: Upward arrows denote positive flows (receipts), downward arrows negative flows (disbursements). Note, too, that the arrows actually represent **net cash flows**: Two or more receipts or disbursements made at the same time are summed and shown as a single arrow. For example, $20,000 received during the same period as a $200 payment would be recorded as an upward arrow of $19,800. Also, the lengths of the arrows can suggest the relative values of particular cash flows.

Cash flow diagrams function in a manner similar to free-body diagrams or circuit diagrams, which most engineers frequently use: Cash flow diagrams give a convenient summary of all the important elements of a problem, as well as a reference point to determine whether the statement of the problem has been converted into its appropriate parameters. The text frequently uses this graphic tool, and you are strongly encouraged to develop the habit of using well-labeled cash flow diagrams as a means to identify and summarize pertinent information in a cash flow problem. Similarly, a table such as Table 3.1 can help you organize information in another summary format.

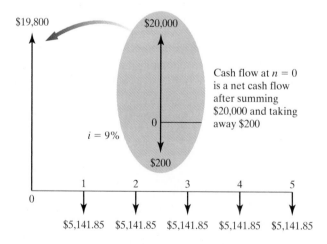

Figure 3.4 A cash flow diagram for Plan 1 of the loan repayment example summarized in Table 3.1.

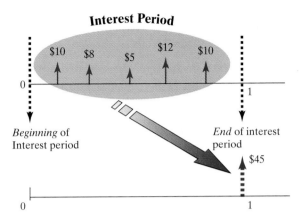

Figure 3.5 Any cash flows occurring during the interest period are summed to a single amount and placed at the end of the interest period.

End-of-Period Convention

In practice, cash flows can occur at the beginning or in the middle of an interest period—or indeed, at practically any point in time. One of the simplifying assumptions we make in engineering economic analysis is the **end-of-period convention**, which is the practice of placing all cash flow transactions at the end of an interest period. (See Figure 3.5.) This assumption relieves us of the responsibility of dealing with the effects of interest within an interest period, which would greatly complicate our calculations.

It is important to be aware of the fact that, like many of the simplifying assumptions and estimates we make in modeling engineering economic problems, the end-of-period convention inevitably leads to some discrepancies between our model and real-world results.

Suppose, for example, that $100,000 is deposited during the first month of the year in an account with an interest period of one year and an interest rate of 10% per year. In such a case, the difference of 1 month would cause an interest income loss of $10,000. This is because, under the end-of-period convention, the $100,000 deposit made during the interest period is viewed as if the deposit were made at the end of the year, as opposed to 11 months earlier. This example gives you a sense of why financial institutions choose interest periods that are less than one year, even though they usually quote their rate as an annual percentage.

Armed with an understanding of the basic elements involved in interest problems, we can now begin to look at the details of calculating interest.

> **End-of-period convention:** Unless otherwise mentioned, all cash flow transactions occur at the end of an interest period.

3.1.3 Methods of Calculating Interest

Money can be lent and repaid in many ways, and, equally, money can earn interest in many different ways. Usually, however, at the end of each interest period, the interest earned on the principal amount is calculated according to a specified interest rate. The two computational schemes for calculating this earned interest are said to yield either **simple interest** or **compound interest**. Engineering economic analysis uses the compound-interest scheme almost exclusively.

> **Simple interest:** The interest rate is applied only to the original principal amount in computing the amount of interest.

Simple Interest

Simple interest is interest earned on only the principal amount during each interest period. In other words, with simple interest, the interest earned during each interest period does not earn additional interest in the remaining periods, *even though you do not withdraw it.*

In general, for a deposit of P dollars at a simple interest rate of i for N periods, the total earned interest would be

$$I = (iP)N. \tag{3.1}$$

The total amount available at the end of N periods thus would be

$$F = P + I = P(1 + iN). \tag{3.2}$$

Simple interest is commonly used with add-on loans or bonds. (See Chapter 4.)

Compound: The ability of an asset to generate *earnings* that are then reinvested and generate their own earnings.

Compound Interest

Under a compound-interest scheme, the interest earned in each period is calculated on the basis of the total amount at the end of the previous period. This total amount includes the original principal plus the accumulated interest that has been left in the account. In this case, you are, in effect, increasing the deposit amount by the amount of interest earned. In general, if you deposited (invested) P dollars at interest rate i, you would have $P + iP = P(1 + i)$ dollars at the end of one period. If the entire amount (principal and interest) is reinvested at the same rate i for another period, at the end of the second period you would have

$$P(1 + i) + i[P(1 + i)] = P(1 + i)(1 + i)$$
$$= P(1 + i)^2.$$

Continuing, we see that the balance after the third period is

$$P(1 + i)^2 + i[P(1 + i)^2] = P(1 + i)^3.$$

This interest-earning process repeats, and after N periods the total accumulated value (balance) F will grow to

$$F = P(1 + i)^N. \tag{3.3}$$

EXAMPLE 3.1 Compound Interest

Suppose you deposit $1,000 in a bank savings account that pays interest at a rate of 10% compounded annually. Assume that you don't withdraw the interest earned at the end of each period (one year), but let it accumulate. How much would you have at the end of year 3?

SOLUTION

Given: $P = \$1,000$, $N = 3$ years, and $i = 10\%$ per year.
Find: F.

Applying Eq. (3.3) to our three-year, 10% case, we obtain

$$F = \$1,000(1 + 0.10)^3 = \$1,331.$$

The total interest earned is $331, which is $31 more than was accumulated under the simple-interest method (Figure 3.6). We can keep track of the interest accruing process more precisely as follows:

Period	Amount at Beginning of Interest Period	Interest Earned for Period	Amount at End of Interest Period
1	$1,000	$1,000(0.10)	$1,100
2	1,100	1,100(0.10)	1,210
3	1,210	1,210(0.10)	1,331

COMMENTS: At the end of the first year, you would have $1,000, plus $100 in interest, or a total of $1,100. In effect, at the beginning of the second year, you would be depositing $1,100, rather than $1,000. Thus, at the end of the second year, the interest earned would be $0.10(\$1,100) = \$110$, and the balance would be $1,100 + \$110 = \$1,210$. This is the amount you would be depositing at the beginning of the third year, and the interest earned for that period would be $0.10(\$1,210) = \$121$. With a beginning principal amount of $1,210 plus the $121 interest, the total balance would be $1,331 at the end of year 3.

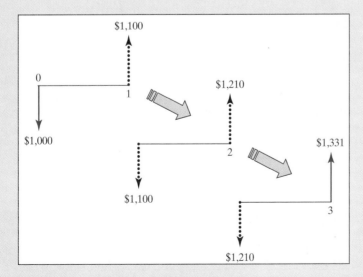

Figure 3.6 The process of computing the balance when $1,000 at 10% is deposited for three years (Example 3.1).

3.1.4 Simple Interest versus Compound Interest

From Eq. (3.3), the total interest earned over N periods is

$$I = F - P = P[(1 + i)^N - 1]. \tag{3.4}$$

Compared with the simple-interest scheme, the additional interest earned with compound interest is

$$\Delta I = P[(1 + i)^N - 1] - (iP)N \tag{3.5}$$

$$= P[(1 + i)^N - (1 + iN)]. \tag{3.6}$$

As either i or N becomes large, the difference in interest earnings also becomes large, so the effect of compounding is further pronounced. Note that, when $N = 1$, compound interest is the same as simple interest.

Using Example 3.1, we can illustrate the difference between compound interest and the simple interest. Under the simple-interest scheme, you earn interest only on the principal amount at the end of each interest period. Under the compounding scheme, you earn interest on the principal, as well as interest on interest.

Figure 3.7 illustrates the fact that compound interest is a sum of simple interests earned on the original principal, as well as periodic simple interests earned on a series of simple interests.

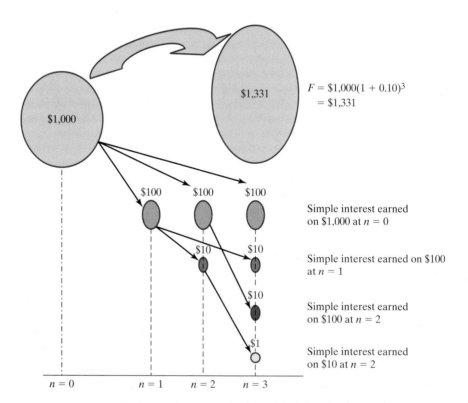

Figure 3.7 The relationship between simple interest and compound interest.

EXAMPLE 3.2 Comparing Simple with Compound Interest

In 1626, Peter Minuit of the Dutch West India Company paid $24 to purchase Manhattan Island in New York from the Indians. In retrospect, if Minuit had invested the $24 in a savings account that earned 8% interest, how much would it be worth in 2007?

SOLUTION

Given: $P = \$24$, $i = 8\%$ per year, and $N = 381$ years.
Find: F, based on (a) 8% simple interest and (b) 8% compound interest.

(a) With 8% simple interest,

$$F = \$24[1 + (0.08)(381)] = \$755.52.$$

(b) With 8% compound interest,

$$F = \$24(1 + 0.08)^{381} = \$130,215,319,909,015.$$

COMMENTS: The significance of compound interest is obvious in this example. Many of us can hardly comprehend the magnitude of $130 trillion. In 2007, the total population in the United States was estimated to be around 300 million. If the money were distributed equally among the population, each individual would receive $434,051. Certainly, there is no way of knowing exactly how much Manhattan Island is worth today, but most real-estate experts would agree that the value of the island is nowhere near $130 trillion. (Note that the U.S. national debt as of December 31, 2007, was estimated to be $9.19 trillion.)

3.2 Economic Equivalence

The observation that money has a time value leads us to an important question: If receiving $100 today is not the same thing as receiving $100 at any future point, how do we measure and compare various cash flows? How do we know, for example, whether we should prefer to have $20,000 today and $50,000 ten years from now, or $8,000 each year for the next ten years? In this section, we describe the basic analytical techniques for making these comparisons. Then, in Section 3.3, we will use these techniques to develop a series of formulas that can greatly simplify our calculations.

3.2.1 Definition and Simple Calculations

The central question in deciding among alternative cash flows involves comparing their economic worth. This would be a simple matter if, in the comparison, we did not need to consider the time value of money: We could simply add the individual payments within a cash flow, treating receipts as positive cash flows and payments (disbursements) as negative cash flows. The fact that money has a time value, however, makes our calculations more complicated. We need to know more than just the size of a payment in order to determine its economic effect completely. In fact, as we will see in this section, we need to know several things:

- The magnitude of the payment.
- The direction of the payment: Is it a receipt or a disbursement?
- The timing of the payment: When is it made?
- The interest rate in operation during the period under consideration.

It follows that, to assess the economic impact of a series of payments, we must consider the impact of each payment individually.

Economic equivalence: The process of comparing two different cash amounts at different points in time.

Calculations for determining the economic effects of one or more cash flows are based on the concept of economic equivalence. **Economic equivalence** exists between cash flows that have the same economic effect and could therefore be traded for one another in the financial marketplace, which we assume to exist.

Economic equivalence refers to the fact that a cash flow—whether a single payment or a series of payments—can be converted to an *equivalent* cash flow at any point in time. For example, we could find the equivalent future value F of a present amount P at interest rate i at period n; or we could determine the equivalent present value P of N equal payments A.

The preceding strict concept of equivalence, which limits us to converting a cash flow into another equivalent cash flow, may be extended to include the comparison of alternatives. For example, we could compare the value of two proposals by finding the equivalent value of each at any common point in time. If financial proposals that appear to be quite different turn out to have the same monetary value, then we can be *economically indifferent* to choosing between them: In terms of economic effect, one would be an even exchange for the other, so no reason exists to prefer one over the other in terms of their economic value.

A way to see the concepts of equivalence and economic indifference at work in the real world is to note the variety of payment plans offered by lending institutions for consumer loans. Table 3.2 extends the example we developed earlier to include three different repayment plans for a loan of $20,000 for five years at 9% interest. You will notice, perhaps to your surprise, that the three plans require significantly different repayment patterns and

TABLE 3.2 Typical Repayment Plans for a Bank Loan of $20,000 (for $N = 5$ years and $i = 9\%$)

	Repayments		
	Plan 1	**Plan 2**	**Plan 3**
Year 1	$ 5,141.85	0	$ 1,800.00
Year 2	5,141.85	0	1,800.00
Year 3	5,141.85	0	1,800.00
Year 4	5,141.85	0	1,800.00
Year 5	5,141.85	$30,772.48	21,800.00
Total of payments	$25,709.25	$30,772.48	$29,000.00
Total interest paid	$ 5,709.25	$10,772.48	$ 9,000.00

Plan 1: Equal annual installments; Plan 2: End-of-loan-period repayment of principal and interest; Plan 3: Annual repayment of interest and end-of-loan repayment of principal

different total amounts of repayment. However, because money has a time value, these plans are equivalent, and economically, the bank is indifferent to a consumer's choice of plan. We will now discuss how such equivalence relationships are established.

Equivalence Calculations: A Simple Example

Equivalence calculations can be viewed as an application of the compound-interest relationships we developed in Section 3.1. Suppose, for example, that we invest $1,000 at 12% annual interest for five years. The formula developed for calculating compound interest, $F = P(1 + i)^N$ (Eq. 3.3), expresses the equivalence between some present amount P and a future amount F, for a given interest rate i and a number of interest periods N. Therefore, at the end of the investment period, our sums grow to

$$\$1,000(1 + 0.12)^5 = \$1,762.34.$$

Thus, we can say that at 12% interest, $1,000 received now is equivalent to $1,762.34 received in five years and that we could trade $1,000 now for the promise of receiving $1,762.34 in five years. Example 3.3 further demonstrates the application of this basic technique.

EXAMPLE 3.3 Equivalence

Suppose you are offered the alternative of receiving either $3,000 at the end of five years or P dollars today. There is no question that the $3,000 will be paid in full (no risk). Because you have no current need for the money, you would deposit the P dollars in an account that pays 8% interest. What value of P would make you indifferent to your choice between P dollars today and the promise of $3,000 at the end of five years?

STRATEGY: Our job is to determine the present amount that is economically equivalent to $3,000 in five years, given the investment potential of 8% per year. Note that the statement of the problem assumes that you would exercise the option of using the earning power of your money by depositing it. The "indifference" ascribed to you refers to economic indifference; that is, in a marketplace where 8% is the applicable interest rate, you could trade one cash flow for the other.

SOLUTION

Given: $F = \$3,000$, $N = 5$ years, and $i = 8\%$ per year.
Find: P.

Equation: Eq. (3.3), $F = P(1 + i)^N$.
Rearranging terms to solve for P gives

$$P = \frac{F}{(1 + i)^N}.$$

Substituting yields

$$P = \frac{\$3,000}{(1 + 0.08)^5} = \$2,042.$$

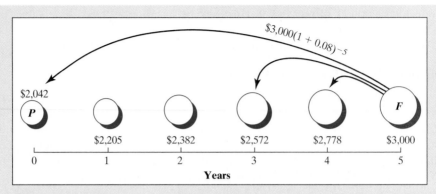

Figure 3.8 Various dollar amounts that will be economically equivalent to $3,000 in five years, given an interest rate of 8% (Example 3.3).

We summarize the problem graphically in Figure 3.8.

COMMENTS: In this example, it is clear that if P is anything less than $2,042, you would prefer the promise of $3,000 in five years to P dollars today; if P is greater than $2,042, you would prefer P. As you may have already guessed, at a lower interest rate, P must be higher to be equivalent to the future amount. For example, at $i = 4\%$, $P = \$2,466$.

3.2.2 Equivalence Calculations: General Principles

In spite of their numerical simplicity, the examples we have developed reflect several important general principles, which we will now explore.

Principle 1: Equivalence Calculations Made to Compare Alternatives Require a Common Time Basis

Common base period: To establish an economic equivalence between two cash flow amounts, a common base period must be selected.

Just as we must convert fractions to common denominators to add them together, we must also convert cash flows to a common basis to compare their value. One aspect of this basis is the choice of a single point in time at which to make our calculations. In Example 3.3, if we had been given the magnitude of each cash flow and had been asked to determine whether they were equivalent, we could have chosen any reference point and used the compound interest formula to find the value of each cash flow at that point. As you can readily see, the choice of $n = 0$ or $n = 5$ would make our problem simpler because we need to make only one set of calculations: At 8% interest, either convert $2,042 at time 0 to its equivalent value at time 5, or convert $3,000 at time 5 to its equivalent value at time 0. (To see how to choose a different reference point, take a look at Example 3.4.)

When selecting a point in time at which to compare the value of alternative cash flows, we commonly use either the present time, which yields what is called the **present worth** of the cash flows, or some point in the future, which yields their **future worth**. The choice of the point in time often depends on the circumstances surrounding a particular decision, or it may be chosen for convenience. For instance, if the present worth is known for the first two of three alternatives, all three may be compared simply by calculating the present worth of the third.

EXAMPLE 3.4 Equivalent Cash Flows Are Equivalent at Any Common Point in Time

In Example 3.3, we determined that, given an interest rate of 8% per year, receiving $2,042 today is equivalent to receiving $3,000 in five years. Are these cash flows also equivalent at the end of year 3?

STRATEGY: This problem is summarized in Figure 3.9. The solution consists of solving two equivalence problems: (1) What is the future value of $2,042 after three years at 8% interest (part (a) of the solution)? (2) Given the sum of $3,000 after five years and an interest rate of 8%, what is the equivalent sum after 3 years (part (b) of the solution)?

SOLUTION

Given:

(a) $P = \$2,042$; $i = 8\%$ per year; $N = 3$ years.
(b) $F = \$3,000$; $i = 8\%$ per year; $N = 5 - 3 = 2$ years.

Find: (1) V_3 for part (a); (2) V_3 for part (b). (3) Are these two values equivalent?

Equation:

(a) $F = P(1 + i)^N$.
(b) $P = F(1 + i)^{-N}$.

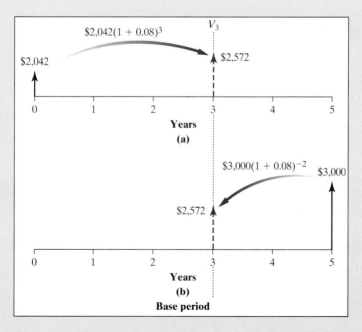

Figure 3.9 Selection of a base period for an equivalence calculation (Example 3.4).

Notation: The usual terminology of F and P is confusing in this example, since the cash flow at $n = 3$ is considered a future sum in part (a) of the solution and a past cash flow in part (b) of the solution. To simplify matters, we are free to arbitrarily designate a reference point $n = 3$ and understand that it need not to be now or the present. Therefore, we assign the equivalent cash flow at $n = 3$ to a single variable, V_3.

1. The equivalent worth of $2,042 after three years is

$$V_3 = 2,042(1 + 0.08)^3$$
$$= \$2,572.$$

2. The equivalent worth of the sum $3,000 two years earlier is

$$V_3 = F(1 + i)^{-N}$$
$$= \$3,000(1 + 0.08)^{-2}$$
$$= \$2,572.$$

(Note that $N = 2$ because that is the number of periods during which discounting is calculated in order to arrive back at year 3.)

3. While our solution doesn't strictly prove that the two cash flows are equivalent at any time, they will be equivalent at any time as long as we use an interest rate of 8%.

Principle 2: Equivalence Depends on Interest Rate

The equivalence between two cash flows is a function of the magnitude and timing of individual cash flows and the interest rate or rates that operate on those flows. This principle is easy to grasp in relation to our simple example: $1,000 received now is equivalent to $1,762.34 received five years from now only at a 12% interest rate. Any change in the interest rate will destroy the equivalence between these two sums, as we will demonstrate in Example 3.5.

EXAMPLE 3.5 Changing the Interest Rate Destroys Equivalence

In Example 3.3, we determined that, given an interest rate of 8% per year, receiving $2,042 today is equivalent to receiving $3,000 in five years. Are these cash flows equivalent at an interest rate of 10%?

SOLUTION

Given: $P = \$2,042$, $i = 10\%$ per year, and $N = 5$ years.
Find: F: Is it equal to $3,000?

We first determine the base period under which an equivalence value is computed. Since we can select any period as the base period, let's select $N = 5$. Then we need to calculate the equivalent value of $2,042 today five years from now.

$$F = \$2,042(1 + 0.10)^5 = \$3,289.$$

Since this amount is greater than $3,000, the change in interest rate destroys the equivalence between the two cash flows.

Principle 3: Equivalence Calculations May Require the Conversion of Multiple Payment Cash Flows to a Single Cash Flow

In all the examples presented thus far, we have limited ourselves to the simplest case of converting a single payment at one time to an equivalent single payment at another time. Part of the task of comparing alternative cash flow series involves moving each individual cash flow in the series to the same single point in time and summing these values to yield a single equivalent cash flow. We perform such a calculation in Example 3.6.

EXAMPLE 3.6 Equivalence Calculations with Multiple Payments

Suppose that you borrow $1,000 from a bank for three years at 10% annual interest. The bank offers two options: (1) repaying the interest charges for each year at the end of that year and repaying the principal at the end of year 3 or (2) repaying the loan all at once (including both interest and principal) at the end of year 3. The repayment schedules for the two options are as follows:

Options	Year 1	Year 2	Year 3
• Option 1: End-of-year repayment of interest, and principal repayment at end of loan	$100	$100	$1,100
• Option 2: One end-of-loan repayment of both principal and interest	0	0	1,331

Determine whether these options are equivalent, assuming that the appropriate interest rate for the comparison is 10%.

STRATEGY: Since we pay the principal after three years in either plan, the repayment of principal can be removed from our analysis. This is an important point: *We can ignore the common elements of alternatives being compared so that we can focus entirely on comparing the interest payments.* Notice that under Option 1, we will pay a total of $300 interest, whereas under Option 2, we will pay a total of $331. Before concluding that we prefer Option 2, remember that a comparison of the two cash flows is based on a *combination of payment amounts and the timing of those payments*. To make our comparison, we must compare the equivalent value of each option at a single point in time. Since Option 2 is already a single payment at $n = 3$ years, it is simplest to convert the cash flow pattern of Option 1 to a single value at $n = 3$. To do this, we must convert the three disbursements of Option 1 to their respective equivalent values at $n = 3$. At that point, since they share a time in common, we can simply sum them in order to compare them with the $331 sum in Option 2.

SOLUTION

Given: Interest payment series; $i = 10\%$ per year.

Find: A single future value F of the flows in Option 1.

Equation: $F = P(1 + i)^N$, applied to each disbursement in the cash flow diagram. N in Eq. (3.3) is the number of interest periods during which interest is in effect, and n is the period number (i.e., for year 1, $n = 1$). We determine the value of F by finding the interest period for each payment. Thus, for each payment in the series, N can be calculated by subtracting n from the total number of years of the loan (3). That is, $N = 3 - n$. Once the value of each payment has been found, we sum the payments:

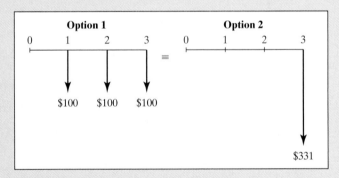

$$F_3 \text{ for } \$100 \text{ at } n = 1 : \$100(1 + .10)^{3-1} = \$121;$$

$$F_3 \text{ for } \$100 \text{ at } n = 2 : \$100(1 + .10)^{3-2} = \$110;$$

$$F_3 \text{ for } \$100 \text{ at } n = 3 : \$100(1 + .10)^{3-3} = \underline{\$100};$$

$$\text{Total} = \$331.$$

By converting the cash flow in Option 1 to a single future payment at year 3, we can compare Options 1 and 2. We see that the two interest payments are equivalent. Thus, the bank would be economically indifferent to a choice between the two plans. Note that the final interest payment in Option 1 does not accrue any compound interest.

Principle 4: Equivalence Is Maintained Regardless of Point of View

As long as we use the same interest rate in equivalence calculations, equivalence can be maintained regardless of point of view. In Example 3.6, the two options were equivalent at an interest rate of 10% from the banker's point of view. What about from a borrower's point of view? Suppose you borrow $1,000 from a bank and deposit it in another bank that pays 10% interest annually. Then you make future loan repayments out of this savings account. Under Option 1, your savings account at the end of year 1 will show a balance of $1,100 after the interest earned during the first period has been credited. Now you withdraw $100 from this savings account (the exact amount required to pay the loan interest during the first year), and you make the first-year interest payment to the bank. This leaves only $1,000 in your savings account. At the end of year 2, your savings account will earn another interest payment in the amount of $1,000(0.10) = 100, making an end-of-year balance of $1,100. Now you withdraw another $100 to make the required loan interest payment. After this payment, your remaining balance will be $1,000. This balance will grow again at 10%, so you will have $1,100 at the end of year 3. After making the last loan payment ($1,100), you will have no money left in either account. For Option 2, you can keep track of the yearly account balances in a similar fashion. You will find that you reach a zero balance after making the lump-sum payment of $1,331. If the borrower had used the same interest rate as the bank, the two options would be equivalent.

3.2.3 Looking Ahead

The preceding examples should have given you some insight into the basic concepts and calculations involved in the concept of economic equivalence. Obviously, the variety of financial arrangements possible for borrowing and investing money is extensive, as is the variety of time-related factors (e.g., maintenance costs over time, increased productivity over time, etc.) in alternative proposals for various engineering projects. It is important to recognize that even the most complex relationships incorporate the basic principles we have introduced in this section.

In the remainder of the chapter, we will represent all cash flow diagrams either in the context of an initial deposit with a subsequent pattern of withdrawals or in an initial borrowed amount with a subsequent pattern of repayments. If we were limited to the methods developed in this section, a comparison between the two payment options would involve a large number of calculations. Fortunately, in the analysis of many transactions, certain cash flow patterns emerge that may be categorized. For many of these patterns, we can derive formulas that can be used to simplify our work. In Section 3.3, we develop these formulas.

3.3 Development of Interest Formulas

Now that we have established some working assumptions and notations and have a preliminary understanding of the concept of equivalence, we will develop a series of interest formulas for use in more complex comparisons of cash flows.

As we begin to compare series of cash flows instead of single payments, the required analysis becomes more complicated. However, when patterns in cash flow transactions can be identified, we can take advantage of these patterns by developing concise expressions for computing either the present or future worth of the series. We will classify five major categories of cash flow transactions, develop interest formulas for them, and present several working examples of each type. Before we give the details, however, we briefly describe the five types of cash flows in the next subsection.

3.3.1 The Five Types of Cash Flows

Whenever we identify patterns in cash flow transactions, we may use those patterns to develop concise expressions for computing either the present or future worth of the series. For this purpose, we will classify cash flow transactions into five categories: (1) a single cash flow, (2) a uniform series, (3) a linear gradient series, (4) a geometric gradient series, and (5) an irregular series. To simplify the description of various interest formulas, we will use the following notation:

1. **Single Cash Flow:** The simplest case involves the equivalence of a single present amount and its future worth. Thus, the single-cash-flow formulas deal with only two amounts: a single present amount P and its future worth F (Figure 3.10a). You have already seen the derivation of one formula for this situation in Section 3.1.3, which gave us Eq. (3.3):

$$F = P(1 + i)^N.$$

2. **Equal (Uniform) Series:** Probably the most familiar category includes transactions arranged as a series of equal cash flows at regular intervals, known as an **equal payment series** (or **uniform series**) (Figure 3.10b). For example, this category describes the cash flows of the common installment loan contract, which arranges the repayment of a loan in equal periodic installments. The equal-cash-flow formulas deal with the equivalence relations P, F, and A (the constant amount of the cash flows in the series).

3. **Linear Gradient Series:** While many transactions involve series of cash flows, the amounts are not always uniform; they may, however, vary in some regular way. One common pattern of variation occurs when each cash flow in a series increases (or decreases) by a fixed amount (Figure 3.10c). A five-year loan repayment plan might specify, for example, a series of annual payments that increase by $500 each year. We call this type of cash flow pattern a **linear gradient series** because its cash flow diagram produces an ascending (or descending) straight line, as you will see in Section 3.3.5. In addition to using P, F, and A, the formulas employed in such problems involve a *constant amount G* of the change in each cash flow.

4. **Geometric Gradient Series:** Another kind of gradient series is formed when the series in a cash flow is determined not by some fixed amount like $500, but by some fixed *rate*, expressed as a percentage. For example, in a five-year financial plan for a project, the cost of a particular raw material might be budgeted to increase at a rate of 4% per year. The curving gradient in the diagram of such a series suggests its name: a **geometric gradient series** (Figure 3.10d). In the formulas dealing with such series, the rate of change is represented by a lowercase g.

5. **Irregular (Mixed) Series:** Finally, a series of cash flows may be irregular, in that it does not exhibit a regular overall pattern. Even in such a series, however, one or more of the patterns already identified may appear over segments of time in the total length of the series. The cash flows may be equal, for example, for 5 consecutive periods in a 10-period series. When such patterns appear, the formulas for dealing with them may be applied and their results included in calculating an equivalent value for the entire series.

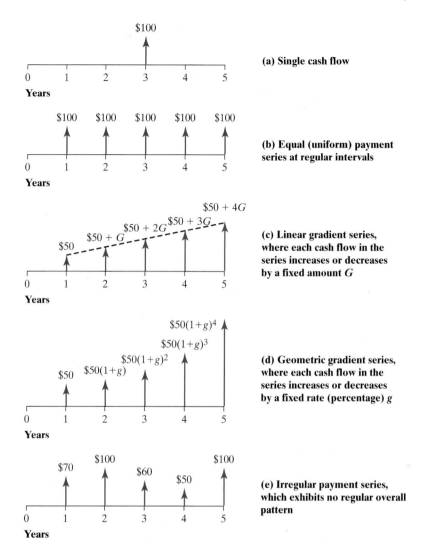

Figure 3.10 Five types of cash flows: (a) Single cash flow, (b) equal (uniform) payment series, (c) linear gradient series, (d) geometric gradient series, and (e) irregular payment series.

3.3.2 Single-Cash-Flow Formulas

We begin our coverage of interest formulas by considering the simplest of cash flows: single cash flows.

Compound Amount Factor

Given a present sum P invested for N interest periods at interest rate i, what sum will have accumulated at the end of the N periods? You probably noticed right away that this description matches the case we first encountered in describing compound interest. To solve for F (the future sum), we use Eq. (3.3):

$$F = P(1 + i)^N.$$

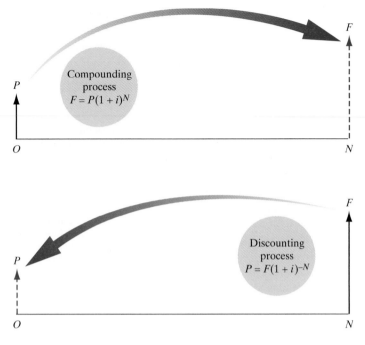

Figure 3.11 Equivalence relation between P and F.

Because of its origin in the compound-interest calculation, the factor $(1 + i)^N$ is known as the **compound-amount factor**. Like the concept of equivalence, this factor is one of the foundations of engineering economic analysis. Given the compound-amount factor, all the other important interest formulas can be derived.

Compounding process: the process of computing the future value of a current sum.

This process of finding F is often called the **compounding process**. The cash flow transaction is illustrated in Figure 3.11. (Note the time-scale convention: The first period begins at $n = 0$ and ends at $n = 1$.) If a calculator is handy, it is easy enough to calculate $(1 + i)^N$ directly.

Interest Tables

Interest formulas such as the one developed in Eq. (3.3), $F = P(1 + i)^N$, allow us to substitute known values from a particular situation into the equation and to solve for the unknown. Before the hand calculator was developed, solving these equations was very tedious. With a large value of N, for example, one might need to solve an equation such as $F = \$20,000(1 + 0.12)^{15}$. More complex formulas required even more involved calculations. To simplify the process, tables of compound-interest factors were developed, and these tables allow us to find the appropriate factor for a given interest rate and the number of interest periods. Even with hand calculators, it is still often convenient to use such tables, and they are included in this text in Appendix A. Take some time now to become familiar with their arrangement and, if you can, locate the compound-interest factor for the example just presented, in which we know P. Remember that, to find F, we need to know the factor by which to multiply $20,000 when the interest rate i is 12% and the number of periods is 15:

$$F = \$20,000 \underbrace{(1 + 0.12)^{15}}_{5.4736} = \$109,472.$$

Factor Notation

As we continue to develop interest formulas in the rest of this chapter, we will express the resulting compound-interest factors in a conventional notation that can be substituted in a formula to indicate precisely which table factor to use in solving an equation. In the preceding example, for instance, the formula derived as Eq. (3.3) is $F = P(1 + i)^N$. In ordinary language, this tells us that, to determine what future amount F is equivalent to a present amount P, we need to multiply P by a factor expressed as 1 plus the interest rate, raised to the power given by the number of interest periods. To specify how the interest tables are to be used, we may also express that factor in functional notation as $(F/P, i, N)$, which is read as "Find F, Given P, i, and N." This is known as the **single-payment compound-amount factor**. When we incorporate the table factor into the formula, it is expressed as

$$F = P(1 + i)^N = P(F/P, i, N).$$

Thus, in the preceding example, where we had $F = \$20,000(1.12)^{15}$, we can write $F = \$20,000(F/P, 12\%, 15)$. The table factor tells us to use the 12% interest table and find the factor in the F/P column for $N = 15$. Because using the interest tables is often the easiest way to solve an equation, this factor notation is included for each of the formulas derived in the sections that follow.

EXAMPLE 3.7 Single Amounts: Find F, Given i, N, and P

If you had $2,000 now and invested it at 10%, how much would it be worth in eight years (Figure 3.12)?

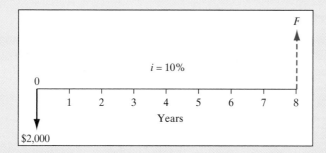

Figure 3.12 A cash flow diagram from the investor's point of view (Example 3.7).

SOLUTION

Given: $P = \$2,000$, $i = 10\%$ per year, and $N = 8$ years.

Find: F.

We can solve this problem in any of three ways:

1. **Using a calculator.** You can simply use a calculator to evaluate the $(1 + i)^N$ term (financial calculators are preprogrammed to solve most future-value problems):

$$F = \$2,000(1 + 0.10)^8$$
$$= \$4,287.18.$$

2. **Using compound-interest tables.** The interest tables can be used to locate the compound-amount factor for $i = 10\%$ and $N = 8$. The number you get can be substituted into the equation. Compound-interest tables are included as Appendix A of this book. From the tables, we obtain

$$F = \$2,000(F/P, 10\%, 8) = \$2,000(2.1436) = \$4,287.20.$$

This is essentially identical to the value obtained by the direct evaluation of the single-cash-flow compound-amount factor. This slight difference is due to rounding errors.

3. **Using Excel.** Many financial software programs for solving compound-interest problems are available for use with personal computers. Excel provides financial functions to evaluate various interest formulas, where the future-worth calculation looks like the following:

$$=FV(10\%,8,0,-2000)$$

Present-Worth Factor

Discounting process: A process of calculating the present value of a future amount.

Finding the present worth of a future sum is simply the reverse of compounding and is known as the **discounting process**. In Eq. (3.3), we can see that if we were to find a present sum P, given a future sum F, we simply solve for P:

$$P = F\left[\frac{1}{(1 + i)^N}\right] = F(P/F, i, N). \tag{3.7}$$

The factor $1/(1 + i)^N$ is known as the **single-payment present-worth factor** and is designated $(P/F, i, N)$. Tables have been constructed for P/F factors and for various values of i and N. The interest rate i and the P/F factor are also referred to as the **discount rate** and **discounting factor**, respectively.

EXAMPLE 3.8 Single Amounts: Find P, Given F, i, and N

Suppose that $1,000 is to be received in five years. At an annual interest rate of 12%, what is the present worth of this amount?

SOLUTION

Given: $F = \$1,000$, $i = 12\%$ per year, and $N = 5$ years.

Find: P.

$$P = \$1,000(1 + 0.12)^{-5} = \$1,000(0.5674) = \$567.40.$$

Using a calculator may be the best way to make this simple calculation. To have $1,000 in your savings account at the end of five years, you must deposit $567.40 now.
   We can also use the interest tables to find that

$$P = \$1,000 \overbrace{(P/F, 12\%, 5)}^{(0.5674)} = \$567.40.$$

Again, you could use a financial calculator or a computer to find the present worth. With Excel, the present-value calculation looks like the following:

$$=PV(12\%,5,0,-1000)$$

Solving for Time and Interest Rates

At this point, you should realize that the compounding and discounting processes are reciprocals of one another and that we have been dealing with one equation in two forms:

$$\text{Future-value form: } F = P(1 + i)^N;$$
$$\text{Present-value form: } P = F(1 + i)^{-N}.$$

There are four variables in these equations: P, F, N, and i. If you know the values of any three, you can find the value of the fourth. Thus far, we have always given you the interest rate i and the number of years N, plus either P or F. In many situations, though, you will need to solve for i or N, as we discuss next.

EXAMPLE 3.9 Solving for i

Suppose you buy a share for $10 and sell it for $20. Then your profit is $10. If that happens within a year, your rate of return is an impressive 100% ($10/$10 = 1). If it takes five years, what would be the average annual rate of return on your investment? (See Figure 3.13.)

SOLUTION

Given: $P = \$10$, $F = \$20$, and $N = 5$.

Find: i.

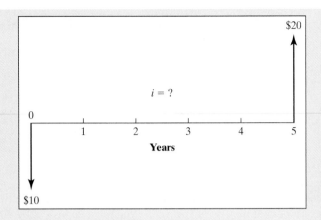

Figure 3.13 Cash flow diagram (Example 3.9).

Here, we know P, F, and N, but we do not know i, the interest rate you will earn on your investment. This type of rate of return is a lot easier to calculate, because you make only a one-time lump-sum investment. Problems such as this are solved as follows:

$$F = P(1 + i)^N;$$
$$\$20 = \$10(1 + i)^5; \text{ solve for } i.$$

- **Method 1.** Go through a trial-and-error process in which you insert different values of i into the equation until you find a value that "works" in the sense that the right-hand side of the equation equals $20. The solution is $i = 14.87\%$. The trial-and-error procedure is extremely tedious and inefficient for most problems, so it is not widely practiced in the real world.
- **Method 2.** You can solve the problem by using the interest tables in Appendix A. Now look across the $N = 5$ row, under the $(F/P, i, 5)$ column, until you can locate the value of 2:

$$\$20 = \$10(1 + i)^5;$$
$$2 = (1 + i)^5 = (F/P, i, 5).$$

This value is close to the 15% interest table with $(F/P, 15\%, 5) = 2.0114$, so the interest rate at which $10 grows to $20 over five years is very close to 15%. This procedure will be very tedious for fractional interest rates or when N is not a whole number, because you may have to approximate the solution by linear interpolation.
- **Method 3.** The most practical approach is to use either a financial calculator or an electronic spreadsheet such as Excel. A financial function such as RATE($N,0,P,F$) allows us to calculate an unknown interest rate. The precise command statement would be

$$= \textbf{RATE}(5,0,-10,20) = 14.87\%$$

Note that, in Excel format, we enter the present value (P) as a negative number, indicating a cash outflow.

	A	B	C
1	P	−10	
2	F	20	
3	N	5	=RATE(5,0,−10,20)
4	i	14.87%	
5			

EXAMPLE 3.10 Single Amounts: Find *N*, Given *P*, *F*, and *i*

You have just purchased 100 shares of General Electric stock at $60 per share. You will sell the stock when its market price has doubled. If you expect the stock price to increase 20% per year, how long do you anticipate waiting before selling the stock (Figure 3.14)?

SOLUTION

Given: $P = \$6,000$, $F = \$12,000$, and $i = 20\%$ per year.
Find: N (years).

Using the single-payment compound-amount factor, we write

$$F = P(1 + i)^N = P(F/P, i, N);$$

$$\$12,000 = \$6,000(1 + 0.20)^N = \$6,000(F/P, 20\%, N);$$

$$2 = (1.20)^N = (F/P, 20\%, N).$$

Again, we could use a calculator or a computer spreadsheet program to find N.

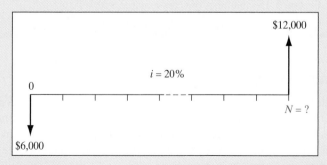

Figure 3.14

1. **Using a calculator.** Solving for N gives

$$\log 2 = N \log 1.20,$$

or

$$N = \frac{\log 2}{\log 1.20}$$

$$= 3.80 \approx 4 \text{ years.}$$

2. **Using Excel.** Within Excel, the financial function $\text{NPER}(i,0,P,F)$ computes the number of compounding periods it will take an investment (P) to grow to a future value (F), earning a fixed interest rate (i) per compounding period. In our example, the Excel command would look like this:

$$=\text{NPER}(20\%,0,-6000,12000)$$

$$= 3.801784.$$

Rule of 72:
Rule giving the approximate number of years that it will take for your investment to double.

COMMENTS: A very handy rule of thumb, called the Rule of 72, estimates approximately how long it will take for a sum of money to double. The rule states that, to find the time it takes for a present sum of money to grow by a factor of two, we divide 72 by the interest rate. In our example, the interest rate is 20%. Therefore, the Rule of 72 indicates $72/20 = 3.60$, or roughly 4 years, for a sum to double. This is, in fact, relatively close to our exact solution. Figure 3.15 illustrates the number of years required to double an investment at various interest rates.

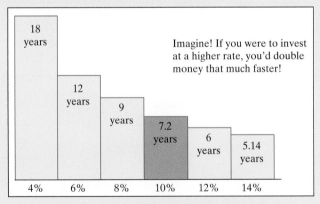

Figure 3.15 Number of years required to double an initial investment at various interest rates.

3.3.3 Uneven Payment Series

A common cash flow transaction involves a series of disbursements or receipts. Familiar examples of series payments are payment of installments on car loans and home mortgage payments. Payments on car loans and home mortgages typically involve identical sums to be paid at regular intervals. However, there is no clear pattern over the series; we call the transaction an uneven cash flow series.

We can find the present worth of any uneven stream of payments by calculating the present value of each individual payment and summing the results. Once the present worth is found, we can make other equivalence calculations (e.g., future worth can be calculated by using the interest factors developed in the previous section).

EXAMPLE 3.11   **Present Values of an Uneven Series by Decomposition into Single Payments**

Wilson Technology, a growing machine shop, wishes to set aside money now to invest over the next four years in automating its customer service department. The company can earn 10% on a lump sum deposited now, and it wishes to withdraw the money in the following increments:

- **Year 1:** $25,000, to purchase a computer and database software designed for customer service use;
- **Year 2:** $3,000, to purchase additional hardware to accommodate anticipated growth in use of the system;
- **Year 3:** No expenses; and
- **Year 4:** $5,000, to purchase software upgrades.

How much money must be deposited now to cover the anticipated payments over the next 4 years?

STRATEGY: This problem is equivalent to asking what value of P would make you indifferent in your choice between P dollars today and the future expense stream of ($25,000, $3,000, $0, $5,000). One way to deal with an uneven series of cash flows is to calculate the equivalent present value of each single cash flow and to sum the present values to find P. In other words, the cash flow is broken into three parts as shown in Figure 3.16.

SOLUTION

Given: Uneven cash flow in Figure 3.16, with $i = 10\%$ per year.
Find: P.

$$P = \$25,000(P/F, 10\%, 1) + \$3,000(P/F, 10\%, 2)$$
$$+ \$5,000(P/F, 10\%, 4)$$
$$= \$28,622.$$

COMMENTS: To see if $28,622 is indeed sufficient, let's calculate the balance at the end of each year. If you deposit $28,622 now, it will grow to (1.10)($28,622), or $31,484, at the end of year 1. From this balance, you pay out $25,000. The remaining balance, $6,484, will again grow to (1.10)($6,484), or $7,132, at the end of year 2. Now you make the second payment ($3,000) out of this balance, which will leave you with only $4,132 at the end of year 2. Since no payment occurs in year 3, the

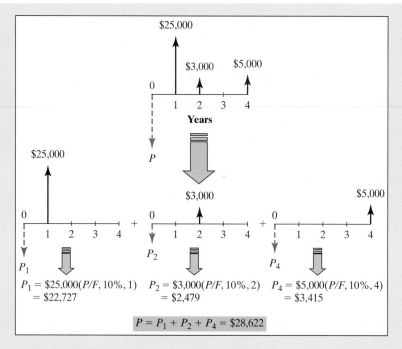

Figure 3.16 Decomposition of uneven cash flow series (Example 3.11).

balance will grow to $\$(1.10)^2(\$4,132)$, or $5,000, at the end of year 4. The final withdrawal in the amount of $5,000 will deplete the balance completely.

EXAMPLE 3.12 Calculating the Actual Worth of a Long-Term Contract of Michael Vick with Atlanta Falcons[2]

On December 23, 2004, Michael Vick became the richest player in the National Football League by agreeing to call Atlanta home for the next decade. The Falcons' quarterback signed a 10-year, $130 million contract extension Thursday that guarantees him an NFL-record $37 million in bonuses.

Base salaries for his new contract are $600,000 (2005), $1.4 million (2006), $6 million (2007), $7 million (2008), $9 million (2009), $10.5 million (2010), $13.5 million (2011), $13 million (2012), $15 million (2013), and $17 million (2014). He received an initial signing bonus of $7.5 million. Vick also received two roster bonuses in the new deal. The first is worth $22.5 million and is due in March 2005. The second is worth $7 million and is due in March 2006. Both roster bonuses will be treated as signing bonuses and prorated annually. Because 2011 is an uncapped year (the league's collective bargaining agreement (CBA) expires after the 2010 season), the initial signing bonus and 2005 roster bonus can be prorated only over the first six years of the contract. If the CBA is extended

[2] *Source:* http://www.falcfans.com/players/michael_vick.html.

prior to March 2006, then the second roster bonus of $7 million can be prorated over the final nine seasons of the contract. If the CBA is extended prior to March 2006, then his cap hits (rounded to nearest thousand) will change to $7.178 million (2006), $11.778 million (2007), $12.778 million (2008), $14.778 million (2009), $16.278 million (2010), $14.278 million (2011), $13.778 million (2012), $15.778 million (2013), and $17.778 million (2014). With the salary and signing bonus paid at the beginning of each season, the net annual payment schedule looks like the following:

Beginning of Season	Base Salary	Prorated Signing Bonus	Total Annual Payment
2005	$ 600,000	$5,000,000	$5,600,000
2006	1,400,000	5,000,000 + 778,000	7,178,000
2007	6,000,000	5,000,000 + 778,000	11,778,000
2008	7,000,000	5,000,000 + 778,000	12,778,000
2009	9,000,000	5,000,000 + 778,000	14,778,000
2010	10,500,000	778,000	16,278,000
2011	13,500,000	778,000	14,278,000
2012	13,000,000	778,000	13,778,000
2013	15,000,000	778,000	15,778,000
2014	17,000,000	778,000	17,778,000

(a) How much is Vick's contract actually worth at the time of signing? Assume that Vick's interest rate is 6% per year.

(b) For the initial signing bonus and the first year's roster bonus, suppose that the Falcons allow Vick to take either the prorated payment option as just described ($30 million over five years) or a lump-sum payment option in the amount of $23 million at the time he signs the contract. Should Vick take the lump-sum option instead of the prorated one?

SOLUTION

Given: Payment series given in Figure 3.17, with $i = 6\%$ per year.
Find: P.

(a) Actual worth of the contract at the time of signing:

$$P_{\text{contract}} = \$5,600,000 + \$7,178,000(P/F, 6\%, 1)$$
$$+ \$11,778,000(P/F, 6\%, 2) + \cdots$$
$$+ \$17,778,000(P/F, 6\%, 9)$$
$$= \$97,102,827.$$

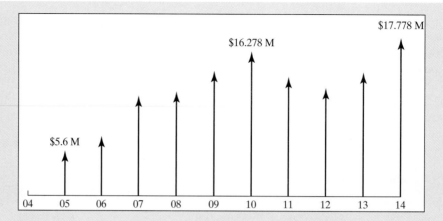

Figure 3.17 Cash flow diagram for Michael Vick's contract with Atlanta Falcons.

(b) Choice between the prorated payment option and the lump-sum payment: The equivalent present worth of the prorated payment option is

$$P_{bonus} = \$5,000,000 + \$5,000,000(P/F, 6\%, 1)$$
$$+ \$5,000,000(P/F, 6\%, 2) + \$5,000,000(P/F, 6\%, 3)$$
$$+ \$5,000,000(P/F, 6\%, 4)$$
$$= \$22,325,528$$

which is smaller than \$23,000,000. Therefore, Vick would be better off taking the lump-sum option if, and only if, his money could be invested at 6% or higher.

COMMENTS: Note that the actual contract is worth less than the published figure of \$130 million. This "brute force" approach of breaking cash flows into single amounts will always work, but it is slow and subject to error because of the many factors that must be included in the calculation. We develop more efficient methods in later sections for cash flows with certain patterns.

3.3.4 Equal Payment Series

As we learned in Example 3.12, the present worth of a stream of future cash flows can always be found by summing the present worth of each of the individual cash flows. However, if cash flow regularities are present within the stream (such as we just saw in the prorated bonus payment series in Example 3.12) then the use of shortcuts, such as finding the present worth of a uniform series, may be possible. We often encounter transactions in which a uniform series of payments exists. Rental payments, bond interest payments, and commercial installment plans are based on uniform payment series.

Compound-Amount Factor: Find F, Given A, i, and N

Suppose we are interested in the future amount F of a fund to which we contribute A dollars each period and on which we earn interest at a rate of i per period. The contributions

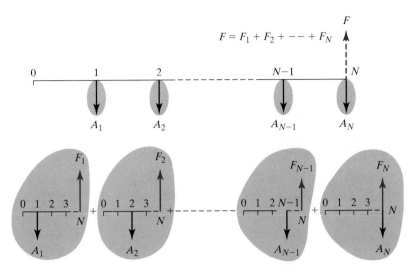

Figure 3.18 The future worth of a cash flow series obtained by summing the future-worth figures of each of the individual flows.

are made at the end of each of N equal periods. These transactions are graphically illustrated in Figure 3.18. Looking at this diagram, we see that if an amount A is invested at the end of each period, for N periods, the total amount F that can be withdrawn at the end of the N periods will be the sum of the compound amounts of the individual deposits.

As shown in Figure 3.18, the A dollars we put into the fund at the end of the first period will be worth $A(1 + i)^{N-1}$ at the end of N periods. The A dollars we put into the fund at the end of the second period will be worth $A(1 + i)^{N-2}$, and so forth. Finally, the last A dollars that we contribute at the end of the Nth period will be worth exactly A dollars at that time. This means that there exists a series of the form

$$F = A(1 + i)^{N-1} + A(1 + i)^{N-2} + \cdots + A(1 + i) + A,$$

or, expressed alternatively,

$$F = A + A(1 + i) + A(1 + i)^2 + \cdots + A(1 + i)^{N-1}. \tag{3.8}$$

Multiplying Eq. (3.8) by $(1 + i)$ results in

$$(1 + i)F = A(1 + i) + A(1 + i)^2 + \cdots + A(1 + i)^N. \tag{3.9}$$

Subtracting Eq. (3.8) from Eq. (3.9) to eliminate common terms gives us

$$F(1 + i) - F = -A + A(1 + i)^N.$$

Solving for F yields

$$F = A\left[\frac{(1 + i)^N - 1}{i}\right] = A(F/A, i, N). \tag{3.10}$$

The bracketed term in Eq. (3.10) is called the **equal payment series compound-amount factor**, or the **uniform series compound-amount factor**; its factor notation is (F/A, i, N). This interest factor has been calculated for various combinations of i and N in the interest tables.

EXAMPLE 3.13 Uniform Series: Find F, Given i, A, and N

Suppose you make an annual contribution of $3,000 to your savings account at the end of each year for 10 years. If the account earns 7% interest annually, how much can be withdrawn at the end of 10 years (Figure 3.19)?

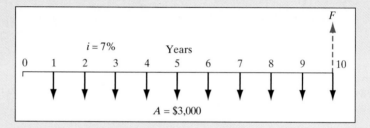

Figure 3.19 Cash flow diagram (Example 3.13).

SOLUTION

Given: $A = \$3,000$, $N = 10$ years, and $i = 7\%$ per year.
Find: F.

$$F = \$3,000(F/A, 7\%, 10)$$
$$= \$3,000(13.8164)$$
$$= \$41,449.20.$$

To obtain the future value of the annuity with the use of Excel, we may use the following financial command:

$$=FV(7\%,10,-3000,0,0)$$

EXAMPLE 3.14 Handling Time Shifts in a Uniform Series

In Example 3.13, the first deposit of the 10-deposit series was made at the end of period 1 and the remaining nine deposits were made at the end of each following period. Suppose that all deposits were made at the beginning of each period instead. How would you compute the balance at the end of period 10?

SOLUTION

Given: Cash flow as shown in Figure 3.20, and $i = 7\%$ per year.
Find: F_{10}.

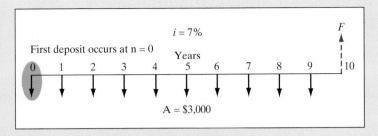

Figure 3.20   Cash Flow diagram (Example 3.14).

Compare Figure 3.20 with Figure 3.19: Each payment has been shifted to one year earlier; thus, each payment would be compounded for one extra year. Note that with the
end-of-year deposit, the ending balance (F) was \$41,449.20. With the beginning-of-year
deposit, the same balance accumulates by the end of period 9. This balance can earn interest for one additional year. Therefore, we can easily calculate the resulting balance as

$$F_{10} = \$41,449.20(1.07) = \$44,350.64.$$

The annuity due can be easily evaluated with the following financial command available
on Excel:

$$=\mathbf{FV}(7\%,10,-3000,0,1)$$

COMMENTS: Another way to determine the ending balance is to compare the two
cash flow patterns. By adding the \$3,000 deposit at period 0 to the original cash flow
and subtracting the \$3,000 deposit at the end of period 10, we obtain the second cash
flow. Therefore, the ending balance can be found by making the following adjustment to the \$41,449.20:

$$F_{10} = \$41,449.20 + \$3,000(F/P, 7\%, 10) - \$3,000 = \$44,350.64.$$

Sinking-Fund Factor: Find *A*, Given *F*, *i*, and *N*

If we solve Eq. (3.10) for A, we obtain

$$A = F\left[\frac{i}{(1 + i)^N - 1}\right] = F(A/F, i, N). \tag{3.11}$$

The term within the brackets is called the **equal payment series sinking-fund
factor**, or **sinking-fund factor**, and is referred to by the notation $(A/F, i, N)$. A sinking
fund is an interest-bearing account into which a fixed sum is deposited each interest
period; it is commonly established for the purpose of replacing fixed assets or retiring
corporate bonds.

Sinking fund:
A means of
repaying funds
advanced
through a bond
issue. This means
that every period,
a company will
pay back a portion
of its bonds.

EXAMPLE 3.15 Combination of a Uniform Series and a Single Present and Future Amount

To help you reach a $5,000 goal five years from now, your father offers to give you $500 now. You plan to get a part-time job and make five additional deposits, one at the end of each year. (The first deposit is made at the end of the first year.) If all your money is deposited in a bank that pays 7% interest, how large must your annual deposit be?

STRATEGY: If your father reneges on his offer, the calculation of the required annual deposit is easy because your five deposits fit the standard end-of-period pattern for a uniform series. All you need to evaluate is

$$A = \$5,000(A/F, 7\%, 5) = \$5,000(0.1739) = \$869.50.$$

If you do receive the $500 contribution from your father at $n = 0$, you may divide the deposit series into two parts: one contributed by your father at $n = 0$ and five equal annual deposit series contributed by yourself. Then you can use the F/P factor to find how much your father's contribution will be worth at the end of year 5 at a 7% interest rate. Let's call this amount F_c. The future value of your five annual deposits must then make up the difference, $\$5,000 - F_c$.

SOLUTION

Given: Cash flow as shown in Figure 3.21, with $i = 7\%$ per year, and $N = 5$ years. Find: A.

$$\begin{aligned}
A &= (\$5,000 - F_C)(A/F, 7\%, 5) \\
&= [\$5,000 - \$500(F/P, 7\%, 5)](A/F, 7\%, 5) \\
&= [\$5,000 - \$500(1.4026)](0.1739) \\
&= \$747.55.
\end{aligned}$$

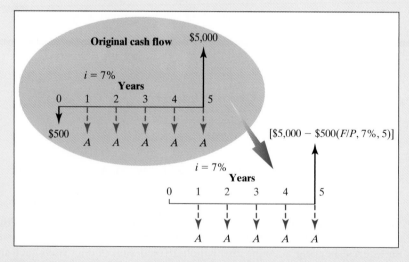

Figure 3.21 Cash flow diagram (Example 3.15).

EXAMPLE 3.16 Comparison of Three Different Investment Plans

Consider three investment plans at an annual interest rate of 9.38% (Figure 3.22):

- **Investor A.** Invest $2,000 per year for the first 10 years of your career. At the end of 10 years, make no further investments, but reinvest the amount accumulated at the end of 10 years for the next 31 years.
- **Investor B.** Do nothing for the first 10 years. Then start investing $2,000 per year for the next 31 years.
- **Investor C.** Invest $2,000 per year for the entire 41 years.

Note that all investments are made at the *beginning* of each year; the first deposit will be made at the beginning of age 25 ($n = 0$), and you want to calculate the balance at the age of 65 ($n = 41$).

STRATEGY: Since the investments are made at the beginning of each year, we need to use the procedure outlined in Example 3.14. In other words, each deposit has one extra interest-earning period.

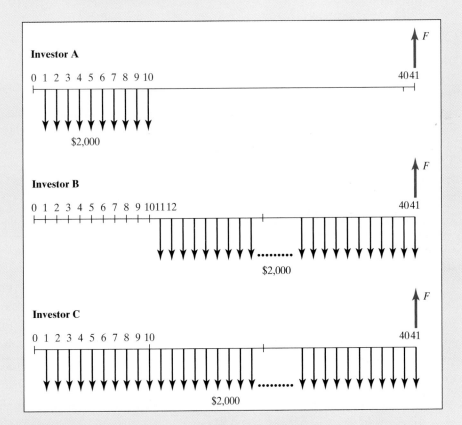

Figure 3.22 Cash flow diagrams for three investment options (Example 3.16).

SOLUTION

Given: Three different deposit scenarios with $i = 9.38\%$ and $N = 41$ years.
Find: Balance at the end of 41 years (or at the age of 65).

- Investor A:

$$F_{65} = \overbrace{\$2,000(F/A, 9.38\%, 10)(1.0938)}^{\text{Balance at the end of 10 years}}(F/P, 9.38\%, 31)$$
$$\underbrace{}_{\$33,845}$$

$$= \$545,216.$$

- Investor B:

$$F_{65} = \$2,000\underbrace{(F/P, 9.38\%, 31)}_{\$322,159}(1.0938)$$

$$= \$352,377.$$

- Investor C:

$$F_{65} = \$2,000\underbrace{(F/P, 9.38\%, 41)}_{\$820,620}(1.0938)$$

$$= \$897,594.$$

If you know how your balance changes at the end of each year, you may want to construct a tableau such as the one shown in Table 3.3. Note that, due to rounding errors, the final balance figures are slightly off from those calculated by interest formulas.

TABLE 3.3 How Time Affects the Value of Money

		Investor A		Investor B		Investor C	
Age	Years	Contribution	Year-End Value	Contribution	Year-End Value	Contribution	Year-End Value
25	1	$2,000	$ 2,188	$0	$0	$2,000	$ 2,188
26	2	$2,000	$ 4,580	$0	$0	$2,000	$ 4,580
27	3	$2,000	$ 7,198	$0	$0	$2,000	$ 7,198
28	4	$2,000	$10,061	$0	$0	$2,000	$10,061
29	5	$2,000	$13,192	$0	$0	$2,000	$13,192
30	6	$2,000	$16,617	$0	$0	$2,000	$16,617
31	7	$2,000	$20,363	$0	$0	$2,000	$20,363
32	8	$2,000	$24,461	$0	$0	$2,000	$24,461
33	9	$2,000	$28,944	$0	$0	$2,000	$28,944
34	10	$2,000	$33,846	$0	$0	$2,000	$33,846

(Continued)

TABLE 3.3 Continued

Age	Years	Investor A Contribution	Investor A Year-End Value	Investor B Contribution	Investor B Year-End Value	Investor C Contribution	Investor C Year-End Value
35	11	$0	$ 37,021	$2,000	$ 2,188	$ 2,000	$ 39,209
36	12	$0	$ 40,494	$2,000	$ 4,580	$ 2,000	$ 45,075
37	13	$0	$ 44,293	$2,000	$ 7,198	$ 2,000	$ 51,490
38	14	$0	$ 48,448	$2,000	$ 10,061	$ 2,000	$ 58,508
39	15	$0	$ 52,992	$2,000	$ 13,192	$ 2,000	$ 66,184
40	16	$0	$ 57,963	$2,000	$ 16,617	$ 2,000	$ 74,580
41	17	$0	$ 63,401	$2,000	$ 20,363	$ 2,000	$ 83,764
42	18	$0	$ 69,348	$2,000	$ 24,461	$ 2,000	$ 93,809
43	19	$0	$ 75,854	$2,000	$ 28,944	$ 2,000	$104,797
44	20	$0	$ 82,969	$2,000	$ 33,846	$ 2,000	$116,815
45	21	$0	$ 90,752	$2,000	$ 39,209	$ 2,000	$129,961
46	22	$0	$ 99,265	$2,000	$ 45,075	$ 2,000	$144,340
47	23	$0	$108,577	$2,000	$ 51,490	$ 2,000	$160,068
48	24	$0	$118,763	$2,000	$ 58,508	$ 2,000	$177,271
49	25	$0	$129,903	$2,000	$ 66,184	$ 2,000	$196,088
50	26	$0	$142,089	$2,000	$ 74,580	$ 2,000	$216,670
51	27	$0	$155,418	$2,000	$ 83,764	$ 2,000	$239,182
52	28	$0	$169,997	$2,000	$ 93,809	$ 2,000	$263,807
53	29	$0	$185,944	$2,000	$104,797	$ 2,000	$290,741
54	30	$0	$203,387	$2,000	$116,815	$ 2,000	$320,202
55	31	$0	$222,466	$2,000	$129,961	$ 2,000	$352,427
56	32	$0	$243,335	$2,000	$144,340	$ 2,000	$387,675
57	33	$0	$266,162	$2,000	$160,068	$ 2,000	$426,229
58	34	$0	$291,129	$2,000	$177,271	$ 2,000	$468,400
59	35	$0	$318,439	$2,000	$196,088	$ 2,000	$514,527
60	36	$0	$348,311	$2,000	$216,670	$ 2,000	$564,981
61	37	$0	$380,985	$2,000	$239,182	$ 2,000	$620,167
62	38	$0	$416,724	$2,000	$263,807	$ 2,000	$680,531
63	39	$0	$455,816	$2,000	$290,741	$ 2,000	$746,557
64	40	$0	$498,574	$2,000	$320,202	$ 2,000	$818,777
65	41	$0	$545,344	$2,000	$352,427	$ 2,000	$897,771
		$20,000		$62,000		$82,000	

	Investor A	Investor B	Investor C
Value at 65	**$545,344**	**$352,427**	**$897,771**
Less Total Contributions	**$20,000**	**$ 62,000**	**$82,000**
Net Earnings	**$525,344**	**$290,427**	**$815,771**

Source: Adapted from *Making Money Work for You*, UNH Cooperative Extension.

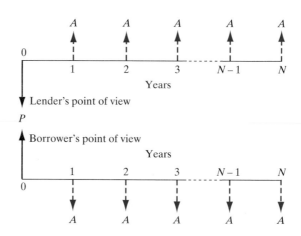

Figure 3.23

Capital Recovery Factor (Annuity Factor): Find *A*, Given *P*, *i*, and *N*

We can determine the amount of a periodic payment *A* if we know *P*, *i*, and *N*. Figure 3.23 illustrates this situation. To relate *P* to *A*, recall the relationship between *P* and *F* in Eq. (3.3), $F = P(1 + i)^N$. Replacing *F* in Eq. (3.11) by $P(1 + i)^N$, we get

$$A = P(1 + i)^N \left[\frac{i}{(1 + i)^N - 1} \right],$$

or

$$A = P \left[\frac{i(1 + i)^N}{(1 + i)^N - 1} \right] = P(A/P, i, N). \tag{3.12}$$

Capital recovery factor: Commonly used to determine the revenue requirements needed to address the up-front capital costs for projects.

Now we have an equation for determining the value of the series of end-of-period payments *A* when the present sum *P* is known. The portion within the brackets is called the **equal payment series capital recovery factor**, or simply **capital recovery factor**, which is designated (*A/P, i, N*). In finance, this *A/P* factor is referred to as the **annuity factor** and indicates a series of payments of a fixed, or constant, amount for a specified number of periods.

Annuity: An annuity is essentially a level stream of *cash flows* for a fixed period of time.

EXAMPLE 3.17 Uniform Series: Find *A*, Given *P*, *i*, and *N*

BioGen Company, a small biotechnology firm, has borrowed $250,000 to purchase laboratory equipment for gene splicing. The loan carries an interest rate of 8% per year and is to be repaid in equal installments over the next six years. Compute the amount of the annual installment (Figure 3.24).

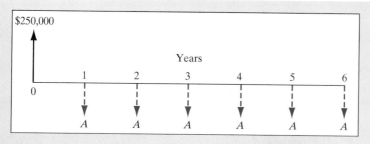

Figure 3.24 A loan cash flow diagram from BioGen's point of view.

SOLUTION

Given: $P = \$250{,}000$, $i = 8\%$ per year, and $N = 6$ years.
Find: A.

$$A = \$250{,}000(A/P, 8\%, 6)$$
$$= \$250{,}000(0.2163)$$
$$= \$54{,}075.$$

Here is an Excel solution using annuity function commands:

$$=\textbf{PMT}(i, N, P)$$
$$=\textbf{PMT}(8\%, 6, -250000)$$
$$=\$54{,}075$$

EXAMPLE 3.18 Deferred Loan Repayment

In Example 3.17, suppose that BioGen wants to negotiate with the bank to defer the first loan repayment until the end of year 2 (but still desires to make six equal install-ments at 8% interest). If the bank wishes to earn the same profit, what should be the annual installment, also known as **deferred annuity** (Figure 3.25)?

SOLUTION

Given: $P = \$250{,}000$, $i = 8\%$ per year, and $N = 6$ years, but the first payment occurs at the end of year 2.
Find: A.

By deferring the loan for year, the bank will add the interest accrued during the first year to the principal. In other words, we need to find the equivalent worth P' of $\$250{,}000$ at the end of year 1:

$$P' = \$250{,}000(F/P, 8\%, 1)$$
$$= \$270{,}000.$$

Deferred annuity:
A type of annuity contract that delays payments of income, installments, or a lump sum until the investor elects to receive them.

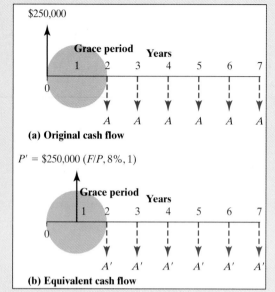

Grace period:
The additional period of time a lender provides for a borrower to make payment on a debt without penalty.

Figure 3.25 A deferred loan cash flow diagram from BioGen's point of view (Example 3.17).

In fact, BioGen is borrowing $270,000 for six years. To retire the loan with six equal installments, the deferred equal annual payment on P' will be

$$A' = \$270{,}000(A/P, 8\%, 6)$$
$$= \$58{,}401.$$

By deferring the first payment for one year, BioGen needs to make additional payments of $4,326 in each year.

Present-Worth Factor: Find *P*, Given *A*, *i*, and *N*

What would you have to invest now in order to withdraw *A* dollars at the end of each of the next *N* periods? In answering this question, we face just the opposite of the equal payment capital recovery factor situation: *A* is known, but *P* has to be determined. With the capital recovery factor given in Eq. (3.12), solving for *P* gives us

$$P = A\left[\frac{(1+i)^N - 1}{i(1+i)^N}\right] = A(P/A, i, N). \tag{3.13}$$

The bracketed term is referred to as the **equal payment series present-worth factor** and is designated $(P/A, i, N)$.

EXAMPLE 3.19 Uniform Series: Find *P*, Given *A*, *i*, and *N*

Let us revisit Louise Outing's lottery problem, introduced in the chapter opening. Suppose that Outing were able to find an investor who was willing to buy her lottery ticket for $2 million. Recall that after an initial gross payment of $283,770, Outing

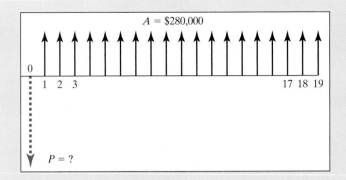

Figure 3.26 A cash flow diagram for Louise Outing's lottery winnings (Example 3.19).

would be paid 19 annual gross checks of $280,000. (See Figure 3.26.) If she could invest her money at 8% interest, what would be the fair amount to trade her 19 future lottery receipts? (Note that she already cashed in $283,770 after winning the lottery, so she is giving up 19 future lottery checks in the amount of $280,000.)

SOLUTION

Given: $i = 8\%$ per year, $A = \$280,000$, and $N = 19$ years.
Find: P.

- Using interest factor:

$$P = \$280,000(P/A, 8\%, 19) = \$280,000(9.6036)$$
$$= \$2,689,008.$$

- Using Excel:

$$=\text{PV}(8\%,19,-280000)=\$2,689,008$$

COMMENTS: Clearly, we can tell Outing that giving up $280,000 a year for 19 years to receive $2 million today is a losing proposition if she can earn only an 8% return on her investment. At this point, we may be interested in knowing at just what rate of return her deal (receiving $2 million) would in fact make sense. Since we know that $P = \$2,000,000$, $N = 19$, and $A = \$280,000$, we solve for i.

If you know the cash flows and the PV (or FV) of a cash flow stream, you can determine the interest rate. In this case, you are looking for the interest rate that caused the P/A factor to equal $(P/A, i, 19) = (\$2,000,000/\$280,000) = 7.1429$. Since we are dealing with an annuity, we could proceed as follows:

- With a financial calculator, enter $N = 19$, $PV = \$2,000,000$, $PMT = -280,000$, and then press the i key to find $i = 12.5086\%$.
- To use the interest tables, first recognize that $\$2,000,000 = \$280,000 \times (P/A, i, 19)$ or $(P/A, i, 19) = 7.1429$. Look up 7.1429 or a close value in

Appendix A. In the P/A column with $N = 19$ in the 12% interest table, you will find that $(P/A, 12\%, 19) = 7.3658$. If you look up the 13% interest table, you find that $(P/A, 12\%, 19) = 6.9380$, indicating that the interest rate should be closer to 12.5%.

- To use Excel's financial command, you simply evaluate the following command to solve the unknown interest rate problem for an annuity:

$$=\textbf{RATE}(N,A,P,F,type,guess)$$
$$=\textbf{RATE}(19,280000,-2000000,0,0,10\%)$$
$$=12.5086\%$$

It is not likely that Outing will find a financial investment which provides this high rate of return. Thus, even though the deal she has been offered is not a good one for economic reasons, she could accept it knowing that she has not much time to enjoy the future benefits.

3.3.5 Linear Gradient Series

Engineers frequently encounter situations involving periodic payments that increase or decrease by a constant amount (G) from period to period. These situations occur often enough to warrant the use of special equivalence factors that relate the arithmetic gradient to other cash flows. Figure 3.27 illustrates a **strict gradient series**, $A_n = (n - 1)G$. Note that the origin of the series is at the end of the first period with a zero value. The gradient G can be either positive or negative. If $G > 0$, the series is referred to as an *increasing* gradient series. If $G < 0$, it is a *decreasing* gradient series.

Unfortunately, the strict form of the increasing or decreasing gradient series does not correspond with the form that most engineering economic problems take. A typical problem involving a linear gradient includes an initial payment during period 1 that increases by G during some number of interest periods, a situation illustrated in Figure 3.28. This contrasts with the strict form illustrated in Figure 3.27, in which no payment is made during period 1 and the gradient is added to the previous payment beginning in period 2.

Gradient Series as Composite Series

In order to utilize the strict gradient series to solve typical problems, we must view cash flows as shown in Figure 3.28 as a **composite series**, or a set of two cash flows, each

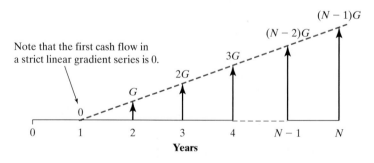

Figure 3.27   A cash flow diagram for a strict gradient series.

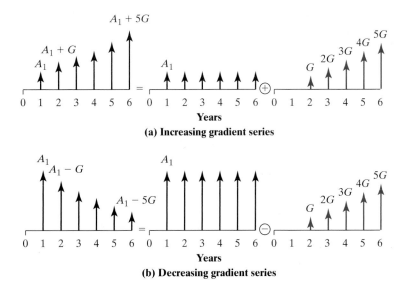

Figure 3.28   Two types of linear gradient series as composites of a uniform series of N payments of A_1 and the gradient series of increments of constant amount G.

corresponding to a form that we can recognize and easily solve: a uniform series of N payments of amount A_1 and a gradient series of increments of constant amount G. The need to view cash flows that involve linear gradient series as composites of two series is very important in solving problems, as we shall now see.

Present-Worth Factor: Linear Gradient: Find *P*, Given *G*, *N*, and *i*

How much would you have to deposit now to withdraw the gradient amounts specified in Figure 3.27? To find an expression for the present amount P, we apply the single-payment present-worth factor to each term of the series and obtain

$$P = 0 + \frac{G}{(1 + i)^2} + \frac{2G}{(1 + i)^3} + \cdots + \frac{(N - 1)G}{(1 + i)^N},$$

or

$$P = \sum_{n=1}^{N} (n - 1)G(1 + i)^{-n}. \tag{3.14}$$

Letting $G = a$ and $1/(1 + i) = x$ yields

$$P = 0 + ax^2 + 2ax^3 + \cdots + (N - 1)ax^N$$

$$= ax[0 + x + 2x^2 + \cdots + (N - 1)x^{N-1}]. \tag{3.15}$$

Since an arithmetic–geometric series $\{0, x, 2x^2, \ldots, (N - 1)x^{N-1}\}$ has the finite sum

$$0 + x + 2x^2 + \cdots + (N - 1)x^{N-1} = x\left[\frac{1 - Nx^{N-1} + (N - 1)x^N}{(1 - x)^2}\right],$$

we can rewrite Eq. (3.15) as

$$P = ax^2 \left[\frac{1 - Nx^{N-1} + (N-1)x^N}{(1-x)^2} \right]. \tag{3.16}$$

Replacing the original values for A and x, we obtain

$$P = G \left[\frac{(1+i)^N - iN - 1}{i^2(1+i)^N} \right] = G(P/G, i, N). \tag{3.17}$$

The resulting factor in brackets is called the **gradient series present-worth factor**, which we denote as $(P/G, i, N)$.

EXAMPLE 3.20 Linear Gradient: Find P, Given A_1, G, i, and N

A textile mill has just purchased a lift truck that has a useful life of five years. The engineer estimates that maintenance costs for the truck during the first year will be $1,000. As the truck ages, maintenance costs are expected to increase at a rate of $250 per year over the remaining life. Assume that the maintenance costs occur at the end of each year. The firm wants to set up a maintenance account that earns 12% annual interest. All future maintenance expenses will be paid out of this account. How much does the firm have to deposit in the account now?

SOLUTION

Given: $A_1 = \$1,000, G = \$250, i = 12\%$ per year, and $N = 5$ years.
Find: P.

Asking how much the firm has to deposit now is equivalent to asking what the equivalent present worth for this maintenance expenditure is if 12% interest is used. The cash flow may be broken into two components as shown in Figure 3.29.

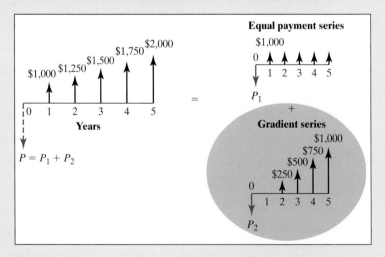

Figure 3.29 Cash flow diagram (Example 3.20).

The first component is an equal payment series (A_1), and the second is a linear gradient series (G). We have

$$P = P_1 + P_2$$

$$P = A_1(P/A, 12\%, 5) + G(P/G, 12\%, 5)$$

$$= \$1,000(3.6048) + \$250(6.397)$$

$$= \$5,204.$$

Note that the value of N in the gradient factor is 5, not 4. This is because, by definition of the series, the first gradient value begins at period 2.

COMMENTS: As a check, we can compute the present worth of the cash flow by using the $(P/F, 12\%, n)$ factors:

Period (n)	Cash Flow	(P/F, 12%, n)	Present Worth
1	$1,000	0.8929	$ 892.90
2	1,250	0.7972	996.50
3	1,500	0.7118	1,067.70
4	1,750	0.6355	1,112.13
5	2,000	0.5674	1,134.80
		Total	$5,204.03

The slight difference is caused by a rounding error.

Gradient-to-Equal-Payment Series Conversion Factor: Find A, Given G, i, and N

We can obtain an equal payment series equivalent to the gradient series, as depicted in Figure 3.30, by substituting Eq. (3.17) for P into Eq. (3.12) to obtain

$$A = G\left[\frac{(1 + i)^N - iN - 1}{i[(1 + i)^N - 1]}\right] = G(A/G, i, N), \qquad (3.18)$$

where the resulting factor in brackets is referred to as the **gradient-to-equal-payment series conversion factor** and is designated $(A/G, i, N)$.

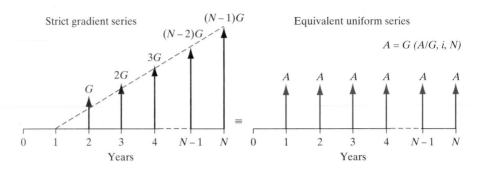

Figure 3.30

EXAMPLE 3.21 Linear Gradient: Find *A*, Given *A₁*, *G*, *i*, and *N*

John and Barbara have just opened two savings accounts at their credit union. The accounts earn 10% annual interest. John wants to deposit $1,000 in his account at the end of the first year and increase this amount by $300 for each of the next five years. Barbara wants to deposit an equal amount each year for the next six years. What should be the size of Barbara's annual deposit so that the two accounts will have equal balances at the end of six years (Figure 3.31)?

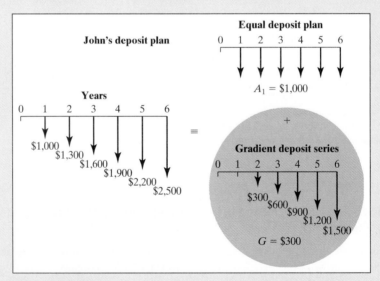

Figure 3.31 John's deposit series viewed as a combination of uniform and gradient series (Example 3.21).

SOLUTION

Given: $A_1 = \$1,000$, $G = \$300$, $i = 10\%$, and $N = 6$.

Find: A.

Since we use the end-of-period convention unless otherwise stated, this series begins at the end of the first year and the last contribution occurs at the end of the sixth year. We can separate the constant portion of $1,000 from the series, leaving the gradient series of 0, 0, 300, 600, ..., 1,500.

To find the equal payment series beginning at the end of year 1 and ending at year 6 that would have the same present worth as that of the gradient series, we may proceed as follows:

$$A = \$1,000 + \$300(A/G, 10\%, 6)$$
$$= \$1,000 + \$300(2.22236)$$
$$= \$1,667.08.$$

Barbara's annual contribution should be $1,667.08.

COMMENTS: Alternatively, we can compute Barbara's annual deposit by first computing the equivalent present worth of John's deposits and then finding the equivalent uniform annual amount. The present worth of this combined series is

$$P = \$1,000(P/A, 10\%, 6) + \$300(P/G, 10\%, 6)$$
$$= \$1,000(4.3553) + \$300(9.6842)$$
$$= \$7,260.56.$$

The equivalent uniform deposit is

$$A = \$7,260.56(A/P, 10\%, 6) = \$1,667.02.$$

(The slight difference in cents is caused by a rounding error.)

EXAMPLE 3.22 Declining Linear Gradient: Find **F**, Given **A₁, G, i,** and **N**

Suppose that you make a series of annual deposits into a bank account that pays 10% interest. The initial deposit at the end of the first year is $1,200. The deposit amounts decline by $200 in each of the next four years. How much would you have immediately after the fifth deposit?

SOLUTION

Given: Cash flow shown in Figure 3.32, $i = 10\%$ per year, and $N = 5$ years.

Find: F.

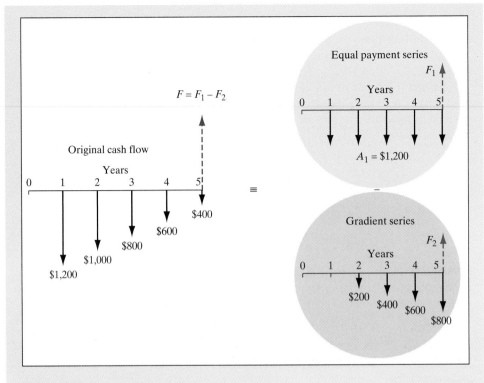

Figure 3.32

The cash flow includes a decreasing gradient series. Recall that we derived the linear gradient factors for an increasing gradient series. For a decreasing gradient series, the solution is most easily obtained by separating the flow into two components: a uniform series and an increasing gradient that is *subtracted* from the uniform series (Figure 3.32). The future value is

$$F = F_1 - F_2$$
$$= A_1(F/A, 10\%, 5) - \$200(P/G, 10\%, 5)(F/P, 10\%, 5)$$
$$= \$1,200(6.105) - \$200(6.862)(1.611)$$
$$= \$5,115.$$

Geometric growth: The year-over-year growth rate of an investment over a specified period of time. Compound growth is an imaginary number that describes the rate at which an investment grew as though it had grown at a steady rate.

3.3.6 Geometric Gradient Series

Many engineering economic problems—particularly those relating to construction costs—involve cash flows that increase or decrease over time, not by a constant *amount* (as with a linear gradient), but rather by a constant percentage (a **geometric gradient**). This kind of cash flow is called **compound growth**. Price changes caused by inflation are a good example of a geometric gradient series. If we use g to designate the percentage change in a payment from one period to the next, the magnitude of the nth payment, A_n, is related to the first payment A_1 by the formula

$$A_n = A_1(1 + g)^{n-1}, n = 1, 2, \ldots, N. \tag{3.19}$$

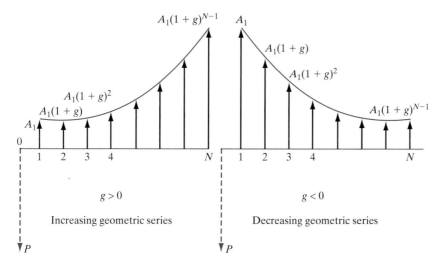

Figure 3.33 A geometrically increasing or decreasing gradient series at a constant rate g.

The variable g can take either a positive or a negative sign, depending on the type of cash flow. If $g > 0$, the series will increase, and if $g < 0$, the series will decrease. Figure 3.33 illustrates the cash flow diagram for this situation.

Present-Worth Factor: Find P, Given A_1, g, i, and N

Notice that the present worth of any cash flow A_n at interest rate i is

$$P_n = A_n(1 + i)^{-n} = A_1(1 + g)^{n-1}(1 + i)^{-n}.$$

To find an expression for the present amount P for the entire series, we apply the **single-payment present-worth factor** to each term of the series:

$$P = \sum_{n=1}^{N} A_1(1 + g)^{n-1}(1 + i)^{-n}. \tag{3.20}$$

Bringing the constant term $A_1(1 + g)^{-1}$ outside the summation yields

$$P = \frac{A_1}{(1 + g)} \sum_{n=1}^{N} \left[\frac{1 + g}{1 + i}\right]^n. \tag{3.21}$$

Let $a = \dfrac{A_1}{1 + g}$ and $x = \dfrac{1 + g}{1 + i}$. Then, rewrite Eq. (3.21) as

$$P = a(x + x^2 + x^3 + \cdots + x^N). \tag{3.22}$$

Since the summation in Eq. (3.22) represents the first N terms of a geometric series, we may obtain the closed-form expression as follows: First, multiply Eq. (3.22) by x to get

$$xP = a(x^2 + x^3 + x^4 + \cdots + x^{N+1}). \tag{3.23}$$

Then, subtract Eq. (3.23) from Eq. (3.22):

$$P - xP = a(x - x^{N+1})$$
$$P(1 - x) = a(x - x^{N+1})$$
$$P = \frac{a(x - x^{N+1})}{1 - x} \qquad (x \neq 1). \qquad (3.24)$$

If we replace the original values for a and x, we obtain

$$P = \begin{cases} A_1\left[\dfrac{1 - (1 + g)^N(1 + i)^{-N}}{i - g}\right] & \text{if } i \neq g \\ NA_1/(1 + i) & \text{if } i = g, \end{cases} \qquad (3.25)$$

or

$$P = A_1(P/A_1, g, i, N).$$

The factor within brackets is called the **geometric-gradient-series present-worth factor** and is designated $(P/A_1, g, i, N)$. In the special case where $i = g$, Eq. (3.21) becomes $P = [A_1/(1 + i)]N$.

EXAMPLE 3.23 Geometric Gradient: Find P, Given A_1, g, i, and N

Ansell, Inc., a medical device manufacturer, uses compressed air in solenoids and pressure switches in its machines to control various mechanical movements. Over the years, the manufacturing floor has changed layouts numerous times. With each new layout, more piping was added to the compressed-air delivery system to accommodate new locations of manufacturing machines. None of the extra, unused old pipe was capped or removed; thus, the current compressed-air delivery system is inefficient and fraught with leaks. Because of the leaks, the compressor is expected to run 70% of the time that the plant will be in operation during the upcoming year. This will require 260 kWh of electricity at a rate of $0.05/kWh. (The plant runs 250 days a year, 24 hours per day.) If Ansell continues to operate the current air delivery system, the compressor run time will increase by 7% per year for the next five years because of ever-worsening leaks. (After five years, the current system will not be able to meet the plant's compressed-air requirement, so it will have to be replaced.) If Ansell decides to replace all of the old piping now, it will cost $28,570.

The compressor will still run the same number of days; however, it will run 23% less (or will have $70\%(1 - 0.23) = 53.9\%$ usage during the day) because of the reduced air pressure loss. If Ansell's interest rate is 12%, is the machine worth fixing now?

SOLUTION

Given: Current power consumption, $g = 7\%$, $i = 12\%$, and $N = 5$ years.
Find: A_1 and P.

Step 1: We need to calculate the cost of power consumption of the current piping system during the first year:

$$\begin{aligned}
\text{Power cost} = \; & \% \text{ of day operating} \\
& \times \text{ days operating per year} \\
& \times \text{ hours per day} \\
& \times \text{ kWh} \times \text{\$/kWh} \\
= \; & (70\%) \times (250 \text{ days/year}) \times (24 \text{ hours/day}) \\
& \times (260 \text{ kWh}) \times (\$0.05/\text{kWh}) \\
= \; & \$54{,}440.
\end{aligned}$$

Step 2: Each year, the annual power cost will increase at the rate of 7% over the previous year's cost. The anticipated power cost over the five-year period is summarized in Figure 3.34. The equivalent present lump-sum cost at 12% for this geometric gradient series is

$$\begin{aligned}
P_{\text{Old}} &= \$54{,}440(P/A_1, 7\%, 12\%, 5) \\
&= \$54{,}440 \left[\frac{1 - (1 + 0.07)^5(1 + 0.12)^{-5}}{0.12 - 0.07} \right] \\
&= \$222{,}283.
\end{aligned}$$

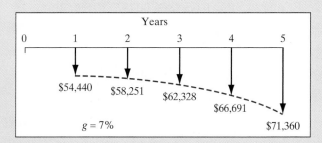

Figure 3.34   Annual power cost series if repair is not performed.

Step 3: If Ansell replaces the current compressed-air system with the new one, the annual power cost will be 23% less during the first year and will remain at that level over the next five years. The equivalent present lump-sum cost at 12% is then

$$\begin{aligned}
P_{\text{New}} &= \$54{,}440(1 - 0.23)(P/A, 12\%, 5) \\
&= \$41{,}918.80(3.6048) \\
&= \$151{,}109.
\end{aligned}$$

Step 4: The net cost for not replacing the old system now is $71,174 (= $222,283 − $151,109). Since the new system costs only $28,570, the replacement should be made now.

COMMENTS: In this example, we assumed that the cost of removing the old system was included in the cost of installing the new system. If the removed system has some salvage value, replacing it will result in even greater savings. We will consider many types of replacement issues in Chapter 14.

EXAMPLE 3.24 Geometric Gradient: Find A_1, Given F, g, i, and N

Jimmy Carpenter, a self-employed individual, is opening a retirement account at a bank. His goal is to accumulate $1,000,000 in the account by the time he retires from work in 20 years' time. A local bank is willing to open a retirement account that pays 8% interest compounded annually throughout the 20 years. Jimmy expects that his annual income will increase 6% yearly during his working career. He wishes to start with a deposit at the end of year 1 (A_1) and increase the deposit at a rate of 6% each year thereafter. What should be the size of his first deposit (A_1)? The first deposit will occur at the end of year 1, and subsequent deposits will be made at the end of each year. The last deposit will be made at the end of year 20.

SOLUTION

Given: F = $1,000,000, g = 6% per year, i = 8% per year, and N = 20 years.
Find: A_1 as in Figure 3.35.

We have

$$F = A_1(P/A_1, 6\%, 8\%, 20)\,(F/P, 8\%, 20)$$

$$= A_1(72.6911).$$

Solving for A_1 yields

$$A_1 = \$1,000,000/72.6911 = \$13,757.$$

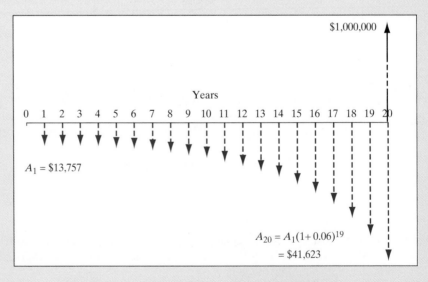

Figure 3.35 Jimmy Carpenter's retirement plan (Example 3.24).

Table 3.4 summarizes the interest formulas developed in this section and the cash flow situations in which they should be used. Recall that these formulas are applicable only to situations where the interest (compounding) period is the same as the payment period (e.g., annual compounding with annual payment). Also, we present some useful Excel financial commands in the table.

3.4 Unconventional Equivalence Calculations

In the preceding section, we occasionally presented two or more methods of attacking problems even though we had standard interest factor equations by which to solve them. It is important that you become adept at examining problems from unusual angles and that you seek out unconventional solution methods, because not all cash flow problems conform to the neat patterns for which we have discovered and developed equations. Two categories of problems that demand unconventional treatment are composite (mixed) cash flows and problems in which we must determine the interest rate implicit in a financial contract. We will begin this section by examining instances of composite cash flows.

3.4.1 Composite Cash Flows

Although many financial decisions do involve constant or systematic changes in cash flows, others contain several components of cash flows that do not exhibit an overall pattern. Consequently, it is necessary to expand our analysis to deal with these mixed types of cash flows.

To illustrate, consider the cash flow stream shown in Figure 3.36. We want to compute the equivalent present worth for this mixed payment series at an interest rate of 15%. Three different methods are presented.

Method 1. A "brute force" approach is to multiply each payment by the appropriate $(P/F, 10\%, n)$ factors and then to sum these products to obtain the present worth of the cash flows, $543.72. Recall that this is exactly the same procedure we used to solve the category of problems called the uneven payment series, described in Section 3.3.3. Figure 3.36 illustrates this computational method.

Method 2. We may group the cash flow components according to the type of cash flow pattern that they fit, such as the single payment, equal payment series, and so forth, as shown in Figure 3.37. Then the solution procedure involves the following steps:

- **Group 1:** Find the present worth of $50 due in year 1:

$$\$50(P/F, 15\%, 1) = \$43.48.$$

- **Group 2:** Find the equivalent worth of a $100 equal payment series at year 1 (V_1), and then bring this equivalent worth at year 0 again:

$$\underbrace{\$100(P/A, 15\%, 3)}_{V_1}(P/F, 15\%, 1) = \$198.54.$$

TABLE 3.4 Summary of Discrete Compounding Formulas with Discrete Payments

Flow Type	Factor Notation	Formula	Excel Command	Cash Flow Diagram
S I N G L E	Compound amount $(F/P, i, N)$	$F = P(1 + i)^N$	=FV$(i, N, P,, 0)$	
	Present worth $(P/F, i, N)$	$P = F(1 + i)^{-N}$	=PV$(i, N, F,, 0)$	
E Q U A L	Compound amount $(F/A, i, N)$	$F = A\left[\dfrac{(1 + i)^N - 1}{i}\right]$	=PV$(i, N, A,, 0)$	
P A Y M E N T	Sinking fund $(A/F, i, N)$	$A = F\left[\dfrac{i}{(1 + i)^N - 1}\right]$	=PMT$(i, N, P, F, 0)$	
S E R I E S	Present worth $(P/A, i, N)$	$P = A\left[\dfrac{(1 + i)^N - 1}{i(1 + i)^N}\right]$	=PV$(i, N, A,, 0)$	
	Capital recovery $(A/P, i, N)$	$A = P\left[\dfrac{i(1 + i)^N}{(1 + i)^N - 1}\right]$	=PMT$(i, N,, P)$	
G R A D I E N T	Linear gradient Present worth $(P/G, i, N)$	$P = G\left[\dfrac{(1 + i)^N - iN - 1}{i^2(1 + i)^N}\right]$		
	Conversion factor $(A/G, i, N)$	$A = G\left[\dfrac{(1 + i)^N - iN - 1}{i[(1 + i)^N - 1]}\right]$		
S E R I E S	Geometric gradient Present worth $(P/A_1, g, i, N)$	$P = \begin{bmatrix} A_1\left[\dfrac{1 - (1 + g)^N(1 + i)^{-N}}{i - g}\right] \\ A_1\left(\dfrac{N}{1 + i}\right)(\text{if } i = g) \end{bmatrix}$		

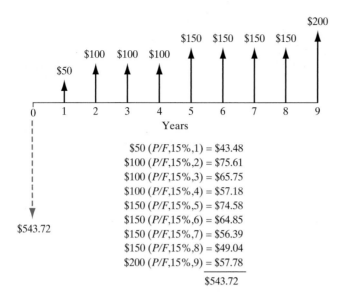

$50 (P/F,15%,1) = \$43.48$
$100 (P/F,15%,2) = \$75.61$
$100 (P/F,15%,3) = \$65.75$
$100 (P/F,15%,4) = \$57.18$
$150 (P/F,15%,5) = \$74.58$
$150 (P/F,15%,6) = \$64.85$
$150 (P/F,15%,7) = \$56.39$
$150 (P/F,15%,8) = \$49.04$
$200 (P/F,15%,9) = \$57.78$
$543.72

Figure 3.36   Equivalent present worth calculation using only
P/F factors (Method 1 "Brute Force Approach").

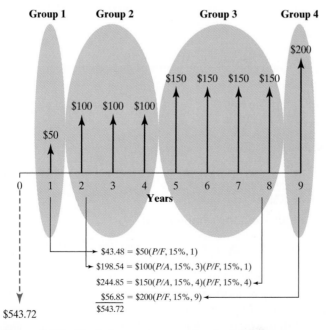

$43.48 = \$50(P/F, 15\%, 1)$
$198.54 = \$100(P/A, 15\%, 3)(P/F, 15\%, 1)$
$244.85 = \$150(P/A, 15\%, 4)(P/F, 15\%, 4)$
$56.85 = \$200(P/F, 15\%, 9)$
$543.72

Figure 3.37   Equivalent present-worth calculation for an
uneven payment series, using *P/F* and *P/A* factors (Method 2:
grouping approach).

- **Group 3:** Find the equivalent worth of a $150 equal payment series at year 4 (V_4), and then bring this equivalent worth at year 0.

$$\underbrace{\$150(P/A, 15\%, 4)}_{V_4}(P/F, 15\%, 4) = \$244.85.$$

- **Group 4:** Find the equivalent present worth of the $200 due in year 9:

$$\$200(P/F, 15\%, 9) = \$56.85.$$

- **Group total—sum the components:**

$$P = \$43.48 + \$198.54 + \$244.85 + \$56.85 = \$543.72.$$

A pictorial view of this computational process is given in Figure 3.37.

Method 3. In computing the present worth of the equal payment series components, we may use an alternative method.

- **Group 1:** Same as in Method 2.
- **Group 2:** Recognize that a $100 equal payment series will be received during years 2 through 4. Thus, we could determine the value of a four-year annuity, subtract the value of a one-year annuity from it, and have remaining the value of a four-year annuity whose first payment is due in year 2. This result is achieved by subtracting the $(P/A, 15\%, 1)$ for a one-year, 15% annuity from that for a four-year annuity and then multiplying the difference by $100:

$$\$100[(P/A, 15\%, 4) - (P/A, 15\%, 1)] = \$100(2.8550 - 0.8696)$$
$$= \$198.54.$$

Thus, the equivalent present worth of the annuity component of the uneven stream is $198.54.

- **Group 3:** We have another equal payment series that starts in year 5 and ends in year 8.

$$\$150[(P/A, 15\%, 8) - (P/A, 15\%, 4)] = \$150(4.4873 - 2.8550)$$
$$= \$244.85.$$

- **Group 4:** Same as Method 2.
- **Group total—sum the components:**

$$P = \$43.48 + \$198.54 + \$244.85 + \$56.85 = \$543.72.$$

Either the "brute force" method of Figure 3.35 or the method utilizing both $(P/A, i, n)$ and $(P/F, i, n)$ factors can be used to solve problems of this type. However, Method 2 or Method 3 is much easier if the annuity component runs for many years. For example, the alternative solution would be clearly superior for finding the equivalent present worth of a stream consisting of $50 in year 1, $200 in years 2 through 19, and $500 in year 20.

Also, note that in some instances we may want to find the equivalent value of a stream of payments at some point other than the present (year 0). In this situation, we proceed as before, but compound and discount to some other point in time—say, year 2, rather than year 0. Example 3.25 illustrates the situation.

EXAMPLE 3.25 Cash Flows with Subpatterns

The two cash flows in Figure 3.38 are equivalent at an interest rate of 12% compounded annually. Determine the unknown value C.

SOLUTION

Given: Cash flows as in Figure 3.38; $i = 12\%$ per year.
Find: C.

- **Method 1.** Compute the present worth of each cash flow at time 0:

$$P_1 = \$100(P/A, 12\%, 2) + \$300(P/A, 12\%, 3)(P/F, 12\%, 2)$$
$$= \$743.42;$$

$$P_2 = C(P/A, 12\%, 5) - C(P/F, 12\%, 3)$$
$$= 2.8930C.$$

  Since the two flows are equivalent, $P_1 = P_2$, and we have

$$743.42 = 2.8930C.$$

  Solving for C, we obtain $C = \$256.97$.

- **Method 2.**   We may select a time point other than 0 for comparison. The best choice of a base period is determined largely by the cash flow patterns. Obviously, we want to select a base period that requires the minimum number of interest factors for the equivalence calculation. Cash flow 1 represents a combined series of two equal payment cash flows, whereas cash flow 2 can be viewed as an equal payment series with the third payment missing. For cash

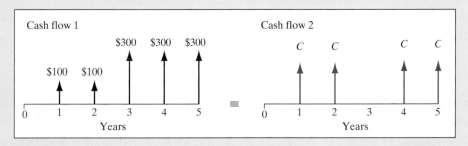

Figure 3.38 Equivalence calculation (Example 3.25).

flow 1, computing the equivalent worth at period 5 will require only two inter-est factors:

$$V_{5,1} = \$100(F/A, 12\%, 5) + \$200(F/A, 12\%, 3)$$
$$= \$1,310.16.$$

For cash flow 2, computing the equivalent worth of the equal payment series at period 5 will also require two interest factors:

$$V_{5,2} = C(F/A, 12\%, 5) - C(F/P, 12\%, 2)$$
$$= 5.0984C.$$

Therefore, the equivalence would be obtained by letting $V_{5,1} = V_{5,2}$:

$$\$1,310.16 = 5.0984C.$$

Solving for C yields $C = \$256.97$, which is the same result obtained from Method 1. The alternative solution of shifting the time point of comparison will require only four interest factors, whereas Method 1 requires five interest factors.

EXAMPLE 3.26 Establishing a College Fund

A couple with a newborn daughter wants to save for their child's college expenses in advance. The couple can establish a college fund that pays 7% annual interest. Assuming that the child enters college at age 18, the parents estimate that an amount of $40,000 per year (actual dollars) will be required to support the child's college expenses for 4 years. Determine the equal annual amounts the couple must save until they send their child to college. (Assume that the first deposit will be made on the child's first birthday and the last deposit on the child's 18th birthday. The first withdrawal will be made at the beginning of the freshman year, which also is the child's 18th birthday.)

SOLUTION

Given: Deposit and withdrawal series shown in Figure 3.39; $i = 7\%$ per year.
Find: Unknown annual deposit amount (X).

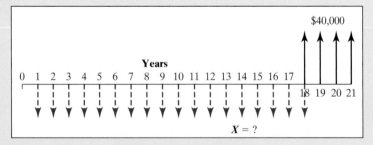

Figure 3.39 Establishing a college fund (Example 3.26). Note that the $40,000 figure represents the actual anticipated expenditures considering the future inflation.

- **Method 1.** Establish economic equivalence at period 0:

  Step 1: Find the equivalent single lump-sum deposit now:

  $$P_{\text{Deposit}} = X(P/A, 7\%, 18)$$
  $$= 10.0591X.$$

  Step 2: Find the equivalent single lump-sum withdrawal now:

  $$P_{\text{Withdrawal}} = \$40{,}000(P/A, 7\%, 4)(P/F, 7\%, 17)$$
  $$= \$42{,}892.$$

  Step 3: Since the two amounts are equivalent, by equating $P_{\text{Deposit}} = P_{\text{Withdrawal}}$, we obtain X:

  $$10.0591X = \$42{,}892$$
  $$X = \$4{,}264.$$

- **Method 2.** Establish the economic equivalence at the child's 18th birthday:

  Step 1: Find the accumulated deposit balance on the child's 18th birthday:

  $$V_{18} = X(F/A, 7\%, 18)$$
  $$= 33.9990X.$$

  Step 2: Find the equivalent lump-sum withdrawal on the child's 18th birthday:

  $$V_{18} = \$40{,}000 + \$40{,}000(P/A, 7\%, 3)$$
  $$= \$144{,}972.$$

  Step 3: Since the two amounts must be the same, we obtain

  $$33.9990X = \$144{,}972$$
  $$X = \$4{,}264.$$

The computational steps are summarized in Figure 3.40. In general, the second method is the more efficient way to obtain an equivalence solution to this type of decision problem.

COMMENTS: To verify whether the annual deposits of $4,264 over 18 years would be sufficient to meet the child's college expenses, we can calculate the actual year-by-year balances: With the 18 annual deposits of $4,264, the balance on the child's 18th birthday is

$$\$4{,}264(F/A, 7\%, 18) = \$144{,}972.$$

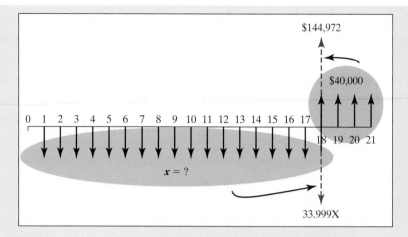

Figure 3.40 An alternative equivalence calculation (Example 3.26).

From this balance, the couple will make four annual tuition payments:

Year N	Beginning Balance	Interest Earned	Tuition Payment	Ending Balance
Freshman	$144,972	$ 0	$40,000	$104,972
Sophomore	104,972	7,348	40,000	72,320
Junior	72,320	5,062	40,000	37,382
Senior	37,382	2,618	40,000	0

3.4.2 Determining an Interest Rate to Establish Economic Equivalence

Thus far, we have assumed that, in equivalence calculations, a typical interest rate is given. Now we can use the same interest formulas that we developed earlier to determine interest rates that are explicit in equivalence problems. For most commercial loans, interest rates are already specified in the contract. However, when making some investments in financial assets, such as stocks, you may want to know the rate of growth (or rate of return) at which your asset is appreciating over the years. (This kind of calculation is the basis of rate-of-return analysis, which is covered in Chapter 7.) Although we can use interest tables to find the rate that is implicit in single payments and annuities, it is more difficult to find the rate that is implicit in an uneven series of payments. In such cases, a trial-and-error procedure or computer software may be used. To illustrate, consider Example 3.27.

EXAMPLE 3.27 Calculating an Unknown Interest Rate with Multiple Factors

You may have already won $2 million! Just peel the game piece off the Instant Winner Sweepstakes ticket, and mail it to us along with your order for subscriptions to your two favorite magazines. As a grand prize winner, you may choose between a $1 million cash prize paid immediately or $100,000 per year for 20 years—that's $2 million! Suppose that, instead of receiving one lump sum of $1 million, you decide to accept the 20 annual installments of $100,000. If you are like most jackpot winners, you will be tempted to spend your winnings to improve your lifestyle during the first several years. Only after you get this type of spending "out of your system" will you save later sums for investment purposes. Suppose that you are considering the following two options:

Option 1: You save your winnings for the first 7 years and then spend every cent of the winnings in the remaining 13 years.

Option 2: You do the reverse, spending for 7 years and then saving for 13 years.

If you can save winnings at 7% interest, how much would you have at the end of 20 years, and what interest rate on your savings will make these two options equivalent? (Cash flows into savings for the two options are shown in Figure 3.41.)

SOLUTION

Given: Cash flows in Figure 3.41.

Find: (a) F and (b) i at which the two flows are equivalent.

(a) In Option 1, the net balance at the end of year 20 can be calculated in two steps: Find the accumulated balance at the end of year 7 (V_7) first; then find the equivalent worth of V_7 at the end of year 20. For Option 2, find the equivalent worth of the 13 equal annual deposits at the end of year 20. We thus have

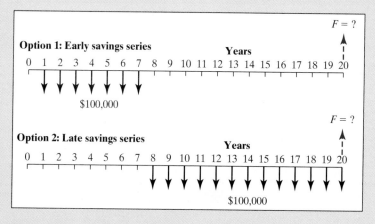

Figure 3.41 Equivalence calculation (Example 3.27).

$$F_{\text{Option 1}} = \$100,000(F/A, 7\%, 7)(F/P, 7\%, 13)$$
$$= \$2,085,485;$$
$$F_{\text{Option 2}} = \$100,000(F/A, 7\%, 13)$$
$$= \$2,014,064.$$

Option 1 accumulates \$71,421 more than Option 2.

(b) To compare the alternatives, we may compute the present worth for each option at period 0. By selecting period 7, however, we can establish the same economic equivalence with fewer interest factors. As shown in Figure 3.42, we calculate the equivalent value V_7 for each option at the end of period 7, remembering that the end of period 7 is also the beginning of period 8. (Recall from Example 3.4 that the choice of the point in time at which to compare two cash flows for equivalence is arbitrary.)

- For Option 1,

$$V_7 = \$100,000(F/A, i, 7).$$

- For Option 2,

$$V_7 = \$100,000(P/A, i, 13).$$

We equate the two values:

$$\$100,000(F/A, i, 7) = \$100,000(P/A, i, 13);$$
$$\frac{(F/A, i, 7)}{(P/A, i, 13)} = 1.$$

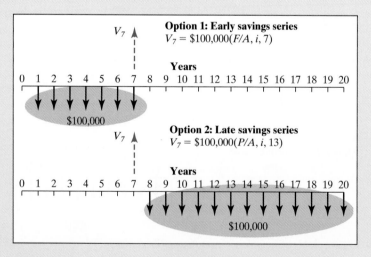

Figure 3.42 Establishing an economic equivalence at period 7 (Example 3.27).

Here, we are looking for an interest rate that gives a ratio of unity. When using the interest tables, we need to resort to a trial-and-error method. Suppose that we guess the interest rate to be 6%. Then

$$\frac{(F/A, 6\%, 7)}{(P/A, 6\%, 13)} = \frac{8.3938}{8.8527} = 0.9482.$$

This is less than unity. To increase the ratio, we need to use a value of i such that it increases the $(F/A, i, 7)$ factor value, but decreases the $(P/A, i, 13)$ value. This will happen if we use a larger interest rate. Let's try $i = 7\%$:

$$\frac{(F/A, 7\%, 7)}{(P/A, 7\%, 13)} = \frac{8.6540}{8.3577} = 1.0355.$$

Now the ratio is greater than unity.

Interest Rate	(F/A, i, 7)/(P/A, i, 13)
6%	0.9482
?	1.0000
7%	1.0355

As a result, we find that the interest rate is between 6% and 7% and may be approximated by **linear interpolation** as shown in Figure 3.43:

$$i = 6\% + (7\% - 6\%)\left[\frac{1 - 0.9482}{1.0355 - 0.9482}\right]$$

$$= 6\% + 1\%\left[\frac{0.0518}{0.0873}\right]$$

$$= 6.5934\%.$$

Interpolation is a method of constructing new data points from a discrete set of known data points.

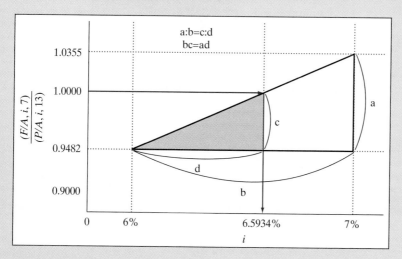

Figure 3.43 Linear interpolation to find an unknown interest rate (Example 3.27).

At 6.5934% interest, the two options are equivalent, and you may decide to indulge your desire to spend like crazy for the first 7 years. However, if you could obtain a higher interest rate, you would be wiser to save for 7 years and spend for the next 13.

COMMENTS: This example demonstrates that finding an interest rate is an iterative process that is more complicated and generally less precise than finding an equivalent worth at a known interest rate. Since computers and financial calculators can speed the process of finding unknown interest rates, such tools are highly recommended for these types of problem solving. With Excel, a more precise break-even value of 6.60219% is found by using the Goal Seek function.[3]

In Figure 3.44, the cell that contains the formula that you want to settle is called the **Set cell (F11 = F7-F9)**. The value you want the formula to change to is called **To value (0)** and the part of the formula that you wish to change is called **By changing cell (F5, interest rate)**. The **Set cell** MUST always contain a formula or a function, whereas the **Changing cell** must contain a value only, not a formula or function.

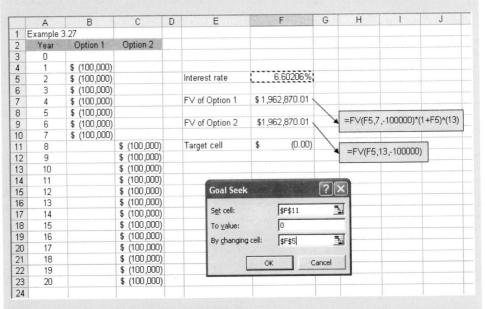

Figure 3.44 Using the Goal Seek function in Excel to find the break-even interest rate (Example 3.27). As soon as you select OK you will see that Goal Seek recalculates your formula. You then have two options, OK or Cancel. If you select OK the new term will be inserted into your worksheet. If you select Cancel, the Goal Seek box will disappear, and your worksheet will be in its original state.

---

[3] Goal Seek can be used when you know the result of a formula, but not the input value required by the formula to decide the result. You can change the value of a specified cell until the formula that is dependent on the changed cell returns the result you want. Goal Seek is found under the Tools menu.

SUMMARY

- Money has a time value because it can earn more money over time. A number of terms involving the time value of money were introduced in this chapter:

  Interest is the cost of money. More specifically, it is a cost to the borrower and an earning to the lender, above and beyond the initial sum borrowed or loaned.

  Interest rate is a percentage periodically applied to a sum of money to determine the amount of interest to be added to that sum.

  Simple interest is the practice of charging an interest rate only to an initial sum.

  Compound interest is the practice of charging an interest rate to an initial sum *and* to any previously accumulated interest that has not been withdrawn from the initial sum. Compound interest is by far the most commonly used system in the real world.

  Economic equivalence exists between individual cash flows and/or patterns of cash flows that have the same value. Even though the amounts and timing of the cash flows may differ, the appropriate interest rate makes them equal.

- The compound-interest formula, perhaps the single most important equation in this text, is

$$F = P(1 + i)^N,$$

where P is a present sum, i is the interest rate, N is the number of periods for which interest is compounded, and F is the resulting future sum. All other important interest formulas are derived from this one.

- **Cash flow diagrams** are visual representations of cash inflows and outflows along a timeline. They are particularly useful for helping us detect which of the following five patterns of cash flow is represented by a particular problem:

  1. **Single payment.** A single present or future cash flow.

  2. **Uniform series.** A series of flows of equal amounts at regular intervals.

  3. **Linear gradient series.** A series of flows increasing or decreasing by a fixed amount at regular intervals.

  4. **Geometric gradient series.** A series of flows increasing or decreasing by a fixed percentage at regular intervals.

  5. **Irregular series.** A series of flows exhibiting no overall pattern. However, patterns might be detected for portions of the series.

- **Cash flow patterns** are significant because they allow us to develop **interest formulas**, which streamline the solution of equivalence problems. Table 3.4 summarizes the important interest formulas that form the foundation for all other analyses you will conduct in engineering economic analysis.

PROBLEMS

Types of Interest

3.1 You deposit $5,000 in a savings account that earns 8% simple interest per year. What is the minimum number of years you must wait to double your balance?

Suppose instead that you deposit the $5,000 in another savings account that earns 7% interest compounded yearly. How many years will it take now to double your balance?

3.2 Compare the interest earned by $1,000 for five years at 8% simple interest with that earned by the same amount for five years at 8% compounded annually.

3.3 You are considering investing $3,000 at an interest rate of 8% compounded annually for five years or investing the $3,000 at 9% per year simple interest for five years. Which option is better?

3.4 You are about to borrow $10,000 from a bank at an interest rate of 9% compounded annually. You are required to make five equal annual repayments in the amount of $2,571 per year, with the first repayment occurring at the end of year 1. Show the interest payment and principal payment in each year.

Equivalence Concept

3.5 Suppose you have the alternative of receiving either $12,000 at the end of five years or P dollars today. Currently you have no need for money, so you would deposit the P dollars in a bank that pays 5% interest. What value of P would make you indifferent in your choice between P dollars today and the promise of $12,000 at the end of five years?

3.6 Suppose that you are obtaining a personal loan from your uncle in the amount of $20,000 (now) to be repaid in two years to cover some of your college expenses. If your uncle usually earns 8% interest (annually) on his money, which is invested in various sources, what minimum lump-sum payment two years from now would make your uncle happy?

Single Payments (Use of F/P or P/F Factors)

3.7 What will be the amount accumulated by each of these present investments?
   (a) $5,000 in 8 years at 5% compounded annually
   (b) $2,250 in 12 years at 3% compounded annually
   (c) $8,000 in 31 years at 7% compounded annually
   (d) $25,000 in 7 years at 9% compounded annually

3.8 What is the present worth of these future payments?
   (a) $5,500 6 years from now at 10% compounded annually
   (b) $8,000 15 years from now at 6% compounded annually
   (c) $30,000 5 years from now at 8% compounded annually
   (d) $15,000 8 years from now at 12% compounded annually

3.9 For an interest rate of 13% compounded annually, find
   (a) How much can be lent now if $10,000 will be repaid at the end of five years?
   (b) How much will be required in four years to repay a $25,000 loan received now?

3.10 How many years will it take an investment to triple itself if the interest rate is 12% compounded annually?

3.11 You bought 300 shares of Microsoft (MSFT) stock at $2,600 on December 31, 2005. Your intention is to keep the stock until it doubles in value. If you expect 15% annual growth for MSFT stock, how many years do you anticipate holding onto the stock? Compare your answer with the solution obtained by the Rule of 72 (discussed in Example 3.10).

3.12 From the interest tables in the text, determine the values of the following factors by interpolation, and compare your answers with those obtained by evaluating the F/P factor or the P/F factor:

(a) The single-payment compound-amount factor for 38 periods at 9.5% interest

(b) The single-payment present-worth factor for 47 periods at 8% interest

Uneven Payment Series

3.13 If you desire to withdraw the following amounts over the next five years from a savings account that earns 8% interest compounded annually, how much do you need to deposit now?

N	Amount
2	$32,000
3	43,000
4	46,000
5	28,000

3.14 If $1,500 is invested now, $1,800 two years from now, and $2,000 four years from now at an interest rate of 6% compounded annually, what will be the total amount in 15 years?

3.15 A local newspaper headline blared, "Bo Smith Signed for $30 Million." A reading of the article revealed that on April 1, 2005, Bo Smith, the former record-breaking running back from Football University, signed a $30 million package with the Dallas Rangers. The terms of the contract were $3 million immediately, $2.4 million per year for the first five years (with the first payment after 1 year) and $3 million per year for the next five years (with the first payment at year 6). If Bo's interest rate is 8% per year, what would his contract be worth at the time he signs it?

3.16 How much invested now at 6% would be just sufficient to provide three payments, with the first payment in the amount of $7,000 occurring two years hence, then $6,000 five years hence, and finally $5,000 seven years hence?

Equal Payment Series

3.17 What is the future worth of a series of equal year-end deposits of $1,000 for 10 years in a savings account that earns 7%, annual interest if

(a) All deposits are made at the *end* of each year?

(b) All deposits are made at the *beginning* of each year?

3.18 What is the future worth of the following series of payments?

(a) $3,000 at the end of each year for 5 years at 7% compounded annually

(b) $4,000 at the end of each year for 12 years at 8.25% compounded annually

(c) $5,000 at the end of each year for 20 years at 9.4% compounded annually

(d) $6,000 at the end of each year for 12 years at 10.75% compounded annually

3.19 What equal annual series of payments must be paid into a sinking fund to accumulate the following amounts?

(a) $22,000 in 13 years at 6% compounded annually

(b) $45,000 in 8 years at 7% compounded annually

(c) $35,000 in 25 years at 8% compounded annually

(d) $18,000 in 8 years at 14% compounded annually

3.20 Part of the income that a machine generates is put into a sinking fund to replace the machine when it wears out. If $1,500 is deposited annually at 7% interest, how many years must the machine be kept before a new machine costing $30,000 can be purchased?

3.21 A no-load (commission-free) mutual fund has grown at a rate of 11% compounded annually since its beginning. If it is anticipated that it will continue to grow at that rate, how much must be invested every year so that $15,000 will be accumulated at the end of five years?

3.22 What equal annual payment series is required to repay the following present amounts?

(a) $10,000 in 5 years at 5% interest compounded annually

(b) $5,500 in 4 years at 9.7% interest compounded annually

(c) $8,500 in 3 years at 2.5% interest compounded annually

(d) $30,000 in 20 years at 8.5% interest compounded annually

3.23 You have borrowed $25,000 at an interest rate of 16%. Equal payments will be made over a three-year period. (The first payment will be made at the end of the first year.) What will the annual payment be, and what will the interest payment be for the second year?

3.24 What is the present worth of the following series of payments?

(a) $800 at the end of each year for 12 years at 5.8% compounded annually

(b) $2,500 at the end of each year for 10 years at 8.5% compounded annually

(c) $900 at the end of each year for 5 years at 7.25% compounded annually

(d) $5,500 at the end of each year for 8 years at 8.75% compounded annually

3.25 From the interest tables in Appendix B, determine the values of the following factors by interpolation and compare your results with those obtained from evaluating the A/P and P/A interest formulas:

(a) The capital recovery factor for 38 periods at 6.25% interest

(b) The equal payment series present-worth factor for 85 periods at 9.25% interest

Linear Gradient Series

3.26 An individual deposits an annual bonus into a savings account that pays 8% interest compounded annually. The size of the bonus increases by $2,000 each year, and the initial bonus amount was $5,000. Determine how much will be in the account immediately after the fifth deposit.

3.27 Five annual deposits in the amounts of $3,000, $2,500, $2,000, $1,500, and $1,000, in that order, are made into a fund that pays interest at a rate of 7% compounded annually. Determine the amount in the fund immediately after the fifth deposit.

3.28 Compute the value of P in the accompanying cash flow diagram, assuming that $i = 9\%$.

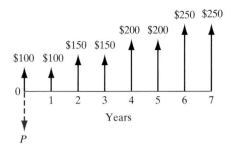

3.29 What is the equal payment series for 12 years that is equivalent to a payment series of $15,000 at the end of the first year, decreasing by $1,000 each year over 12 years? Interest is 8% compounded annually.

Geometric Gradient Series

3.30 Suppose that an oil well is expected to produce 100,000 barrels of oil during its first year in production. However, its subsequent production (yield) is expected to decrease by 10% over the previous year's production. The oil well has a proven reserve of 1,000,000 barrels.

(a) Suppose that the price of oil is expected to be $60 per barrel for the next several years. What would be the present worth of the anticipated revenue stream at an interest rate of 12% compounded annually over the next seven years?

(b) Suppose that the price of oil is expected to start at $60 per barrel during the first year, but to increase at the rate of 5% over the previous year's price. What would be the present worth of the anticipated revenue stream at an interest rate of 12% compounded annually over the next seven years?

(c) Consider part (b) again. After three years' production, you decide to sell the oil well. What would be a fair price?

3.31 A city engineer has estimated the annual toll revenues from a newly proposed highway construction over 20 years as follows:

$$A_n = (\$2{,}000{,}000)(n)(1.06)^{n-1},$$

$$n = 1, 2, \ldots, 20.$$

To validate the bond, the engineer was asked to present the estimated total present value of toll revenue at an interest rate of 6%. Assuming annual compounding, find the present value of the estimated toll revenue.

3.32 What is the amount of 10 equal annual deposits that can provide five annual withdrawals when a first withdrawal of $5,000 is made at the end of year 11 and subsequent withdrawals increase at the rate of 8% per year over the previous year's withdrawal if

(a) The interest rate is 9% compounded annually?

(b) The interest rate is 6% compounded annually?

Various Interest Factor Relationships

3.33 By using only those factors given in interest tables, find the values of the factors that follow, which are not given in your tables. Show the relationship between the factors by using factor notation, and calculate the value of the factor. Then compare the solution you obtained by using the factor formulas with a direct calculation of the factor values.

Example: $(F/P, 8\%, 38) = (F/P, 8\%, 30)(F/P, 8\%, 8) = 18.6253$

(a) $(P/F, 8\%, 67)$

(b) $(A/P, 8\%, 42)$

(c) $(P/A, 8\%, 135)$

3.34 Prove the following relationships among interest factors:

(a) $(F/P, i, N) = i(F/A, i, N) + 1$

(b) $(P/F, i, N) = 1 - (P/A, i, N)i$

(c) $(A/F, i, N) = (A/P, i, N) - i$

(d) $(A/P, i, N) = i/[1 - (P/F, i, N)]$

(e) $(P/A, i, N \rightarrow \infty) = 1/i$

(f) $(A/P, i, N \rightarrow \infty) = i$

(g) $(P/G, i, N \rightarrow \infty) = 1/i^2$

Equivalence Calculations

3.35 Find the present worth of the cash receipts where $i = 12\%$ compounded annually with only four interest factors.

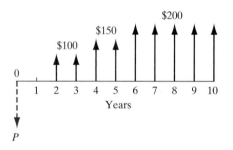

3.36 Find the equivalent present worth of the cash receipts where $i = 8\%$. In other words, how much do you have to deposit now (with the second deposit in the amount of $200 at the end of the first year) so that you will be able to withdraw $200 at the end of second year, $120 at the end of third year, and so forth if the bank pays you a 8% annual interest on your balance?

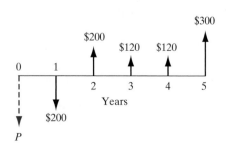

3.37 What value of *A* makes two annual cash flows equivalent at 13% interest compounded annually?

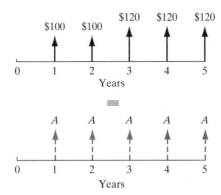

3.38 The two cash flow transactions shown in the accompanying cash flow diagram are said to be equivalent at 6% interest compounded annually. Find the unknown value of *X* that satisfies the equivalence.

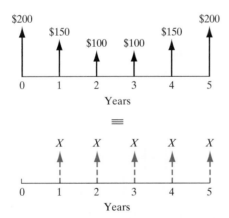

3.39 Solve for the present worth of this cash flow using at most three interest factors at 10% interest compounded annually.

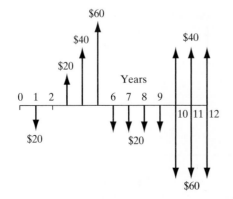

3.40 From the accompanying cash flow diagram, find the value of C that will establish the economic equivalence between the deposit series and the withdrawal series at an interest rate of 8% compounded annually.

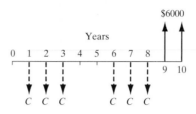

(a) $1,335   (b) $862

(c) $1,283   (d) $828

3.41 The following equation describes the conversion of a cash flow into an equivalent equal payment series with $N = 10$:

$$A = [800 + 20(A/G, 6\%, 7)]$$
$$\times\ (P/A, 6\%, 7)(A/P, 6\%, 10)$$
$$+ [300(F/A, 6\%, 3) - 500](A/F, 6\%, 10).$$

Reconstruct the original cash flow diagram.

3.42 Consider the cash flow shown in the accompanying diagram. What value of C makes the inflow series equivalent to the outflow series at an interest rate of 10%?

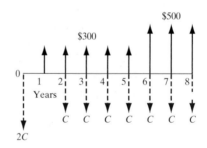

3.43 Find the value of X so that the two cash flows shown in the diagram are equivalent for an interest rate of 8%.

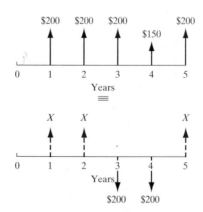

3.44 What single amount at the end of the fifth year is equivalent to a uniform annual series of $3,000 per year for 10 years if the interest rate is 9% compounded annually?

3.45 From the following list, identify all the correct equations used in computing either the equivalent present worth (P) or future worth (F) for the cash flow shown at $i = 10\%$.

Years

(1) $P = R(P/A, 10\%, 6)$
(2) $P = R + R(P/A, 10\%, 5)$
(3) $P = R(P/F, 10\%, 5) + R(P/A, 10\%, 5)$
(4) $F = R(F/A, 10\%, 5) + R(F/P, 10\%, 5)$
(5) $F = R + R(F/A, 10\%, 5)$
(6) $F = R(F/A, 10\%, 6)$
(7) $F = R(F/A, 10\%, 6) - R$

3.46 On the day his baby was born, a father decided to establish a savings account for the child's college education. Any money that is put into the account will earn an interest rate of 8% compounded annually. The father will make a series of annual deposits in equal amounts on each of his child's birthdays from the 1st through the 18th, so that the child can make four annual withdrawals from the account in the amount of $30,000 on each birthday. Assuming that the first withdrawal will be made on the child's 18th birthday, which of the following equations are correctly used to calculate the required annual deposit?

(1) $A = (\$30,000 \times 4)/18$
(2) $A = \$30,000(F/A, 8\%, 4) \times (P/F, 8\%, 21)(A/P, 8\%, 18)$
(3) $A = \$30,000(P/A, 8\%, 18) \times (F/P, 8\%, 21)(A/F, 8\%, 4)$
(4) $A = [\$30,000(P/A, 8\%, 3) + \$30,000](A/F, 8\%, 18)$
(5) $A = \$30,000[(P/F, 8\%, 18) + (P/F, 8\%, 19) +$
$(P/F, 8\%, 20) + (P/F, 8\%, 21)](A/P, 8\%, 18)$

3.47 Find the equivalent equal payment series (A) using an A/G factor such that the two cash flows are equivalent at 10% compounded annually.

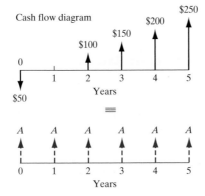

3.48 Consider the following cash flow:

Year End	Payment
0	$500
1–5	$1,000

In computing F at the end of year 5 at an interest rate of 12%, which of the following equations is *incorrect*?

(a) $F = \$1,000(F/A, 12\%, 5) - \$500(F/P, 12\%, 5)$
(b) $F = \$500(F/A, 12\%, 6) + \$500(F/A, 12\%, 5)$
(c) $F = [\$500 + \$1,000(P/A, 12\%, 5)] \times (F/P, 12\%, 5)$
(d) $F = [\$500(A/P, 12\%, 5) + \$1,000] \times (F/A, 12\%, 5)$

3.49 Consider the cash flow series given. In computing the equivalent worth at $n = 4$, which of the following equations is *incorrect*?

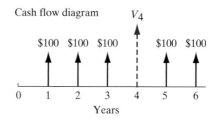

Cash flow diagram

(a) $V_4 = [\$100(P/A, i, 6) - \$100(P/F, i, 4)](F/P, i, 4)$
(b) $V_4 = \$100(F/A, i, 3) + \$100(P/A, i, 2)$
(c) $V_4 = \$100(F/A, i, 4) - \$100 + \$100(P/A, i, 2)$
(d) $V_4 = [\$100(F/A, i, 6) - \$100(F/P, i, 2)](P/F, i, 2)$

3.50 Henry Cisco is planning to make two deposits: $25,000 now and $30,000 at the end of year 6. He wants to withdraw C each year for the first six years and $(C + \$1,000)$ each year for the next six years. Determine the value of C if the deposits earn 10% interest compounded annually.

(a) $7,711
(b) $5,794
(c) $6,934
(d) $6,522

Solving for an Unknown Interest Rate or Unknown Interest Periods

3.51 At what rate of interest compounded annually will an investment double itself in five years?

3.52 Determine the interest rate (i) that makes the pairs of cash flows shown economically equivalent.

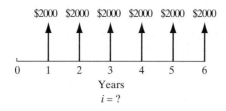

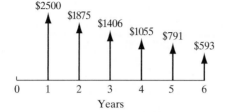

3.53 You have $10,000 available for investment in stock. You are looking for a growth stock whose value can grow to $35,000 over five years. What kind of growth rate are you looking for?

3.54 How long will it take to save $1 million if you invest $2,000 each year at 6%?

Short Case Studies

ST3.1 Read the following letter from a magazine publisher:

Dear Parent:

Currently your *Growing Child/Growing Parent* subscription will expire with your 24-month issue. To renew on an annual basis until your child reaches 72 months would cost you a total of $63.84 ($15.96 per year). We feel it is so important for you to continue receiving this material until the 72nd month, that we offer you an opportunity to renew now for $57.12. Not only is this a savings of 10% over the regular rate, but it is an excellent inflation hedge for you against increasing rates in the future. Please act now by sending $57.12.

(a) If your money is worth 6% per year, determine whether this offer can be of any value.

(b) What rate of interest would make you indifferent between the two renewal options?

ST3.2 The State of Florida sold a total of 36.1 million lottery tickets at $1 each during the first week of January 2006. As prize money, a total of $41 million will be distributed ($1,952,381 at the *beginning* of each year) over the next 21 years. The distribution of the first-year prize money occurs now, and the remaining lottery proceeds will be put into the state's educational reserve fund, which earns interest at the rate of 6% compounded annually. After making the last prize distribution (at the beginning of year 21), how much will be left over in the reserve account?

ST3.3 A local newspaper carried the following story:

Texas Cowboys wide receiver John Young will earn either $11,406,000 over 12 years or $8,600,000 over 6 years. Young must declare which plan he prefers. The $11 million package is deferred through the year 2017, while the nondeferred arrangement ends after the 2011 season. Regardless of which plan is chosen, Young will be playing through the 2011 season. Here are the details of the two plans:

Deferred Plan		Nondeferred Plan	
2006	$2,000,000	2006	$2,000,000
2007	566,000	2007	900,000
2008	920,000	2008	1,000,000
2009	930,000	2009	1,225,000
2010	740,000	2010	1,500,000
2011	740,000	2011	1,975,000
2012	740,000		
2013	790,000		
2014	540,000		
2015	1,040,000		
2016	1,140,000		
2017	1,260,000		
Total	$11,406,000	Total	$8,600,000

(a) As it happened, Young ended up with the nondeferred plan. In retrospect, if Young's interest rate were 6%, did he make a wise decision in 2006?

(b) At what interest rate would the two plans be economically equivalent?

ST3.4 Fairmont Textile has a plant in which employees have been having trouble with carpal tunnel syndrome (CTS, an inflammation of the nerves that pass through the carpal tunnel, a tight space at the base of the palm), resulting from long-term repetitive activities, such as years of sewing. It seems as if 15 of the employees working in this facility developed signs of CTS over the last five years. Deep-South, the company's insurance firm, has been increasing Fairmont's liability insurance steadily because of this problem. DeepSouth is willing to lower the insurance premiums to $16,000 a year (from the current $30,000 a year) for the next five years if Fairmont implements an acceptable CTS-prevention program that includes making the employees aware of CTS and how to reduce the chances of it developing. What would be the maximum amount that Fairmont should invest in the program to make it worthwhile? The firm's interest rate is 12% compounded annually.

ST3.5 Kersey Manufacturing Co., a small fabricator of plastics, needs to purchase an extrusion molding machine for $120,000. Kersey will borrow money from a bank at an interest rate of 9% over five years. Kersey expects its product sales to be slow during the first year, but to increase subsequently at an annual rate of 10%. Kersey therefore arranges with the bank to pay off the loan on a "balloon scale," which results in the lowest payment at the end of the first year and each subsequent payment being just 10% over the previous one. Determine the five annual payments.

ST3.6 Adidas will put on sale what it bills as the world's first computerized "smart shoe." But consumers will decide whether to accept the bionic running shoe's $250 price tag—four times the average shoe price at stores such as Foot Locker. Adidas uses a sensor, a microprocessor, and a motorized cable system to automatically adjust the shoe's cushioning. The sensor under the heel measures compression and decides whether the shoe is too soft or firm. That information is sent to the microprocessor and, while the shoe is in the air, the cable adjusts the heel cushion. The whole system weighs less than 40 grams. Adidas's computer-driven shoe—three years in the making—is the latest innovation in the $16.4 billion U.S. sneaker industry. The top-end running shoe from New Balance lists for $199.99. With runners typically replacing shoes by 500 miles, the $250 Adidas could push costs to 50 cents per mile. Adidas is spending an estimated $20 million on the rollout.[4]

The investment required to develop a full-scale commercial rollout cost Adidas $70 million (including the $20 million ad campaign), which will be financed at an interest rate of 10%. With a price tag of $250, Adidas will have about $100 net cash profit from each sale. The product will have a five-year market life. Assuming that the annual demand for the product remains constant over the market life, how many units does Adidas have to sell each year to pay off the initial investment and interest?

ST3.7 *Millionaire Babies: How to Save Our Social Security System.* It sounds a little wild, but that is probably the point. Former Senator Bob Kerrey, D-Nebraska, had proposed giving every newborn baby a $1,000 government savings account at birth, followed by five annual contributions of $500 each. If the money is then left untouched in an investment account, Kerrey said, by the time the baby reaches age 65, the $3,500 contribution per child would grow to $600,000, even at medium returns for a thrift savings plan. At about 9.4%, the balance would grow to be $1,005,132. (How would you calculate this number?) With about 4 million babies born each year, the proposal would cost the federal government $4 billion annually. Kerrey offered this idea in a speech devoted to tackling Social Security reform. About 90% of the total annual Social Security tax collections of more than $300 billion are used to pay current beneficiaries in the largest federal program. The remaining 10% is invested in interest-bearing government bonds that finance the day-to-day expenses of the federal government. Discuss the economics of Kerrey's Social Security savings plan.

---

[4] Source: "Adidas puts computer on new footing," by Michael McCarthy, USA Today—Thursday, March 3, 2006, section 5B.

ST3.8 Recently an NFL quarterback agreed to an eight-year, $50 million contract that at the time made him one of the highest paid players in professional football history. The contract included a signing bonus of $11 million. The agreement called for annual salaries of $2.5 million in 2005, $1.75 million in 2006, $4.15 million in 2007, $4.90 million in 2008, $5.25 million in 2009, $6.2 million in 2010, $6.75 million in 2011, and $7.5 million in 2012. The $11 million signing bonus was prorated over the course of the contract, so that an additional $1.375 million was paid each year over the eight-year contract period. Table ST3.8 shows the net annual payment schedule, with the salary paid at the beginning of each season.

(a) How much was the quarterback's contract actually worth at the time of signing?

(b) For the signing bonus portion, suppose that the quarterback was allowed to take either the prorated payment option as just described or a lump-sum payment option in the amount of $8 million at the time he signed the contract. Should he have taken the lump-sum option instead of the prorated one? Assume that his interest rate is 6%.

TABLE ST3.8 Net Annual Payment Schedule

Beginning of Season	Prorated Contract Salary	Actual Signing Bonus	Annual Payment
2005	$2,500,000	$1,375,000	$3,875,000
2006	1,750,000	1,375,000	3,125,000
2007	4,150,000	1,375,000	5,525,000
2008	4,900,000	1,375,000	6,275,000
2009	5,250,000	1,375,000	6,625,000
2010	6,200,000	1,375,000	7,575,000
2011	6,750,000	1,375,000	8,125,000
2012	7,500,000	1,375,000	8,875,000

ST3.9 Yuma, Arizona, resident Rosalind Setchfield won $1.3 million in a 1987 Arizona lottery drawing, to be paid in 20 annual installments of $65,277. However, in 1989, her husband, a construction worker, suffered serious injuries when a crane dropped a heavy beam on him. The couple's medical expenses and debt mounted. Six years later, in early 1995, a prize broker from Singer Asset Finance Co. telephoned Mrs. Setchfield with a promising offer. Singer would immediately give her $140,000 for one-half of each of her next 9 prize checks, an amount equal to 48% of the $293,746 she had coming over that period. A big lump sum had obvious appeal to Mrs. Setchfield at that time, and she ended up signing a contract with Singer. Did she make the right decision? The following table gives the details of the two options:

Year	Installment	Year	Installment	Reduced Payment
1988	$65,277	1995	$65,277	$32,639
1989	65,277	1996	65,277	32,639
1990	65,277	1997	65,277	32,639
1991	65,277	1998	65,277	32,639
1992	65,277	1999	65,277	32,639
1993	65,277	2000	65,277	32,639
1994	65,277	2001	65,277	32,639
		2002	65,277	32,639
		2003	65,277	32,639
		2004	65,277	
		2005	65,277	
		2006	65,277	
		2007	65,277	

FOUR

Understanding Money and Its Management

Hybrid Mortgages Can Cause Pain Should Rates Start to Rise[1] Are you shopping for a mortgage to finance a home that you expect to own for no more than a few years? If so, you should know about a hybrid mortgage. Hybrid loans give prospective home buyers the ability to buy a lot more home than they can afford, thanks to the initially lower interest rate. But with such flexibility comes greater risk. Since lenders are free to design loans to fit borrowers' needs, the terms and fees vary widely and homeowners can get burned if rates turn higher.

Hybrid mortgages allow homeowners to benefit from the best aspects of both fixed-rate and adjustable-rate mortgages (ARMs). With hybrids, borrowers choose to accept a fixed interest rate over a number of years—usually, 3, 5, 7, or 10 years—and afterward the loan

Rates Rising Mortgages are more costly than they were a year ago.		
	2006	2005
30-year fixed-rate mortgage	6.48%	5.85%
Hybrid ARM with fixed rate for first 10 years	6.32	5.58
Hybrid ARM with fixed rate for first 5 years	6.02	5.01
ARM with rate that adjusts annually	5.29	4.21

[1] "Hybrid mortgages can cause pain should rates start to rise," by Terri Cullen, *Wall Street Journal Online*, December 5, 2002. @2002 Dow Jones & Company, Inc.

converts to an ARM. But therein lies the danger: While you're getting an extraordinarily low rate up front for a few years, when the fixed-rate period expires you could very well end up paying more than double your current rate of interest.

At today's rate of 6.16% for a 30-year mortgage, a person borrowing $200,000 would pay $1,220 a month. With a 7-year hybrid, more commonly called a 7/1 loan, at the going rate of 5.61% that monthly payment drops to $1,150. By the end of the seventh year, the homeowner would save $7,700 in interest charges by going with a 7-year hybrid.

To say that there are drawbacks is an understatement. Despite the surge in popularity, a hybrid loan can be a ticking time bomb for borrowers who plan on holding the loan for the long term.

Let's say a borrower takes out a 30-year $200,000 hybrid loan that will remain at a fixed rate of 5.19% for 5 years and then will switch to an adjustable-rate mortgage for the remaining period. The lender agrees to set a cap on the adjustable-rate portion of the loan, so that the rate will climb no more than 5 percentage points. Conceivably, then, if rates are sharply higher after that initial 7-year period, a borrower could be looking at a mortgage rate of more than 10%. Under this scenario, the homeowner's monthly payment on a $200,000 mortgage would jump to $1,698 from $1,097 after the 5-year term expires—a 55% increase! But if there's a very real chance you'll be looking to sell your home over the next 10 years, hybrids can make a lot of sense, since shorter term loans usually carry the lowest rates.

In this chapter, we will consider several concepts crucial to managing money. In Chapter 3, we examined how time affects the value of money, and we developed various interest formulas for that purpose. Using these basic formulas, we will now extend the concept of equivalence to determine interest rates that are implicit in many financial contracts. To this end, we will introduce several examples in the area of loan transactions. For example, many commercial loans require that interest compound more frequently than once a year—for instance, monthly or quarterly. To consider the effect of more frequent compounding, we must begin with an understanding of the concepts of nominal and effective interest.

CHAPTER LEARNING OBJECTIVES

After completing this chapter, you should understand the following concepts:

- The difference between the nominal interest rate and the effective interest rate.
- The procedure for computing the effective interest rate, based on a payment period.
- How commercial loans and mortgages are structured in terms of interest and principal payments.
- The basics of investing in financial assets.

4.1 Nominal and Effective Interest Rates

In all the examples in Chapter 3, the implicit assumption was that payments are received *once a year*, or *annually*. However, some of the most familiar financial transactions, both personal and in engineering economic analysis, involve payments that are not based on one annual payment—for example, monthly mortgage payments and quarterly earnings on savings accounts. Thus, if we are to compare different cash flows with different compounding periods, we need to evaluate them on a common basis. This need has led to the development of the concepts of the **nominal interest rate** and the **effective interest rate**.

4.1.1 Nominal Interest Rates

Take a closer look at the billing statement from any of your credit cards. Or if you financed a new car recently, examine the loan contract. You will typically find the interest that the bank charges on your unpaid balance. Even if a financial institution uses a unit of time other than a year—say, a month or a quarter (e.g., when calculating interest payments)—the institution usually quotes the interest rate on an *annual basis*. Many banks, for example, state the interest arrangement for credit cards in this way:

18% compounded monthly.

Annual percentage rate (APR) is the yearly cost of a loan, including interest, insurance, and the origination fee, expressed as a percentage.

This statement simply means that each month the bank will charge 1.5% interest on an unpaid balance. We say that 18% is the **nominal interest rate** or **annual percentage rate** (APR), and the compounding frequency is monthly (12). As shown in Figure 4.1, to obtain the interest rate per compounding period, we divide, for example, 18% by 12, to get 1.5% per month.

Although the annual percentage rate, or APR, is commonly used by financial institutions and is familiar to many customers, the APR does not explain precisely the amount of interest that will accumulate in a year. To explain the true effect of more frequent compounding on annual interest amounts, we will introduce the term *effective interest rate*, commonly known as *annual effective yield*, or *annual percentage yield* (APY).

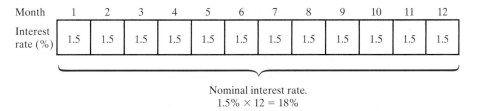

Figure 4.1 shows:

Month	1	2	3	4	5	6	7	8	9	10	11	12
Interest rate (%)	1.5	1.5	1.5	1.5	1.5	1.5	1.5	1.5	1.5	1.5	1.5	1.5

Nominal interest rate.
$1.5\% \times 12 = 18\%$

Figure 4.1 The nominal interest rate is determined by summing the individual interest rates per period.

4.1.2 Effective Annual Interest Rates

The **effective annual interest rate** is the one rate that truly represents the interest earned in a year. For instance, in our credit card example, the bank will charge 1.5% interest on any unpaid balance at the end of each month. Therefore, the 1.5% rate represents the effective interest rate per month. On a yearly basis, you are looking for a cumulative rate—1.5% each month for 12 months. This cumulative rate predicts the actual interest payment on your outstanding credit card balance.

Suppose you deposit $10,000 in a savings account that pays you at an interest rate of 9% compounded quarterly. Here, 9% represents the nominal interest rate, and the interest rate per quarter is 2.25% (9%/4). The following is an example of how interest is compounded when it is paid quarterly:

End of Period	Base amount	Interest Earned 2.25% × (Base amount)	New Base
First quarter	$10,000.00	2.25% × $10,000.00 = $225.00	$10,225.00
Second quarter	$10,225.00	2.25% × $10,225.00 = $230.06	$10,455.06
Third quarter	$10,455.06	2.25% × $10,455.06 = $225.24	$10,690.30
Fourth quarter	$10,690.30	2.25% × $10,690.30 = $240.53	$10,930.83

Clearly, you are earning more than 9% on your original deposit. In fact, you are earning 9.3083% ($930.83/$10,000). We could calculate the total annual interest payment for a principal amount of $10,000 with the formula given in Eq. (3.3). If $P = \$10,000$, $i = 2.25\%$, and $N = 4$, we obtain

$$F = P(1 + i)^N$$

$$= \$10,000(1 + 0.0225)^4$$

$$= \$10,930.83.$$

The implication is that, for each dollar deposited, you are earning an equivalent annual interest of 9.38 cents. In terms of an effective annual interest rate (i_a), the interest payment can be rewritten as a percentage of the principal amount:

$$i_a = \$930.83/\$10,000 = 0.093083, \text{ or } 9.3083\%.$$

In other words, earning 2.25% interest per quarter for four quarters is equivalent to earning 9.3083% interest just one time each year.

Table 4.1 shows effective interest rates at various compounding intervals for 4%–12% APR. As you can see, depending on the frequency of compounding, the effective interest earned or paid by the borrower can differ significantly from the APR. Therefore, truth-in-lending laws require that financial institutions quote both the nominal interest rate and the compounding frequency (i.e., the effective interest) when you deposit or borrow money.

Certainly, more frequent compounding increases the amount of interest paid over a year at the same nominal interest rate. Assuming that the nominal interest rate is r, and M compounding periods occur during the year, we can calculate the effective annual interest rate

$$i_a = \left(1 + \frac{r}{M}\right)^M - 1. \tag{4.1}$$

When $M = 1$, we have the special case of annual compounding. Substituting $M = 1$ in Eq. (4.1) reduces it to $i_a = r$. That is, when compounding takes place once annually, the effective interest is equal to the nominal interest. Thus, in the examples in Chapter 3, in which only annual interest was considered, we were, by definition, using effective interest rates.

TABLE 4.1 Nominal and Effective Interest Rates with Different Compounding Periods

	Effective Rates				
Nominal Rate	Compounding Annually	Compounding Semiannually	Compounding Quarterly	Compounding Monthly	Compounding Daily
4%	4.00%	4.04%	4.06%	4.07%	4.08%
5	5.00	5.06	5.09	5.12	5.13
6	6.00	6.09	6.14	6.17	6.18
7	7.00	7.12	7.19	7.23	7.25
8	8.00	8.16	8.24	8.30	8.33
9	9.00	9.20	9.31	9.38	9.42
10	10.00	10.25	10.38	10.47	10.52
11	11.00	11.30	11.46	11.57	11.62
12	12.00	12.36	12.55	12.68	12.74

EXAMPLE 4.1 Determining a Compounding Period

The following table summarizes interest rates on certificates of deposit (CDs) offered by various lending institutions during November 2005:

Product	Bank	Minimum	Rate	APY*
3-Month CD	Imperial Capital Bank	$2,000	4.03%	4.10%
6-Month Jumbo CD	IndyMac Bank	$5,000	4.21%	4.30%
1-Year Jumbo CD	VirtualBank	$10,000	4.50%	4.60%
1.5-Year CD	AmTrust Bank	$1,000	4.50%	4.60%
2-Year CD	Ohio Savings Bank	$1,000	4.59%	4.70%
2.5-Year Jumbo CD	Countrywide Bank	$98,000	4.66%	4.77%
3-Year CD	ING Direct	$0	4.70%	4.70%
5-Year CD	Citizens & Northern Bank	$500	4.70%	4.78%

Annual percentage yield (APY) is the rate actually earned or paid in one year, taking into account the affect of compounding.

*Annual percentage yield = effective annual interest rate (i_a)

In the table, no mention is made of specific interest compounding frequencies. (a) Find the interest periods assumed for the 2.5-year Jumbo CD offered by Countrywide Bank. (b) Find the total balance for a deposit amount of $100,000 at the end of 2.5 years.

SOLUTION

Given: $r = 4.66\%$ per year, i_a (APY) $= 4.77\%$, $P = \$100,000$, and $N = 2.5$ years. Find: M and the balance at the end of 2.5 years.

(a) The nominal interest rate is 4.66% per year, and the effective annual interest rate (yield) is 4.77%. Using Eq. (4.1), we obtain the expression

$$0.0477 = \left(1 + \frac{0.0466}{M}\right)^M - 1,$$

or

$$1.0477 = \left(1 + \frac{0.0466}{M}\right)^M.$$

By trial and error, we find that $M = 365$, which indicates daily compounding. Thus, the 2.5-year Jumbo CD earns 4.66% interest compounded daily.

Normally, if the CD is not cashed at maturity, it will be renewed automatically at the original interest rate. Similarly, we can find the interest periods for the other CDs.

(b) If you purchase the 2.5-year Jumbo CD, it will earn 4.66% interest compounded daily. This means that your CD earns an effective annual interest of 4.77%:

$$F = P(1 + i_a)^N$$
$$= \$100,000(1 + 0.0477)^{2.5}$$
$$= \$100,000\left(1 + \frac{0.0466}{365}\right)^{365 \times 2.5}$$
$$= \$112,355.$$

4.1.3 Effective Interest Rates per Payment Period

We can generalize the result of Eq. (4.1) to compute the effective interest rate for periods of *any duration*. As you will see later, the effective interest rate is usually computed on the basis of the payment (transaction) period. For example, if cash flow transactions occur quarterly, but interest is compounded monthly, we may wish to calculate the effective interest rate on a quarterly basis. To do this, we may redefine Eq. (4.1) as

$$i = \left(1 + \frac{r}{M}\right)^C - 1$$
$$= \left(1 + \frac{r}{CK}\right)^C - 1, \tag{4.2}$$

where

$$M = \text{the number of interest periods per year,}$$
$$C = \text{the number of interest periods per payment period, and}$$
$$K = \text{the number of payment periods per year.}$$

Note that $M = CK$ in Eq. (4.2). For the special case of annual payments with annual compounding, we obtain $i = i_a$ with $C = M$ and $K = 1$.

EXAMPLE 4.2 Effective Rate per Payment Period

Suppose that you make quarterly deposits in a savings account that earns 9% interest compounded monthly. Compute the effective interest rate per quarter.

SOLUTION

Given: $r = 9\%$, $C =$ three interest periods per quarter, $K =$ four quarterly payments per year, and $M = 12$ interest periods per year.
Find: i.

Using Eq. (4.2), we compute the effective interest rate per quarter as

$$i = \left(1 + \frac{0.09}{12}\right)^3 - 1$$

$$= 2.27\%.$$

COMMENTS: Figure 4.2 illustrates the relationship between the nominal and effective interest rates per payment period.

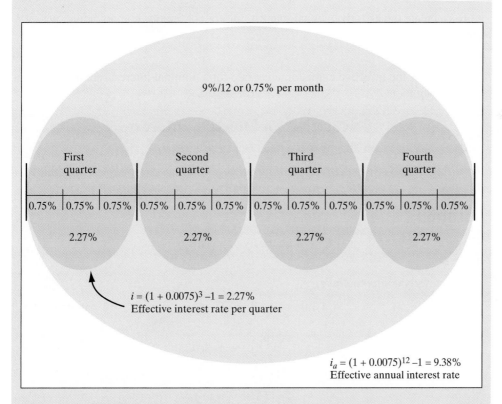

9%/12 or 0.75% per month

| First quarter | Second quarter | Third quarter | Fourth quarter |

0.75% 0.75% 0.75% 0.75% 0.75% 0.75% 0.75% 0.75% 0.75% 0.75% 0.75% 0.75%

2.27%   2.27%   2.27%   2.27%

$i = (1 + 0.0075)^3 - 1 = 2.27\%$
Effective interest rate per quarter

$i_a = (1 + 0.0075)^{12} - 1 = 9.38\%$
Effective annual interest rate

Figure 4.2   Functional relationships among r, i, and i_a, where interest is calculated based on 9% compounded monthly and payments occur quarterly (Example 4.2).

4.1.4 Continuous Compounding

To be competitive on the financial market or to entice potential depositors, some financial institutions offer frequent compounding. As the number of compounding periods (M) becomes very large, the interest rate per compounding period (r/M) becomes very small. As M approaches infinity and r/M approaches zero, we approximate the situation of **continuous compounding**.

By taking limits on both sides of Eq. (4.2), we obtain the effective interest rate per payment period as

$$
i = \lim_{CK \to \infty}\left[\left(1 + \frac{r}{CK}\right)^{C} - 1\right]
$$

$$
= \lim_{CK \to \infty}\left(1 + \frac{r}{CK}\right)^{C} - 1
$$

$$
= (e^{r})^{1/K} - 1.
$$

Continuous compounding: The process of calculating interest and adding it to existing principal and interest at infinitely short time intervals.

Therefore, the effective interest rate per payment period is

$$
i = e^{r/K} - 1. \tag{4.3}
$$

To calculate the effective annual interest rate for continuous compounding, we set K equal to unity, resulting in

$$
i_a = e^{r} - 1. \tag{4.4}
$$

As an example, the effective annual interest rate for a nominal interest rate of 12% compounded continuously is $i_a = e^{0.12} - 1 = 12.7497\%$.

EXAMPLE 4.3 Calculating an Effective Interest Rate with Quarterly Payment

Find the effective interest rate per *quarter* at a nominal rate of 8% compounded (a) quarterly, (b) monthly, (c) weekly, (d) daily, and (e) continuously.

SOLUTION

Given: $r = 8\%$, M, C, and $K = 4$ quarterly payments per year.
Find: i.

(a) Quarterly compounding:

$r = 8\%$, $M = 4$, $C = 1$ interest period per quarter, and $K = 4$ payments per year. Then

$$
i = \left(1 + \frac{0.08}{4}\right)^{1} - 1 = 2.00\%.
$$

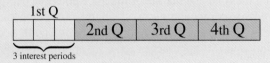

(b) Monthly compounding:

$r = 8\%$, $M = 12$, $C = 3$ interest periods per quarter, and $K = 4$ payments per year. Then

$$
i = \left(1 + \frac{0.08}{12}\right)^{3} - 1 = 2.013\%.
$$

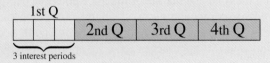

(c) Weekly compounding:

$r = 8\%, M = 52, C = 13$ interest periods per quarter, and $K = 4$ payments per year. Then

$$i = \left(1 + \frac{0.08}{52}\right)^{13} - 1 = 2.0186\%.$$

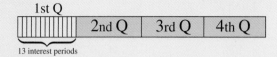

13 interest periods

(d) Daily compounding:

$r = 8\%, M = 365, C = 91.25$ days per quarter, and $K = 4$. Then

$$i = \left(1 + \frac{0.08}{365}\right)^{91.25} - 1 = 2.0199\%.$$

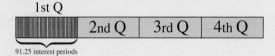

91.25 interest periods

(e) Continuous compounding:

$r = 8\%, M \rightarrow \infty, C \rightarrow \infty$, and $K = 4$. Then, from Eq. (4.3),

$$i = e^{0.08/4} - 1 = 2.0201\%.$$

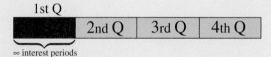

∞ interest periods

COMMENTS: Note that the difference between daily compounding and continuous compounding is often negligible. Many banks offer continuous compounding to entice deposit customers, but the extra benefits are small.

4.2 Equivalence Calculations with Effective Interest Rates

All the examples in Chapter 3 assumed annual payments and annual compounding. However, a number of situations involve cash flows that occur at intervals that are not the same as the compounding intervals often used in practice. Whenever payment and compounding periods differ from each other, *one or the other must be transformed so that both conform to the same unit of time.* For example, if payments occur quarterly and compounding occurs monthly, the most logical procedure is to calculate the effective interest rate per quarter. By contrast, if payments occur monthly and compounding occurs quarterly, we may be able to find the equivalent monthly interest rate. The bottom line is that, to proceed with equivalency analysis, the compounding and payment periods must be in the same order.

4.2.1 When Payment Period Is Equal to Compounding Period

Whenever the compounding and payment periods are equal ($M = K$), whether the interest is compounded annually or at some other interval, the following solution method can be used:

1. Identify the number of compounding periods (M) per year.
2. Compute the effective interest rate per payment period—that is, using Eq. (4.2). Then, with $C = 1$ and $K = M$, we have

$$i = \frac{r}{M}.$$

3. Determine the number of compounding periods:

$$N = M \times (\text{number of years}).$$

EXAMPLE 4.4 Calculating Auto Loan Payments

Suppose you want to buy a car. You have surveyed the dealers' newspaper advertisements, and the following one has caught your attention:

College Graduate Special: New 2006 Nissan Altima 2.55 with automatic transmission, A/C, power package, and cruise control

MSRP:	$20,870
Dealer's discount:	$1,143
Manufacturer rebate	$800
College graduate cash:	$500
Sale price:	$18,427

MSRP: Manufacturer's Suggested Retail Price.

Price and payment is plus tax, title, customer service fee, with approved credit for 72 months at 6.25% APR.

You can afford to make a down payment of $3,427 (and taxes and insurance as well), so the net amount to be financed is $15,000. What would the monthly payment be (Figure 4.3)?

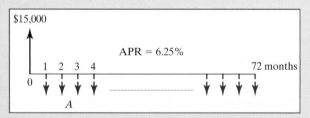

Figure 4.3 A car loan cash transaction (Example 4.4).

DISCUSSION: The advertisement does not specify a compounding period, but in automobile financing, the interest and the payment periods are almost always monthly. Thus, the 6.25% APR means 6.25% compounded monthly.

SOLUTION

Given: $P = \$25{,}000$, $r = 6.25\%$ per year, K = 12 payments per year, $N = 72$ months, and $M = 12$ interest periods per year.

Find: A.

In this situation, we can easily compute the monthly payment with Eq. (3.12):

$$i = 6.25\%/12 = 0.5208\% \text{ per month,}$$

$$N = (12)(6) = 72 \text{ months,}$$

$$A = \$15{,}000(A/P, 0.5208\%, 72) = \$250.37.$$

4.2.2 Compounding Occurs at a Different Rate than That at Which Payments Are Made

We will consider two situations: (1) compounding is more frequent than payments and (2) compounding is less frequent than payments.

Compounding Is More Frequent than Payments

The computational procedure for compounding periods and payment periods that cannot be compared is as follows:

1. Identify the number of compounding periods per year (M), the number of payment periods per year (K), and the number of interest periods per payment period (C).
2. Compute the effective interest rate per payment period:
   - For discrete compounding, compute

$$i = \left(1 + \frac{r}{M}\right) - 1.$$

   - For continuous compounding, compute

$$i = e^{r/K} - 1.$$

3. Find the total number of payment periods:

$$N = K \times (\text{number of years}).$$

4. Use i and N in the appropriate formulas in Table 3.4.

EXAMPLE 4.5 Compounding Occurs More Frequently than Payments Are Made (Discrete-Compounding Case)

Suppose you make equal quarterly deposits of $1,500 into a fund that pays interest at a rate of 6% compounded monthly, as shown in Figure 4.4. Find the balance at the end of year 2.

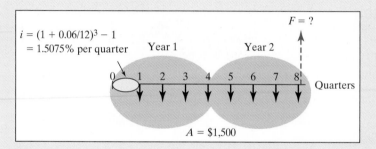

Figure 4.4 Quarterly deposits with monthly compounding (Example 4.5).

SOLUTION

Given: $A = \$1,500$ per quarter, $r = 6\%$ per year, $M = 12$ compounding periods per year, and $N = 8$ quarters.

Find: F.

We follow the aforementioned procedure for noncomparable compounding and payment periods:

1. Identify the parameter values for M, K, and C, where

$$M = 12 \text{ compounding periods per year,}$$

$$K = 4 \text{ payment periods per year,}$$

$$C = 3 \text{ interest periods per payment period.}$$

2. Use Eq. (4.2) to compute the effective interest:

$$i = \left(1 + \frac{0.06}{12}\right)^3 - 1$$

$$= 1.5075\% \text{ per quarter.}$$

3. Find the total number of payment periods, N:

$$N = K(\text{number of years}) = 4(2) = 8 \text{ quarters.}$$

4. Use i and N in the appropriate equivalence formulas:

$$F = \$1,500(F/A, 1.5075\%, 8) = \$12,652.60.$$

COMMENT: No 1.5075% interest table appears in Appendix A, but the interest factor can still be evaluated by $F = \$1,500(A/F, 0.5\%, 3)(F/A, 0.5\%, 24)$, where the first interest factor finds its equivalent monthly payment and the second interest factor converts the monthly payment series to an equivalent lump-sum future payment.

EXAMPLE 4.6 Compounding Occurs More Frequently than Payments Are Made (Continuous-Compounding Case)

A series of equal quarterly receipts of $500 extends over a period of five years as shown in Figure 4.5. What is the present worth of this quarterly payment series at 8% interest compounded continuously?

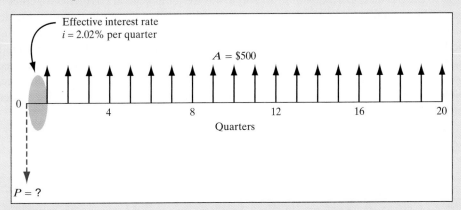

Figure 4.5   A present-worth calculation for an equal payment series with an interest rate of 8% compounded continuously (Example 4.6).

Discussion: A question that is equivalent to the preceding one is "How much do you need to deposit now in a savings account that earns 8% interest compounded continuously so that you can withdraw $500 at the end of each quarter for five years?" Since the payments are quarterly, we need to compute i per quarter for the equivalence calculations:

$$i = e^{r/K} - 1 = e^{0.08/4} - 1$$
$$= 2.02\% \text{ per quarter}$$

$$N = (4 \text{ payment periods per year})(5 \text{ years})$$
$$= 20 \text{ quarterly periods.}$$

SOLUTION

Given: $i = 2.02\%$ per quarter, $N = 20$ quarters, and $A = \$500$ per quarter.
Find: P.

Using the $(P/A, i, N)$ factor with $i = 2.02\%$ and $N = 20$, we find that

$$P = \$500(P/A, 2.02\%, 20)$$
$$= \$500(16.3199)$$
$$= \$8,159.96.$$

Compounding Is Less Frequent than Payments

The next two examples contain identical parameters for savings situations in which compounding occurs less frequently than payments. However, two different underlying assumptions govern how interest is calculated. In Example 4.7, the assumption is that, whenever a deposit is made, it starts to earn interest. In Example 4.8, the assumption is that the deposits made within a quarter do not earn interest until the end of that quarter. As a result, in Example 4.7 we transform the compounding period to conform to the payment period, and in Example 4.8 we lump several payments together to match the compounding period. In the real world, which assumption is applicable depends on the transactions and the financial institutions involved. The accounting methods used by many firms record cash transactions that occur within a compounding period as if they had occurred at the end of that period. For example, when cash flows occur daily, but the compounding period is monthly, the cash flows within each month are summed (ignoring interest) and treated as a single payment on which interest is calculated.

Note: *In this textbook, we assume that whenever the time point of a cash flow is specified, one cannot move it to another time point without considering the time value of money (i.e., the practice demonstrated in Example 4.7 should be followed).*

EXAMPLE 4.7 Compounding Is Less Frequent than Payments: Effective Interest Rate per Payment Period

Suppose you make $500 monthly deposits to a tax-deferred retirement plan that pays interest at a rate of 10% compounded quarterly. Compute the balance at the end of 10 years.

SOLUTION

Given: $r = 10\%$ per year, $M = 4$ quarterly compounding periods per year, $K = 12$ payment periods per year, $A = \$500$ per month, $N = 120$ months, and interest is accrued on deposits made during the compounding period.
Find: i, F.

As in the case of Example 4.5, the procedure for noncomparable compounding and payment periods is followed:

1. The parameter values for M, K, and C are

$$M = 4 \text{ compounding periods per year,}$$
$$K = 12 \text{ payment periods per year,}$$
$$C = \tfrac{1}{3} \text{ interest period per payment period.}$$

2. As shown in Figure 4.6, the effective interest rate per payment period is calculated with Eq. (4.2):

$$i = 0.826\% \text{ per month.}$$

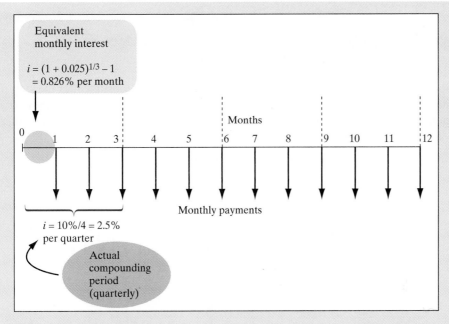

Figure 4.6 Calculation of equivalent monthly interest when the quarterly interest rate is specified (Example 4.7).

3. Find N:

$$N = (12)(10) = 120 \text{ payment periods.}$$

4. Use i and N in the appropriate equivalence formulas (Figure 4.7):

$$F = \$500(F/A, 0.826\%, 120)$$
$$= \$101,907.89.$$

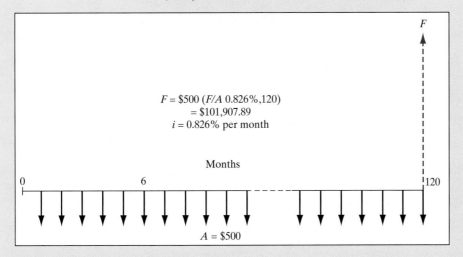

Figure 4.7 Cash flow diagram (Example 4.7).

EXAMPLE 4.8 Compounding Is Less Frequent than Payment: Summing Cash Flows to the End of the Compounding Period

Some financial institutions will not pay interest on funds deposited after the start of the compounding period. To illustrate, consider Example 4.7 again. Suppose that money deposited during a quarter (the compounding period) will not earn any interest (Figure 4.8). Compute what F will be at the end of 10 years.

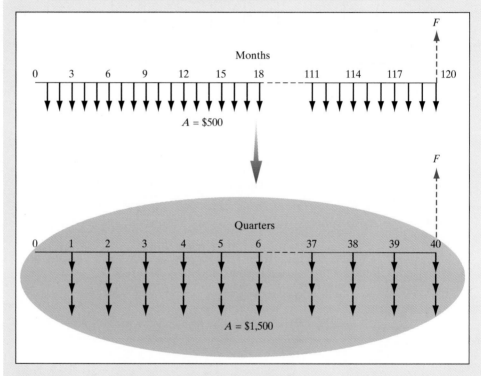

Figure 4.8 Transformed cash flow diagram created by summing monthly cash flows to the end of the quarterly compounding period (Example 4.8).

SOLUTION

Given: Same as for Example 4.7; however, no interest on flow during the compounding period.
Find: F.

In this case, the three monthly deposits during each quarterly period will be placed at the end of each quarter. Then the payment period coincides with the interest period, and we have

$$i = \frac{10\%}{4} = 2.5\% \text{ per quarter,}$$

$$A = 3(\$500) = \$1,500 \text{ per quarter,}$$

$$N = 4(10) = 40 \text{ payment periods,}$$
$$F = \$1,500(F/A, 2.5\%, 40) = \$101,103.83.$$

COMMENTS: In Example 4.8, the balance will be $804.06 less than in Example 4.7, a fact that is consistent with our understanding that increasing the frequency of compounding increases the future value of money. Some financial institutions follow the practice illustrated in Example 4.7. As an investor, you should reasonably ask yourself whether it makes sense to make deposits in an interest-bearing account more frequently than interest is paid. In the interim between interests compounding, you may be tying up your funds prematurely and forgoing other opportunities to earn interest.

Figure 4.9 is a decision chart that allows you to sum up how you can proceed to find the effective interest rate per payment period, given the various possible compounding and interest arrangements.

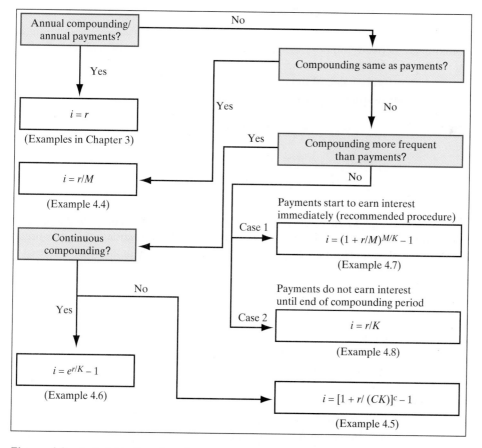

Figure 4.9 A decision flowchart demonstrating how to compute the effective interest rate i per payment period.

4.3 Equivalence Calculations with Continuous Payments

As we have seen so far, interest can be compounded annually, semiannually, monthly, or even continuously. Discrete compounding is appropriate for many financial transactions; mortgages, bonds, and installment loans, which require payments or receipts at discrete times, are good examples. In most businesses, however, transactions occur continuously throughout the year. In these circumstances, we may describe the financial transactions as having a continuous flow of money, for which continuous compounding and discounting are more realistic. This section illustrates how one establishes economic equivalence between cash flows under continuous compounding.

Continuous cash flows represent situations in which money flows continuously and at a known rate throughout a given period. In business, many daily cash flow transactions can be viewed as continuous. An advantage of the continuous-flow approach is that it more closely models the realities of business transactions. Costs for labor, for carrying inventory, and for operating and maintaining equipment are typical examples. Others include capital improvement projects that conserve energy or water or that process steam. Savings on these projects can occur continuously.

4.3.1 Single-Payment Transactions

First we will illustrate how single-payment formulas for continuous compounding and discounting are derived. Suppose that you invested P dollars at a nominal rate of r % interest for N years. If interest is compounded continuously, the effective annual interest is $i = e^r - 1$. The future value of the investment at the end of N years is obtained with the F/P factor by substituting $e^r - 1$ for i:

$$F = P(1 + i)^N$$
$$= P(1 + e^r - 1)^N$$
$$= Pe^{rN}.$$

This implies that $1 invested now at an interest rate of r% compounded continuously accumulates to e^{rN} dollars at the end of N years. Correspondingly, the present value of F due N years from now and discounted continuously at an interest rate of r% is equal to

$$P = Fe^{-rN}.$$

We can say that the present value of $1 due N years from now and discounted continuously at an annual interest rate of r% is equal to e^{-rN} dollars.

4.3.2 Continuous-Funds Flow

Suppose that an investment's future cash flow per unit of time (e.g., per year) can be expressed by a continuous function ($f(t)$) that can take any shape. Suppose also that the

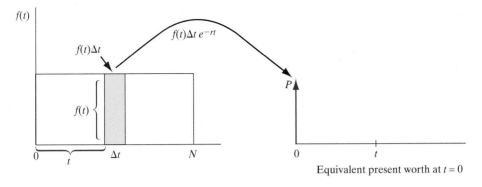

Figure 4.10   Finding an equivalent present worth of a continuous-flow payment function $f(t)$ at a nominal rate of $r\%$.

investment promises to generate cash of $f(t)\Delta t$ dollars between t and $t + \Delta t$, where t is a point in the time interval $0 \leq t \leq N$ (Figure 4.10). If the nominal interest rate is a constant r during this interval, the present value of the cash stream is given approximately by the expression

$$\Sigma(f(t)\Delta t)e^{-rt},$$

where e^{-rt} is the discounting factor that converts future dollars into present dollars. With the project's life extending from 0 to N, we take the summation over all subperiods (compounding periods) in the interval from 0 to N. As the interval is divided into smaller and smaller segments (i.e., as Δt approaches zero), we obtain the expression for the present value by the integral

$$P = \int_0^N f(t)e^{-rt}\,dt. \tag{4.5}$$

Similarly, the expression for the future value of the cash flow stream is given by the equation

$$F = Pe^{rN} = \int_0^N f(t)e^{r(N-t)}\,dt, \tag{4.6}$$

where $e^{r(N-t)}$ is the compounding factor that converts present dollars into future dollars. It is important to observe that the time unit is the *year*, because the effective interest rate is expressed in terms of a year. Therefore, all time units in equivalence calculations must be converted into years. Table 4.2 summarizes some typical continuous cash functions that can facilitate equivalence calculations.[2]

---

[2] Chan S. Park and Gunter P. Sharp-Bette, *Advanced Engineering Economics*. New York: John Wiley & Sons, 1990. (Reprinted by permission of John Wiley & Sons, Inc.)

TABLE 4.2 Summary of Interest Factors for Typical Continuous Cash Flows with Continuous Compounding

Type of Cash Flow	Cash Flow Function	Parameters Find	Parameters Given	Algebraic Notation	Factor Notation
Uniform (step)	$f(t) = \bar{A}$	P	$\bar{A}$	$\bar{A}\left[\dfrac{e^{rN} - 1}{re^{rN}}\right]$	$(P/\bar{A}, r, N)$
		$\bar{A}$	P	$P\left[\dfrac{re^{rN}}{e^{rN} - 1}\right]$	$(\bar{A}/P, r, N)$
		F	$\bar{A}$	$\bar{A}\left[\dfrac{e^{rN} - 1}{r}\right]$	$(F/\bar{A}, r, N)$
		$\bar{A}$	F	$F\left[\dfrac{r}{e^{rN} - 1}\right]$	$(\bar{A}/P, r, N)$
Gradient (ramp)	$f(t) = Gt$	P	G	$\dfrac{G}{r^2}(1 - e^{-rN}) - \dfrac{G}{r}(Ne^{-rN})$	
Decay	$f(t) = ce^{-jt}$ j^t = decay rate with time	P	c, j	$\dfrac{c}{r + j}(1 - e^{-(r+j)N})$	

EXAMPLE 4.9 Comparison of Daily Flows and Daily Compounding with Continuous Flows and Continuous Compounding

Consider a situation in which money flows daily. Suppose you own a retail shop and generate $200 cash each day. You establish a special business account and deposit your daily cash flows in an account for 15 months. The account earns an interest rate of 6%. Compare the accumulated cash values at the end of 15 months, assuming

(a) Daily compounding and

(b) Continuous compounding.

SOLUTION

(a) With daily compounding:

Given: $A = \$200$ per day $r = 6\%$ per year, $M = 365$ compounding periods per year, and $N = 455$ days.
Find: F.

Assuming that there are 455 days in the 15-month period, we find that

$$i = 6\%/365$$
$$= 0.01644\% \text{ per day,}$$
$$N = 455 \text{ days.}$$

The balance at the end of 15 months will be

$$F = \$200(F/A, 0.01644\%, 455)$$
$$= \$200(472.4095)$$
$$= \$94,482.$$

(b) With continuous compounding:

Now we approximate this discrete cash flow series by a uniform continuous cash flow function as shown in Figure 4.11. In this situation, an amount flows at the rate of $\overline{A}$ per year for N years.

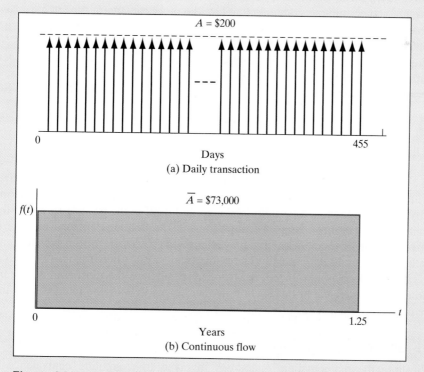

$A = \$200$

Days
(a) Daily transaction

$\overline{A} = \$73,000$

$f(t)$

Years
(b) Continuous flow

Figure 4.11 Comparison between daily transaction and continuous-funds flow transaction (Example 4.9).

Note: *Our time unit is a year.* Thus, a 15-month period is 1.25 years. Then the cash flow function is expressed as

$$f(t) = \overline{A}, 0 \le t \le 1.25$$

$$= \$200(365)$$

$$= \$73,000 \text{ per year.}$$

Given: $\overline{A} = \$73,000$ per year, $r = 6\%$ per year, compounded continuously, and $N = 1.25$ years.

Find: F.

Substituting these values back into Eq. (4.6) yields

$$F = \int_0^{1.25} 73,000 e^{0.06(1.25-t)} \, dt$$

$$= \$73,000 \left[\frac{e^{0.075} - 1}{0.06} \right]$$

$$= \$94,759.$$

The factor in the bracket is known as the **funds flow compound amount factor** and is designated $(F/\overline{A}, r, N)$ as shown in Table 4.2. Notice that the difference between the two methods is only $277 (less than 0.3%).

COMMENTS: As shown in this example, the differences between discrete daily compounding and continuous compounding have no practical significance in most cases. Consequently, as a mathematical convenience, instead of assuming that money flows in discrete increments at the end of each day, we could assume that money flows continuously at a uniform rate during the period in question. This type of cash flow assumption is common practice in the chemical industry.

4.4 Changing Interest Rates

Up to this point, we have assumed a constant interest rate in our equivalence calculations. When an equivalence calculation extends over several years, more than one interest rate may be applicable to properly account for the time value of money. That is to say, over time, interest rates available in the financial marketplace fluctuate, and a financial institution that is committed to a long-term loan may find itself in the position of losing the opportunity to earn higher interest because some of its holdings are tied up in a lower interest loan. The financial institution may attempt to protect itself from such lost earning opportunities by building gradually increasing interest rates into a long-term loan at the outset. Adjustable-rate mortgage (ARM) loans are perhaps the most common examples of variable interest rates. In this section, we will consider variable interest rates in both single payments and a series of cash flows.

4.4.1 Single Sums of Money

To illustrate the mathematical operations involved in computing equivalence under changing interest rates, first consider the investment of a single sum of money, P, in a savings account for N interest periods. If i_n denotes the interest rate appropriate during period n, then the future worth equivalent for a single sum of money can be expressed as

$$F = P(1 + i_1)(1 + i_2)\cdots(1 + i_{N-1})(1 + i_N), \qquad (4.7)$$

and solving for P yields the inverse relation

$$P = F[(1 + i_1)(1 + i_2)\cdots(1 + i_{N-1})(1 + i_N)]^{-1}. \qquad (4.8)$$

EXAMPLE 4.10 Changing Interest Rates with a Lump-Sum Amount

Suppose you deposit $2,000 in an individual retirement account (IRA) that pays interest at 6% compounded monthly for the first two years and 9% compounded monthly for the next three years. Determine the balance at the end of five years (Figure 4.12).

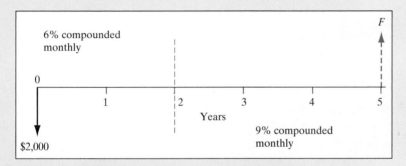

Figure 4.12   Changing interest rates (Example 4.10).

SOLUTION

Given: $P = \$2,000$, $r = 6\%$ per year for first two years, 9% per year for last three years, $M = 4$ compounding periods per year, $N = 20$ quarters.
Find: F.

We will compute the value of F in two steps. First we will compute the balance B_2 in the account at the end of two years. With 6% compounded quarterly, we have

$$i = 6\%/12 = 0.5\%$$

$$N = 12(2) = 24 \text{ months}$$

$$B_2 = \$2,000(F/P, 0.5\%, 24)$$

$$= \$2,000(1.12716)$$

$$= \$2,254.$$

Since the fund is not withdrawn, but reinvested at 9% compounded monthly, as a second step we compute the final balance as follows:

$$i = 9\%/12 = 0.75\%$$

$$N = 12(3) = 36 \text{ months}$$

$$F = B_2(F/P, 0.75\%, 36)$$

$$= \$2,254(1.3086)$$

$$= \$2,950.$$

4.4.2 Series of Cash Flows

The phenomenon of changing interest rates can easily be extended to a series of cash flows. In this case, the present worth of a series of cash flows can be represented as

$$P = A_1(1 + i_1)^{-1} + A_2[(1 + i_1)^{-1}(1 + i_2)^{-1}] + \ldots$$
$$+ A_N[(1 + i_1)^{-1}(1 + i_2)^{-1} \ldots (1 + i_N)^{-1}]. \qquad (4.9)$$

The future worth of a series of cash flows is given by the inverse of Eq. (4.9):

$$F = A_1[(1 + i_2)(1 + i_3) \ldots (1 + i_N)]$$
$$+ A_2[(1 + i_3)(1 + i_4) \ldots (1 + i_N)] + \ldots + A_N. \qquad (4.10)$$

The uniform series equivalent is obtained in two steps. First, the present-worth equivalent of the series is found from Eq. (4.9). Then A is obtained after establishing the following equivalence equation:

$$P = A(1 + i_1)^{-1} + A[(1 + i_1)^{-1}(1 + i_2)^{-1}] + \ldots$$
$$+ A[(1 + i_1)^{-1}(1 + i_2)^{-1} \ldots (1 + i_N)^{-1}]. \qquad (4.11)$$

EXAMPLE 4.11 Changing Interest Rates with Uneven Cash Flow Series

Consider the cash flow in Figure 4.13 with the interest rates indicated, and determine the uniform series equivalent of the cash flow series.

DISCUSSION: In this problem and many others, the easiest approach involves collapsing the original flow into a single equivalent amount, for example, at time zero, and then converting the single amount into the final desired form.

SOLUTION

Given: Cash flows and interest rates as shown in Figure 4.13; $N = 3$.
Find: A.

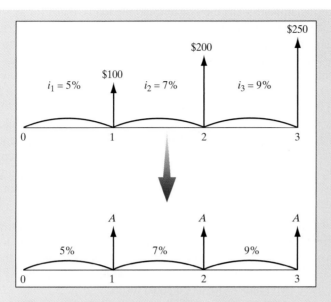

Figure 4.13 Equivalence calculation with changing interest rates (Example 4.11).

Using Eq. (4.9), we find the present worth:

$$P = \$100(P/F, 5\%, 1) + \$200(P/F, 5\%, 1)(P/F, 7\%, 1)$$
$$+ \$250(P/F, 5\%, 1)(P/F, 7\%, 1)(P/F, 9\%, 1)$$
$$= \$477.41.$$

Then we obtain the uniform series equivalent as follows:

$$\$477.41 = A(P/F, 5\%, 1) + A(P/F, 5\%, 1)(P/F, 7\%, 1)$$
$$+ A(P/F, 5\%, 1)(P/F, 7\%, 1)(P/F, 9\%, 1)$$
$$= 2.6591A$$
$$A = \$179.54.$$

4.5 Debt Management

Credit card debt and commercial loans are among the most significant financial transactions involving interest. Many types of loans are available, but here we will focus on those most frequently used by individuals and in business.

4.5.1 Commercial Loans

One of the most important applications of compound interest involves loans that are paid off in **installments** over time. If the loan is to be repaid in equal periodic amounts (weekly, monthly, quarterly, or annually), it is said to be an **amortized loan**.

Examples of installment loans include automobile loans, loans for appliances, home mortgage loans, and the majority of business debts other than very short-term loans. Most commercial loans have interest that is compounded monthly. With an auto loan, a local bank or a dealer advances you the money to pay for the car, and you repay the principal plus interest in monthly installments, usually over a period of three to five years. The car is your collateral. If you don't keep up with your payments, the lender can repossess the car and keep all the payments you have made.

Two things determine what borrowing will cost you: the finance charge and the length of the loan. The cheapest loan is not the one with the lowest payments or even the one with the lowest interest rate. Instead, you have to look at (1) the total cost of borrowing, which depends on the interest rate plus fees, and (2) the term, or length of time it takes you to repay the loan. While you probably cannot influence the rate or the fees, you may be able to arrange for a shorter term.

- **The annual percentage rate** (APR) is set by lenders, who are required to tell you what a loan will actually cost per year, expressed as an APR. Some lenders charge lower interest, but add high fees; others do the reverse. Combining the fees with a year of interest charges to give you the true annual interest rate, the APR allows you to compare these two kinds of loans on equal terms.
- **Fees** are the expenses the lender will charge to lend the money. The application fee covers processing expenses. Attorney fees pay the lender's attorney. Credit search fees cover researching your credit history. Origination fees cover administrative costs and sometimes appraisal fees. All these fees add up very quickly and can substantially increase the cost of your loan.
- **Finance charges** are the cost of borrowing. For most loans, they include all the interest, fees, service charges, points, credit-related insurance premiums, and any other charges.
- **The periodic interest rate** is the interest the lender will charge on the amount you borrow. If lender also charges fees, the periodic interest rate will not be the true interest rate.
- **The term of your loan** is crucial in determining its cost. Shorter terms mean squeezing larger amounts into fewer payments. However, they also mean paying interest for fewer years, saving a lot of money.

Amortized Installment Loans

So far, we have considered many instances of amortized loans in which we calculated present or future values of the loans or the amounts of the installment payments. An additional aspect of amortized loans that will be of great interest to us is calculating the amount of interest versus the portion of the principal that is paid off in each installment. As we shall explore more fully in Chapter 10, the interest paid on a loan is an important element in calculating taxable income and has repercussions for both personal and business loan transactions. For now, we will focus on several methods of calculating interest and principal paid at any point in the life of the loan.

In calculating the size of a monthly installment, lending institutions may use two types of schemes. The first is the conventional amortized loan, based on the compound-interest method, and the other is the add-on loan, based on the simple-interest concept. We explain each method in what follows, but it should be understood that the amortized loan is the most common in various commercial lending. The add-on loan is common in financing appliances as well as furniture.

In a typical amortized loan, the amount of interest owed for a specified period is calculated on the basis of the remaining balance on the loan at the beginning of the period. A set of formulas has been developed to compute the remaining loan balance, interest payment, and principal payment for a specified period. Suppose we borrow an amount P at an interest rate i and agree to repay this principal sum P, including interest, in equal payments A over N periods. Then the size of the payment is $A = P(A/P, i, N)$, and each payment is divided into an amount that is interest and a remaining amount that goes toward paying the principal.

Let

B_n = Remaining balance at the end of period n, with $B_0 = P$,

I_n = Interest payment in period n, where $I_n = B_{n-1}i$,

P_n = Principal payment in period n.

Then each payment can be defined as

$$A = P_n + I_n. \tag{4.12}$$

The interest and principal payments for an amortized loan can be determined in several ways; two are presented here. No clear-cut reason is available to prefer one method over the other. Method 1, however, may be easier to adopt when the computational process is automated through a spreadsheet application, whereas Method 2 may be more suitable for obtaining a quick solution when a period is specified. You should become comfortable with at least one of these methods; pick the one that comes most naturally to you.

Method 1: Tabular Method. The first method is tabular. The interest charge for a given period is computed progressively on the basis of the remaining balance at the beginning of that period. Example 4.12 illustrates the process of creating a loan repayment schedule based on an iterative approach.

EXAMPLE 4.12 Loan Balance, Principal, and Interest: Tabular Method

Suppose you secure a home improvement loan in the amount of $5,000 from a local bank. The loan officer computes your monthly payment as follows:

Contract amount = $5,000,

Contract period = 24 months,

Annual percentage rate = 12%,

Monthly installments = $235.37.

Figure 4.14 is the cash flow diagram. Construct the loan payment schedule by showing the remaining balance, interest payment, and principal payment at the end of each period over the life of the loan.

SOLUTION

Given: $P = \$5,000$, $A = \$235.37$ per month, $r = 12\%$ per year, $M = 12$ compounding periods per year, and $N = 24$ months.

Find: B_n and I_n for $n = 1$ to 24.

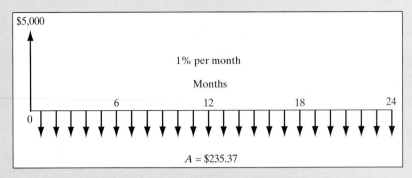

Figure 4.14 Cash flow diagram of the home improvement loan with an APR of 12% (Example 4.12).

TABLE 4.3 Creating a Loan Repayment Schedule with Excel (Example 4.12)

Payment No.	Size of Payment	Principal Payment	Interest Payment	Loan Balance
1	$235.37	$185.37	$50.00	$4,814.63
2	235.37	187.22	48.15	4,627.41
3	235.37	189.09	46.27	4,438.32
4	235.37	190.98	44.38	4,247.33
5	235.37	192.89	42.47	4,054.44
6	235.37	194.83	40.54	3,859.62
7	235.37	196.77	38.60	3,662.85
8	235.37	198.74	36.63	3,464.11
9	235.37	200.73	34.64	3,263.38
10	235.37	202.73	32.63	3,060.65
11	235.37	204.76	30.61	2,855.89
12	235.37	206.81	28.56	2,649.08
13	235.37	208.88	26.49	2,440.20
14	235.37	210.97	24.40	2,229.24
15	235.37	213.08	22.29	2,016.16
16	235.37	215.21	20.16	1,800.96
17	235.37	217.36	18.01	1,583.60
18	235.37	219.53	15.84	1,364.07
19	235.37	221.73	13.64	1,142.34
20	235.37	223.94	11.42	918.40
21	235.37	226.18	9.18	692.21
22	235.37	228.45	6.92	463.77
23	235.37	230.73	4.64	233.04
24	235.37	233.04	2.33	0.00

We can easily see how the bank calculated the monthly payment of $235.37. Since the effective interest rate per payment period on this loan is 1% per month, we establish the following equivalence relationship:

$$\$235.37(P/A, 1\%, 24) = \$235.37(21.2431) = \$5,000.$$

The loan payment schedule can be constructed as in Table 4.3. The interest due at $n = 1$ is $50.00, 1% of the $5,000 outstanding during the first month. The $185.37 left over is applied to the principal, reducing the amount outstanding in the second month to $4,814.63. The interest due in the second month is 1% of $4,814.63, or $48.15, leaving $187.22 for repayment of the principal. At $n = 24$, the last $235.37 payment is just sufficient to pay the interest on the unpaid principal of the loan and to repay the remaining principal. Figure 4.15 illustrates the ratios between the interest and principal payments over the life of the loan.

COMMENTS: Certainly, constructing a loan repayment schedule such as that in Table 4.3 can be tedious and time consuming, unless a computer is used. As you can see in the website for this book, you can download an Excel file that creates the loan repayment schedule, on which you can make any adjustment to solve a typical loan problem of your choice.

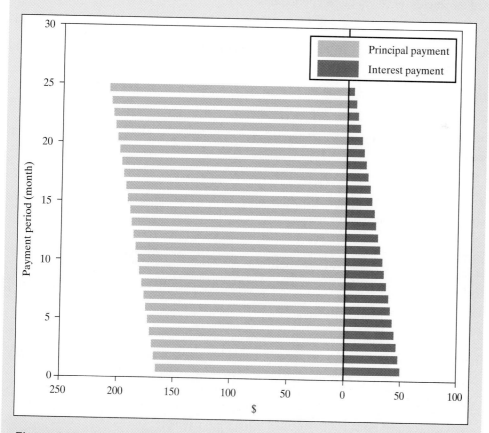

Figure 4.15 The proportions of principal and interest payments over the life of the loan (monthly payment = $235.37) (Example 4.12).

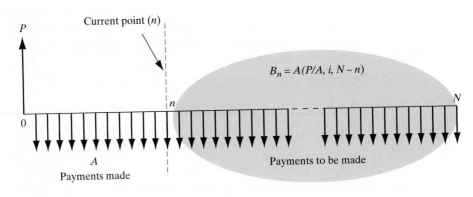

Figure 4.16 Calculating the remaining loan balance on the basis of Method 2.

Method 2: Remaining-Balance Method. Alternatively, we can derive B_n by computing the equivalent payments remaining after the nth payment. Thus, the balance with $N - n$ payments remaining is

$$B_n = A(P/A, i, N - n),\qquad(4.13)$$

and the interest payment during period n is

$$I_n = (B_{n-1})i = A(P/A, i, N - n + 1)i,\qquad(4.14)$$

where $A(P/A, i, N - n + 1)$ is the balance remaining at the end of period $n - 1$ and

$$P_n = A - I_n = A - A(P/A, i, N - n + 1)i$$
$$= A[1 - (P/A, i, N - n + 1)i].$$

Knowing the interest factor relationship $(P/F, i, n) = 1 - (P/A, i, n)i$ from Table 3.4, we obtain

$$P_n = A(P/F, i, N - n + 1).\qquad(4.15)$$

As we can see in Figure 4.16, this method provides more concise expressions for computing the balance of the loan.

EXAMPLE 4.13 Loan Balances, Principal, and Interest: Remaining-Balance Method

Consider the home improvement loan in Example 4.12, and

(a) For the sixth payment, compute both the interest and principal payments.
(b) Immediately after making the sixth monthly payment, you would like to pay off the remainder of the loan in a lump sum. What is the required amount?

SOLUTION

(a) Interest and principal payments for the sixth payment.
   Given: (as for Example 4.12)

Find: I_6 and P_6.

Using Eqs. (4.14) and (4.15), we compute

$$I_6 = \$235.37(P/A, 1\%, 19)(0.01)$$
$$= (\$4,054.44)(0.01)$$
$$= \$40.54.$$
$$P_6 = \$235.37(P/F, 1\%, 19) = \$194.83,$$

or we simply subtract the interest payment from the monthly payment:

$$P_6 = \$235.37 - \$40.54 = \$194.83.$$

(b) Remaining balance after the sixth payment.

The lower half of Figure 4.17 shows the cash flow diagram that applies to this part of the problem. We can compute the amount you owe after you make the sixth payment by calculating the equivalent worth of the remaining 18 payments at the end of the sixth month, with the time scale shifted by 6:

Given: $A = \$235.37$, $i = 1\%$ per month, and $N = 18$ months.

Find: Balance remaining after six months (B_6).

$$B_6 = \$235.37(P/A, 1\%, 18) = \$3,859.62.$$

If you desire to pay off the remainder of the loan at the end of the sixth payment, you must come up with $3,859.62. To verify our results, compare this answer with the value given in Table 4.3.

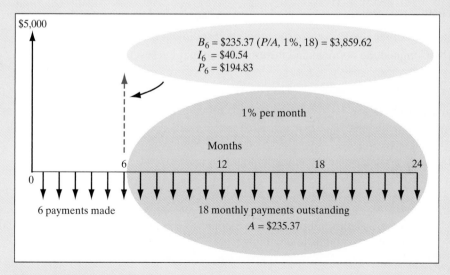

$$B_6 = \$235.37 \, (P/A, 1\%, 18) = \$3,859.62$$
$$I_6 = \$40.54$$
$$P_6 = \$194.83$$

1% per month

Months

6 payments made

18 monthly payments outstanding
$A = \$235.37$

Figure 4.17 Computing the outstanding loan balance after making the sixth payment on the home improvement loan (Example 4.13).

Add-On Interest Loans

Add-on Interest:
A method of computing interest whereby interest charges are made for the entire principal amount for the entire term, regardless of any repayments of principal made.

The add-on loan is totally different from the popular amortized loan. In the add-on loan, the total interest to be paid is precalculated and added to the principal. The principal and this precalculated interest amount are then paid together in equal installments. In such a case, the interest rate quoted is not the effective interest rate, but what is known as **add-on interest**. If you borrow P for N years at an add-on rate of i, with equal payments due at the end of each month, a typical financial institution might compute the monthly installment payments as follows:

$$\text{Total add-on interest} = P(i)(N),$$

$$\text{Principal plus add-on interest} = P + P(i)(N) = P(1 + iN)$$

$$\text{Monthly installments} = \frac{P(1 + iN)}{(12 \times N)}. \qquad (4.16)$$

Notice that the add-on interest is *simple interest*. Once the monthly payment is determined, the financial institution computes the APR on the basis of this payment, and you will be told what the value will be. Even though the add-on interest is specified along with the APR, many ill-informed borrowers think that they are actually paying the add-on rate quoted for this installment loan. To see how much interest you actually pay under a typical add-on loan arrangement, consider Example 4.14.

EXAMPLE 4.14 Effective Interest Rate for an Add-On Interest Loan

Consider again the home improvement loan problem in Example 4.12. Suppose that you borrow $5,000 with an add-on rate of 12% for two years. You will make 24 equal monthly payments.

(a) Determine the amount of the monthly installment.

(b) Compute the nominal and the effective annual interest rate on the loan.

SOLUTION

Given: Add-on rate = 12% per year, loan amount $(P) = \$5,000$, and $N = 2$ years.
Find: (a) A and (b) i_a and i.

(a) First we determine the amount of add-on interest:

$$iPN = (0.12)(\$5,000)(2) = \$1,200.$$

Then we add this simple-interest amount to the principal and divide the total amount by 24 months to obtain A:

$$A = \frac{(\$5,000 + \$1,200)}{24} = \$258.33.$$

(b) Putting yourself in the lender's position, compute the APR value of the loan just described. Since you are making monthly payments with monthly compounding, you need to find the effective interest rate that makes the present $5,000 sum equivalent to 24 future monthly payments of $258.33. In this situation, we are solving for i in the equation

$$\$258.33 = \$5,000(A/P, i, 24),$$

or

$$(A/P, i, 24) = 0.0517.$$

You know the value of the A/P factor, but you do not know the interest rate i. As a result, you need to look through several interest tables and determine i by interpolation. A more effective approach is to use Excel's RATE function with the following parameters:

$$=\text{RATE}(\text{N,A,P,F,type,guess})$$
$$=\text{RATE}(24,258.33,-5000,0,0,1\%) \rightarrow 1.7975\%$$

The nominal interest rate for this add-on loan is $1.7975 \times 12 = 21.57\%$, and the effective annual interest rate is $(1 + 0.01975)^{12} - 1 = 26.45\%$, rather than the 12% quoted add-on interest. When you take out a loan, you should not confuse the add-on interest rate stated by the lender with the actual interest cost of the loan.

COMMENTS: In the real world, truth-in-lending laws require that APR information always be provided in mortgage and other loan situations, so you would not have to calculate nominal interest as a prospective borrower (although you might be interested in calculating the actual or effective interest). However, in later engineering economic analyses, you will discover that solving for implicit interest rates, or rates of return on investment, is performed regularly. Our purpose in this text is to periodically give you some practice with this type of problem, even though the scenario described does not exactly model the real-world information you would be given.

4.5.2 Loan versus Lease Financing

When, for example, you choose a car, you also choose how to pay for it. If you do not have the cash on hand to buy a new car outright—and most of us don't—you can consider taking out a loan or leasing the car to spread the payments over time. Deciding whether to pay cash, take a loan, or sign a lease depends on a number of personal as well as economic factors. Leasing is an option that lets you pay for the portion of a vehicle you expect to use over a specified term, plus a charge for rent, taxes, and fees. For instance, you might want a $20,000 vehicle. Suppose that vehicle might be worth about $9,000 at the end of your three-year lease. (This is called the residual value.)

- If you have enough money to buy the car, you could purchase it in cash. If you pay cash, however, you will lose the opportunity to earn interest on the amount you spend. That could be substantial if you know of an investment that is paying a good return.
- If you purchase the vehicle using debt financing, your monthly payments will be based on the entire $20,000 value of the vehicle. You will continue to own the vehicle

at the end of your financing term, but the interest you will pay on the loan will drive up the real cost of the car.

- If you lease the same vehicle, your monthly payments will be based on the amount of the vehicle you expect to "use up" over the term of the lease. This value ($11,000 in our example) is the difference between the original cost ($20,000) and the estimated value at the end of the lease ($9,000). With leasing, the length of your lease agreement, the monthly payments, and the yearly mileage allowance can be tailored to your driving needs. The greatest financial appeal for leasing is low initial outlay costs: Usually you pay a leasing administrative fee, one month's lease payment, and a refundable security deposit. The terms of your lease will include a specific mileage allowance; if you put additional miles on your car, you will have to pay more for each extra mile.

Discount Rate to Use in Comparing Different Financing Options

In calculating the net cost of financing a car, we need to decide which interest rate to use in discounting the loan repayment series. The dealer's (bank's) interest rate is supposed to reflect the time value of money of the dealer (or the bank) and is factored into the required payments. However, the correct interest rate to use in comparing financing options is the interest rate that reflects *your* time value of money. For most individuals, this interest rate might be equivalent to the savings rate from their deposits. To illustrate, consider Example 4.15, in which we compare three auto financing options.

EXAMPLE 4.15 Financing your Vehicle: Paying Cash, Taking a Loan, or Leasing

Suppose you intend to own or lease a vehicle for 42 months. Consider the following three ways of financing the vehicle—say, a 2006 BMW 325 Ci 2-D coupe:

- **Option A:** Purchase the vehicle at the normal price of $32,508, and pay for the vehicle over 42 months with equal monthly payments at 5.65% APR financing.

- **Option B:** Purchase the vehicle at a discount price of $31,020 to be paid immediately.

- **Option C:** Lease the vehicle for 42 months.

The accompanying chart lists the items of interest under each option. For each option, license, title, and registration fees, as well as taxes and insurance, are extra.

For the lease option, the lessee must come up with $1,507.76 at signing. This cash due at signing includes the first month's lease payment of $513.76 and a $994 administrative fee. The lease rate is based on 60,000 miles over the life of the contract. There will be a surcharge at the rate of 18 cents per mile for any additional miles driven over 60,000. No security deposit is required; however, a $395 disposition fee is due at the end of the lease, at which time the lessee has the option to purchase the car for $17,817. The lessee is also responsible for excessive wear and

Item	Option A Debt Financing	Option B Paying Cash	Option C Lease Financing
Price	$32,508	$32,508	$32,508
Down payment	$4,500	$0	$0
APR(%)	5.65%		
Monthly payment	$736.53		$513.76
Length	42 months		42 months
Fees			$994
Cash due at end of lease			$395
Purchase option at end of lease			$17,817
Cash due at signing	$4,500	$31,020	$1,507.76

use. If the funds that would be used to purchase the vehicle are presently earning 4.5% annual interest compounded monthly, which financing option is a better choice?

DISCUSSION: With a lease payment, you pay for the portion of the vehicle you expect to use. At the end of the lease, you simply return the vehicle to the dealer and pay the agreed-upon disposal fee. With traditional financing, your monthly payment is based on the entire $32,508 value of the vehicle, and you will own the vehicle at the end of your financing terms. Since you are comparing the options over 42 months, you must explicitly consider the unused portion (resale value) of the vehicle at the end of the term. In other words, you must consider the resale value of the vehicle in order to figure out the net cost of owning it. *As the resale value, you could use the $17,817 quoted by the dealer in the lease option.* Then you have to ask yourself if you can get that kind of resale value after 42 months' ownership.

Note that the 5.65% APR represents the dealer's interest rate used in calculating the loan payments. With 5.65% interest, your monthly payments will be $A = (\$32,508 - \$4,500)(A/P, 5.65\%/12, 42) = \$736.53$. Note also, however, that the 4.5% APR represents your earning opportunity rate. In other words, if you do not buy the car, your money continues to earn 4.5% APR. Therefore, 4.5% represents your opportunity cost of purchasing the car. So which interest rate do you use in your analysis? Clearly, the 4.5% rate is the appropriate one to use.

SOLUTION

Given: Financial facts shown in Figure 4.18, $r = 4.5\%$, payment period = monthly, and compounding period = monthly.

Find: The most economical financing option, under the assumption that you will be able to sell the vehicle for $17,817 at the end of 42 months.

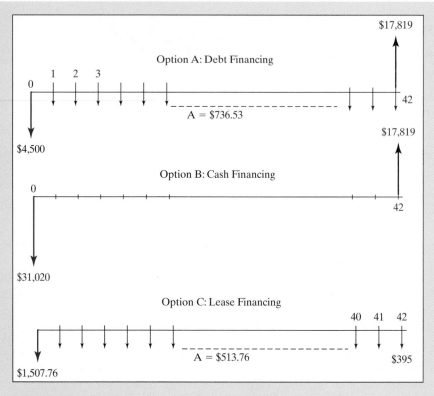

Figure 4.18 Comparing different financing options.

For each option, we will calculate the net equivalent total cost at $n = 0$. Since the loan payments occur monthly, we need to determine the effective interest rate per month, which is 4.5%/12.

- **Option A: Conventional debt financing**

    The equivalent present cost of the total loan payments is

    $$P_1 = \$4,500 + \$736.53(P/A, 4.5\%/12, 42)$$
    $$= \$33,071.77.$$

    The equivalent present worth of the resale value of the car is

    $$P_2 = \$17,817(P/F, 4.5\%/12, 42) = \$15,225.13.$$

    The equivalent net financing cost is

    $$P_{\text{Option A}} = \$33,071.77 - \$15,225.13$$
    $$= \$17,846.64.$$

- **Option B: Cash financing**

    $$P_{\text{Option B}} = \$31,020 - \$17,817(P/F, 4.5\%/12, 42)$$
    $$= \$31,020 - \$15,225.13$$
    $$= \$15,844.87.$$

> - **Option C: Lease financing**
>
>   The equivalent present cost of the total lease payments is
>
> $$P_{\text{Option C}} = \$1,507.76 + \$513.76(P/A, 4.5\%/12, 41)$$
> $$+ \ \$395(P/F, 4.5\%/12, 42)$$
> $$= \$1,507.76 + \$19,490.96 + \$337.54$$
> $$= \$21,336.26$$
>
>   It appears that, at 4.5% interest compounded monthly, the cash financing option is the most economical one.

4.5.3 Home Mortgage

The term **mortgage** refers to a special type of loan for buying a piece of property, such as a house. The most common mortgages are fixed-rate amortized loans, adjustable-rate loans, and graduated-payment loans. As with many of the amortized loans we have considered so far, most home mortgages involve monthly payments and monthly compounding. However, as the names of these types of mortgages suggest, a number of ways determine how monthly payments and interest rates can be structured over the life of the mortgage.

Mortgage: A loan to finance the purchase of real estate, usually with specified payment periods and interest rates.

The Cost of a Mortgage

The cost of a mortgage depends on the amount you borrow, the interest you pay, and how long you take to repay the loan. Since monthly payments spread the cost of a mortgage over a long period, it is easy to forget the total expense. For example, if you borrow $100,000 for 30 years at 8.5% interest, your total repayment will be around $277,000, more than two-and-a-half times the original loan! Minor differences in the interest rate—8.5% versus 8%—can add up to a lot of money over 30 years. At 8%, the total repaid would be $264,240, almost $13,000 less than at the 8.5% rate. Other than the interest rate, any of the following factors will increase the overall cost, but a higher interest rate and longer term will have the greatest impact:

ARM: A mortgage with an interest rate that may change, usually in response to changes in the Treasury Bill rate or the prime rate.

- **Loan amount.** This is the amount you actually borrow after fees and points are deducted. It is the basis for figuring the real interest, or APR, on the money you are borrowing.
- **Loan term.** With a shorter term, you will pay less interest overall, and your monthly payments will be somewhat larger. A 15-year mortgage, as opposed to a 30-year mortgage for the same amount, can cut your costs by more than 55%.
- **Payment frequency.** You can pay your mortgage biweekly instead of monthly, or you can make an additional payment each month. With biweekly payments, you make 26 regular payments instead of 12 every year. The mortgage is paid off in a little more than half the time, and you pay a little more than half the interest. With an additional payment each month, you can reduce your principal. With a fixed-rate mortgage, you pay off the loan more quickly, but regular monthly payments remain the same.
- **Points (prepaid interest).** Points are interest that you prepay at the closing on your home. Each point is 1% of the loan amount. For example, on a $100,000

Origination fee:
A fee charged
by a lender for
processing a loan
application,
expressed as a
percentage of
the mortgage
amount.

loan, three points represents a prepayment of $3,000. This is equivalent to financing $97,000, but your payments are based on a $100,000 loan.

- **Fees.** Fees include application fees, loan origination fees, and other initial costs imposed by the lender.

Lenders might be willing to raise the rate by a fraction (say, $\frac{1}{8}\%$ or $\frac{1}{4}\%$) and lower the points—or the reverse—as long as they make the same profit. The advantages of fewer points are lower closing costs and laying out less money when you are apt to need it most. However, if you plan to keep the house longer than five to seven years, paying more points to get a lower interest rate will reduce your long-term cost. Example 4.16 examines the effect of points on the cost of borrowing.

EXAMPLE 4.16 Points or No Points?

When you are shopping for a home mortgage loan, you frequently encounter various types of borrowing options, including paying points up front and paying no points but accepting a slightly higher interest rate. Suppose you want to finance a home mortgage of $100,000 at a 15-year fixed interest rate. Countrywide, a leading independent home lender offers the following two options with no origination fees:

- **Option 1:** Pay one point with 6.375% interest.
- **Option 2:** Pay no points with 6.75% interest.

A point is equivalent to 1% of the face value of the mortgage. In other words, with Option 1, you are borrowing only $99,000, but your lender will calculate your monthly payments on the basis of $100,000. Compute the APR for Option 1.

DISCUSSION: Discount fees or points are a fact of life with mortgages. A point is a fee charged by a lender to increase the lender's effective yield on the money borrowed. Points are charged in lieu of interest; the more points paid, the lower is the rate of interest required on the loan to provide the same yield or return to the lender. One point equals 1% of the loan amount. Origination fees are the fees charged by a lender to prepare a loan document, make credit checks, and inspect and sometimes appraise a property. These fees are usually computed as a percentage of the face value of the mortgage.

SOLUTION

Given: $P = \$100,000$, $r = 6.375\%$, $N = 180$ months, and discount point $= 1$ point. Find: APR.

We first need to find out how much the lender will calculate your monthly payments to be. Since the mortgage payments will be based on the face value of the loan, your monthly payment will be

$$A = \$100,000(A/P, 6.375\%/12, 180) = \$863.98.$$

Because you have to pay $1,000 to borrow $100,000, you are actually borrowing $99,000. Therefore, you need to find out what kind of interest you are actually paying. In other words, you borrow $99,000 and you make $863.98 monthly payments over 15 years. To find the interest rate, you can set up the following equivalence equation and solve for the unknown interest rate.

$$\$99,000 = \$863.98(P/A, i, 180),$$

$$i = 0.545\% \text{ per month,}$$

$$\text{APR} = 0.545\% \times 12 = 6.54\%.$$

Note that with a one-point fee, the lender was able to raise its effective yield from 6.375% to 6.54%. However, the lender is still earning less than Option 2 affords.

Variable-Rate Mortgages

Mortgages can have either fixed or adjustable rates (or both, in which case they are known as hybrid mortgages). As we mentioned earlier, interest rates in the financial marketplace rise and fall over time. If there is a possibility that market rates will rise above the fixed rate, some lenders may be reluctant to lock a loan into a fixed interest rate over a long term. If they did, they would have to forgo the opportunity to earn better rates because their assets are tied up in a loan that earns a lower rate of interest. The variable-rate mortgage is meant to encourage lending institutions to commit funds over long periods. Typically, the rate rises gradually each year for the first several years of the mortgage and then settles into a single rate for the remainder of the loan. By contrast, a hybrid loan offers a plan with a fixed interest rate for the first few years and then converts to a variable-rate schedule for the remaining periods. Example 4.17 illustrates how a lender would calculate the monthly payments with varying interest rates.

EXAMPLE 4.17 A 5/1 Hybrid Mortgage Plan

Consider again the hybrid mortgage issue discussed in the chapter opening story. Suppose that you finance a home on the basis of a 5/1 hybrid mortgage (five-year fixed/adjustable) plan over 30 years. The $100,000 hybrid loan plan offers an initial rate of 6.02% fixed for 60 months. The loan rate would be adjusted thereafter every 12 months to the lowest of three options: the then-current rate on one-year Treasury bills plus 2.75 percent, the previous rate plus a maximum annual cap of 2.0 percent, or a lifetime cap of 11.02 percent. There is no prepayment penalty for this type of loan. The projected interest rates by the lender after 5 years are as follows:

Period	Projected APR
Years 1–5	6.02%
Year 6	6.45%
Year 7	6.60%
Year 8	6.80%
Year 9	7.15%
Year 10	7.30%

(a) Develop the payment schedule for the first 10 years.

(b) Determine the total interest paid over a 10-year ownership.

SOLUTION

Given: Varying annual mortgage rates and $N = 30$ years.

Find: (a) the monthly payment; (b) the total interest payment over the 10-year ownership of the home.

(a) *Monthly payment calculation.* During the first 5 years, you borrow $100,000 for 30 years at 6.02%. To compute the monthly payment, use $i = 6.02\%/12 = 0.5017\%$ per month and $N = 360$ months:

$$A_{1-60} = \$100,000(A/P, 0.5017\%, 360) = \$600.84.$$

The balance remaining on the loan after you make the 60th payment will be

$$B_{60} = \$600.84(P/A, 0.5017\%, 300) = \$93,074.$$

During the 6th year, the interest rate changes to 6.15%, or 0.5125% per month, but the remaining term is only 300 months. Therefore, the new monthly payment would be

$$A_{60-72} = \$93,074(A/P, 0.5375\%, 300) = \$625.54.$$

After you make the 72nd payment, the balance remaining on the mortgage is

$$B_{72} = \$625.54(P/A, 0.5375\%, 288) = \$91,526.$$

During the 7th year, the interest rate changes to 6.60%, or 0.5500% per month. The new monthly payment and the remaining balance after you make the 84th payment are then

$$A_{73-84} = \$91,526(A/P, 0.5500\%, 288) = \$634.03$$

and

$$B_{84} = \$634.03(P/A, 0.5500\%, 276) = \$89,908.$$

You can compute the monthly payments in the same fashion for the remaining years. The accompanying table gives the details over the life of the loan.

(b) To determine the total interest paid over 10 years, we first determine the total monthly mortgage payments over 10 years. Since we know the ending balance at the end of 10 years, we can easily calculate the interest payments during this home ownership period:

$$\text{Total mortgage payment} = 60 \times \$600.84$$

$$+ 12\left(\frac{\$625.54 + \$634.03 + \$645.09}{+ \$664.09 + \$672.05}\right)$$

$$= \$74,940.$$

	Year	Month	Forecast Rate	Monthly Payment	Ending Loan Balance
	1	1–12	6.02%	$600.84	$98,777
	2	13–24	6.02%	$600.84	$97,477
Fixed rate	3	25–36	6.02%	$600.84	$96,098
	4	37–48	6.02%	$600.84	$94,633
	5	49–60	6.02%	$600.84	$93,704
	6	61–72	6.45%	$625.54	$91,526
	7	73–84	6.60%	$634.03	$89,908
Variable rate	8	85–96	6.80%	$645.09	$88,229
	9	97–108	7.15%	$664.09	$86,513
	10	109–120	7.30%	$672.05	$84,704

$$\text{Interest payment} = \text{Ending balance} + \text{Total mortgage payment}$$
$$- \$100,000$$
$$= \$84,704 + \$74,940 - \$100,000$$
$$= \$59,644.$$

4.6 Investing in Financial Assets

Most individual investors have three basic investment opportunities in financial assets: stocks, bonds, and cash. Cash investments include money in bank accounts, certificates of deposit (CDs), and U.S. Treasury bills. You can invest directly in any or all of the three, or indirectly, by buying mutual funds that pool your money with money from other people and then invest it. If you want to invest in financial assets, you have plenty of opportunities. In the United States alone, there are more than 9,000 stocks, 7,500 mutual funds, and thousands of corporate and government bonds to choose from. Even though we will discuss investment basics in the framework of financial assets in this chapter, the same economic concepts are applicable to any business assets examined in later chapters.

4.6.1 Investment Basics

Selecting the best investment for you depends on your personal circumstances as well as general market conditions. For example, a good investment strategy for a long-term retirement plan may not be a good strategy for a short-term college savings plan. In each case, the right investment is a balance of three things: liquidity, safety, and return.

- **Liquidity: How accessible is your money?** If your investment money must be available to cover financial emergencies, you will be concerned about liquidity: how easily you can convert it to cash. Money-market funds and savings accounts are very liquid; so are investments with short maturity dates, such as CDs. However, if you are investing for longer term goals, liquidity is not a critical issue. What you are after in that case is growth, or building your assets. We normally consider certain stocks and stock mutual funds as growth investments.

- **Risk: How safe is your money?** Risk is the chance you take of making or losing money on your investment. For most investors, the biggest risk is losing money, so they look for investments they consider safe. Usually, that means putting money into bank accounts and U.S. Treasury bills, as these investments are either insured or default free. The opposite, but equally important, risk is that your investments will not provide enough growth or income to offset the impact of inflation, the gradual increase in the cost of living. There are additional risks as well, including how the economy is doing. However, *the biggest risk is not investing at all.*

- **Return: How much profit will you be able to expect from your investment?** Safe investments often promise a specific, though limited, return. Those which involve more risk offer the opportunity to make—or lose—a lot of money. Both risk and reward are time dependent. On the one hand, as time progresses, low-yielding investments become more risky because of inflation. On the other hand, the returns associated with higher risk investments could become more stable and predictable over time, thereby reducing the perceived level of risk.

4.6.2 How to Determine Your Expected Return

Return is what you get back in relation to the amount you invested. Return is one way to evaluate how your investments in financial assets are doing in relation to each other and to the performance of investments in general. Let us look first at how we may derive rates of return.

Risk-free return: A theoretical interest rate that would be returned on an investment which was completely free of risk. The 3-month Treasury Bill is a close approximation, since it is virtually risk-free.

Basic Concepts

Conceptually, the rate of return that we realistically expect to earn on any investment is a function of three components: (1) risk-free real return, (2) an inflation factor, and (3) a risk premium.

Suppose you want to invest in stock. First, you should expect to be rewarded in some way for not being able to use your money while you are holding the stock. Then, you would be compensated for decreases in purchasing power between the time you invest the money and the time it is returned to you. Finally, you would demand additional rewards for any chance that you would not get your money back or that it will have declined in value while invested.

Risk premium: The reward for holding a risky investment rather that a risk-free one.

For example, if you were to invest $1,000 in risk-free U.S. Treasury bills for a year, you would expect a real rate of return of about 2%. Your risk premium would be also zero. You probably think that the 2% does not sound like much. However, to that you have to add an allowance for inflation. If you expect inflation to be about 4% during the investment period, you should realistically expect to earn 6% during that interval (2% real return + 4% inflation factor + 0% for risk premium). Here is what the situation looks like in tabular form:

Real return	2%
Inflation (loss of purchasing power)	4%
Risk premium (U.S. Treasury Bills)	0%
Total expected return	6%

How would it work out for a riskier investment, say, in an Internet stock such as Google.com? Because you consider this stock to be a very volatile one, you would increase the risk premium to something like the following:

Real return	2%
Inflation (loss of purchasing power)	4%
Risk premium (Google.com)	20%
Total expected return	26%

So you will not invest your money in Google.com unless you are reasonably confident of having it grow at an annual rate of 26%. Again, the risk premium of 20% is a perceived value that can vary from one investor to another.

Return on Investment over Time

If you start out with $1,000 and end up with $2,000, your return is $1,000 on that investment, or 100%. If a similar investment grows to $1,500, your return is $500, or 50%. However, unless you held those investments for the same period, you cannot determine which has a better performance. What you need in order to compare your return on one investment with the return on another investment is the **compound annual return**, the average percentage that you have gained on each investment over a series of one-year periods. For example, if you buy a share for $15 and sell it for $20, your profit is $5. If that happens within a year, your rate of return is an impressive 33.33%. If it takes five years, your return (compounded) will be closer to 5.92%, since the profit is spread over a five-year period. Mathematically, you are solving the following equivalence equation for i:

Compound annual return: The year-over-year growth rate of an investment over a specified period of time.

$$\$20 = \$15(1 + i)^5,$$
$$i = 5.92\%.$$

Figuring out the actual return on your portfolio investment is not always that simple. There are several reasons:

1. The amount of your investment changes. Most investment portfolios are active, with money moving in and out.
2. The method of computing the return can vary. For example, the performance of a stock can be averaged or compounded, which changes the rate of return significantly, as we will demonstrate in Example 4.18.
3. The time you hold specific investments varies. When you buy or sell can have a dramatic effect on the overall return.

EXAMPLE 4.18 Figuring Average versus Compound Return

Consider the following six different cases of the performance of a $1,000 investment over a three-year holding period:

	Annual Investment Yield					
Investment	**Case 1**	**Case 2**	**Case 3**	**Case 4**	**Case 5**	**Case 6**
Year 1	9%	5%	0%	0%	−1%	−5%
Year 2	9%	10%	7%	0%	−1%	−8%
Year 3	9%	12%	20%	27%	29%	40%

Compute the average versus compound return for each case.

SOLUTION

Given: Three years' worth of annual investment yield data.

Find: Compound versus average rate of return.

Average annual return: A figure used when reporting the historical return of a mutual fund.

As an illustration, consider Case 6 for an investment of $1,000. At the end of the first year, the value of the investment decreases to $950; at the end of second year, it decreases again, to $950(1 − 0.08) = 874; at the end of third year, it increases to $874(1 + 0.40) = $1,223.60$. Therefore, one way you can look at the investment is to ask, "At what annual interest rate would the initial $1,000 investment grow to $1,223.60 over three years?" This is equivalent to solving the following equivalence problem:

$$\$1,223.60 = \$1,000(1 + i)^3,$$

$$i = 6.96\%.$$

If someone evaluates the investment on the basis of the average annual rate of return, he or she might proceed as follows:

$$i = \frac{(-5\% - 8\% + 40\%)}{3} = 9\%.$$

If you calculate the remaining cases, you will observe that all six cases have the same average annual rate of return, although their compound rates of return vary from 6.96% to 9%:

	Compound versus Average Rate of Return					
	Case 1	**Case 2**	**Case 3**	**Case 4**	**Case 5**	**Case 6**
Average return	9%	9%	9%	9%	9%	9%
Balance at end of three years	$1,295	$1,294	$1,284	$1,270	$1,264	$1,224
Compound rate of return	9.00%	8.96%	8.69%	8.29%	8.13%	6.96%

Your immediate question is "Are they the same indeed?" Certainly not: You will have the most money with Case 1, which also has the highest compound rate of return. The average rate of return is easy to calculate, but it ignores the basic principle of the time value of money. In other words, according to the average-rate-of-return concept, we may view all six cases as indifferent. However, the amount of money available at the end of year 3 would be different in each case. Although the average rate of return is popular for comparing investments in terms of their yearly performance, it is not a correct measure in comparing the performance for investments over a multiyear period.

COMMENTS: You can evaluate the performance of your portfolio by comparing it against standard indexes and averages that are widely reported in the financial press. If you own stocks, you can compare their performance with the performance of the Dow Jones Industrial Average (DJIA), perhaps one of the best-known measures of stock market performance in the world. If you own bonds, you can identify an index that tracks the type you own: corporate, government, or agency bonds. If your investments are in cash, you can follow the movement of interest rates on Treasury bills, CDs, and similar investments. In addition, total-return figures for the performance of mutual funds are reported regularly. You can compare how well your investments are doing against those numbers. Another factor to take into account in evaluating your return is the current inflation rate. Certainly, your return needs to be higher than the inflation rate if your investments are going to have real growth.

4.6.3 Investing in Bonds

Bonds are loans that investors make to corporations and governments. As shown in Figure 4.19, the borrowers get the cash they need while the lenders earn interest. Americans have more money invested in bonds than in stocks, mutual funds, or any other types of securities. One of the major appeals is that bonds pay a set amount of interest on a regular basis. That is why they are called *fixed-income securities*. Another attraction is that the issuer promises to repay the loan in full and on time.

Bond versus Loan

A bond is similar to a loan. For example, say you lend out $1,000 for 10 years in return for a yearly payment of 7% interest. Here is how that arrangement translates into bond terminology. You did not make a loan; you bought a bond. The $1,000 of principal is the **face value** of the bond, the yearly interest payment is its **coupon**, and the length of the loan, 10 years, is the bond's **maturity**. If you buy a bond at face value, or **par**, and hold it until it matures, you will earn interest at the stated, or coupon, rate. For example, if you buy a 20-year $1,000 bond paying 8%, you will earn $80 a year for 20 years. The yield, or return on your investment, will also be 8%, and you get your $1,000 back. You can also buy and sell bonds through a broker after their date of issue. This is known as the **secondary market**. There the price fluctuates, with a bond sometimes selling at more than par value, at a premium price (premium bonds), and sometimes below, at a discount. Changes in price are tied directly to the interest rate of the bond. If its rate is higher than the rate being paid on similar bonds, buyers are willing to pay more to get the higher interest. If its rate is lower, the bond will sell for less in order to attract buyers. However, as the price goes up, the yield goes down, and when the price goes down, the yield goes up.

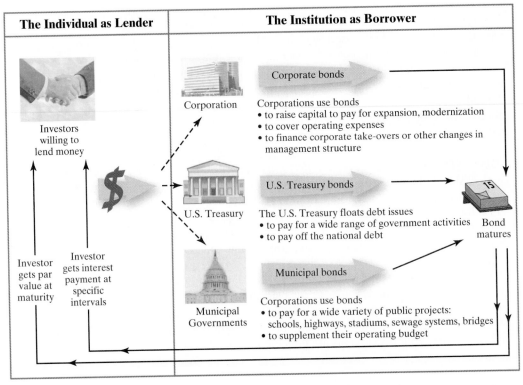

The Individual as Lender	The Institution as Borrower

Figure 4.19 Types of bonds and how they are issued in the financial market.
(*Source*: Kenneth M. Morris and Alan M. Stegel, *The Wall Street Guide to Understanding Money and Investing*, © 1993 by Lightbulb Press, Inc.)

Types of Bonds

You can choose different types of bonds to fit your financial needs—be they investing for college, finding tax-free income, or operating over a range of other possibilities. That is why it is important to have a sense of how the various types work. Several types of bonds are available in the financial market:

- **Corporate bonds.** Bonds are the major source of corporate borrowing. *Debentures* are backed by the general credit of the corporation. *Specific corporate assets, such as property or equipment, may back bonds.*
- **Municipal bonds.** State and local governments issue bonds to finance many projects that could not be funded through tax revenues. *General-obligation* bonds are backed by the full faith and credit of the issuer, *revenue bonds* by the earnings of the particular project being financed.
- **Treasury notes and bonds.** The U.S. Treasury floats debt issues to pay for a wide range of government activities, including financing the national debt. Intermediate (2 to 10 years) and long-term (10 to 30 years) government bonds are a major source of government funding.
- **Treasury bills.** These are the largest components of the money market, where short-term (13 weeks to 52 weeks) securities are bought and sold. Investors use T-bills for part of their cash reserve or as an interim holding place. Interest is the difference between the buying price and the amount paid at maturity.

- **Zero-coupon bonds.** Corporations and governments sell zero-coupon bonds at a deep discount. Investors do not collect interest; instead, the value of the bond increases to its full value when it matures. In this way, zero-coupon bonds are similar to old Series savings bonds that you bought for $37.50 and could cash in for $50 after seven years. Organizations like to issue zeros because they can continue to use the loan money without paying periodic interest. Investors like zeros because they can buy more bonds for their money and then time the maturities to coincide with anticipated expenses.
- **Callable bonds.** Callable bonds do not run their full term. In a process called redemption, the issuer may call the bond—pay off the debt—before the maturity date. Issuers will sometimes call bonds when interest rates drop, so that they can reduce their debt. If they pay off their outstanding bonds, they can float another bond at the lower rate. It is the same idea as that of refinancing a mortgage to get a lower interest rate and make lower monthly payments. Callable bonds are more risky for investors than noncallable bonds, because the investors are often faced with reinvesting the money at a lower, less attractive rate. To protect bondholders who expect long-term steady income, call provisions often specify that a bond cannot be called before a certain number of years, usually 5 or 10 years.
- **Floating-rate bonds.** These bonds promise periodic adjustments of the interest rate to persuade investors that they are not locked into what seems like an unattractively low rate.

Under the U.S. Constitution, any type of government bond must include a tax break. Because federal and local governments cannot interfere with each other's affairs, income from local government bonds (municipals, or simply **munis**) is immune from federal taxes, and income from U.S. Treasury bonds is free from local taxes.

Understanding Bond Prices

Corporate bond prices are quoted either by a percentage of the bond's face value or in increments of points and eight fractions of a point, with a par of $1,000 as the base. The value of each point is $10 and of each fraction $1.25. For example, a bond quoted at $86\frac{1}{2}$ would be selling for $865, and one quoted at $100\frac{3}{4}$ would be selling for $1,007.50. Treasury bonds are measured in thirty-seconds, rather than hundredths, of a point. Each $\frac{1}{32}$ equals 31.25 cents, and we normally drop the fractional part of the cent when stating a price. For example, if a bond is quoted at 100.2, or $100 + \frac{2}{32}$, the price translates to $1,000.62.

Are Bonds Safe?

Just because bonds have a reputation as a conservative investment does not mean that they are always safe. There are sources of risk in holding or trading bonds:

- To begin with, not all loans are paid back. Companies, cities, and counties occasionally do go bankrupt. Only U.S. Treasury bonds are considered rock solid.
- Another source of risk for certain bonds is that your loan may be called, or paid back early. While that is certainly better than its not being paid back at all, it forces you to find another, possibly less lucrative, place to put your money.
- The main danger for buy-and-hold investors, however, is a rising inflation rate. Since the dollar amount they earn on a bond investment does not change, the value of that money can be eroded by inflation. Worse yet, with your money locked away in the bond, you will not be able to take advantage of the higher interest rates that are usually available in an inflationary economy. Now you know why

bond investors cringe at cheerful headlines about full employment and strong economic growth: These traditional signs of inflation hint that bondholders may soon lose their shirts.

How Do Prices and Yields Work?

Yield to maturity (YTM): Yield that would be realized on a bond or other fixed income security if the bond was held until the maturity date.

You can trade bonds on the market just as you do stocks. Once a bond is purchased, you may keep it until it matures or for a variable number of interest periods before selling it. You can purchase or sell bonds at prices other than face value, depending on the economic environment. Furthermore, bond prices change over time because of the risk of nonpayment of interest or par value, supply and demand, and the economic outlook. These factors affect the **yield to maturity** (or **return on investment**) of the bond:

- The **yield to maturity** represents the actual interest earned from a bond over the holding period. In other words, the yield to maturity on a bond is the interest rate that establishes the equivalence between all future interest and face-value receipts and the market price of the bond.

Current yield: Annual income (interest) divided by the current market price of the bond.

- The **current yield** of a bond is the annual interest earned, as a percentage of the current market price. The current yield provides an indication of the annual return realized from investment in the bond. To illustrate, we will explain these values with numerical examples shortly.

Bond quotes follow a few unique conventions. As an example, let's take a look at a corporate bond with a face value of $1,000 (issued by Ford Motor Company) and traded on November 18, 2005:

Price:	94.50
Coupon (%):	6.625
Maturity Date:	16-Jun-2008
Yield to Maturity (%):	9.079
Current Yield (%):	7.011
Debt Rating:	BBB
Coupon Payment Frequency:	Semiannual
First Coupon Date:	16-Dec-2005
Type:	Corporate
Industry:	Industrial

- Prices are given as percentages of face value, with the last digits not decimals, but in eighths. The bond on the list, for instance, has just fallen to sell for 94.50, or 94.5% of its $1,000 face value. In other words, an investor who bought the bond when it was issued (at 100) could now sell it for a 5.5% discount.
- The discount over face value is explained by examining the bond's coupon rate, 6.625%, and its current yield, 7.011%. The current yield will be higher than the coupon rate whenever the bond is selling for less than its par value.
- The coupon rate is 6.625% paid semiannually, meaning that, for every six-month period, you will receive $66.25/2 = $33.13.

- The bond rating system helps investors determine a company's credit risk. Think of a bond rating as the report card on a company's credit rating. Blue-chip firms, which are safer investments, have a high rating, while risky companies have a low rating. The following chart illustrates the different bond rating scales from Moody's, Standard and Poor's (S&P), and Fitch Ratings—the major rating agencies in the United States.

Bond Rating			
Moody's	**S&P/Fitch**	**Grade**	**Risk**
Aaa	AAA	Investment	Highest quality
Aa	AA	Investment	High quality
A	A	Investment	Strong
Baa	BBB	Investment	Medium grade
Ba, B	BB, B	Junk	Speculative
Caa/Ca/C	CCC/CC/C	Junk	Highly speculative
C	D	Junk	In default

Notice that if the company falls below a certain credit rating, its grade changes from investment quality to junk status. Junk bonds are aptly named: They are the debt of companies that are in some sort of financial difficulty. Because they are so risky, they have to offer much higher yields than any other debt. This brings up an important point: Not all bonds are inherently safer than stocks. Certain types of bonds can be just as risky, if not riskier, than stocks.

- If you buy a bond at face value, its rate of return, or yield, is just the coupon rate. However, a glance at a table of bond quotes (like the preceding one) will tell you that after they are first issued, bonds rarely sell for exactly face value. So how much is the yield then? In our example, if you can purchase the bond for $945, you are getting two bonuses. First, you have effectively bought a bond with a 6.625% coupon, since the $66.25 coupon is 7.011% of your $945 purchase price. (Recall that the coupon rate, adjusted for the current price, is the current yield of the bond.) However, there's more: Although you paid $945, in 2008 you will receive the full $1,000 face value.

Example 4.19 illustrates how you calculate the yield to maturity and the current yield, considering both the purchase price and the capital gain for a new issue.

EXAMPLE 4.19 Yield to Maturity and Current Yield

Consider buying a $1,000 corporate (Delta Corporation) bond at the market price of $996.25. The interest will be paid semiannually, the interest rate per payment period will be simply 4.8125%, and 20 interest payments over 10 years are required. We show the resulting cash flow to the investor in Figure 4.20. Find (a) the yield to maturity and (b) the current yield.

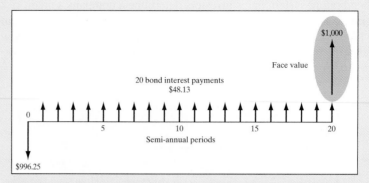

Figure 4.20 A typical cash flow transaction associated with an investment in Delta's corporate bond.

DISCUSSION

- **Debenture bond:** Delta wants to issue bonds totaling $100 million, and the company will back the bonds with specific pieces of property, such as buildings.
- **Par value:** Delta's bond has a par value of $1,000.
- **Maturity date:** Delta's bonds, which were issued on January 30, 2006, will mature on January 31, 2016; thus, they have a 10-year maturity at the time of their issue.
- **Coupon rate:** Delta's coupon rate is 9.625%, and interest is payable semiannually. For example, Delta's bonds have a $1,000 par value, and they pay $96.25 in simple interest $\left(9\frac{5}{8}\%\right)$ each year ($48.13 every six months).
- **Discount bond:** At 99.625%, or a 0.375% discount, Delta's bonds are offered at less than their par value.

SOLUTION

Given: Initial purchase price = $996.25, coupon rate = 9.625% per year paid semi-annually, and 10-year maturity with a par value of $1,000.

Find: (a) Yield to maturity and (b) current yield.

(a) *Yield to maturity.* We find the yield to maturity by determining the interest rate that makes the present worth of the receipts equal to the market price of the bond:

$$\$996.25 = \$48.13(P/A, i, 20) + \$1,000(P/F, i, 20).$$

The value of i that makes the present worth of the receipts equal to $996.25 lies between 4.5% and 5%. We could use a linear interpolation to find the yield to maturity, but it is much easier to obtain with Excel's **Goal Seek** function. As shown in Figure 4.21, solving for i yields $i = 4.8422\%$.

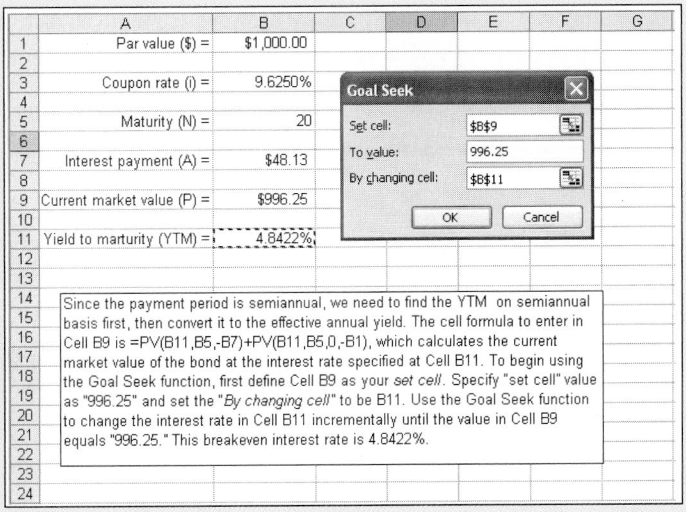

	A	B	C	D	E	F	G
1	Par value ($) =	$1,000.00					
2							
3	Coupon rate (i) =	9.6250%					
4							
5	Maturity (N) =	20					
6							
7	Interest payment (A) =	$48.13					
8							
9	Current market value (P) =	$996.25					
10							
11	Yield to maturity (YTM) =	4.8422%					
12							
13							

Goal Seek dialog box:
Set cell: B9
To value: 996.25
By changing cell: B11
OK   Cancel

Since the payment period is semiannual, we need to find the YTM on semiannual basis first, then convert it to the effective annual yield. The cell formula to enter in Cell B9 is =PV(B11,B5,-B7)+PV(B11,B5,0,-B1), which calculates the current market value of the bond at the interest rate specified at Cell B11. To begin using the Goal Seek function, first define Cell B9 as your *set cell*. Specify "set cell" value as "996.25" and set the "*By changing cell*" to be B11. Use the Goal Seek function to change the interest rate in Cell B11 incrementally until the value in Cell B9 equals "996.25." This breakeven interest rate is 4.8422%.

Figure 4.21   Finding the yield to maturity of the Delta bond with Excel's Goal Seek function.

Note that this 4.8422% is the yield to maturity *per semiannual period*. The nominal (annual) yield is $2(4.8422) = 9.6844\%$, compounded semiannually. Compared with the coupon rate of $\left(9\frac{5}{8}\%\right)$ (or 9.625%), purchasing the bond with the price discounted at 0.375% brings about an additional 0.0594% yield. The effective annual interest rate is then

$$i_a = (1 + 0.048422)^2 - 1 = 9.92\%.$$

The 9.92% represents the **effective annual yield** to maturity on the bond. Notice that when you purchase a bond at par value and sell at par value, the yield to maturity will be the *same* as the coupon rate of the bond.

Until now, we have observed differences in nominal and effective interest rates because of the frequency of compounding. In the case of bonds, the reason is different: The stated (par) value of the bond and the actual price for which it is sold are not the same. We normally state the nominal interest as a percentage of par value. However, when the bond is sold at a discount, the same nominal interest on a smaller initial investment is earned; hence, your effective interest earnings are greater than the stated nominal rate.

(b) *Current yield*. For our example of Delta, we compute the current yield as follows:

$$\frac{\$48.13}{996.25} = 4.83\% \text{ per semiannual period,}$$

$$4.83\% \times 2 = 9.66\% \text{ per year (nominal current yield),}$$

$$i_a = (1 + 0.0483)^2 - 1 = 9.90\%.$$

This effective current yield is 0.02% lower than the 9.92% yield to maturity we just computed. If the bond is selling at a discount, the current yield is smaller than the yield to maturity. If the bond is selling at a premium, the current yield is larger than the yield to maturity. A significant difference between the yield to maturity and the current yield of a bond can exist because the market price of a bond may be more or less than its face value. Moreover, both the current yield and the yield to maturity may differ considerably from the stated coupon value of the bond.

EXAMPLE 4.20 Bond Value over Time

Consider again the Delta bond investment introduced in Example 4.19. If the yield to maturity remains constant at 9.68%, (a) what will be the value of the bond one year after it was purchased? (b) If the market interest rate drops to 9% a year later, what would be the market price of the bond?

SOLUTION

Given: The same data as in Example 4.19.
Find: (a) The value of the bond one year later and (b) the market price of the bond a year later at the going rate of 9% interest.

(a) We can find the value of the bond one year later by using the same valuation procedure as in Example 4.19, but now the term to maturity is only nine years:

$$\$48.13(P/A, 4.84\%, 18) + \$1,000(P/F, 4.84\%, 18) = \$996.80.$$

The value of the bond will remain at $996.80 as long as the yield to maturity remains constant at 9.68% over nine years.

(b) Now suppose interest rates in the economy have fallen since the Delta bonds were issued, and consequently, the going rate of interest is 9%. Then both the coupon interest payments and the maturity value remain constant, but now 9% values have to be used to calculate the value of the bond. The value of the bond at the end of the first year would be

$$\$48.13(P/A, 4.5\%, 18) + \$1,000(P/F, 4.5\%, 18) = \$1,038.06.$$

Thus, the bond would sell at a premium over its par value.

COMMENTS: The arithmetic of the bond price increase should be clear, but what is the logic behind it? We can explain the reason for the increase as follows: Because the going market interest rate for the bond has fallen to 9%, if we had $1,000 to invest, we could buy new bonds with a coupon rate of 9%. These would pay $90 interest each year, rather than $96.80. We would prefer $96.80 to $90; therefore, we would be willing to pay more than $1,000 for Delta's bonds to obtain higher coupons. All investors would recognize these facts, and hence the Delta's bonds would be bid up in price to $1,038.06. At that point, they would provide the same yield to maturity (rate of return) to a potential investor as would the new bonds, namely, 9%.

SUMMARY

- Interest is most frequently quoted by financial institutions as an **annual percentage rate**, or **APR**. However, compounding frequently occurs more often than once annually, and the APR does not account for the effect of this more frequent compounding. The situation leads to the distinction between nominal and effective interest:

- **Nominal interest** is a stated rate of interest for a given period (usually a year).

- **Effective interest** is the actual rate of interest, which accounts for the interest amount accumulated over a given period. The **effective rate** is related to the APR by the equation

$$i = \left(1 + \frac{r}{M}\right)^M - 1,$$

where r is the APR, M is the number of compounding periods, and i is the effective interest rate.

   In any equivalence problem, the interest rate to use is the effective interest rate per payment period, or

$$i = \left[1 + \frac{r}{CK}\right]^C - 1,$$

where C is the number of interest periods per payment period, K is the number of payment periods per year, and r/K is the nominal interest rate per payment period. Figure 4.9 outlines the possible relationships between compounding and payment periods and indicates which version of the effective-interest formula to use.

- The equation for determining the effective interest of continuous compounding is

$$i = e^{r/K} - 1.$$

The difference in accumulated interest between continuous compounding and very frequent compounding ($M > 50$) is minimal.

- Cash flows, as well as compounding, can be continuous. Table 4.2 shows the interest factors to use for continuous cash flows with continuous compounding.

- Nominal (and hence effective) interest rates may fluctuate over the life of a cash flow series. Some forms of home mortgages and bond yields are typical examples.

- **Amortized loans** are paid off in equal installments over time, and most of these loans have interest that is compounded monthly.

- Under a typical **add-on loan**, the lender precalculates the total simple interest amount and adds it to the principal. The principal and this precalculated interest amount are then paid together in equal installments.

- The term **mortgage** refers to a special type of loan for buying a piece of property, such as a house or a commercial building. The cost of the mortgage will depend on many factors, including the amount and term of the loan and the frequency of payments, as well as points and fees.

- Two types of mortgages are common: fixed-rate mortgages and variable-rate mortgages. Fixed-rate mortgages offer loans whose interest rates are fixed over the period of the contract, whereas variable-rate mortgages offer interest rates that fluctuate with market conditions. In general, the initial interest rate is lower for variable-rate mortgages, as the lenders have the flexibility to adjust the cost of the loans over the period of the contract.

- Allocating one's assets is simply a matter of answering the following question: "Given my personal tolerance for risk and my investment objectives, what percentage of my assets should be allocated for **growth**, what percentage for **income**, and what percentage for **liquidity**?"

- You can determine the **expected rate of return** on a portfolio by computing the weighted average of the returns on each investment.

- You can determine the **expected risk** of a portfolio by computing the weighted average of the volatility of each investment.

- All other things being equal, if the expected returns are approximately the same, choose the portfolio with the lowest expected risk.

- All other things being equal, if the expected risk is about the same, choose the portfolio with the highest expected return.

- **Asset-backed bonds:** If a company backs its bonds with specific pieces of property, such as buildings, we call these types of bonds **mortgage bonds**, which indicate the terms of repayment and the particular assets pledged to the bondholders in case of default. It is much more common, however, for a corporation simply to pledge its overall assets. A **debenture bond** represents such a promise.

- **Par value:** Individual bonds are normally issued in even denominations of $1,000 or multiples of $1,000. The stated face value of an individual bond is termed the **par value**.

- **Maturity date:** Bonds generally have a specified **maturity** date on which the par value is to be repaid.

- **Coupon rate:** We call the interest paid on the par value of a bond the **annual coupon rate**. The time interval between interest payments could be of any duration, but a semiannual period is the most common.

- **Discount or premium bond:** A bond that sells below its par value is called a **discount bond**. When a bond sells above its par value, it is called a **premium bond**.

PROBLEMS

Nominal and Effective Interest Rates

4.1 If your credit card calculates interest based on 12.5% APR,
   (a) What are your monthly interest rate and annual effective interest rate?
   (b) If you current outstanding balance is $2,000 and you skip payments for two months, what would be the total balance two months from now?

4.2 A department store has offered you a credit card that charges interest at 1.05% per month, compounded monthly. What is the nominal interest (annual percentage) rate for this credit card? What is the effective annual interest rate?

4.3 A local bank advertised the following information: Interest 6.89%—effective annual yield 7.128%. No mention was made of the interest period in the advertisement. Can you figure out the compounding scheme used by the bank?

4.4 College Financial Sources, which makes small loans to college students, offers to lend $500. The borrower is required to pay $400 at the end of each week for 16 weeks. Find the interest rate per week. What is the nominal interest rate per year? What is the effective interest rate per year?

4.5 A financial institution is willing to lend you $40. However, $450 is repaid at the end of one week.
  (a) What is the nominal interest rate?
  (b) What is the effective annual interest rate?

4.6 The Cadillac Motor Car Company is advertising a 24-month lease of a Cadillac Deville for $520, payable at the beginning of each month. The lease requires a $2,500 down payment, plus a $500 refundable security deposit. As an alternative, the company offers a 24-month lease with a single up-front payment of $12,780, plus a $500 refundable security deposit. The security deposit will be refunded at the end of the 24-month lease. Assuming an interest rate of 6%, compounded monthly, which lease is the preferred one?

4.7 As a typical middle-class consumer, you are making monthly payments on your home mortgage (9% annual interest rate), car loan (12%), home improvement loan (14%), and past-due charge accounts (18%). Immediately after getting a $100 monthly raise, your friendly mutual fund broker tries to sell you some investment funds with a guaranteed return of 10% per year. Assuming that your only other investment alternative is a savings account, should you buy?

Compounding More Frequent than Annually

4.8 A loan company offers money at 1.8% per month, compounded monthly.
  (a) What is the nominal interest rate?
  (b) What is the effective annual interest rate?
  (c) How many years will it take an investment to triple if interest is compounded monthly?
  (d) How many years will it take an investment to triple if the nominal rate is compounded continuously?

4.9 Suppose your savings account pays 9% interest compounded quarterly. If you deposit $10,000 for one year, how much would you have?

4.10 What will be the amount accumulated by each of these present investments?
  (a) $5,635 in 10 years at 5% compounded semiannually.
  (b) $7,500 in 15 years at 6% compounded quarterly.
  (c) $38,300 in 7 years at 9% compounded monthly.

4.11 How many years will it take an investment to triple if the interest rate is 9% compounded
  (a) Quarterly?　　　(b) Monthly?　　　(c) Continuously?

4.12 A series of equal quarterly payments of $5,000 for 12 years is equivalent to what present amount at an interest rate of 9% compounded
  (a) Quarterly?　　　(b) Monthly?　　　(c) Continuously?

4.13 What is the future worth of an equal payment series of $3,000 each quarter for five years if the interest rate is 8% compounded continuously?

4.14 Suppose that $2,000 is placed in a bank account at the end of each quarter over the next 15 years. What is the future worth at the end of 15 years when the interest rate is 6% compounded

(a) Quarterly?   (b) Monthly?   (c) Continuously?

4.15 A series of equal quarterly deposits of $1,000 extends over a period of three years. It is desired to compute the future worth of this quarterly deposit series at 12% compounded monthly. Which of the following equations is correct?

(a) $F = 4(\$1,000)(F/A, 12\%, 3)$.   (b) $F = \$1,000(F/A, 3\%, 12)$.

(c) $F = \$1,000(F/A, 1\%, 12)$.   (d) $F = \$1,000(F/A, 3.03\%, 12)$.

4.16 If the interest rate is 8.5% compounded continuously, what is the required quarterly payment to repay a loan of $12,000 in five years?

4.17 What is the future worth of a series of equal monthly payments of $2,500 if the series extends over a period of eight years at 12% interest compounded

(a) Quarterly?   (b) Monthly?   (c) Continuously?

4.18 Suppose you deposit $500 at the end of each quarter for five years at an interest rate of 8% compounded monthly. What equal end-of-year deposit over the five years would accumulate the same amount at the end of the five years under the same interest compounding? To answer the question, which of the following is correct?

(a) $A = [\$500(F/A, 2\%, 20)] \times (A/F, 8\%, 5)$.

(b) $A = \$500(F/A, 2.013\%, 4)$.

(c) $A = \$500\left(F/A, \dfrac{8\%}{12}, 20 \right) \times (A/F, 8\%, 5)$.

(d) None of the above.

4.19 A series of equal quarterly payments of $2,000 for 15 years is equivalent to what future lump-sum amount at the end of 10 years at an interest rate of 8% compounded continuously?

4.20 What will be the required quarterly payment to repay a loan of $32,000 in five years, if the interest rate is 7.8% compounded continuously?

4.21 A series of equal quarterly payments of $4,000 extends over a period of three years. What is the present worth of this quarterly payment series at 8.75% interest compounded continuously?

4.22 What is the future worth of the following series of payments?

(a) $6,000 at the end of each six-month period for 6 years at 6% compounded semiannually.

(b) $42,000 at the end of each quarter for 12 years at 8% compounded quarterly.

(c) $75,000 at the end of each month for 8 years at 9% compounded monthly.

4.23 What equal series of payments must be paid into a sinking fund to accumulate the following amount?

(a) $21,000 in 10 years at 6.45% compounded semiannually when payments are semiannual.

(b) $9,000 in 15 years at 9.35% compounded quarterly when payments are quarterly.

(c) $24,000 in 5 years at 6.55% compounded monthly when payments are monthly.

4.24 You have a habit of drinking a cup of Starbucks coffee ($2.50 a cup) on the way to work every morning. If, instead, you put the money in the bank for 30 years, how much would you have at the end of that time, assuming that your account earns

5% interest compounded *daily*? Assume also that you drink a cup of coffee every day, including weekends.

4.25 John Jay is purchasing a $24,000 automobile, which is to be paid for in 48 monthly installments of $543.35. What effective annual interest is he paying for this financing arrangement?

4.26 A loan of $12,000 is to be financed to assist in buying an automobile. On the basis of monthly compounding for 42 months, the end-of-the-month equal payment is quoted as $445. What nominal interest rate in percentage is being charged?

4.27 Suppose a young newlywed couple is planning to buy a home two years from now. To save the down payment required at the time of purchasing a home worth $220,000 (let's assume that the down payment is 10% of the sales price, or $22,000), the couple decides to set aside some money from each of their salaries at the end of every month. If each of them can earn 6% interest (compounded monthly) on his or her savings, determine the equal amount this couple must deposit each month until the point is reached where the couple can buy the home.

4.28 What is the present worth of the following series of payments?
   (a) $1,500 at the end of each six-month period for 12 years at 8% compounded semiannually.
   (b) $2,500 at the end of each quarter for 8 years at 8% compounded quarterly.
   (c) $3,800 at the end of each month for 5 years at 9% compounded monthly.

4.29 What is the amount of the quarterly deposits *A* such that you will be able to withdraw the amounts shown in the cash flow diagram if the interest rate is 8% compounded quarterly?

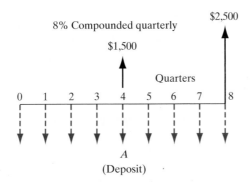

4.30 Georgi Rostov deposits $15,000 in a savings account that pays 6% interest compounded monthly. Three years later, he deposits $14,000. Two years after the $14,000 deposit, he makes another deposit in the amount of $12,500. Four years after the $12,500 deposit, half of the accumulated funds is transferred to a fund that pays 8% interest compounded quarterly. How much money will be in each account six years after the transfer?

4.31 A man is planning to retire in 25 years. He wishes to deposit a regular amount every three months until he retires, so that, beginning one year following his retirement, he will receive annual payments of $60,000 for the next 10 years. How much must he deposit if the interest rate is 6% compounded quarterly?

4.32 You borrowed $15,000 for buying a new car from a bank at an interest rate of 12% compounded monthly. This loan will be repaid in 48 equal monthly installments over four years. Immediately after the 20th payment, you desire to pay the remainder of the loan in a single payment. Compute this lump-sum amount of that time.

4.33 A building is priced at $125,000. If a down payment of $25,000 is made and a payment of $1,000 every month thereafter is required, how many months will it take to pay for the building? Interest is charged at a rate of 9% compounded monthly.

4.34 You obtained a loan of $20,000 to finance an automobile. Based on monthly compounding over 24 months, the end-of-the-month equal payment was figured to be $922.90. What APR was used for this loan?

4.35 *The Engineering Economist* (a professional journal) offers three types of subscriptions, payable in advance: one year at $66, two years at $120, and three years at $160. If money can earn 6% interest compounded monthly, which subscription should you take? (Assume that you plan to subscribe to the journal over the next three years.)

4.36 A couple is planning to finance its three-year-old son's college education. Money can be deposited at 6% compounded quarterly. What quarterly deposit must be made from the son's 3rd birthday to his 18th birthday to provide $50,000 on each birthday from the 18th to the 21st? (Note that the last deposit is made on the date of the first withdrawal.)

4.37 Sam Salvetti is planning to retire in 15 years. Money can be deposited at 8% compounded quarterly. What quarterly deposit must be made at the end of each quarter until Sam retires so that he can make a withdrawal of $25,000 semiannually over the first five years of his retirement? Assume that his first withdrawal occurs at the end of six months after his retirement.

4.38 Michelle Hunter received $250,000 from an insurance company after her husband's death. Michelle wants to deposit this amount in a savings account that earns interest at a rate of 6% compounded monthly. Then she would like to make 120 equal monthly withdrawals over the 10-year period such that, when she makes the last withdrawal, the savings account will have a balance of zero. How much can she withdraw each month?

4.39 Anita Tahani, who owns a travel agency, bought an old house to use as her business office. She found that the ceiling was poorly insulated and that the heat loss could be cut significantly if 6 inches of foam insulation were installed. She estimated that with the insulation, she could cut the heating bill by $40 per month and the air-conditioning cost by $25 per month. Assuming that the summer season is three months (June, July, and August) of the year and that the winter season is another three months (December, January, and February) of the year, how much can Anita spend on insulation if she expects to keep the property for five years? Assume that neither heating nor air-conditioning would be required during the fall and spring seasons. If she decides to install the insulation, it will be done at the beginning of May. Anita's interest rate is 9% compounded monthly.

Continuous Payments with Continuous Compounding

4.40 A new chemical production facility that is under construction is expected to be in full commercial operation 1 year from now. Once in full operation, the facility will generate $63,000 cash profit daily over the plant's service life of 12 years.

Determine the equivalent present worth of the future cash flows generated by the facility at the beginning of commercial operation, assuming

(a) 12% interest compounded daily, with the daily flows.

(b) 12% interest compounded continuously, with the daily flow series approximated by a uniform continuous cash flow function.

Also, compare the difference between (a) discrete (daily) and (b) continuous compounding.

4.41 Income from a project is expected to decline at a constant rate from an initial value of $500,000 at time 0 to a final value of $40,000 at the end of year 3. If interest is compounded continuously at a nominal annual rate of 11%, determine the present value of this continuous cash flow.

4.42 A sum of $80,000 will be received uniformly over a five-year period beginning two years from today. What is the present value of this deferred-funds flow if interest is compounded continuously at a nominal rate of 9%?

4.43 A small chemical company that produces an epoxy resin expects its production volume to decay exponentially according to the relationship

$$y_t = 5e^{-0.25t},$$

where y_t is the production rate at time t. Simultaneously, the unit price is expected to increase linearly over time at the rate

$$u_t = \$55(1 + 0.09t).$$

What is the expression for the present worth of sales revenues from $t = 0$ to $t = 20$ at 12% interest compounded continuously?

Changing Interest Rates

4.44 Consider the accompanying cash flow diagram, which represents three different interest rates applicable over the five-year time span shown.

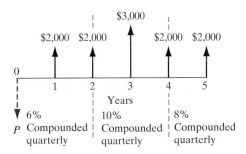

(a) Calculate the equivalent amount P at the present time.

(b) Calculate the single-payment equivalent to F at $n = 5$.

(c) Calculate the equal-payment-series cash flow A that runs from $n = 1$ to $n = 5$.

4.45 Consider the cash flow transactions depicted in the accompanying cash flow diagram, with the changing interest rates specified.

(a) What is the equivalent present worth? (In other words, how much do you have to deposit now so that you can withdraw $300 at the end of year 1, $300 at the end of year 2, $500 at the end of year 3, and $500 at the end of year 4?)

(b) What is the single effective annual interest rate over four years?

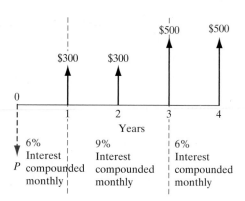

4.46 Compute the future worth of the cash flows with the different interest rates specified. The cash flows occur at the end of each year over four years.

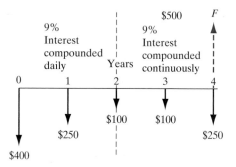

Amortized Loans

4.47 An automobile loan of $20,000 at a nominal rate of 9% compounded monthly for 48 months requires equal end-of-month payments of $497.70. Complete the following table for the first six payments, as you would expect a bank to calculate the values:

End of Month (n)	Interest Payment	Repayment of Principal	Remaining Loan Balance
1			$19,652.30
2			
3		$352.94	
4	$142.12		
5	$139.45		
6			$17,874.28

4.48 Mr. Smith wants to buy a new car that will cost $18,000. He will make a down payment in the amount of $8,000. He would like to borrow the remainder from a bank at an interest rate of 9% compounded monthly. He agrees to pay off the loan monthly for a period of two years. Select the correct answer for the following questions:

(a) What is the amount of the monthly payment A?

   (i) $A = \$10,000(A/P, 0.75\%, 24)$.

   (ii) $A = \$10,000(A/P, 9\%, 2)/12$.

   (iii) $A = \$10,000(A/F, 0.75\%, 24)$.

   (iv) $A = \$12,500(A/F, 9\%, 2)/12$.

(b) Mr. Smith has made 12 payments and wants to figure out the balance remaining immediately after 12th payment. What is that balance?

   (i) $B_{12} = 12A$.

   (ii) $B_{12} = A(P/A, 9\%, 1)/12$.

   (iii) $B_{12} = A(P/A, 0.75\%, 12)$.

   (iv) $B_{12} = 10,000 - 12A$.

4.49 Tony Wu is considering purchasing a used automobile. The price, including the title and taxes, is $12,345. Tony is able to make a $2,345 down payment. The balance, $10,000, will be borrowed from his credit union at an interest rate of 8.48% compounded daily. The loan should be paid in 36 equal monthly payments. Compute the monthly payment. What is the total amount of interest Tony has to pay over the life of the loan?

4.50 Suppose you are in the market for a new car worth $18,000. You are offered a deal to make a $1,800 down payment now and to pay the balance in equal end-of-month payments of $421.85 over a 48-month period. Consider the following situations:

(a) Instead of going through the dealer's financing, you want to make a down payment of $1,800 and take out an auto loan from a bank at 11.75% compounded monthly. What would be your monthly payment to pay off the loan in four years?

(b) If you were to accept the dealer's offer, what would be the effective rate of interest per month the dealer charges on your financing?

4.51 Bob Pearson borrowed $25,000 from a bank at an interest rate of 10% compounded monthly. The loan will be repaid in 36 equal monthly installments over three years. Immediately after his 20th payment, Bob desires to pay the remainder of the loan in a single payment. Compute the total amount he must pay.

4.52 You plan to buy a $200,000 home with a 10% down payment. The bank you want to finance the loan suggests two options: a 20-year mortgage at 9% APR and a 30-year mortgage at 10% APR. What is the difference in monthly payments (for the first 20 years) between these two options?

4.53 David Kapamagian borrowed money from a bank to finance a small fishing boat. The bank's terms allowed him to defer payments (including interest) on the loan for six months and to make 36 equal end-of-month payments thereafter. The original bank note was for $4,800, with an interest rate of 12% compounded monthly. After 16 monthly payments, David found himself in a financial bind and went to a loan company for assistance in lowering his monthly payments. Fortunately, the loan company offered to pay his debts in one lump sum if he would pay the company $104 per month for the next 36 months. What monthly rate of interest is the loan company charging on this transaction?

4.54 You are buying a home for $250,000.

(a) If you make a down payment of $50,000 and take out a mortgage on the rest of the money at 8.5% compounded monthly, what will be your monthly payment to retire the mortgage in 15 years?

(b) Consider the seventh payment. How much will the interest and principal payments be?

4.55 With a $350,000 home mortgage loan with a 20-year term at 9% APR compounded monthly, compute the total payments on principal and interest over the first 5 years of ownership.

4.56 A lender requires that monthly mortgage payments be no more than 25% of gross monthly income with a maximum term of 30 years. If you can make only a 15% down payment, what is the minimum monthly income needed to purchase a $400,000 house when the interest rate is 9% compounded monthly?

4.57 To buy a $150,000 house, you take out a 9% (APR) mortgage for $120,000. Five years later, you sell the house for $185,000 (after all other selling expenses). What equity (the amount that you can keep before tax) would you realize with a 30-year repayment term?

4.58 Just before their 15th payment,
- Family A had a balance of $80,000 on a 9%, 30-year mortgage;
- Family B had a balance of $80,000 on a 9%, 15-year mortgage; and
- Family C had a balance of $80,000 on a 9%, 20-year mortgage.

How much interest did each family pay on the 15th payment?

4.59 Home mortgage lenders usually charge points on a loan to avoid exceeding a legal limit on interest rates or to be competitive with other lenders. As an example, for a two-point loan, the lender would lend only $98 for each $100 borrowed. The borrower would receive only $98, but would have to make payments just as if he or she had received $100. Suppose that you receive a loan of $130,000, payable at the end of each month for 30 years with an interest rate of 9% compounded monthly, but you have been charged three points. What is the effective interest rate on this home mortgage loan?

4.60 A restaurant is considering purchasing a lot adjacent to its business to provide adequate parking space for its customers. The restaurant needs to borrow $35,000 to secure the lot. A deal has been made between a local bank and the restaurant so that the restaurant would pay the loan back over a five-year period with the following payment terms: 15%, 20%, 25%, 30%, and 35% of the initial loan at the end of first, second, third, fourth, and fifth years, respectively.

(a) What rate of interest is the bank earning from this loan?

(b) What would be the total interest paid by the restaurant over the five-year period?

4.61 Don Harrison's current salary is $60,000 per year, and he is planning to retire 25 years from now. He anticipates that his annual salary will increase by $3,000 each year (to $60,000 the first year, $63,000 the second year, $66,000 the third year, and so forth), and he plans to deposit 5% of his yearly salary into a retirement fund that earns 7% interest compounded daily. What will be the amount accumulated at the time of Don's retirement?

4.62 Consider the following two options for financing a car:

- **Option A.** Purchase the vehicle at the normal price of $26,200 and pay for the vehicle over three years with equal monthly payments at 1.9% APR financing.

- **Option B.** Purchase the vehicle for a discount price of $24,048, to be paid immediately. The funds that would be used to purchase the vehicle are presently earning 5% annual interest compounded monthly.

(a) What is the meaning of the APR of 1.9% quoted by the dealer?

(b) Under what circumstances would you prefer to go with the dealer's financing?

(c) Which interest rate (the dealer's interest rate or the savings rate) would you use in comparing the two options?

Add-On Loans

4.63 Katerina Unger wants to purchase a set of furniture worth $3,000. She plans to finance the furniture for two years. The furniture store tells Katerina that the interest rate is only 1% per month, and her monthly payment is computed as follows:

- Installment period = 24 months.
- Interest = $24(0.01)(\$3,000) = \$720$.
- Loan processing fee = $25.
- Total amount owed = $\$3,000 + \$720 + \$25 = \$3,745$.
- Monthly payment = $\$3,745/24 = \$156.04$ per month.

(a) What is the annual effective interest rate that Katerina is paying for her loan transaction? What is the nominal interest (annual percentage rate) for the loan?

(b) Katerina bought the furniture and made 12 monthly payments. Now she wants to pay off the remaining installments in one lump sum (at the end of 12 months). How much does she owe the furniture store?

4.64 You purchase a piece of furniture worth $5,000 on credit through a local furniture store. You are told that your monthly payment will be $146.35, including an acquisition fee of $25, at a 10% add-on interest rate over 48 months. After making 15 payments, you decide to pay off the balance. Compute the remaining balance, based on the conventional amortized loan.

Loans with Variable Payments

4.65 Kathy Stonewall bought a new car for $15,458. A dealer's financing was available through a local bank at an interest rate of 11.5% compounded monthly. Dealer financing required a 10% down payment and 60 equal monthly payments. Because the interest rate was rather high, Kathy checked her credit union for possible financing. The loan officer at the credit union quoted a 9.8% interest rate for a new-car loan and 10.5% for a used car. But to be eligible for the loan, Kathy has to be a member of the union for at least six months. Since she joined the union two months ago, she has to wait four more months to apply for the loan. Consequently, she decided to go ahead with the dealer's financing, and four months later she refinanced the balance through the credit union at an interest rate of 10.5%.

(a) Compute the monthly payment to the dealer.

(b) Compute the monthly payment to the union.

(c) What is the total interest payment on each loan?

4.66 A house can be purchased for $155,000, and you have $25,000 cash for a down payment. You are considering the following two financing options:

- **Option 1.** Getting a new standard mortgage with a 7.5% (APR) interest and a 30-year term.

- **Option 2.** Assuming the seller's old mortgage, which has an interest rate of 5.5% (APR), a remaining term of 25 years (the original term was 30 years), a remaining balance of $97,218, and payments of $597 per month. You can obtain a second mortgage for the remaining balance ($32,782) from your credit union at 9% (APR) with a 10-year repayment period.

(a) What is the effective interest rate of the combined mortgage?

(b) Compute the monthly payments for each option over the life of the mortgage.

(c) Compute the total interest payment for each option.

(d) What homeowner's interest rate makes the two financing options equivalent?

Loans with Variable Interest Rates

4.67 A loan of $10,000 is to be financed over a period of 24 months. The agency quotes a nominal rate of 8% for the first 12 months and a nominal rate of 9% for any remaining unpaid balance after 12 months, compounded monthly. Based on these rates, what equal end-of-the-month payment for 24 months would be required to repay the loan with interest?

4.68 Emily Wang financed her office furniture from a furniture dealer. The dealer's terms allowed her to defer payments (including interest) for six months and to make 36 equal end-of-month payments thereafter. The original note was for $15,000, with interest at 9% compounded monthly. After 26 monthly payments, Emily found herself in a financial bind and went to a loan company for assistance. The loan company offered to pay her debts in one lump sum if she would pay the company $186 per month for the next 30 months.

(a) Determine the original monthly payment made to the furniture store.

(b) Determine the lump-sum payoff amount the loan company will make.

(c) What monthly rate of interest is the loan company charging on this loan?

4.69 If you borrow $120,000 with a 30-year term at a 9% (APR) variable rate and the interest rate can be changed every five years,

(a) What is the initial monthly payment?

(b) If the lender's interest rate is 9.75% (APR) at the end of five years, what will the new monthly payments be?

Investment in Bonds

4.70 The Jimmy Corporation issued a new series of bonds on January 1, 1996. The bonds were sold at par ($1,000), have a 12% coupon rate, and mature in 30 years, on December 31, 2025. Coupon interest payments are made semiannually (on June 30 and December 31).

(a) What was the yield to maturity (YTM) of the bond on January 1, 1996?

(b) Assuming that the level of interest rates had fallen to 9%, what was the price of the bond on January 1, 2001, five years later?

(c) On July 1, 2001, the bonds sold for $922.38. What was the YTM at that date? What was the current yield at that date?

4.71 A $1,000, 9.50% semiannual bond is purchased for $1,010. If the bond is sold at the end of three years and six interest payments, what should the selling price be to yield a 10% return on the investment?

4.72 Mr. Gonzalez wishes to sell a bond that has a face value of $1,000. The bond bears an interest rate of 8%, with bond interests payable semiannually. Four years ago, $920 was paid for the bond. At least a 9% return (yield) on the investment is desired. What must be the minimum selling price?

4.73 Suppose you have the choice of investing in (1) a zero-coupon bond, which costs $513.60 today, pays nothing during its life, and then pays $1,000 after five years, or (2) a bond that costs $1,000 today, pays $113 in interest semiannually, and matures at the end of five years. Which bond would provide the higher yield?

4.74 Suppose you were offered a 12-year, 15% coupon, $1,000 par value bond at a price of $1,298.68. What rate of interest (yield to maturity) would you earn if you bought the bond and held it to maturity (at semiannual interest)?

4.75 The Diversified Products Company has two bond issues outstanding. Both bonds pay $100 semiannual interest, plus $1,000 at maturity. Bond A has a remaining maturity of 15 years, bond B a maturity of 1 year. What is the value of each of these bonds now, when the going rate of interest is 9%?

4.76 The AirJet Service Company's bonds have four years remaining to maturity. Interest is paid annually, the bonds have a $1,000 par value, and the coupon interest rate is 8.75%.

(a) What is the yield to maturity at a current market price of $1,108?

(b) Would you pay $935 for one of these bonds if you thought that the market rate of interest was 9.5%?

4.77 Suppose Ford sold an issue of bonds with a 15-year maturity, a $1,000 par value, a 12% coupon rate, and semiannual interest payments.

(a) Two years after the bonds were issued, the going rate of interest on bonds such as these fell to 9%. At what price would the bonds sell?

(b) Suppose that, two years after the bonds' issue, the going interest rate had risen to 13%. At what price would the bonds sell?

(c) Today, the closing price of the bond is $783.58. What is the current yield?

4.78 Suppose you purchased a corporate bond with a 10-year maturity, a $1,000 par value, a 10% coupon rate, and semiannual interest payments. All this means that you receive a $50 interest payment at the end of each six-month period for 10 years (20 times). Then, when the bond matures, you will receive the principal amount (the face value) in a lump sum. Three years after the bonds were purchased, the going rate of interest on new bonds fell to 6% (or 6% compounded semiannually). What is the current market value (P) of the bond (three years after its purchase)?

Short Case Studies

ST4.1 Jim Norton, an engineering junior, was mailed two guaranteed line-of-credit applications from two different banks. Each bank offered a different annual fee and finance charge.

Jim expects his average monthly balance after payment to the bank to be $300 and plans to keep the credit card he chooses for only 24 months. (After graduation, he will apply for a new card.) Jim's interest rate on his savings account is 6% compounded daily. The following table lists the terms of each bank:

Terms	Bank A	Bank B
Annual fee	$20	$30
Finance charge	1.55%	16.5%
	monthly interest rate	annual percentage rate

(a) Compute the effective annual interest rate for each card.

(b) Which bank's credit card should Jim choose?

(c) Suppose Jim decided to go with Bank B and used the card for one year. The balance after one year is $1,500. If he makes just a minimum payment each month (say, 5% of the unpaid balance), how long will it take to pay off the card debt? Assume that he will not make any new purchases on the card until he pays off the debt.

ST4.2 The following is an actual promotional pamphlet prepared by Trust Company Bank in Atlanta, Georgia:

"Lower your monthly car payments as much as 48%." Now you can buy the car you want and keep the monthly payments as much as 48% lower than they would be if you financed with a conventional auto loan. Trust Company's *Alternative Auto Loan* (AAL)[SM] makes the difference. It combines the lower monthly payment advantages of leasing with tax and ownership of a conventional loan. And if you have your monthly payment deducted automatically from your Trust Company checking account, you will save $\frac{1}{2}$% on your loan interest rate. Your monthly payments can be spread over 24, 36 or 48 months.

Amount Financed	Financing Period (months)	Monthly Alternative Auto Loan	Payment Conventional Auto Loan
	24	$249	$477
$10,000	36	211	339
	48	191	270
	24	498	955
$20,000	36	422	678
	48	382	541

The amount of the final payment will be based on the residual value of the car at the end of the loan. Your monthly payments are kept low because you make principal payments on only a portion of the loan and not on the residual value of the car. Interest is computed on the full amount of the loan. At the end of the loan period you may:

1. Make the final payment and keep the car.
2. Sell the car yourself, repay the note (remaining balance), and keep any profit you make.

3. Refinance the car.

4. Return the car to Trust Company in good working condition and pay only a return fee.

So, if you've been wanting a special car, but not the high monthly payments that could go with it, consider the *Alternative Auto Loan*. For details, ask at any Trust Company branch.

Note 1: The chart above is based on the following assumptions. Conventional auto loan 13.4% annual percentage rate. *Alternative Auto Loan* 13.4% annual percentage rate.

Note 2: The residual value is assumed to be 50% of sticker price for 24 months; 45% for 36 months. The amount financed is 80% of sticker price.

Note 3: Monthly payments are based on principal payments equal to the depreciation amount on the car and interest in the amount of the loan.

Note 4: The residual value of the automobile is determined by a published residual value guide in effect at the time your Trust Company's *Alternative Auto Loan* is originated.

Note 5: The minimum loan amount is $10,000 (Trust Company will lend up to 80% of the sticker price). Annual household income requirement is $50,000.

Note 6: Trust Company reserves the right of final approval based on customer's credit history. Offer may vary at other Trust Company banks in Georgia.

(a) Show how the monthly payments were computed for the *Alternative Auto Loan* by the bank.

(b) Suppose that you have decided to finance a new car for 36 months from Trust Company. Suppose also that you are interested in owning the car (not leasing it). If you decided to go with the *Alternative Auto Loan*, you would make the final payment and keep the car at the end of 36 months. Assume that your opportunity cost rate (personal interest rate) is an interest rate of 8% compounded monthly. (You may view this opportunity cost rate as an interest rate at which you can invest your money in some financial instrument, such as a savings account.) Compare Trust Company's alternative option with the conventional option and make a choice between them.

ST4.3 In 1988, the Michigan legislature enacted the nation's first state-run program, the *Pay-Now, Learn-Later Plan*, to guarantee college tuition for students whose families invested in a special tax-free trust fund. The minimum deposit now is $1,689 for each year of tuition that sponsors of a newborn want to prepay. The yearly amount to buy into the plan increases with the age of the child: Parents of infants pay the least, and parents of high school seniors pay the most—$8,800 this year. This is because high school seniors will go to college sooner. Michigan State Treasurer Robert A. Bowman contends that the educational trust is a better deal than putting money into a certificate of deposit (CD) or a tuition prepayment plan at a bank, because the state promises to stand behind the investment. "Regardless of how high tuition goes, you know it's paid for," he said. "The disadvantage of a CD or a savings account is you have to hope and cross your fingers that tuition won't outpace the amount you save." At the newborns' rate, $6,756 will prepay four years of college, which is 25% less than the statewide average public-college cost of $9,000 for four years in 1988. In 2006, when a child born in 1988 will be old enough for college, four years of college could cost $94,360 at a private institution and $36,560 at a state school if costs continue to rise the expected average of at least 7% a year. The Internal Revenue Service issued its opinion, ruling that the person who sets aside the money would not be taxed on the amount paid into the fund. The agency said that the student would be subject to federal tax on the difference between the amount paid in and the amount paid out. Assuming that you are interested in the program for a newborn, would you join it?

ST4.4 Suppose you are going to buy a home worth $110,000 and you make a down payment in the amount of $50,000. The balance will be borrowed from the Capital Savings and Loan Bank. The loan officer offers the following two financing plans for the property:

- **Option 1.** A conventional fixed loan at an interest rate of 13% over 30 years, with 360 equal monthly payments.
- **Option 2.** A graduated payment schedule (FHA 235 plan) at 11.5% interest, with the following monthly payment schedule:

Year (n)	Monthly Payment	Monthly Mortgage Insurance
1	$497.76	$25.19
2	522.65	25.56
3	548.78	25.84
4	576.22	26.01
5	605.03	26.06
6–30	635.28	25.96

For the FHA 235 plan, mortgage insurance is a must.
(a) Compute the monthly payment under Option 1.
(b) What is the effective annual interest rate you are paying under Option 2?
(c) Compute the outstanding balance at the end of five years under each option.
(d) Compute the total interest payment under each option.
(e) Assuming that your only investment alternative is a savings account that earns an interest rate of 6% compounded monthly, which option is a better deal?

ST4.5 Consider the following advertisement seeking to sell a beachfront condominium at SunDestin, Florida:

$$95\% \text{ Financing } 8\tfrac{1}{8}\% \text{ interest!!}$$

5% Down Payment. Own a Gulf-Front Condominium for only $100,000 with a 30-year variable-rate mortgage. We're providing incredible terms: $95,000 mortgage (30 years), year 1 at 8.125%, year 2 at 10.125%, year 3 at 12.125%, and years 4 through 30 at 13.125%.
(a) Compute the monthly payments for each year.
(b) Calculate the total interest paid over the life of the loan.
(c) Determine the equivalent single-effect annual interest rate for the loan.

Evaluation
of Business
and Engineering
Assets

FIVE

Present-Worth Analysis

Parking Meters Get Smarter[1]—Wireless Technology Turns Old-Fashioned Coin-Operated Device into a Sophisticated Tool for Catching Scofflaws and Raising Cash Technology is taking much of the fun out of finding a place to park the car:

- In Pacific Grove, California, parking meters "know" when a car pulls out of the spot and quickly reset to zero—eliminating drivers' little joy of parking for free on someone else's quarters.
- In Montreal, when cars stay past their time limit, meters send real-time alerts to an enforcement officer's handheld device, reducing the number of people needed to monitor parking spaces—not to mention drivers' chances of getting away with violations.
- In Aspen, Colorado, wireless "in-car" meters may eliminate the need for curbside parking meters altogether: They dangle from the rearview mirror inside the car, ticking off a prepaid time.

Now, in cities from New York to Seattle, the door is open to a host of wireless technologies seeking to improve the parking meter even further. Chicago and Sacramento, California, among others, are equipping enforcement vehicles with infrared cameras capable of scanning license plates even at 30 miles an hour. Using a global positioning system, the cameras can tell which individual cars have parked too long in a two-hour parking zone. At a cost of $75,000 a camera, the system is an expensive upgrade of the old method of chalking tires and then coming back two hours later to see if the car has moved.

The camera system, supplied by Canada's Autovu Technologies, also helps identify scofflaws and stolen vehicles, by linking to a database of unpaid tickets and auto thefts. Sacramento bought three cameras in August, and since then its practice of "booting," or immobilizing, cars

---

[1] Christopher Conkey, staff reporter of *The Wall Street Journal*, June 30, 2005, p. B1.

with a lot of unpaid tickets has increased sharply. Revenue is soaring, too. According to Howard Chan, Sacramento's parking director, Sacramento booted 189 cars and took in parking revenue of $169,000 for the fiscal year ended in June 2004; for fiscal 2005, the city expects to boot 805 cars and take in more than $475,000.

In downtown Montreal, more than 400 "pay-by-space" meters, each covering 10 to 15 spaces, are a twist on regular multispace meters. Motorists park, then go to the meter to type in the parking-space number, and pay by card or coin. These meters, which cost about $9,000 each, identify violators in real time for enforcement officers carrying handheld devices: A likeness of the block emerges on the screen, and cars parked illegally show up in red.

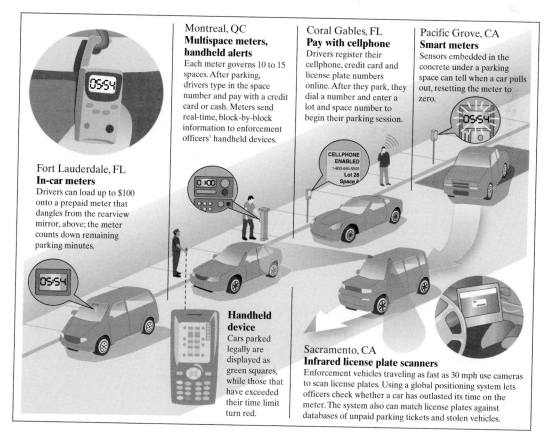

Montreal, QC
Multispace meters, handheld alerts
Each meter governs 10 to 15 spaces. After parking, drivers type in the space number and pay with a credit card or cash. Meters send real-time, block-by-block information to enforcement officers' handheld devices.

Coral Gables, FL
Pay with cellphone
Drivers register their cellphone, credit card and license plate numbers online. After they park, they dial a number and enter a lot and space number to begin their parking session.

Pacific Grove, CA
Smart meters
Sensors embedded in the concrete under a parking space can tell when a car pulls out, resetting the meter to zero.

CELLPHONE ENABLED
1-800-555-5555
Lot 28
Space 6

Fort Lauderdale, FL
In-car meters
Drivers can load up to $100 onto a prepaid meter that dangles from the rearview mirror, above; the meter counts down remaining parking minutes.

Handheld device
Cars parked legally are displayed as green squares, while those that have exceeded their time limit turn red.

Sacramento, CA
Infrared license plate scanners
Enforcement vehicles traveling as fast as 30 mph use cameras to scan license plates. Using a global positioning system lets officers check whether a car has outlasted its time on the meter. The system also can match license plates against databases of unpaid parking tickets and stolen vehicles.

Parking czars in municipalities across the country are starting to realize parking meters' original goals: generating revenue and creating a continuous turnover of parking spaces on city streets. Clearly, their main question is "Would there be enough new revenues from installing the expensive parking monitoring devices?" or "How many devices could be installed to maximize the revenue streams?" From the device manufacturer's point of view, the question is "Would there be enough demand for their products to justify the investment required in new facilities and marketing?" If the manufacturer decides to go ahead and market the products, but the actual demand is far less than its forecast or the adoption of the technology is too slow, what would be the potential financial risk?

In Chapters 3 and 4, we presented the concept of the time value of money and developed techniques for establishing cash flow equivalence with compound-interest factors. That background provides a foundation for accepting or rejecting a capital investment: the economic evaluation of a project's desirability. The forthcoming coverage of investment worth in this chapter will allow us to go a step beyond merely accepting or rejecting an investment to making comparisons of alternative investments. We will learn how to compare alternatives on an equal basis and select the wisest alternative from an economic standpoint.

The three common measures based on cash flow equivalence are (1) equivalent present worth (PW), (2) equivalent future worth (FW), and (3) equivalent annual worth (AE). Present worth represents a measure of future cash flow relative to the time point "now," with provisions that account for earning opportunities. Future worth is a measure of cash flow at some future planning horizon and offers a consideration of the earning opportunities of intermediate cash flows. Annual worth is a measure of cash flow in terms of equivalent equal payments made on an annual basis.

Our treatment of measures of investment worth is divided into three chapters. Chapter 5 begins with a consideration of the payback period, a project screening tool that was the first formal method used to evaluate investment projects. Then it introduces two measures based on fundamental cash flow equivalence techniques: present-worth and future-worth analysis. Because the annual-worth approach has many useful engineering applications related to estimating the unit cost, Chapter 6 is devoted to annual cash flow analysis. Chapter 7 presents measures of investment worth based on yield—measures known as rate-of-return analysis.

We must also recognize that one of the most important parts of the capital budgeting process is the estimation of relevant cash flows. For all examples in this chapter, and those in Chapters 6 and 7, however, net cash flows can be viewed as before-tax values or after-tax values for which tax effects have been recalculated. Since some organizations (e.g., governments and nonprofit organizations) are not subject to tax, the before-tax situation provides a valid base for this type of economic evaluation. Taking this after-tax view will allow us to focus on our main area of concern: the economic evaluation of investment projects. The procedures for determining after-tax net cash flows in taxable situations are developed in Chapter 10.

CHAPTER LEARNING OBJECTIVES

After completing this chapter, you should understand the following concepts:

■ How firms screen potential investment opportunities.

■ How firms evaluate the profitability of an investment project by considering the time value of money.

■ How firms compare mutually exclusive investment opportunities.

COMMENTS: If the company purchases the computer process control system for $650,000 now, it can expect an annual savings of $162,500 for eight years. (Note that these savings occur in discrete lumps at the ends of years.) We also considered only the benefits associated with the production of the demulsification chemical. We could also have quantified some benefits attributable to the production of the other chemicals from this plant. Suppose that the demulsification chemical benefits alone justify the acquisition of the new system. Then it is obvious that, had we considered the benefits deriving from the other chemicals as well, the acquisition of the system would have been even more clearly justified.

We draw a cash flow diagram of this situation in Figure 5.2. Assuming that these cost savings and cash flow estimates are correct, should management give the go-ahead for installation of the system? If management decides not to purchase the computer control system, what should it do with the $650,000 (assuming that it has this amount in the first place)? The company could buy $650,000 of Treasury bonds, or it could invest the amount in other cost-saving projects. How would the company compare cash flows that differ both in timing and amount for the alternatives it is considering? This is an extremely important question, because virtually every engineering investment decision involves a comparison of alternatives. Indeed, these are the types of questions this chapter is designed to help you answer.

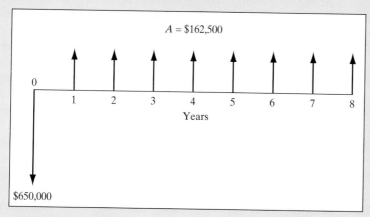

Figure 5.2   Cash flow diagram for the computer process control project described in Example 5.1.

5.1.2 Independent versus Mutually Exclusive Investment Projects

Most firms have a number of unrelated investment opportunities available. For example, in the case of XL Chemicals, other projects being considered in addition to the computer process control project in Example 5.1 are a new waste heat recovery boiler, a CAD system for the engineering department, and a new warehouse. The economic attractiveness of each of these projects can be measured, and a decision to accept or reject the project can be made without reference to any of the other projects. In other words, the decision regarding any one project has no effect on the decision to accept or reject another project. Such projects are said to be **independent**.

In Section 5.5, we will see that in many engineering situations we are faced with selecting the most economically attractive project from a number of alternative projects, all of which solve the same problem or meet the same need. It is unnecessary to choose more than one project in this situation, and the acceptance of one automatically entails the rejection of all of the others. Such projects are said to be **mutually exclusive**.

As long as the total cost of all the independent projects found to be economically attractive is less than the investment funds available to a firm, all of these projects could proceed. However, this is rarely the case. The selection of those projects which should proceed when investment funds are limited is the subject of capital budgeting. Apart from Chapter 15, which deals with capital budgeting, the availability of funds will not be a consideration in accepting or rejecting projects dealt with in this book.

5.2 Initial Project Screening Method

Let's suppose that you are in the market for a new punch press for your company's machine shop, and you visit an equipment dealer. As you take a serious look at one of the punch press models in the display room, an observant equipment salesperson approaches you and says, "That press you are looking at is the state of the art in its category. If you buy that top-of-the-line model, it will cost a little bit more, but it will pay for itself in less than two years." Before studying the four measures of investment attractiveness, we will review a simple, but nonrigorous, method commonly used to screen capital investments. One of the primary concerns of most businesspeople is whether and when the money invested in a project can be recovered. The **payback method** screens projects on the basis of how long it takes for net receipts to equal investment outlays. This calculation can take one of two forms: either ignore time-value-of-money considerations or include them. The former case is usually designated the **conventional payback method**, the latter case the **discounted payback method**.

Payback period: The length of time required to recover the cost of an investment.

A common standard used to determine whether to pursue a project is that the project does not merit consideration unless its payback period is shorter than some specified period. (This time limit is determined largely by management policy. For example, a high-tech firm, such as a computer chip manufacturer, would set a short time limit for any new investment, because high-tech products rapidly become obsolete.) If the payback period is within the acceptable range, a formal project evaluation (such as a present-worth analysis) may begin. It is important to remember that **payback screening** is not an *end* in itself, but rather a method of screening out certain obviously unacceptable investment alternatives before progressing to an analysis of potentially acceptable ones.

5.2.1 Payback Period: The Time It Takes to Pay Back

Determining the relative worth of new production machinery by calculating the time it will take to pay back what it cost is the single most popular method of project screening. If a company makes investment decisions solely on the basis of the payback period, it considers only those projects with a payback period *shorter* than the maximum acceptable payback period. (However, because of shortcomings of the payback screening method, which we will discuss later, it is rarely used as the only decision criterion.)

What does the payback period tell us? One consequence of insisting that each proposed investment have a short payback period is that investors can assure themselves of being restored to their initial position within a short span of time. By restoring their initial position, investors can take advantage of additional, perhaps better, investment possibilities that may come along.

EXAMPLE 5.2 Conventional Payback Period for the Computer Process Control System Project

Consider the cash flows given in Example 5.1. Determine the payback period for this computer process control system project.

SOLUTION

Given: Initial cost = $650,000 and annual net benefits = $162,500.
Find: Conventional payback period.

Given a uniform stream of receipts, we can easily calculate the payback period by dividing the initial cash outlay by the annual receipts:

$$\text{Payback period} = \frac{\text{Initial cost}}{\text{Uniform annual benefit}} = \frac{\$650,000}{\$162,500}$$

$$= 4 \text{ years.}$$

If the company's policy is to consider only projects with a payback period of five years or less, this computer process control system project passes the initial screening.

In Example 5.2, dividing the initial payment by annual receipts to determine the payback period is a simplification we can make because the annual receipts are uniform. Whenever the expected cash flows vary from year to year, however, the payback period must be determined by adding the expected cash flows for each year until the sum is equal to or greater than zero. The significance of this procedure is easily explained. The cumulative cash flow equals zero at the point where cash inflows exactly match, or pay back, the cash outflows; thus, the project has reached the payback point. Similarly, if the cumulative cash flows are greater than zero, then the cash inflows exceed the cash outflows, and the project has begun to generate a profit, thus surpassing its payback point. To illustrate, consider Example 5.3.

EXAMPLE 5.3 Conventional Payback Period with Salvage Value

Autonumerics Company has just bought a new spindle machine at a cost of $105,000 to replace one that had a salvage value of $20,000. The projected annual after-tax savings via improved efficiency, which will exceed the investment cost, are as follows:

Period	Cash Flow	Cumulative Cash Flow
0	−$105,000 + $20,000	−$85,000
1	15,000	−70,000
2	25,000	−45,000
3	35,000	−10,000
4	45,000	35,000
5	45,000	80,000
6	35,000	115,000

SOLUTION

Given: Cash flow series as shown in Figure 5.3(a).

Find: Conventional payback period.

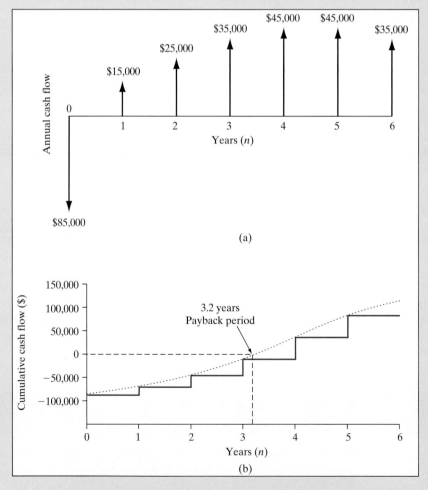

Figure 5.3 Illustration of conventional payback period (Example 5.3).

The salvage value of retired equipment becomes a major consideration in most justification analysis. (In this example, the salvage value of the old machine should be taken into account, as the company already had decided to replace the old machine.) When used, the salvage value of the retired equipment is subtracted from the purchase price of new equipment, revealing a closer true cost of the investment. As we see from the cumulative cash flow in Figure 5.3(b), the total investment is recovered during year 4. If the firm's stated maximum payback period is three years, the project will not pass the initial screening stage.

COMMENTS: In Example 5.2, we assumed that cash flows occur only in discrete lumps at the ends of years. If instead cash flows occur continuously throughout the year, the payback period calculation needs adjustment. A negative balance of $10,000 remains at the start of year 4. If $45,000 is expected to be received as a more or less continuous flow during year 4, the total investment will be recovered two-tenths ($10,000/$45,000) of the way through the fourth year. Thus, in this situation, the payback period is 3.2 years.

5.2.2 Benefits and Flaws of Payback Screening

The simplicity of the payback method is one of its most appealing qualities. Initial project screening by the method reduces the information search by focusing on that time at which the firm expects to recover the initial investment. The method may also eliminate some alternatives, thus reducing the firm's time spent analyzing. But the much-used payback method of equipment screening has a number of serious drawbacks. The principal objection to the method is that it fails to measure profitability (i.e., no "profit" is made during the payback period). Simply measuring how long it will take to recover the initial investment outlay contributes little to gauging the earning power of a project. (In other words, you already know that the money you borrowed for the drill press is costing you 12% per year; the payback method can't tell you how much your invested money is contributing toward the interest expense.) Also, because payback period analysis ignores differences in the timing of cash flows, it fails to recognize the difference between the present and future value of money. For example, although the payback on two investments can be the same in terms of numbers of years, a front-loaded investment is better because money available today is worth more than that to be gained later. Finally, because payback screening ignores all proceeds after the payback period, it does not allow for the possible advantages of a project with a longer economic life.

By way of illustration, consider the two investment projects listed in Table 5.1. Each requires an initial investment outlay of $90,000. Project 1, with expected annual cash proceeds of $30,000 for the first 3 years, has a payback period of 3 years. Project 2 is expected to generate annual cash proceeds of $25,000 for 6 years; hence, its payback period is 3.6 years. If the company's maximum payback period is set to 3 years, then project 1 would pass the initial project screening, whereas project 2 would fail even though it is clearly the more profitable investment.

TABLE 5.1 Investment Cash Flows for Two Competing Projects

n	Project 1	Project 2
0	−$90,000	−$90,000
1	30,000	25,000
2	30,000	25,000
3	30,000	25,000
4	1,000	25,000
5	1,000	25,000
6	1,000	25,000
	$ 3,000	$60,000

5.2.3 Discounted Payback Period

To remedy one of the shortcomings of the conventional payback period, we may modify the procedure so that it takes into account the time value of money—that is, the cost of funds (interest) used to support a project. This modified payback period is often referred to as the **discounted payback period**. In other words, we may define the discounted payback period as the number of years required to recover the investment from *discounted* cash flows.

For the project in Example 5.3, suppose the company requires a rate of return of 15%. To determine the period necessary to recover both the capital investment and the cost of funds required to support the investment, we may construct Table 5.2, showing cash flows and costs of funds to be recovered over the life of the project.

To illustrate, let's consider the cost of funds during the first year: With $85,000 committed at the beginning of the year, the interest in year 1 would be $12,750 ($85,000 × 0.15). Therefore, the total commitment grows to $97,750, but the $15,000 cash flow in year 1

> **Discounted payback period:** The length of time required to recover the cost of an investment based on discounted cash flows.

TABLE 5.2 Payback Period Calculation Taking into Account the Cost of Funds (Example 5.3)

Period	Cash Flow	Cost of Funds (15%)*		Cumulative Cash Flow
0	−$85,000	0		−$85,000
1	15,000	−$85,000(0.15) =	−$12,750	−82,750
2	25,000	−$82,750(0.15) =	−12,413	−70,163
3	35,000	−$70,163(0.15) =	−10,524	−45,687
4	45,000	−$45,687(0.15) =	−6,853	−7,540
5	45,000	−$7,540(0.15) =	−1,131	36,329
6	35,000	$36,329(0.15) =	5,449	76,778

*Cost of funds = Unrecovered beginning balance × interest rate.

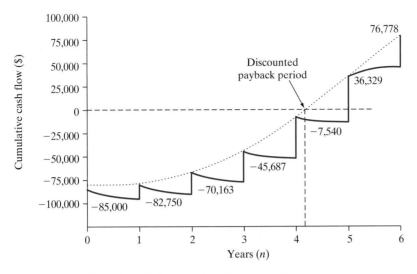

Figure 5.4 Illustration of discounted payback period.

leaves a net commitment of $82,750. The cost of funds during the second year would be $12,413 ($82,750 × 0.15), but with the $25,000 receipt from the project, the net commitment drops to $70,163. When this process repeats for the remaining years of the project's life, we find that the net commitment to the project ends during year 5. Depending on which cash flow assumption we adopt, the project must remain in use about 4.2 years (continuous cash flows) or 5 years (year-end cash flows) in order for the company to cover its cost of capital and also recover the funds it has invested. Figure 5.4 illustrates this relationship.

The inclusion of effects stemming from time value of money has increased the payback period calculated for this example by a year. Certainly, this modified measure is an improved one, but it does not show the complete picture of the project's profitability either.

5.2.4 Where Do We Go from Here?

Should we abandon the payback method? Certainly not, but if you use payback screening exclusively to analyze capital investments, look again. You may be missing something that another method can help you spot. Therefore, it is illogical to claim that payback is either a good or bad method of justification. Clearly, it is not a measure of profitability. But when it is used to supplement other methods of analysis, it can provide useful information. For example, payback can be useful when a company needs a measure of the speed of cash recovery, when the company has a cash flow problem, when a product is built to last only for a short time, and when the machine the company is contemplating buying itself is known to have a short market life.

5.3 Discounted Cash Flow Analysis

Until the 1950s, the payback method was widely used as a means of making investment decisions. As flaws in this method were recognized, however, businesspeople began to search for methods to improve project evaluations. The result was the development of **discounted**

cash flow techniques (DCFs), which take into account the time value of money. One of the DCFs is the net-present-worth, or net-present-value, method. A capital investment problem is essentially a problem of determining whether the anticipated cash inflows from a proposed project are sufficient to attract investors to invest funds in the project. In developing the NPW criterion, we will use the concept of cash flow equivalence discussed in Chapter 3.

Discounted cash flow analysis (DCF): A method of evaluating an investment by estimating future cash flows and taking into consideration the time value of money.

As we observed, the most convenient point at which to calculate the equivalent values is often at time 0. Under the NPW criterion, the present worth of all cash inflows is compared against the present worth of all cash outflows associated with an investment project. The difference between the present worth of these cash flows, referred to as the **net present worth** (NPW), **net present value** (NPV) determines whether the project is an acceptable investment. When two or more projects are under consideration, NPW analysis further allows us to select the best project by comparing their NPW figures.

5.3.1 Net-Present-Worth Criterion

Net present worth: The difference between the present value of cash inflows and the present value of cash outflows.

We will first summarize the basic procedure for applying the net-present-worth criterion to a typical investment project:

Investment pool operates like a mutual fund to earn a targeted return by investing the firm's money in various investment assets.

- Determine the interest rate that the firm wishes to earn on its investments. The interest rate you determine represents the rate at which the firm can always invest the money in its **investment pool**. This interest rate is often referred to as either a **required rate of return** or a **minimum attractive rate of return** (MARR). Usually, selection of the MARR is a policy decision made by top management. It is possible for the MARR to change over the life of a project, as we saw in Section 4.4, but for now we will use a single rate of interest in calculating the NPW.
- Estimate the service life of the project.
- Estimate the cash inflow for each period over the service life.
- Estimate the cash outflow over each service period.
- Determine the net cash flows (net cash flow = cash inflow − cash outflow).
- Find the present worth of each net cash flow at the MARR. Add up all the present-worth figures; their sum is defined as the project's NPW, given by

$$PW(i) = \text{NPW calculated at } i = \frac{A_0}{(1 + i)^0} + \frac{A_1}{(1 + i)^1} + \frac{A_2}{(1 + i)^2} + \cdots$$

$$+ \frac{A_N}{(1 + i)^N}$$

$$= \sum_{n=0}^{N} \frac{A_n}{(1 + i)^n}$$

$$= \sum_{n=0}^{N} A_n(P/F, i, n), \tag{5.1}$$

where

$$A_n = \text{Net cash flow at end of period } n,$$
$$i = \text{MARR (or cost of capital)},$$
$$N = \text{Service life of the project.}$$

A_n will be positive if the corresponding period has a net cash inflow and negative if there is a net cash outflow.

- **Single Project Evaluation.** In this context, a positive NPW means that the equivalent worth of the inflows is greater than the equivalent worth of outflows, so the project makes a profit. Therefore, if the PW(i) is positive for a single project, the project should be accepted; if the PW(i) is negative, the project should be rejected.[2] The decision rule is

$$\text{If } PW(i) > 0, \text{ accept the investment.}$$

$$\text{If } PW(i) = 0, \text{ remain indifferent.}$$

$$\text{If } PW(i) < 0, \text{ reject the investment.}$$

- **Comparing Multiple Alternatives.** Compute the PW(i) for each alternative and select the one with the largest PW(i). As you will learn in Section 5.5, when you compare mutually exclusive alternatives with the *same revenues*, they are compared on a *cost-only basis*. In this situation (because you are minimizing costs, rather than maximizing profits), you should accept the project that results in the *smallest*, or *least negative*, NPW.

EXAMPLE 5.4 Net Present Worth: Uniform Flows

Consider the investment cash flows associated with the computer process control project discussed in Example 5.1. If the firm's MARR is 15%, compute the NPW of this project. Is the project acceptable?

SOLUTION

Given: Cash flows in Figure 5.2 and MARR = 15% per year.
Find: NPW.

Since the computer process control project requires an initial investment of $650,000 at $n = 0$, followed by the eight equal annual savings of $162,000, we can easily determine the NPW as follows:

$$PW(15\%)_{\text{Outflow}} = \$650,000;$$
$$PW(15\%)_{\text{Inflow}} = \$162,500(P/A, 15\%, 8)$$
$$= \$729,190.$$

Then the NPW of the project is

$$PW(15\%) = PW(15\%)_{\text{Inflow}} - PW(15\%)_{\text{Outflow}}$$
$$= \$729,190 - \$650,000$$
$$= \$79,190,$$

or, from Eq. (5.1),

$$PW(15\%) = -\$650,000 + \$162,500(P/A, 15\%, 8)$$
$$= \$79,190.$$

Since PW(15%) > 0, the project is acceptable.

---

[2] Some projects (e.g., the installation of pollution control equipment) cannot be avoided. In a case such as this, the project would be accepted even though its NPW $\leq$ 0. This type of project will be discussed in Chapter 12.

Now let's consider an example in which the investment cash flows are not uniform over the service life of the project.

EXAMPLE 5.5 Net Present Worth: Uneven Flows

Tiger Machine Tool Company is considering acquiring a new metal-cutting machine. The required initial investment of $75,000 and the projected cash benefits[3] over the project's three-year life are as follows:

End of Year	Net Cash Flow
0	−$75,000
1	24,400
2	27,340
3	55,760

You have been asked by the president of the company to evaluate the economic merit of the acquisition. The firm's MARR is known to be 15%.

SOLUTION

Given: Cash flows as tabulated and MARR = 15% per year.

Find: NPW.

If we bring each flow to its equivalent at time zero, we find that

$$PW(15\%) = -\$75,000 + \$24,000(P/F, 15\%, 1) + \$27,340(P/F, 15\%, 2)$$
$$+ \$55,760(P/F, 15\%, 3)$$
$$= \$3,553.$$

Since the project results in a surplus of $3,553, the project is acceptable.

In Example 5.5, we computed the NPW of a project at a fixed interest rate of 15%. If we compute the NPW at varying interest rates, we obtain the data in Table 5.3. Plotting the NPW as a function of interest rate gives Figure 5.5, the present-worth profile. (You may use a spreadsheet program such as Excel to generate Table 5.3 or Figure 5.5.)

Figure 5.5 indicates that the investment project has a positive NPW if the interest rate is below 17.45% and a negative NPW if the interest rate is above 17.45%. As we will see in Chapter 7, this **break-even interest rate** is known as the **internal rate of return**. If the

---

[3] As we stated at the beginning of this chapter, we treat net cash flows as before-tax values or as having their tax effects precalculated. Explaining the process of obtaining cash flows requires an understanding of income taxes and the role of depreciation, which are discussed in Chapter 9.

TABLE 5.3 Present-Worth Amounts at Varying Interest Rates (Example 5.5)

i(%)	PW(i)	i(%)	PW(i)
0	$32,500	20	−$3,412
2	27,743	22	−5,924
4	23,309	24	−8,296
6	19,169	26	−10,539
8	15,296	28	−12,662
10	11,670	30	−14,673
12	8,270	32	−16,580
14	5,077	34	−18,360
16	2,076	36	−20,110
17.45*	0	38	−21,745
18	−750	40	−23,302

*Break-even interest rate (also known as the rate of return).

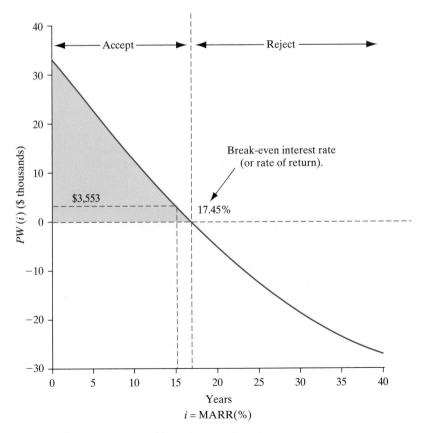

Figure 5.5 Present-worth profile described in Example 5.5.

firm's MARR is 15%, then the project has an NPW of $3,553 and so may be accepted. The $3,553 figure measures the equivalent immediate gain in present worth to the firm following acceptance of the project. By contrast, at $i = 20\%, \mathrm{PW}(20\%) = -\$3,412$, and the firm should reject the project. (Note that either accepting or rejecting an investment is influenced by the choice of a MARR, so it is crucial to estimate the MARR correctly. We will defer this important issue until Section 5.3.3. For now, we will assume that the firm has an accurate MARR estimate available for use in investment analysis.)

5.3.2 Meaning of Net Present Worth

In present-worth analysis, we assume that all the funds in a firm's treasury can be placed in investments that yield a return equal to the MARR. We may view these funds as an **investment pool**. Alternatively, if no funds are available for investment, we assume that the firm can borrow them at the MARR (or cost of capital) from the capital market. In this section, we will examine these two views as we explain the meaning of the MARR in NPW calculations.

Investment Pool Concept

An investment pool is equivalent to a firm's treasury. All fund transactions are administered and managed by the firm's comptroller. The firm may withdraw funds from this investment pool for other investment purposes, but if left in the pool, these funds will earn at the MARR. Thus, in investment analysis, net cash flows will be net cash flows relative to an investment pool. To illustrate the investment pool concept, we consider again the project in Example 5.5 that required an investment of $75,000.

  If the firm did not invest in the project and left $75,000 in the investment pool for three years, these funds would grow as follows:

$$\$75,000(F/P, 15\%, 3) = \$114,066.$$

Suppose the company decided instead to invest $75,000 in the project described in Example 5.5. Then the firm would receive a stream of cash inflows during the project's life of three years in the following amounts:

Period (n)	Net Cash Flow (A_n)
1	$24,400
2	27,340
3	55,760

Since the funds that return to the investment pool earn interest at a rate of 15%, it is worthwhile to see how much the firm would benefit from its $75,000 investment. For this alternative, the returns after reinvestment are

$$\$24,400(F/P, 15\%, 2) = \$32,269,$$
$$\$27,340(F/P, 15\%, 1) = \$31,441,$$
$$\$55,760(F/P, 15\%, 0) = \$55,760.$$

These returns total $119,470. At the end of three years, the additional cash accumulation from investing in the project is

$$\$119,470 - \$114,066 = \$5,404.$$

If we compute the equivalent present worth of this net cash surplus at time 0, we obtain

$$\$5,404(P/F, 15\%, 3) = \$3,553,$$

which is exactly what we get when we compute the NPW of the project with Eq. (5.1). Clearly, on the basis of its positive NPW, the alternative of purchasing a new machine should be preferred to that of simply leaving the funds in the investment pool at the MARR. Thus, in PW analysis, any investment is assumed to be returned at the MARR. If a surplus exists at the end of the project, then $PW(MARR) > 0$. Figure 5.6 illustrates the reinvestment concept as it relates to the firm's investment pool.

Borrowed-Funds Concept

Suppose that the firm does not have $75,000 at the outset. In fact, the firm doesn't have to have an investment pool at all. Suppose further that the firm borrows all of its capital from a bank at an interest rate of 15%, invests in the project, and uses the proceeds from the investment to pay off the principal and interest on the bank loan. How much is left over for the firm at the end of the project's life?

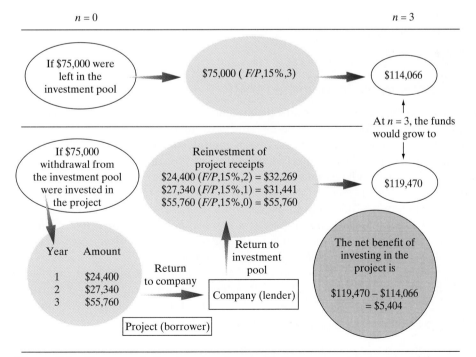

Figure 5.6 The concept of an investment pool with the company as a lender and the project as a borrower.

At the end of first year, the interest on the project's use of the bank loan would be $75,000(0.15) = \$11,250$. Therefore, the total loan balance grows to $\$75,000(1.15) = \$86,250$. Then, the firm receives \$24,400 from the project and applies the entire amount to repay the loan portion, leaving a balance due of

$$\$75,000(1 + 0.15) - \$24,400 = \$61,850.$$

This amount becomes the net amount the project is borrowing at the beginning of year 2, which is also known as the **project balance**. At the end of year 2, the bank debt grows to $\$61,850(1.15) = \$71,128$, but with the receipt of \$27,340, the project balance is reduced to

$$\$71,128 - \$27,340 = \$43,788.$$

Similarly, at the end of year 3, the project balance becomes

$$\$43,788(1.15) = \$50,356.$$

But with the receipt of \$55,760 from the project, the firm should be able to pay off the remaining balance and come out with a surplus in the amount of \$5,404. This terminal project balance is also known as the **net future worth** of the project. In other words, the firm repays its initial bank loan and interest at the end of year 3, with a resulting profit of \$5,404. If we compute the equivalent present worth of this net profit at time 0, we obtain

$$\text{PW}(15\%) = \$5,404(P/F, 15\%, 3) = \$3,553.$$

The result is identical to the case in which we directly computed the NPW of the project at $i = 15\%$, shown in Example 5.5. Figure 5.7 illustrates the project balance as a function of time.[4]

5.3.3 Basis for Selecting the MARR

Cost of capital:
The required
return
necessary
to make an
investment
project
worthwhile.

The basic principle used to determine the discount rate in project evaluations is similar to the concept of the required return on investment for financial assets discussed in Section 4.6.2. The first element to cover is the **cost of capital**, which is the required return necessary to make an investment project worthwhile. The cost of capital would include both the *cost of debt* (the interest rate associated with borrowing) and the *cost of equity* (the return that stockholders require for a company). Both the cost of debt and the cost of equity reflect the presence of inflation in the economy. *The cost of capital determines how a company can raise money (through issuing a stock, borrowing, or a mix of the two). Therefore, this is normally considered as the rate of return that a firm would receive if it invested its money someplace else with a similar risk.*

The second element is a consideration of any additional risk associated with the project. If the project belongs to the normal risk category, the cost of capital may already reflect the risk premium. However, if you are dealing with a project with higher risk, the additional risk premium may be added onto the cost of capital.

---

[4] Note that the sign of the project balance changes from negative to positive during year 3. The time at which the project balance becomes zero is known as the **discounted payback period**.

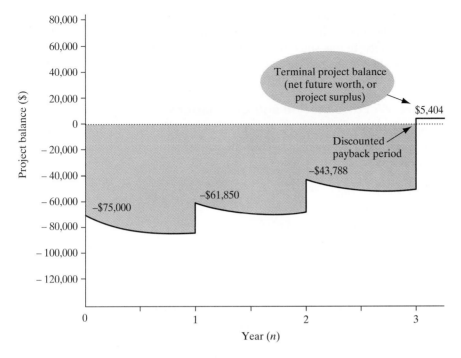

Figure 5.7 Project balance diagram as a function of time. (A negative project balance indicates the amount of the loan remaining to be paid off or the amount of investment to be recovered.)

In sum, the discount rate (**MARR**) to use for project evaluation would be equivalent to the firm's cost of capital for a project of normal risk, but could be much higher if you are dealing with a risky project. Chapter 15 will detail the analytical process of determining this discount rate. For now, we assume that such a rate is already known to us, and we will focus on the evaluation of the investment project.

MARR: this is based on the firm's cost of capital plus or minus a risk premium to reflect the project's specific risk characteristics.

5.4 Variations of Present-Worth Analysis

As variations of present-worth analysis, we will consider two additional measures of investment worth: **future-worth analysis** and **capitalized equivalent-worth analysis**. (The equivalent annual worth measure is another variation of the present-worth measure, but we present it in Chapter 6.) Future-worth analysis calculates the future worth of an investment undertaken. Capitalized equivalent-worth analysis calculates the present worth of a project with a perpetual life span.

5.4.1 Future-Worth Analysis

Net present worth measures the surplus in an investment project at time 0. **Net future worth (NFW)** measures this surplus at a time other than 0. Net-future-worth analysis is particularly useful in an investment situation in which we need to compute the equivalent worth of a project at the end of its investment period, rather than at its beginning. For example, it may take 7 to 10 years to build a nuclear power plant because of the complexities of engineering design and the many time-consuming regulatory procedures that must be followed to ensure public safety. In this situation, it is more common to measure the worth of the investment at the time of the project's commercialization (i.e., we conduct an NFW analysis at the end of the investment period).

Net future worth: The value of an asset or cash at a specified date in the future that is equivalent in value to a specified sum today.

Net-Future-Worth Criterion and Calculations

Let A_n represent the cash flow at time n for $n = 0, 1, 2, \ldots, N$ for a typical investment project that extends over N periods. Then the net-future-worth (NFW) expression at the end of period N is

$$\begin{aligned} \text{FW}(i) &= A_0(1 + i)^N + A_1(1 + i)^{N-1} + A_2(1 + i)^{N-2} + \cdots + A_N \\ &= \sum_{n=0}^{N} A_n(1 + i)^{N-n} \\ &= \sum_{n=0}^{N} A_n(F/P, i, N - n). \end{aligned} \tag{5.2}$$

As you might expect, the decision rule for the NFW criterion is the same as that for the NPW criterion: For a single project evaluation,

If $\text{FW}(i) > 0$, accept the investment.

If $\text{FW}(i) = 0$, remain indifferent to the investment.

If $\text{FW}(i) < 0$, reject the investment.

EXAMPLE 5.6 Net Future Worth: At the End of the Project

Consider the project cash flows in Example 5.5. Compute the NFW at the end of year 3 at $i = 15\%$.

SOLUTION

Given: Cash flows in Example 5.5 and MARR $= 15\%$ per year.
Find: NFW.

As seen in Figure 5.8, the NFW of this project at an interest rate of 15% would be

$$\begin{aligned} \text{FW}(15\%) &= -\$75{,}000(F/P, 15\%, 3) + \$24{,}400(F/P, 15\%, 2) \\ &\quad + \$27{,}340(F/P, 15\%, 1) + \$55{,}760 \\ &= \$5{,}404. \end{aligned}$$

Note that the net future worth of the project is equivalent to the terminal project balance as calculated in Section 5.3.2. Since $\text{FW}(15\%) > 0$, the project is acceptable. We reach the same conclusion under present-worth analysis.

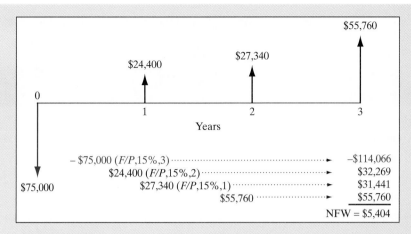

Figure 5.8 Future-worth calculation at the end of year 3 (Example 5.6).

EXAMPLE 5.7 Future Equivalent: At an Intermediate Time

Higgins Corporation (HC), a Detroit-based robot-manufacturing company, has developed a new advanced-technology robot called Helpmate, which incorporates advanced technology such as vision systems, tactile sensing, and voice recognition. These features allow the robot to roam the corridors of a hospital or office building without following a predetermined track or bumping into objects. HC's marketing department plans to target sales of the robot toward major hospitals. The robots will ease nurses' workloads by performing low-level duties such as delivering medicines and meals to patients.

• The firm would need a new plant to manufacture the Helpmates; this plant could be built and made ready for production in two years. It would require a 30-acre site, which can be purchased for $1.5 million in year 0. Building construction would begin early in year 1 and continue throughout year 2. The building would cost an estimated $10 million, with a $4 million payment due to the contractor at the end of year 1, and with another $6 million payable at the end of year 2.

• The necessary manufacturing equipment would be installed late in year 2 and would be paid for at the end of year 2. The equipment would cost $13 million, including transportation and installation. When the project terminates, the land is expected to have an after-tax market value of $2 million, the building an after-tax value of $3 million, and the equipment an after-tax value of $3 million.

For capital budgeting purposes, assume that the cash flows occur at the end of each year. Because the plant would begin operations at the beginning of year 3, the first operating cash flows would occur at the end of year 3. The Helpmate plant's estimated economic life is six years after completion, with the following expected after-tax operating cash flows in millions:

Calendar Year	'06	'07	'08	'09	'10	'11	'12	'13	'14
End of Year	0	1	2	3	4	5	6	7	8
After-tax cash flows									
A. Operating revenue				$6	$8	$13	$18	$14	$8
B. Investment									
Land	−1.5								+2
Building		−4	−6						+3
Equipment			−13						+3
Net cash flow	−$1.5	−$4	−$19	$6	$8	$13	$18	$14	$16

Compute the equivalent worth of this investment at the start of operations. Assume that HC's MARR is 15%.

SOLUTION

Given: Preceding cash flows and MARR = 15% per year.
Find: Equivalent worth of project at the end of calendar year 2.

One easily understood method involves calculating the present worth and then transforming it to the equivalent worth at the end of year 2. First, we can compute PW(15%) at time 0 of the project:

$$PW(15\%) = -\$1.5 - \$4(P/F, 15\%, 1) - \$19(P/F, 15\%, 2)$$
$$+ \$6(P/F, 15\%, 3) + \$8(P/F, 15\%, 4) + \$13(P/F, 15\%, 5)$$
$$+ \$18(P/F, 15\%, 6) + \$14(P/F, 15\%, 7) + \$16(P/F, 15\%, 8)$$
$$= \$13.91 \text{ million.}$$

Then, the equivalent project worth at the start of operation is

$$FW(15\%) = PW(15\%)(F/P, 15\%, 2)$$
$$= \$18.40 \text{ million.}$$

A second method brings all flows prior to year 2 up to that point and discounts future flows back to year 2. The equivalent worth of the earlier investment, when the plant begins full operation, is

$$-\$1.5(F/P, 15\%, 2) - \$4(F/P, 15\%, 1) - \$19 = -\$25.58 \text{ million,}$$

which produces an equivalent flow as shown in Figure 5.9. If we discount the future flows to the start of operation, we obtain

$$FW(15\%) = -\$25.58 + \$6(P/F, 15\%, 1) + \$8(P/F, 15\%, 2) + \cdots$$
$$+ \$16(F/P, 15\%, 6)$$
$$= \$18.40 \text{ million.}$$

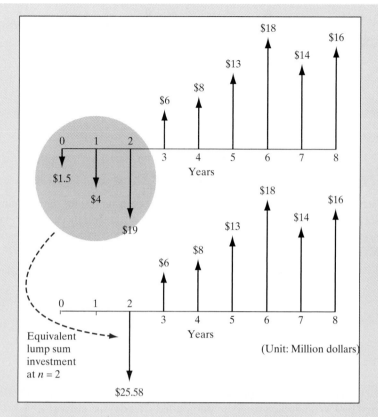

Figure 5.9 Cash flow diagram for the Helpmate project (Example 5.7).

COMMENTS: If another company is willing to purchase the plant and the right to manufacture the robots immediately after completion of the plant (year 2), HC would set the price of the plant at $43.98 million ($18.40 + $25.58) at a minimum.

5.4.2 Capitalized Equivalent Method

Another special case of the PW criterion is useful when the life of a proposed project is **perpetual** or the planning horizon is extremely long (say, 40 years or more). Many public projects, such as bridges, waterway structures, irrigation systems, and hydro-electric dams, are expected to generate benefits over an extended period (or forever). In this section, we will examine the **capitalized equivalent** (CE(i)) method for evaluating such projects.

Perpetual Service Life

Consider the cash flow series shown in Figure 5.10. How do we determine the PW for an infinite (or almost infinite) uniform series of cash flows or a repeated cycle of cash flows? The process of computing the PW cost for this infinite series is referred to as the **capitalization**

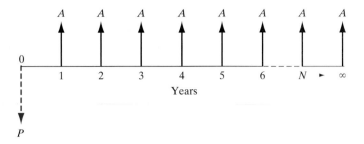

Figure 5.10 Equivalent present worth of an infinite cash flow series.

Capitalized cost
related to car
leasing, means
the amount that
is being financed.

of the project cost. The cost, known as the **capitalized cost**, represents the amount of money that must be invested today to yield a certain return A at the end of each and every period forever, assuming an interest rate of i. Observe the limit of the uniform series present-worth factor as N approaches infinity:

$$\lim_{N \to \infty} (P/A, i, N) = \lim_{N \to \infty} \left[\frac{(1 + i)^N - 1}{i(1 + i)^N} \right] = \frac{1}{i}.$$

Thus,

$$\text{PW}(i) = A(P/A, i, N \to \infty) = \frac{A}{i}. \tag{5.3}$$

Another way of looking at this problem is to ask what constant income stream could be generated by $\text{PW}(i)$ dollars today in perpetuity. Clearly, the answer is $A = i\text{PW}(i)$. If withdrawals were greater than A, you would be eating into the principal, which would eventually reduce it to 0.

EXAMPLE 5.8 Capitalized Equivalent Cost

An engineering school has just completed a new engineering complex worth $50 million. A campaign targeting alumni is planned to raise funds for future maintenance costs, which are estimated at $2 million per year. Any unforeseen costs above $2 million per year would be obtained by raising tuition. Assuming that the school can create a trust fund that earns 8% interest annually, how much has to be raised now to cover the perpetual string of $2 million in annual costs?

SOLUTION

Given: $A = \$2$ million, $i = 8\%$ per year, and $N = \infty$.
Find: CE(8%).

The capitalized cost equation is

$$\text{CE}(i) = \frac{A}{i},$$

so

$$CE(8\%) = \$2,000,000/0.08$$
$$= \$25,000,000.$$

COMMENTS: It is easy to see that this lump-sum amount should be sufficient to pay maintenance expenses for the school forever. Suppose the school deposited $25 million in a bank that paid 8% interest annually. Then at the end of the first year, the $25 million would earn 8%($25 million) = $2 million interest. If this interest were withdrawn, the $25 million would remain in the account. At the end of the second year, the $25 million balance would again earn 8%($25 million) = $2 million. This annual withdrawal could be continued forever, and the endowment (gift funds) would always remain at $25 million.

Project's Service Life Is Extremely Long

The benefits of typical civil engineering projects, such as bridge and highway construction, although not perpetual, can last for many years. In this section, we will examine the use of the CE(i) criterion to approximate the NPW of engineering projects with long lives.

EXAMPLE 5.9 Comparison of Present Worth for Long Life and Infinite Life

Mr. Gaynor L. Bracewell amassed a small fortune developing real estate in Florida and Georgia over the past 30 years. He sold more than 700 acres of timber and farmland to raise $800,000, with which he built a small hydroelectric plant, known as High Shoals Hydro. The plant was a decade in the making. The design for Mr. Bracewell's plant, which he developed using his Army training as a civil engineer, is relatively simple. A 22-foot-deep canal, blasted out of solid rock just above the higher of two dams on his property, carries water 1,000 feet along the river to a "trash rack," where leaves and other debris are caught. A 6-foot-wide pipeline capable of holding 3 million pounds of liquid then funnels the water into the powerhouse at 7.5 feet per second, thereby creating 33,000 pounds of thrust against the turbines. Under a 1978 federal law designed to encourage alternative power sources, Georgia Power Company is required to purchase any electricity Mr. Bracewell can supply. Mr. Bracewell estimates that his plant can generate 6 million kilowatt-hours per year.

Suppose that, after paying income taxes and operating expenses, Mr. Bracewell's annual income from the hydroelectric plant will be $120,000. With normal maintenance, the plant is expected to provide service for at least 50 years. Figure 5.11 illustrates when and in what quantities Mr. Bracewell spent his $800,000 (not taking into account the time value of money) during the last 10 years. Was Mr. Bracewell's $800,000 investment a wise one? How long will he have to wait to recover his initial investment, and will he ever make a profit? Examine the situation by computing the project worth at varying interest rates.

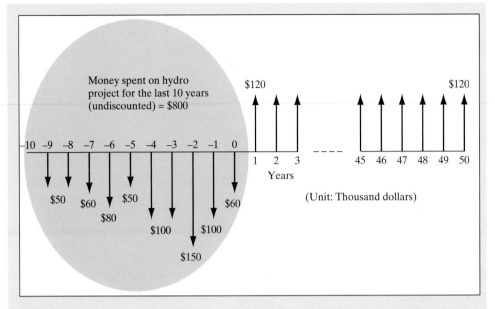

Figure 5.11 Net cash flow diagram for Mr. Bracewell's hydroelectric project (Example 5.9).

(a) If Mr. Bracewell's interest rate is 8%, compute the NPW (at time 0 in Figure 5.11) of this project with a 50-year service life and infinite service, respectively.

(b) Repeat part (a), assuming an interest rate of 12%.

SOLUTION

Given: Cash flow in Figure 5.11 (to 50 years or ∞) and $i = 8\%$ or 12%.
Find: NPW at time 0.

One of the main questions is whether Mr. Bracewell's plant will be profitable. Now we will compute the equivalent total investment and the equivalent worth of receiving future revenues at the start of power generation (i.e., at time 0).

(a) Let $i = 8\%$. Then

- with a plant service life of 50 years, we can make use of single-payment compound-amount factors in the invested cash flow to help us find the equivalent total investment at the start of power generation. Using K to indicate thousands, we obtain

$$V_1 = -\$50K(F/P, 8\%, 9) - \$50K(F/P, 8\%, 8)$$
$$- \$60K(F/P, 8\%, 7) \cdots - \$100K(F/P, 8\%, 1) - \$60K$$
$$= -\$1,101K.$$

The equivalent total benefit at the start of generation is

$$V_2 = \$120K(P/A, 8\%, 50) = \$1,468K.$$

Summing, we find the net equivalent worth at the start of power generation:

$$V_1 + V_2 = -\$1,101K + \$1,468K$$
$$= \$367K.$$

- With an infinite service life, the net equivalent worth is called the capitalized equivalent worth. The investment portion prior to time 0 is identical, so the capitalized equivalent worth is

$$CE(8\%) = -\$1,101K + \$120K/(0.08)$$
$$= \$399K.$$

Note that the difference between the infinite situation and the planning horizon of 50 years is only \$32,000.

(b) Let $i = 12\%$. Then
- With a service life of 50 years, proceeding as we did in part (a), we find that the equivalent total investment at the start of power generation is

$$V_1 = -\$50K(F/P, 12\%, 9) - \$50K(F/P, 12\%, 8)$$
$$- \$60K(F/P, 12\%, 7) \cdots - \$100K(F/P, 12\%, 1) - 60K$$
$$= -\$1,299K.$$

Equivalent total benefits at the start of power generation are

$$V_2 = \$120K(P/A, 12\%, 50) = \$997K.$$

The net equivalent worth at the start of power generation is

$$V_1 + V_2 = -\$1,299K + \$997K$$
$$= -\$302K.$$

- With infinite cash flows, the capitalized equivalent worth at the current time is

$$CE(12\%) = -\$1,299K + \$120K/(0.12)$$
$$= -\$299K.$$

Note that the difference between the infinite situation and a planning horizon of 50 years is merely \$3,000, which demonstrates that we may approximate the present worth of long cash flows (i.e., 50 years or more) by using the capitalized equivalent value. The accuracy of the approximation improves as the interest rate increases (or the number of years is greater).

COMMENTS: At $i = 12\%$, Mr. Bracewell's investment is not a profitable one, but at 8% it is. This outcome indicates the importance of using the appropriate i in investment analysis. The issue of selecting an appropriate i will be presented again in Chapter 15.

5.5 Comparing Mutually Exclusive Alternatives

Until now, we have considered situations involving either a single project alone or projects that were independent of each other. In both cases, we made the decision to reject or accept each project individually according to whether it met the MARR requirements, evaluated with either the PW or FW criterion.

In the real world of engineering practice, however, it is typical for us to have two or more choices of projects that are not independent of one another in seeking to accomplish a business objective. (As we shall see, even when it appears that we have only one project to consider, the implicit "do-nothing" alternative must be factored into the decision-making process.) In this section, we extend our evaluation techniques to multiple projects that are mutually exclusive. Other dependencies between projects will be considered in Chapter 15.

Often, various projects or investments under consideration do not have the same duration or do not match the desired study period. Adjustments must then be made to account for the differences. In this section, we explain the concept of an analysis period and the process of accommodating for different lifetimes, two important considerations that apply in selecting among several alternatives. Up to now in this chapter, all available options in a decision problem were assumed to have equal lifetimes. In the current section, this restriction is also relaxed.

5.5.1 Meaning of Mutually Exclusive and "Do Nothing"

As we briefly mentioned in Section 5.1.2, several alternatives are **mutually exclusive** when any one of them will fulfill the same need and the selection of one of them implies that the others will be excluded. Take, for example, buying versus leasing an automobile for business use; when one alternative is accepted, the other is excluded. We use the terms **alternative** and **project** interchangeably to mean "decision option."

"Do Nothing" Is a Decision Option

When considering an investment, we are in one of two situations: Either the project is aimed at replacing an existing asset or system, or it is a new endeavor. In either case, a do-nothing alternative may exist. On the one hand, if a process or system already in place to accomplish our business objectives is still adequate, then we must determine which, if any, new proposals are economical replacements. If none are feasible, then we do nothing. On the other hand, if the existing system has failed, then the choice among proposed alternatives is mandatory (i.e., do nothing is not an option).

New endeavors occur as alternatives to the "green fields" do-nothing situation, which has zero revenues and zero costs (i.e., nothing currently exists). For most new endeavors, do nothing is generally an alternative, as we won't proceed unless at least one of the proposed alternatives is economically sound. In fact, undertaking even a single project entails making a decision between two alternatives, because the do-nothing alternative is implicitly included. Occasionally, a new initiative must be undertaken, cost notwithstanding, and in this case the goal is to choose the most economical alternative, since "do nothing" is not an option.

When the option of retaining an existing asset or system is available, there are two ways to incorporate it into the evaluation of the new proposals. One way is to treat the do-nothing option as a distinct alternative; we cover this approach primarily in Chapter 14, where methodologies specific to replacement analysis are presented. The second approach, used mostly in this chapter, is to generate the cash flows of the new proposals relative to that of the do-nothing alternative. That is, for each new alternative, the **incremental costs** (and incremental savings or revenues if applicable) relative to "do nothing" are used in the economic evaluation. For a replacement-type problem, these costs are calculated by subtracting the do-nothing cash flows from those of each new alternative. For new endeavors, the incremental cash flows are the same as the absolute amounts associated with each alternative, since the do-nothing values are all zero.

Because the main purpose of this chapter is to illustrate how to choose among mutually exclusive alternatives, most of the problems are structured so that one of the options presented must be selected. Therefore, unless otherwise stated, it is assumed that "do nothing" is not an option, and costs and revenues can be viewed as incremental to "do nothing."

Service Projects versus Revenue Projects

When comparing mutually exclusive alternatives, we need to classify investment projects into either service or revenue projects. **Service projects** are projects whose revenues do not depend on the choice of project; rather, such projects *must produce the same amount of output (revenue).* In this situation, we certainly want to choose an alternative with the least input (or cost). For example, suppose an electric utility is considering building a new power plant to meet the peak-load demand during either hot summer or cold winter days. Two alternative service projects could meet this demand: a combustion turbine plant and a fuel-cell power plant. No matter which type of plant is selected, the firm will collect the same amount of revenue from its customers. The only difference is how much it will cost to generate electricity from each plant. If we were to compare these service projects, we would be interested in knowing which plant could provide the cheaper power (lower production cost). Further, if we were to use the NPW criterion to compare alternatives so as to minimize expenditures, we would choose the alternative with the **lower present-value** production cost over the service life of the plant.

Revenue projects, by contrast, are projects whose revenues depend on the choice of alternative. With revenue projects, we are not limiting the amount of input going into the project or the amount of output that the project would generate. Then our decision is to select the alternative with the largest net gains (output – input). For example, a TV manufacturer is considering marketing two types of high-resolution monitors. With its present production capacity, the firm can market only one of them. Distinct production processes for the two models could incur very different manufacturing costs, so the revenues from each model would be expected to differ due to divergent market prices and potentially different sales volumes. In this situation, if we were to use the NPW criterion, we would select the model that promises to bring in the higher net present worth.

Total-Investment Approach

Applying an evaluation criterion to each mutually exclusive alternative individually and then comparing the results to make a decision is referred to as the **total-investment approach**. We compute the PW for each individual alternative, as we did in Section 5.3, and select the one with the highest PW. Note that this approach guarantees valid results only when PW, FW, and AE criteria are used. (As you will see in Chapters 7 and 16, the total-investment approach does not work for any decision criterion based on either a percentage (rate of return) or a ratio (e.g., a benefit–cost ratio). With percentages or ratios, you need to use the incremental investment approach, which also works with any decision criterion, including PW, FW, and AE.) The incremental investment approach will be discussed in detail in Chapter 7.

Scale of Investment

Frequently, mutually exclusive investment projects may require different levels of investments. At first, it seems unfair to compare a project requiring a smaller investment with one requiring a larger investment. However, the disparity in scale of investment should not be of concern in comparing mutually exclusive alternatives, as long as you understand the basic assumption: *Funds not invested in the project will continue to earn interest at the MARR.* We will look at mutually exclusive alternatives that require different levels of investments for both service projects and revenue projects:

- **Service projects.** Typically, what you are asking yourself to do here is to decide whether the higher initial investment can be justified by additional savings that will occur in the future. More efficient machines are usually more expensive to acquire initially, but they will reduce future operating costs, thereby generating more savings.
- **Revenue projects.** If two mutually exclusive revenue projects require different levels of investments with varying future revenue streams, then your question is what to do with the difference in investment funds if you decide to go with the project that requires the smaller investment. To illustrate, consider the two mutually exclusive revenue projects illustrated in Figure 5.12. Our objective is to compare these two projects at a MARR of 10%.

Suppose you have exactly $4,000 to invest. If you choose Project B, you do not have any leftover funds. However, if you go with Project A, you will have $3,000 in unused

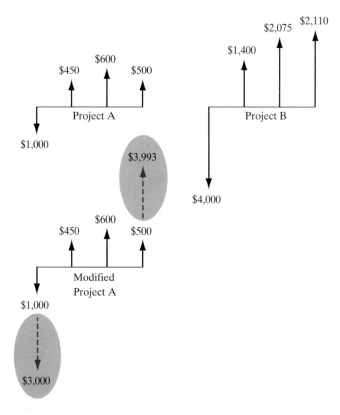

Figure 5.12 Comparing mutually exclusive revenue projects requiring different levels of investment.

funds. Our assumption is that these unused funds will continue to earn an interest rate that is the MARR. Therefore, the unused funds will grow at 10%, or $3,933, at the end of the project term, or three years from now. Consequently, selecting Project A is equivalent to having a modified project cash flow as shown in Figure 5.12.

Let's calculate the net present worth for each option at 10%:

- **Project A:**

$$PW(10\%)_A = -\$1,000 + \$450(P/F, 10\%, 1) + \$600(P/F, 10\%, 2)$$
$$+ \$500(P/F, 10\%, 3)$$
$$= \$283.$$

- **Project B:**

$$PW(10\%)_B = -\$4,000 + \$1,400(P/A, 10\%, 1)$$
$$+ \$2,075(P/F, 10\%, 2) + \$2,110(P/F, 10\%, 3)$$
$$= \$579.$$

Clearly, Project B is the better choice. But how about the modified Project A? If we calculate the present worth for the modified Project A, we have the following:

- **Modified Project A:**

$$PW(10\%)_A = -\$4,000 + \$450(P/A, 10\%, 1) + \$600(P/F, 10\%, 2)$$
$$+ \$4,493(P/F, 10\%, 3)$$
$$= \$283.$$

This is exactly the same as the net present worth without including the investment consequence of the unused funds. It is not a surprising result, as the return on investment in the unused funds will be exactly 10%. If we also discount the funds at 10%, there will be no surplus. So, what is the conclusion? It is this: If there is any disparity in investment scale for mutually exclusive revenue projects, go ahead and calculate the net present worth for each option without worrying about the investment differentials.

5.5.2 Analysis Period

The **analysis period** is the time span over which the economic effects of an investment will be evaluated. The analysis period may also be called the **study period** or **planning horizon**. The length of the analysis period may be determined in several ways: It may be a predetermined amount of time set by company policy, or it may be either implied or explicit in the need the company is trying to fulfill. (For example, a diaper manufacturer sees the need to dramatically increase production over a 10-year period in response to an anticipated "baby boom.") In either of these situations, we consider the analysis period to be the **required service period**.

When the required service period is not stated at the outset, the analyst must choose an appropriate analysis period over which to study the alternative investment projects. In such a case, one convenient choice of analysis period is the period of the useful life of the investment project.

When the useful life of an investment project does not match the analysis or required service period, we must make adjustments in our analysis. A further complication in a consideration of two or more mutually exclusive projects is that the investments themselves may have different useful lives. Accordingly, we must compare projects with different useful lives over an **equal time span**, which may require further adjustments in our analysis.

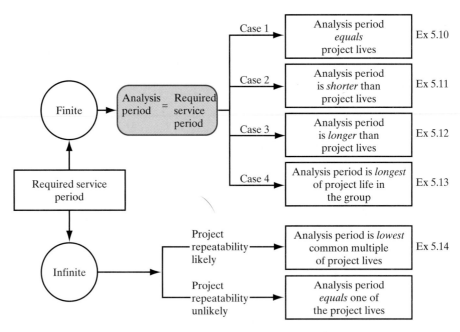

Figure 5.13 Analysis period implied in comparing mutually exclusive alternatives.

(Figure 5.13 is a flowchart showing the possible combinations of the analysis period and the useful life of an investment.) In the sections that follow, we will explore in more detail how to handle situations in which project lives differ from the analysis period and from each other. But we begin with the most straightforward situation: when the project lives and the analysis period coincide.

5.5.3 Analysis Period Equals Project Lives

When the project lives equal the analysis period, we compute the NPW for each project and select the one with the highest NPW. Example 5.10 illustrates this point.

EXAMPLE 5.10 Present-Worth Comparison (Revenue Projects with Equal Lives): Three Alternatives

Bullard Company (BC) is considering expanding its range of industrial machinery products by manufacturing machine tables, saddles, machine bases, and other similar parts. Several combinations of new equipment and personnel could serve to fulfill this new function:

- **Method 1 (M1):** new machining center with three operators.
- **Method 2 (M2):** new machining center with an automatic pallet changer and three operators.
- **Method 3 (M3):** new machining center with an automatic pallet changer and two task-sharing operators.

Each of these arrangements incurs different costs and revenues. The time taken to load and unload parts is reduced in the pallet-changer cases. Certainly, it costs more to acquire, install, and tool-fit a pallet changer, but because the device is more efficient and versatile, it can generate larger annual revenues. Although saving on labor costs, task-sharing operators take longer to train and are more inefficient initially. As the operators become more experienced at their tasks and get used to collaborating with each other, it is expected that the annual benefits will increase by 13% per year over the five-year study period. BC has estimated the investment costs and additional revenues as follows:

| | Machining Center Methods | | |
	M1	M2	M3
Investment:			
Machine tool purchase	$121,000	$121,000	$121,000
Automatic pallet changer		$ 66,600	$ 66,600
Installation	$ 30,000	$ 42,000	$ 42,000
Tooling expense	$ 58,000	$ 65,000	$ 65,000
Total investment	$209,000	$294,600	$294,600
Annual benefits: Year 1			
Additional revenues	$ 55,000	$ 69,300	$ 36,000
Direct labor savings			$ 17,300
Setup savings		$ 4,700	$ 4,700
Year 1: Net revenues	$ 55,000	$ 74,000	$ 58,000
Years 2–5: Net revenues	constant	constant	$g = 13\%$/year
Salvage value in year 5	$ 80,000	$120,000	$120,000

All cash flows include all tax effects. "Do nothing" is obviously an option, since BC will not undertake this expansion if none of the proposed methods is economically viable. If a method is chosen, BC expects to operate the machining center over the next five years. On the basis of the use of the PW measure at $i = 12\%$, which option would be selected?

SOLUTION

Given: Cash flows for three revenue projects and $i = 12\%$ per year.
Find: NPW for each project and which project to select.

For these revenue projects, the net-present-worth figures at $i = 12\%$ would be as follows:

- For Option M1,

$$PW(12\%)_{M1} = -\$209,000 + \$55,000(P/A, 12\%, 5)$$
$$+ \$80,000(P/F, 12\%, 5)$$
$$= \$34,657.$$

- For Option M2,

$$PW(12\%)_{M2} = -\$294,600 + \$74,000(P/A, 12\%, 5)$$
$$+ \$120,000(P/F, 12\%, 5)$$
$$= \$40,245.$$

- For Option M3,

$$PW(12\%)_{M3} = -\$294,600 + \$58,000(P/A_1, 13\%, 12\%, 5)$$
$$+ \$120,000(P/F, 12\%, 5)$$
$$= \$37,085.$$

Clearly, Option M2 is the most profitable. Given the nature of BC parts and shop orders, management decides that the best way to expand would be with an automatic pallet changer, but without task sharing.

5.5.4 Analysis Period Differs from Project Lives

In Example 5.10, we assumed the simplest scenario possible when analyzing mutually exclusive projects: The projects had useful lives equal to each other and to the required service period. In practice, this is seldom the case. Often, project lives do not match the required analysis period or do not match each other (or both). For example, two machines may perform exactly the same function, but one lasts longer than the other, and both of them last longer than the analysis period over which they are being considered. In the sections and examples that follow, we will develop some techniques for dealing with these complications.

Project's Life Is Longer than Analysis Period

Project lives rarely conveniently coincide with a firm's predetermined required analysis period; they are often too long or too short. The case of project lives that are too long is the easier one to address.

    Consider the case of a firm that undertakes a five-year production project when all of the alternative equipment choices have useful lives of seven years. In such a case, we analyze each project for only as long as the required service period (in this case, five years). We are then left with some unused portion of the equipment (in this case, two years' worth), which we include as salvage value in our analysis. **Salvage value** is the amount of money for which the equipment could be sold after its service to the project has been rendered. Alternatively, salvage value is the dollar measure of the remaining usefulness of the equipment.

    A common instance of project lives that are longer than the analysis period occurs in the construction industry: A building project may have a relatively short completion time, but the equipment that is purchased has a much longer useful life.

Salvage value: The estimated value that an asset will realize upon its sale at the end of its useful life.

EXAMPLE 5.11 Present-Worth Comparison: Project Lives Longer than the Analysis Period

Waste Management Company (WMC) has won a contract that requires the firm to re-move radioactive material from government-owned property and transport it to a des-ignated dumping site. This task requires a specially made ripper–bulldozer to dig and load the material onto a transportation vehicle. Approximately 400,000 tons of waste must be moved in a period of two years.

- Model A costs $150,000 and has a life of 6,000 hours before it will require any major overhaul. Two units of model A would be required to remove the material within two years, and the operating cost for each unit would run to $40,000/year for 2,000 hours of operation. At this operational rate, the model would be operable for three years, at the end of which time it is estimated that the salvage value will be $25,000 for each machine.
- A more efficient model B costs $240,000 each, has a life of 12,000 hours without any major overhaul, and costs $22,500 to operate for 2,000 hours per year to com-plete the job within two years. The estimated salvage value of model B at the end of six years is $30,000. Once again, two units of model B would be required to re-move the material within two years.

Since the lifetime of either model exceeds the required service period of two years (Figure 5.14), WMC has to assume some things about the used equipment at the end of that time. Therefore, the engineers at WMC estimate that, after two years, the model A units could be sold for $45,000 each and the model B units for $125,000 each. After considering all tax effects, WMC summarized the resulting cash flows (in thousand of dollars) for each project as follows:

Period	Model A		Model B	
0	−$300		−$480	
1	−80		−45	
2	−80	+90	−45	+250
3	−80	+50	−45	
4			−45	
5			−45	
6			−45	+60

Here, the figures in the boxes represent the estimated salvage values at the end of the analysis period (the end of year 2). Assuming that the firm's MARR is 15%, which option would be acceptable?

SOLUTION

Given: Cash flows for two alternatives as shown in Figure 5.14 and $i = 15\%$ per year.
Find: NPW for each alternative and which alternative is preferred.

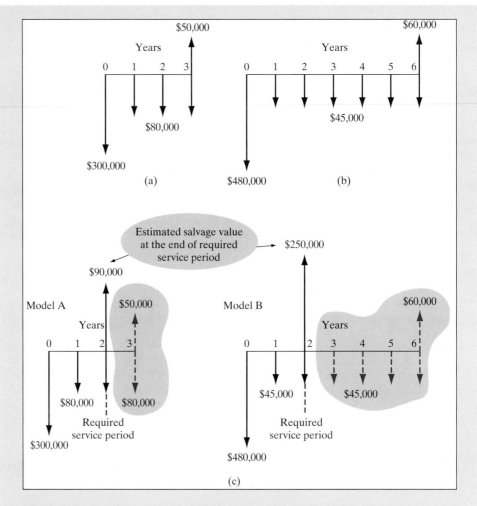

Figure 5.14 (a) Cash flow for model A; (b) cash flow for model B; (c) comparison of service projects with unequal lives when the required service period is shorter than the individual project life (Example 5.11).

First, note that these are service projects, so we can assume the same revenues for both configurations. Since the firm explicitly estimated the market values of the assets at the end of the analysis period (two years), we can compare the two models directly. Because the benefits (removal of the waste) are equal, we can concentrate on the costs:

$$\text{PW}(15\%)_\text{A} = -\$300 - \$80(P/A, 15\%, 2) + \$90(P/F, 15\%, 2)$$
$$= -\$362;$$
$$\text{PW}(15\%)_\text{B} = -\$480 - \$45(P/A, 15\%, 2) + \$250(P/F, 15\%, 2)$$
$$= -\$364.$$

Model A has the least negative PW costs and thus would be preferred.

Project's Life Is Shorter than Analysis Period

When project lives are shorter than the required service period, we must consider how, at the end of the project lives, we will satisfy the rest of the required service period. Replacement projects—additional projects to be implemented when the initial project has reached the limits of its useful life—are needed in such a case. A sufficient number of replacement projects that match or exceed the required service period must be analyzed.

To simplify our analysis, we could assume that the replacement project will be exactly the same as the initial project, with the same costs and benefits. However, this assumption is not necessary. For example, depending on our forecasting skills, we may decide that a different kind of technology—in the form of equipment, materials, or processes—is a preferable replacement. Whether we select exactly the same alternative or a new technology as the replacement project, we are ultimately likely to have some unused portion of the equipment to consider as salvage value, just as in the case when the project lives are longer than the analysis period. Of course, we may instead decide to lease the necessary equipment or subcontract the remaining work for the duration of the analysis period. In this case, we can probably match our analysis period and not worry about salvage values.

In any event, at the outset of the analysis period, we must make some initial guess concerning the method of completing the analysis. Later, when the initial project life is closer to its expiration, we may revise our analysis with a different replacement project. This is only reasonable, since economic analysis is an ongoing activity in the life of a company and an investment project, and we should always use the most reliable, up-to-date data we can reasonably acquire.

EXAMPLE 5.12 Present-Worth Comparison: Project Lives Shorter than the Analysis Period

The Smith Novelty Company, a mail-order firm, wants to install an automatic mailing system to handle product announcements and invoices. The firm has a choice between two different types of machines. The two machines are designed differently, but have identical capacities and do exactly the same job. The $12,500 semiautomatic model A will last three years, while the fully automatic model B will cost $15,000 and last four years. The expected cash flows for the two machines, including maintenance, salvage value, and tax effects, are as follows:

n	Model A	Model B
0	−$12,500	−$15,000
1	−5,000	−4,000
2	−5,000	−4,000
3	−5,000 + 2,000	−4,000
4		−4,000 + 1,500
5		

As business grows to a certain level, neither of the models may be able to handle the expanded volume at the end of year 5. If that happens, a fully computerized mail-order system will need to be installed to handle the increased business volume. In the scenario just presented, which model should the firm select at MARR = 15%?

SOLUTION

Given: Cash flows for two alternatives as shown in Figure 5.15, analysis period of five years, and $i = 15\%$.

Find: NPW of each alternative and which alternative to select.

Since both models have a shorter life than the required service period of 5 years, we need to make an explicit assumption of how the service requirement is to be met. Suppose that the company considers leasing equipment comparable to model A at an annual payment of $6,000 (after taxes) and with an annual operating cost of $5,000

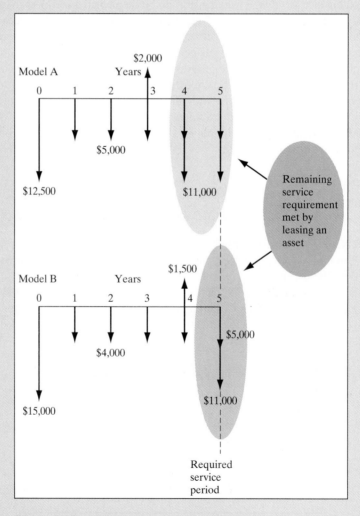

Figure 5.15 Comparison for service projects with unequal lives when the required service period is longer than the individual project life (Example 5.12).

for the remaining required service period. In this case, the cash flow would look like that shown in Figure 5.15:

n	Model A		Model B	
0	−$12,500		−$15,000	
1	−5,000		−4,000	
2	−5,000		−4,000	
3	−5,000 +	2,000	−4,000	
4	−5,000 −	6,000	−4,000 +	1,500
5	−5,000 −	6,000	−5,000 −	6,000

Here, the boxed figures represent the annual lease payments. (It costs $6,000 to lease the equipment and $5,000 to operate it annually. Other maintenance costs will be paid by the leasing company.) Note that both alternatives now have the same required service period of five years. Therefore, we can use NPW analysis:

$$PW(15\%)_A = -\$12,500 - \$5,000(P/A, 15\%, 2) - \$3,000(P/F, 15\%, 3)$$
$$- \$11,000(P/A, 15\%, 2)(P/F, 15\%, 3)$$
$$= -\$34,359.$$

$$PW(15\%)_B = -\$15,000 - \$4,000(P/A, 15\%, 3) - \$2,500(P/F, 15\%, 4)$$
$$- \$11,000(P/F, 15\%, 5)$$
$$= -\$31,031.$$

Since these are service projects, model B is the better choice.

Analysis Period Coincides with Longest Project Life

As seen in the preceding pages, equal future periods are generally necessary to achieve comparability of alternatives. In some situations, however, revenue projects with different lives can be compared if they require only a one-time investment because the task or need within the firm is a one-time task or need. An example of this situation is the extraction of a fixed quantity of a natural resource such as oil or coal.

Consider two mutually exclusive processes: One requires 10 years to recover some coal, and the other can accomplish the task in only 8 years. There is no need to continue the project if the short-lived process is used and all the coal has been retrieved. In this example, the two processes can be compared over an analysis period of 10 years (the longest project life of the two being considered), assuming that no cash flows after 8 years for the shorter lived project. Because of the time value of money, the revenues must be included in the analysis even if the price of coal is constant. Even if the total (undiscounted) revenue is equal for either process, that for the faster process has a larger present worth. Therefore, the two projects could be compared by using the NPW of each over its own life. Note that in this case the analysis period is determined by, and coincides with, the longest project life in the group. (Here we are still, in effect, assuming an analysis period of 10 years.)

EXAMPLE 5.13 Present-Worth Comparison: A Case where the Analysis Period Coincides with the Project with the Longest Life in the Mutually Exclusive Group

The family-operated Foothills Ranching Company (FRC) owns the mineral rights to land used for growing grain and grazing cattle. Recently, oil was discovered on this property. The family has decided to extract the oil, sell the land, and retire. The company can lease the necessary equipment and extract and sell the oil itself, or it can lease the land to an oil-drilling company:

- **Drill option.** If the company chooses to drill, it will require $300,000 leasing expenses up front, but the net annual cash flow after taxes from drilling operations will be $600,000 at the end of each year for the next five years. The company can sell the land for a net cash flow of $1,000,000 in five years, when the oil is depleted.
- **Lease option.** If the company chooses to lease, the drilling company can extract all the oil in only three years, and FRC can sell the land for a net cash flow of $800,000 at that time. (The difference in resale value of the land is due to the increasing rate of land appreciation anticipated for this property.) The net cash flow from the lease payments to FRC will be $630,000 at the *beginning* of each of the next three years.

All benefits and costs associated with the two alternatives have been accounted for in the figures listed. Which option should the firm select at $i = 15\%$?

SOLUTION

Given: Cash flows shown in Figure 5.16 and $i = 15\%$ per year.

Find: NPW of each alternative and which alternative to select.

As illustrated in Figure 5.16, the cash flows associated with each option look like this:

n	Drill	Lease
0	−$300,000	$630,000
1	600,000	630,000
2	600,000	630,000
3	600,000	800,000
4	600,000	
5	1,600,000	

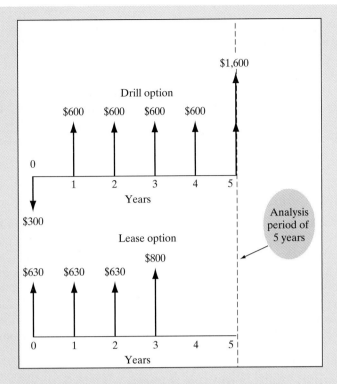

Figure 5.16 Comparison of revenue projects with unequal lives when the analysis period coincides with the project with the longest life in the mutually exclusive group (Example 5.13). In our example, the analysis period is five years, assuming no cash flow in years 4 and 5 for the lease option.

After depletion of the oil, the project will terminate. The present worth of each of the two options is as follows:

$$PW(15\%)_{Drill} = -\$300,000 + \$600,000(P/A, 15\%, 4)$$
$$+ \$1,600,000(P/F, 15\%, 5)$$
$$= \$2,208,470.$$

$$PW(15\%)_{Lease} = \$630,000 + \$630,000(P/A, 15\%, 2)$$
$$+ \$800,000(P/F, 15\%, 3)$$
$$= \$2,180,210.$$

Note that these are revenue projects; therefore, the drill option appears to be the marginally better option.

COMMENTS: The relatively small difference between the two NPW amounts ($28,260) suggests that the actual decision between drilling and leasing might be based on noneconomic issues. Even if the drilling option were slightly better, the company might prefer to forgo the small amount of additional income and select the lease option, rather than undertake an entirely new business venture and do its own drilling. A variable that might also have a critical effect on this decision is the sales value of the land in each alternative. The value of land is often difficult to forecast over any long period, and the firm may feel some uncertainty about the accuracy of its guesses. In Chapter 12, we will discuss sensitivity analysis, a method by which we can factor uncertainty about the accuracy of project cash flows into our analysis.

5.5.5 Analysis Period Is Not Specified

Our coverage so far has focused on situations in which an analysis period is known. When an analysis period is not specified, either explicitly by company policy or practice or implicitly by the projected life of the investment project, it is up to the analyst to choose an appropriate one. In such a case, the most convenient procedure is to choose an analysis on the basis of the useful lives of the alternatives. When the alternatives have equal lives, this is an easy selection. When the lives of at least some of the alternatives differ, we must select an analysis period that allows us to compare projects with different lifetimes on an equal time basis—that is, a **common service period**.

Lowest Common Multiple of Project Lives

A required service period of infinity may be assumed if we anticipate that an investment project will be proceeding at roughly the same level of production for some indefinite period. It is certainly possible to make this assumption mathematically, although the analysis is likely to be complicated and tedious. Therefore, in the case of an indefinitely ongoing investment project, we typically select a finite analysis period by using the **lowest common multiple** of project lives. For example, if alternative A has a 3-year useful life and alternative B has a 4-year life, we may select 12 years as the analysis or common service period. We would consider alternative A through four life cycles and alternative B through three life cycles; in each case, we would use the alternatives completely. We then accept the finite model's results as a good prediction of what will be the economically wisest course of action for the foreseeable future. The next example is a case in which we conveniently use the lowest common multiple of project lives as our analysis period.

Least common multiple of two numbers is the lowest number that can be divided by both.

EXAMPLE 5.14 Present-Worth Comparison: Unequal Lives, Lowest-Common-Multiple Method

Consider Example 5.12. Suppose that models A and B can each handle the increased future volume and that the system is not going to be phased out at the end of five years. Instead, the current mode of operation is expected to continue indefinitely. Suppose also that the two models will be available in the future without significant changes in price and operating costs. At MARR = 15%, which model should the firm select?

SOLUTION

Given: Cash flows for two alternatives as shown in Figure 5.17, $i = 15\%$ per year, and an indefinite period of need.

Find: NPW of each alternative and which alternative to select.

Recall that the two mutually exclusive alternatives have different lives, but provide identical annual benefits. In such a case, we ignore the common benefits and can make the decision solely on the basis of costs, as long as a common analysis period is used for both alternatives.

To make the two projects comparable, let's assume that, after either the 3- or 4-year period, the system would be reinstalled repeatedly, using the same model, and that the same costs would apply. The lowest common multiple of 3 and 4 is 12, so we will use 12 years as the common analysis period. Note that any cash flow difference between the alternatives will be revealed during the first 12 years. After that, the same

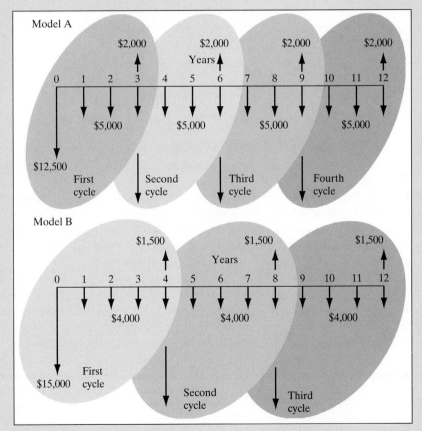

Figure 5.17 Comparison of projects with unequal lives when the required service period is infinite and the project is likely to be repeatable with the same investment and operations and maintenance costs in all future replacement cycles (Example 5.14).

cash flow pattern repeats every 12 years for an indefinite period. The replacement cycles and cash flows are shown in Figure 5.17. Here is our analysis:

- **Model A.** Four replacements occur in a 12-year period. The PW for the first investment cycle is

$$PW(15\%) = -\$12,500 - \$5,000(P/A, 15\%, 3)$$
$$+ \$2,000(P/F, 15\%, 3)$$
$$= -\$22,601.$$

With four replacement cycles, the total PW is

$$PW(15\%) = -\$22,601[1 + (P/F, 15\%, 3)$$
$$+ (P/F, 15\%, 6) + (P/F, 15\%, 9)]$$
$$= -\$53,657.$$

- **Model B.** Three replacements occur in a 12-year period. The PW for the first investment cycle is

$$PW(15\%) = -\$15,000 - \$4,000(P/A, 15\%, 4)$$
$$+ \$1,500(P/F, 15\%, 4)$$
$$= -\$25,562.$$

With three replacement cycles in 12 years, the total PW is

$$PW(15\%) = -\$25,562[1 + (P/F, 15\%, 4) + (P/F, 15\%, 8)]$$
$$= -\$48,534.$$

Clearly, model B is a better choice, as before.

COMMENTS: In Example 5.14, an analysis period of 12 years seems reasonable. The number of actual reinvestment cycles needed with each type of system will depend on the technology of the future system, so we may or may not actually need the four reinvestment cycles (model A) or three (model B) we used in our analysis. The validity of the analysis also depends on the costs of the system and labor remaining constant. If we assume **constant-dollar prices** (see Chapter 11), this analysis would provide us with a reasonable result. (As you will see in Example 6.3, the annual-worth approach makes it mathematically easier to solve this type of comparison.) If we cannot assume constant-dollar prices in future replacements, we need to estimate the costs for each replacement over the analysis period. This will certainly complicate the problem significantly.

Other Common Analysis Periods

In some cases, the lowest common multiple of project lives is an unwieldy analysis period to consider. Suppose, for example, that you were considering two alternatives with lives of 7 and 12 years, respectively. Besides making for tedious calculations, an 84-year analysis period may lead to inaccuracies, since, over such a long time, we can be less and less confident about the ability to install identical replacement projects with identical costs and benefits. In a case like this, it would be reasonable to use the useful life of one of the alternatives by either factoring in a replacement project or salvaging the remaining useful life, as the case may be. The important idea is that we must compare both projects on the same time basis.

SUMMARY

In this chapter, we presented the concept of present-worth analysis, based on cash flow equivalence along with the payback period. We observed the following important results:

- Present worth is an equivalence method of analysis in which a project's cash flows are discounted to a single present value. It is perhaps the most efficient analysis method we can use in determining the acceptability of a project on an economic basis. Other analysis methods, which we will study in Chapters 6 and 7, are built on a sound understanding of present worth.

- The minimum attractive rate of return (MARR) is the interest rate at which a firm can always earn or borrow money under a normal operating environment. It is generally dictated by management and is the rate at which NPW analysis should be conducted.

- Revenue projects are those for which the income generated depends on the choice of project. Service projects are those for which the income remains the same, regardless of which project is selected.

- Several alternatives that meet the same need are **mutually exclusive** if, whenever one of them is selected, the others will be rejected.

- When not specified by management or company policy, the analysis period to use in a comparison of mutually exclusive projects may be chosen by an individual analyst. Several efficiencies can be applied when an analysis period is selected. In general, the analysis period should be chosen to cover the required service period, as highlighted in Figure 5.13.

PROBLEMS

Note: *Unless otherwise stated, all cash flows are after-tax cash flows. The interest rate (MARR) is also given on an after-tax basis.*

Identifying Cash Inflows and Outflows

5.1 Camptown Togs, Inc., a children's clothing manufacturer, has always found payroll processing to be costly because it must be done by a clerk so that the number of piece-goods coupons received by each employee can be collected and the types of tasks performed by each employee can be calculated. Not long ago, an industrial engineer designed a system that partially automates the process by means of a scanner that reads the piece-goods coupons. Management is enthusiastic about this system because it utilizes some personal computer systems that were purchased recently. It is expected that this new automated system will save $45,000 per year in labor. The new system will cost about $30,000 to build and test prior to operation. It is expected that operating costs, including income taxes, will be about $5,000 per year. The system will have a five-year useful life. The expected net salvage value of the system is estimated to be $3,000.

(a) Identify the cash inflows over the life of the project.

(b) Identify the cash outflows over the life of the project.

(c) Determine the net cash flows over the life of the project.

Payback Period

5.2 Refer to Problem 5.1, and answer the following questions:

(a) How long does it take to recover the investment?

(b) If the firm's interest rate is 15% after taxes, what would be the discounted payback period for this project?

5.3 Consider the following cash flows:

	Project's Cash Flow ($)			
n	A	B	C	D
0	-$2,500	-$3,000	-$5,500	-$4,000
1	300	2,000	2,000	5,000
2	300	1,500	2,000	-3,000
3	300	1,500	2,000	-2,500
4	300	500	5,000	1,000
5	300	500	5,000	1,000
6	300	1,500		2,000
7	300			3,000
8	300			

(a) Calculate the payback period for each project.

(b) Determine whether it is meaningful to calculate a payback period for project D.

(c) Assuming that $i = 10\%$, calculate the discounted payback period for each project.

NPW Criterion

5.4 Consider the following sets of investment projects, all of which have a three-year investment life:

	Project's Cash Flow ($)			
n	A	B	C	D
0	-$1,500	-$1,200	-$1,600	-$3,100
1	0	600	-1,800	800
2	0	800	800	1,900
3	3,000	1,500	2,500	2,300

(a) Compute the net present worth of each project at $i = 10\%$.

(b) Plot the present worth as a function of the interest rate (from 0% to 30%) for project B.

5.5 You need to know whether the building of a new warehouse is justified under the following conditions:

The proposal is for a warehouse costing $200,000. The warehouse has an expected useful life of 35 years and a net salvage value (net proceeds from sale after tax adjustments) of $35,000. Annual receipts of $37,000 are expected, annual maintenance and administrative costs will be $8,000/year, and annual income taxes are $5,000.

Given the foregoing data, which of the following statements are correct?

(a) The proposal is justified for a MARR of 9%.

(b) The proposal has a net present worth of $152,512 when 6% is used as the interest rate.

(c) The proposal is acceptable, as long as MARR $\leq 11.81\%$.

(d) All of the preceding are correct.

5.6 Your firm is considering purchasing an old office building with an estimated remaining service life of 25 years. Recently, the tenants signed long-term leases, which leads you to believe that the current rental income of $150,000 per year will remain constant for the first 5 years. Then the rental income will increase by 10% for every 5-year interval over the remaining life of the asset. For example, the annual rental income would be $165,000 for years 6 through 10, $181,500 for years 11 through 15, $199,650 for years 16 through 20, and $219,615 for years 21 through 25. You estimate that operating expenses, including income taxes, will be $45,000 for the first year and that they will increase by $3,000 each year thereafter. You also estimate that razing the building and selling the lot on which it stands will realize a net amount of $50,000 at the end of the 25-year period. If you had the opportunity to invest your money elsewhere and thereby earn interest at the rate of 12% per annum, what would be the maximum amount you would be willing to pay for the building and lot at the present time?

5.7 Consider the following investment project:

n	A_n	i
0	−$42,000	10%
1	32,400	11
2	33,400	13
3	32,500	15
4	32,500	12
5	33,000	10

Suppose the company's reinvestment opportunities change over the life of the project as shown in the preceding table (i.e., the firm's MARR changes over the life of the project). For example, the company can invest funds available now at 10% for the first year, 11% for the second year, and so forth. Calculate the net present worth of this investment and determine the acceptability of the investment.

5.8 Cable television companies and their equipment suppliers are on the verge of installing new technology that will pack many more channels into cable networks, thereby creating a potential programming revolution with implications for broad-casters, telephone companies, and the consumer electronics industry.

Digital compression uses computer techniques to squeeze 3 to 10 programs into a single channel. A cable system fully using digital compression technology would be able to offer well over 100 channels, compared with about 35 for the average cable television system now used. If the new technology is combined with the increased use of optical fibers, it might be possible to offer as many as 300 channels.

A cable company is considering installing this new technology to increase sub-scription sales and save on satellite time. The company estimates that the installa-tion will take place over 2 years. The system is expected to have an 8-year service life and produce the following savings and expenditures:

Digital Compression	
Investment	
Now	$500,000
First year	$3,200,000
Second year	$4,000,000
Annual savings in satellite time	$2,000,000
Incremental annual revenues due to new subscriptions	$4,000,000
Incremental annual expenses	$1,500,000
Incremental annual income taxes	$1,300,000
Economic service life	8 years
Net salvage value	$1,200,000

Note that the project has a 2-year investment period, followed by an 8-year serv-ice life (a total 10-year life for the project). This implies that the first annual sav-ings will occur at the end of year 3 and the last will occur at the end of year 10. If the firm's MARR is 15%, use the NPW method to justify the economic worth of the project.

5.9 A large food-processing corporation is considering using laser technology to speed up and eliminate waste in the potato-peeling process. To implement the sys-tem, the company anticipates needing $3.5 million to purchase the industrial-strength lasers. The system will save $1,550,000 per year in labor and materials. However, it will require an additional operating and maintenance cost of $350,000. Annual income taxes will also increase by $150,000. The system is ex-pected to have a 10-year service life and will have a salvage value of about $200,000. If the company's MARR is 18%, use the NPW method to justify the economics of the project.

Future Worth and Project Balance

5.10 Consider the following sets of investment projects, all of which have a three-year investment life:

Period	Project's Cash Flow			
(n)	A	B	C	D
0	−$12,500	−11,000	12,500	−13,000
1	5,400	−3,000	−7,000	5,500
2	14,400	21,000	−2,000	5,500
3	7,200	13,000	4,000	8,500

(a) Compute the net present worth of each project at $i = 15\%$.

(b) Compute the net future worth of each project at $i = 15\%$.

Which project or projects are acceptable?

5.11 Consider the following project balances for a typical investment project with a service life of four years:

n	A_n	Project Balance
0	−$1,000	−$1,000
1	( )	−1,100
2	( )	−800
3	460	−500
4	( )	0

(a) Construct the original cash flows of the project.

(b) Determine the interest rate used in computing the project balance.

(c) Would the project be acceptable at $i = 15\%$?

5.12 Your R&D group has developed and tested a computer software package that assists engineers to control the proper chemical mix for the various process-manufacturing industries. If you decide to market the software, your first-year operating net cash flow is estimated to be $1,000,000. Because of market competition, product life will be about 4 years, and the product's market share will decrease by 25% each year over the previous year's share. You are approached by a big software house which wants to purchase the right to manufacture and distribute the product. Assuming that your interest rate is 15%, for what minimum price would you be willing to sell the software?

5.13 Consider the accompanying project balance diagram for a typical investment project with a service life of five years. The numbers in the figure indicate the beginning project balances.

(a) From the project balance diagram, construct the project's original cash flows.

(b) What is the project's conventional payback period (without interest)?

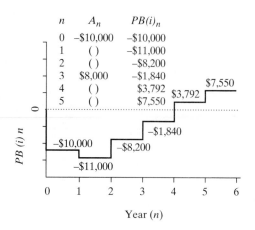

n	A_n	$PB(i)_n$
0	-$10,000	-$10,000
1	( )	-$11,000
2	( )	-$8,200
3	$8,000	-$1,840
4	( )	$3,792
5	( )	$7,550

5.14 Consider the following cash flows and present-worth profile:

	Net Cash Flows ($)	
Year	Project 1	Project 2
0	-$1,000	-$1,000
1	400	300
2	800	Y
3	X	800

(a) Determine the values of X and Y.
(b) Calculate the terminal project balance of project 1 at MARR = 24%.
(c) Find the values of a, b, and c in the NPW plot.

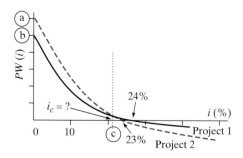

5.15 Consider the project balances for a typical investment project with a service life of five years, as shown in Table P5.15.

(a) Construct the original cash flows of the project and the terminal balance, and fill in the blanks in the preceding table.
(b) Determine the interest rate used in the project balance calculation, and compute the present worth of this project at the computed interest rate.

TABLE P5.15 Investment Project Balances

n	A_n	Project Balance
0	−$3,000	−$3,000
1		−2,700
2	1,470	−1,500
3		0
4		−300
5	600	

5.16 Refer to Problem 5.3, and answer the following questions:
   (a) Graph the project balances (at $i = 10\%$) of each project as a function of n.
   (b) By examining the graphical results in part (a), determine which project appears to be the safest to undertake if there is some possibility of premature termination of the projects at the end of year 2.

5.17 Consider the following investment projects:

n	A	B	Project's Cash Flow C	D	E
0	−$1,800	−$5,200	−$3,800	−$4,000	−$6,500
1	−500	2,500	0	500	1,000
2	900	−4,000	0	2,000	3,600
3	1,300	5,000	4,000	3,000	2,400
4	2,200	6,000	7,000	4,000	
5	−700	3,000	12,000	1,250	

   (a) Compute the future worth at the end of life for each project at $i = 12\%$.
   (b) Determine the acceptability of each project.

5.18 Refer to Problem 5.17, and answer the following questions:
   (a) Plot the future worth for each project as a function of the interest rate (0%–50%).
   (b) Compute the project balance of each project at $i = 12\%$.
   (c) Compare the terminal project balances calculated in (b) with the results obtained in Problem 5.17(a). Without using the interest factor tables, compute the future worth on the basis of the project balance concept.

5.19 Covington Corporation purchased a vibratory finishing machine for $20,000 in year 0. The useful life of the machine is 10 years, at the end of which the machine is estimated to have a salvage value of zero. The machine generates net annual revenues of $6,000. The annual operating and maintenance expenses are estimated to be $1,000. If Covington's MARR is 15%, how many years will it take before this machine becomes profitable?

5.20 Gene Research, Inc., just finished a 4-year program of R&D and clinical trial. It expects a quick approval from the Food and Drug Administration. If Gene markets the product its own, the company will require $30 million immediately ($n = 0$) to build a new manufacturing facility, and it is expected to have a 10-year product life. The R&D expenditure in the previous years and the anticipated revenues that the company can generate over the next 10 years are summarized as follows:

Period (n)	Cash Flow (Unit: $ million)
−4	−$10
−3	−10
−2	−10
−1	−10
0	−10 − 30
1–10	100

Merck, a large drug company is interested, in purchasing the R&D project and the right to commercialize the product from Gene Research, Inc.; it wants to do so immediately ($n = 0$). What would be a starting negotiating price for the project from Merck? Assume that Gene's MARR = 20%.

5.21 Consider the following independent investment projects:

	Project Cash Flows		
n	A	B	C
0	−$400	−$300	$100
1	150	140	−40
2	150	140	−40
3	350	140	−40
4	−200	110	
5	400	110	
6	300		

Assume that MARR = 10%, and answer the following questions:

(a) Compute the net present worth for each project, and determine the acceptability of each.

(b) Compute the net future worth of each project at the end of each project period, and determine the acceptability of each project.

(c) Compute the project worth of each project at the end of six years with variable MARRs as follows: 10% for $n = 0$ to $n = 3$ and 15% for $n = 4$ to $n = 6$.

5.22 Consider the project balance profiles shown in Table P5.22 for proposed investment projects. Project balance figures are rounded to nearest dollars.

(a) Compute the net present worth of projects A and C.

(b) Determine the cash flows for project A.

(c) Identify the net future worth of project C.

(d) What interest rate would be used in the project balance calculations for project B?

TABLE P5.22 Profiles for Proposed Investment Projects

	Project Balances		
n	**A**	**B**	**C**
0	−$1,000	−$1,000	−$1,000
1	−1,000	−650	−1,200
2	−900	−348	−1,440
3	−690	−100	−1,328
4	−359	85	−1,194
5	105	198	−1,000
Interest rate used	10%	?	20%
NPW	?	$79.57	?

5.23 Consider the following project balance profiles for proposed investment projects:

	Project Balances		
n	**A**	**B**	**C**
0	−$1,000	−$1,000	−$1,000
1	−800	−680	−530
2	−600	−302	*X*
3	−400	−57	−211
4	−200	233	−89
5	0	575	0
Interest rate used	10%	18%	12%

Project balance figures are rounded to the nearest dollar.

(a) Compute the net present worth of each investment.

(b) Determine the project balance for project C at the end of period 2 if $A_2 = \$500$.

(c) Determine the cash flows for each project.

(d) Identify the net future worth of each project.

Capitalized Equivalent Worth

5.24 Maintenance money for a new building has been sought. Mr. Kendall would like to make a donation to cover all future expected maintenance costs for the building. These maintenance costs are expected to be $50,000 each year for the first 5 years, $70,000 each year for years 6 through 10, and $90,000 each year after that. (The building has an indefinite service life.)

(a) If the money is placed in an account that will pay 13% interest compounded annually, how large should the gift be?

(b) What is the equivalent annual maintenance cost over the infinite service life of the building?

5.25 Consider an investment project, the cash flow pattern of which repeats itself every five years forever as shown in the accompanying diagram. At an interest rate of 14%, compute the capitalized equivalent amount for this project.

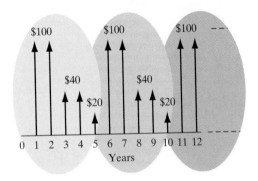

5.26 A group of concerned citizens has established a trust fund that pays 6% interest, compounded monthly, to preserve a historical building by providing annual maintenance funds of $30,000 forever. Compute the capitalized equivalent amount for these building maintenance expenses.

5.27 A newly constructed bridge costs $5,000,000. The same bridge is estimated to need renovation every 15 years at a cost of $1,000,000. Annual repairs and maintenance are estimated to be $100,000 per year.

(a) If the interest rate is 5%, determine the capitalized cost of the bridge.

(b) Suppose that, in (a), the bridge must be renovated every 20 years, not every 15 years. What is the capitalized cost of the bridge?

(c) Repeat (a) and (b) with an interest rate of 10%. What have you to say about the effect of interest on the results?

5.28 To decrease the costs of operating a lock in a large river, a new system of operation is proposed. The system will cost $650,000 to design and build. It is estimated that it will have to be reworked every 10 years at a cost of $100,000. In addition, an expenditure of $50,000 will have to be made at the end of the fifth year for a new type of gear that will not be available until then. Annual operating costs are expected to be $30,000 for the first 15 years and $35,000 a year thereafter. Compute the capitalized cost of perpetual service at $i = 8\%$.

Mutually Exclusive Alternatives

5.29 Consider the following cash flow data for two competing investment projects:

	Cash Flow Data (Unit: $ thousand)	
n	**Project A**	**Project B**
0	$-$800	$-$2,635
1	$-1,500$	-565
2	-435	820
3	775	820
4	775	1,080
5	1,275	1,880
6	1,275	1,500
7	975	980
8	675	580
9	375	380
10	660	840

At $i = 12\%$, which of the two projects would be a better choice?

5.30 Consider the cash flows for the following investment projects.

	Project's Cash Flow				
n	**A**	**B**	**C**	**D**	**E**
0	$-$1,500	$-$1,500	$-$3,000	$1,500	$-$1,800
1	1,350	1,000	1,000	-450	600
2	800	800	X	-450	600
3	200	800	1,500	-450	600
4	100	150	X	-450	600

(a) Suppose projects A and B are mutually exclusive. On the basis of the NPW criterion, which project would be selected? Assume that MARR $= 15\%$.

(b) Repeat (a), using the NFW criterion.

(c) Find the minimum value of X that makes project C acceptable.

(d) Would you accept project D at $i = 18\%$?

(e) Assume that projects D and E are mutually exclusive. On the basis of the NPW criterion, which project would you select?

5.31 Consider two mutually exclusive investment projects, each with MARR $= 12\%$, as shown in Table 5.31.

(a) On the basis of the NPW criterion, which alternative would be selected?

(b) On the basis of the NFW criterion, which alternative would be selected?

TABLE P5.31 Two Mutually Exclusive Investment Projects

	Project's Cash Flow	
n	A	B
0	$-\$14,500$	$-\$12,900$
1	12,610	11,210
2	12,930	11,720
3	12,300	11,500

5.32 Consider the following two mutually exclusive investment projects, each with MARR = 15%:

	Project's Cash Flow	
n	A	B
0	$-\$6,000$	$-\$8,000$
1	800	11,500
2	14,000	400

(a) On the basis of the NPW criterion, which project would be selected?

(b) Sketch the PW(i) function for each alternative on the same chart between 0% and 50%. For what range of i would you prefer project B?

5.33 Two methods of carrying away surface runoff water from a new subdivision are being evaluated:

Method A. Dig a ditch. The first cost would be $60,000, and $25,000 of redigging and shaping would be required at five-year intervals forever.

Method B. Lay concrete pipe. The first cost would be $150,000, and a replacement would be required at 50-year intervals at a net cost of $180,000 forever.

At i = 12%, which method is the better one?

5.34 A local car dealer is advertising a standard 24-month lease of $1,150 per month for its new XT 3000 series sports car. The standard lease requires a down payment of $4,500, plus a $1,000 refundable initial deposit *now*. The first lease payment is due at the end of month 1. In addition, the company offers a 24-month lease plan that has a single up-front payment of $30,500, plus a refundable initial deposit of $1,000. Under both options, the initial deposit will be refunded at the end of month 24. Assume an interest rate of 6% compounded monthly. With the present-worth criterion, which option is preferred?

5.35 Two machines are being considered for a manufacturing process. Machine A has a first cost of $75,200, and its salvage value at the end of six years of estimated service life is $21,000. The operating costs of this machine are estimated to be $6,800 per year. Extra income taxes are estimated at $2,400 per year. Machine B has a first cost of $44,000, and its salvage value at the end of six years' service is estimated to be negligible. The annual operating costs will be $11,500. Compare these two mutually exclusive alternatives by the present-worth method at i = 13%.

5.36 An electric motor is rated at 10 horsepower (HP) and costs $800. Its full load effi-
ciency is specified to be 85%. A newly designed high-efficiency motor of the
same size has an efficiency of 90%, but costs $1,200. It is estimated that the
motors will operate at a rated 10 HP output for 1,500 hours a year, and the cost of
energy will be $0.07 per kilowatt-hour. Each motor is expected to have a 15-year
life. At the end of 15 years, the first motor will have a salvage value of $50 and the
second motor will have a salvage value of $100. Consider the MARR to be 8%.
(Note: 1 HP = 0.7457 kW.)

(a) Use the NPW criterion to determine which motor should be installed.

(b) In (a), what if the motors operated 2,500 hours a year instead of 1,500 hours a
year? Would the motor you chose in (a) still be the choice?

5.37 Consider the following two mutually exclusive investment projects:

	Project's Cash Flow	
n	A	B
0	−$20,000	−$25,000
1	17,500	25,500
2	17,000	18,000
3	15,000	

On the basis of the NPW criterion, which project would be selected if you use an
infinite planning horizon with project repeatability (the same costs and benefits)
likely? Assume that $i = 12\%$.

5.38 Consider the following two mutually exclusive investment projects, which have
unequal service lives:

	Project's Cash Flow	
n	A1	A2
0	−$900	−$1,800
1	−400	−300
2	−400	−300
3	−400 + 200	−300
4		−300
5		−300
6		−300
7		−300
8		−300 + 500

(a) What assumption(s) do you need in order to compare a set of mutually exclusive investments with unequal service lives?

(b) With the assumption(s) defined in (a) and using $i = 10\%$, determine which
project should be selected.

(c) If your analysis period (study period) is just three years, what should be the salvage value of project A2 at the end of year 3 to make the two alternatives economically indifferent?

5.39 Consider the following two mutually exclusive projects:

| | B1 | | B2 | |
| | Cash Flow | Salvage Value | Cash Flow | Salvage Value |
n				
0	−$18,000		−$15,000	
1	−2,000	6,000	−2,100	6,000
2	−2,000	4,000	−2,100	3,000
3	−2,000	3,000	−2,100	1,000
4	−2,000	2,000		
5	−2,000	2,000		

Salvage values represent the net proceeds (after tax) from disposal of the assets if they are sold at the end of each year. Both B1 and B2 will be available (or can be repeated) with the same costs and salvage values for an indefinite period.

(a) Assuming an infinite planning horizon, which project is a better choice at MARR = 12%?

(b) With a 10-year planning horizon, which project is a better choice at MARR = 12%?

5.40 Consider the following cash flows for two types of models:

| | Project's Cash Flow | |
n	Model A	Model B
0	−$6,000	−$15,000
1	3,500	10,000
2	3,500	10,000
3	3,500	

Both models will have no salvage value upon their disposal (at the end of their respective service lives). The firm's MARR is known to be 15%.

(a) Notice that the models have different service lives. However, model A will be available in the future with the same cash flows. Model B is available at one time only. If you select model B now, you will have to replace it with model A at the end of year 2. If your firm uses the present worth as a decision criterion, which model should be selected, assuming that the firm will need either model for an indefinite period?

(b) Suppose that your firm will need either model for only two years. Determine the salvage value of model A at the end of year 2 that makes both models indifferent (equally likely).

5.41 An electric utility is taking bids on the purchase, installation, and operation of microwave towers. Following are some details associated with the two bids that were received:

	Cost per Tower	
	Bid A	**Bid B**
Equipment cost	$112,000	$98,000
Installation cost	$25,000	$30,000
Annual maintenance and inspection fee	$2,000	$2,500
Annual extra income taxes		$800
Life	40 years	35 years
Salvage value	$0	$0

Which is the most economical bid if the interest rate is considered to be 11%? Either tower will have no salvage value after 20 years of use.

Use the NPW method to compare these two mutually exclusive plans.

5.42 Consider the following two investment alternatives:

	Project's Cash Flow	
n	**A1**	**A2**
0	−$15,000	−$25,000
1	9,500	0
2	12,500	X
3	7,500	X
PW(15%)	?	9,300

The firm's MARR is known to be 15%.

(a) Compute PW(15%) for A1.

(b) Compute the unknown cash flow X in years 2 and 3 for A2.

(c) Compute the project balance (at 15%) of A1 at the end of period 3.

(d) If these two projects are mutually exclusive alternatives, which one would you select?

5.43 Consider each of the after-tax cash flows shown in Table P5.43.

(a) Compute the project balances for projects A and D as a function of the project year at $i = 10\%$.

(b) Compute the net future-worth values for projects A and D at $i = 10\%$.

(c) Suppose that projects B and C are mutually exclusive. Suppose also that the required service period is eight years and that the company is considering leasing comparable equipment with an annual lease expense of $3,000 for the remaining years of the required service period. Which project is a better choice?

TABLE P5.43 After-Tax Cash Flows

	Project's Cash Flow			
n	**A**	**B**	**C**	**D**
0	−$2,500	−$7,000	−$5,000	−$5,000
1	650	−2,500	−2,000	−500
2	650	−2,000	−2,000	−500
3	650	−1,500	−2,000	4,000
4	600	−1,500	−2,000	3,000
5	600	−1,500	−2,000	3,000
6	600	−1,500	−2,000	2,000
7	300		−2,000	3,000
8	300			

5.44 A mall with two levels is under construction. The plan is to install only 9 escalators at the start, although the ultimate design calls for 16. The question arises as to whether to provide necessary facilities (stair supports, wiring conduits, motor foundations, etc.) that would permit the installation of the additional escalators at the mere cost of their purchase and installation or to defer investment in these facilities until the escalators need to be installed.

- **Option 1.** Provide these facilities now for all 7 future escalators at $200,000.
- **Option 2.** Defer the investment in the facility as needed. Install 2 more escalators in two years, 3 more in five years, and the last 2 in eight years. The installation of these facilities at the time they are required is estimated to cost $100,000 in year 2, $160,000 in year 5, and $140,000 in year 8.

Additional annual expenses are estimated at $3,000 for each escalator facility installed. At an interest rate of 12%, compare the net present worth of each option over eight years.

5.45 An electrical utility is experiencing a sharp power demand that continues to grow at a high rate in a certain local area.

Two alternatives are under consideration. Each is designed to provide enough capacity during the next 25 years, and both will consume the same amount of fuel, so fuel cost is not considered in the analysis.

- **Alternative A.** Increase the generating capacity now so that the ultimate demand can be met with additional expenditures later. An initial investment of $30 million would be required, and it is estimated that this plant facility would be in service for 25 years and have a salvage value of $0.85 million. The annual operating and maintenance costs (including income taxes) would be $0.4 million.
- **Alternative B.** Spend $10 million now and follow this expenditure with future additions during the 10th year and the 15th year. These additions would cost $18 million and $12 million, respectively. The facility would be sold 25 years from now, with a salvage value of $1.5 million. The annual operating and maintenance costs (including income taxes) initially will be $250,000 and will increase to $0.35 million after the second addition (from the 11th year to the 15th year)

and to $0.45 million during the final 10 years. (Assume that these costs begin 1 year subsequent to the actual addition.)

On the basis of the present-worth criterion, if the firm uses 15% as a MARR, which alternative should be undertaken?

5.46 A large refinery–petrochemical complex is to manufacture caustic soda, which will use feedwater of 10,000 gallons per day. Two types of feedwater storage installation are being considered over the 40 years of their useful life.

Option 1. Build a 20,000-gallon tank on a tower. The cost of installing the tank and tower is estimated to be $164,000. The salvage value is estimated to be negligible.

Option 2. Place a tank of 20,000-gallon capacity on a hill, which is 150 yards away from the refinery. The cost of installing the tank on the hill, including the extra length of service lines, is estimated to be $120,000, with negligible salvage value. Because of the tank's location on the hill, an additional investment of $12,000 in pumping equipment is required. The pumping equipment is expected to have a service life of 20 years, with a salvage value of $1,000 at the end of that time. The annual operating and maintenance cost (including any income tax effects) for the pumping operation is estimated at $1,000.

If the firm's MARR is known to be 12%, which option is better, on the basis of the present-worth criterion?

Short Case Studies

ST5.1 Apex Corporation requires a chemical finishing process for a product under contract for a period of six years. Three options are available. Neither Option 1 nor Option 2 can be repeated after its process life. However, Option 3 will always be available from H&H Chemical Corporation at the same cost during the period that the contract is operative. Here are the options:

- **Option 1.** Process device A, which costs $100,000, has annual operating and labor costs of $60,000 and a useful service life of four years with an estimated salvage value of $10,000.
- **Option 2.** Process device B, which costs $150,000, has annual operating and labor costs of $50,000 and a useful service life of six years with an estimated salvage value of $30,000.
- **Option 3.** Subcontract out the process at a cost of $100,000 per year.

According to the present-worth criterion, which option would you recommend at $i = 12\%$?

ST5.2 Tampa Electric Company, an investor-owned electric utility serving approximately 2,000 square miles in west central Florida, was faced with providing electricity to a newly developed industrial park complex. The distribution engineering department needs to develop guidelines for the design of the distribution circuit. The "main feeder," which is the backbone of each 13-kV distribution circuit, represents a substantial investment by the company.[5]

---

[5] Example provided by Andrew Hanson from Tampa Electric Company.

Tampa Electric has four approved main feeder construction configurations—(1) crossarm, (2) vertical (horizontal line post), (3) vertical (standoff pin), and (4) triangular, as illustrated in the accompanying figure. The width of the easement sought depends on the planned construction configuration. If crossarm construction is planned, a 15-foot easement is sought. A 10-foot wide easement is sought for vertical and triangular configurations.

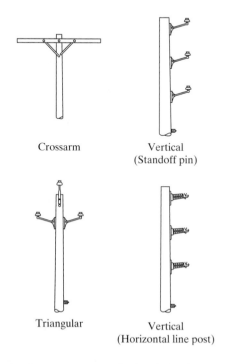

Crossarm

Vertical
(Standoff pin)

Triangular

Vertical
(Horizontal line post)

Once the required easements are obtained, the line clearance department clears any foliage that would impede the construction of the line. The clearance cost is dictated by the typical tree densities along road rights-of-way. The average cost to trim 1 tree is estimated at $20, and the average tree density in the service area is estimated to be 75 trees per mile. The costs of each type of construction are as follows:

	Design Configurations			
Factors	**Crossarm**	**Triangular**	**Horizontal Line**	**Standoff**
Easements	$487,000	$388,000	$388,000	$388,000
Line clearance	$613	$1,188	$1,188	$1,188
Line construction	$7,630	$7,625	$12,828	$8,812

Additional factors to consider in selecting the best main feeder configuration are as follows: In certain sections of Tampa Electric's service territory, ospreys often nest on transmission and distribution poles. The nests reduce the structural and electrical integrity of the poles. Crossarm construction is most vulnerable to osprey nesting, since the crossarm and braces provide a secure area for construction of the nest. Vertical and triangular construction do not provide such spaces. Furthermore, in areas where ospreys are known to nest, vertical and triangular configuration have added advantages. The insulation strength of a construction configuration may favorably or adversely affect the reliability of the line for which the configuration is used. A common measure of line insulation strength is the critical flashover (CFO) voltage. The greater the CFO, the less susceptible the line is to nuisance flashovers from lightning and other electrical phenomena.

The utility's existing inventory of crossarms is used primarily for main feeder construction and maintenance. The use of another configuration for main feeder construction would result in a substantial reduction in the inventory of crossarms. The line crews complain that line spacing on vertical and triangular construction is too restrictive for safe live line work. Each accident would cost $65,000 in lost work and other medical expenses. The average cost of each flashover repair would be $3,000. The following table lists the values of the factors involved in the four design configurations:

	Design Configurations			
Factors	**Crossarm**	**Triangular**	**Horizontal Line**	**Standoff**
Nesting	Severe	None	None	None
Insulation strength, CFO (kV)	387	474	476	462
Annual flashover occurrence (n)	2	1	1	1
Annual inventory savings		$4,521	$4,521	$4,521
Safety	OK	Problem	Problem	Problem

All configurations would last about 20 years, with no salvage value. It appears that noncrossarm designs are better, but engineers need to consider other design factors, such as safety, rather than just monetary factors when implementing the project. It is true that the line spacing on triangular construction is restrictive. However, with a better clearance design between phases for vertical construction, the hazard would be minimized. In the utility industry, the typical opposition to new types of construction is caused by the confidence acquired from constructing lines in the crossarm configuration for many years. As more vertical and triangular lines are built, opposition to these configurations should decrease. Which of the four designs described in the preceding table would you recommend to the management? Assume Tampa Electric's interest rate to be 12%.

CHAPTER SIX

Annual Equivalent-Worth Analysis

Thermally Activated Technologies: Absorption Chillers for Buildings[1] Absorption chillers provide cooling to buildings by using heat. This seemingly paradoxical, but highly efficient, technology is most cost effective in large facilities with significant heat loads. Not only do absorption chillers use less energy than conventional equipment does, but they also cool buildings without the use of ozone-depleting chlorofluorocarbons (CFCs). Unlike conventional electric chillers, which use mechanical energy in a vapor compression process to provide refrigeration, absorption chillers primarily use heat energy, with limited mechanical energy for pumping. Absorption chillers can be powered by natural gas, steam, or waste heat.

- The most promising markets for absorption chillers are in commercial buildings, government facilities, college campuses, hospital complexes, industrial parks, and municipalities.
- Absorption chillers generally become economically attractive when there is a source of inexpensive thermal energy at temperatures between 212°F and 392°F.

An absorption chiller transfers thermal energy from the heat source to the heat sink through an absorbent fluid and a refrigerant. The absorption chiller accomplishes its refrigerative effect by absorbing and then releasing water vapor into and out of a lithium bromide solution. Absorption chiller systems are classified by single-, double-, or triple-stage effects, which indicate the number of generators in the given system. The greater the number of stages, the higher is the overall efficiency of the system. Double-effect absorption chillers typically have a higher first cost, but a significantly lower energy cost, than single-effect chillers, resulting in a lower net present worth.

---

[1] Tech Brief, Office of Power Technology, U.S. Department of Energy, Washington, DC.

Single-Effect Absorption Chiller

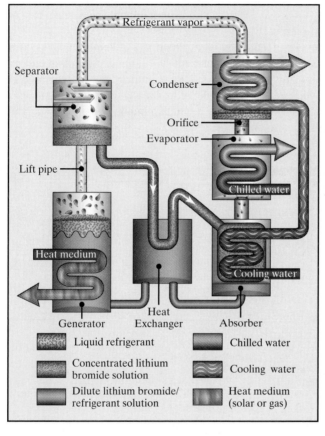

Some of the known economic benefits of the absorption chiller over the conventional mechanical chiller are as follows: [2]

- In a plant where low-pressure steam is currently being vented to the atmosphere, a mechanical chiller with a Coefficient of Performance (COP) of 4.0 is used 4,000 hours a year to produce an average of 300 tons of refrigeration.
- The plant's cost of electricity is $0.05 a kilowatt-hour. An absorption unit requiring 5,400 lb/hr of 15-psig steam could replace the mechanical chiller, providing the following annual electrical cost savings:

$$\text{Annual Savings} = 300 \text{ tons} \times (12,000 \text{ Btu/ton/}4.0) \times 4,000 \text{ hrs/yr}$$
$$\times \ \$0.05/\text{kWh} \times \text{kWh}/3,413 \text{ Btu} = \$52,740.$$

[2] *Source*: *EcoGeneration Solutions™, LLC, Companies,* 12615 Jones Rd., Suite 209, Houston, Texas 77070 (http://www.cogeneration.net/Absorption_Chillers.htm).

Suppose you plan to install the chiller and expect to operate it continuously for 10 years. How would you calculate the operating cost per hour? Suppose you are considering buying a new car. If you expect to drive 12,000 miles per year, can you figure out how much the car costs per mile? You would have good reason to want to know the cost if you were being reimbursed by your employer on a per mile basis for the business use of your car. Or consider a real-estate developer who is planning to build a 500,000-square-foot shopping center. What would be the minimum annual rental fee per square foot required to recover the initial investment?

Annual equivalence analysis is the method by which these and other unit costs (or profits) are calculated. Along with present-worth analysis, annual equivalence analysis is the second major equivalence technique for putting alternatives on a common basis of comparison. In this chapter, we develop the annual equivalent-worth criterion and demonstrate a number of situations in which annual equivalence analysis is preferable to other methods of comparison.

CHAPTER LEARNING OBJECTIVES

After completing this chapter, you should understand the following concepts:

■ How to determine the equivalent annual worth (cost) for a given project.

■ Why the annual equivalent approach facilitates the comparison of unequal service life problems.

■ How to determine the capital cost (or ownership cost) when you purchase an asset.

■ How to determine the unit cost or unit profit.

■ How to conduct a life-cycle cost analysis.

■ How to optimize design parameters in engineering design.

6.1 Annual Equivalent-Worth Criterion

In this section, we set forth a fundamental decision rule based on annual equivalent worth by considering both revenue and cost streams of a project. If revenue streams are irrelevant, then we make a decision solely on the basis of cost. This leads to a popular decision tool known as "life-cycle cost analysis," which we will discuss in Section 6.4.

6.1.1 Fundamental Decision Rule

The **annual equivalent worth (AE)** criterion provides a basis for measuring the worth of an investment by determining equal payments on an annual basis. Knowing that any lump-sum cash amount can be converted into a series of equal annual payments, we may first find the net present worth (NPW) of the original series and then multiply this amount by the capital recovery factor:

$$AE(i) = PW(i)(A/P, i, N). \tag{6.1}$$

- **Single-project evaluation:** The accept–reject selection rule for a single *revenue* project is as follows:

$$\text{If } AE(i) > 0, \text{ accept the investment.}$$
$$\text{If } AE(i) = 0, \text{ remain indifferent to the investment.}$$
$$\text{If } AE(i) < 0, \text{ reject the investment.}$$

Notice that the factor $(A/P, i, N)$ in Eq. (6.1) is positive for $-1 < i < \infty$, which indicates that the value of $AE(i)$ will be positive if, and only if, $PW(i)$ is positive. In other words, accepting a project that has a positive $AE(i)$ is equivalent to accepting a project that has a positive $PW(i)$. Therefore, the AE criterion for evaluating a project is consistent with the NPW criterion.

- **Comparing mutually exclusive alternatives:** As with present-worth analysis, when you compare mutually exclusive *service* projects whose revenues are the same, you may compare them on a *cost*-only basis. In this situation, the alternative with the minimum *annual equivalent cost* (or least negative annual equivalent worth) is selected.

Example 6.1 illustrates how to find the equivalent annual worth for a proposed energy-savings project. As you will see, you first calculate the net present worth of the project and then convert this present worth into an equivalent annual basis.

EXAMPLE 6.1 Annual Equivalent Worth: A Single-Project Evaluation

A utility company is considering adding a second feedwater heater to its existing system unit to increase the efficiency of the system and thereby reduce fuel costs. The 150-MW unit will cost $1,650,000 and has a service life of 25 years. The expected salvage value of the unit is considered negligible. With the second unit installed, the efficiency of the system will improve from 55% to 56%. The fuel cost to run the feedwater is estimated at $0.05 kWh. The system unit will have a load factor of 85%, meaning that the system will run 85% of the year.

(a) Determine the equivalent annual worth of adding the second unit with an interest rate of 12%.

(b) If the fuel cost increases at the annual rate of 4% after first year, what is the equivalent annual worth of having the second feedwater unit at $i = 12\%$?

DISCUSSION: Whenever we compare machines with different efficiency ratings, we need to determine the input powers required to operate the machines. Since the percent efficiency is equal to the ratio of the output power to the input power, we can determine the input power by dividing the output power by the motor's percent efficiency:

$$\text{Input power} = \frac{\text{output power}}{\text{percent efficiency}}.$$

A feedwater heater is a power-plant component used to preheat water delivered to a boiler. Preheating the feedwater reduces the amount of energy needed to make steam and thus reduces plant operation costs.

For example, a 30-HP motor with 90% efficiency will require an input power of

$$\text{Input power} = \frac{(30 \text{ HP} \times 0.746 \text{ kW/HP})}{0.90}$$

$$= 24.87 \text{ kW}.$$

Therefore, energy consumption with and without the second unit can be calculated as follows:

- Before adding the second unit, $\dfrac{150,000 \text{ kW}}{0.55} = 272,727 \text{ kW}$
- After adding the second unit, $\dfrac{150,000 \text{ kW}}{0.56} = 267,857 \text{ kW}$

So the reduction in energy consumption is 4,871 kW.

Since the system unit will operate only 85% of the year, the total annual operating hours are calculated as follows:

$$\text{Annual operating hours} = (365)(24)(0.85) = 7,446 \text{ hours/year.}$$

SOLUTION

Given: $I = \$1,650,000$, $N = 25$ years, $S = 0$, annual fuel savings, and $i = 12\%$. Find: AE of fuel savings due to improved efficiency.

(a) With the assumption of constant fuel cost over the service life of the second heater,

$$A_{\text{fuel savings}} = (\text{reduction in fuel requirement}) \times (\text{fuel cost})$$

$$\times (\text{operating hours per year})$$

$$= \left(\frac{150,000 \text{ kW}}{0.55} - \frac{150,000 \text{ kW}}{0.56} \right) \times (\$0.05/\text{kWh})$$

$$\times ((8,760)(0.85) \text{ hours/year})$$

$$= (4,871 \text{ kW}) \times (\$0.05/\text{kWh}) \times (7,446 \text{ hours/year})$$

$$= \$1,813,473/\text{year;}$$

$$\text{PW}(12\%) = -\$1,650,000 + \$1,813,473(P/A, 12\%, 25)$$

$$= \$12,573,321;$$

$$\text{AE}(12\%) = \$12,573,321(A/P, 12\%, 25)$$

$$= \$1,603,098.$$

(b) With the assumption of escalating energy cost at the annual rate of 4%, since the first year's fuel savings is already calculated in (a), we use it as A_1 in the geometric-gradient-series present-worth factor $(P/A_1, g, i, N)$:

$$A_1 = \$1,813,473$$

$$PW(12\%) = -\$1,650,000 + \$1,813,473(P/A_1, 4\%, 12\%, 25)$$

$$= \$17,463,697$$

$$AE(12\%) = \$17,463,697(A/P, 12\%, 25)$$

$$= \$2,226,621$$

Clearly, either situation generates enough fuel savings to justify adding the second unit of the feedwater. Figure 6.1 illustrates the cash flow series associated with the required investment and fuel savings over the heater's service life of 25 years.

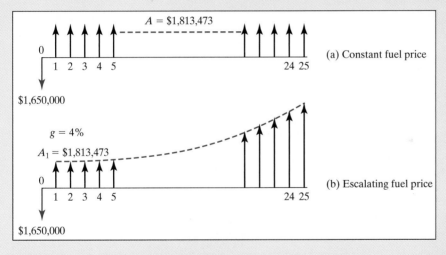

Figure 6.1   Cash flow diagram (Example 6.1).

6.1.2 Annual-Worth Calculation with Repeating Cash Flow Cycles

In some situations, a **cyclic cash flow pattern** may be observed over the life of the project. Unlike the situation in Example 6.1, where we first computed the NPW of the entire cash flow and then calculated the AE, we can compute the AE by examining the first cash flow cycle. Then we can calculate the NPW for the first cash flow cycle and derive the AE over that cycle. This shortcut method gives the same solution when the NPW of the entire project is calculated, and then the AE can be computed from this NPW.

EXAMPLE 6.2 Annual Equivalent Worth: Repeating Cash Flow Cycles

SOLEX Company is producing electricity directly from a solar source by using a large array of solar cells and selling the power to the local utility company. SOLEX decided to use amorphous silicon cells because of their low initial cost, but these cells degrade over time, thereby resulting in lower conversion efficiency and power output. The cells must be replaced every four years, which results in a particular cash flow pattern that repeats itself as shown in Figure 6.2. Determine the annual equivalent cash flows at $i = 12\%$.

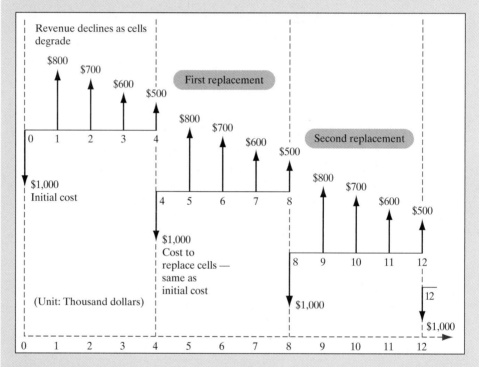

Figure 6.2 Conversion of repeating cash flow cycles into an equivalent annual payment (Example 6.2).

SOLUTION

Given: Cash flows in Figure 6.2 and $i = 12\%$.

Find: Annual equivalent benefit.

To calculate the AE, we need only consider one cycle over the four-year replacement period of the cells. For $i = 12\%$, we first obtain the NPW for the first cycle as follows:

$$\text{PW}(12\%) = -\$1,000,000$$
$$+ [(\$800,000 - \$100,000(A/G, 12\%, 4)](P/A, 12\%, 4)$$

$$= -\$1,000,000 + \$2,017,150.$$

$$= \$1,017,150.$$

Then we calculate the AE over the four-year life cycle:

$$AE(12\%) = \$1,017,150(A/P, 12\%, 4)$$

$$= \$334,880.$$

We can now say that the two cash flow series are equivalent:

Original Cash Flows		Annual Equivalent Flows	
n	A_n	n	A_n
0	−$1,000,000	0	0
1	800,000	1	$334,880
2	700,000	2	334,880
3	600,000	3	334,880
4	500,000	4	334,880

We can extend this equivalency over the remaining cycles of the cash flow. The reasoning is that each similar set of five values (one disbursement and four receipts) is equivalent to four annual receipts of $334,880 each. In other words, the $1 million investment in the solar project will recover the entire investment and generate equivalent annual savings of $334,880 over a four-year life cycle.

6.1.3 Comparing Mutually Exclusive Alternatives

In this section, we consider a situation in which two or more mutually exclusive alternatives need to be compared on the basis of annual equivalent worth. In Section 5.5, we discussed the general principles that should be applied when mutually exclusive alternatives with unequal service lives were compared. The same general principles should be applied in comparing mutually exclusive alternatives on the basis of annual equivalent worth: Mutually exclusive alternatives in equal time spans must be compared. Therefore, we must give careful consideration to the period covered by the analysis: the **analysis period**. We will consider two situations: (1) The analysis period equals project lives and (2) the analysis period differs from project lives.

With situation (1), we compute AE for each project and select the project that has the least negative AE for service projects (or the largest AE for revenue projects). With situation (2), we need to consider the issue of unequal project lives. As we saw in Chapter 5, comparing projects with unequal service lives is complicated by the need to determine the lowest common multiple life. For the special situation of an indefinite service period and replacement with identical projects, we can avoid this complication by the use of AE analysis, provided that the following criteria are met:

1. The service of the selected alternative is required on a continuous basis.
2. Each alternative will be replaced by an *identical* asset that has the same costs and performance.

When these two criteria are met, we may solve for the AE of each project on the basis of its initial life span, rather than on that of the lowest common multiple of the projects' lives. Example 6.3 illustrates the process of comparing unequal service projects.

EXAMPLE 6.3 Annual Equivalent Cost Comparison: Unequal Project Lives

Consider again Example 5.14, in which we compared two types of equipment with unequal service lives. Apply the annual equivalent approach to select the most economical equipment.

SOLUTION

Given: Cost cash flows shown in Figure 6.3 and $i = 15\%$ per year.
Find: AE cost and which alternative is the preferred one.

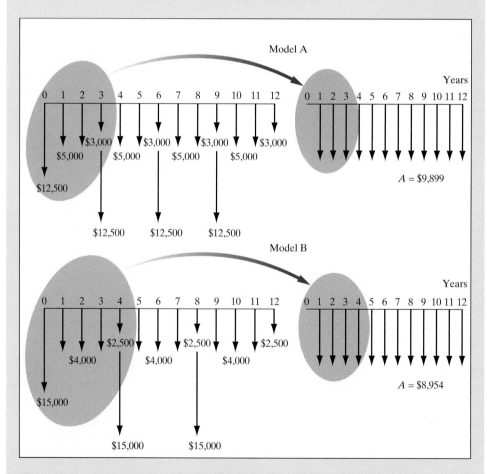

Figure 6.3 Comparison of projects with unequal lives and an indefinite analysis period using the annual equivalent-worth criterion (Example 6.3).

An alternative procedure for solving Example 5.14 is to compute the annual equivalent cost of an outlay of $12,500 for model A every 3 years and the annual equivalent cost of an outlay of $15,000 for model B every 4 years. Notice that the AE of each 12-year cash flow is the same as that of the corresponding 3- or 4-year cash flow (Figure 6.3). From Example 5.14, we calculate

- **Model A:**   For a 3-year life,

$$PW(15\%) = \$22,601$$
$$AE(15\%) = 22,601(A/P, 15\%, 3)$$
$$= \$9,899.$$

  For a 12-year period (the entire analysis period),

$$PW(15\%) = \$53,657$$
$$AE(15\%) = 53,657(A/P, 15\%, 12)$$
$$= \$9,899.$$

- **Model B:**   For a 4-year life,

$$PW(15\%) = \$25,562$$
$$AE(15\%) = \$25,562(A/P, 15\%, 4)$$
$$= \$8,954.$$

  For a 12-year period (the entire analysis period),

$$PW(15\%) = \$48,534$$
$$AE(15\%) = \$48,534(A/P, 15\%, 12)$$
$$= \$8,954.$$

Notice that the annual equivalent values that were calculated on the basis of the common service period are the same as those which were obtained over their initial life spans. Thus, for alternatives with unequal lives, we will obtain the same selection by comparing the NPW over a common service period using repeated projects or by comparing the AE for initial lives.

6.2 Capital Costs versus Operating Costs

When only costs are involved, the AE method is sometimes called the **annual equivalent cost (AEC)** method. In this case, revenues must cover two kinds of costs: **operating costs** and **capital costs**. Operating costs are incurred through the operation of physical plant or equipment needed to provide service; examples include items such as labor and

Capital cost:
the amount of
net investment.

raw materials. Capital costs are incurred by purchasing assets to be used in production and service. Normally, capital costs are nonrecurring (i.e., one-time) costs, whereas operating costs recur for as long as an asset is owned.

Because operating costs recur over the life of a project, they tend to be estimated on an annual basis anyway, so, for the purposes of annual equivalent cost analysis, no special calculation is required. However, because capital costs tend to be one-time costs, in conducting an annual equivalent cost analysis we must translate this one-time cost into its annual equivalent over the life of the project. The annual equivalent of a capital cost is given a special name: **capital recovery cost**, designated CR(i).

Capital recovery cost: The annual payment that will repay the cost of a fixed asset over the useful life of the asset and will provide an economic rate of return on the investment.

Two general monetary transactions are associated with the purchase and eventual retirement of a capital asset: its initial cost (I) and its salvage value (S). Taking into account these sums, we calculate the capital recovery factor as follows:

$$CR(i) = I(A/P, i, N) - S(A/F, i, N). \tag{6.2}$$

Now, recall algebraic relationships between factors in Table 3.4, and notice that the factor $(A/F, i, N)$ can be expressed as

$$(A/F, i, N) = (A/P, i, N) - i.$$

Then we may rewrite CR(i) as

$$CR(i) = I(A/P, i, N) - S[(A/P, i, N) - i]$$
$$= (I - S)(A/P, i, N) + iS. \tag{6.3}$$

Since we are calculating the equivalent annual costs, we treat cost items with a positive sign. Then the salvage value is treated as having a negative sign in Eq. (6.3). We may interpret this situation thus: To obtain the machine, one borrows a total of I dollars, S dollars of which are returned at the end of the Nth year. The first term, $(I - S)(A/P, i, N)$, implies that the balance $(I - S)$ will be paid back in equal installments over the N-year period at a rate of i. The second term, iS, implies that simple interest in the amount iS is paid on S until it is repaid (Figure 6.4). Thus, the amount to be financed is $I - S$ ($P/F, i, N$), and the installments of this loan over the N-year period are

$$AE(i) = [I - S(P/F, i, N)](A/P, i, N)$$
$$= I(A/P, i, N) - S(P/F, i, N)(A/P, i, N)$$
$$= [I(A/P, i, N) - S(A/F, i, N)]$$
$$= CR(i). \tag{6.4}$$

Therefore, CR(i) tells us what the bank would charge each year. Many auto leases are based on this arrangement, in that most require a guarantee of S dollars in salvage. From an industry viewpoint, CR(i) is the annual cost to the firm of owning the asset.

With this information, the amount of annual savings required to recover the capital and operating costs associated with a project can be determined, as Example 6.4 illustrates.

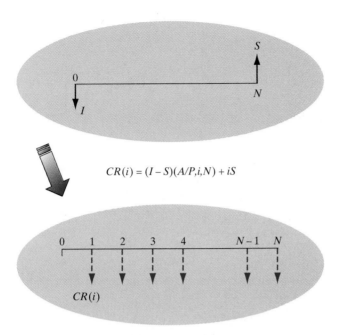

$$CR(i) = (I - S)(A/P, i, N) + iS$$

Figure 6.4 Capital recovery (ownership) cost calculation.

EXAMPLE 6.4 Capital Recovery Cost

Consider a machine that costs $20,000 and has a five-year useful life. At the end of the five years, it can be sold for $4,000 after tax adjustment. The annual operating and maintenance (O&M) costs are about $500. If the firm could earn an after-tax revenue of $5,000 per year with this machine, should it be purchased at an interest rate of 10%? (All benefits and costs associated with the machine are accounted for in these figures.)

SOLUTION

Given: I = $20,000, S = $4,000, O&M = $500, A = $5,000, N = 5 years, and i = 10% per year.

Find: AE, and determine whether to purchase the machine.

The first task is to separate cash flows associated with acquisition and disposal of the asset from the normal operating cash flows. Since the operating cash flows—the $5,000 yearly revenue—are already given in equivalent annual flows, we need to convert only the cash flows associated with acquisition and disposal of the asset into equivalent annual flows (Figure 6.5). Using Eq. (6.3), we obtain

$$\begin{aligned}
CR(i) &= (I - S)(A/P, i, N) + iS \\
&= (\$20{,}000 - \$4{,}000)(A/P, 10\%, 5) + (0.10)\ \$4{,}000 \\
&= \$4{,}620.76,
\end{aligned}$$

$$O\&M(i) = \$500,$$

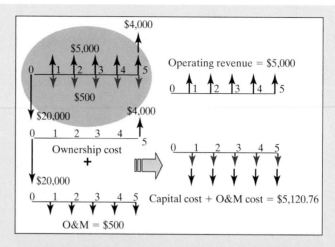

Figure 6.5 Separating ownership cost (capital cost) and operating cost from operating revenue, which must exceed the annual equivalent cost to make the project acceptable.

$$\text{AEC}(10\%) = \text{CR}(10\%) + \text{O\&M}(10\%)$$

$$= \$4,620.76 + \$500$$

$$= \$5,120.76,$$

$$\text{AE}(10\%) = \$5,000 - \$5,120.76$$

$$= -\$120.76.$$

This negative AE value indicates that the machine does not generate sufficient revenue to recover the original investment, so we must reject the project. In fact, there will be an equivalent loss of $120.76 per year over the life of the machine.

COMMENTS: We may interpret the value found for the annual equivalent cost as asserting that the annual operating revenues must be at least $5,120.76 in order to recover the cost of owning and operating the asset. However, the annual operating revenues actually amount to only $5,000, resulting in a loss of $120.76 per year. Therefore, the project is not worth undertaking.

6.3 Applying Annual-Worth Analysis

In general, most engineering economic analysis problems can be solved by the present-worth methods that were introduced in Chapter 5. However, some economic analysis problems can be solved more efficiently by annual-worth analysis. In this section, we introduce several applications that call for such an analysis.

6.3.1 Benefits of AE Analysis

Example 6.1 should look familiar to you: It is exactly the situation we encountered in Chapter 4 when we converted a mixed cash flow into a single present value and then into a series of equivalent cash flows. In the case of Example 6.1, you may wonder why we bother to convert NPW to AE at all, since we already know that the project is acceptable from NPW analysis. In fact, the example was mainly an exercise to familiarize you with the AE calculation.

However, in the real world, a number of situations can occur in which AE analysis is preferred, or even demanded, over NPW analysis. For example, corporations issue annual reports and develop yearly budgets. For these purposes, a company may find it more useful to present the annual cost or benefit of an ongoing project, rather than its overall cost or benefit. Following are some additional situations in which AE analysis is preferred:

1. **Consistency of report formats.** Financial managers commonly work with annual rather than overall costs in any number of internal and external reports. Engineering managers may be required to submit project analyses on an annual basis for consistency and ease of use by other members of the corporation and stockholders.

2. **Need for unit costs or profits.** In many situations, projects must be broken into unit costs (or profits) for ease of comparison with alternatives. Make-or-buy and reimbursement analyses are key examples, and these will be discussed in the chapter.

3. **Life-cycle cost analysis.** When there is no need for estimating the revenue stream for a proposed project, we can consider only the cost streams of the project. In that case, it is common to convert this life-cycle cost (LCC) into an equivalent annual cost for purposes of comparison. Industry has used the LCC to help determine which project will cost less over the life of a product. LCC analysis has had a long tradition in the Department of Defense, having been applied to virtually every new weapon system proposed or under development.

6.3.2 Unit Profit or Cost Calculation

In many situations, we need to know the unit profit (or cost) of operating an asset. To obtain this quantity, we may proceed as follows:

- Determine the number of units to be produced (or serviced) each year over the life of the asset.
- Identify the cash flow series associated with production or service over the life of the asset.
- Calculate the net present worth of the project cash flow series at a given interest rate, and then determine the equivalent annual worth.
- Divide the equivalent annual worth by the number of units to be produced or serviced during each year. When the number of units varies each year, you may need to convert them into equivalent annual units.

To illustrate the procedure, Example 6.5 uses the annual equivalent concept in estimating the savings per machine hour for the proposed acquisition of a machine.

EXAMPLE 6.5 Unit Profit per Machine Hour When Annual Operating Hours Remain Constant

Consider the investment in the metal-cutting machine of Example 5.5. Recall that this three-year investment was expected to generate an NPW of $3,553. Suppose that the machine will be operated for 2,000 hours per year. Compute the equivalent savings per machine hour at $i = 15\%$.

SOLUTION

Given: NPW = $3,553, N = 3 years, i = 15% per year, and 2,000 machine hours per year.
Find: Equivalent savings per machine hour.

We first compute the annual equivalent savings from the use of the machine. Since we already know the NPW of the project, we obtain the AE by the formula

$$AE(15\%) = \$3,553(A/P, 15\%, 3) = \$1,556.$$

With an annual usage of 2,000 hours, the equivalent savings per machine hour would be

$$\text{Savings per machine hour} = \$1,556/2,000 \text{ hours} = \$0.78/\text{hour.}$$

COMMENTS: Note that we cannot simply divide the NPW ($3,553) by the total number of machine hours over the three-year period (6,000 hours) to obtain $0.59/hour. This $0.59 figure represents the instant savings in present worth for each hourly use of the equipment, but does not consider the time over which the savings occur. Once we have the annual equivalent worth, we can divide by the desired time unit if the compounding period is one year. If the compounding period is shorter, then the equivalent worth should be calculated for the compounding period.

EXAMPLE 6.6 Unit Profit per Machine Hour When Annual Operating Hours Fluctuate

Consider again Example 6.5, and suppose that the metal-cutting machine will be operated according to varying hours: 1,500 hours the first year, 2,500 hours the second year, and 2,000 hours the third year. The total operating hours still remain at 6,000 over three years. Compute the equivalent savings per machine hour at $i = 15\%$.

SOLUTION

Given: NPW = $3,553, N = 3 years, i = 15% per year, and operating hours of 1,500 the first year, 2,500 the second year, and 2,000 the third year.
Find: Equivalent savings per machine hour.

As calculated in Example 6.5, the annual equivalent savings is $1,556. Let C denote the equivalent annual savings per machine hour, which we need to determine. Now, with varying annual usages of the machine, we can set up the equivalent annual savings as a function of C:

$$
\begin{aligned}
\text{Equivalent annual savings} = {}& [C(1{,}500)(P/F, 15\%, 1) \\
& + C(2{,}500)(P/F, 15\%, 2) \\
& + C(2{,}000)(P/F, 15\%, 3)](A/P, 15\%, 3) \\
= {}& 1{,}975.16C.
\end{aligned}
$$

We can equate this amount to the $1,556 we calculated in Example 6.5 and solve for C. This gives us

$$
C = \$1{,}556/1{,}975.16 = \$0.79/\text{hour},
$$

which is about a penny more than the $0.78 we found in Example 6.5.

6.3.3 Make-or-Buy Decision

Make-or-buy problems are among the most common of business decisions. At any given time, a firm may have the option of either buying an item or producing it. Unlike the make-or-buy situation we will consider in Chapter 8, *if either the "make" or the "buy" alternative requires the acquisition of machinery or equipment, then it becomes an investment decision.* Since the cost of an outside service (the "buy" alternative) is usually quoted in terms of dollars per unit, it is easier to compare the two alternatives if the differential costs of the "make" alternative are also given in dollars per unit. This unit cost comparison requires the use of annual-worth analysis. The specific procedure is as follows:

Step 1. Determine the time span (planning horizon) for which the part (or product) will be needed.

Step 2. Determine the annual quantity of the part (or product).

Step 3. Obtain the unit cost of purchasing the part (or product) from the outside firm.

Step 4. Determine the equipment, manpower, and all other resources required to make the part (or product) in-house.

Step 5. Estimate the net cash flows associated with the "make" option over the planning horizon.

Step 6. Compute the annual equivalent cost of producing the part (or product).

Step 7. Compute the unit cost of making the part (or product) by dividing the annual equivalent cost by the required annual volume.

Step 8. Choose the option with the minimum unit cost.

EXAMPLE 6.7 Equivalent Worth: Outsourcing the Manufacture of Cassettes and Tapes

Ampex Corporation currently produces both videocassette cases and metal particle magnetic tape for commercial use. An increased demand for metal particle tapes is projected, and Ampex is deciding between increasing the internal production of empty cassette cases and magnetic tape or purchasing empty cassette cases from an outside vendor. If Ampex purchases the cases from a vendor, the company must also buy specialized equipment to load the magnetic tapes, since its current loading machine is not compatible with the cassette cases produced by the vendor under consideration. The projected production rate of cassettes is 79,815 units per week for 48 weeks of operation per year. The planning horizon is seven years. After considering the effects of income taxes, the accounting department has itemized the annual costs associated with each option as follows:

- **Make option (annual costs):**

Labor	$1,445,633
Materials	$2,048,511
Incremental overhead	$1,088,110
Total annual cost	$4,582,254

- **Buy option:**

Capital expenditure	
Acquisition of a new loading machine	$ 405,000
Salvage value at end of seven years	$ 45,000
Annual Operating Costs	
Labor	$ 251,956
Purchasing empty cassette ($0.85/unit)	$3,256,452
Incremental overhead	$ 822,719
Total annual operating costs	$4,331,127

(Note the conventional assumption that cash flows occur in discrete lumps at the ends of years, as shown in Figure 6.6.) Assuming that Ampex's MARR is 14%, calculate the unit cost under each option.

SOLUTION

Given: Cash flows for two options and $i = 14\%$.
Find: Unit cost for each option and which option is preferred.

The required annual production volume is

$$79,815 \text{ units/week} \times 48 \text{ weeks} = 3,831,120 \text{ units per year.}$$

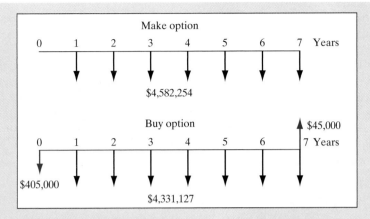

Figure 6.6 Make-or-buy analysis.

We now need to calculate the annual equivalent cost under each option:

- **Make option.** Since the "make option" is already given on an annual basis, the equivalent annual cost will be

$$AEC(14\%)_{Make} = \$4,582,254.$$

- **Buy option.** The two cost components are capital cost and operating cost.
  Capital cost:

$$CR(14\%) = (\$405,000 - \$45,000)(A/P, 14\%, 7)$$
$$+ (0.14)(\$45,000)$$
$$= \$90,249$$
$$AEC(14\%)_1 = CR(14\%) = \$90,249.$$

  Operating cost:

$$AEC(14\%)_2 = \$4,331,127.$$

  Total annual equivalent cost:

$$AEC(14\%)_{Buy} = AEC(14\%)_1 + AEC(14\%)_2 = \$4,421,376.$$

Obviously, this annual equivalent calculation indicates that Ampex would be better off buying cassette cases from the outside vendor. However, Ampex wants to know the unit costs in order to set a price for the product. In such a situation, we need to calculate the unit cost of producing the cassette tapes under each option. We do this by dividing the magnitude of the annual equivalent cost for each option by the annual quantity required:

- **Make option:**

$$\text{Unit cost} = \$4,582,254/3,831,120 = \$1.20/\text{unit}.$$

- **Buy option:**

$$\text{Unit cost} = \$4,421,376/3,831,120 = \$1.15/\text{unit}.$$

Buying the empty cassette cases from the outside vendor and loading the tape in-house will save Ampex 5 cents per cassette before any consideration of taxes.

COMMENTS: Two important noneconomic factors should also be considered. The first is the question of whether the quality of the supplier's component is better than, equal to, or worse than the component the firm is presently manufacturing. The second is the reliability of the supplier in terms of providing the needed quantities of the cassette cases on a timely basis. A reduction in quality or reliability should virtually always rule out buying.

6.3.4 Pricing the Use of an Asset

Companies often need to calculate the cost of equipment that corresponds to a **unit of use** of that equipment. For example, if you own an asset such as a building, you would be interested in knowing the cost per square foot of owning and operating the asset. This information will be the basis for determining the rental fee for the asset. A familiar example is an employer's reimbursement of costs for the use of an employee's personal car for business purposes. If an employee's job depends on obtaining and using a personal vehicle on the employer's behalf, reimbursement on the basis of the employee's overall costs per mile seems fair.

EXAMPLE 6.8 Pricing an Apartment Rental Fee

Sunbelt Corporation, an investment company, is considering building a 50-unit apartment complex in a growing area near Tucson, Arizona. Since the long-term growth potential of the town is excellent, it is believed that the company could average 85% full occupancy for the complex each year. If the following financial data are reasonably accurate estimates, determine the minimum monthly rent that should be charged if a 15% rate of return is desired:

- Land investment cost = $1,000,000
- Building investment cost = $2,500,000
- Annual upkeep cost = $150,000
- Property taxes and insurance = 5% of total initial investment
- Study period = 25 years
- Salvage value = Only land cost can be recovered in full.

SOLUTION

Given: Preceding financial data, study period = 25 years, and i = 15%.
Find: Minimum monthly rental charge.

First we need to determine the capital cost associated with ownership of the property:

$$\text{Total investment required} = \text{land cost} + \text{building cost} = \$3,500,000,$$
$$\text{Salvage value} = \$1,000,000 \text{ at the end of 25 years,}$$

$$CR(15\%) = (\$3,500,000 - \$1,000,000)(A/P, 15\%, 25)$$
$$+ (\$1,000,000)(0.15)$$
$$= \$536,749.$$

Second, the annual operating cost has two elements: (1) property taxes and insurance and (2) annual upkeep cost. Thus,

$$O\&M \text{ cost} = (0.05)(\$3,500,000) + \$150,000$$
$$= \$325,000.$$

So the total annual equivalent cost is

$$AEC(15\%) = \$536,749 + \$325,000$$
$$= \$861,749,$$

which is the minimum annual rental required to achieve a 15% rate of return. Therefore, with annual compounding, the monthly rental amount is

$$\text{Required monthly charge} = \frac{\$861,749}{(12 \times 50)(0.85)}$$
$$= \$1,690.$$

COMMENTS: The rental charge that is exactly equal to the cost of owning and operating the building is known as the **break-even point**.

6.4 Life-Cycle Cost Analysis

Because 80% of the total life-cycle cost of a system occurs after the system has entered service, the best long-term system acquisition and support decisions are based on a full understanding of the total cost of acquiring, operating, and supporting the system (Figure 6.7). **Life-cycle cost analysis** (LCCA) enables the analyst to make sure that the selection of a design alternative is not based solely on the lowest initial costs, but also takes into account all the future costs over the project's usable life. Some of the unique features of LCCA are as follows:

Life-cycle cost analysis is useful when project alternatives that fulfill the same performance requirements, but differ with respect to initial costs and operating costs.

- LCCA is used appropriately only to select from among design alternatives that would yield the same level of performance or benefits to the project's users during normal operations. If benefits vary among the design alternatives, then the alternatives cannot be compared solely on the basis of cost. Rather, the analyst would need to employ present-worth analysis or benefit–cost analysis (BCA), which measures the monetary value of life-cycle benefits as well as costs. BCA is discussed in Chapter 16.
- LCCA is a way to predict the most cost-effective solution; it does not guarantee a particular result, but allows the plant designer or manager to make a reasonable comparison among alternative solutions within the limits of the available data.
- To make a fair comparison, the plant designer or manager might need to consider the measure used. For example, the same process output volume should be considered, and if the two items being examined cannot give the same output volume, it may be appropriate to express the figures in cost per unit of output (e.g., $/ton, or euro/kg), which requires a calculation of annual equivalent dollars generated. This calculation is based on the annual output.

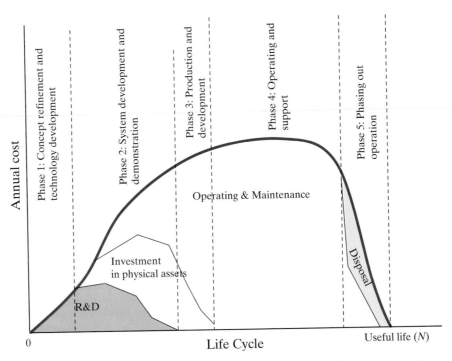

Figure 6.7 Stages of life-cycle cost. These include the concept refinement and technology development phase, the system development and demonstration phase, the production and deployment phase, the operating phase, and the disposal phase.

In many situations, we need to compare a set of different design alternatives, each of which would produce the same number of units (constant revenues), but would require different amounts of investment and operating costs (because of different degrees of mechanization). Example 6.9 illustrates the use of the annual equivalent-cost concept to compare the cost of operating an existing pumping system with that of an improved system.

EXAMPLE 6.9 Pumping System with a Problem Valve[3]

Consider a single-pump circuit that transports a process fluid containing some solids from a storage tank to a pressurized tank. A heat exchanger heats the fluid, and a control valve regulates the rate of flow into the pressurized tank to 80 cubic meters per hour (m³/h), or 350 gallons per minute (gpm). The process is depicted in Figure 6.8.

The plant engineer is experiencing problems with a fluid control valve (FCV) that fails due to erosion caused by cavitations. The valve fails every 10 to 12 months at a cost of $4,000 per repair. A change in the control valve is being considered: Replace the existing valve with one that can resist cavitations. Before the control valve

---

[3] *Pump Life Cycle Costs: A Guide to LCC Analysis for Pumping Systems.* DOE/GO-102001-1190. December 2000, Office of Industrial Technologies, U.S. Department of Energy and Hydraulic Institute.

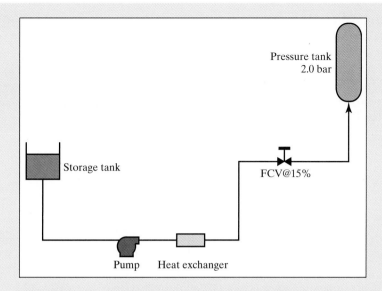

Figure 6.8 Sketch of a pumping system in which the control valve fails.

is repaired again, the project engineer wants to look at other options and perform an LCCA on alternative solutions.

Engineering Solution Alternatives

The first step is to determine how the system is currently operating and why the control valve fails. Then the engineer can see what can be done to correct the problem. The control valve currently operates between 15 and 20% open and with considerable cavitation noise from the valve. It appears that the valve was not sized properly for the application. After reviewing the original design calculations, it was discovered that the pump was oversized: 110 m³/h (485 gpm) instead of 80 m³/h (350 gpm). This resulted in a larger pressure drop across the control valve than was originally intended.

As a result of the large differential pressure at the operating rate of flow, and because the valve is showing cavitation damage at regular intervals, the engineer determines that the control valve is not suitable for this process.

The following four options are suggested:

- **Option A.** A new control valve can be installed to accommodate the high pressure differential.
- **Option B.** The pump impeller can be trimmed so that the pump does not develop as much head, resulting in a lower pressure drop across the current valve.
- **Option C.** A variable-frequency drive (VFD) can be installed and the flow control valve removed. The VFD can vary the pump speed and thus achieve the desired process flow.
- **Option D.** The system can be left as it is, with a yearly repair of the flow control valve to be expected.

Cost Summary:

Pumping systems often have a life span of 15 to 20 years. Some cost elements will be incurred at the outset, and others will be incurred at different times throughout the lives of the different solutions evaluated. Therefore, you need to calculate a present value of the LCC to accurately assess the different solutions.

Some of the major LCC elements related to a typical pumping system are summarized as follows:

$$LCC = C_{ic} + C_{in} + C_e + C_o + C_m + C_s + C_{env} + C_d$$

LCC = life-cycle cost

C_{ic} = initial costs, purchase price (costs of pump, system, pipe, auxiliary services)

C_{in} = installation and commissioning cost (including cost of training)

C_e = energy costs (predicted cost for system operation, including costs of pump driver, controls, and any auxiliary services)

C_o = operation cost (labor cost of normal system supervision)

C_m = maintenance and repair costs (costs of routine and predicted repairs)

C_s = downtime costs (cost of loss of production)

C_{env} = environmental costs (costs due to contamination from pumped liquid and auxiliary equipment)

C_d = decommissioning and disposal costs (including the cost of of restoration of the local environment and disposal of auxiliary services)

For each option, the major cost elements identified are as follows:

- **Option A.** The cost of a new control valve that is properly sized is $5,000. The cost of modifying the pump's performance by reducing the diameter of the impeller is $2,250. The process operates at 80 m³/h for 6,000 h/year. The energy cost is $0.08 per kWh and the motor efficiency is 90%.

- **Option B.** By trimming the impeller to 375 mm, the pump's total head is reduced to 42.0 m (138 ft) at 80 m³/h. This drop in pressure reduces the differential pressure across the control valve to less than 10 m (33 ft), which better matches the valve's original design intent. The resulting annual energy cost with the smaller impeller is $6,720 per year. It costs $2,250 to trim the impeller. This cost includes the machining cost as well as the cost to disassemble and reassemble the pump.

- **Option C.** A 30-kW VFD costs $20,000 and an additional $1,500 to install. The VFD will cost $500 to maintain each year. It is assumed that it will not need any repairs over the project's eight-year life.

- **Option D.** The option to leave the system unchanged will result in a yearly cost of $4,000 for repairs to the cavitating flow control value.

Table 6.1 summarizes financial as well as technical data related to the various options.

TABLE 6.1 Cost Comparison for Options A through D in the System with a Failing Control Valve

Cost	Change Control Valve (A)	Trim Impeller (B)	VFD (C)	Repair Control Valve (D)
Pump Cost Data				
Impeller diameter	430 mm	375 mm	430 mm	430 mm
Pump head	71.7 m (235 ft)	42.0 m (138 ft)	34.5 m (113 ft)	71.7 m (235 ft)
Pump efficiency	75.1%	72.1%	77%	75.1%
Rate of flow	80 m³/h (350 gpm)	80 m³/h (350 gpm)	80 m³/h (350 gpm)	80 m³/h (350 gpm)
Power consumed	23.1 kW	14.0 kW	11.6 kW	23.1 kW
Energy cost/year	$11,088	$6,720	$ 5,568	$11,088
New valve	$ 5,000	0	0	0
Modify impeller	0	$2,250	0	0
VFD	0	0	$20,000	0
Installation of VFD	0	0	$ 1,500	0
Valve repair/year	0	0	0	$ 4,000

SOLUTION

Given: Financial data as summarized in Table 6.1.

Find: Which design option to choose.

Assumptions:

- The current energy price is $0.08/kWh.
- The process is operated for 6,000 hours/year.
- The company has a cost of $500 per year for routine maintenance of pumps of this size, with a repair cost of $2,500 every second year.
- There is no decommissioning cost or environmental disposal cost associated with this project.
- The project has an eight-year life.
- The interest rate for new capital projects is 8%, and an inflation rate of 4% is expected.

A sample LCC calculation for Option A is shown in Table 6.2. Note that the energy cost and other cost data are escalated at the annual rate of 4%. For example, the current estimate of energy cost is $11,088. To find the cost at the end of year 1, we

TABLE 6.2 LCC Calculation for Option A

	A	B	C	D	E	F	G	H	I	J
1	**Calculation Chart for LCC: Option A**									
2										
3	Year No	0	1	2	3	4	5	6	7	8
4	Initial investment cost:	$5,000								
5	Installation and commissioning cost:	$0								
6										
7	Energy cost/year (calculated)		$11,532	$11,993	$12,472	$12,971	$13,490	$14,030	$14,591	$15,175
8	Routine maintenance		$520	$541	$562	$585	$608	$633	$658	$684
9	Repair cost every 2nd year			$2,704		$2,925		$3,163		$3,421
10	Operating costs		$0	$0	$0	$0	$0	$0	$0	$0
11	Other costs		$0	$0	$0	$0	$0	$0	$0	$0
12										
13	Downtime costs		$0	$0	$0	$0	$0	$0	$0	$0
14										
15	Decommissioning costs									$0
16	Environmental & disposal costs									$0
17										
18	**Sum costs**	$5,000	$12,052	$15,238	$13,035	$16,481	$14,099	$17,826	$15,249	$19,280
19										
20	**Present costs**	$5,000	$11,159	$13,064	$10,348	$12,114	$9,595	$11,233	$8,898	$10,417
21	Present costs, energy		$10,677	$10,282	$9,901	$9,534	$9,181	$8,841	$8,514	$8,198
22	Present costs, routine maintenance		$481	$464	$446	$430	$414	$399	$384	$370
23										
24	**Sum of present costs**	**$91,827**								
25	Sum of energy costs	$75,129								
26	Sum of routine maintenance costs	$3,388								
27										
28	Annual equivalent cost (8%)	$15,979								
29	Cost per operating hour	**$2.66**								

TABLE 6.3 Comparison of LCC for Options A through D

	Option A Change Control Valve	Option B Trim Impeller	Option C VFD and Remove Control Valve	Option D Repair Control Valve
Input				
Initial investment cost:	$5,000	$2,250	$21,500	0
Energy price (present) per kWh:	0.080	0.080	0.080	0.080
Weighted average power of equipment in kW:	23.1	14.0	11.6	23.1
Average operating hours/year:	6,000	6,000	6,000	6,000
Energy cost/year (calculated) + energy price × weighted average power × average operating hours/year:	11,088	6,720	5,568	11,088
Maintenance cost (routine maintenance/year:	500	500	1,000	500
Repair every second year:	2,500	2,500	2,500	2,500
Other yearly costs:	0	0	0	4,000
Downtime cost/year:	0	0	0	0
Environmental cost:	0	0	0	0
Decommissioning/disposal (salvage) cost:	0	0	0	0
Lifetime in years:	8	8	8	8
Interest rate (%):	8.0%	8.0%	8.0%	8.0%
Inflation rate (%):	4.0%	4.0%	4.0%	4.0%
Output				
Present LCC value:	$91,827	$59,481	$74,313	$113,930
Cost per operating hour	**$2.66**	**$1.73**	**$2.16**	**$3.30**

multiply $11,088 by $(1 + 0.04)$, yielding $11,532. For year 2, we multiply $11,532 by $(1 + 0.04)$ to obtain $11,993. Once we calculate the LCC in present worth ($91,827), we find the equivalent annual value ($15,979) at 8% interest. Finally, to calculate the cost per hour, we divide $15,979 by 6,000 hours, resulting in $2.66. We can calculate the unit costs for other options in a similar fashion. (See Table 6.3.)

Option B, trimming the impeller, has the lowest life-cycle cost ($1.73 per hour) and is the preferred option for this example.

6.5 Design Economics

Engineers are frequently involved in making design decisions that provide the required functional quality at the lowest cost. Another valuable extension of AE analysis is minimum-cost analysis, which is used to determine optimal engineering designs. The AE analysis method is useful when two or more cost components are affected differently by the same design element (i.e., for a single design variable, some costs may increase while others decrease). When the equivalent annual total cost of a design variable is a function of increasing and decreasing cost components, we can usually find the optimal value that will minimize the cost of the design with the formula

$$\text{AEC}(i) = a + bx + \frac{c}{x},\tag{6.5}$$

where x is a common design variable and a, b, and c are constants.

To find the value of the common design variable that minimizes AE(i), we need to take the first derivative, equate the result to zero, and solve for x:

$$\frac{d\text{AEC}(i)}{dx} = b - \frac{c}{x^2}$$

$$= 0$$

$$x = x^* = \sqrt{\frac{c}{b}}.\tag{6.6}$$

(It is common engineering practice to denote the optimal solution with an asterisk.)

The logic of the first-order requirement is that an activity should, if possible, be carried to a point where its **marginal yield** $d\text{AEC}(i)/dx$ is zero. However, to be sure whether we have found a maximum or a minimum when we have located a point whose marginal yield is zero, we must examine it in terms of what are called the **second-order conditions**. Second-order conditions are equivalent to the usual requirements that the second derivative be negative in the case of a maximum and positive for a minimum. In our situation,

$$\frac{d^2\text{AEC}(i)}{dx^2} = \frac{2C}{x^3}.$$

As long as $C > 0$, the second derivative will be positive, indicating that the value x^* is the minimum-cost point for the design alternative. To illustrate the optimization concept, two examples are given, the first having to do with designing the optimal cross-sectional area of a conductor and the second dealing with selecting an optimal size for a pipe.

EXAMPLE 6.10 Optimal Cross-Sectional Area

A constant electric current of 5,000 amps is to be transmitted a distance of 1,000 feet from a power plant to a substation for 24 hours a day, 365 days a year. A copper conductor can be installed for $8.25 per pound. The conductor will have an estimated life of 25 years and a salvage value of $0.75 per pound. The power loss from a conductor is inversely proportional to the cross-sectional area A of the conductor.

It is known that the resistance of the conductor sought is 0.8145×10^{-5} ohm for 1 square inch per foot of cross section. The cost of energy is \$0.05 per kilowatt-hour, the interest rate is 9%, and the density of copper is 555 lb/ft^3. For the given data, calculate the optimum cross-sectional area A of the conductor.

DISCUSSION: The resistance of the conductor is the most important cause of power loss in a transmission line. The resistance of an electrical conductor varies directly with its length and inversely with its cross-sectional area according to the equation

$$R = \rho\left(\frac{L}{A}\right), \qquad (6.7)$$

where R is the resistance of the conductor, L is the length of the conductor, A is the cross-sectional area of the conductor, and ρ is the resistivity of the conductor material.

Any consistent set of units may be used. In power work in the United States, L is usually given in feet, A in circular mils (cmil), and ρ in ohm-circular mils per foot.

A circular mil is the area of a circle that has a diameter of 1 mil, equal to 1×10^{-3} in. The cross-sectional area of a wire in square inches equals its area in circular mils multiplied by 0.7854×10^{-6}. More specifically, one unit relates to another as follows:

$$1 \text{ linear mil} = 0.001 \text{ in.}$$
$$= 0.0254 \text{ millimeter.}$$
$$1 \text{ circular mil} = \text{area of circle 1 linear mil in diameter}$$
$$= (0.5)^2 \pi \text{ mil}^2$$
$$= 0.7854 \times 10^{-6} \text{ in.}^2.$$
$$1 \text{ in.}^2 = 1/(0.7854 \times 10^{-6})$$
$$= 1.27324 \times 10^6 \text{ cmil.}$$

In SI units (the official designation for the Système International d'Unités), L is in meters, A in square meters, and ρ in ohm-meters. In terms of SI units, the copper conductor has a ρ value of 1.7241×10^{-8} Ω-meter. We can easily convert this value into units of Ω-in.^{2} per foot. From Eq. (6.7), solving for ρ yields

$$\rho = RA/L = 1.7241 \times 10^{-8} \ \Omega(1 \text{ meter})^2/1 \text{ meter}$$
$$= 1.7241 \times 10^{-8} \ \Omega(39.37 \text{ in.})^2/3.2808 \text{ ft}$$
$$= 0.8145421 \times 10^{-5} \ \Omega \text{ in.}^2/\text{ft.}$$

When current flows through a circuit, power is used to overcome resistance. The unit of electrical work is the kilowatt hour (kWh), which is equal to the power in kilowatts, multiplied by the hours during which work is performed. If the current I is steady, then the charge passing through the wire in time T is equivalent to the power that is converted to heat (known as energy loss) and is equal to

$$\text{Power} = I^2 \frac{RT}{1{,}000 \text{ kWh}}, \tag{6.8}$$

where I is the current in amperes, R is the resistance in ohms, and T is the duration, in hours, during which work is performed.

SOLUTION

Given: Cost components as a function of cross-sectional area (A), $N = 25$ years, and $i = 9\%$.
Find: Optimal value of A.

Step 1: This classic minimum-cost example, the design of the cross-sectional area of an electrical conductor, involves increasing and decreasing cost components. Since the resistance of a conductor is inversely proportional to the size of the conductor, energy loss will decrease as the size of the conductor increases. More specifically, the energy loss in kilowatt-hours (kWh) in a conductor due to resistance is equal to

$$\begin{aligned}
\text{Energy loss in kilowatt-hours} &= \frac{I^2 R}{1{,}000 A} T \\
&= \frac{(5{,}000^2)(0.008145)}{1{,}000 A}(24 \times 365) \\
&= \frac{1{,}783{,}755}{A} \text{kWh.}
\end{aligned}$$

Step 2: Again, since the electrical resistance is inversely proportional to the cross-sectional area A of the conductor, the total energy loss in dollars per year for a specified conductor material is

$$\begin{aligned}
\text{Cost of energy loss} &= \frac{1{,}783{,}755}{A}(\phi) \\
&= \frac{1{,}783{,}755}{A}(\$0.05) \\
&= \frac{\$89{,}188}{A},
\end{aligned}$$

where ϕ is the cost of energy in dollars per kWh.

Step 3: As we increase the size of the conductor, however, it costs more to build. First, we need to calculate the total amount of conductor material in pounds.

Since the cross-sectional area is given in square inches, we need to convert the total length in feet to inches before finding the weight of the material:

$$\text{Weight of material in pounds} = \frac{(1,000)(12)(555)A}{12^3}$$
$$= 3,854(A)$$
$$\text{Total material cost} = 3,854(A)(\$8.25)$$
$$= \$31,797(A).$$

Here, we are looking for the trade-off between the cost of installation and the cost of energy loss.

Step 4: Since, at the end of 25 years, the copper material will be salvaged at the rate of $0.75 per pound, we can compute the capital recovery cost as

$$CR(9\%) = [31,797A - 0.75(3,854A)](A/P, 9\%, 25) + 0.75(3,854A)(0.09)$$
$$= 2,943A + 260A$$
$$= 3,203A.$$

Step 5: Using Eq. (6.5), we express the total annual equivalent cost as a function of the design variable A:

$$AEC(9\%) = \overbrace{3,203A}^{\text{Capital cost}} + \underbrace{\frac{89,188}{A}}_{\text{Operating cost}}.$$

To find the minimum annual equivalent cost, we use Eq. (6.6):

$$\frac{d\,AEC(9\%)}{dA} = 3,203 - \frac{89,188}{A^2} = 0,$$
$$A^* = \sqrt{\frac{89,188}{3,203}}$$
$$= 5.276 \text{ in.}^2.$$

The minimum annual equivalent total cost is

$$AEC(9\%) = 3,203(5.276) + \frac{89,188}{5.276}$$
$$= \$33,802.$$

Figure 6.9 illustrates the nature of this design trade-off problem.

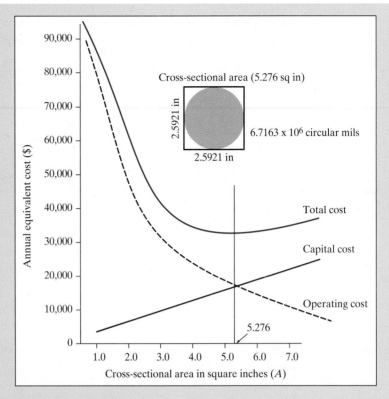

Figure 6.9 Optimal cross-sectional areas for a copper conductor. Note that the minimum point almost coincides with the crossing point of the capital-cost and operating-cost lines. This is not always true. Since the cost components can have a variety of cost patterns, the minimum point does not in general occur at the crossing point (Example 6.10).

EXAMPLE 6.11 Economical Pipe Size

As a result of the 1990 conflict in the Persian Gulf, Kuwait is studying the feasibility of running a steel pipeline across the Arabian Peninsula to the Red Sea. The pipeline will be designed to handle 3 million barrels of crude oil per day under optimum conditions. The length of the line will be 600 miles. Calculate the optimum pipeline diameter that will be used for 20 years for the following data at $i = 10\%$:

$$\text{Pumping power} = \frac{1.333 Q \Delta P}{1,980,000} \text{HP}$$

$$Q = \text{volume flow rate, ft}^3/\text{hour}$$

$$\Delta P = \frac{128 Q \mu L}{g \pi D^4}, \text{pressure drop, lb/ft}^2$$

$$L = \text{pipe length, ft}$$

$D =$ inside pipe diameter, ft

$t = 0.01D$, pipeline wall thickness, ft

$\mu = 8{,}500$ lb/hour ft, oil viscosity

$g = 32.3 \times 12{,}960{,}000$ ft/hour^2

Power cost, $0.015 per HP hour

Oil cost, $50 per barrel

Pipeline cost, $1.00 per pound of steel

Pump and motor costs, $195 per HP

The salvage value of the steel after 20 years is assumed to be zero because removal costs exhaust scrap profits from steel. (See Figure 6.10 for the relationship between D and t.)

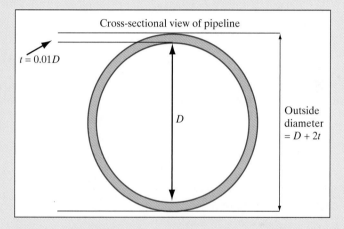

Figure 6.10 Designing economical pipe size to handle 3 million barrels of crude oil per day (Example 6.11).

DISCUSSION: In general, when a progressively larger size of pipe is used to carry a given fluid at a given volume flow rate, the energy required to move the fluid will progressively decrease. However, as we increase the pipe size, the cost of its construction will increase. In practice, to obtain the best pipe size for a particular situation, you may choose a reasonable, but small, starting size. Compute the energy cost of pumping fluid through this size of pipe and the total construction cost, and then compare the difference in energy cost with the difference in construction cost. When the savings in energy cost exceeds the added construction cost, the process may be repeated with progressively larger pipe sizes until the added construction cost exceeds the savings in energy cost. As soon as this happens, the best pipe size to use in the particular application is identified. However, this search process can be simplified by using the minimum-cost concept encompassed by Eqs. (6.5) and (6.6).

SOLUTION

Given: Cost components as a function of pipe diameter (D), $N = 20$ years, and $i = 10\%$.

Find: Optimal pipe size (D).

Step 1: Several different units are introduced; however, we need to work with common units. We will assume the following conversions:

$$1 \text{ mile} = 5,280 \text{ ft}$$
$$600 \text{ miles} = 600 \times 5,280 = 3,168,000 \text{ ft}$$
$$1 \text{ barrel} = 42 \text{ U.S. gallons}$$
$$1 \text{ barrel} = 42 \text{ gallons} \times 231 \text{ in.}^3/\text{gallon} = 9,702 \text{ in.}^3$$
$$1 \text{ barrel} = 9,702 \text{ in.}^3/12^3 = 5.6146 \text{ ft}^3$$
$$\text{Density of steel} = 490.75 \text{ lb/ft}^3$$

Step 2: Power required to pump oil:

It is well known that, for any given set of operating conditions involving the flow of a noncompressible fluid, such as oil, through a pipe of constant diameter, a small-diameter pipe will have a high fluid velocity and a high fluid pressure drop through the pipe. This will require a pump that will deliver a high discharge pressure and a motor with large energy consumption. To determine the pumping power required to boost the pressure drop, we need to determine both the volume flow rate and the amount of pressure drop. Then we can calculate the cost of the power required to pump oil.

- Volume flow rate per hour:

$$Q = 3,000,000 \text{ barrels/day} \times 5.6146 \text{ ft}^3/\text{barrel}$$
$$= 16,843,800 \text{ ft}^3/\text{day}$$
$$= 701,825 \text{ ft}^3/\text{hour.}$$

- Pressure drop:

$$\Delta P = \frac{128 Q \mu L}{g \pi D^4}$$
$$= \frac{128 \times 701,825 \times 8,500 \times 3,168,000}{32.3 \times 12,960,000 \times 3.14159 D^4}$$
$$= \frac{1,845,153,595}{D^4} \text{ lb/ft}^2.$$

- Pumping power required to boost the pressure drop:

$$\text{Power} = \frac{1.333 Q \Delta P}{1,980,000}$$
$$= \frac{1.333 \times 701,825 \times \dfrac{1,845,153,595}{D^4}}{1,980,000}$$
$$= \frac{871,818,975}{D^4} \text{ HP.}$$

- Power cost to pump oil:

$$\text{Power cost} = \frac{871,818,975}{D^4}\text{HP} \times \$0.015/\text{HP. hour}$$

$$\times 24 \text{ hours/day} \times 365 \text{ days/year}$$

$$= \frac{\$114,557,013,315}{D^4}/\text{year.}$$

Step 3: Pump and motor cost calculation:

Once we determine the required pumping power, we can find the size of the pump and the motor costs. This is because the capacity of the pump and motor is proportional to the required pumping power. Thus,

$$\text{Pump and motor cost} = \frac{871,818,975}{D^4} \times \$195/\text{HP}$$

$$= \frac{\$170,004,700,125}{D^4}.$$

Step 4: Required amount and cost of steel:

The pumping cost will be counterbalanced by the lower costs for the smaller pipe, valves, and fittings. If the pipe diameter is made larger, the fluid velocity drops markedly and the pumping costs become substantially lower. Conversely, the capital cost for the larger pipe, fittings, and valves becomes greater. For a given cross-sectional area of the pipe, we can determine the total volume of the pipe as well as the weight. Once the total weight of the pipe is determined, we can easily convert it into the required investment cost. The calculations are as follows:

$$\text{Cross-sectional area} = 3.14159[(0.51D)^2 - (0.50D)^2]$$

$$= 0.032D^2 \text{ ft}^2,$$

$$\text{Total volume of pipe} = 0.032D^2 \text{ ft}^2 \times 3,168,000 \text{ ft}$$

$$= 101,376D^2 \text{ ft}^3,$$

$$\text{Total weight of steel} = 101,376D^2 \text{ ft}^3 \times 490.75 \text{ lb/ft}^3$$

$$= 49,750,272D^2 \text{ lb,}$$

$$\text{Total pipeline cost} = \$1.00/\text{lb} \times 49,750,272D^2 \text{ lb}$$

$$= \$49,750,272D^2.$$

Step 5: Annual equivalent cost calculation:

For a given total pipeline cost and its salvage value at the end of 20 years of service life, we can find the equivalent annual capital cost by using the formula for the capital recovery factor with return:

$$\text{Capital cost} = \left(\$49,750,272D^2 + \frac{\$170,004,700,125}{D^4} \right)(A/P, 10\%, 20)$$

$$= 5,843,648D^2 + \frac{19,968,752,076}{D^4}$$

$$\text{Annual power cost} = \frac{\$114,557,013,315}{D^4}.$$

Step 6: Economical pipe size:

Now that we have determined the annual pumping and motor costs and the equivalent annual capital cost, we can express the total equivalent annual cost as a fraction of the pipe diameter (D):

$$\text{AEC}(10\%) = 5,843,648D^2 + \frac{19,968,752,076}{D^4} + \frac{114,557,013,315}{D^4}.$$

To find the optimal pipe size (D) that results in the minimum annual equivalent cost, we take the first derivative of AEC(10%) with respect to D, equate the result to zero, and solve for D:

$$\frac{d\text{AEC}(10\%)}{dD} = 11,687,297D - \frac{538,103,061,567}{D^5}$$

$$= 0;$$

$$11,687,297D^6 = 538,103,061,567,$$

$$D^6 = 46,041.70,$$

$$D^* = 5.9868 \text{ ft.}$$

Note that, ideally, the velocity in a pipe should be no more than approximately 10 ft/sec, because friction wears the pipe. To check whether the preceding answer is reasonable, we may compute

$$Q = \text{velocity} \times \text{pipe inner area,}$$

$$701,825 \text{ ft}^3/\text{hr} \times \frac{1}{3,600}\text{hr/sec} = V\frac{3.14159(5.9868)^2}{4},$$

$$V = 6.93 \text{ ft/sec,}$$

which is less than 10 ft/sec. Therefore, the optimal answer as calculated is practical.

Step 7: Equivalent annual cost at optimal pipe size:

$$\text{Capital cost} = \left[\$49{,}750{,}272(5.9868)^2 + \frac{\$170{,}004{,}700{,}125}{5.9868^4} \right](A/P, 10\%, 20)$$

$$= 5{,}843{,}648(5.9868)^2 + \frac{19{,}968{,}752{,}076}{5.9868^4}$$

$$= \$224{,}991{,}039;$$

$$\text{Annual power cost} = \frac{114{,}557{,}013{,}315}{5.9868^4}$$

$$= \$89{,}174{,}911;$$

$$\text{Total annual cost} = \$224{,}991{,}039 + \$89{,}174{,}911$$

$$= \$314{,}165{,}950.$$

Step 8: Total annual oil revenue:

$$\text{Annual oil revenue} = \$50/\text{bbl} \times 3{,}000{,}000 \text{ bbl/day}$$

$$= \times 365 \text{ day/year}$$

$$= \$54{,}750{,}000{,}000/\text{year}.$$

Enough revenues are available to offset the capital cost as well as the operating cost.

COMMENTS: A number of other criteria exist for choosing the pipe size for a particular fluid transfer application. For example, low velocity may be required when erosion or corrosion concerns must be considered. Alternatively, higher velocities may be desirable for slurries when settling is a concern. Ease of construction may also weigh significantly in choosing a pipe size. On the one hand, a small pipe size may not accommodate the head and flow requirements efficiently; on the other, space limitations may prohibit the selection of large pipe sizes.

SUMMARY

■ Annual equivalent worth analysis, or AE, is—along with present-worth analysis—one of the two main analysis techniques based on the concept of equivalence. The equation for AE is

$$\text{AE}(i) = \text{PW}(i)(A/P, i, N).$$

AE analysis yields the same decision result as PW analysis.

- The capital recovery cost factor, or CR(i), is one of the most important applications of AE analysis, in that it allows managers to calculate an annual equivalent cost of capital for ease of itemization with annual operating costs. The equation for CR(i) is

$$CR(i) = (I - S)(A/P, i, N) + iS,$$

where I = initial cost and S = salvage value.

- AE analysis is recommended over NPW analysis in many key real-world situations for the following reasons:

   1. In many financial reports, an annual equivalent value is preferred to a present-worth value.

   2. Unit costs often must be calculated to determine reasonable pricing for items that are on sale.

   3. The cost per unit of use of an item must be calculated in order to reimburse employees for the business use of personal cars.

   4. Make-or-buy decisions usually require the development of unit costs for the various alternatives.

   5. Minimum-cost analysis is easy to do when it is based on annual equivalent cost.

- LCCA is a way to predict the most cost-effective solution by allowing engineers to make a reasonable comparison among alternatives within the limit of the available data.

PROBLEMS

Note: *Unless otherwise stated, all cash flows given in the problems that follow represent after-tax cash flows.*

Annual Equivalent-Worth Calculation

6.1   Consider the following cash flows and compute the equivalent annual worth at $i = 10\%$:

n	A_n Investment	Revenue
0	−$5,000	
1		$2,000
2		2,000
3		3,000
4		3,000
5		1,000
6	2,000	500

6.2 Consider the accompanying cash flow diagram. Compute the equivalent annual worth at $i = 12\%$.

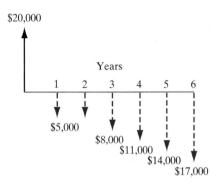

6.3 Consider the accompanying cash flow diagram. Compute the equivalent annual worth at $i = 10\%$.

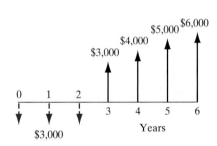

6.4 Consider the accompanying cash flow diagram. Compute the equivalent annual worth at $i = 13\%$.

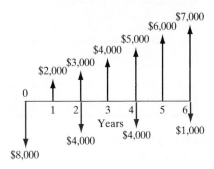

6.5 Consider the accompanying cash flow diagram. Compute the equivalent annual worth at $i = 8\%$.

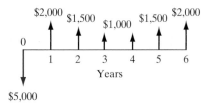

6.6 Consider the following sets of investment projects:

	Project's Cash Flow ($)			
n	A	B	C	D
0	−$2,500	−$4,500	−$8,000	−$12,000
1	400	3,000	−2,000	2,000
2	500	2,000	6,000	4,000
3	600	1,000	2,000	8,000
4	700	500	4,000	8,000
5	800	500	2,000	4,000

Compute the equivalent annual worth of each project at $i = 10\%$, and determine the acceptability of each project.

6.7 Sun-Devil Company is producing electricity directly from a solar source by using a large array of solar cells and selling the power to the local utility company. Because these cells degrade over time, thereby resulting in lower conversion efficiency and power output, the cells must be replaced every four years, which results in a particular cash flow pattern that repeats itself as follows: $n = 0$, −$500,000; $n = 1$, $600,000; $n = 2$, $400,000; $n = 3$, $300,000$, and $n = 4$, $200,000$. Determine the annual equivalent cash flows at $i = 12\%$.

6.8 Consider the following sets of investment projects:

	Project's Cash Flow			
n	A	B	C	D
0	−$7,500	−$4,000	−$5,000	−$6,600
1	0	1,500	4,000	3,800
2	0	1,800	3,000	3,800
3	15,500	2,100	2,000	3,800

Compute the equivalent annual worth of each project at $i = 13\%$, and determine the acceptability of each project.

6.9   The cash flows for a certain project are as follows:

n	Investment	Operating Income	
		Net Cash Flow	
0	−$800		1st cycle
1		$900	
2		700	
3	−800	500	2nd cycle
4		900	
5		700	
6	−800	500	3rd cycle
7		900	
8		700	
9		500	

Find the equivalent annual worth for this project at $i = 10\%$, and determine the acceptability of the project.

6.10 Beginning next year, a foundation will support an annual seminar on campus by the earnings of a $100,000 gift it received this year. It is felt that 8% interest will be realized for the first 10 years, but that plans should be made to anticipate an interest rate of 6% after that time. What amount should be added to the foundation now to fund the seminar at the $10,000 level into infinity?

Capital (Recovery) Cost/Annual Equivalent Cost

6.11 The owner of a business is considering investing $55,000 in new equipment. He estimates that the net cash flows will be $5,000 during the first year, but will increase by $2,500 per year the next year and each year thereafter. The equipment is estimated to have a 10-year service life and a net salvage value of $6,000 at that time. The firm's interest rate is 12%.
   (a) Determine the annual capital cost (ownership cost) for the equipment.
   (b) Determine the equivalent annual savings (revenues).
   (c) Determine whether this is a wise investment.

6.12 You are considering purchasing a dump truck. The truck will cost $45,000 and have an operating and maintenance cost that starts at $15,000 the first year and increases by $2,000 per year. Assume that the salvage value at the end of five years is $9,000 and interest rate is 12%. What is the equivalent annual cost of owning and operating the truck?

6.13 Emerson Electronics Company just purchased a soldering machine to be used in its assembly cell for flexible disk drives. The soldering machine cost $250,000. Because of the specialized function it performs, its useful life is estimated to be five years. It is also estimated that at that time its salvage value will be $40,000. What is the capital cost for this investment if the firm's interest rate is 18%?

6.14 The present price (year 0) of kerosene is $1.80 per gallon, and its cost is expected to increase by $.15 per year. (At the end of year 1, kerosene will cost $1.95 per gallon.) Mr. Garcia uses about 800 gallons of kerosene for space heating during a winter season. He has an opportunity to buy a storage tank for $600, and at the end of four years he can sell the storage tank for $100. The tank has a capacity to supply four years of Mr. Garcia's heating needs, so he can buy four years' worth of kerosene at its present price ($1.80), or he can invest his money elsewhere at 8%. Should he purchase the storage tank? Assume that kerosene purchased on a pay-as-you-go basis is paid for at the end of the year. (However, kerosene purchased for the storage tank is purchased now.)

6.15 Consider the following advertisement, which appeared in a local paper.

Pools-Spas-Hot Tubs—Pure Water without Toxic Chemicals: The comparative costs between the conventional chemical system (chlorine) and the IONETICS systems are as follows:

Item	Conventional System	IONETICS System
Annual costs		
Chemical	$471	
IONETICS		$ 85
Pump	$576	$ 100
($0.667/kWh)		
Capital investment		$1,200

Note that the IONETICS system pays for itself in less than 2 years.

Assume that the IONETICS system has a 12-year service life and the interest rate is 6%. What is the equivalent annual cost of operating the IONETICS system?

6.16 The cash flows for two investment projects are as follows:

	Project's Cash Flow	
n	A	B
0	-$4,000	$5,500
1	1,000	-1,400
2	X	-1,400
3	1,000	-1,400
4	1,000	-1,400

(a) For project A, find the value of X that makes the equivalent annual receipts equal the equivalent annual disbursement at $i = 13\%$.

(b) Would you accept project B at $i = 15\%$, based on an AE criterion?

6.17 An industrial firm can purchase a special machine for $50,000. A down payment of $5,000 is required, and the unpaid balance can be paid off in five equal year-end installments at 7% interest. As an alternative, the machine can be purchased for $46,000 in cash. If the firm's MARR is 10%, use the annual equivalent method to determine which alternative should be accepted.

6.18 An industrial firm is considering purchasing several programmable controllers and automating the company's manufacturing operations. It is estimated that the equipment will initially cost $100,000 and the labor to install it will cost $35,000. A service contract to maintain the equipment will cost $5,000 per year. Trained service personnel will have to be hired at an annual salary of $30,000. Also estimated is an approximate $10,000 annual income-tax savings (cash inflow). How much will this investment in equipment and services have to increase the annual revenues after taxes in order to break even? The equipment is estimated to have an operating life of 10 years, with no salvage value because of obsolescence. The firm's MARR is 10%.

6.19 A construction firm is considering establishing an engineering computing center. The center will be equipped with three engineering workstations that cost $35,000 each, and each has a service life of five years. The expected salvage value of each workstation is $2,000. The annual operating and maintenance cost would be $15,000 for each workstation. At a MARR of 15%, determine the equivalent annual cost for operating the engineering center.

Unit-Cost Profit Calculation

6.20 You have purchased a machine costing $20,000. The machine will be used for two years, at the end of which time its salvage value is expected to be $10,000. The machine will be used 6,000 hours during the first year and 8,000 hours during the second year. The expected annual net savings will be $30,000 during the first year and $40,000 during the second year. If your interest rate is 10%, what would be the equivalent net savings per machine hour?

6.21 The engineering department of a large firm is overly crowded. In many cases, several engineers share one office. It is evident that the distraction caused by the crowded conditions reduces the productive capacity of the engineers considerably. Management is considering the possibility of providing new facilities for the department, which could result in fewer engineers per office and a private office for some. For an office presently occupied by five engineers, what minimum individual increase in effectiveness must result to warrant the assignment of only three engineers to an office if the following data apply?

- The office size is 16 × 20 feet.
- The average annual salary of each engineer is $80,000.
- The cost of the building is $110 per square foot.
- The estimated life of the building is 25 years.
- The estimated salvage value of the building is 10% of the initial cost.
- The annual taxes, insurance, and maintenance are 6% of the initial cost.
- The cost of janitorial service, heating, and illumination is $5.00 per square foot per year.
- The interest rate is 12%.

Assume that engineers reassigned to other office space will maintain their present productive capability as a minimum.

6.22 Sam Tucker is a sales engineer at Buford Chemical Engineering Company. Sam owns two vehicles, and one of them is entirely dedicated to business use. His business car is a used small pickup truck, which he purchased with $11,000 of personal savings. On the basis of his own records and with data compiled by the U.S. Department of Transportation, Sam has estimated the costs of owning and operating his business vehicle for the first three years as follows:

	First Year	Second Year	Third Year
Depreciation	$2,879	$1,776	$1,545
Scheduled maintenance	100	153	220
Insurance	635	635	635
Registration and taxes	78	57	50
Total ownership cost	$3,692	$2,621	$2,450
Nonscheduled repairs	35	85	200
Replacement tires	35	30	27
Accessories	15	13	12
Gasoline and taxes	688	650	522
Oil	80	100	100
Parking and tolls	135	125	110
Total operating costs	$988	$1,003	$971
Total of all costs	$4,680	$3,624	$3,421
Expected miles driven	14,500	13,000	11,500

If his interest rate is 6%, what should be Sam's reimbursement rate per mile so that he can break even?

6.23 Two 150-horsepower (HP) motors are being considered for installation at a municipal sewage-treatment plant. The first costs $4,500 and has an operating efficiency of 83%. The second costs $3,600 and has an efficiency of 80%. Both motors are projected to have zero salvage value after a life of 10 years. If all the annual charges, such as insurance, maintenance, etc., amount to a total of 15% of the original cost of each motor, and if power costs are a flat 5 cents per kilowatt-hour, how many minimum hours of full-load operation per year are necessary to justify purchasing the more expensive motor at $i = 6\%$? (A conversion factor you might find useful is 1 HP $= 746$ watts $= 0.746$ kilowatts.)

6.24 Danford Company, a manufacturer of farm equipment, currently produces 20,000 units of gas filters per year for use in its lawn-mower production. The costs, based on the previous year's production, are reported in Table P6.24.

It is anticipated that gas-filter production will last five years. If the company continues to produce the product in-house, annual direct material costs will increase at the rate of 5%. (For example, annual material costs during the first production

TABLE P6.24 Production costs

Item	Expense ($)
Direct materials	$ 60,000
Direct labor	180,000
Variable overhead (power and water)	135,000
Fixed overhead (light and heat)	70,000
Total cost	$445,000

year will be $63,000.) Direct labor will also increase at the rate of 6% per year. However, variable overhead costs will increase at the rate of 3%, but the fixed overhead will remain at its current level over the next five years. John Holland Company has offered to sell Danford 20,000 units of gas filters for $25 per unit. If Danford accepts the offer, some of the manufacturing facilities currently used to manufacture the filter could be rented to a third party for $35,000 per year. In addition, $3.5 per unit of the fixed overheard applied to the production of gas filters would be eliminated. The firm's interest rate is known to be 15%. What is the unit cost of buying the gas filter from the outside source? Should Danford accept John Holland's offer, and why?

6.25 Southern Environmental Consulting (SEC), Inc., designs plans and specifications for asbestos abatement (removal) projects in public, private, and governmental buildings. Currently, SEC must conduct an air test before allowing the reoccupancy of a building from which asbestos has been removed. SEC subcontracts air-test samples to a laboratory for analysis by transmission electron microscopy (TEM). To offset the cost of TEM analysis, SEC charges its clients $100 more than the subcontractor's fee. The only expenses in this system are the costs of shipping the air-test samples to the subcontractor and the labor involved in shipping the samples. With the growth of the business, SEC is having to consider either continuing to subcontract the TEM analysis to outside companies or developing its own TEM laboratory. Because of the passage of the Asbestos Hazard Emergency Response Act (AHERA) by the U.S. Congress, SEC expects about 1,000 air-sample testings per year over eight years. The firm's MARR is known to be 15%.

- **Subcontract option.** The client is charged $400 per sample, which is $100 above the subcontracting fee of $300. Labor expenses are $1,500 per year, and shipping expenses are estimated to be $0.50 per sample.

- **TEM purchase option.** The purchase and installation cost for the TEM is $415,000. The equipment would last for eight years, at which time it should have no salvage value. The design and renovation cost is estimated to be $9,500. The client is charged $300 per sample, based on the current market price. One full-time manager and two part-time technicians are needed to operate the laboratory. Their combined annual salaries will be $50,000. Material required to operate the lab includes carbon rods, copper grids, filter equipment, and acetone. The costs of these materials are estimated at $6,000 per year. Utility costs, operating and maintenance costs, and the indirect labor needed to maintain the lab are estimated at $18,000 per year. The extra income-tax expenses would be $20,000.

(a) Determine the cost of an air-sample test by the TEM (in-house).

(b) What is the required number of air samples per year to make the two options equivalent?

6.26 A company is currently paying its employees $0.38 per mile to drive their own cars on company business. The company is considering supplying employees with cars, which would involve purchasing at $25,000, with an estimated three-year life, a net salvage value of $8,000, taxes and insurance at a cost of $900 per year, and operating and maintenance expenses of $0.22 per mile. If the interest rate is 10% and the company anticipates an employee's annual travel to be 22,000 miles, what is the equivalent cost per mile (neglecting income taxes)?

6.27 An automobile that runs on electricity can be purchased for $25,000. The automobile is estimated to have a life of 12 years with annual travel of 20,000 miles. Every 3 years, a new set of batteries will have to be purchased at a cost of $3,000. Annual maintenance of the vehicle is estimated to cost $700. The cost of recharging the batteries is estimated at $0.015 per mile. The salvage value of the batteries and the vehicle at the end of 12 years is estimated to be $2,000. Suppose the MARR is 7%. What is the cost per mile to own and operate this vehicle, based on the preceding estimates? The $3,000 cost of the batteries is a net value, with the old batteries traded in for the new ones.

6.28 The estimated cost of a completely installed and ready-to-operate 40-kilowatt generator is $30,000. Its annual maintenance costs are estimated at $500. The energy that can be generated annually at full load is estimated to be 100,000 kilowatt-hours. If the value of the energy generated is $0.08 per kilowatt-hour, how long will it take before this machine becomes profitable? Take the MARR to be 9% and the salvage value of the machine to be $2,000 at the end of its estimated life of 15 years.

6.29 A large land-grant university that is currently facing severe parking problems on its campus is considering constructing parking decks off campus. A shuttle service could pick up students at the off-campus parking deck and transport them to various locations on campus. The university would charge a small fee for each shuttle ride, and the students could be quickly and economically transported to their classes. The funds raised by the shuttle would be used to pay for trolleys, which cost about $150,000 each. Each trolley has a 12-year service life, with an estimated salvage value of $3,000. To operate each trolley, the following additional expenses will be incurred:

Item	Annual Expenses ($)
Driver	$50,000
Maintenance	10,000
Insurance	3,000

If students pay 10 cents for each ride, determine the annual ridership per trolley (number of shuttle rides per year) required to justify the shuttle project, assuming an interest rate of 6%.

6.30 Eradicator Food Prep, Inc., has invested $7 million to construct a food irradiation plant. This technology destroys organisms that cause spoilage and disease, thus extending the shelf life of fresh foods and the distances over which they can be shipped. The plant can handle about 200,000 pounds of produce in an hour, and it will be operated for 3,600 hours a year. The net expected operating and maintenance costs (taking into account income-tax effects) would be $4 million per year. The plant is expected to have a useful life of 15 years, with a net salvage value of $700,000. The firm's interest rate is 15%.

(a) If investors in the company want to recover the plant investment within 6 years of operation (rather than 15 years), what would be the equivalent after-tax annual revenues that must be generated?

(b) To generate annual revenues determined in part (a), what minimum processing fee per pound should the company charge to its producers?

6.31 The local government of Santa Catalina Island, off the coast of Long Beach, California, is completing plans to build a desalination plant to help ease a critical drought on the island. The drought has combined with new construction on Catalina to leave the island with an urgent need for a new water source. A modern desalination plant could produce fresh water from seawater for $1,000 an acre-foot (326,000 gallons), or enough to supply two households for 1 year. On Catalina, the cost of acquiring water from natural sources is about the same as that for desalting. The $3 million plant, with a daily desalting capacity of 0.4 acre-foot, can produce 132,000 gallons of fresh water a day (enough to supply 295 households daily), more than a quarter of the island's total needs. The desalination plant has an estimated service life of 20 years, with no appreciable salvage value. The annual operating and maintenance costs would be about $250,000. Assuming an interest rate of 10%, what should be the minimum monthly water bill for each household?

6.32 A California utility firm is considering building a 50-megawatt geothermal plant that generates electricity from naturally occurring underground heat. The binary geothermal system will cost $85 million to build and $6 million (including any income-tax effect) to operate per year. (Virtually no fuel costs will accrue compared with fuel costs related to a conventional fossil-fuel plant.) The geothermal plant is to last for 25 years. At that time, its expected salvage value will be about the same as the cost to remove the plant. The plant will be in operation for 70% (plant utilization factor) of the year (or 70% of 8,760 hours per year). If the firm's MARR is 14% per year, determine the cost per kilowatt-hour of generating electricity.

6.33 A corporate executive jet with a seating capacity of 20 has the following cost factors:

Item	Cost
Initial cost	$12,000,000
Service life	15 years
Salvage value	$2,000,000
Crew costs per year	$225,000
Fuel cost per mile	$1.10
Landing fee	$250
Maintenance per year	$237,500
Insurance cost per year	$166,000
Catering per passenger trip	$75

The company flies three round trips from Boston to London per week, a distance of 3,280 miles one way. How many passengers must be carried on an average trip in order to justify the use of the jet if the first-class round-trip fare is $3,400? The firm's MARR is 15%. (Ignore income-tax consequences.)

Comparing Mutually Exclusive Alternatives by the AE Method

6.34 The following cash flows represent the potential annual savings associated with two different types of production processes, each of which requires an investment of $12,000:

n	Process A	Process B
0	−$12,000	−$12,000
1	9,120	6,350
2	6,840	6,350
3	4,560	6,350
4	2,280	6,350

Assuming an interest rate of 15%,

(a) Determine the equivalent annual savings for each process.

(b) Determine the hourly savings for each process if it is in operation 2,000 hours per year.

(c) Which process should be selected?

6.35 Birmingham Steel, Inc., is considering replacing 20 conventional 25-HP, 230-V, 60-Hz, 1800-rpm induction motors in its plant with modern premium efficiency (PE) motors. Both types of motors have a power output of 18.650 kW per motor (25 HP × 0.746 kW/HP). Conventional motors have a published efficiency of 89.5%, while the PE motors are 93% efficient. The initial cost of the conventional motors is $13,000, whereas the initial cost of the proposed PE motors is $15,600. The motors are operated 12 hours per day, five days per week, 52 weeks per year, with a local utility cost of $0.07 per kilowatt-hour (kWh). The motors are to be operated at 75% load, and the life cycle of both the conventional and PE motor is 20 years, with no appreciable salvage value.

(a) At an interest rate of 13%, what are the savings per kWh achieved by switching from the conventional motors to the PE motors?

(b) At what operating hours are the two motors equally economical?

6.36 A certain factory building has an old lighting system, and lighting the building costs, on average, $20,000 a year. A lighting consultant tells the factory supervisor that the lighting bill can be reduced to $8,000 a year if $50,000 were invested in relighting the building. If the new lighting system is installed, an incremental maintenance cost of $3,000 per year must be taken into account. The new lighting system has zero salvage value at the end of its life. If the old lighting system also has zero salvage value, and the new lighting system is estimated to have a life of 20 years, what is the net annual benefit for this investment in new lighting? Take the MARR to be 12%.

6.37 Travis Wenzel has $2,000 to invest. Usually, he would deposit the money in his savings account, which earns 6% interest compounded monthly. However, he is considering three alternative investment opportunities:

- **Option 1.** Purchasing a bond for $2,000. The bond has a face value of $2,000 and pays $100 every six months for three years, after which time the bond matures.
- **Option 2.** Buying and holding a stock that grows 11% per year for three years.
- **Option 3.** Making a personal loan of $2,000 to a friend and receiving $150 per year for three years.

Determine the equivalent annual cash flows for each option, and select the best option.

6.38 A chemical company is considering two types of incinerators to burn solid waste generated by a chemical operation. Both incinerators have a burning capacity of 20 tons per day. The following data have been compiled for comparison:

	Incinerator A	Incinerator B
Installed cost	$1,200,000	$750,000
Annual O&M costs	$ 50,000	$ 80,000
Service life	20 years	10 years
Salvage value	$ 60,000	$ 30,000
Income taxes	$ 40,000	$ 30,000

If the firm's MARR is known to be 13%, determine the processing cost per ton of solid waste incurred by each incinerator. Assume that incinerator B will be available in the future at the same cost.

6.39 Consider the cash flows for the following investment projects (MARR = 15%):

	Project's Cash Flow		
n	A	B	C
0	− $2,500	− $4,000	− $5,000
1	1,000	1,600	1,800
2	1,800	1,500	1,800
3	1,000	1,500	2,000
4	400	1,500	2,000

(a) Suppose that projects A and B are mutually exclusive. Which project would you select, based on the AE criterion?

(b) Assume that projects B and C are mutually exclusive. Which project would you select, based on the AE criterion?

Life-Cycle Cost Analysis

6.40 An airline is considering two types of engine systems for use in its planes. Each has the same life and the same maintenance and repair record.

- **System A** costs $100,000 and uses 40,000 gallons per 1,000 hours of operation at the average load encountered in passenger service.
- **System B** costs $200,000 and uses 32,000 gallons per 1,000 hours of operation at the same level.

Both engine systems have three-year lives before any major overhaul is required. On the basis of the initial investment, the systems have 10% salvage values. If jet fuel currently costs $2.10 a gallon and fuel consumption is expected to increase at the rate of 6% per year because of degrading engine efficiency, which engine system should the firm install? Assume 2,000 hours of operation per year and a MARR of 10%. Use the AE criterion. What is the equivalent operating cost per hour for each engine?

6.41 Mustang Auto-Parts, Inc., is considering one of two forklift trucks for its assembly plant. Truck A costs $15,000 and requires $3,000 annually in operating expenses. It will have a $5,000 salvage value at the end of its three-year service life. Truck B costs $20,000, but requires only $2,000 annually in operating expenses; its service life is four years, at which time its expected salvage value will be $8,000. The firm's MARR is 12%. Assuming that the trucks are needed for 12 years and that no significant changes are expected in the future price and functional capacity of each truck, select the most economical truck, on the basis of AE analysis.

6.42 A small manufacturing firm is considering purchasing a new machine to modernize one of its current production lines. Two types of machines are available on the market. The lives of machine A and machine B are four years and six years, respectively, but the firm does not expect to need the service of either machine for more than five years. The machines have the following expected receipts and disbursements:

Item	Machine A	Machine B
First cost	$6,500	$8,500
Service life	4 years	6 years
Estimated salvage value	$600	$1,000
Annual O&M costs	$800	$520
Change oil filter every other year	$100	None
Engine overhaul	$200 (every 3 years)	$280 (every 4 years)

The firm always has another option: leasing a machine at $3,000 per year, fully maintained by the leasing company. After four years of use, the salvage value for machine B will remain at $1,000.

(a) How many decision alternatives are there?

(b) Which decision appears to be the best at $i = 10\%$?

6.43 A plastic-manufacturing company owns and operates a polypropylene production facility that converts the propylene from one of its cracking facilities to polypropylene plastics for outside sale. The polypropylene production facility is currently forced to operate at less than capacity due to an insufficiency of propylene production capacity in its hydrocarbon cracking facility. The chemical engineers are considering alternatives for supplying additional propylene to the polypropylene production facility. Two feasible alternatives are to build a pipeline to the nearest outside supply source and to provide additional propylene by truck from an outside source. The engineers also gathered the following projected cost estimates:

- Future costs for purchased propylene excluding delivery: $0.215 per lb.
- Cost of pipeline construction: $200,000 per pipeline mile.
- Estimated length of pipeline: 180 miles.
- Transportation costs by tank truck: $0.05 per lb, utilizing a common carrier.
- Pipeline operating costs: $0.005 per lb, excluding capital costs.
- Projected additional propylene needs: 180 million lb per year.
- Projected project life: 20 years.
- Estimated salvage value of the pipeline: 8% of the installed costs.

Determine the propylene cost per pound under each option if the firm's MARR is 18%. Which option is more economical?

6.44 The City of Prattsville is comparing two plans for supplying water to a newly developed subdivision:

- **Plan A** will take care of requirements for the next 15 years, at the end of which time the initial cost of $400,000 will have to be duplicated to meet the requirements of subsequent years. The facilities installed at dates 0 and 15 may be considered permanent; however, certain supporting equipment will have to be replaced every 30 years from the installation dates, at a cost of $75,000. Operating costs are $31,000 a year for the first 15 years and $62,000 thereafter, although they are expected to increase by $1,000 a year beginning in the 21st year.

- **Plan B** will supply all requirements for water indefinitely into the future, although it will be operated only at half capacity for the first 15 years. Annual costs over this period will be $35,000 and will increase to $55,000 beginning in the 16th year. The initial cost of Plan B is $550,000; the facilities can be considered permanent, although it will be necessary to replace $150,000 worth of equipment every 30 years after the initial installation.

The city will charge the subdivision for the use of water on the basis of the equivalent annual cost. At an interest rate of 10%, determine the equivalent annual cost for each plan, and make a recommendation to the city.

Minimum-Cost Analysis

6.45 A continuous electric current of 2,000 amps is to be transmitted from a generator to a transformer located 200 feet away. A copper conductor can be installed for $6 per pound, will have an estimated life of 25 years, and can be salvaged for $1 per pound. Power loss from the conductor will be inversely proportional to the cross-sectional area of the conductor and may be expressed as $6.516/A$ kilowatt, where

A is in square inches. The cost of energy is $0.0825 per kilowatt-hour, the interest rate is 11%, and the density of copper is 555 pounds per cubic foot.

(a) Calculate the optimum cross-sectional area of the conductor.

(b) Calculate the annual equivalent total cost for the value you obtained in part (a).

(c) Graph the two individual cost factors (capital cost and power-loss cost) and the total cost as a function of the cross-sectional area A, and discuss the impact of increasing energy cost on the optimum obtained in part (a).

Short Case Studies

ST6.1 Automotive engineers at Ford are considering the laser blank welding (LBW) technique to produce a windshield frame rail blank. The engineers believe that, compared with the conventional sheet metal blanks, LBW would result in a significant savings as follows:

1. Scrap reduction through more efficient blank nesting on coil.

2. Scrap reclamation (weld scrap offal into a larger usable blank).

The use of a laser welded blank provides a reduction in engineered scrap for the production of a window frame rail blank.

On the basis of an annual volume of 3,000 blanks, Ford engineers have estimated the following financial data:

| | Blanking Method | |
| | Conventional | Laser Blank Welding |
Description		
Weight per blank (lb/part)	63.764	34.870
Steel cost/part	$ 14.98	$ 8.19
Transportation/part	$ 0.67	$ 0.42
Blanking/part	$ 0.50	$ 0.40
Die investment	$106,480	$83,000

The LBW technique appears to achieve significant savings, so Ford's engineers are leaning toward adopting it. Since the engineers have had no previous experience with LBW, they are not sure whether producing the windshield frames in-house at this time is a good strategy. For this windshield frame, it may be cheaper to use the services of a supplier that has both the experience with, and the machinery for, laser blanking. Ford's lack of skill in laser blanking may mean that it will take six months to get up to the required production volume. If, however, Ford relies on a supplier, it can only assume that supplier labor problems will not halt the production of Ford's parts. The make-or-buy decision depends on two factors: the amount of new investment that is required in laser welding and whether additional machinery will be required for future products. Assuming a lifetime of 10 years and an interest rate of 16%, recommend the best course of action. Assume also that the salvage value at the end of 10 years is estimated to be insignificant for either system. If Ford considers the subcontracting option, what would be the acceptable range of contract bid (unit cost per part)?

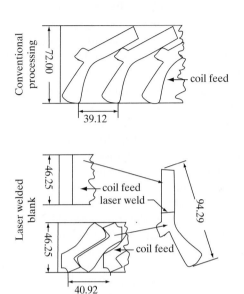

ST6.2 The proliferation of computers into all aspects of business has created an ever-increasing need for data capture systems that are fast, reliable, and cost effective. One technology that has been adopted by many manufacturers, distributors, and retailers is a bar-coding system. Hermes Electronics, a leading manufacturer of underwater surveillance equipment, evaluated the economic benefits of installing an automated data acquisition system into its plant. The company could use the system on a lim-ited scale, such as for tracking parts and assemblies for inventory management, or it could opt for a broader implementation by recording information that is useful for quality control, operator efficiency, attendance, and other functions. All these aspects are currently monitored, but although computers are used to manage the information, the recording is conducted primarily manually. The advantages of an automated data collection system, which include faster and more accurate data capture, quicker analysis of and response to production changes, and savings due to tighter control over operations, could easily outweigh the cost of the new system. Two alternative systems from competing suppliers are under consideration:

- **System 1** relies on handheld bar-code scanners that transmit radio frequencies. The hub of this wireless network can then be connected to the company's existing LAN and integrated with its current MRP II system and other management software.

- **System 2** consists primarily of specialized data terminals installed at every collection point, with connected bar-code scanners where required. This system is configured in such a way as to facilitate phasing in the components over two stages or installing the system all at once. The former would allow Hermes to defer some of the capital investment, while becoming thoroughly familiar with the functions introduced in the first stage.

Either of these systems would satisfy Hermes's data collection needs. They each have some unique elements, and the company needs to compare the relative benefits of the features offered by each system. From the point of view of engineering economics, the two systems have different capital costs, and their operating and maintenance costs are not identical. One system may also be rated to last longer than the other before replacement is required, particularly if the option of acquiring System 2 in phases is selected. Discuss many issues to be considered before making the best choice of those types of technology in manufacturing.

ST6.3 A Veterans Administration (VA) hospital is to decide which type of boiler fuel system will most efficiently provide the required steam energy output for heating, laundry, and sterilization purposes. The current boilers were installed in the early 1950s and are now obsolete. Much of the auxiliary equipment is also old and in need of repair. Because of these general conditions, an engineering recommendation was made to replace the entire plant with a new boiler plant building that would house modern equipment. The cost of demolishing the old boiler plant would be almost a complete loss, as the salvage value of the scrap steel and used brick was estimated to be only about $1,000. The VA hospital's engineer finally selected two alternative proposals as being worthy of more intensive analysis. The hospital's annual energy requirement, measured in terms of steam output, is approximately 145,000,000 pounds of steam. As a rule of thumb for analysis, 1 pound of steam is approximately 1,000 Btu, and 1 cubic foot of natural gas is also approximately 1,000 Btu. The two alternatives are as follows:

- **Proposal 1.** Replace the old plant with a new coal-fired boiler plant that costs $1,770,300. To meet the requirements for particulate emission as set by the Environmental Protection Agency, this coal-fired boiler, even if it burned low-sulfur coal, would need an electrostatic precipitator, which would cost approximately $100,000. The plant would last for 20 years. One pound of dry coal yields about 14,300 Btu. To convert the 145,000,000 pounds of steam energy to the common denominator of Btu, it is necessary to multiply by 1,000. To find the Btu input requirements, it is necessary to divide by the relative boiler efficiency for the type of fuel. The boiler efficiency for coal is 0.75. The price of coal is estimated to be $55.50 per ton.

- **Proposal 2.** Build a gas-fired boiler plant with No. 2 fuel oil, and use the new plant as a standby. This system would cost $889,200 and have an expected service life of 20 years. Since small household or commercial gas users that are entirely dependent on gas have priority, large plants must have an oil switchover capability. It has been estimated that 6% of 145,000,000 pounds of steam energy (or 8,700,000 pounds) would come about as a result of the switch to oil. The boiler efficiency with each fuel would be 0.78 for gas and 0.81 for oil, respectively. The heat value of natural gas is approximately 1,000,000 Btu/MCF (thousand cubic feet), and for No. 2 fuel oil it is 139,400 Btu/gal. The estimated gas price is $9.50/MCF, and the price of No. 2 fuel oil is $1.35 per gallon.

(a) Calculate the annual fuel costs for each proposal.

(b) Determine the unit cost per steam pound for each proposal. Assume that $i = 10\%$.

(c) Which proposal is the more economical?

ST6.4 The following is a letter that I received from a local city engineer:

Dear Professor Park:

Thank you for taking the time to assist with this problem. I'm really embarrassed at not being able to solve it myself, since it seems rather straightforward. The situation is as follows:

A citizen of Opelika paid for concrete drainage pipe approximately 20 years ago to be installed on his property. (We have a policy that if drainage trouble exists on private property and the owner agrees to pay for the material, city crews will install it.) That was the case in this problem. Therefore, we are dealing with only material costs, disregarding labor.

However, this past year, we removed the pipe purchased by the citizen, due to a larger area drainage project. He thinks, and we agree, that he is due some refund for salvage value of the pipe due to its remaining life.

Problem:

- Known: 80′ of 48″ pipe purchased 20 years ago. Current quoted price of 48″ pipe = $52.60/foot, times 80 feet = $4,208 total cost in today's dollars.
- Unknown: Original purchase price.
- Assumptions: 50-year life; therefore, assume 30 years of life remaining at removal after 20 years. A 4% price increase per year, average, over 20 years.

Thus, we wish to calculate the cost of the pipe 20 years ago. Then we will calculate, in today's dollars, the present salvage value after 20 years, use with 30 years of life remaining. Thank you again for your help. We look forward to your reply.

Charlie Thomas, P.E.
Director of Engineering
City of Opelika

After reading this letter, recommend a reasonable amount of compensation to the citizen for the replaced drainage pipe.

CHAPTER SEVEN

Rate-of-Return Analysis

Will That be Cash, Credit—or Fingertip?[1] Have you ever found yourself short of cash or without a wallet when you want to buy something? Consider the following two types of technologies available in retail stores to speed up checkouts:

- **Pay By Touch** takes fingerprints when customers enroll in the program. The image is then converted to about 40 unique points of the finger. Those points are stored in a computer system with "military-level encryption." They want this to be your cash replacement because of the time savings, and a lot of customers who are paying cash will find it more convenient now to use these cards.

- **A contactless card** allows the shopper to pay in seconds by waving his or her contactless card in front of a reader, which lights up and beeps to tell the shopper the transaction is done.

  A contactless payment is twice as fast as a no-signature credit card purchase and three times as fast as using cash. That's why it's catching on at fast-food restaurants and convenience stores.

These stores' profits depend, in part, on how quickly they get customers—typically with small purchases—through the line. These new technologies being rolled out at convenience stores, supermarkets, and gas stations could some day make it passé to carry bulky wallets. Without the need to dig for cash and checks at the register, the quick stop-and-go payments promise speedier transactions for consumers—and perhaps fatter profits for retailers. The appeal is that there's no need to run them through a machine. A contactless-card transaction is usually more expensive for a retailer to process than a cash payment. But retailers that adopt contactless payments hope they'll bring in more customers, offsetting higher costs. If that turns out to be false, then some could turn their backs on the new technology.

[1] "Will that be cash, credit—or fingertip?" Kathy Chu, *USA Today*, Section B1, Friday, December 2, 2005.

Getting Through the Checkout Line Faster
Contactless Payment

1. Hold the card 1 to 2 inches from electronic reader.

2. Microchip in card helps encrypt data and sends account number.

3. Unique transaction number generated to deter fraud.

4. No signing, no swiping. Transaction processed normally.

1. Place your finger on the electronic reader.

2. Unique points from your finger identify you.

3. Enter phone number and choose payment method.

4. No signing, no swiping. Transaction processed normally.

One retailer who just installed a Pay By Touch™ system hopes to increase its customer traffic so that a 10% return on investment can be attained. The Pay By Touch™ scanners cost about $50 each, the monthly service fee ranges between $38 and $45, and each transaction fee costs 10 cents. In a society driven by convenience, anything that speeds up the payment process attracts consumers. But technology providers will need to convince consumers of the safety of their information before the technologies can become a staple in the checkout line.

What does the 10% rate of return for the retailer really represent? How do we compute the figure from the projected additional retail revenues? And once computed, how do we use the figure when evaluating an investment project? Our consideration of the concept of rate of return in this chapter will answer these and other questions.

Along with the NPW and the AE criteria, the third primary measure of investment worth is **rate of return**. As shown in Chapter 5, the NPW measure is easy to calculate and apply. Nevertheless, many engineers and financial managers prefer rate-of-return analysis to the NPW method because they find it intuitively more appealing to analyze investments in terms of percentage rates of return rather than dollars of NPW. Consider the following statements regarding an investment's profitability:

- This project will bring in a 15% rate of return on the investment.
- This project will result in a net surplus of $10,000 in the NPW.

Neither statement describes the nature of the investment project in any complete sense. However, the rate of return is somewhat easier to understand because many of us are so familiar with savings-and-loan interest rates, which are in fact rates of return.

In this chapter, we will examine four aspects of rate-of-return analysis: (1) the concept of return on investment, (2) the calculation of a rate of return, (3) the development of an internal rate-of-return criterion, and (4) the comparison of mutually exclusive alternatives based on a rate of return.

CHAPTER LEARNING OBJECTIVES

After completing this chapter, you should understand the following concepts:

- The meaning of the rate of return.
- The various methods to compute the rate of return.
- How you make an accept and reject decision with the rate of return.
- How to resolve the multiple rates of return problem.
- How you conduct an incremental analysis with the rate of return.

7.1 Rate of Return

Yield: The annual rate of return on an investment, expressed as a percentage.

Many different terms are used to refer to **rate of return**, including **yield** (i.e., the yield to maturity, commonly used in bond valuation), **internal rate of return**, and **marginal efficiency of capital**. First we will review three common definitions of rate of return. Then we will use the definition of internal rate of return as a measure of profitability for a single investment project throughout the text.

7.1.1 Return on Investment

There are several ways of defining the concept of a rate of return on investment. The first is based on a typical loan transaction, the second on the mathematical expression of the present-worth function, and the third on the project cash flow series.

Definition 1

The rate of return is the interest rate earned on the unpaid balance of an amortized loan.

Suppose that a bank lends $10,000 and is repaid $4,021 at the end of each year for three years. How would you determine the interest rate that the bank charges on this transaction? As we learned in Chapter 3, you would set up the equivalence equation

$$\$10,000 = \$4,021(P/A, i, 3)$$

and solve for i. It turns out that $i = 10\%$. In this situation, the bank will earn a return of 10% on its investment of $10,000. The bank calculates the balances over the life of the loan as follows:

Year	Unpaid Balance at Beginning of Year	Return on Unpaid Balance (10%)	Payment Received	Unpaid Balance at End of Year
0	−$10,000	$0	$0	−$10,000
1	−10,000	−1,000	+4,021	−6,979
2	−6,979	−698	+4,021	−3,656
3	−3,656	−366	+4,021	0

A negative balance indicates an unpaid balance. In other words, the customer still owes money to the bank.

Observe that, for the repayment schedule shown, the 10% interest is calculated only on each year's outstanding balance. In this situation, only part of the $4,021 annual payment represents interest; the remainder goes toward repaying the principal. Thus, the three annual payments repay the loan itself and additionally provide a return of 10% on the *amount still outstanding each year*.

Note that when the last payment is made, the outstanding principal is eventually reduced to zero.[2] If we calculate the NPW of the loan transaction at its rate of return (10%), we see that

$$PW(10\%) = -\$10,000 + \$4,021(P/A, 10\%, 3) = 0,$$

which indicates that the bank can break even at a 10% rate of interest. In other words, the rate of return becomes the rate of interest that equates the present value of future cash repayments to the amount of the loan. This observation prompts the second definition of rate of return.

Definition 2

The rate of return is the break-even interest rate i^* that equates the present worth of a project's cash outflows to the present worth of its cash inflows, or

$$PW(i^*) = PW_{\text{Cash inflows}} - PW_{\text{Cash outflows}}$$
$$= 0.$$

Note that the expression for the NPW is equivalent to

$$PW(i^*) = \frac{A_0}{(1 + i^*)^0} + \frac{A_1}{(1 + i^*)^1} + \ldots + \frac{A_N}{(1 + i^*)^N} = 0. \qquad (7.1)$$

Here we know the value of A_n for each period, but not the value of i^*. Since it is the only unknown, however, we can solve for i^*. (Inevitably, there will be N values of i^* that satisfy this equation. In most project cash flows, you would be able to find a unique positive i^* that satisfies Eq. (7.1). However, you may encounter some cash flows that cannot be solved for a single rate of return greater than 100%. By the nature of the NPW function in

---

[2] As we learned in Section 5.3.2, this terminal balance is equivalent to the net future worth of the investment. If the net future worth of the investment is zero, its NPW should also be zero.

Eq. (7.1), it is possible to have more than one rate of return for certain types of cash flows. For some cash flows, we may not find a specific rate of return at all.)[3]

Note that the formula in Eq. (7.1) is simply the NPW formula solved for the particular interest rate $(i*)$ at which PW(i) is equal to zero. By multiplying both sides of Eq. (7.1) by $(1 + i*)^N$, we obtain

$$PW(i*)(1 + i*)^N = FW(i*) = 0.$$

If we multiply both sides of Eq. (7.1) by the capital recovery factor $(A/P, i*, N)$, we obtain the relationship $AE(i*) = 0$. Therefore, the $i*$ of a project may be defined as the rate of interest that equates the present worth, future worth, and annual equivalent worth of the entire series of cash flows to zero.

7.1.2 Return on Invested Capital

Investment projects can be viewed as analogous to bank loans. We will now introduce the concept of rate of return based on the return on invested capital in terms of a project investment. A project's return is referred to as the internal rate of return (IRR) or the **yield** promised by an **investment project** over its **useful life**.

Definition 3

The internal rate of return is the interest rate charged on the unrecovered project balance of the investment such that, when the project terminates, the unrecovered project balance will be zero.

Suppose a company invests $10,000 in a computer with a three-year useful life and equivalent annual labor savings of $4,021. Here, we may view the investing firm as the lender and the project as the borrower. The cash flow transaction between them would be identical to the amortized loan transaction described under Definition 1:

n	Beginning Project Balance	Return on Invested Capital	Ending Cash Payment	Project Balance
0	$0	$0	−$10,000	−$10,000
1	−10,000	−1,000	4,021	6,979
2	−6,979	−697	4,021	3,656
3	−3,656	−365	4,021	0

Internal rate of return: This is the return that a company would earn if it invested in itself, rather than investing that money elsewhere.

In our project balance calculation, we see that 10% is earned (or charged) on $10,000 during year 1, 10% is earned on $6,979 during year 2, and 10% is earned on $3,656 during year 3. This indicates that the firm earns a 10% rate of return on funds that remain *internally* invested in the project. Since it is a return that is *internal* to the project, we refer to it as the **internal rate of return**, or IRR. This means that the computer project under consideration brings in enough cash to pay for itself in three years and also to provide the firm with a return

---

[3] You will always have N rates of return. The issue is whether they are real or imaginary. If they are real, the question "Are they in the $(-100\%, \infty)$ interval?" should be asked. A **negative rate of return** implies that you never recover your initial investment.

of 10% on its invested capital. Put differently, if the computer is financed with funds costing 10% annually, the cash generated by the investment will be exactly sufficient to repay the principal and the annual interest charge on the fund in three years.

Notice that only one cash outflow occurs at time 0, and the present worth of this outflow is simply $10,000. There are three equal receipts, and the present worth of these inflows is $4,021(P/A, 10%, 3) = $10,000. Since the NPW = $PW_{Inflow} − PW_{Outflow}$ = $10,000 − $10,000 = 0, 10% also satisfies Definition 2 of the rate of return. Even though the preceding simple example implies that $i*$ coincides with IRR, only Definitions 1 and 3 correctly describe the true meaning of the internal rate of return. As we will see later, if the cash expenditures of an investment are not restricted to the initial period, several break-even interest rates ($i*$'s) may exist that satisfy Eq. (7.1). However, there may not be a rate of return that is *internal* to the project.

7.2 Methods for Finding the Rate of Return

We may find $i*$ by several procedures, each of which has its advantages and disadvantages. To facilitate the process of finding the rate of return for an investment project, we will first classify various types of investment cash flow.

7.2.1 Simple versus Nonsimple Investments

We can classify an investment project by counting the number of sign changes in its net cash flow sequence. A change from either "+" to "−" or "−" to "+" is counted as one sign change. (We ignore a zero cash flow.) Then,

- A **simple investment** is an investment in which the initial cash flows are negative and only one sign change occurs in the remaining net cash flow series. If the initial flows are positive and only one sign change occurs in the subsequent net cash flows, they are referred to as **simple borrowing** cash flows.
- A **nonsimple investment** is an investment in which more than one sign change occurs in the cash flow series.

Simple investment: The project with only one sign change in the net cash flow series.

Multiple $i*$'s, as we will see later, occur only in nonsimple investments. Three different types of investment possibilities are illustrated in Example 7.1.

EXAMPLE 7.1 Investment Classification

Consider the following three cash flow series and classify them into either simple or nonsimple investments:

Period	Net Cash Flow		
n	Project A	Project B	Project C
0	−$1,000	−$1,000	$1,000
1	−500	3,900	−450
2	800	−5,030	−450
3	1,500	2,145	−450
4	2,000		

SOLUTION

Given: Preceding cash flow sequences.

Find: Classify the investments shown into either simple and nonsimple investments.

- Project A represents many common simple investments. This situation reveals the NPW profile shown in Figure 7.1(a). The curve crosses the i-axis only once.
- Project B represents a nonsimple investment. The NPW profile for this investment has the shape shown in Figure 7.1(b). The i-axis is crossed at 10%, 30%, and 50%.
- Project C represents neither a simple nor a nonsimple investment, even though only one sign change occurs in the cash flow sequence. Since the first cash flow is positive, this is a **simple borrowing** cash flow, not an investment flow. Figure 7.1(c) depicts the NPW profile for this type of investment.

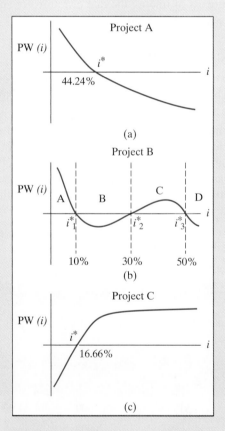

Figure 7.1 Present-worth profiles:
(a) Simple investment, (b) nonsimple investment with multiple rates of return, and (c) simple borrowing cash flows.

COMMENTS: Not all NPW profiles for nonsimple investments have multiple crossings of the *i*-axis. Clearly, then, we should place a high priority on discovering this situation early in our analysis of a project's cash flows. The quickest way to predict multiple *i**'s is to generate an NPW profile on a computer and check whether it crosses the horizontal axis more than once. In the next section, we illustrate when to expect such multiple crossings by examining types of cash flows.

7.2.2 Predicting Multiple *i**'s

As hinted at in Example 7.1, for certain series of project cash flows, we may uncover the complication of multiple *i** values that satisfy Eq. (7.1). By analyzing and classifying cash flows, we may anticipate this difficulty and adjust our analysis approach later. Here we will focus on the initial problem of whether we can predict a unique *i** for a project by examining its cash flow pattern. Two useful rules allow us to focus on sign changes (1) in net cash flows and (2) in accounting net profit (accumulated net cash flows).

Net Cash Flow Rule of Signs

One useful method for predicting an upper limit on the number of positive *i**'s of a cash flow stream is to apply the rule of signs: *The number of real i*'s that are greater than* -100% *for a project with N periods is never greater than the number of sign changes in the sequence of the A_n's. A zero cash flow is ignored.*

  An example is

Period	A_n	Sign Change
0	$-\$100$	
1	-20	
2	$+50$	1
3	0	
4	$+60$	
5	-30	1
6	$+100$	1

Three sign changes occur in the cash flow sequence, so three or fewer real positive *i**'s exist.

  It must be emphasized that the rule of signs provides an indication only of the *possibility* of multiple rates of return: The rule predicts only the *maximum* number of possible *i**'s. Many projects have multiple sign changes in their cash flow sequence, but still possess a unique real *i** in the $(-100\%, \infty)$ range.

Accumulated Cash Flow Sign Test

The accumulated cash flow is the sum of the net cash flows up to and including a given time. If the rule of cash flow signs indicates multiple $i*$'s, we should proceed to the **accumulated cash flow sign test** to eliminate some possibility of multiple rates of return.

If we let A_n represent the net cash flow in period n and S_n represent the accumulated cash flow (the accounting sum) up to period n, we have the following:

Period (n)	Cash Flow (A_n)	Accumulated Cash Flow (S_n)
0	A_0	$S_0 = A_0$
1	A_1	$S_1 = S_0 + A_1$
2	A_2	$S_2 = S_1 + A_2$
$\vdots$	$\vdots$	$\vdots$
N	A_N	$S_N = S_{N-1} + A_N$

We then examine the sequence of accumulated cash flows $(S_0, S_1, S_2, S_3, \ldots, S_N)$ to determine the number of sign changes. *If the series S_n starts negatively and changes sign only once, then a unique positive $i*$ exists.* This cumulative cash flow sign rule is a more discriminating test for identifying the uniqueness of $i*$ than the previously described method.

EXAMPLE 7.2 Predicting the Number of $i*$'s

Predict the number of real positive rates of return for each of the following cash flow series:

Period	A	B	C	D
0	−$100	−$100	$0	−$100
1	−200	+50	−50	+50
2	+200	−100	+115	0
3	+200	+60	−66	+200
4	+200	−100		−50

SOLUTION

Given: Four cash flow series and cumulative flow series.

Find: The upper limit on number of $i*$'s for each series.

The cash flow rule of signs indicates the following possibilities for the positive values of $i*$:

Project	Number of Sign Changes in Net Cash Flows	Possible Number of Positive Values of i^*
A	1	1 or 0
B	4	4, 3, 2, 1, or 0
C	2	2, 1, or 0
D	2	2, 1, or 0

For cash flows B, C, and D, we would like to apply the more discriminating cumulative cash flow test to see if we can specify a smaller number of possible values of i^*. Accordingly, we write

	Project B		Project C		Project D	
n	A_n	S_n	A_n	S_n	A_n	S_n
0	−$100	−$100	$ 0	$ 0	−$100	−$100
1	+50	−50	−50	−50	+50	−50
2	−100	−150	+115	+65	0	−50
3	+60	−90	−66	−1	+200	+150
4	−100	−190			−50	+100

Recall the test: If the series starts *negatively* and changes sign only once, a unique positive i^* exists.

- Only project D begins negatively and passes the test; therefore, we may predict a unique i^* value, rather than 2, 1, or 0 as predicted by the cash flow rule of signs. ($i_1^* = -75.16\%$ and $i_2^* = 35.05\%$)
- Project B, with no sign change in the cumulative cash flow series, has no rate of return.
- Project C fails the test, and we cannot eliminate the possibility of multiple i^*'s. ($i_1^* = 10\%$ and $i_2^* = 20\%$)

7.2.3 Computational Methods

Once we identify the type of an investment cash flow, several ways to determine its rate of return are available. Some of the most practical methods are as follows:

- Direct solution method,
- Trial-and-error method, and
- Computer solution method.

Direct Solution Method

For the special case of a project with only a two-flow transaction (an investment followed by a single future payment) or a project with a service life of two years of return, we can seek a direct mathematical solution for determining the rate of return. These two cases are examined in Example 7.3.

> ## EXAMPLE 7.3 Finding $i*$ by Direct Solution: Two Flows and Two Periods

Consider two investment projects with the following cash flow transactions:

n	Project 1	Project 2
0	−$2,000	−$2,000
1	0	1,300
2	0	1,500
3	0	
4	3,500	

Compute the rate of return for each project.

SOLUTION

Given: Cash flows for two projects.
Find: $i*$ for each project.

Project 1: Solving for $i*$ in $PW(i*) = 0$ is identical to solving $FW(i*) = 0$, because FW equals PW times a constant. We could do either here, but we will set $FW(i*) = 0$ to demonstrate the latter. Using the single-payment future-worth relationship, we obtain

$$FW(i*) = -\$2,000(F/P, i*, 4) + \$3,500 = 0,$$
$$\$3,500 = \$2,000(F/P, i*, 4) = \$2,000(1 + i*)^4,$$
$$1.75 = (1 + i*)^4.$$

Solving for $i*$ yields

$$i* = \sqrt[4]{1.75} - 1$$
$$= 0.1502 \text{ or } 15.02\%.$$

Project 2: We may write the NPW expression for this project as

$$PW(i) = -\$2,000 + \frac{\$1,300}{(1 + i)} + \frac{\$1,500}{(1 + i)^2} = 0.$$

Let $X = 1/(1 + i)$. We may then rewrite $PW(i)$ as a function of X as follows:

$$PW(X) = -\$2,000 + \$1,300X + \$1,500X^2 = 0.$$

This is a quadratic equation that has the following solution:[4]

---

[4] The solution of the quadratic equation $aX^2 + bX + c = 0$ is $X = \dfrac{-b \pm \sqrt{b^2 - 4ac}}{2a}$.

$$X = \frac{-1,300 \pm \sqrt{1,300^2 - 4(1,500)(-2,000)}}{2(1,500)}$$

$$= \frac{-1,300 \pm 3,700}{3,000}$$

$$= 0.8 \text{ or } -1.667.$$

Replacing X values and solving for i gives us

$$0.8 = \frac{1}{(1+i)} \rightarrow i = 25\%,$$

$$-1.667 = \frac{1}{(1+i)} \rightarrow i = -160\%.$$

Since an interest rate less than -100% has no economic significance, we find that the project's i^* is 25%.

COMMENTS: In both projects, one sign change occurred in the net cash flow series, so we expected a unique i^*. Also, these projects had very simple cash flows. When cash flows are more complicated, generally we use a trial-and-error method or a computer to find i^*.

Trial-and-Error Method

The first step in the trial-and-error method is to make an estimated **guess**[5] at the value of i^*. For a simple investment, we use the "guessed" interest rate to compute the present worth of net cash flows and observe whether it is positive, negative, or zero. Suppose, then, that PW(i) is negative.

Since we are aiming for a value of i that makes PW(i) = 0, we must raise the present worth of the cash flow. To do this, we lower the interest rate and repeat the process. If PW(i) is positive, however, we raise the interest rate in order to lower PW(i). The process is continued until PW(i) is approximately equal to zero. Whenever we reach the point where PW(i) is bounded by one negative and one positive value, we use **linear interpolation** to approximate i^*. This process is somewhat tedious and inefficient. (The trial-and-error method does not work for nonsimple investments in which the NPW function is not, in general, a monotonically decreasing function of the interest rate.)

EXAMPLE 7.4 Finding i^* by Trial and Error

The Imperial Chemical Company is considering purchasing a chemical analysis machine worth $13,000. Although the purchase of this machine will not produce any

---

[5] As we shall see later in this chapter, the ultimate objective of finding i^* is to compare it against the MARR. Therefore, it is a good idea to use the MARR as the initial guess.

increase in sales revenues, it will result in a reduction of labor costs. In order to operate the machine properly, it must be calibrated each year. The machine has an expected life of six years, after which it will have no salvage value. The following table summarizes the annual savings in labor cost and the annual maintenance cost in calibration over six years:

Year (n)	Costs ($)	Savings ($)	Net Cash Flow ($)
0	13,000		−13,000
1	2,300	6,000	3,700
2	2,300	7,000	4,700
3	2,300	9,000	6,700
4	2,300	9,000	6,700
5	2,300	9,000	6,700
6	2,300	9,000	6,700

Find the rate of return for this project.

SOLUTION

Given: Cash flows over six years as shown in Figure 7.2.
Find: i^*.

We start with a guessed interest rate of 25%. The present worth of the cash flows is

$$PW(25\%) = -\$13,000 + \$3,700(P/F, 25\%, 1) + \$4,700(P/F, 25\%, 2)$$
$$+ \$6,700(P/A, 25\%, 4)(P/F, 25\%, 2)$$
$$= \$3,095.$$

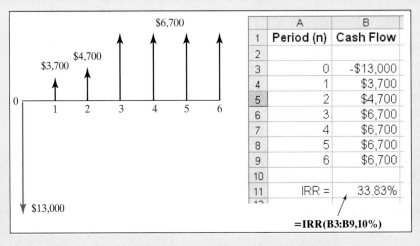

Figure 7.2 Cash flow diagram for a simple investment (Example 7.4).

Since this present worth is positive, we must raise the interest rate to bring PW toward zero. When we use an interest rate of 35%, we find that

$$PW(35\%) = -\$13,000 + \$3,700(P/F, 35\%, 1) + \$4,700(P/F, 35\%, 2)$$
$$+ \$6,700(P/A, 35\%, 4)(P/F, 35\%, 2)$$
$$= -\$339.$$

We have now bracketed the solution: PW(i) will be zero at i somewhere between 25% and 35%. Using straight-line interpolation, we approximate

$$i^* \cong 25\% + (35\% - 25\%)\left[\frac{3,095 - 0}{3,095 - (-339)}\right]$$
$$= 25\% + 10\%(0.9013)$$
$$= 34.01\%.$$

Now we will check to see how close this value is to the precise value of i^*. If we compute the present worth at this interpolated value, we obtain

$$PW(34\%) = -\$13,000 + \$3,700(P/F, 34\%, 1) + \$4,700(P/F, 34\%, 2)$$
$$+ \$6,700(P/A, 34\%, 4)(P/F, 34\%, 2)$$
$$= -\$50.58.$$

As this is not zero, we may recompute i^* at a lower interest rate, say, 33%:

$$PW(33\%) = -\$13,000 + \$3,700(P/F, 33\%, 1) + \$4,700(P/F, 33\%, 2)$$
$$+ \$6,700(P/A, 33\%, 4)(P/F, 33\%, 2)$$
$$= \$248.56.$$

With another round of linear interpolation, we approximate

$$i^* \cong 33\% + (34\% - 33\%)\left[\frac{248.56 - 0}{248.56 - (-50.58)}\right]$$
$$= 33\% + 1\%(0.8309)$$
$$= 33.83\%.$$

At this interest rate,

$$PW(33.83\%) = -\$13,000 + \$3,700(P/F, 33.83\%, 1)$$
$$+ \$4,700(P/F, 33.83\%, 2)$$
$$+ \$6,700(P/A, 33.83\%, 4)(P/F, 33.83\%, 2)$$
$$= -\$0.49,$$

which is practically zero, so we may stop here. In fact, there is no need to be more precise about these interpolations, because the final result can be no more accurate than the basic data, which ordinarily are only rough estimates.

COMMENT: With Excel, you can evaluate the IRR for the project as $=$ IRR (range,guess), where you specify the cell range for the cash flow (e.g., B3:B9) and the initial guess, such as 25%. Computing $i*$ for this problem in Excel, incidentally, gives us 33.8283%. Instead of using the factor notations, you may attempt use a tabular approach as follows:

		Internal Rate of Return: What It Looks Like			
		Discount Rate: 25%		Discount Rate: 35%	
Year	Cash Flow	Factor	Amount	Factor	Amount
0	−$13,000	1.0000	−$13,000	1.0000	−$13,000
1	3,700	0.8000	2,960	0.7407	2,741
2	4,700	0.6400	3,008	0.5487	2,579
3	6,700	0.5120	3,430	0.4064	2,723
4	6,700	0.4096	2,744	0.3011	2,017
5	6,700	0.3277	2,196	0.2230	1,494
6	6,700	0.2621	1,756	0.1652	1,107
Total	+$22,200	**NPW = +$3,095**		**NPW = −$339**	
			IRR = close to 34%		

Graphical Method

We don't need to do laborious manual calculations to find $i*$. Many financial calculators have built-in functions for calculating $i*$. It is worth noting that many online financial calculators or spreadsheet packages have $i*$ functions, which solve Eq. (7.1) very rapidly,[6] usually with the user entering the cash flows via a computer keyboard or by reading a cash flow data file. (For example, Microsoft Excel has an IRR financial function that analyzes investment cash flows, as illustrated in Example 7.4.)

The most easily generated and understandable graphic method of solving for $i*$ is to create the **NPW profile** on a computer. On the graph, the horizontal axis indicates the interest rate and the vertical axis indicates the NPW. For a given project's cash flows, the NPW is calculated at an interest rate of zero (which gives the vertical-axis intercept) and several other interest rates. Points are plotted and a curve is sketched. Since $i*$ is defined as the interest rate at which $PW(i*) = 0$, the point at which the curve crosses the horizontal axis closely approximates $i*$. The graphical approach works for both simple and nonsimple investments.

[6] An alternative method of solving for $i*$ is to use a computer-aided economic analysis program. Cash Flow Analyzer (CFA) finds $i*$ visually by specifying the lower and upper bounds of the interest search limit and generates NPW profiles when given a cash flow series. In addition to the savings in calculation time, the advantage of computer-generated profiles is their precision. CFA can be found from the book's website.

EXAMPLE 7.5 Graphical Approach to Estimate i^*

Consider the cash flow series shown in Figure 7.3(a). Estimate the rate of return by generating the NPW profile on a computer.

SOLUTION

Given: Cash flow series in Figure 7.3.

Find: (a) i^* by plotting the NPW profile and (b) i^* by using Excel.

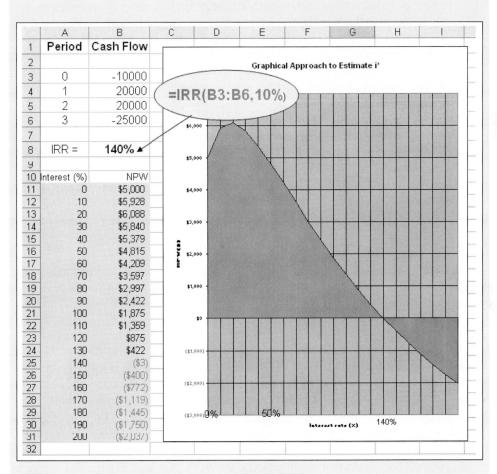

Figure 7.3 Graphical solution to rate-of-return problem for a typical nonsimple investment (Example 7.5).

(a) The present-worth function for the project cash flow series is

$$PW(i) = -\$10{,}000 + \$20{,}000(P/A, i, 2) - \$25{,}000(P/F, i, 3).$$

First we use $i = 0$ in this equation to obtain NPW $= \$5{,}000$, which is the vertical-axis intercept. Then we substitute several other interest rates—10%, 20%, ..., 140%—and plot these values of PW(i) as well. The result is Figure 7.3, which shows the curve crossing the horizontal axis at roughly 140%. This value can be verified by other methods if we desire to do so. Note that, in addition to establishing the interest rate that makes NPW $= 0$, the NPW profile indicates where positive and negative NPW values fall, thus giving us a broad picture of those interest rates for which the project is acceptable or unacceptable. (Note also that a trial-and-error method would lead to some confusion: As you increase the interest rate from 0% to 20%, the NPW value also keeps increasing, instead of decreasing.) Even though the project is a nonsimple investment, the curve crosses the horizontal axis only once. As mentioned in the previous section, however, most nonsimple projects have more than one value of i^* that makes NPW $= 0$ (i.e., more than one i^* per project). In such a case, the NPW profile would cross the horizontal axis more than once.[7]

(b) With Excel, you can evaluate the IRR for the project with the function

$$=\text{IRR}(\text{range,guess})$$

in which you specify the cell range for the cash flow and the initial guess, such as 10%.

7.3 Internal-Rate-of-Return Criterion

Now that we have classified investment projects and learned methods for determining the i^* value for a given project's cash flows, our objective is to develop an accept–reject decision rule that gives results consistent with those obtained from NPW analysis.

7.3.1 Relationship to PW Analysis

As we already observed in Chapter 5, NPW analysis depends on the rate of interest used for the computation of NPW. A different rate may change a project from being considered acceptable to being unacceptable, or it may change the ranking of several projects:

- Consider again the NPW profile as drawn for the simple project in Figure 7.1(a). For interest rates below i^*, this project should be accepted because NPW > 0; for interest rates above i^*, it should be rejected.
- By contrast, for certain nonsimple projects, the NPW may look like the one shown in Figure 7.1(b). NPW analysis would lead you to accept the projects in regions **A** and **C**, but reject those in regions **B** and **D**. Of course, this result goes against intuition: A higher interest rate would change an unacceptable project into an acceptable one. The situation graphed in Figure 7.1(b) is one of the cases of multiple i^*'s mentioned in Definition 2.

---

[7] In Section 7.2.2, we discuss methods of predicting the number of i^* values by looking at cash flows. However, generating an NPW profile to discover multiple i^*'s is as practical and informative as any other method.

Therefore, for the simple investment situation in Figure 7.1(a), $i*$ can serve as an appropriate index for either accepting or rejecting the investment. However, for the nonsimple investment of Figure 7.1(b), it is not clear which $i*$ to use to make an accept–reject decision. Therefore, the $i*$ value fails to provide an appropriate measure of profitability for an investment project with multiple rates of return.

7.3.2 Net-Investment Test: Pure versus Mixed Investments

To develop a consistent accept–reject decision rule with the NPW, we need to further classify a project into either a pure or a mixed investment:

- A project is said to be a **net investment** when the project balances computed at the project's $i*$ values, $PB(i*)_n$, are either less than or equal to zero throughout the life of the investment, with the first cash flow being negative $(A_0 < 0)$. The investment is *net* in the sense that the firm does not overdraw on its return at any point and hence investment is *not indebted* to the project. This type of project is called a **pure investment**. In contrast, **pure borrowing** is defined as the situation in which $PB(i*)_n$ values are positive or zero throughout the life of the loan, with $A_0 > 0$. *Simple investments will always be pure investments.*
- If any of the project balances calculated at the project's $i*$ is positive, the project is not a pure investment. A positive project balance indicates that, at some time during the project life, the firm acts as a borrower $[PB(i*)_n > 0]$ rather than an investor in the project $[PB(i*)_n < 0]$. This type of investment is called a **mixed investment**.

Net investment test: A process to determine whether or not a firm borrows money from a project during the investment period.

Pure investment: An investment in which a firm never borrows money from the project.

Mixed investment: An investment in which a firm borrows money from the project during the investment period.

EXAMPLE 7.6 Pure versus Mixed Investments

Consider the following four investment projects with known $i*$ values:

	Project Cash Flows			
n	**A**	**B**	**C**	**D**
0	−$1,000	−$1,000	−$1,000	−$1,000
1	1,000	1,600	500	3,900
2	2,000	−300	−500	−5,030
3	1,500	−200	2,000	2,145
$i*$	33.64%	21.95%	29.95%	(10%, 30%, 50%)

Determine which projects are pure investments.

SOLUTION

Given: Four projects with cash flows and $i*$'s as shown.
Find: Which projects are pure investments?

We will first compute the project balances at the projects' respective i^*'s. If multiple rates of return exist, we may use the largest value of i^* greater than zero.[8]

Project A:

$$PB(33.64\%)_0 = -\$1,000,$$

$$PB(33.64\%)_1 = -\$1,000(1 + 0.3364) + (-\$1,000) = -\$2,336.40,$$

$$PB(33.64\%)_2 = -\$2,336.40(1 + 0.3364) + \$2,000 = -\$1,122.36,$$

$$PB(33.64\%)_3 = -\$1,122.36(1 + 0.3364) + \$1,500 = 0.$$

$(-, -, -, 0)$: passes the net-investment test (pure investment).

Project B:

$$PB(21.95\%)_0 = -\$1,000,$$

$$PB(21.95\%)_1 = -\$1,000(1 + 0.2195) + \$1,600 = \$380.50,$$

$$PB(21.95\%)_2 = +\$380.50(1 + 0.2195) - \$300 = \$164.02,$$

$$PB(21.95\%)_3 = +\$164.02(1 + 0.2195) - \$200 = 0.$$

$(-, +, +, 0)$: fails the net-investment test (mixed investment).

Project C:

$$PB(29.95\%)_0 = -\$1,000,$$

$$PB(29.95\%)_1 = -\$1,000(1 + 0.2995) + \$500 = -\$799.50,$$

$$PB(29.95\%)_2 = -\$799.50(1 + 0.2995) - \$500 = -\$1,538.95,$$

$$PB(29.95\%)_3 = -\$1,538.95(1 + 0.2995) + \$2,000 = 0.$$

$(-, -, -, 0)$: passes the net-investment test (pure investment).

Project D: There are three rates of return. We can use any of them for the net investment test. Thus,

$$PB(50\%)_0 = -\$1,000,$$

$$PB(50\%)_1 = -\$1,000(1 + 0.50) + \$3,900 = \$2,400,$$

$$PB(50\%)_2 = +\$2,400(1 + 0.50) - \$5,030 = -\$1,430,$$

$$PB(50\%)_3 = -\$1,430(1 + 0.50) + \$2,145 = 0.$$

$(-, +, -, 0)$: fails the net-investment test (mixed investment).

COMMENTS: As shown in Figure 7.4, projects A and C are the only pure investments. Project B demonstrates that the existence of a unique i^* is a necessary but not sufficient condition for a pure investment.

---

[8] In fact, it does not matter which rate we use in applying the net-investment test. If one value passes the test, they will all pass. If one value fails, they will all fail.

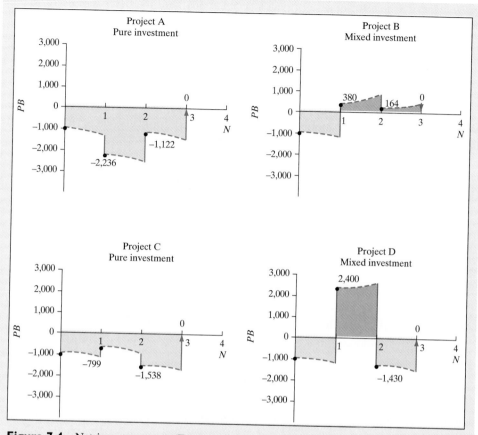

Figure 7.4 Net-investment test (Example 7.6).

7.3.3 Decision Rule for Pure Investments

Suppose we have a pure investment. (Recall that all simple investments are pure investments as well.) Why are we interested in finding the particular interest rate that equates a project's cost with the present worth of its receipts? Again, we may easily answer this question by examining Figure 7.1(a). In this figure, we notice two important characteristics of the NPW profile. First, as we compute the project's PW(i) at a varying interest rate i, we see that the NPW is positive for $i < i^*$, indicating that the project would be acceptable under PW analysis for those values of i. Second, the NPW is negative for $i > i^*$, indicating that the project is unacceptable for those values of i. Therefore, i^* serves as a **benchmark** interest rate, knowledge of which will enable us to make an accept–reject decision consistent with NPW analysis.

Note that, for a pure investment, i^* is indeed the IRR of the investment. (See Section 7.1.2.) Merely knowing i^*, however, is not enough to apply this method. Because firms typically wish to do better than break even (recall that at NPW = 0 we were indifferent to the project), a minimum acceptable rate of return (MARR) is indicated by company policy, management, or the project decision maker. If the IRR exceeds this MARR, we are assured that the company will more than break even. Thus, the IRR becomes a useful gauge against which to judge a project's acceptability, and the decision rule for a pure project is as follows:

If IRR > MARR, accept the project.

If IRR = MARR, remain indifferent.

If IRR < MARR, reject the project.

Note that this decision rule is designed to be applied for a single project evaluation. When we have to compare mutually exclusive investment projects, we need to apply the **incremental analysis approach**, as we shall see in Section 7.4.2.

EXAMPLE 7.7 Investment Decision for a Pure Investment

Merco, Inc., a machinery builder in Louisville, Kentucky, is considering investing $1,250,000 in a complete structural beam-fabrication system. The increased productivity resulting from the installation of the drilling system is central to the project's justification. Merco estimates the following figures as a basis for calculating productivity:

- Increased fabricated steel production: 2,000 tons/year.
- Average sales price/ton fabricated steel: $2,566.50/ton.
- Labor rate: $10.50/hour.
- Tons of steel produced in a year: 15,000 tons.
- Cost of steel per ton (2,205 lb): $1,950/ton.
- Number of workers on layout, hole making, sawing, and material handling: 17.
- Additional maintenance cost: $128,500/year.

With the cost of steel at $1,950 per ton and the direct labor cost of fabricating 1 lb at 10 cents, the cost of producing a ton of fabricated steel is about $2,170.50. With a selling price of $2,566.50 per ton, the resulting contribution to overhead and profit becomes $396 per ton. Assuming that Merco will be able to sustain an increased production of 2,000 tons per year by purchasing the system, engineers have estimated the projected additional contribution to be 2,000 tons × $396 = $792,000.

Since the drilling system has the capacity to fabricate the full range of structural steel, two workers can run the system, one on the saw and the other on the drill. A third operator is required to operate a crane for loading and unloading materials. Merco estimates that, to do the equivalent work of these three workers with conventional manufacturing techniques would require, on the average, an additional 14 people for center punching, hole making with a radial or magnetic drill, and material handling. This translates into a labor savings in the amount of $294,000 per year (14 × $10.50 × 40 hours/week × 50 weeks/year). The system can last for 15 years, with an estimated after-tax salvage value of $80,000. However, after an annual deduction of $226,000 in corporate income taxes, the net investment costs, as well as savings, are as follows:

- Project investment cost: $1,250,000.
- Projected annual net savings:
  ($792,000 + $294,000) − $128,500 − $226,000 = $731,500.

- Projected after-tax salvage value at the end of year 15: $80,000.
  - (a) What is the projected IRR on this fabrication investment?
  - (b) If Merco's MARR is known to be 18%, is this investment justifiable?

SOLUTION

Given: Projected cash flows as shown in Figure 7.5 and MARR $= 18\%$.
Find: (a) The IRR and (b) whether to accept or reject the investment.

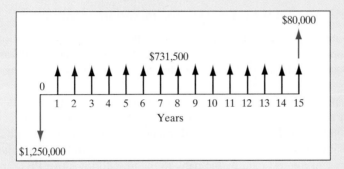

Figure 7.5 Cash flow diagram (Example 7.7).

(a) Since only one sign change occurs in the net cash flow series, the fabrication project is a simple investment. This indicates that there will be a unique rate of return that is internal to the project:

$$PW(i) = -\$1,250,000 + \$731,500(P/A, i, 15)$$

$$+ \$80,000(P/F, i, 15)$$

$$= 0$$

$$i^* = 58.71\%.$$

With Excel, you will also find that the IRR is about 58.71% for the net investment of $1,250,000.

(b) The IRR figure far exceeds Merco's MARR, indicating that the fabrication system project is an economically attractive one. Merco's management believes that, over a broad base of structural products, there is no doubt that the installation of the fabricating system would result in a significant savings, even after considering some potential deviations from the estimates used in the analysis.

7.3.4 Decision Rule for Mixed Investments

Applied to pure projects, $i*$ provides an unambiguous criterion for measuring profitability. However, when multiple rates of return occur, none of them is an accurate portrayal of a project's acceptability or profitability. However, there is a correct method, which uses an **external interest rate**, for refining our analysis when we do discover multiple $i*$'s. An external rate of return allows us to calculate a single accurate rate of return; if you choose to avoid these more complicated applications of rate-of-return techniques, you must be able to predict multiple $i*$'s via the NPW profile and, when they occur, select an alternative method such as NPW or AE analysis for determining the project's acceptability.

Need for an External Interest Rate for Mixed Investments

In the case of a mixed investment, we can extend the economic interpretation of the IRR to the return-on-invested-capital measure if we are willing to make an assumption about what happens to the extra cash that the investor gets from the project during the intermediate years.

Project balance:
The amount
of money
committed to
a project at a
specific period.

First, the **project balance** (PB), or investment balance, can also be interpreted from the viewpoint of a financial institution that borrows money from an investor and then pays interest on the PB. Thus, a negative PB means that the investor has money in a bank account; a positive PB means that the investor has borrowed money from the bank. Negative PBs represent interest paid by the bank to the investor; positive PBs represent interest paid by the investor to the bank.

Now, can we assume that the interest paid by the bank and the interest received from the investor are the same for the same amount of balance? In our banking experience, we know that is not the case. Normally, the borrowing rate (interest paid by the investor) is higher than the interest rate on your deposit (interest paid by the bank).

However, when we calculate the project balance at an $i*$ for mixed investments, we notice an important point: Cash borrowed (released) from the project is assumed to earn the same interest rate through external investment as money that remains internally invested. In other words, in solving a cash flow for an unknown interest rate, it is assumed that money released from a project can be reinvested to yield a rate of return equal to that received from the project. In fact, we have been making this assumption regardless of whether a cash flow does or does not produce a unique positive $i*$. Note that money is borrowed only when $\text{PB}(i*) > 0$, and the magnitude of the borrowed amount is the project balance. When $\text{PB}(i*) < 0$, no money is borrowed, even though the cash flow may be positive at that time.

In reality, it is not always possible for cash borrowed (released) from a project to be reinvested to yield a rate of return equal to that received from the project. Instead, it is likely that the rate of return available on a capital investment in the business is much different—usually higher—from the rate of return available on other external investments. Thus, it may be necessary to compute the project balances for a project's cash flow at two rates of interest—one on the internal investment and one on the external investments. As we will see later, by separating the interest rates, we can measure the **true rate of return** of any internal portion of an investment project.

Calculation of Return on Invested Capital for Mixed Investments

For a mixed investment, we must calculate a rate of return on the portion of capital that remains invested internally. This rate is defined as the **true IRR** for the mixed investment and is commonly known as the **return on invested capital (RIC)**. Then, what interest rate should we assume for the portion of external investment? Insofar as a project is not a net investment, one or more periods when the project has a net outflow of money (a positive project balance) must later be returned to the project. This money can be put into the firm's investment pool until such time as it is needed in the project. The interest rate of this investment pool is the interest rate at which the money can in fact be invested outside the project.

Return on invested capital (RIC): The amount that a company earns on the total investment it has made in its project.

Recall that the NPW method assumed that the interest rate charged to any funds withdrawn from a firm's investment pool would be equal to the MARR. In this book, *we will use the MARR as an established external interest rate* (i.e., the rate earned by money invested outside of the project). We can then compute the RIC as a function of the MARR by finding the value of the RIC that will make the terminal project balance equal to zero. (This implies that the firm wants to fully recover any investment made in the project and pays off any borrowed funds at the end of the project life.) This way of computing the rate of return is an accurate measure of the profitability of the project as represented by the cash flow. The following procedure outlines the steps for determining the IRR for a mixed investment:

Step 1. Identify the MARR (or external interest rate).

Step 2. Calculate $PB(i, MARR)_n$ (or simply PB_n) according to the rule

$$PB(i, MARR)_0 = A_0.$$

$$PB(i, MARR)_1 = \begin{cases} PB_0(1 + i) + A_1, \text{ if } PB_0 < 0 \\ PB_0(1 + MARR) + A_1, \text{ if } PB_0 > 0 \end{cases}$$

$$\vdots$$

$$PB(i, MARR)_n = \begin{cases} PB_{n-1}(1 + i) + A_n, \text{ if } PB_{n-1} < 0 \\ PB_{n-1}(1 + MARR) + A_n, \text{ if } PB_{n-1} > 0 \end{cases}$$

(As defined in the text, A_n stands for the net cash flow at the end of period n. Note that the terminal project balance must be zero.)

Step 3. Determine the value of i by solving the terminal project balance equation

$$PB(i, MARR)_N = 0.$$

The interest rate i is the RIC (or IRR) for the mixed investment.

Using the MARR as an external interest rate, we may accept a project if the IRR exceeds the MARR, and we should reject the project otherwise. Figure 7.6 summarizes the IRR computation for a mixed investment.

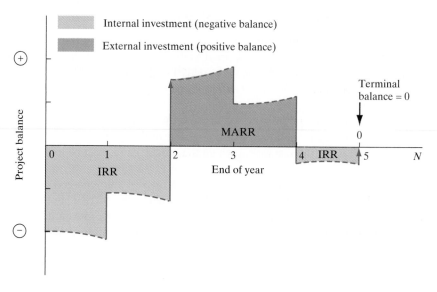

Figure 7.6 Computational logic for IRR (mixed investment).

EXAMPLE 7.8 IRR for a Mixed Investment

By outbidding its competitors, Trane Image Processing (TIP), a defense contractor, received a contract worth $7,300,000 to build navy flight simulators for U.S. Navy pilot training over two years. With some defense contracts, the U.S. government makes an advance payment when the contract is signed, but in this case the government will make two progressive payments: $4,300,000 at the end of the first year and the $3,000,000 balance at the end of the second year. The expected cash outflows required to produce the simulators are estimated to be $1,000,000 now, $2,000,000 during the first year, and $4,320,000 during the second year. The expected net cash flows from this project are summarized as follows:

Year	Cash Inflow	Cash Outflow	Net Cash Flow
0		$1,000,000	−$1,000,000
1	$4,300,000	2,000,000	2,300,000
2	3,000,000	4,320,000	−1,320,000

In normal situations, TIP would not even consider a marginal project such as this one. However, hoping that the company can establish itself as a technology leader in the field, management felt that it was worth outbidding its competitors. Financially, what is the economic worth of outbidding the competitors for this project? That is,

(a) Compute the values of i^*'s for this project.

(b) Make an accept–reject decision based on the results in part (a). Assume that the contractor's MARR is 15%.

SOLUTION

Given: Cash flow shown and MARR = 15%.

Find: (a) Compute the NPW, (b) i^*, and (c) RIC at MARR = 15%, and determine whether to accept the project.

(a)

$$PW(15\%) = -\$1,000,000 + \$2,300,000(P/F, 15\%, 1)$$
$$= -\$1,320,000(P/F, 15\%, 2)$$
$$= \$1,890 > 0.$$

(b) Since the project has a two-year life, we may solve the net-present-worth equation directly via the quadratic formula:

$$-\$1,000,000 + \$2,300,000/(1 + i^*) - \$1,320,000/(1 + i^*)^2 = 0.$$

If we let $X = 1/(1 + i^*)$, we can rewrite the preceding expression as

$$-1,000,000 + 2,300,000X - 1,320,000X^2 = 0.$$

Solving for X gives $X = 10/11$ and $10/12$, or $i^* = 10\%$ and 20%. As shown in Figure 7.7, the NPW profile intersects the horizontal axis twice, once at 10% and again at 20%. The investment is obviously not a simple one; thus, neither 10% nor 20% represents the true internal rate of return of this government project.

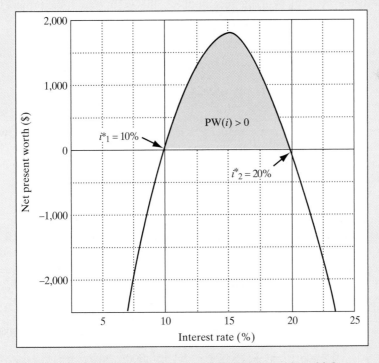

Figure 7.7 NPW plot for a nonsimple investment with multiple rates of return (Example 7.8).

(c) As calculated in (b), the project has multiple rates of return. This is obviously not a net investment, as the following table shows:

Net Investment Test N	Using $i^* = 10\%$ 0	1	2	Using $i^* = 20\%$ 0	1	2
Beginning balance	$0	−$1,000	$1,200	$0	−$1,000	$1,100
Return on investment	0	−100	120	0	−200	220
Payment	−1,000	2,300	−1,320	−1,000	2,300	−1,320
Ending balance	−$1,000	$1,200	0	−$1,000	$1,100	0

(Unit: $1,000)

At $n = 0$, there is a net investment to the firm, so the project balance expression becomes

$$PB(i, 15\%)_0 = -\$1,000,000.$$

The net investment of $1,000,000 that remains invested internally grows at the interest rate i for the next period. With the receipt of $2,300,000 in year 1, the project balance becomes

$$PB(i, 15\%)_1 = -\$1,000,000(1 + i) + \$2,300,000$$
$$= \$1,300,000 - \$1,000,000i$$
$$= \$1,000,000(1.3 - i).$$

At this point, we do not know whether $PB(i, 15\%)_1$ is positive or negative; we want to know this in order to test for net investment and the presence of a unique i^*. It depends on the value of i, which we want to determine. Therefore, we need to consider two situations: (1) $i < 1.3$ and (2) $i > 1.3$.

- **Case 1:** $i < 1.3 \rightarrow PB(i, 15\%)_1 > 0$.
  Since this indicates a positive balance, the cash released from the project would be returned to the firm's investment pool to grow at the MARR until it is required back in the project. By the end of year 2, the cash placed in the investment pool would have grown at the rate of 15% [to $\$1,000,000(1.3 - i)$ $(1 + 0.15)$] and must equal the investment into the project of $1,320,000 required at that time. Then the terminal balance must be

$$PB(i, 15\%)_2 = \$1,000,000(1.3 - i)(1 + 0.15) - \$1,320,000$$
$$= \$175,000 - \$1,150,000i$$
$$= 0.$$

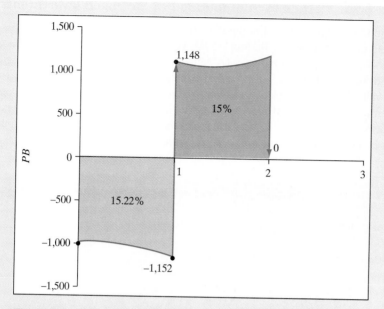

Figure 7.8 Calculation of the IRR for a mixed investment (Example 7.8).

Solving for i yields

$$\text{RIC} = \text{IRR} = 0.1522, \text{ or } 15.22\% > 15\%.$$

The computational process is shown graphically in Figure 7.8.

- **Case 2:** $i > 1.3 \rightarrow \text{PB}(i, 15\%)_1 < 0$.

The firm is still in an investment mode. Therefore, the balance at the end of year 1 that remains invested will grow at the rate i for the next period. With the investment of \$1,320,000 required in year 2 and the fact that the net investment must be zero at the end of the project life, the balance at the end of year 2 should be

$$\begin{aligned} \text{PB}(i, 15\%)_2 &= \$1,000,000(1.3 - i)(1 + i) - \$1,320,000 \\ &= -\$20,000 + \$300,000i - \$1,000,000i^2 \\ &= 0. \end{aligned}$$

Solving for i gives

$$\text{IRR} = 0.1 \text{ or } 0.2 < 1.3,$$

which violates the initial assumption that $i > 1.3$. Therefore, Case 1 is the only correct situation. Since it indicates that IRR > MARR, the project is acceptable, resulting in the same decision as obtained in (a) by applying the NPW criterion.

COMMENTS: In this example, we could have seen by inspection that Case 1 was correct. Since the project required an investment as the final cash flow, the project balance at the end of the previous period (year 1) had to be positive in order for the final balance to equal zero. Inspection does not generally work with more complicated cash flows.

Trial-and-Error Method for Computing IRR for Mixed Investments

The trial-and-error approach to finding the IRR (RIC) for a mixed investment is similar to the trial-and-error approach to finding i^*. We begin with a given MARR and a guess for IRR and solve for the project balance. (A value of IRR close to the MARR is a good starting point for most problems.) Since we desire the project balance to approach zero, we can adjust the value of IRR as needed after seeing the result of the initial guess. For example, for a given pair of interest rates (IRRguess, MARR), if the terminal project balance is positive, the IRRguess value is too low, so we raise it and recalculate. We can continue adjusting our IRR guesses in this way until we obtain a project balance equal or close to zero.

EXAMPLE 7.9 IRR for a Mixed Investment by Trial and Error

Consider project D in Example 7.6. The project has the following cash flow:

n	A_n
0	$-\$1,000$
1	3,900
2	$-5,030$
3	2,145

We know from an earlier calculation that this is a mixed investment. Compute the IRR for this project. Assume that MARR = 6%.

SOLUTION

Given: Cash flow as stated for mixed investment and MARR = 6%.
Find: IRR.

For MARR = 6%, we must compute i by trial and error. Suppose we guess $i = 8\%$:

$$PB(8\%, 6\%)_0 = -\$1,000,$$

$$PB(8\%, 6\%)_1 = -\$1,000(1 + 0.08) + \$3,900 = \$2,820.$$

$$PB(8\%, 6\%)_2 = +\$2,820(1 + 0.06) - \$5,030 = -\$2,040.80,$$

$$PB(8\%, 6\%)_3 = -\$2,040.80(1 + 0.08) + \$2,145 = -\$59.06.$$

The net investment is negative at the end of the project, indicating that our trial $i = 8\%$ is in error. After several trials, we conclude that, for MARR = 6%, the IRR is approximately 6.13%. To verify the results, we write

$$PB(6.13\%, 6\%)_0 = -\$1,000,$$

$$PB(6.13\%, 6\%)_1 = -\$1,000.00(1 + 0.0613) + \$3,900 = \$2,838.66.$$

$$PB(6.13\%, 6\%)_2 = +\$2,838.66(1 + 0.0600) - \$5,030 = -\$2,021.02,$$

$$PB(6.13\%, 6\%)_3 = -\$2,021.02(1 + 0.0613) + \$2,145 = 0.$$

The positive balance at the end of year 1 indicates the need to borrow from the project during year 2. However, note that the net investment becomes zero at the end of the project life, confirming that 6.13% is the IRR for the cash flow. Since IRR > MARR, the investment is acceptable.

COMMENTS: On the basis of the NPW criterion, the investment would be acceptable if the MARR was between zero and 10% or between 30% and 50%. The rejection region is $10\% < i < 30\%$ and $i > 50\%$. This can be verified in Figure 7.1(b). Note that the project also would be marginally accepted under the NPW analysis at MARR = i = 6%:

$$PW(6\%) = -\$1,000 + 3,900(P/F, 6\%, 1)$$
$$= -\$5,030(P/F, 6\%, 2) + 2,145(P/F, 6\%, 3)$$
$$= \$3.55 > 0.$$

The flowchart in Figure 7.9 summarizes how you should proceed to apply the net cash flow sign test, accumulated cash flow sign test, and net-investment test to calculate an IRR and make an accept–reject decision for a single project. Given the complications

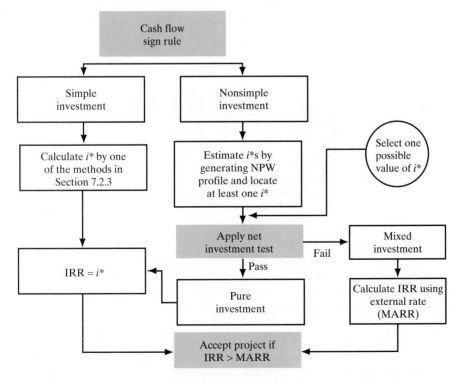

Figure 7.9 Summary of IRR criterion: A flowchart that summarizes how you may proceed to apply the net cash flow sign rule and net-investment test to calculate IRR for a pure as well as a mixed investment.

involved in using IRR analysis to compare alternative projects, it is usually more desirable to use one of the other equivalence techniques for this purpose. As an engineering manager, you should keep in mind the intuitive appeal of the rate-of-return measure. Once you have selected a project on the basis of NPW or AE analysis, you may also wish to express its worth as a rate of return, for the benefit of your associates.

7.4 Mutually Exclusive Alternatives

In this section, we present the decision procedures that should be used in comparing two or more mutually exclusive projects on the basis of the rate-of-return measure. We will consider two situations: (1) alternatives that have the same economic service life and (2) alternatives that have unequal service lives.

7.4.1 Flaws in Project Ranking by IRR

Under NPW or AE analysis, the mutually exclusive project with the highest worth was preferred. (This is known as the "total investment approach.") Unfortunately, the analogy does not carry over to IRR analysis: The project with the highest IRR may *not* be the preferred alternative. To illustrate the flaws inherent in comparing IRRs in order to choose from mutually exclusive projects, suppose you have two mutually exclusive alternatives, each with a 1-year service life: One requires an investment of $1,000 with a return of $2,000, and the other requires $5,000 with a return of $7,000. You already obtained the IRRs and NPWs at MARR = 10% as follows:

n	A1	A2
0	−$1,000	−$5,000
1	2,000	7,000
IRR	100%	40%
PW(10%)	$818	$1,364

Assuming that you have enough money in your investment pool to select either alternative, would you prefer the first project simply because you expect a higher rate of return?

On the one hand, we can see that A2 is preferred over A1 by the NPW measure. On the other hand, the IRR measure gives a numerically higher rating for A1. This inconsistency in ranking occurs because the NPW, NFW, and AE are **absolute (dollar)** measures of investment worth, whereas the IRR is a **relative (percentage)** measure and cannot be applied in the same way. That is, the IRR measure ignores the **scale** of the investment. Therefore, the answer to our question in the previous paragraph is no; instead, you would prefer the second project, with the lower rate of return, but higher NPW. Either the NPW or the AE measure would lead to that choice, but a comparison of IRRs would rank the smaller project higher. Another approach, referred to as **incremental analysis**, is needed.

7.4.2 Incremental Investment Analysis

In the previous example, the more costly option requires an incremental investment of $4,000 at an incremental return of $5,000. Let's assume that you have exactly $5,000 in your investment pool.

- If you decide to invest in option A1, you will need to withdraw only $1,000 from your investment pool. The remaining $4,000 will continue to earn 10% interest. One year later, you will have $2,000 from the outside investment and $4,400 from the investment pool. With an investment of $5,000, in one year you will have $6,400. The equivalent present worth of this change in wealth is PW(10%) = −$5,000 + $6,400(P/F, 10%, 1) = $818.
- If you decide to invest in option A2, you will need to withdraw $5,000 from your investment pool, leaving no money in the pool, but you will have $7,000 from your outside investment. Your total wealth changes from $5,000 to $7,000 in a year. The equivalent present worth of this change in wealth is PW(10%) = −$5,000 + $7,000(P/F, 10%, 1) = $1,364.

In other words, if you decide to take the more costly option, certainly you would be interested in knowing that this additional investment can be justified at the MARR. The 10%-of-MARR value implies that you can always earn that rate from other investment sources (i.e., $4,400 at the end of 1 year for a $4,000 investment). However, in the second option, by investing the additional $4,000, you would make an additional $5,000, which is equivalent to earning at the rate of 25%. Therefore, the incremental investment can be justified.

Now we can generalize the decision rule for comparing mutually exclusive projects. For a pair of mutually exclusive projects (A and B, with B defined as the more costly option), we may rewrite B as

$$B = A + (B - A).$$

In other words, B has two cash flow components: (1) the same cash flow as A and (2) the incremental component $(B - A)$. Therefore, the only situation in which B is preferred to A is when the rate of return on the incremental component $(B - A)$ exceeds the MARR. Therefore, for two mutually exclusive projects, rate-of-return analysis is done by computing the *internal rate of return on the incremental investment* $(IRR\Delta)$ between the projects. Since we want to consider increments of investment, we compute the cash flow for the difference between the projects by subtracting the cash flow for the lower investment-cost project (A) from that of the higher investment-cost project (B). Then the decision rule is

Incremental IRR: IRR on the incremental investment from choosing a large project instead of a smaller project.

If IRR_{B-A} > MARR, select B,

If IRR_{B-A} = MARR, select either project,

If IRR_{B-A} < MARR, select A,

where $B - A$ is an investment increment (negative cash flow). *If a "do-nothing" alternative is allowed, the smaller cost option must be profitable (its IRR must be greater than the MARR) at first.* This means that you compute the rate of return for each alternative in the mutually exclusive group and then eliminate the alternatives whose IRRs are less than the MARR before applying the incremental analysis.

It may seem odd to you how this simple rule allows us to select the right project. Example 7.10 illustrates the incremental investment decision rule.

EXAMPLE 7.10 IRR on Incremental Investment: Two Alternatives

John Covington, a college student, wants to start a small-scale painting business during his off-school hours. To economize the start-up business, he decides to purchase some used painting equipment. He has two mutually exclusive options: Do most of the painting by himself by limiting his business to only residential painting jobs (B1) or purchase more painting equipment and hire some helpers to do both residential and commercial painting jobs that he expects will have a higher equipment cost, but provide higher revenues as well (B2). In either case, John expects to fold up the business in three years, when he graduates from college.

The cash flows for the two mutually exclusive alternatives are as follows:

n	B1	B2	B2 − B1
0	−$3,000	−$12,000	−$9,000
1	1,350	4,200	2,850
2	1,800	6,225	4,425
3	1,500	6,330	4,830
IRR	25%	17.43%	

Knowing that both alternatives are revenue projects, which project would John select at MARR = 10%? (Note that both projects are also profitable at 10%.)

SOLUTION

Given: Incremental cash flow between two alternatives and MARR = 10%.
Find: (a) IRR on the increment and (b) which alternative is preferable.

(a) To choose the best project, we compute the incremental cash flow B2 − B1. Then we compute the IRR on this increment of investment by solving the equation

$$-\$9,000 + \$2,850(P/F, i, 1) + \$4,425(P/F, i, 2) + \$4,830(P/F, i, 3) = 0.$$

(b) We obtain $i^*_{B2-B1} = 15\%$, as plotted in Figure 7.10. By inspection of the incremental cash flow, we know that it is a simple investment, so $IRR_{B2-B1} = i^*_{B2-B1}$. Since $IRR_{B2-B1} >$ MARR, we select B2, which is consistent with the NPW analysis. Note that, at MARR > 25%, neither project would be acceptable.

COMMENTS: Why did we choose to look at the increment B2 − B1 instead of B1 − B2? Because we want the first flow of the incremental cash flow series to be negative (an investment flow), so that we can calculate an IRR. By subtracting the lower initial investment project from the higher, we guarantee that the first increment will be an investment flow. If we ignore the investment ranking, we might end up

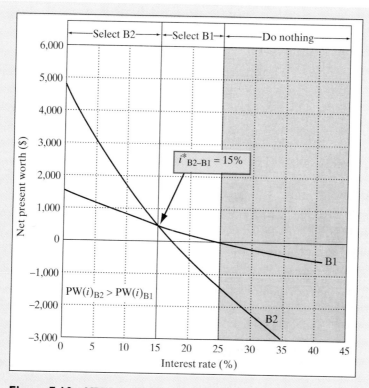

Figure 7.10   NPW profiles for B1 and B2 (Example 7. 10).

with an increment that involves *borrowing cash flow* and has no internal rate of return. This is indeed the case for B1 − B2. (i^*_{B1-B2} is also 15%, not −15%, but it has a different meaning: it is a borrowing rate, not a rate of return on your investment.) If, erroneously, we had compared this i^* with the MARR, we might have accepted project B1 over B2. This undoubtedly would have damaged our credibility with management!

The next example indicates that the inconsistency in ranking between NPW and IRR can also occur when differences in the timing of a project's future cash flows exist, even if their initial investments are the same.

EXAMPLE 7.11   IRR on Incremental Investment When Initial Flows Are Equal

Consider the following two mutually exclusive investment projects that require the same amount of investment:

Which project would you select on the basis of the rate of return on incremental investment, assuming that MARR = 12%? (Once again, both projects are profitable at 12%.)

n	C1	C2
0	−$9,000	−$9,000
1	480	5,800
2	3,700	3,250
3	6,550	2,000
4	3,780	1,561
IRR	18%	20%

SOLUTION

Given: Cash flows for two mutually exclusive alternatives as shown and MARR = 12%.

Find: (a) IRR on incremental investment and (b) which alternative is preferable.

(a) When the initial investments are equal, we progress through the cash flows until we find the first difference and then set up the increment so that this first nonzero flow is negative (i.e., an investment). Thus, we set up the incremental investment by taking (C1 − C2):

n	C1 − C2
0	$ 0
1	−5,320
2	450
3	4,550
4	2,219

We next set the PW equation equal to zero:

$$-\$5,320 + \$450(P/F, i, 1) + \$4,550(P/F, i, 2) + \$2,219(P/F, i, 3) = 0.$$

(b) Solving for i yields $i^* = 14.71\%$, which is also an IRR, since the increment is a simple investment. Since $IRR_{C1-C2} = 14.71\% > MARR$, we would select C1. If we used NPW analysis, we would obtain $PW(12\%)_{C1} = \$1,443$ and $PW(12\%)_{C2} = \$1,185$, indicating that C1 is preferred over C2.

When you have more than two mutually exclusive alternatives, they can be compared in pairs by successive examination. Example 7.12 illustrates how to compare three mutually exclusive alternatives. (In Chapter 15, we will examine some multiple-alternative problems in the context of capital budgeting.)

EXAMPLE 7.12 IRR on Incremental Investment: Three Alternatives

Consider the following three sets of mutually exclusive alternatives:

n	DI	D2	D3
0	−$2,000	−$1,000	−$3,000
1	1,500	800	1,500
2	1,000	500	2,000
3	800	500	1,000
IRR	34.37%	40.76%	24.81%

Which project would you select on the basis of the rate of return on incremental investment, assuming that MARR $= 15\%$?

SOLUTION

Given: Preceding cash flows and MARR $= 15\%$.

Find: IRR on incremental investment and which alternative is preferable.

Step 1: Examine the IRR of each alternative. At this point, we can eliminate any alternative that fails to meet the MARR. In this example, all three alternatives exceed the MARR.

Step 2: Compare D1 and D2 in pairs.[9] Because D2 has a lower initial cost, compute the rate of return on the increment (D1−D2), which represents an increment of investment.

n	DI − D2
0	−$1,000
1	700
2	500
3	300

The incremental cash flow represents a simple investment. To find the incremental rate of return, we write

$$-\$1,000 + \$700(P/F, i, 1) + \$500(P/F, i, 2) + \$300(P/F, i, 3) = 0.$$

Solving for i^*_{D1-D2} yields 27.61%, which exceeds the MARR; therefore, D1 is preferred over D2. Now you eliminate D2 from further consideration.

Step 3: Compare D1 and D3. Once again, D1 has a lower initial cost. Examine the increment (D3 − D1):

---

[9] When faced with many alternatives, you may arrange them in order of increasing initial cost. This is not a required step, but it makes the comparison more tractable.

n	D3 − D1
0	−$1,000
1	0
2	1,000
3	200

Here, the incremental cash flow represents another simple investment. The increment (D3 − D1) has an unsatisfactory 8.8% rate of return; therefore, D1 is preferred over D3. Accordingly, we conclude that D1 is the best alternative.

EXAMPLE 7.13 Incremental Analysis for Cost-Only Projects

Falk Corporation is considering two types of manufacturing systems to produce its shaft couplings over six years: (1) a cellular manufacturing system (CMS) and (2) a flexible manufacturing system (FMS). The average number of pieces to be produced with either system would be 544,000 per year. The operating cost, initial investment, and salvage value for each alternative are estimated as follows:

Items	CMS Option	FMS Option
Annual O&M costs:		
Annual labor cost	$1,169,600	$707,200
Annual material cost	832,320	598,400
Annual overhead cost	3,150,000	1,950,000
Annual tooling cost	470,000	300,000
Annual inventory cost	141,000	31,500
Annual income taxes	1,650,000	1,917,000
Total annual costs	$7,412,920	$5,504,100
Investment	$4,500,000	$12,500,000
Net salvage value	$500,000	$1,000,000

Figure 7.11 illustrates the cash flows associated with each alternative. The firm's MARR is 15%. On the basis of the IRR criterion, which alternative would be a better choice?

DISCUSSION: Since we can assume that both manufacturing systems would provide the same level of revenues over the analysis period, we can compare the two alternatives on the basis of cost only. (These are service projects.) Although we cannot compute the IRR for each option without knowing the revenue figures, we can still

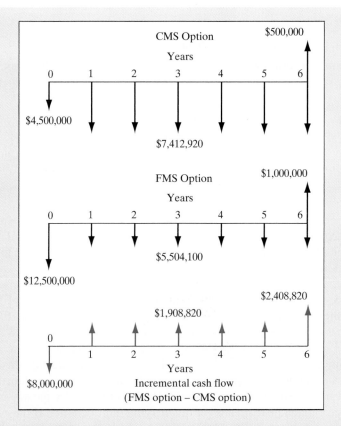

Figure 7.11 Comparison of mutually exclusive alternatives with equal revenues (cost only) (Example 7.13).

calculate the IRR on incremental cash flows. Since the FMS option requires a higher initial investment than that for the CMS, the incremental cash flow is the difference (FMS − CMS).

n	CMS Option	FMS Option	Incremental (FMS − CMS)
0	−$4,500,000	−$12,500,000	−$8,000,000
1	−7,412,920	−5,504,100	1,908,820
2	−7,412,920	−5,504,100	1,908,820
3	−7,412,920	−5,504,100	1,908,820
4	−7,412,920	−5,504,100	1,908,820
5	−7,412,920	−5,504,100	1,908,820
6	−7,412,920	−5,504,100	
	+ $500,000	+ $1,000,000	$2,408,820

SOLUTION

Given: Cash flows shown in Figure 7.11 and $i = 15\%$ per year.
Find: Incremental cash flows, and select the better alternative on the basis of the IRR criterion.

First, we have

$$PW(i)_{\text{FMS-CMS}} = -\$8,000,000 + \$1,908,820(P/A, i, 5)$$
$$= +\$2,408,820(P/F, i, 6)$$
$$= 0.$$

Solving for i by trial and error yields 12.43%. Since $\text{IRR}_{\text{FMS-CMS}} = 12.43\% < 15\%$, we would select CMS. Although the FMS would provide an incremental annual savings of \$1,908,820 in operating costs, the savings are not large enough to justify the incremental investment of \$8,000,000.

COMMENTS: Note that the CMS option is marginally preferred to the FMS option. However, there are dangers in relying solely on the easily quantified savings in input factors—such as labor, energy, and materials—from the FMS and in not considering gains from improved manufacturing performance that are more difficult and subjective to quantify. Factors such as improved product quality, increased manufacturing flexibility (rapid response to customer demand), reduced inventory levels, and increased capacity for product innovation are frequently ignored because we have inadequate means for quantifying their benefits. If these intangible benefits were considered, the FMS option could come out better than the CMS option.

7.4.3 Handling Unequal Service Lives

In Chapters 5 and 6, we discussed the use of the NPW and AE criteria as bases for comparing projects with unequal lives. The IRR measure can also be used to compare projects with unequal lives, as long as we can establish a common analysis period. The decision procedure is then exactly the same as for projects with equal lives. It is likely, however, that we will have a multiple-root problem, which creates a substantial computational burden. For example, suppose we apply the IRR measure to a case in which one project has a 5-year life and the other project has an 8-year life, resulting in a least common multiple of 40 years. Then when we determine the incremental cash flows over the analysis period, we are bound to observe many sign changes. This leads to the possibility of having many i^*'s. Example 7.14 uses i^* to compare mutually exclusive projects, in which the incremental cash flows reveal several sign changes. (Our purpose is not to encourage you to use the IRR approach to compare projects with unequal lives; rather, it is to show the correct way to compare them if the IRR approach *must* be used.)

EXAMPLE 7.14 IRR Analysis for Projects with Different Lives in Which the Increment is a Nonsimple Investment

Consider Example 5.14, in which a mail-order firm wants to install an automatic mailing system to handle product announcements and invoices. The firm had a choice between two different types of machines. Using the IRR as a decision criterion, select the best machine. Assume a MARR of 15%, as before.

SOLUTION

Given: Cash flows for two projects with unequal lives, as shown in Figure 5.11, and MARR = 15%.

Find: The alternative that is preferable.

Since the analysis period is equal to the least common multiple of 12 years, we may compute the incremental cash flow over this 12-year period. As shown in Figure 7.12, we subtract the cash flows of model A from those of model B to form the increment of investment. (Recall that we want the first cash flow difference to be a negative value.) We can then compute the IRR on this incremental cash flow.

Five sign changes occur in the incremental cash flows, indicating a nonsimple incremental investment:

n	Model A		Model B		Model B — Model A
0	−$12,500		−$15,000		−$2,500
1		−5,000		−4,000	1,000
2		−5,000		−4,000	1,000
3	−12,500	−3,000		−4,000	11,500
4		−5,000	−15,000	−2,500	−12,500
5		−5,000		−4,000	1,000
6	−12,500	−3,000		−4,000	11,500
7		−5,000		−4,000	1,000
8		−5,000	−15,000	−2,500	−12,500
9	−12,500	−3,000		−4,000	11,500
10		−5,000		−4,000	1,000
11		−5,000		−4,000	1,000
12		−3,000		−2,500	500
	Four replacement cycles		Three replacement cycles		Incremental cash flows

Even though there are five sign changes in the cash flow, there is only one positive i^* for this problem: 63.12%. Unfortunately, however, the investment is not a pure

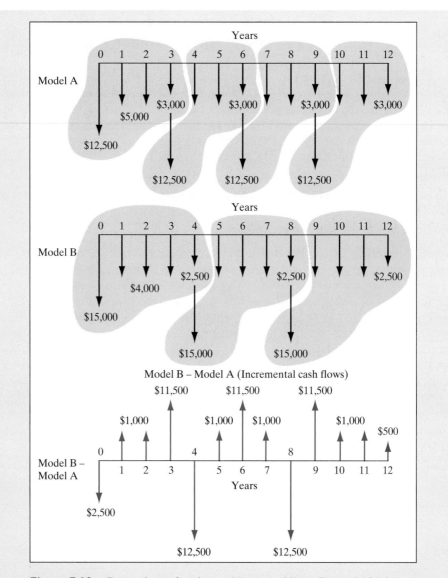

Figure 7.12 Comparison of projects with unequal lives (Example 7.14).

investment. We need to employ an external rate to compute the IRR in order to make a proper accept–reject decision. Assuming that the firm's MARR is 15%, we will use a trial-and-error approach. Try $i = 20\%$:

$$PB(20\%, 15\%)_0 = -\$2,500,$$

$$PB(20\%, 15\%)_1 = -\$2,500(1.20) + \$1,000 = -\$2,000.$$

$$PB(20\%, 15\%)_2 = -\$2,000(1.20) + \$1,000 = -\$1,400,$$

$$PB(20\%, 15\%)_3 = -\$1,400(1.20) + \$11,500 = \$9,820.$$

$$PB(20\%, 15\%)_4 = \$9,820(1.15) - \$12,500 = -\$1,207,$$

$$PB(20\%, 15\%)_5 = -\$1,207(1.20) + \$1,000 = -\$448.40.$$

$$PB(20\%, 15\%)_6 = -\$448.40(1.20) + \$11,500 = \$10,961.92,$$

$$PB(20\%, 15\%)_7 = \$10,961.92(1.15) + \$1,000 = \$13,606.21.$$

$$PB(20\%, 15\%)_8 = \$13,606.21(1.15) - \$12,500 = \$3,147.14,$$

$$PB(20\%, 15\%)_9 = \$3,147.14(1.15) + \$11,500 = \$15,119.21.$$

$$PB(20\%, 15\%)_{10} = \$15,119.21(1.15) + \$1,000 = \$18,387.09,$$

$$PB(20\%, 15\%)_{11} = \$18,387.09(1.15) + \$1,000 = \$22,145.16,$$

$$PB(20\%, 15\%)_{12} = \$22,145.16(1.15) + \$500 = \$25,966.93.$$

Since $PB(20\%, 15\%)_{12} > 0$, the guessed 20% is not the RIC. We may increase the value of i and repeat the calculations. After several trials, we find that the RIC is 50.68%.[10] Since $IRR_{B-A} > MARR$, model B would be selected, which is consistent with the NPW analysis. In other words, the additional investment over the years to obtain model B ($-\$2,500$ at $n = 0$, $-\$12,500$ at $n = 4$, and $-\$12,500$ at $n = 8$) yields a satisfactory rate of return.

COMMENTS: Given the complications inherent in IRR analysis in comparing alternative projects, it is usually more desirable to employ one of the other equivalence techniques for this purpose. As an engineering manager, you should keep in mind the intuitive appeal of the rate-of-return measure. Once you have selected a project on the basis of NPW or AE analysis, you may also wish to express its worth as a rate of return, for the benefit of your associates.

SUMMARY

■ The **rate of return (ROR)** is the interest rate earned on unrecovered project balances such that an investment's cash receipts make the terminal project balance equal to zero. The rate of return is an intuitively familiar and understandable measure of project profitability that many managers prefer to NPW or other equivalence measures.

■ Mathematically, we can determine the rate of return for a given project cash flow series by locating an interest rate that equates the net present worth of the project's cash flows to zero. This break-even interest rate is denoted by the symbol i^*.

■ The **internal rate of return (IRR)** is another term for ROR which stresses the fact that we are concerned with the interest earned on the portion of the project that is internally invested, not those portions released by (borrowed from) the project.

---

[10] It is tedious to solve this type of problem by a trial-and-error method on your calculator. The problem can be solved quickly by using the Cash Flow Analyzer, which can be found from the book's website.

▪ To apply rate-of-return analysis correctly, we need to classify an investment as either simple or nonsimple. A **simple investment** is defined as an investment in which the initial cash flows are negative and only one sign change in the net cash flow occurs, whereas a **nonsimple investment** is an investment for which more than one sign change in the cash flow series occurs. Multiple i^*'s occur only in nonsimple investments. However, not all nonsimple investments will have multiple i^*'s. In this regard,

1. The possible presence of multiple i^*'s (rates of return) can be predicted by

   - The net cash flow sign test.
   - The accumulated cash flow sign test.

   When multiple rates of return cannot be ruled out by the two methods, it is useful to generate an NPW profile to approximate the value of i^*.

2. All i^* values should be exposed to the **net investment test**. Passing this test indicates that i^* is an internal rate of return and is therefore a suitable measure of project profitability. Failing to pass the test indicates project borrowing, a situation that requires further analysis with the use of an **external interest rate**.

3. **Return-on-invested-capital** analysis uses one rate (the firm's MARR) on externally invested balances and solves for another rate (i^*) on internally invested balances.

▪ For a pure investment, i^* is the rate of return that is internal to the project. For a mixed investment, the RIC calculated with the use of the external interest rate (or MARR) is the true IRR; so the decision rule is as follows:

$$\text{If IRR} > \text{MARR, accept the project.}$$

$$\text{If IRR} = \text{MARR, remain indifferent.}$$

$$\text{If IRR} < \text{MARR, reject the project.}$$

IRR analysis yields results consistent with NPW and other equivalence methods.

▪ In properly selecting among alternative projects by IRR analysis, **incremental investment** must be used. In creating an incremental investment, we always subtract the lower cost investment from the higher cost one. Basically, you want to know that the extra investment required can be justified on the basis of the extra benefits generated in the future.

PROBLEMS

Note: *The symbol i^* represents the interest rate that makes the net present value of the project in question equal to zero. The symbol IRR represents the internal rate of return of the investment. For a simple investment, $IRR = i^*$. For a nonsimple investment, i^* is generally not equal to IRR.*

Concept of Rate of Return

7.1 You are going to buy a new car worth $14,500. The dealer computes your monthly payment to be $267 for 72 months' financing. What is the dealer's rate of return on this loan transaction?

7.2 You wish to sell a bond that has a face value of $1,000. The bond bears an interest rate of 6%, payable semiannually. Four years ago, the bond was purchased at $900. At least an 8% annual return on the investment is desired. What must be the minimum selling price of the bond now in order to make the desired return on the investment?

7.3 In 1970, Wal-Mart offered 300,000 shares of its common stock to the public at a price of $16.50 per share. Since that time, Wal-Mart has had 11 two-for-one stock splits. On a purchase of 100 shares at $16.50 per share on the company's first offering, the number of shares has grown to 204,800 shares worth $10,649,600 on January 2006. What is the return on investment for investors who purchased the stock in 1970 (say, over a 36-year ownership period)? Assume that the dividends received during that period were not reinvested.

7.4 Johnson Controls spent more than $2.5 million retrofitting a government complex and installing a computerized energy-management system for the State of Massachusetts. As a result, the state's energy bill dropped from an average of $6 million a year to $3.5 million. Moreover, both parties will benefit from the 10-year life of the contract. Johnson recovers half the money it saved in reduced utility costs (about $1.2 million a year over 10 years); Massachusetts has its half to spend on other things. What is the rate of return realized by Johnson Controls in this energy-control system?

7.5 Pablo Picasso's 1905 portrait *Boy with a Pipe* sold for $104.2 million in an auction at Sotheby's Holdings, Inc., on June 24, 2004, shattering the existing record for art and ushering in a new era in pricing for 20th-century paintings. The Picasso, sold by the philanthropic Greentree Foundation, cost Mr. Whitney about $30,000 in 1950. Determine the annual rate of appreciation of the artwork over 54 years.

Investment Classification and Calculation of $i*$

7.6 Consider four investments with the following sequences of cash flows:

	Net Cash Flow			
n	**Project A**	**Project B**	**Project C**	**Project D**
0	−$18,000	−$30,000	$34,578	−$56,500
1	30,000	32,000	−18,000	2,500
2	20,000	32,000	−18,000	6,459
3	10,000	−22,000	−18,000	−78,345

(a) Identify all the simple investments.
(b) Identify all the nonsimple investments.
(c) Compute $i*$ for each investment.
(d) Which project has no rate of return?

7.7   Consider the following infinite cash flow series with repeated cash flow patterns:

n	A_n
0	-$1,000
1	400
2	800
3	500
4	500
5	400
6	800
7	500
8	500
⋮	⋮

Determine $i*$ for this infinite cash flow series.

7.8   Consider the following investment projects:

	Project Cash Flow				
n	A	B	C	D	E
0	-$100	-$100	-$200	-$50	-$50
1	60	70	$20	120	-100
2	150	70	10	40	-50
3		40	5	40	0
4		40	-180	-20	150
5			60	40	150
6			50	30	100
7			400		100

(a) Classify each project as either simple or nonsimple.
(b) Use the quadratic equation to compute $i*$ for project A.
(c) Obtain the rate(s) of return for each project by plotting the NPW as a function of the interest rate.

7.9   Consider the projects in Table P7.9.
(a) Classify each project as either simple or nonsimple.
(b) Identify all positive $i*$'s for each project.
(c) For each project, plot the present worth as a function of the interest rate (i).

TABLE P7.9 Net Cash Flow for Four Projects

	Net Cash Flow			
n	A	B	C	D
0	−$2,000	−$1,500	−$1,800	−$1,500
1	500	800	5,600	−360
2	100	600	4,900	4,675
3	100	500	−3,500	2,288
4	2,000	700	7,000	
5			−1,400	
6			2,100	
7			900	

7.10 Consider the following financial data for a project:

Initial investment	$50,000
Project life	8 years
Salvage value	$10,000
Annual revenue	$25,000
Annual expenses (including income taxes)	$ 9,000

(a) What is i^* for this project?
(b) If the annual expense increases at a 7% rate over the previous year's expenses, but the annual income is unchanged, what is the new i^*?
(c) In part (b), at what annual rate will the annual income have to increase to maintain the same i^* obtained in part (a)?

7.11 Consider two investments, A and B, with the following sequences of cash flows:

	Net Cash Flow	
n	Project A	Project B
0	−$25,000	−$25,000
1	2,000	10,000
2	6,000	10,000
3	12,000	10,000
4	24,000	10,000
5	28,000	5,000

(a) Compute i^* for each investment.

(b) Plot the present-worth curve for each project on the same chart, and find the interest rate that makes the two projects equivalent.

7.12 Consider the following investment projects:

			Project Cash Flows			
N	**A**	**B**	**C**	**D**	**E**	**F**
0	−$100	−$100	−$100	−$100	−$100	−$100
1	200	470	−200	0	300	300
2	300	720	200	0	250	100
3	400	360	250	500	−40	400

(a) For each project, apply the sign rule to predict the number of possible i^*'s.

(b) For each project, plot the NPW profile as a function of i between 0 and 200%.

(c) For each project, compute the value(s) of i^*.

7.13 Consider an investment project with the following cash flows:

n	Net Cash Flow
0	−$120,000
1	94,000
2	144,000
3	72,000

(a) Find the IRR for this investment.

(b) Plot the present worth of the cash flow as a function of i.

(c) On the basis of the IRR criterion, should the project be accepted at MARR = 15%?

Mixed Investments

7.14 Consider the following investment projects:

	Net Cash Flow		
n	**Project 1**	**Project 2**	**Project 3**
0	−$1,000	−$2,000	−$1,000
1	500	1,560	1,400
2	840	944	−100
IRR	?	?	?

Assume that MARR = 12% in the following questions:
(a) Compute i^* for each investment. If the problem has more than one i^*, identify all of them.
(b) Compute IRR(true) for each project.
(c) Determine the acceptability of each investment.

7.15 Consider the following investment projects:

	Project Cash Flow				
n	A	B	C	D	E
0	−$100	−$100	−$5	−$100	$200
1	100	30	10	30	100
2	24	30	30	30	−500
3		70	−40	30	−500
4		70		30	200
5				30	600

(a) Use the quadratic equation to compute i^* for A.
(b) Classify each project as either simple or nonsimple.
(c) Apply the cash flow sign rules to each project, and determine the number of possible positive i^*'s. Identify all projects having a unique i^*.
(d) Compute the IRRs for projects B through E.
(e) Apply the net-investment test to each project.

7.16 Consider the following investment projects:

	Net Cash Flow		
n	**Project 1**	**Project 2**	**Project 3**
0	−$1,600	−$5,000	−$1,000
1	10,000	10,000	4,000
2	10,000	30,000	−4,000
3		−40,000	

Assume that MARR = 12% in the following questions:
(a) Identify the i^*('s) for each investment. If the project has more than one i^*, identify all of them.
(b) Which project(s) is (are) a mixed investment?

(c) Compute the IRR for each project.

(d) Determine the acceptability of each project.

7.17 Consider the following investment projects:

	Net Cash Flow		
n	**Project A**	**Project B**	**Project C**
0	−$100	−$150	−$100
1	30	50	410
2	50	50	−558
3	80	50	252
4		100	
IRR	(23.24%)	(21.11%)	(20%, 40%, 50%)

Assume that MARR = 12% for the following questions:

(a) Identify the pure investment(s).

(b) Identify the mixed investment(s).

(c) Determine the IRR for each investment.

(d) Which project would be acceptable?

7.18 The Boeing Company has received a NASA contract worth $460 million to build rocket boosters for future space missions. NASA will pay $50 million when the contract is signed, another $360 million at the end of the first year, and the $50 million balance at the end of second year. The expected cash outflows required to produce these rocket boosters are estimated to be $150 million now, $100 million during the first year, and $218 million during the second year. The firm's MARR is 12%. The cash flow is as follows:

n	Outflow	Inflow	Net Cash Flow
0	$150	$50	−$100
1	100	360	260
2	218	50	−168

(a) Show whether this project is or is not a mixed investment.

(b) Compute the IRR for this investment.

(c) Should Boeing accept the project?

7.19 Consider the following investment projects:

	Net Cash Flow		
n	**Project A**	**Project B**	**Project C**
0	−$100		−$100
1	216	−150	50
2	−116	100	−50
3		50	200
4		40	
i*	?	15.51%	29.95%

(a) Compute $i*$ for project A. If there is more than one $i*$, identify all of them.
(b) Identify the mixed investment(s).
(c) Assuming that MARR = 10%, determine the acceptability of each project on the basis of the IRR criterion.

7.20 Consider the following investment projects:

	Net Cash Flow				
n	**A**	**B**	**C**	**D**	**E**
0	−$1,000	−$5,000	−$2,000	−$2,000	−$1,000
1	3,100	20,000	1,560	2,800	3,600
2	−2,200	12,000	944	−200	−5,700
3		−3,000			3,600
i*	?	?	18%	32.45%	35.39%

Assume that MARR = 12% in the following questions:
(a) Compute $i*$ for projects A and B. If the project has more than one $i*$, identify all of them.
(b) Classify each project as either a pure or a mixed investment.
(c) Compute the IRR for each investment.
(d) Determine the acceptability of each project.

7.21 Consider an investment project whose cash flows are as follows:

n	Net Cash Flow
0	−$5,000
1	10,000
2	30,000
3	−40,000

(a) Plot the present-worth curve by varying i from 0% to 250%.

(b) Is this a mixed investment?

(c) Should the investment be accepted at MARR $= 18\%$?

7.22 Consider the following two mutually exclusive investment projects:

	Net Cash Flow	
n	**Project A**	**Project B**
0	$-\$300$	$-\$800$
1	0	1,150
2	690	40
$i*$	51.66%	46.31%

Assume that MARR $= 15\%$.

(a) According to the IRR criterion, which project would be selected?

(b) Sketch the PW(i) function on the incremental investment $(B - A)$.

7.23 Consider the following cash flows of a certain project:

n	Net Cash Flow
0	$-\$100,000$
1	310,000
2	$-220,000$

The project's $i*$'s are computed as 10% and 100%, respectively. The firm's MARR is 8%.

(a) Show why this investment project fails the net-investment test.

(b) Compute the IRR, and determine the acceptability of this project.

7.24 Consider the following investment projects:

	Net Cash Flow		
n	**Project 1**	**Project 2**	**Project 3**
0	$-\$1,000$	$-\$1,000$	$-\$1,000$
1	$-1,000$	1,600	1,500
2	2,000	-300	-500
3	3,000	-200	2,000

Which of the following statements is correct?

(a) All projects are nonsimple investments.

(b) Project 3 should have three real rates of return.

(c) All projects will have a unique positive real rate of return.

(d) None of the above.

IRR Analysis

7.25 Agdist Corporation distributes agricultural equipment. The board of directors is considering a proposal to establish a facility to manufacture an electronically controlled "intelligent" crop sprayer invented by a professor at a local university. This crop sprayer project would require an investment of $10 million in assets and would produce an annual after-tax net benefit of $1.8 million over a service life of eight years. All costs and benefits are included in these figures. When the project terminates, the net proceeds from the sale of the assets will be $1 million. Compute the rate of return of this project. Is this a good project at MARR = 10%?

7.26 Consider an investment project with the following cash flows:

n	Cash Flow
0	− $5,000
1	0
2	4,840
3	1,331

Compute the IRR for this investment. Is the project acceptable at MARR = 10%?

7.27 Consider the following cash flow of a certain project:

n	Net Cash Flow
0	− $2,000
1	800
2	900
3	X

If the project's IRR is 10%,
(a) Find the value of X.
(b) Is this project acceptable at MARR = 8%?

7.28 You are considering a luxury apartment building project that requires an investment of $12,500,000. The building has 50 units. You expect the maintenance cost for the apartment building to be $250,000 the first year and $300,000 the second year. The maintenance cost will continue to increase by $50,000 in subsequent years. The cost to hire a manager for the building is estimated to be $80,000 per year. After five years of operation, the apartment building can be sold for $14,000,000. What is the annual rent per apartment unit that will provide a return on investment of 15%? Assume that the building will remain fully occupied during its five years of operation.

7.29 A machine costing $25,000 to buy and $3,000 per year to operate will save mainly labor expenses in packaging over six years. The anticipated salvage value of the machine at the end of the six years is $5,000. To receive a 10% return on investment (rate of return), what is the minimum required annual savings in labor from this machine?

7.30 Champion Chemical Corporation is planning to expand one of its propylene-manufacturing facilities. At $n = 0$, a piece of property costing $1.5 million must be purchased to build a plant. The building, which needs to be expanded during the first year, costs $3 million. At the end of the first year, the company needs to spend about $4 million on equipment and other start-up costs. Once the building becomes operational, it will generate revenue in the amount of $3.5 million during the first operating year. This will increase at the annual rate of 5% over the previous year's revenue for the next 9 years. After 10 years, the sales revenue will stay constant for another 3 years before the operation is phased out. (It will have a project life of 13 years after construction.) The expected salvage value of the land at the end of the project's life would be about $2 million, the building about $1.4 million, and the equipment about $500,000. The annual operating and maintenance costs are estimated to be approximately 40% of the sales revenue each year. What is the IRR for this investment? If the company's MARR is 15%, determine whether the investment is a good one. (Assume that all figures represent the effect of the income tax.)

7.31 Recent technology has made possible a computerized vending machine that can grind coffee beans and brew fresh coffee on demand. The computer also makes possible such complicated functions as changing $5 and $10 bills, tracking the age of an item, and moving the oldest stock to the front of the line, thus cutting down on spoilage. With a price tag of $4,500 for each unit, Easy Snack has estimated the cash flows in millions of dollars over the product's six-year useful life, including the initial investment, as follows:

n	Net Cash Flow
0	−$20
1	8
2	17
3	19
4	18
5	10
6	3

(a) On the basis of the IRR criterion, if the firm's MARR is 18%, is this product worth marketing?

(b) If the required investment remains unchanged, but the future cash flows are expected to be 10% higher than the original estimates, how much of an increase in IRR do you expect?

(c) If the required investment has increased from $20 million to $22 million, but the expected future cash flows are projected to be 10% smaller than the original estimates, how much of a decrease in IRR do you expect?

Comparing Alternatives

7.32 Consider two investments A and B with the following sequences of cash flows:

	Net Cash Flow	
n	**Project A**	**Project B**
0	−$120,000	−$100,000
1	20,000	15,000
2	20,000	15,000
3	120,000	130,000

(a) Compute the IRR for each investment.

(b) At MARR = 15%, determine the acceptability of each project.

(c) If A and B are mutually exclusive projects, which project would you select, based on the rate of return on incremental investment?

7.33 With $10,000 available, you have two investment options. The first is to buy a certificate of deposit from a bank at an interest rate of 10% annually for five years. The second choice is to purchase a bond for $10,000 and invest the bond's interest in the bank at an interest rate of 9%. The bond pays 10% interest annually and will mature to its face value of $10,000 in five years. Which option is better? Assume that your MARR is 9% per year.

7.34 A manufacturing firm is considering the following mutually exclusive alternatives:

	Net Cash Flow	
n	**Project A1**	**Project A2**
0	−$2,000	−$3,000
1	1,400	2,400
2	1,640	2,000

Determine which project is a better choice at a MARR = 15%, based on the IRR criterion.

7.35 Consider the following two mutually exclusive alternatives:

	Net Cash Flow	
n	**Project A1**	**Project A2**
0	−$10,000	−$12,000
1	5,000	6,100
2	5,000	6,100
3	5,000	6,100

(a) Determine the IRR on the incremental investment in the amount of $2,000.

(b) If the firm's MARR is 10%, which alternative is the better choice?

7.36 Consider the following two mutually exclusive investment alternatives:

	Net Cash Flow	
n	**Project A1**	**Project A2**
0	−$15,000	−$20,000
1	7,500	8,000
2	7,500	15,000
3	7,500	5,000
IRR	23.5%	20%

(a) Determine the IRR on the incremental investment in the amount of $5,000. (Assume that MARR = 10%.)

(b) If the firm's MARR is 10%, which alternative is the better choice?

7.37 You are considering two types of automobiles. Model A costs $18,000 and model B costs $15,624. Although the two models are essentially the same, after four years of use model A can be sold for $9,000, while model B can be sold for $6,500. Model A commands a better resale value because its styling is popular among young college students. Determine the rate of return on the incremental investment of $2,376. For what range of values of your MARR is model A preferable?

7.38 A plant engineer is considering two types of solar water heating system:

	Model A	Model B
Initial cost	$7,000	$10,000
Annual savings	$700	$1,000
Annual maintenance	$100	$50
Expected life	20 years	20 years
Salvage value	$400	$500

The firm's MARR is 12%. On the basis of the IRR criterion, which system is the better choice?

7.39 Consider the following investment projects:

	Net Cash Flow					
n	**A**	**B**	**C**	**D**	**E**	**F**
0	−$100	−$200	−$4,000	−$2,000	−$2,000	−$3,000
1	60	120	2,410	1,400	3,700	2,500
2	50	150	2,930	1,720	1,640	1,500
3	50					
*i**	28.89%	21.65%	21.86%	31.10%	121.95%	23.74%

Assume that MARR = 15%.

(a) Projects A and B are mutually exclusive. Assuming that both projects can be repeated for an indefinite period, which one would you select on the basis of the IRR criterion?

(b) Suppose projects C and D are mutually exclusive. According to the IRR criterion, which project would be selected?

(c) Suppose projects E and F are mutually exclusive. Which project is better according to the IRR criterion?

7.40 Fulton National Hospital is reviewing ways of cutting the cost of stocking medical supplies. Two new stockless systems are being considered, to lower the hospital's holding and handling costs. The hospital's industrial engineer has compiled the relevant financial data for each system as follows (dollar values are in millions):

	Current Practice	Just-in-Time System	Stockless Supply System
Start-up cost	$0	$2.5	$5
Annual stock holding cost	$3	$1.4	$0.2
Annual operating cost	$2	$1.5	$1.2
System life	8 years	8 years	8 years

The system life of eight years represents the period that the contract with the medical suppliers is in force. If the hospital's MARR is 10%, which system is more economical?

7.41 Consider the cash flows for the following investment projects:

	Project Cash Flow				
n	A	B	C	D	E
0	−$1,000	−$1,000	−$2,000	$1,000	−$1,200
1	900	600	900	−300	400
2	500	500	900	−300	400
3	100	500	900	−300	400
4	50	100	900	−300	400

Assume that the MARR = 12%.

(a) Suppose A, B, and C are mutually exclusive projects. Which project would be selected on the basis of the IRR criterion?

(b) What is the borrowing rate of return (BRR) for project D?

(c) Would you accept project D at MARR $= 20\%$?

(d) Assume that projects C and E are mutually exclusive. Using the IRR criterion, which project would you select?

7.42 Consider the following investment projects:

	Net Cash Flow		
n	Project 1	Project 2	Project 3
0	−$1,000	−$5,000	−$2,000
1	500	7,500	1,500
2	2,500	600	2,000

Assume that MARR $= 15\%$.

(a) Compute the IRR for each project.

(b) On the basis of the IRR criterion, if the three projects are mutually exclusive investments, which project should be selected?

7.43 Consider the following two investment alternatives:

	Net Cash Flow	
N	Project A	Project B
0	−$10,000	−$20,000
1	5,500	0
2	5,500	0
3	5,500	40,000
IRR	30%	?
PW(15%)	?	6300

The firm's MARR is known to be 15%.

(a) Compute the IRR of project B.

(b) Compute the NPW of project A.

(c) Suppose that projects A and B are mutually exclusive. Using the IRR, which project would you select?

7.44 The E. F. Fedele Company is considering acquiring an automatic screwing machine for its assembly operation of a personal computer. Three different models with varying automatic features are under consideration. The required investments are $360,000 for model A, $380,000 for model B, and $405,000 for model C. All three models are expected to have the same service life of eight years. The following financial information, in which model $(B - A)$ represents the incremental cash flow determined by subtracting model A's cash flow from model B's, is available:

Model	IRR (%)
A	30%
B	15
C	25

Model	Incremental IRR (%)
(B − A)	5%
(C − B)	40
(C − A)	15

If the firm's MARR is known to be 12%, which model should be selected?

7.45 The GeoStar Company, a leading manufacturer of wireless communication devices, is considering three cost-reduction proposals in its batch job-shop manufacturing operations. The company has already calculated rates of return for the three projects, along with some incremental rates of return:

Incremental Investment	Incremental Rate of Return (%)
$A_1 - A_0$	18%
$A_2 - A_0$	20
$A_3 - A_0$	25
$A_2 - A_1$	10
$A_3 - A_1$	18
$A_3 - A_2$	23

A_0 denotes the do-nothing alternative. The required investments are $420,000 for A_1, $550,000 for A_2, and $720,000 for A_3. If the MARR is 15%, what system should be selected?

7.46 A manufacturer of electronic circuit boards is considering six mutually exclusive cost-reduction projects for its PC-board manufacturing plant. All have lives of 10 years and zero salvage values. The required investment and the estimated after-tax reduction in annual disbursements for each alternative are as follows, along with computed rates of return on incremental investments:

Proposal A_j	Required After-Tax Investment	Savings	Rate of Return (%)
A_1	$60,000	$22,000	35.0%
A_2	100,000	28,200	25.2
A_3	110,000	32,600	27.0
A_4	120,000	33,600	25.0
A_5	140,000	38,400	24.0
A_6	150,000	42,200	25.1

Incremental Investment	Incremental Rate of Return (%)
$A_2 - A_1$	9.0%
$A_3 - A_2$	42.8
$A_4 - A_3$	0.0
$A_5 - A_4$	20.2
$A_6 - A_5$	36.3

If the MARR is 15%, which project would you select, based on the rate of return on incremental investment?

7.47 Baby Doll Shop manufactures wooden parts for dollhouses. The worker is paid $8.10 an hour and, using a handsaw, can produce a year's required production (1,600 parts) in just eight 40-hour weeks. That is, the worker averages five parts per hour when working by hand. The shop is considering purchasing of a power band saw with associated fixtures, to improve the productivity of this operation. Three models of power saw could be purchased: Model A (the economy version), model B (the high-powered version), and model C (the deluxe high-end version). The major operating difference between these models is their speed of operation. The investment costs, including the required fixtures and other operating characteristics, are summarized as follows:

Category	By Hand	Model A	Model B	Model C
Production rate (parts/hour)	5	10	15	20
Labor hours required (hours/year)	320	160	107	80
Annual labor cost (@ $8.10/hour)	2,592	1,296	867	648
Annual power cost ($)		400	420	480
Initial investment ($)		4,000	6,000	7,000
Salvage value ($)		400	600	700
Service life (years)		20	20	20

Assume that MARR = 10%. Are there enough savings to purchase any of the power band saws? Which model is most economical, based on the rate-of-return principle? (Assume that any effect of income tax has been already considered in the dollar estimates.) (*Source*: This problem is adapted with the permission of Professor Peter Jackson of Cornell University.)

Unequal Service Lives

7.48 Consider the following two mutually exclusive investment projects for which MARR = 15%:

	Net Cash Flow	
n	**Project A**	**Project B**
0	−$100	−$200
1	60	120
2	50	150
3	50	
IRR	28.89%	21.65%

On the basis of the IRR criterion, which project would be selected under an infinite planning horizon with project repeatability likely?

7.49 Consider the following two mutually exclusive investment projects:

	Net Cash Flow	
n	**Project A1**	**Project A2**
0	−$10,000	−$15,000
1	5,000	20,000
2	5,000	
3	5,000	

(a) To use the IRR criterion, what assumption must be made in comparing a set of mutually exclusive investments with unequal service lives?

(b) With the assumption defined in part (a), determine the range of MARRs which will indicate that project A1 should be selected.

Short Case Studies

ST7.1 Critics have charged that, in carrying out an economic analysis, the commercial nuclear power industry does not consider the cost of decommissioning, or "mothballing," a nuclear power plant and that the analysis is therefore unduly optimistic. As an example, consider the Tennessee Valley Authority's Bellefont twin nuclear generating facility under construction at Scottsboro, in northern Alabama: The initial cost is $1.5 billion (present worth at the start of operations), the estimated life is 40 years, the annual operating and maintenance costs the first year are assumed to be 4.6% of the initial cost and are expected to increase at the fixed rate of 0.05%

of the initial cost each year, and annual revenues are estimated to be three times the annual operating and maintenance costs throughout the life of the plant.

(a) The criticism that the economic analysis is overoptimistic because it omits "mothballing" costs is not justified, since the addition of a cost of 50% of the initial cost to "mothball" the plant decreases the 10% rate of return only to approximately 9.9%.

(b) If the estimated life of the plants is more realistically taken to be 25 years instead of 40 years, then the criticism is justified. By reducing the life to 25 years, the rate of return of approximately 9% without a "mothballing" cost drops to approximately 7.7% when a cost to "mothball" the plant equal to 50% of the initial cost is added to the analysis.

Comment on these statements.

ST7.2 The B&E Cooling Technology Company, a maker of automobile air-conditioners, faces an uncertain, but impending, deadline to phase out the traditional chilling technique, which uses chlorofluorocarbons (CFCs), a family of refrigerant chemicals believed to attack the earth's protective ozone layer. B&E has been pursuing other means of cooling and refrigeration. As a near-term solution, the engineers recommend a cold technology known as absorption chiller, which uses plain water as a refrigerant and semiconductors that cool down when charged with electricity. B&E is considering two options:

- **Option 1.** Retrofit the plant now to adapt the absorption chiller and continue to be a market leader in cooling technology. Because of untested technology on a large scale, it may cost more to operate the new facility while personnel are learning the new system.

- **Option 2.** Defer the retrofitting until the federal deadline, which is three years away. With expected improvement in cooling technology and technical know-how, the retrofitting cost will be cheaper, but there will be tough market competition, and the revenue would be less than that of Option 1.

The financial data for the two options are as follows:

	Option 1	Option 2
Investment timing	Now	3 years from now
Initial investment	$6 million	$5 million
System life	8 years	8 years
Salvage value	$1 million	$2 million
Annual revenue	$15 million	$11 million
Annual O&M costs	$6 million	$7 million

(a) What assumptions must be made to compare these two options?

(b) If B&E's MARR is 15%, which option is the better choice, based on the IRR criterion?

ST7.3 An oil company is considering changing the size of a small pump that is currently operational in wells in an oil field. If this pump is kept, it will extract 50% of the known crude-oil reserve in the first year of its operation and the remaining 50% in the second year. A pump larger than the current pump will cost $1.6 million, but it will extract 100% of the known reserve in the first year. The total oil revenues over the two years are the same for both pumps, namely, $20 million. The advantage of the large pump is that it allows 50% of the revenues to be realized a year earlier than with the small pump.

	Current Pump	Larger Pump
Investment, year 0	0	$1.6 million
Revenue, year 1	$10 million	$20 million
Revenue, year 2	$10 million	0

If the firm's MARR is known to be 20%, what do you recommend, based on the IRR criterion?

ST7.4 You have been asked by the president of the company you work for to evaluate the proposed acquisition of a new injection molding machine for the firm's manufacturing plant. Two types of injection molding machines have been identified, with the following estimated cash flows:

	Net Cash Flow	
n	**Project 1**	**Project 2**
0	−$30,000	−$40,000
1	20,000	43,000
2	18,200	5,000
IRR	18.1%	18.1%

You return to your office, quickly retrieve your old engineering economics text, and then begin to smile: Aha—this is a classic rate-of-return problem! Now, using a calculator, you find out that both projects have about the same rate of return: 18.1%. This figure seems to be high enough to justify accepting the project, but you recall that the ultimate justification should be done with reference to the firm's MARR. You call the accounting department to find out the current MARR the firm should use in justifying a project. "Oh boy, I wish I could tell you, but my boss will be back next week, and he can tell you what to use," says the accounting clerk.

A fellow engineer approaches you and says, "I couldn't help overhearing you talking to the clerk. I think I can help you. You see, both projects have the same IRR, and on top of that, project 1 requires less investment, but returns more cash flows (−$30,000 + $20,000 + $18,200 = $8,200, and −$40,000 + $43,000 + $5,000 = $8,000); thus, project 1 dominates project 2. For this type of decision problem, you don't need to know a MARR!"

(a) Comment on your fellow engineer's statement.

(b) At what range of MARRs would you recommend the selection of project 2?

Analysis of Project
Cash Flows

EIGHT

Cost Concepts Relevant to Decision Making

High Hopes for Beer Bottles[1] Three hundred billion beer bottles a year worldwide is a mighty tempting target for the plastics industry. Brewers generally say they need a bottle that provides shelf life of over 120 days with less than 15% loss of CO_2 and admittance of no more than 1 ppm of oxygen. Internal or external coatings, and three- or five-layer polyethylene terephthalate (PET) structures using barrier materials are being evaluated to reach that performance.

What is the least expensive way to make a 0.5L PET barrier bottle? Summit International LLC, Smyrna, GA, which specializes in preform and container development and market research, compared the manufacturing costs of five different barrier technologies against a standard monolayer PET bottle. Summit looked at three-layer and five-layer and at internally and externally coated containers, all of them reheat stretch blow molded. It found the bottle with an external coating to be least costly, while the internally coated bottle was the most expensive.

The firm compared a five-layer structure with an oxygen-scavenger material, a three-layer bottle with a $2.50/lb barrier material, and a second three-layer structure with a $6/lb barrier. Also compared were a bottle coated inside using Sidel's new Actis plasma technology and a bottle coated on the outside.

Capital investment (preform and bottle machines, utilities, downstream equipment, quality control, spare parts, and installation) for producing 20,000 bottles/hr is $10.8 million for the five-layer bottle, $9.9 million for both three-layer structures, $9.2 million for internal coating, $7.5 for external coating, and $6.8 million with no barrier.

---

[1] "Blow Molding Close-Up: Prospects Brighten for PET Beer Bottles," Mikell Knights, Plastic Technology Online, © copyright 2005, Gardner Publications, Inc.

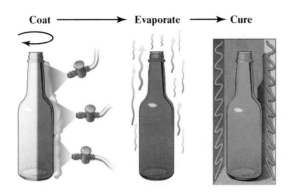

Coat ⟶ Evaporate ⟶ Cure

Spray coating of external PET bottles

Direct manufacturing cost per 1,000 (materials, energy, labor, maintenance, and scrap) amounts to $66.57 for three layers with the expensive barrier, $59.35 for five layers, $55.34 for the external coating, $54.63 for three layers with the low-cost barrier, $46.90 for internal coating, and $44.63 without barrier.

You may be curious how all of those cost data were estimated in the chapter opening story. Before we study the different kinds of engineering economic decision problems, we need to understand the concept of various costs. At the level of plant operations, engineers must make decisions involving materials, plant facilities, and the in-house capabilities of company personnel. Consider, for example, the manufacture of food processors. In terms of selecting materials, several of the parts could be made of plastic, whereas others must be made of metal. Once materials have been chosen, engineers must consider the production methods, the shipping weight, and the method of packaging necessary to protect the different types of materials. In terms of actual production, parts may be made in-house or purchased from an outside vendor, depending on the availability of machinery and labor.

Present economic studies: Various economic analyses for short operating decisions.

All these operational decisions (commonly known as **present economic studies** in traditional engineering economic texts) require estimating the costs associated with various production or manufacturing activities. Because these costs also provide the basis for developing successful business strategies and planning future operations, it is important to understand how various costs respond to changes in levels of business activity. In this chapter, we discuss many of the possible uses of cost data. We also discuss how to define, classify, and estimate costs for each use. Our ultimate task in doing so is to explain how costs are classified in making various engineering economic decisions.

CHAPTER LEARNING OBJECTIVES

After completing this chapter, you should understand the following concepts:

- Various cost terminologies that are common in cost accounting and engineering economic studies.

- How a cost item reacts or responds to changes in the level of production or business activities.

- The types of cost data that management needs in making choice between alternative courses of action.

- The types of present economic studies frequently performed by engineers in manufacturing and business environment.

- How to develop a production budget related to operating activities.

8.1 General Cost Terms

In engineering economics, the term **cost** is used in many different ways. Because there are many types of costs, each is classified differently according to the immediate needs of management. For example, engineers may want cost data to prepare external reports, to prepare planning budgets, or to make decisions. Also, each different use of cost data demands a different classification and definition of *cost*. For example, the preparation of external financial reports requires the use of historical cost data, whereas decision making may require current cost data or estimated future cost data.

Our initial focus in this chapter is on manufacturing companies, because their basic activities (acquiring raw materials, producing finished goods, marketing, etc.) are commonly found in most other businesses. Therefore, the understanding of costs in a manufacturing company can be helpful in understanding costs in other types of business organizations.

8.1.1 Manufacturing Costs

Several types of manufacturing costs incurred by a typical manufacturer are illustrated in Figure 8.1. In converting raw materials into finished goods, a manufacturer incurs various costs associated with operating a factory. Most manufacturing companies divide manufacturing costs into three broad categories: direct raw material costs, direct labor costs, and manufacturing overhead.

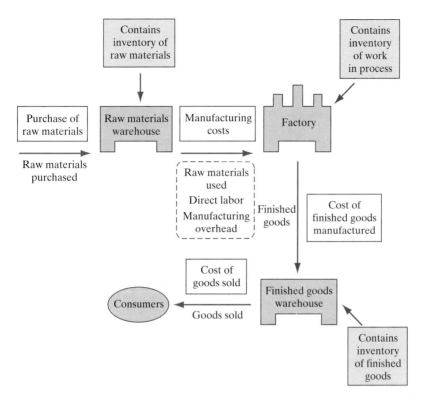

Figure 8.1   Various types of manufacturing costs incurred by a manufacturer.

Direct Raw Materials

Direct raw materials are any materials that are used in the final product and that can be easily traced to it. Some examples are wood in furniture, steel in bridge construction, paper in printing firms, and fabric for clothing manufacturers. It is important to note that the finished product of one company can become the raw materials of another company. For example, the computer chips produced by Intel are a raw material used by Dell in its personal computers.

Direct cost: A cost that can be directly traced to producing specific goods or services.

Direct Labor

Like direct raw materials, direct labor incurs costs that go into the production of a product. The labor costs of assembly-line workers, for example, would be direct labor costs, as would the labor costs of welders in metal-fabricating industries, carpenters or bricklayers in home building, and machine operators in various manufacturing operations.

Manufacturing Overhead

Manufacturing overhead, the third element of manufacturing cost, includes all costs of manufacturing except the costs of direct materials and direct labor. In particular, manufacturing overhead includes such items as the costs of indirect materials; indirect labor; maintenance and repairs on production equipment; heat and light, property taxes, depreciation, and insurance on manufacturing facilities; and overtime premiums. The most

Overhead: A reference in accounting to all costs not including or related to direct labor, materials, or administration costs.

important thing to note about manufacturing overhead is the fact that, unlike direct materials and direct labor, it is not easily traceable to specific units of output. In addition, many manufacturing overhead costs do not change as output changes, as long as the production volume stays within the capacity of the plant. For example, depreciation of factory buildings is unaffected by the amount of production during any particular period. If, however, a new building is required to meet any increased production, manufacturing overhead will certainly increase.

- Sometimes it may not be worth the effort to trace the costs of materials that are relatively insignificant in the finished products. Such minor items include the solder used to make electrical connections in a computer circuit board and the glue used to bind this textbook. Materials such as solder and glue are called *indirect materials* and are included as part of manufacturing overhead.
- Sometimes we may not be able to trace some of the labor costs to the creation of a product. We treat this type of labor cost as a part of manufacturing overhead, along with indirect materials. *Indirect labor* includes the wages of janitors, supervisors, material handlers, and night security guards. Although the efforts of these workers are essential to production, it would be either impractical or impossible to trace their costs to specific units of product. Therefore, we treat such labor costs as indirect labor costs.

8.1.2 Nonmanufacturing Costs

Two additional costs incurred in supporting any manufacturing operation are (1) marketing or selling costs and (2) administrative costs. Marketing or selling costs include all costs necessary to secure customer orders and get the finished product or service into the hands of the customer. Breakdowns of these types of costs provide data for control over selling and administrative functions in the same way that manufacturing cost breakdowns provide data for control over manufacturing functions. For example, a company incurs costs for

- **Overhead.** Heat and light, property taxes, and depreciation or similar items associated with the company's selling and administrative functions.
- **Marketing.** Advertising, shipping, sales travel, sales commissions, and sales salaries. Marketing costs include all executive, organizational, and clerical costs associated with sales activities.
- **Administrative functions.** Executive compensation, general accounting, public relations, and secretarial support, associated with the general management of an organization.

Matching concept: The accounting principle that requires the recognition of all costs that are associated with the generation of the revenue reported in the income statement.

8.2 Classifying Costs for Financial Statements

For purposes of preparing financial statements, we often classify costs as either period costs or product costs. To understand the difference between them, we must introduce the matching concept essential to any accounting studies. In financial accounting, the **matching principle** states that *the costs incurred in generating a certain amount of revenue should be recognized as expenses in the same period that the revenue is recognized.* This matching principle is the key to distinguishing between period costs and product costs. Some costs are matched against periods and become expenses immediately. Other costs are matched against products and do not become expenses until the products are sold, which may be in the next accounting period.

8.2.1 Period Costs

Period costs are costs charged to expenses in the period in which they are incurred. The underlying assumption is that the associated benefits are received in the same period the cost is incurred. Some specific examples are all general and administrative expenses, selling expenses, and insurance and income tax expenses. Therefore, advertising costs, executives' salaries, sales commissions, public-relations costs, and other nonmanufacturing costs discussed earlier would all be period costs. Such costs are not related to the production and flow of manufactured goods, but are deducted from revenue in the income statement. In other words, period costs will appear on the income statement as expenses during the time in which they occur.

8.2.2 Product Costs

Some costs are better matched against products than they are against periods. Costs of this type—called **product costs**—consist of the costs involved in the purchase or manufacture of goods. In the case of manufactured goods, product costs are the costs of direct materials, direct labor costs, and manufacturing overhead. Product costs are not viewed as expenses; rather, they are the cost of creating inventory. Thus, product costs are considered an asset until the associated goods are sold. At the time they are sold, the costs are released from inventory as expenses (typically called cost of goods sold) and matched against sales revenue. Since product costs are assigned to inventories, they are also known as *inventory costs*. In theory, product costs include all manufacturing costs—that is, all costs relating to the manufacturing process. As shown in Figure 8.2, product costs appear on financial statements when the inventory, or final goods, is sold, not when the product is manufactured.

To understand product costs more fully, let us look briefly at the flow of costs in a manufacturing company. By doing so, we will be able to see how product costs move

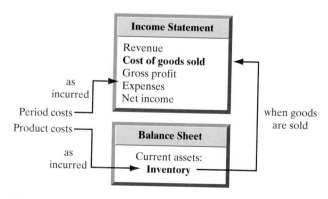

Figure 8.2 How the period costs and product costs flow through financial statements from the manufacturing floor to sales.

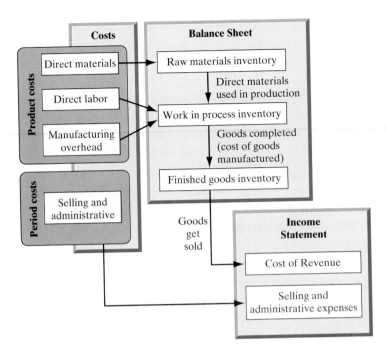

Figure 8.3 Cost flows and classifications in a manufacturing company.

through the various accounts and affect the balance sheet and the income statement in the course of the manufacture and sale of goods. The flows of period costs and product costs through the financial statements are illustrated in Figure 8.3. All product costs filter through the balance-sheet statement as "inventory cost." If the product gets sold, the inventory costs in the balance-sheet statement are transferred to the income statement under the head "cost of goods sold."

- **Raw-materials inventory.** This account in the balance-sheet statement represents the unused portion of the raw materials on hand at the end of the fiscal year.
- **Work-in-process inventory.** This balance-sheet entry consists of the partially completed goods on hand in the factory at year-end. When raw materials are used in production, their costs are transferred to the work-in-process inventory account as direct materials. Note that direct labor cost and manufacturing overhead cost are also added directly to work-in-process, which can be viewed as the assembly line in a manufacturing plant, where workers are stationed and where products slowly take shape as they move from one end of the line to the other.
- **Finished-goods inventory.** This account shows the cost of finished goods that are on hand and awaiting sale to customers at year-end. As the goods are completed, accountants transfer the cost from the work-in-process account into the finished-goods account. At this stage, the goods await sale to a customer. As the goods are sold, their cost is transferred from the finished-goods account into the cost-of-goods-sold (or cost-of-revenue) account. At this point, we finally treat the various material, labor, and overhead costs that were involved in the manufacture of the units being sold as *expenses* in the income statement.

Example 8.1 serves to explain the classification scheme for financial statements.

EXAMPLE 8.1 Classifying Costs for Uptown Ice Cream Shop

Here is a look at why it costs $2.50 for a single-dip ice cream cone at a typical store in Washington, DC. The annual sales volume (the number of ice cream cones sold) averages around 185,000 cones, bringing in revenue of $462,500. This is equivalent to selling more than 500 cones a day, assuming a seven-day operation. The following table shows the unit price of an ice cream cone and the costs that go into producing the product:

Items	Total Cost	Unit Price*	% of Price
Ice cream (cream, sugar, milk, and milk solids)	$120,250	$0.65	26%
Cone	9,250	0.05	2
Rent	112,850	0.61	24
Wages	46,250	0.25	10
Payroll taxes	9,250	0.05	2
Sales taxes	42,550	0.23	9
Business taxes	14,800	0.08	3
Debt service	42,550	0.23	9
Supplies	16,650	0.09	4
Utilities	14,800	0.08	3
Other expenses (insurance, advertising, fees, and heating and lighting for shop)	9,250	0.05	2
Profit	24,050	0.13	5
Total	**$462,500**	**$2.50**	**100**

*Based on an annual volume of 185,000 cones.

If you were to classify the operating costs into either product costs or period costs, how would you do it?

SOLUTION:

Given: Financial data just described.

Find: Classify the cost elements into product costs and period costs.

The following is a breakdown of the two kinds of costs:

- **Product costs:** Costs incurred in preparing 185,000 ice cream cones per year.

Raw materials:	
Ice cream @ $0.65	$120,250
Cone @ $0.05	9,250
Labor:	
Wages @ $0.25	46,250
Overhead:	
Supplies @ $0.09	16,650
Utilities @ $0.08	14,800
Total product cost	$207,200

• **Period costs:** Costs incurred in running the shop regardless of sales volume.

Business taxes:	
Payroll taxes @ $0.05	$ 9,250
Sales taxes @ $0.23	42,550
Business taxes @ $0.08	14,800
Operating expenses:	
Rent @ $0.61	112,850
Debt service @ $0.23	42,550
Other @ $0.05	9,250
Total period cost	$231,250

8.3 Cost Classification for Predicting Cost Behavior

In engineering economic analysis, we need to predict how a certain cost will behave in response to a change in activity. For example, a manager will want to estimate the impact a 5% increase in production will have on the company's total wages before he or she decides whether to alter production. **Cost behavior** describes how a cost item will react or respond to changes in the level of business activity.

8.3.1 Volume Index

In general, the operating costs of any company are likely to respond in some way to changes in the company's operating volume. In studying cost behavior, we need to determine some measurable volume or activity that has a strong influence on the amount of cost incurred. The unit of measure used to define volume is called a **volume index**. A volume index may be based on production inputs, such as tons of coal processed, direct labor hours used, or machine-hours worked; or it may be based on production outputs, such as the number of kilowatt-hours generated. For a vehicle, the number of miles driven per year may be used as a volume index. Once we identify a volume index, we try to find out how costs change in response to changes in the index.

8.3.2 Cost Behaviors

Accounting systems typically record the cost of resources acquired and track their subsequent usage. Fixed costs and variable costs are the two most common cost behavior patterns. An additional category known as "mixed costs" contains two parts, the first of which is fixed and the other of which varies with the volume of output.

Fixed Costs

The costs of providing a company's basic operating capacity are known as the company's **fixed cost** or **capacity cost**. For a cost item to be classified as fixed, it must have a relatively wide span of output over which costs are expected to remain constant. This span is called the **relevant range**. In other words, fixed costs do not change within a given period, although volume may change. In the case of an automobile, for example, the annual insurance premium, property tax, and license fee are fixed costs, since they are independent of the number of miles driven per year. Some other typical examples are building rents; depreciation of buildings, machinery, and equipment; and salaries of administrative and production personnel. In our Uptown Scoop Ice Cream Store example, we may classify expenses such as rent, business taxes, debt service, and other (insurance, advertising, professional fees) as fixed costs (costs that are fixed in total for a given period of time and for given production levels).

Fixed cost: A cost that remains constant, regardless of any change in a company's activity.

Variable Costs

In contrast to fixed operating costs, **variable operating costs** have a close relationship to the level of volume of a business. If, for example, volume increases 10%, a total variable cost will also increase by approximately 10%. Gasoline is a good example of a variable automobile cost, because fuel consumption is directly related to miles driven. Similarly, the cost of replacing tires will increase as a vehicle is driven more.

In a typical manufacturing environment, direct labor and material costs are major variable costs. In our Uptown Scoop example, the variable costs would include the cost of the ice cream and cone (direct materials), wages, payroll taxes, sales taxes, and supplies. Both payroll and sales taxes are related to sales volume. In other words, if the store becomes busy, more servers are needed, which will increase the payroll as well as taxes.

Variable cost: A cost that changes in proportion to a change in a company's activity or business.

Mixed Costs

Some costs do not fall precisely into either the fixed or the variable category, but contain elements of both. We refer to these as mixed costs (or **semivariable costs**). In our automobile example, **depreciation** (loss of value) is a mixed cost. On the one hand, some depreciation occurs simply from the passage of time, regardless of how many miles a car is driven, and this represents the fixed portion of depreciation. On the other hand, the more miles an automobile is driven a year, the faster it loses its market value, and this represents the variable portion of depreciation. A typical example of a mixed cost in manufacturing is the cost of electric power. Some components of power consumption, such as lighting, are independent of the operating volume, while other components (e.g., the number of machine-hours equipment is operated) may vary directly with volume. In our Uptown Scoop example, the utility cost can be a mixed cost item: Some lighting and heating requirements might stay the same, but the use of power mixers will be in proportion to sales volume.

Mixed cost: Costs are fixed for a set level of production or consumption, becoming variable after the level is exceeded.

Average Unit Cost

The foregoing description of fixed, variable, and mixed costs was expressed in terms of total volume over a given period. We often use the term **average cost** to express activity cost on a per unit basis. In terms of unit costs, the description of cost is quite different:

- The variable cost per unit of volume is a constant.
- Fixed cost per unit varies with changes in volume: As the volume increases, the fixed cost per unit decreases.
- The mixed cost per unit also changes as volume changes, but the amount of change is smaller than that for fixed costs.

To explain the behavior of the fixed, variable, mixed, and average costs in relation to volume, let's consider a medium-size car, say, the 2005 Ford Taurus SEL Deluxe six-cylinder (3.0-liter) four-door sedan. In calculating the vehicle's operating costs, we may use the procedure outlined by the American Automobile Association (AAA) as shown in Table 8.1.

On the basis of the assumptions in Table 8.1, the operating and ownership costs of our Ford vehicle are estimated as follows:

Operating costs:

- Gas and oil   8.5 cents
- Maintenance   5.8 cents
- Tires   0.7 cent

  Cost per mile   15.0 cents

TABLE 8.1 **Assumptions Used in Calculating the Average Cost of Owning and Operating a New Vehicle**

What's Covered	Costs Base
Fuel	U.S. price of unleaded gasoline from AAA's *Fuel Gauge Report*, weighted 60% city and 40% highway driving.
Maintenance	Costs of retail parts and labor for normal, routine maintenance, as specified by the vehicle manufacturer.
Tires	Costs are based on the price of one set of replacement tires of the same quality, size, and ratings as those which came with the vehicle.
Insurance	A full-coverage policy for a married 47-year-old male with a good driving record living a small city and commuting 3 to 10 miles daily to work.
License, Registration, and Taxes	All government taxes and fees payable at time of purchase, as well as fees due each year to keep the vehicle licensed and registered.
Depreciation	Based on the difference between the new-vehicle purchase price and the estimated trade-in value at the end of five years.
Finance	Based on a five-year loan at 6% interest with a 10% down payment.

Source: American Automobile Association (AAA).

Ownership costs:

- Comprehensive insurance ($250 deductible) $1,195
- License, registration, taxes $390
- Depreciation (15,000 miles annually) $4,005
- Finance charge (20% down, loan @8.5%/4 years) $740
  Cost per year $6,330
  Cost per day $17.34
  Added depreciation costs (per 1,000 miles over
     15,000 miles annually) $185

Total cost per mile:

- Cost per mile × 15,000 miles $2,250
- Cost per day × 365 days per year $6,330
  Total cost per year $8,580
  Average cost per mile ($8,580/15,000)   57.2 cents

Now, if you drive the same vehicle for 20,000 miles instead of 15,000 miles, you may be interested in knowing how the average cost per mile would change. Example 8.2 illustrates how you determine the average cost as a function of mileage.

EXAMPLE 8.2 Calculating Average Cost per Mile as a Function of Mileage

Table 8.2 itemizes the operating and ownership costs associated with driving a passenger car by fixed, variable, and mixed classes. Note that the only change from the preceding list is in the depreciation amount. Using the given data, develop a cost–volume chart and calculate the average cost per mile as a function of the annual mileage.

DISCUSSION: First we may examine the effect of driving an additional 1,000 miles over the 15,000 allotted miles. Since the loss in the car's value due to driving an additional 1,000 miles over 15,000 miles is estimated to be $185, the total cost per year and the average cost per mile, based on 16,000 miles, can be recalculated as follows:

- Added depreciation cost: $185.
- Added operating cost: 1,000 miles × 15 cents = $150.
- Total cost per year: $8,580 + $185 + $150 = $8,915.
- Average cost per mile ($8,915/16,000 miles): 55.72 cents.

Note that the average cost comes down as you drive more, as the ownership cost per mile is further reduced.

SOLUTION

Given: Financial data.

Find: The average cost per mile at an annual operating volume between 15,000 and 20,000 miles.

TABLE 8.2 Cost Classification of Owning and Operating a Passenger Car

Cost Classification	Reference	Cost
Variable costs:		
Standard miles per gallon	20 miles/gallon	
Average fuel price per gallon	$1.939/gallon	
Fuel and oil per mile		$0.085
Maintenance per mile		$0.058
Tires per mile		$0.007
Annual fixed costs:		
Insurance (comprehensive)		$1,195
License, registration, taxes		$390
Finance charge		$740
Mixed costs: Depreciation		
Fixed portion per year (15,000 miles)		$4,005
Variable portion per mile (above 15,000 miles)		$0.185

In Table 8.3, we summarize the costs of owning and operating the automobile at various annual operating volumes from 15,000 to 20,000 miles. Once the total cost figures are available at specific volumes, we can calculate the effect of volume on unit (per mile) costs by converting the total cost figures to average unit costs.

TABLE 8.3 Operating Costs as a Function of Mileage Driven

Volume Index (miles)	15,000	16,000	17,000	18,000	19,000	20,000
Variable costs (15¢ per mile)	$ 2,250	$ 2,400	$ 2,550	$ 2,700	$ 2,850	$ 3,000
Fixed costs	2,325	2,325	2,325	2,325	2,325	2,325
Mixed costs (depreciation):						
Variable portion	0	185	370	555	740	925
Fixed portion	4,005	4,005	4,005	4,005	4,005	4,005
Total variable cost	2,250	2,585	2,920	3,255	3,590	3,925
Total fixed cost	6,330	6,330	6,330	6,330	6,330	6,330
Total costs	$ 8,580	$ 8,915	$ 9,250	$ 9,585	$ 9,920	$10,255
Cost per mile	$0.5720	$0.5572	$0.5441	$0.5325	$0.5221	$0.5128

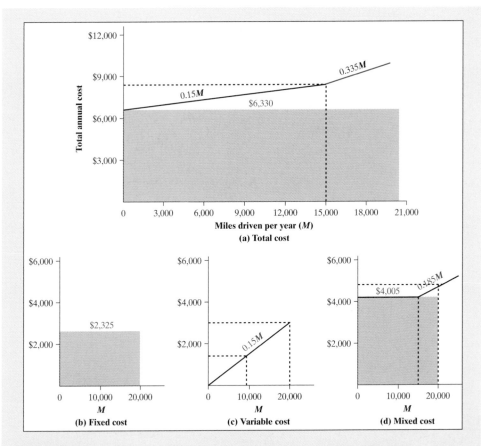

Figure 8.4 Cost–volume relationships pertaining to annual automobile costs (Example 8.2).

To estimate annual costs for any assumed mileage, we construct a **cost–volume diagram** as shown in Figure 8.4(a). Further, we can show the relation between volume (miles driven per year) and the three cost classes separately, as in Figure 8.4(b) through (d). We can use these cost–volume graphs to estimate both the separate and combined costs of operating the car at other possible volumes. For example, an owner who expects to drive 16,500 miles in a given year may estimate the total cost at $9,082.50, or 55.05 cents per mile. In Figure 8.4, all costs more than those necessary to operate at the zero level are known as variable costs. Since the fixed cost is $6,330 a year, the remaining $2,752.50 is variable cost. By combining all the fixed and variable elements of cost, we can state simply that the cost of owning and operating an automobile is $6,330 per year, plus 16.68 cents per mile driven during the year.

Figure 8.5 illustrates graphically the average unit cost of operating the automobile. The average fixed unit cost, represented by the height of the middle curve in the figure, declines steadily as the volume increases. The average unit cost is high when volume is low because the total fixed cost is spread over a relatively few units

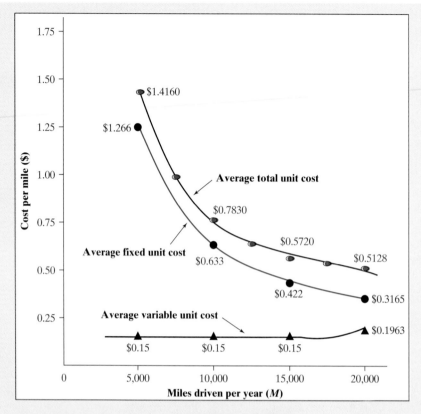

Figure 8.5 Average cost per mile of owning and operating a car (Example 8.2).

of volume. In other words, the total fixed costs remain the same regardless of the number of miles driven, but the average fixed cost decreases on a per mile basis as the number of miles driven increases.

8.4 Future Costs for Business Decisions

In the previous sections, our focus has been on classifying cost data that serve management's need to control and evaluate the operations of a firm. However, these data are historical in nature; they may not be suitable for management's planning future business operations. We are not saying that historical cost data are of no use in making decisions about the future. In fact, they serve primarily as the first step in predicting the uncertain future. However, the types of cost data that management needs in making choices between alternative courses of action are different from historical cost data.

8.4.1 Differential Cost and Revenue

As we have seen throughout the text, decisions involve choosing among alternatives. In business decisions, each alternative has certain costs and benefits that must be compared with the costs and benefits of the other available alternatives. A difference in cost between any two alternatives is known as a **differential cost**. Similarly, a difference in revenue between any two alternatives is known as **differential revenue**. A differential cost is also known as an incremental cost, although, technically, an incremental cost should refer only to an increase in cost from one alternative to another.

Cost–volume relationships based on differential costs find many engineering applications. In particular, they are useful in making a variety of short-term operational decisions. Many short-run problems have the following characteristics:

- The base case is the status quo (the current operation or existing method), and we propose an alternative to the base case. If we find the alternative to have lower costs than the base case, we accept the alternative, assuming that nonquantitative factors do not offset the cost advantage. The **differential (incremental) cost** is the difference in total cost that results from selecting one alternative instead of another. If several alternatives are possible, we select the one with the maximum savings from the base. Problems of this type are often called trade-off problems, because one type of cost is traded off for another.
- New investments in physical assets are not required.
- The planning horizon is relatively short (a week or a month—certainly less than a year).
- Relatively few cost items are subject to change by management decision.

> **Incremental cost** is the overall change that a company experiences by producing one additional unit of good.

Some common examples of short-run problems are method changes, operations planning, and make-or-buy decisions.

Method Changes

Often, we may derive the best information about future costs from an analysis of historical costs. Suppose the proposed alternative is to consider some new method of performing an activity. Then, as Example 8.3 shows, if the differential costs of the proposed method are significantly lower than the current method, we adopt the new method.

EXAMPLE 8.3 Differential Cost Associated with Adopting a New Production Method

The engineering department at an auto-parts manufacturer recommends that the current dies (the base case) be replaced with higher quality dies (the alternative), which would result in substantial savings in manufacturing one of the company's products. The higher cost of materials would be more than offset by the savings in machining time and electricity. If estimated monthly costs of the two alternatives are as shown in the accompanying table, what is the differential cost for going with better dies?

SOLUTION

Given: Financial data.

Find: Which production method is preferred.

In this problem, the differential cost is −$5,000 a month. The differential cost's being negative indicates a saving, rather than an addition to total cost. An important point to remember is that differential costs usually include variable costs, but fixed costs are affected only if the decision involves going outside of the relevant range. Although the production volume remains unchanged in our example, the slight increase

	Current Dies	**Better Dies**	**Differential Cost**
Variable costs:			
Materials	$150,000	$170,000	+$20,000
Machining labor	85,000	64,000	−21,000
Electricity	73,000	66,000	−7,000
Fixed costs:			
Supervision	25,000	25,000	0
Taxes	16,000	16,000	0
Depreciation	40,000	43,000	+3,000
Total	$392,000	$387,000	−$5,000

in depreciation expense is due to the acquisition of new machine tools (dies). All other items of fixed cost remained within their relevant ranges.

Operations Planning

In a typical manufacturing environment, when demand is high, managers are interested in whether to use a one-shift-plus-overtime operation or to add a second shift. When demand is low, it is equally possible to explore whether to operate temporarily at very low volume or to shut down until operations at normal volume become economical. In a chemical plant, several routes exist for scheduling products through the plant. The problem is which route provides the lowest cost. Example 8.4 illustrates how engineers may use cost–volume relationships in a typical operational analysis.

Break-even analysis: An analysis of the level of sales at which a project would make zero profit.

EXAMPLE 8.4 Break-Even Volume Analysis

Sandstone Corporation has one of its manufacturing plants operating on a single-shift five-day week. The plant is operating at its full capacity (24,000 units of output per week) without the use of overtime or extra shifts. Fixed costs for single-shift operation amount to $90,000 per week. The average variable cost is a constant $30 per unit, at all output rates, up to 24,000 units per week. The company has received an order to produce an extra 4,000 units per week beyond the current single-shift maximum capacity. Two options are being considered to fill the new order:

- **Option 1.** Increase the plant's output to 36,000 units a week by adding overtime, by adding Saturday operations, or both. No increase in fixed costs is entailed, but the variable cost is $36 per unit for any output in excess of 24,000 units per week, up to a 36,000-unit capacity.
- **Option 2.** Operate a second shift. The maximum capacity of the second shift is 21,000 units per week. The variable cost of the second shift is $31.50 per unit, and the operation of a second shift entails additional fixed costs of $13,500 per week.

Determine the range of operating volume that will make Option 2 profitable.

SOLUTION

Given: Financial data.

Find: The break-even volume that will make both options indifferent.

In this example, the operating costs related to the first-shift operation will remain unchanged if one alternative is chosen instead of another. Therefore, those costs are irrelevant to the current decision and can safely be left out of the analysis. Consequently, we need to examine only the increased total cost due to the additional operating volume under each option. (This kind of study is known as *incremental analysis*.) Let Q denote the additional operating volume. Then we have

- **Option 1.** Overtime and Saturday operation: $36Q$.
- **Option 2.** Second-shift operation: $13,500 + \$31.50Q$.

We can find the break-even volume (Q_b) by equating the incremental cost functions and solving for Q:

$$36Q = 13,500 + 31.5Q,$$
$$4.5Q = 13,500,$$
$$Q_b = 3,000 \text{ units.}$$

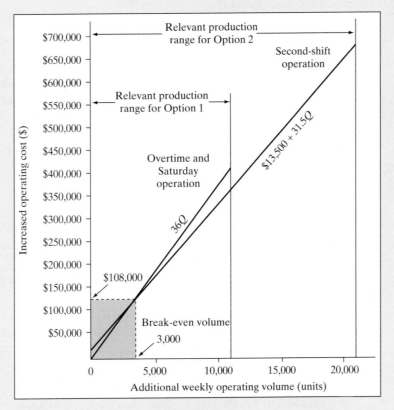

Figure 8.6 Cost–volume relationships of operating overtime and a Saturday operation versus second-shift operation beyond 24,000 units (Example 8.4).

If the additional volume exceeds 3,000 units, the second-shift operation becomes more efficient than overtime or Saturday operation. A break-even (or cost–volume) graph based on the foregoing data is shown in Figure 8.6. The horizontal scale represents additional volume per week. The upper limit of the relevant volume range for Option 1 is 12,000 units, whereas the upper limit for Option 2 is 21,000 units. The vertical scale is in dollars of cost. The operating savings expected at any volume may be read from the cost–volume graph. For example, the break-even point (zero savings) is 3,000 units per week. Option 2 is a better choice, since the additional weekly volume from the new order exceeds 3,000 units.

Make-or-Buy (Outsourcing) Decision

In business, the make-or-buy decision arises on a fairly frequent basis. Many firms perform certain activities using their own resources and pay outside firms to perform certain other activities. It is a good policy to constantly seek to improve the balance between these two types of activities by asking whether we should outsource some function that we are now performing ourselves or vice versa. Make-or-buy decisions are often grounded in the concept of *opportunity cost*.

8.4.2 Opportunity Cost

Opportunity cost: The benefits you could have received by taking an alternative action.

Opportunity cost may be defined as the potential benefit that is given up as you seek an alternative course of action. In fact, virtually every alternative has some opportunity cost associated with it. For example, suppose you have a part-time job while attending college that pays you $200 per week. You would like to spend a week at the beach during spring break, and your employer has agreed to give you the week off. What would be the opportunity cost of taking the time off to be at the beach? The $200 in lost wages would be an opportunity cost.

In an economic sense, opportunity cost could mean the contribution to income that is forgone by not using a limited resource in the best way possible. Or we may view opportunity costs as cash flows that could be generated from an asset the firm already owns, provided that such flows are not used for the alternative in question. In general, *accountants do not post opportunity cost in the accounting records of an organization. However, this cost must be explicitly considered in every decision.* In sum,

- An opportunity cost arises when a project uses a resource that may already have been paid for by the firm.
- When a resource that is already owned by a firm is being considered for use in a project, that resource has to be priced on its next-best alternative use, which may be
  1. A sale of the asset, in which case the opportunity cost is the expected proceeds from the sale, net of any taxes on gains.
  2. Renting or leasing the asset out, in which case the opportunity cost is the expected present value of the after-tax revenue from the rental or lease.
  3. Some use elsewhere in the business, in which case the opportunity cost is the cost of replacing the resource.
  4. That the asset has been abandoned or is of no use. Then the opportunity cost is zero.

EXAMPLE 8.5 Opportunity Cost: Lost Rental Income (Opportunity Cost)

Benson Company is a farm equipment manufacturer that currently produces 20,000 units of gas filters annually for use in its lawn-mower production. The expected annual production cost of the gas filters is summarized as follows:

Variable costs:	
Direct materials	$100,000
Direct labor	190,000
Power and water	35,000
Fixed costs:	
Heating and light	20,000
Depreciation	100,000
Total cost	**$445,000**

Tompkins Company has offered to sell Benson 20,000 units of gas filters for $17.00 per unit. If Benson accepts the offer, some of the manufacturing facilities currently used to manufacture the filters could be rented to a third party at an annual rent of $35,000. Should Benson accept Tompkins's offer, and why?

SOLUTION

Given: Financial data; production volume = 20,000 units.
Find: Whether Benson should outsource the gas filter operation.

	Make Option	Buy Option	Differential Cost (Make-Buy)
Variable costs:			
Direct materials	$100,000		$100,000
Direct labor	190,000		190,000
Power and water	35,000		35,000
Gas filters		340,000	−340,000
Fixed costs:			
Heating and light	20,000	20,000	0
Depreciation	100,000	100,000	0
Rental income lost	35,000		35,000
Total cost	$480,000	$460,000	$20,000
Unit cost	$24.00	$23.00	$1.00

This problem is unusual in the sense that the buy option would generate a rental fee of $35,000. In other words, Benson could rent out the current manufacturing facilities if it were to purchase the gas filters from Tompkins. To compare the two options, we need to examine the cost of each option.

The buy option has a lower unit cost and saves $1 for each use of a gas filter. If the lost rental income (opportunity cost) were not considered, however, the decision would favor the make option.

Sunk cost: A cost that has been incurred and cannot be reversed

8.4.3 Sunk Costs

A sunk cost is a cost that has already been incurred by past actions. Sunk costs are not relevant to decisions, because they cannot be changed regardless of what decision is made now or in the future. The only costs relevant to a decision are costs that vary among the alternative courses of action being considered. To illustrate a sunk cost, suppose you have a very old car that requires frequent repairs. You want to sell the car, and you figure that the current market value would be about $1,200 at best. While you are in the process of advertising the car, you find that the car's water pump is leaking. You decided to have the pump repaired, which cost you $200. A friend of yours is interested in buying your car and has offered $1,300 for it. Would you take the offer, or would you decline it simply because you cannot recoup the repair cost with that offer? In this example, the $200 repair cost is a sunk cost. You cannot change this repair cost, regardless of whether you keep or sell the car. Since your friend's offer is $100 more than the best market value, it would be better to accept the offer.

8.4.4 Marginal Cost

We make decisions every day, as entrepreneurs, professionals, executives, investors, and consumers, with little thought as to where our motivations come from or how our assumptions of logic fit a particular academic regimen. As you have seen, the engineering economic decisions that we describe throughout this text owe much to English economist Alfred Marshall (1842–1924) and his concepts of **marginalism**. In our daily quest for material gain, rarely do we recall the precepts of microeconomics or marginal utility, yet the ideas articulated by Marshall remain some of the most useful core principles guiding economic decision making.

Definition

Marginal cost: The cost associated with one additional unit of production.

Another cost term useful in cost–volume analysis is marginal cost. We define **marginal cost** as the added cost that would result from increasing the rate of output by a single unit. The accountant's differential-cost concept can be compared to the economist's marginal-cost concept. In speaking of changes in cost and revenue, the economist employs the terms *marginal cost* and *marginal revenue*. The revenue that can be obtained from selling one more unit of product is called **marginal revenue**. The cost involved in producing one more unit of product is called **marginal cost**.

EXAMPLE 8.6 Marginal Costs versus Average Costs

Consider a company that has an available electric load of 37 horsepower and that purchases its electricity at the following rates:

kWh/Month	@$/kWh	Average Cost ($/kWh)
First 1,500	$0.050	$0.050
Next 1,250	0.035	$\dfrac{\$75 + 0.0350(X - 1{,}500)}{X}$
Next 3,000	0.020	$\dfrac{\$118.75 + 0.020(X - 2{,}750)}{X}$
All over 5,750	0.010	$\dfrac{\$178.25 + 0.010(X - 5{,}750)}{X}$

According to this rate schedule, the unit variable cost in each rate class represents the marginal cost per kilowatt-hours (kWh). Alternatively, we may determine the average costs in the third column by finding the cumulative total cost and dividing it by the total number of kWh (X). Suppose that the current monthly consumption of electric power averages 3,200 kWh. On the basis of this rate schedule, determine the marginal cost of adding one more kWh and, for a given operating volume (3,200 kWh), the average cost per kWh.

SOLUTION

Given: Marginal cost schedule for electricity; operating volume $= 3,200$ kWh.
Find: Marginal and average cost per kWh.

The marginal cost of adding one more kWh is $0.020. The average variable cost per kWh is calculated as follows:

kWh	Rate ($/kWh)	Cost
First 1,500	0.050	$75.00
Next 1,250	0.035	43.75
Remaining 450	0.020	9.00
Total		$127.75

The average variable cost per kWh is $127.75/3,200 kWh $= \$0.0399$ kWh. Or we can find the value by using the formulas in the third column of the rate schedule:

$$\frac{\$118.75 + 0.020(3{,}200 - 2{,}750)}{3{,}200} = \frac{\$127.75}{3{,}200 \text{ kWh}} = \$0.0399 \text{ kWh}.$$

Changes in the average variable cost per unit are the result of changes in the marginal cost. As shown in Figure 8.7, the average variable cost continues to fall because the marginal cost is lower than the average variable cost over the entire volume.

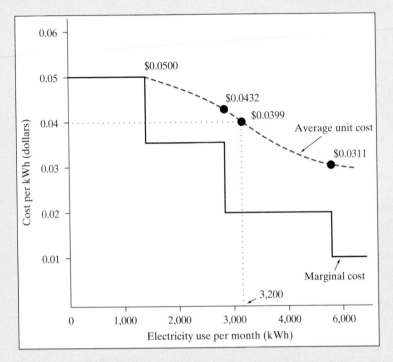

Figure 8.7   Marginal versus average cost per kWh (Example 8.6).

Marginal Analysis

Marginal analysis: A technique used in microeconomics by which very small changes in specific variables are studied in terms of the effect on related variables and the system as a whole.

The fundamental concept of marginal thinking is that you are where you are, and what is past is irrelevant. (This means that we should ignore the sunk cost.) The point is whether you move forward, and you will do so if the benefits outweigh the costs, even if they do so by a smaller margin than before. You will continue to produce a product or service until the cost of doing so equals the revenue derived. Microeconomics suggests that businesspersons and consumers measure progress not in great leaps, but in small incremental steps. Rational persons will reevaluate strategies and take new actions if benefits exceed costs on a marginal basis. If there is more than one alternative to choose, do so always with the alternative with greatest marginal benefit.

As we have mentioned, the economist's marginal-cost concept is the same as the accountant's differential-cost concept. For a business problem on maximization of profit, economists place a great deal of emphasis on the importance of marginal analysis: Rational individuals and institutions should perform the most cost-effective activities first. Specifically,

- When you examine marginal costs, you want to use the least expensive methods first and the most expensive last (if at all).

- Similarly, when you examine marginal revenue (or demand) the most revenue-enhancing methods should be used first, the least revenue enhancing last.
- If we need to consider the revenue along with the cost, the volume at which the total revenue and total cost are the same is known as the *break-even point*. If the cost and revenue function are assumed to be linear, this point may also be calculated, by means of the following formula:

$$\text{Break-even volume} = \frac{\text{Fixed costs}}{\text{Sales price per unit} - \text{Variable cost per unit}}$$

$$= \frac{\text{Fixed costs}}{\text{Marginal contribution per unit}} \qquad (8.1)$$

The difference between the unit sales price and the unit variable cost is the producer's **marginal contribution**, also known as **marginal income**. This means that each unit sold contributes toward absorbing the company's fixed costs.

EXAMPLE 8.7 Profit-Maximization Problem: Marginal Analysis[2]

Suppose you are a chief executive officer (CEO) of a small pharmaceutical company that manufactures generic aspirin. You want the company to maximize its profits. You can sell as many aspirins as you make at the prevailing market price. You have only one manufacturing plant, which is the constraint. You have the plant working at full capacity Monday through Saturday, but you close the plant on Sunday because on Sundays you have to pay workers overtime rates, and it is not worth it. The marginal costs of production are constant Monday through Saturday. Marginal costs are higher on Sundays, only because labor costs are higher.

Now you obtain a long-term contract to manufacture a brand-name aspirin. The costs of making the generic aspirin or the brand-name aspirin are identical. In fact, there is no cost or time involved in switching from the manufacture of one to the other. You will make much larger profits from the brand-name aspirin, but the demand is limited. One day of manufacturing each week will permit you to fulfill the contract. You can manufacture both the brand-name and the generic aspirin. Compared with the situation before you obtained the contract, your profits will be much higher if you now begin to manufacture on Sundays—even if you have to pay overtime wages.

- *Generic aspirin.* Each day, you can make 1,000 cases of generic aspirin. You can sell as many as you make, for the market price of $10 per case. Every week you have fixed costs of $5,000 (land tax and insurance). No matter how many cases you manufacture, the cost of materials and supplies is $2 per case; the cost of labor is $5 per case, except on Sundays, when it is $10 per case.
- *Brand-name aspirin.* Your order for the brand-name aspirin requires that you manufacture 1,000 cases per week, which you sell for $30 per case. The cost for the brand-name aspirin is identical to the cost of the generic aspirin.

What do you do?

[2] *Source:* "Profit Maximization Problem," by David Hemenway and Elon Kohlberg, *Economic Inquiry,* October 1, 1997, Page 862, Copyright 1997 Western Economic Association International.

SOLUTION

Given: Sales price, $10 per case for generic aspirin, $30 per case for band aspirin; fixed cost, $5,000; variable cost, $7 per case during weekdays, $12 per case on Sunday operation; weekly production, 6,000 cases of generic aspirin, 1,000 cases of brand-name aspirin.

Find: (a) Optimal production mix and (b) break-even volume.

(a) *Optimal production mix*: The marginal costs of manufacturing brand-name aspirin are constant Monday through Saturday; they rise substantially on Sunday and are above the marginal revenue from manufacturing generic aspirin. In other words, your company should manufacture the brand-name aspirin first. Your marginal revenue is the highest the "first" day, when you manufacture the brand-name aspirin. It then falls and remains constant for the rest of the week. On the seventh day (Sunday), the marginal revenue from manufacturing the generic aspirin is still below the marginal cost. You should manufacture brand-name aspirin one day a week and generic aspirin five days a week. On Sundays, the plant should close.

- The marginal revenue from manufacturing on Sunday is $10,000 (1,000 cases times $10 per case).

- The marginal cost from manufacturing on Sunday is $12,000 (1,000 cases times $12 per case—$10 labor + $2 materials).

- Profits will be $2,000 lower than revenue if the plant operates on Sunday.

(b) *Break-even volume*: The total revenue and cost functions can be represented as follows:

$$\text{Total revenue function:} \begin{cases} 30Q & \text{for } 0 \le Q \le 1,000 \\ 30,000 + 10Q & \text{for } 1,000 < Q \le 6,000, \end{cases}$$

$$\text{Total cost function:} \begin{cases} 5,000 + 7Q & \text{for } 0 \le Q \le 6,000 \\ 47,000 + 12Q & \text{for } 6,000 \le Q \le 7,000. \end{cases}$$

Table 8.4 shows the various factors involved in the production of the brand-name and the generic aspirin.

- If you produce the brand-name aspirin first, the break-even volume is

$$30Q - 7Q - 5,000 = 0,$$
$$Q_b = 217.39.$$

- If you produce the generic aspirin first, the break-even volume is

$$10Q - 7Q - 5,000 = 0,$$
$$Q_b = 1,666.67.$$

Clearly, scheduling the production of the brand-name aspirin first is the better strategy, as you can recover the fixed cost ($5,000) much faster by selling just 217.39 cases of brand-name aspirin. Also, Sunday operation is not economical, as the marginal cost exceeds the marginal revenue by $2,000, as shown in Figure 8.8.

TABLE 8.4 Net Profit Calculation as a Function of Production Volume

	Production Volume (Q)	Product Mix	Revenue	Total Revenue	Variable Cost	Fixed Cost	Total Cost	Net Profit
	0		0	0	0	$5,000	$5,000	−$5,000
Mon	1,000	Brand name	$30,000	$30,000	$7,000	0	12,000	18,000
Tue	2,000	Generic	10,000	40,000	7,000	0	19,000	21,000
Wed	3,000	Generic	10,000	50,000	7,000	0	26,000	24,000
Thu	4,000	Generic	10,000	60,000	7,000	0	33,000	27,000
Fri	5,000	Generic	10,000	70,000	7,000	0	40,000	30,000
Sat	6,000	Generic	10,000	80,000	7,000	0	47,000	33,000
Sun	7,000	Generic	10,000	90,000	12,000	0	59,000	31,000

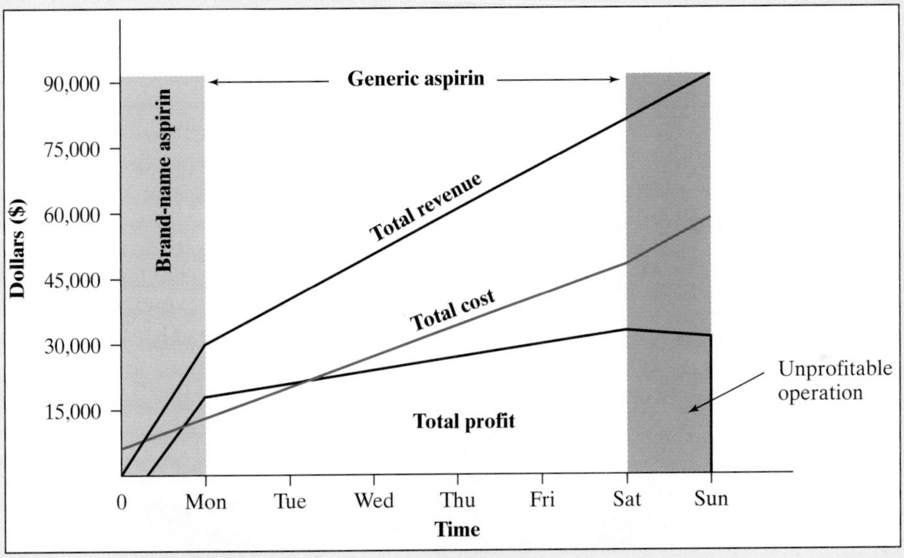

Figure 8.8 Weekly profits as a function of time. Sunday operation becomes unprofitable, because the marginal revenue stays at $10 per case whereas the marginal cost increases to $12 per case (Example 8.7).

8.5 Estimating Profit from Production

Up to this point, we have defined various cost elements and their behaviors in a manufacturing environment. Our ultimate objective is to develop a project's cash flows; we do so in Chapter 10, where the first step is to formulate the budget associated with the sales and production of the product. As we will explain in that chapter, the income statement developed in the current chapter will be the focal point for laying out the project's cash flows.

8.5.1 Calculation of Operating Income

Accountants measure the net income of a specified operating period by subtracting expenses from revenues for that period. These terms can be defined as follows:

1. The **project revenue** is the income earned[3] by a business as a result of providing products or services to customers. Revenue comes from sales of merchandise to customers and from fees earned by services performed for clients or others.

2. The **project expenses** that are incurred[4] are the cost of doing business to generate the revenues of the specified operating period. Some common expenses are the cost of the goods sold (labor, material, inventory, and supplies), depreciation, the cost of employees' salaries, the business's operating costs (such as the cost of renting buildings and the cost of insurance coverage), and income taxes.

The business expenses just listed are accounted for in a straightforward fashion on a company's income statement and balance sheet: The amount paid by the organization for each item would translate, dollar for dollar, into expenses in financial reports for the period. One additional category of expenses—the purchase of new assets—is treated by depreciating the total cost gradually over time. Because capital goods are given this unique accounting treatment, depreciation is accounted for as a separate expense in financial reports. Because of its significance in engineering economic analysis, we will treat depreciation accounting in a separate chapter.

Figure 8.9 will be used as a road map to construct the income statement related to the proposed manufacturing activities.

8.5.2 Sales Budget for a Manufacturing Business

The first step in laying out a sales budget for a manufacturing business is to predict the dollar sales, which is the key to the entire process. The estimates are typically expressed in both dollars and units of the product. Using the expected selling price, we can easily extend unit sales

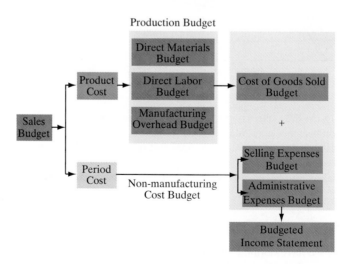

Figure 8.9 Process of creating a master production budget.

[3] Note that the cash may be received in a different accounting period.
[4] Note that the cash may be paid in a different accounting period.

to revenues. Often, the process is initiated by individual salespersons or managers predicting sales in their own area for the budget period. Of course, an aggregate sales budget is influenced by economic conditions, pricing decisions, competition, marketing programs, etc. To illustrate, suppose that you want to create a sales budget schedule on a quarterly basis. Projected sales units are 1,000 units during the first quarter, 1,200 units during the second quarter, 1,300 units during the third quarter, and 1,500 units during the fourth quarter. The projected sales price is $15 per unit. Then an estimated sales schedule looks like the following:

Sales Budget Schedule (Year 2006)—Product X					
	1Q	2Q	3Q	4Q	Annual Total
Budgeted units	1,000	1,200	1,300	1,500	5,000
Sales price	$ 15	$ 15	$ 15	$ 15	$ 15
Estimated sales	$ 15,000	$ 18,000	$ 19,500	$ 22,500	$ 75,000

8.5.3 Preparing the Production Budget

Once the sales budget is known, we can prepare the production activities. Using the sales budget for the number of project units needed, we prepare an estimate of the number of units to be produced in the budget period. Note that the production budget is the basis for projecting the cost-of-goods-manufactured budget, which we will discuss subsequently. To illustrate, we may use the following steps:

1. From the sales budget, record the projected number of units to be sold.
2. Determine the desired number of units to carry in the ending inventory; usually, that number is a percentage of next quarter's needs. In our example, we will assume that 20% of the budgeted units will be the desirable ending inventory position in each quarter.
3. *Add* to determine the total number of units needed each period.
4. Determine the projected number of units in the beginning inventory. Note that the beginning inventory is last quarter's ending inventory. In our example, we will assume that the first quarter's beginning inventory is 100 units.
5. *Subtract* to determine the projected production budget.

Then a typical production budget may look like the following:

Production Budget (Year 2006)—Product X					
	1Q	2Q	3Q	4Q	Annual Total
Budgeted units to be produced	1,000	1,200	1,300	1,500	5,000
Desired ending inventory	200	240	260	300	1,000
Total units needed	1,200	1,440	1,560	1,800	6,000
Less beginning inventory	100	200	240	260	800
Units to produce	1,100	1,240	1,320	1,540	5,200

Materials Budgets

Once we know how many units we need to produce, we are ready to develop direct materials budgets. We may use the following steps:

1. From the production budget, copy the projected number of units to be produced.
2. *Multiply* by the amount of raw materials needed per unit to calculate the amount of materials needed. In our example, we will assume that each production unit consumes $4 of materials.
3. Calculate the desired ending inventory (the number of units required in the ending inventory × $4).
4. *Add* to calculate the total amount of materials needed.
5. *Subtract* the beginning inventory, which is last quarter's ending inventory, to calculate the amount of raw materials needed to be purchased.
6. Calculate the net cost of raw materials.

Then a typical direct materials budget may look like the following:

Direct Materials Budget (Year 2006)—Product X					
	1Q	2Q	3Q	4Q	Annual Total
Units to produce	1,100	1,240	1,320	1,540	5,200
Unit cost of materials	$ 4	$ 4	$ 4	$ 4	
Cost of materials for units to be produced	$ 4,400	$ 4,960	$ 5,280	$ 6,160	$ 20,800
Plus cost of materials in ending inventory	$ 800	$ 960	$ 1,040	$ 1,200	$ 4,000
Total cost of materials needed	$ 5,200	$ 5,920	$ 6,320	$ 7,360	$ 24,800
Less cost of materials in beginning inventory	$ 400	$ 800	$ 960	$ 1,040	$ 3,200
Cost of materials to purchase	$ 4,800	$ 5,120	$ 5,360	$ 6,320	$ 21,600

Direct Labor Budget for a Manufacturing Business

As with the materials budgets, once the production budget has been completed, we can easily prepare the direct labor budget. This budget allows the firm to estimate labor requirements—both labor hours and dollars—in advance. To illustrate, use the following steps:

1. From the production budget, copy the projected number of units to be produced.
2. *Multiply* by the direct labor cost per unit to calculate the total direct labor cost. In our example, we will assume that the direct labor cost is $3 per unit.

Then a typical direct labor budget may look like the following:

Direct Labor Budget (Year 2006)—Product X					
	1Q	2Q	3Q	4Q	Annual Total
Units to produce	1,100	1,240	1,320	1,540	5,200
× Direct labor cost per unit	$ 3.00	$ 3.00	$ 3.00	$ 3.00	
Total direct labor cost ($)	$ 3,300	$ 3,720	$ 3,960	$ 4,620	$ 15,600

Overhead Budget for a Manufacturing Business

The overhead budget should provide a schedule of all costs of production other than direct materials and direct labor. Typically, the overhead budget is expressed in dollars, based on a predetermined overhead rate. In preparing a manufacturing overhead budget, we may take the following steps:

1. From the production budget, copy the projected number of units to be produced.
2. *Multiply* by the variable overhead rate to calculate the budgeted variable overhead. In our example, we will assume the variable overhead rate to be $1.50 per unit. There are several ways to determine this overhead rate. One common approach (known as traditional standard costing) is to divide the expected total overhead cost by the budgeted number of direct labor hours (or units). Another approach is to adopt an **activity-based costing** concept, allocating indirect costs against the activities that caused them. We will not review this accounting method here, but it can more accurately reflect indirect cost improvement than traditional standard costing can.
3. *Add* any budgeted fixed overhead to calculate the total budgeted overhead. In our example, we will assume the fixed overhead to be $230 each quarter.

> **Activity-based costing (ABC)** identifies opportunities to improve business process effectiveness and efficiency by determining the "true" cost of a product or service.

Then a typical manufacturing overhead budget may look like the following:

Manufacturing Overhead Budget (Year 2006)—Product X					
	1Q	2Q	3Q	4Q	Annual Total
Units to produce	1,100	1,240	1,320	1,540	5,200
Variable mfg overhead rate per unit ($1.50)	$ 1,650	$ 1,860	$ 1,980	$ 2,310	$ 7,800
Fixed mfg overhead	$ 230	$ 230	$ 230	$ 230	$ 920
Total overhead	$ 1,880	$ 2,090	$ 2,210	$ 2,540	$ 8,720

8.5.4 Preparing the Cost-of-Goods-Sold Budget

The production budget developed in the previous section shows how much it would cost to produce the required production volume. Note that the number of units to be produced includes both the anticipated number of sales units and the number of units in inventory.

The cost-of-goods-sold budget is different from the production budget, because we do not count the costs incurred to carry the inventory. Therefore, we need to prepare a budget that details the costs related to the sales, not the inventory. Typical steps in preparing a cost-of-goods-sold budget are as follows:

1. From the sales budget, and not the production budget, copy the budgeted number of sales units.
2. *Multiply* by the direct material cost per unit to estimate the amount of direct materials.
3. *Multiply* by the direct labor cost per unit to estimate the direct labor.
4. *Multiply* by the manufacturing overhead per unit to estimate the overhead.

Then a typical cost-of-goods-sold budget may look like the following:

Cost of goods sold: A figure reflecting the cost of the product or good that a company sells to generate revenue.

Cost of Goods Sold (Year 2006)—Product X					
	1Q	2Q	3Q	4Q	Annual Total
Budgeted sales units	1,000	1,200	1,300	1,500	5,000
Direct materials ($4/unit)	$ 4,000	$ 4,800	$ 5,200	$ 6,000	$ 20,000
Direct labor ($3/unit)	$ 3,000	$ 3,600	$ 3,900	$ 4,500	$ 15,000
Mfg overhead:					
Variable ($1.50 per unit)	$ 1,500	$ 1,800	$ 1,950	$ 2,250	$ 7,500
Fixed	$ 230	$ 230	$ 230	$ 230	$ 920
Cost of goods sold	$ 7,000	$ 8,400	$ 9,100	$ 10,500	$ 35,000

8.5.5 Preparing the Nonmanufacturing Cost Budget

To complete the entire production budget, we need to add two more items related to production: the selling expenses and the administrative expenses.

Selling Expenses Budget for a Manufacturing Business

Since we know the sales volume, we can develop a selling expenses budget by considering the budgets of various individuals involved in marketing the products. The budget includes both variable cost items (shipping, handling, and sales commission) and fixed items, such as advertising and salaries for marketing personnel. To prepare the selling expenses budget, we may take the following steps:

1. From the sales budget, copy the projected number of unit sales.
2. List the variable selling expenses, such as the sales commission. In our example, we will assume that the sales commission is calculated at 5% of unit sales.
3. List the fixed selling expenses, typically rent, depreciation expenses, advertising, and other office expenses. In our example, we will assume the following: rent, $500 per quarter; advertising, $300 per quarter; office expense, $200 per quarter.
4. *Add* to calculate the total budgeted selling expenses.

Then a typical selling expenses budget may look like the following:

Selling Expenses (Year 2006)—Product X					
	1Q	2Q	3Q	4Q	Annual Total
Budgeted unit sales ($)	$ 15,000	$ 18,000	$ 19,500	$ 22,500	$ 75,000
Variable expenses:					
Commission	$ 750	$ 900	$ 975	$ 1,125	$ 3,750
Fixed expenses:					
Rent	$ 500	$ 500	$ 500	$ 500	$ 2,000
Advertising	$ 300	$ 300	$ 300	$ 300	$ 1,200
Office expenses	$ 200	$ 200	$ 200	$ 200	$ 800
Total selling expenses	$ 1,750	$ 1,900	$ 1,975	$ 2,125	$ 7,750

Administrative Expenses Budget for a Manufacturing Business

Another nonmanufacturing expense category to consider is administrative expenses, a category that contains mostly fixed-cost items such as executive salaries and the depreciation of administrative buildings and office furniture. To prepare an administrative expense budget, we may take the following steps:

1. List the variable administrative expenses, if any.
2. List the fixed administrative expenses, which include salaries, insurance, office supplies, and utilities. We will assume the following quarterly expenses: salaries, $1,400; insurance, $135; office supplies, $300; other office expenses, $150.
3. *Add* to calculate the total administrative expenses.

Then a typical administrative expenses budget may look like the following:

Administrative Expenses (Year 2006)—Product X					
	1Q	2Q	3Q	4Q	Annual Total
Variable expenses:					
Fixed expenses:					
Salaries	$ 1,400	$ 1,400	$ 1,400	$ 1,400	$ 5,600
Insurance	$ 135	$ 135	$ 135	$ 135	$ 540
Office supplies	$ 300	$ 300	$ 300	$ 300	$ 1,200
Utilities (phone, power, water, etc.)	$ 500	$ 500	$ 500	$ 500	$ 2,000
Other office expenses	$ 150	$ 150	$ 150	$ 150	$ 600
Total administrative expenses	$ 2,485	$ 2,485	$ 2,485	$ 2,485	$ 9,940

8.5.6 Putting It All Together: The Budgeted Income Statement

Now we are ready to develop a consolidated budget that details a company's projected sales, costs, expenses, and net income. To determine the net income, we need to include any other operating expenses, such as interest expenses and renting or leasing expenses. Once we calculate the taxable income, we can determine the federal taxes to pay. The procedure to follow is

1. Copy the net sales from the sales budget.
2. Copy the cost of goods sold.
3. Copy both the selling and administrative expenses.
4. Gross income = Sales − Cost of goods sold.
5. Operating income = Sales − (Cost of goods sold + Operating expenses).
6. Determine any interest expense incurred during the budget period.
7. Calculate the federal income tax at a percentage of taxable income. In our example, let's assume a 35% tax rate.
8. Net income = Income from operations − Federal income tax, which summarizes the projected profit from the budgeted sales units.

Then the budgeted income statement would look like the format shown in Table 8.5. Once you have developed a budgeted income statement, you may be able to examine the profitability of the manufacturing operation. Three margin figures are commonly used to quickly assess the profitability of the operation: gross margin, operating margin, and net profit margin.

TABLE 8.5 Budgeted Income Statement

Budgeted Income Statement (Year 2006)—Product X					
	1Q	2Q	3Q	4Q	Annual Total
Sales	$ 15,000	$ 18,000	$ 19,500	$ 22,500	$ 75,000
Cost of goods sold	$ 7,000	$ 8,400	$ 9,100	$ 10,500	$ 35,000
Gross income	$ 8,000	$ 9,600	$ 10,400	$ 12,000	$ 40,000
Operating expenses:					
Selling expenses	$ 1,750	$ 1,900	$ 1,975	$ 2,125	$ 7,750
Administrative expenses	$ 2,485	$ 2,485	$ 2,485	$ 2,485	$ 9,940
Operating income	$ 3,765	$ 5,215	$ 5,940	$ 7,390	$ 22,310
Interest expenses	$ —	$ —	$ —	$ —	$ —
Net income before taxes	$ 3,765	$ 5,215	$ 5,940	$ 7,390	$ 22,310
Income taxes (35%)	$ 1,318	$ 1,825	$ 2,079	$ 2,587	$ 7,809
Net income	$ 2,447	$ 3,390	$ 3,861	$ 4,804	$ 14,502

Gross Margin

The gross margin reveals how much a company earns, taking into consideration the costs that it incurs for producing its products or services. In other words, gross margin is equal to gross income divided by net sales and is expressed as a percentage:

$$\text{Gross margin} = \frac{\text{Gross income}}{\text{Net sales}}. \qquad (8.2)$$

In our example, gross margin = $40,000/$75,000 = 53%. Gross margin indicates how profitable a company is at the most fundamental level. Companies with higher gross margins will have more money left over to spend on other business operations, such as research and development or marketing.

Operating Margin

The operating margin is defined as the operating profit for a certain period, divided by revenues for that period:

$$\text{Operating margin} = \frac{\text{Operating income}}{\text{Net sales}}. \qquad (8.3)$$

In our example, operating margin = $22,310/$75,000 = 30%. Operating *profit margin* indicates how effective a company is at controlling the costs and expenses associated with its normal business operations.

> **Operating margin** gives analysts an idea of how much a company makes (before interest and taxes) on each dollar of sales.

Net Profit Margin

As we defined the term in Chapter 2, the net profit margin is obtained by dividing net profit by net revenues, often expressed as a percentage:

$$\text{Net profit margin} = \frac{\text{Net income}}{\text{Net sales}}. \qquad (8.4)$$

This number is an indication of how effective a company is at cost control. The higher the net profit margin, the more effective the company is at converting revenue into actual profit. In our example, the net profit margin = $14,502/$75,000 = 19%. The net profit margin is a good way of comparing companies in the same industry, since such companies are generally subject to similar business conditions.

8.5.7 Looking Ahead

The operational decision problem described in this chapter tends to have a relatively short-term horizon (weekly or monthly). That is, such decisions do not commit a firm to a certain course of action over a relatively long period. If operational decision problems significantly affect the amount of funds that must be invested in a firm, fixed costs will have to increase. In Example 8.7, if the daily aspirin production increases to 10,000 cases per day, exceeding the current production capacity, the firm must make new investments in plant and equipment. Since the benefits of the expansion will continue to occur over an extended period, we need to consider the economic effects of the fixed costs over the life of the assets. Also, we have not discussed how we actually calculate the depreciation amount related to production or the amount of income taxes to be paid from profits generated from the operation. Doing so requires an understanding the concepts of **capital investment, depreciation, and income taxes**, which we will discuss in the next chapter.

SUMMARY

In this chapter, we examined several ways in which managers classify costs. How the costs will be used dictates how they will be classified:

- Most manufacturing companies divide **manufacturing costs** into three broad categories: *direct materials*, *direct labor*, and *manufacturing overhead*. **Nonmanufacturing costs** are classified into two categories: *marketing* or *selling costs* and *administrative costs*.

- For the purpose of valuing inventories and determining expenses for the balance sheet and income statement, costs are classified as either **product costs** or **period costs**.

- For the purpose of predicting **cost behavior**—how costs will react to changes in activity—managers commonly classify costs into two categories: variable costs and fixed costs.

- An understanding of the following **cost–volume relationships** is essential to developing successful business strategies and to planning future operations:

  Fixed operating costs: Costs that do not vary with production volume.

  Variable operating costs: Costs that vary with the level of production or sales.

  Average costs: Costs expressed in terms of units obtained by dividing total costs by total volumes.

  Differential (incremental) costs: Costs that represent differences in total costs, which result from selecting one alternative instead of another.

  Opportunity costs: Benefits that could have been obtained by taking an alternative action.

  Sunk costs: Past costs not relevant to decisions because they cannot be changed no matter what actions are taken.

  Marginal costs: Added costs that result from increasing rates of outputs, usually by single units.

  Marginal analysis: In economic analysis, we need to answer the apparently trivial question, "Is it worthwhile?"—whether the action in question will add sufficiently to the benefits enjoyed by the decision maker to make performing the action worth the cost. This is the heart of marginal-decision making—the statement that an action merits performance if, and only if, as a result, we can expect to be better off than we were before.

- At the level of plant operations, engineers must make decisions involving materials, production processes, and the in-house capabilities of company personnel. Most of these operating decisions do not require any investments in physical assets; therefore, they depend solely on the cost and volume of business activity, without any consideration of the time value of money.

- Engineers are often asked to prepare the production budgets related to their operating division. Doing this requires a knowledge of budgeting scarce resources, such as labor and materials, and an understanding of the overhead cost. The same budgeting practice will be needed in preparing the estimates of costs and revenues associated with undertaking a new project.

PROBLEMS

Classifying Costs

8.1 Identify which of the following transactions and events are product costs and which are period costs:
- Storage and material handling costs for raw materials.
- Gains or losses on the disposal of factory equipment.
- Lubricants for machinery and equipment used in production.
- Depreciation of a factory building.
- Depreciation of manufacturing equipment.
- Depreciation of the company president's automobile.
- Leasehold costs for land on which factory buildings stand.
- Inspection costs of finished goods.
- Direct labor cost.
- Raw-materials cost.
- Advertising expenses.

Cost Behavior

8.2 Identify which of the following costs are fixed and which are variable:
- Wages paid to temporary workers.
- Property taxes on a factory building.
- Property taxes on an administrative building.
- Sales commission.
- Electricity for machinery and equipment in the plant.
- Heating and air-conditioning for the plant.
- Salaries paid to design engineers.
- Regular maintenance on machinery and equipment.
- Basic raw materials used in production.
- Factory fire insurance.

8.3 The accompanying figures are a number of cost behavior patterns that might be found in a company's cost structure. The vertical axis on each graph represents total cost, and the horizontal axis on each graph represents level of activity (volume). For each of the situations that follow, identify the graph that illustrates the cost pattern involved. Any graph may be used more than once[5].

(a) Electricity bill—a flat-rate fixed charge, plus a variable cost after a certain number of kilowatt-hours are used.

(b) City water bill, which is computed as follows:

First 1,000,000 gallons $1,000 flat or less rate

Next 10,000 gallons $0.003 per gallon used

Next 10,000 gallons $0.006 per gallon used

Next 10,000 gallons $0.009 per gallon used

Etc. etc.

---

[5] Adapted originally from a CPA exam, and the same materials are also found in R. H. Garrison and E. W. Noreen, *Managerial Accounting*, 8th ed. Irwin, 1997, copyright © Richard D. Irwin, p. 271.

(c) Depreciation of equipment, where the amount is computed by the straight-line method. When the depreciation rate was established, it was anticipated that the obsolescence factor would be greater than the wear-and-tear factor.

(d) Rent on a factory building donated by the city, where the agreement calls for a fixed fee payment unless 200,000 labor-hours or more are worked, in which case no rent need be paid.

(e) Cost of raw materials, where the cost decreases by 5 cents per unit for each of the first 100 units purchased, after which it remains constant at $2.50 per unit.

(f) Salaries of maintenance workers, where one maintenance worker is needed for every 1,000 machine-hours or less (that is, 0 to 1,000 hours requires one maintenance worker, 1,001 to 2,000 hours requires two maintenance workers, etc.).

(g) Cost of raw materials used.

(h) Rent on a factory building donated by the county, where the agreement calls for rent of $100,000, less $1 for each direct labor-hour worked in excess of 200,000 hours, but a minimum rental payment of $20,000 must be paid.

(i) Use of a machine under a lease, where a minimum charge of $1,000 must be paid for up to 400 hours of machine time. After 400 hours of machine time, an additional charge of $2 per hour is paid, up to a maximum charge of $2,000 per period.

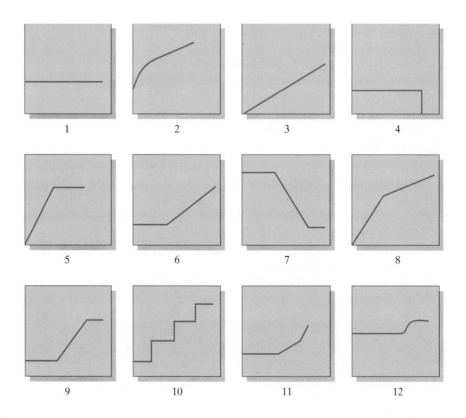

8.4 Harris Company manufactures a single product. Costs for the year 2001 for output levels of 1,000 and 2,000 units are as follows:

Units produced	1,000	2,000
Direct labor	$30,000	$30,000
Direct materials	20,000	40,000
Overhead:		
Variable portion	12,000	24,000
Fixed portion	36,000	36,000
Selling and administrative costs:		
Variable portion	5,000	10,000
Fixed portion	22,000	22,000

At each level of output, compute the following:
(a) Total manufacturing costs.
(b) Manufacturing costs per unit.
(c) Total variable costs.
(d) Total variable costs per unit.
(e) Total costs that have to be recovered if the firm is to make a profit.

Cost–Volume–Profit Relationships

8.5 Bragg & Stratton Company manufactures a specialized motor for chain saws. The company expects to manufacture and sell 30,000 motors in year 2001. It can manufacture an additional 10,000 motors without adding new machinery and equipment. Bragg & Stratton's projected total costs for the 30,000 units are as follows:

Direct Materials	$150,000
Direct labor	300,000
Manufacturing overhead:	
Variable portion	100,000
Fixed portion	80,000
Selling and administrative costs:	
Variable portion	180,000
Fixed portion	70,000

The selling price for the motor is $80.

(a) What is the total manufacturing cost per unit if 30,000 motors are produced?

(b) What is the total manufacturing cost per unit if 40,000 motors are produced?

(c) What is the break-even price on the motors?

8.6 The accompanying chart shows the expected monthly profit or loss of Cypress Manufacturing Company within the range of its monthly practical operating capacity. Using the information provided in the chart, answer the following questions:

(a) What is the company's break-even sales volume?

(b) What is the company's marginal contribution rate?

(c) What effect would a 5% decrease in selling price have on the break-even point in (a)?

(d) What effect would a 10% increase in fixed costs have on the marginal contribution rate in (b)?

(e) What effect would a 6% increase in variable costs have on the break-even point in (a)?

(f) If the chart also reflects $20,000 monthly depreciation expenses, compute the sale at the break-even point for cash costs.

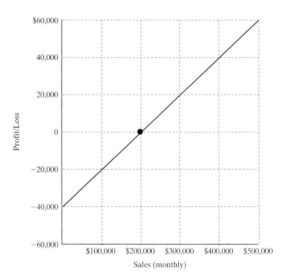

8.7 The accompanying graph is a cost–volume–profit graph. In the graph, identify the following line segments or points:

(a) Line EF represents _____.

(b) The horizontal axis AB represents _____, and the vertical axis AD represents _____.

(c) Point V represents _____.

(d) The distance CB divided by the distance AB is _____.

(e) The point V_b is a break-even _____.

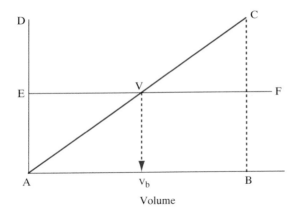

8.8 A cost–volume–profit (CVP) graph is a useful technique for showing relation-ships between costs, volume, and profits in an organization.

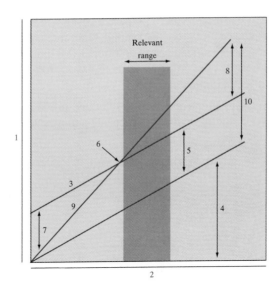

(a) Identify the numbered components in the accompanying CVP graph.

No.	Description	No.	Description
1		6	
2		7	
3		8	
4		9	
5		10	

(b) Using the typical CVP relationship shown, fill in the missing amounts in each of the following four situations (each case is independent of the others):

Case	Units Sold	Sales	Variable Expenses	Contribution Margin per Unit	Fixed Expenses	Net Income (Loss)
A	9,000	$270,000	$162,000		$ 90,000	
B		$350,000		$15	$170,000	$ 40,000
C	20,000		$280,000	$ 6		$ 35,000
D	5,000	$100,000			$ 82,000	($12,000)

Cost Concepts Relevant to Decision Making

8.9 An executive from a large merchandising firm has called your vice-president for production to get a price quote for an additional 100 units of a given product. The vice-president has asked you to prepare a cost estimate. The number of hours required to produce a unit is 5. The average labor rate is $12 per hour. The materials cost $14 per unit. Overhead for an additional 100 units is estimated at 50% of the direct labor cost. If the company wants to have a 30% profit margin, what should be the unit price to quote?

8.10 The Morton Company produces and sells two products: A and B. Financial data related to producing these two products are summarized as follows:

	Product A	Product B
Selling price	$ 10.00	$12.00
Variable costs	$ 5.00	$10.00
Fixed costs	$ 2,000	$ 600

(a) If these products are sold in the ratio of 4A for 3B, what is the break-even point?
(b) If the product mix has changed to 5A for 5B, what would happen to the break-even point?
(c) In order to maximize the profit, which product mix should be pushed?
(d) If both products must go through the same manufacturing machine and there are only 30,000 machine hours available per period, which product should be pushed? Assume that product A requires 0.5 hour per unit and B requires 0.25 hour per unit.

8.11 Pearson Company manufactures a variety of electronic printed circuit boards (PCBs) that go into cellular phones. The company has just received an offer from an outside supplier to provide the electrical soldering for Pearson's Motorola product line (Z-7 PCB, slimline). The quoted price is $4.80 per unit. Pearson is interested in this offer, since its own soldering operation of the PCB is at its peak capacity.

- **Outsourcing option.** The company estimates that if the supplier's offer were accepted, the direct labor and variable overhead costs of the Z-7 slimline would be reduced by 15% and the direct material cost would be reduced by 20%.
- **In-house production option.** Under the present operations, Pearson manufactures all of its own PCBs from start to finish. The Z-7 slimlines are sold through Motorola at $20 per unit. Fixed overhead charges to the Z-7 slimline total $20,000 each year. The further breakdown of producing one unit is as follows:

Direct materials	$ 7.50
Direct labor	5.00
Manufacturing overhead	4.00
Total cost	$16.00

The manufacturing overhead of $4.00 per unit includes both variable and fixed manufacturing overhead, based on a production of 100,000 units each year.

(a) Should Pearson Company accept the outside supplier's offer?

(b) What is the maximum unit price that Pearson Company should be willing to pay the outside supplier?

Short Case Study

ST8.1 The Hamilton Flour Company is currently operating its mill six days a week, 24 hours a day, on three shifts. At current prices, the company could easily obtain a sufficient volume of sales to take the entire output of a seventh day of operation each week. The mill's practical capacity is 6,000 hundredweight of flour per day. Note that

- Flour sells for $12.40 a hundredweight (cwt.) and the price of wheat is $4.34 a bushel. About 2.35 bushels of wheat are required per cwt. of flour. Fixed costs now average $4,200 a day, or $0.70 per cwt. The average variable cost of mill operation, almost entirely wages, is $0.34 per cwt.
- With Sunday operation, wages would be doubled for Sunday work, which would bring the variable cost of Sunday operation to $0.66 per cwt. Total fixed costs per week would increase by $420 (or $29,820) if the mill were to operate on Sunday.

(a) Using the information provided, compute the break-even volumes for six-day and seven-day operation.

(b) What are the marginal contribution rates for six-day and seven-day operation?

(c) Compute the average total cost per cwt. for six-day operation and the net profit margin per cwt. before taxes.

(d) Would it be economical for the mill to operate on Sundays? (Justify your answer numerically.)

CHAPTER NINE

Depreciation and Corporate Taxes

***Know What It Costs to Own a Piece of Equipment:
A Hospital Pharmacy Gets a Robotic Helper[1]*** When most
patients at Kirkland's Evergreen Hospital Medical Center are asleep,
the robot comes to life. Its machinery thumps like a heart beating
as it moves around the hospital pharmacy, preparing prescriptions.
In a little more than an hour, he'll ready 1,500 doses of medication.

Ernie, or Evergreen Robot Noticeably Improving Efficiency, is a
new $3 million addition to [the] pharmacy staff. The robot uses bar
codes to match each drug dispensed with an electronic patient profile,
helping prevent errors, said Bob Blanchard, pharmacy director.

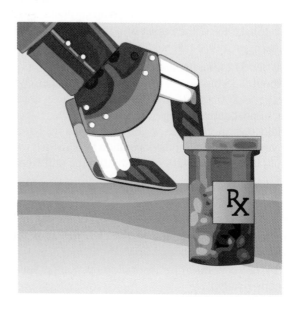

[1] "Know What It Costs to Own a Piece of Equipment: A Hospital Pharmacy Gets a Robotic Helper,"
Katherine Sather, *Seattle Times Eastside bureau,* Copyright © 2004 The Seattle Times Company.

"It's the future," he said. "Safety is the main benefit." Efficiency is another plus. The robot can prepare a 24-hour medication supply for in-house patients, about 1,500 doses, in a little more than an hour. The same task used to take three people about three hours to complete, Blanchard said.

Now, staff [members] only label the medications with bar codes. Ernie does the rest.

The machine looks like a mini space ship. A door opens into an octagon-shaped room, 12 feet long diagonally, stocked with more than 400 racks of medicine. Each dose is labeled with a bar code that tells Ernie what it is and where it should be stored. Affixed to the center of the room is a mechanical arm that scans the bar-coded medicine and places it on its designated rack.

When staff [members] give Ernie a computerized order, his arm buzzes to the correct row of medication, grabs it with suction cups and drops it into an envelope that is bar coded with the patient's profile.

"Research shows using it decreases certain predictable errors," Blanchard said. "We're very excited about it—we've really led the drive to move to automation."

At Evergreen, [the] pharmacy staff hope[s] for Ernie to eventually dispense 93 percent of the medication that is distributed to patients in the 244-bed hospital. Medication that needs care such as refrigeration is prepared by staff.

"The technology is such that it's been tested and it's reliable. Given the volume of patients we see, it makes sense," said Amy Gepner, a spokesperson for the hospital. "It's a safety initiative." Ernie is being purchased with a seven-year lease along with 23 automated medical cabinets placed throughout the hospital. But [the] staff [is] fonder of the robot.

Now ask yourself, How does the cost of this robot ($3 million) affect the financial position of the hospital? In the long run, the system promises to create greater cost savings for the hospital by enhancing productivity, improving safety, and cutting down lead time in filling orders. In the short run, however, the high initial cost of the robot will adversely affect the organization's "bottom line," because that cost is only gradually rewarded by the benefits the robot offers.

Another consideration should come to mind as well: This state-of-the-art robot must inevitably wear out over time, and even if its productive service extends over many years, the cost of maintaining its high level of functioning will increase as the individual pieces of hardware wear out and need to be replaced. Of even greater concern is the question of how long this robot will be the state of the art. When will the competitive advantage the hospital has just acquired become a competitive disadvantage through obsolescence?

One of the facts of life that organizations must deal with and account for is that fixed assets lose their value—even as they continue to function and contribute to the engineering projects that use them. This loss of value, called **depreciation**, can involve deterioration and obsolescence.

The main function of **depreciation accounting** is to account for the cost of fixed assets in a pattern that matches their decline in value over time. The cost of the robot we have just described, for example, will be allocated over several years in the hospital's financial statements, so that the pattern of the robot's costs roughly matches its pattern of service. In this way, as we shall see, depreciation accounting enables the firm to stabilize the statements about its financial position that it distributes to stockholders and the outside world.

On a project level, engineers must be able to assess how the practice of depreciating fixed assets influences the investment value of a given project. To do this, the engineers need to estimate the allocation of capital costs over the life of the project, which requires an understanding of the conventions and techniques that accountants use to depreciate assets. In this chapter, we will review the conventions and techniques of asset depreciation and income taxes.

We begin by discussing the nature and significance of depreciation, distinguishing its general economic definition from the related, but different, accounting view of depreciation. We then focus our attention almost exclusively on the rules and laws that govern asset depreciation and the methods that accountants use to allocate depreciation expenses. Knowledge of these rules will prepare you to apply them in assessing the depreciation of assets acquired in engineering projects. Then we turn our attention to the subject of depletion, which utilizes similar ideas, but specialized techniques, to allocate the cost of the depletion of natural-resource assets.

Once we understand the effect of depreciation at the project level, we need to address the effect of corporate taxes on project cash flows. There are many forms of government taxation, including sales taxes, property taxes, user taxes, and state and federal income taxes. In this chapter, we will focus on federal income taxes. When you are operating a business, any profits or losses you incur are subject to income tax consequences. Therefore, we cannot ignore the impact of income taxes in project evaluation. The chapter will give you a good idea of how the U.S. tax system operates and of how federal income taxes affect economic analysis. Although tax law is subject to frequent changes, the analytical procedures presented here provide a basis for tax analysis that can be adapted to reflect future changes in tax law. Thus, while we present many examples based on current tax rates, in a larger context we present a general approach to the analysis of *any* tax law.

CHAPTER LEARNING OBJECTIVES

After completing this chapter, you should understand the following concepts:

- How to account for the loss of value of an asset in business.
- The meaning and types of depreciation.
- The difference between book depreciation and tax depreciation.
- The effects of depreciation on net income calculation.
- The general scheme of U.S. corporate taxes.
- How to determine ordinary gains and capital gains.
- How to determine the appropriate tax rate to use in project analysis.
- The relationship between net income and net cash flow.

9.1 Asset Depreciation

Fixed assets, such as equipment and real estate, are economic resources that are acquired to provide future cash flows. Generally, **depreciation** can be defined as a gradual decrease in the utility of fixed assets with use and time. While this general definition does not capture the subtleties inherent in a more specific definition of depreciation, it does provide us with a starting point for examining the variety of underlying ideas and practices that are discussed in this chapter. Figure 9.1 will serve as a road map for understanding the different types of depreciation that we will explore here.

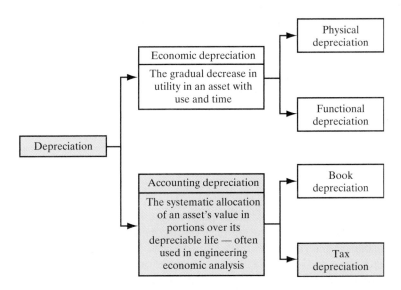

Figure 9.1   Classification of types of depreciation.

We can classify depreciation into the categories of physical or functional depreciation. **Physical depreciation** can be defined as a reduction in an asset's capacity to perform its intended service due to physical impairment. Physical depreciation can occur in any fixed asset in the form of (1) deterioration from interaction with the environment, including such agents as corrosion, rotting, and other chemical changes, and (2) wear and tear from use. Physical depreciation leads to a decline in performance and high maintenance costs.

Functional depreciation occurs as a result of changes in the organization or in technology that decrease or eliminate the need for an asset. Examples of functional depreciation include obsolescence attributable to advances in technology, a declining need for the services performed by an asset, and the inability to meet increased quantity or quality demands.

9.1.1 Economic Depreciation

This chapter is concerned primarily with accounting depreciation, which is the form of depreciation that provides an organization with the information it uses to assess its financial position. It would also be useful, however, to discuss briefly the economic ideas upon which accounting depreciation is based. In the course of the discussion, we will develop a precise definition of economic depreciation that will help us distinguish between various conceptions of depreciation.

If you have ever owned a car, you are probably familiar with the term *depreciation* as it is used to describe the decreasing value of your vehicle. Because a car's reliability and appearance usually decline with age, the vehicle is worth less with each passing year. You can calculate the economic depreciation accumulated for your car by subtracting the current market value, or "blue book" value, of the car from the price you originally paid for it. We can define **economic depreciation** as follows:

$$\text{Economic depreciation} = \text{Purchase price} - \text{market value.}$$

Physical and functional depreciation are categories of economic depreciation. The measurement of economic depreciation does not require that an asset be sold: The market value of the asset can be closely estimated without actually testing it in the marketplace. The need to have a precise scheme for recording the ongoing decline in the value of an asset as a part of the accounting process leads us to an exploration of how organizations account for depreciation.

9.1.2 Accounting Depreciation

The acquisition of fixed assets is an important activity for a business organization, whether the organization is starting up or acquiring new assets to remain competitive. Like other disbursements, the cost of these fixed assets must be recorded as expenses on a firm's balance sheet and income statement. However, unlike costs such as maintenance, material, and labor costs, the costs of fixed assets are not treated simply as expenses to be accounted for in the year that they are acquired. Rather, these assets are **capitalized**; that is, their costs are distributed by subtracting them as expenses from gross income, one part at a time over a number of periods. The systematic allocation of the initial cost of an asset in parts over a time, known as the asset's depreciable life, is what we mean by **accounting depreciation**. Because accounting depreciation is the standard of the business world, we sometimes refer to it more generally as **asset depreciation**.

Accounting depreciation is based on the **matching concept**: A fraction of the cost of the asset is chargeable as an expense in each of the accounting periods in which the asset provides service to the firm, and each charge is meant to be a percentage of the whole cost that "matches" the percentage of the value utilized in the given period. The matching concept suggests that the accounting depreciation allowance generally reflects, at least to some extent, the actual economic depreciation of the asset. *In engineering economic analysis, we use the concept of accounting depreciation exclusively.* This is because accounting depreciation provides a basis for determining the income taxes associated with any project undertaken.

> **Accounting depreciation:** Amount allocated during the period to amortize the cost of acquiring long term assets over the useful life of the assets.

9.2 Factors Inherent in Asset Depreciation

The process of depreciating an asset requires that we make several preliminary determinations: (1) What is the cost of the asset? (2) What is the asset's value at the end of its useful life? (3) What is the depreciable life of the asset? and, finally, (4) What method of depreciation do we choose? In this section, we will examine each of these questions.

9.2.1 Depreciable Property

As a starting point, it is important to recognize what constitutes a **depreciable asset**—that is, a property for which a firm may take depreciation deductions against income. For the purposes of U.S. tax law, any depreciable property has the following characteristics:

1. It must be used in business or must be held for the production of income.
2. It must have a definite service life, and that life must be longer than 1 year.
3. It must be something that wears out, decays, gets used up, becomes obsolete, or loses value from natural causes.

Depreciable property includes buildings, machinery, equipment, and vehicles. Inventories are not depreciable property, because they are held primarily for sale to customers in the ordinary course of business. If an asset has no definite service life, the asset cannot be depreciated. For example, *you can never depreciate land.*[2]

As a side note, we mention the fact that, while we have been focusing on depreciation within firms, individuals may also depreciate assets, as long as they meet the conditions we have listed. For example, an individual may depreciate an automobile if the vehicle is used exclusively for business purposes.

---

[2] This also means that you cannot depreciate the cost of clearing, grading, planting, and landscaping. All four expenses are considered part of the cost of the land.

9.2.2 Cost Basis

The **cost basis** of an asset represents the total cost that is claimed as an expense over the asset's life (i.e., the sum of the annual depreciation expenses). The cost basis generally includes the actual cost of the asset and all other incidental expenses, such as freight, site preparation, and installation. This total cost, rather than the cost of the asset only, must be the depreciation basis charged as an expense over the asset's life.

Besides being used in figuring depreciation deductions, an asset's cost basis is used in calculating the gain or loss to the firm if the asset is ever sold or salvaged. (We will discuss these topics in Section 9.8.)

Cost basis: The cost of an asset used to determine the depreciation base.

EXAMPLE 9.1 Cost Basis

Lanier Corporation purchased an automatic hole-punching machine priced at $62,500. The vendor's invoice included a sales tax of $3,263. Lanier also paid the inbound transportation charges of $725 on the new machine, as well as the labor cost of $2,150 to install the machine in the factory. In addition, Lanier had to prepare the site at a cost of $3,500 before installation. Determine the cost basis for the new machine for depreciation purposes.

SOLUTION

Book value: The value of an asset as it appears on a balance sheet, equal to cost minus accumulated depreciation.

Given: Invoice price = $62,500, freight = $725, installation cost = $2,150, and site preparation = $3,500.
Find: The cost basis.

The cost of machine that is applicable for depreciation is computed as follows:

Cost of new hole-punching machine	$62,500
Freight	725
Installation labor	2,150
Site preparation	3,500
Cost of machine (cost basis)	$68,875

COMMENTS: Why do we include all the incidental charges relating to the acquisition of a machine in its cost? Why not treat these incidental charges as expenses incurred during the period in which the machine is acquired? The matching of costs and revenue is the basic accounting principle. Consequently, the total costs of the machine should be viewed as an asset and allocated against the future revenue that the machine will generate. All costs incurred in acquiring the machine are costs of the services to be received from using the machine.

If the asset is purchased by trading in a similar asset, the difference between the book value (the cost basis minus the total accumulated depreciation) and the trade-in allowance must be considered in determining the cost basis for the new asset. If the trade-in allowance exceeds the book value, the difference (known as **unrecognized gain**) needs to be subtracted from the cost basis of the new asset. If the opposite is true (**unrecognized loss**), the difference should be added to the cost basis for the new asset.

EXAMPLE 9.2 Cost Basis with Trade-In Allowance

In Example 9.1, suppose Lanier purchased the hole-punching press by trading in a similar machine and paying cash for the remainder. The trade-in allowance is $5,000, and the book value of the hole-punching machine that was traded in is $4,000. Determine the cost basis for this hole-punching press.

SOLUTION

Given: Accounting data from Example 9.1; trade allowance = $5,000.
Find: The revised cost basis.

Old hole-punching machine (**book value**)	$ 4,000
Less: Trade-in allowance	5,000
Unrecognized gains	$ 1,000
Cost of new hole-punching machine	$62,500
Less: Unrecognized gains	(1,000)
Freight	725
Installation labor	2,150
Site preparation	3,500
Cost of machine (cost basis)	$67,875

9.2.3 Useful Life and Salvage Value

Over how many periods will an asset be useful to a company? What do published statutes allow you to choose as the life of an asset? These are the central questions to be answered in determining an asset's depreciable life (i.e., the number of years over which the asset is to be depreciated).

Historically, depreciation accounting included choosing a depreciable life that was based on the service life of an asset. Determining the service life of the asset, however, was often very difficult, and the uncertainty of these estimates often led to disputes between taxpayers and the Internal Revenue Service (IRS). To alleviate the problems, the IRS published guidelines on lives for categories of assets. The guidelines, known as **asset depreciation ranges**, or **ADRs**, specified a range of lives for classes of assets based on historical data, and taxpayers were free to choose a depreciable life within the specified range for a given asset. An example of ADRs for some assets is given in Table 9.1.

TABLE 9.1 Asset Depreciation: Some Selected Asset Guideline Classes

Assets Used	Asset Depreciation Range (Years)		
	Lower Limit	Midpoint Life	Upper Limit
Office furniture, fixtures, and equipment	8	10	12
Information systems (computers)	5	6	7
Airplanes	5	6	7
Automobiles, taxis	2.5	3	3.5
Buses	7	9	11
Light trucks	3	4	5
Heavy trucks (concrete ready-mixer)	5	6	7
Railroad cars and locomotives	12	15	18
Tractor units	5	6	7
Vessels, barges, tugs, and water transportation systems	14.5	18	21.5
Industrial steam and electrical generation and/or distribution systems	17.5	22	26.5
Manufacturer of electrical and nonelectrical machinery	8	10	12
Manufacturer of electronic components, products, and systems	5	6	7
Manufacturer of motor vehicles	9.5	12	14.5
Telephone distribution plant	28	35	42

Source: IRS Publication 534. *Depreciation*. Washington, DC: U.S. Government Printing Office, 1995.

The **salvage value** is an asset's estimated value at the end of its life—the amount eventually recovered through sale, trade-in, or salvage. The eventual salvage value of an asset must be estimated when the depreciation schedule for the asset is established. If this estimate subsequently proves to be inaccurate, then an adjustment must be made. We will discuss these specific issues in Section 9.6.

9.2.4 Depreciation Methods: Book and Tax Depreciation

Most firms calculate depreciation in two different ways, depending on whether the calculation is (1) intended for financial reports (the **book depreciation method**), such as for the balance sheet or income statement, or (2) for the Internal Revenue Service (IRS), for the purpose of determining taxes (the **tax depreciation method**). In the United States, this distinction is totally legitimate under IRS regulations, as it is in many other countries.

Calculating depreciation differently for financial reports and for tax purposes allows the following benefits:

- It enables firms to report depreciation to stockholders and other significant outsiders on the basis of the matching concept. Therefore, the actual loss in value of the assets is generally reflected.
- It allows firms to benefit from the tax advantages of depreciating assets more quickly than would be possible with the matching concept. In many cases, tax depreciation allows firms to defer paying income taxes. This does not mean that they pay less tax overall, because the total depreciation expense accounted for over time is the same in either case. However, because tax depreciation methods usually permit a higher depreciation in earlier years than do book depreciation methods, the tax benefit of depreciation is enjoyed earlier, and firms generally pay lower taxes in the initial years of an investment project. Typically, this leads to a better cash position in early years, and the added cash leads to greater future wealth because of the time value of the funds.

As we proceed through the chapter, we will make increasing use of the distinction between depreciation accounting for financial reporting and depreciation accounting for income tax calculation. Now that we have established the context for our interest in both tax and book depreciation, we can survey the different methods with an accurate perspective.

9.3 Book Depreciation Methods

Three different methods can be used to calculate the periodic depreciation allowances: (1) the straight-line method, (2) accelerated methods, and (3) the unit-of-production method. In engineering economic analysis, we are interested primarily in depreciation in the context of income tax computation. Nonetheless, a number of reasons make the study of book depreciation methods useful. First, tax depreciation methods are based largely on the same principles that are used in book depreciation methods. Second, firms continue to use book depreciation methods for financial reporting to stockholders and outside parties. Third, book depreciation methods are still used for state income tax purposes in many states and even for federal income tax purposes for assets that were put into service before 1981. Finally, our discussion of depletion in Section 9.5 is based largely on one of these three book depreciation methods.

9.3.1 Straight-Line Method

The **straight-line (SL) method** of depreciation interprets a fixed asset as an asset that offers its services in a uniform fashion. The asset provides an equal amount of service in each year of its useful life.

Straight-line depreciation: An equal dollar amount of depreciation in each accounting period.

The straight-line method charges, as an expense, an equal fraction of the net cost of the asset each year, as expressed by the relation

$$D_n = \frac{(I - S)}{N}, \tag{9.1}$$

where D_n = Depreciation charge during year n,

I = Cost of the asset, including installation expenses,

S = Salvage value at the end of the asset's useful life,

N = Useful life.

The book value of the asset at the end of n years is then defined as

Book value in a given year = Cost basis − total depreciation charges made to date

or

$$B_n = I - (D_1 + D_2 + D_3 + \cdots + D_n).$$ (9.2)

EXAMPLE 9.3 Straight-Line Depreciation

Consider the following data on an automobile:

Cost basis of the asset, I = $10,000,

Useful life, N = 5 years,

Estimated salvage value, S = $2,000.

Use the straight-line depreciation method to compute the annual depreciation allowances and the resulting book values.

SOLUTION

Given: I = $10,000, S = $2,000, and N = 5 years.
Find: D_n and B_n for n = 1 to 5.

The straight-line depreciation rate is $\frac{1}{5}$, or 20%. Therefore, the annual depreciation charge is

$$D_n = (0.20)(\$10,000 - \$2,000) = \$1,600.$$

The asset would then have the following book values during its useful life:

n	B_{n-1}	D_n	B_n
1	$10,000	$1,600	$8,400
2	8,400	1,600	6,800
3	6,800	1,600	5,200
4	5,200	1,600	3,600
5	3,600	1,600	2,000

Here, B_{n-1} represents the book value before the depreciation charge for year n. This situation is illustrated in Figure 9.2.

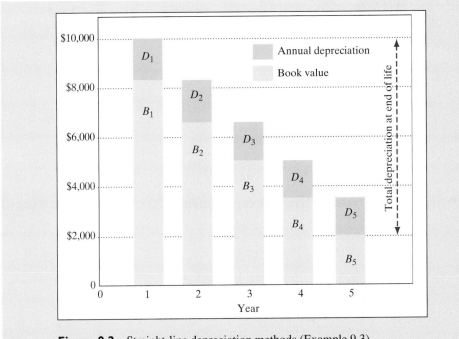

Figure 9.2   Straight-line depreciation methods (Example 9.3).

9.3.2 Accelerated Methods

The second concept of depreciation recognizes that the stream of services provided by a fixed asset may decrease over time; in other words, the stream may be greatest in the first year of an asset's service life and least in its last year. This pattern may occur because the mechanical efficiency of an asset tends to decline with age, because maintenance costs tend to increase with age, or because of the increasing likelihood that better equipment will become available and make the original asset obsolete. This kind of reasoning leads to a method that charges a larger fraction of the cost as an expense of the early years than of the later years. Any such method is called an **accelerated method**. The most widely used accelerated method is the **double-declining-balance method**.

Accelerated method: Any depreciation method that produces larger deductions for depreciation in the early years of a project's life.

Declining-Balance Method

The **declining-balance (DB) method** of calculating depreciation allocates a fixed fraction of the beginning book balance each year. The fraction, α, is obtained as follows:

$$\alpha = \left(\frac{1}{N}\right)(\text{multiplier}).\qquad(9.3)$$

The most commonly used multipliers in the United States are 1.5 (called 150% DB) and 2.0 (called 200%, or double-declining balance, DDB). As N increases, α decreases, resulting in a situation in which depreciation is highest in the first year and then decreases over the asset's depreciable life.

Double-declining balance method: A depreciation method, in which double the straight-line depreciation amount is taken the first year, and then that same percentage is applied to the undepreciated amount in subsequent years.

The fractional factor can be utilized to determine depreciation charges for a given year, D_n, as follows:

$$D_1 = \alpha I,$$

$$D_2 = \alpha(I - D_1) = \alpha I(1 - \alpha),$$

$$D_3 = \alpha(I - D_1 - D_2) = \alpha I(1 - \alpha)^2,$$

Thus, for any year n, we have a depreciation charge

$$D_n = \alpha I(1 - \alpha)^{n-1}. \tag{9.4}$$

We can also compute the total DB depreciation (TDB) at the end of n years as follows:

$$
\begin{aligned}
\text{TDB} &= D_1 + D_2 + \cdots + D_n \\
&= \alpha I + \alpha I(1 - \alpha) + \alpha I(1 - \alpha)^2 + \cdots + \alpha I(1 - \alpha)^{n-1} \\
&= \alpha I[1 + (1 - \alpha) + (1 - \alpha)^2 + \cdots + (1 - \alpha)^{n-1}] \\
&= I[1 - (1 - \alpha)^n]. \tag{9.5}
\end{aligned}
$$

The book value B_n at the end of n years will be the cost I of the asset minus the total depreciation at the end of n years:

$$
\begin{aligned}
B_n &= I - \text{TDB} \\
&= I - I[1 - (1 - \alpha)^n] \\
B_n &= I(1 - \alpha)^n. \tag{9.6}
\end{aligned}
$$

EXAMPLE 9.4 Declining-Balance Depreciation

Consider the following accounting information for a computer system:

Cost basis of the asset, $I = \$10,000$,

Useful life, $N = 5$ years,

Estimated salvage value, $S = \$778$.

Use the double-declining-depreciation method to compute the annual depreciation allowances and the resulting book values (Figure 9.3).

SOLUTION

Given: $I = \$10,000$, $S = \$778$, $N = 5$ years
Find: D_n and B_n for $n = 1$ to 5

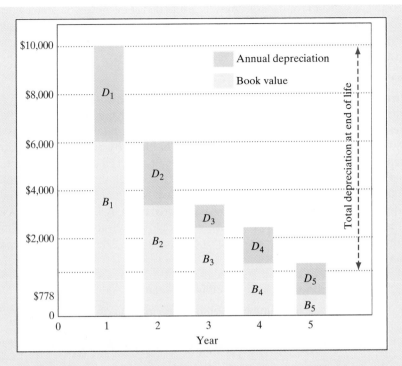

Figure 9.3 Double-declining-balance method (Example 9.4).

The book value at the beginning of the first year is $10,000, and the declining-balance rate (α) is $\left(\frac{1}{5}\right)(2) = 40\%$. Then the depreciation deduction for the first year will be $4,000 $(40\% \times \$10,000 = \$4,000)$. To figure the depreciation deduction in the second year, we must first adjust the book value for the amount of depreciation we deducted in the first year. The first year's depreciation from the beginning book value is subtracted ($10,000 - $4,000 = $6,000), and the resulting amount is multiplied by the rate of depreciation ($6,000 \times 40\% = $2,400). By continuing the process, we obtain the following table:

n	B_n	D_n	B_n
1	10,000	4,000	6,000
2	6,000	2,400	3,600
3	3,600	1,440	2,160
4	2,160	864	1,296
5	1,296	518	778

The declining balance is illustrated in terms of the book value of time in Figure 9.3.

The salvage value (*S*) of the asset must be estimated at the outset of depreciation analysis. In Example 9.4, the final book value (B_N) conveniently equals the estimated salvage value of $778, a coincidence that is rather unusual in the real world. When $B_N \neq S$, we would want to make adjustments in our depreciation methods.

- **Case 1: $B_N > S$**

  When $B_N > S$, we are faced with a situation in which we have not depreciated the entire cost of the asset and thus have not taken full advantage of depreciation's tax-deferring benefits. If you would prefer to reduce the book value of an asset to its salvage value as quickly as possible, it can be done by switching from DB to SL whenever SL depreciation results in larger depreciation charges and therefore a more rapid reduction in the book value of the asset. The switch from DB to SL depreciation can take place in any of the *n* years, the objective being to identify the optimal year to switch. The switching rule is as follows: If depreciation by DB in any year is less than (or equal to) what it would be by SL, we should switch to and remain with the SL method for the duration of the project's depreciable life. The straight-line depreciation in any year *n* is calculated by

$$D_n = \frac{\text{Book value at beginning of year } n - \text{salvage value}}{\text{Remaining useful life at beginning of year } n} \qquad (9.7)$$

EXAMPLE 9.5 Declining Balance with Conversion to Straight-Line Depreciation ($B_N > S$)

Suppose the asset given in Example 9.4 has a zero salvage value instead of $778; that is,

$$\text{Cost basis of the asset, } I = \$10,000$$
$$\text{Useful life, } N = 5 \text{ years,}$$
$$\text{Salvage value, } S = \$0,$$
$$a = (1/5)(2) = 40\%.$$

Determine the optimal time to switch from DB to SL depreciation and the resulting depreciation schedule.

SOLUTION

Given: $I = \$10,000$, $S = 0$, $N = 5$ years, and $\alpha = 40\%$.
Find: Optimal conversion time, D_n and B_n for $n = 1$ to 5.

We will first proceed by computing the DDB depreciation for each year, as before:

Year	D_n	B_n
1	$4,000	$6,000
2	2,400	3,600
3	1,440	2,160
4	864	1,296
5	518	778

Then, using Eq. (9.7), we compute the SL depreciation for each year. We compare SL with DDB depreciation for each year and use the aforementioned decision rule for when to change:

If Switch to SL at Beginning of Year	SL Depreciation	DDB Depreciation	Switching Decision
2	($6,000 − 0)/4 = $1,500	<$2,400	Do not switch
3	(3,600 − 0)/3 = 1,200	< 1,440	Do not switch
4	(2,160 − 0)/2 = 1,080	> 864	Switch to SL

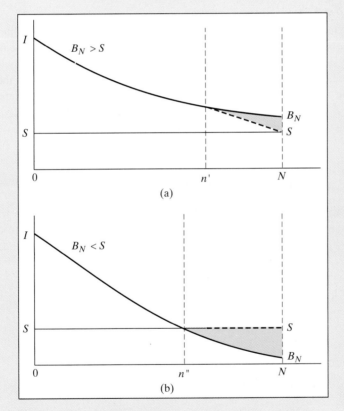

Figure 9.4   Adjustments to the declining-balance method:
(a) Switch from declining balance to straight line after n';
(b) no further depreciation allowances are available after n''
(Examples 9.5 and 9.6).

The optimal time (year 4) in this situation corresponds to n' in Figure 9.4(a). The resulting depreciation schedule is

Year	DDB with Switch to SL	End-of-Year Book Value
1	$ 4,000	$6,000
2	2,400	3,600
3	1,440	2,160
4	1,080	1,080
5	1,080	0
	$10,000	

- **Case 2: $B_N < S$**

  With a relatively high salvage value, it is possible that the book value of the asset could decline below the estimated salvage value. When $B_n < S$, we must readjust our analysis because tax law does not permit us to depreciate assets below their salvage value. To avoid deducting depreciation charges that would drop the book value below the salvage value, you simply stop depreciating the asset whenever you get down to $B_n = S$. In other words, if, at any period, the implied book value is lower than S, then the depreciation amounts are adjusted so that $B_n = S$.

EXAMPLE 9.6 Declining Balance, $B_N < S$

Compute the double-declining-balance (DDB) depreciation schedule for the data used in Example 9.5, this time with the asset having a salvage value of $2,000. Then

$$\text{Cost basis of the asset, } I = \$10,000,$$

$$\text{Useful life, } N = 5 \text{ years,}$$

$$\text{Salvage value, } S = \$2,000,$$

$$\alpha = \left(\tfrac{1}{5}\right)(2) = 40\%.$$

SOLUTION

Given: $I = \$10,000$, $S = \$2,000$, $N = 5$ years, and $\alpha = 40\%$.
Find: D_n and B_n for $n = 1$ to 5.

The given data result in the accompanying table.
Note that B_4 would be less than $S = \$2,000$ if the full deduction ($864) had been taken. Therefore, we adjust D_4 to $160, making $B_4 = \$2,000$. D_5 is zero and B_5 remains at $2,000. Year 4 is equivalent to n'' in Figure 9.4(b).

End of Year	D_n	B_n
1	$0.4(\$10,000) = \$4,000$	$\$10,000 - \$4,000 = \$6,000$
2	$0.4(6,000) = 2,400$	$6,000 - 2,400 = 3,600$
3	$0.4(3,600) = 1,440$	$3,600 - 1,440 = 2,160$
4	$0.4(2,160) = 864 > \boxed{160}$	$2,160 - 160 = 2,000$
5	0	$2,000 - 0 = 2,000$
	Total $= \$8,000$	

9.3.3 Units-of-Production Method

Straight-line depreciation can be defended only if the machine is used for exactly the same amount of time each year. What happens when a punch press machine runs 1,670 hours one year and 780 the next or when some of its output is shifted to a new machining center? This leads us to a consideration of another depreciation method that views the asset as consisting of a bundle of service units; unlike the SL and accelerated methods, however, this one does not assume that the service units will be consumed in a time-phased pattern. Rather, the cost of each service unit is the net cost of the asset divided by the total number of such units. The depreciation charge for a period is then related to the number of service units consumed in that period. The result is the **units-of-production method**, according to which the depreciation in any year is given by

$$D_n = \frac{\text{Service units consumed during year } n}{\text{Total service units}} (I - S). \qquad (9.8)$$

When the units-of-production method is used, depreciation charges are made proportional to the ratio of the actual output to the total expected output. Usually, this ratio is figured in machine hours. The advantages of using the units-of-production method include the fact that depreciation varies with production volume, so the method gives a more accurate picture of machine usage. A disadvantage of the method is that collecting data on machine usage is somewhat tedious, as are the accounting methods.

This method can be useful in depreciating equipment used to exploit natural resources if the resources will be depleted before the equipment wears out. It is not, however, considered a practical method for general use in depreciating industrial equipment.

EXAMPLE 9.7 Units-of-Production Depreciation

A truck for hauling coal has an estimated net cost of $55,000 and is expected to give service for 250,000 miles, resulting in a $5,000 salvage value. Compute the allowed depreciation amount for a truck usage of 30,000 miles.

SOLUTION

Given: $I = \$55,000$, $S = \$5,000$, total service units = 250,000 miles, and usage for this year = 30,000 miles.

Find: Depreciation amount in this year.

The depreciation expense in a year in which the truck traveled 30,000 miles would be

$$\frac{30,000 \text{ miles}}{250,000 \text{ miles}} (\$55,000 - \$5,000) = \left(\frac{3}{25}\right)(\$50,000)$$

$$= \$6,000.$$

9.4 Tax Depreciation Methods

Prior to the Economic Recovery Act of 1981, taxpayers could choose among several methods when depreciating assets for tax purposes. The most widely used methods were the straight-line method and the declining-balance method. The subsequent imposition of the Accelerated Cost Recovery System (ACRS) and the Modified Accelerated Cost Recovery System (MACRS) superseded these methods for use in tax purposes.

9.4.1 MACRS Depreciation

MACRS method allows taxpayers to deduct greater amounts during the first few years of an asset's life.

From 1954 to 1981, congressional changes in tax law evolved fairly consistently toward simpler, more rapid depreciation methods. Prior to 1954, the straight-line method was required for tax purposes, but that year accelerated methods such as double-declining balance and sum-of-years'-digits were permitted. In 1981, these conventional accelerated methods were replaced by the simpler ACRS. In 1986, Congress modified the ACRS and introduced the MACRS, sharply reducing depreciation allowances that were enacted in the Economic Recovery Tax Act of 1981. This section will present some of the primary features of MACRS tax depreciation.

MACRS Recovery Periods

Historically, for both tax and accounting purposes, an asset's depreciable life was determined by its estimated useful life; it was intended that the asset be fully depreciated at approximately the end of its useful life. The MACRS scheme, however, totally abandoned this practice, and simpler guidelines were established that created several classes of assets, each with a more or less arbitrary life called a **recovery period**. (Note: *Recovery periods do not necessarily bear any relationship to expected useful lives*.)

A major effect of the original ACRS method of 1981 was to shorten the depreciable lives of assets, thus giving businesses larger depreciation deductions that, in early years, resulted in lower taxes and increased cash flows available for reinvestment. As shown in Table 9.2, the MACRS method of 1986 reclassified certain assets on the basis of the

TABLE 9.2 MACRS Property Classifications

Recovery Period	ADR* Midpoint Class	Applicable Property
3 years	ADR ≤ 4	Special tools for the manufacture of plastic products, fabricated metal products, and motor vehicles
5 years	4 < ADR ≤ 10	Automobiles,[†] light trucks, high-tech equipment, equipment used for research and development, computerized telephone switching systems
7 years	10 < ADR ≤ 16	Manufacturing equipment, office furniture, fixtures
10 years	16 < ADR ≤ 20	Vessels, barges, tugs, railroad cars
15 years	20 < ADR ≤ 25	Wastewater plants, telephone-distribution plants, similar utility property
20 years	25 < ADR	Municipal sewers, electrical power plant
$27\frac{1}{2}$ years		Residential rental property
39 years		Nonresidential real property, including elevators and escalators

* ADR = Asset depreciation range; guidelines are published by the IRS.
[†] Automobiles have a midpoint life of 3 years in the ADR guidelines, but are classified into a 5-year property class.

MACRS: Depreciation methods applied to assets placed in service after 1986; less favorable than the earlier ACRS system.

midpoint of their lives under the ADR system. The MACRS scheme includes eight categories of assets, with lives of 3, 5, 7, 10, 15, 20, 27.5, and 39 years:

- Investments in some short-lived assets are depreciated over 3 years by using 200% DB and then switching to SL depreciation.
- Computers, automobiles, and light trucks are written off over 5 years by using 200% DB and then switching to SL depreciation.
- Most types of manufacturing equipment are depreciated over 7 years, but some long-lived assets are written off over 10 years. Most equipment write-offs are calculated by the 200% DB method, followed by a switch to SL depreciation, which allows faster write-offs in the first few years after an investment is made.
- Sewage-treatment plants and telephone-distribution plants are written off over 15 years by using 150% DB and then switching to SL depreciation.
- Sewer pipes and certain other very long-lived equipment are written off over 20 years by using 150% DB and then switching to SL depreciation.
- Investments in residential rental property are written off in straight-line fashion over $27\frac{1}{2}$ years. Nonresidential real estate (commercial buildings), by contrast, is written off by the SL method over 39 years.

Under the MACRS, *the salvage value of property is always treated as zero.*

9.4.2 MACRS Depreciation Rules

Under earlier depreciation methods, the rate at which the value of an asset declined was estimated and was then used as the basis for tax depreciation. Thus, different assets were depreciated along different paths over time. The MACRS method, however, established prescribed depreciation rates, called **recovery allowance percentages**, for all assets within each class. These rates, as set forth in 1986 and 1993, are shown in Table 9.3. The yearly recovery, or depreciation expense, is determined by multiplying the asset's depreciation base by the applicable recovery allowance percentage.

Half-Year Convention

The MACRS recovery percentages shown in Table 9.3 use the **half-year convention**; that is, it is assumed that all assets are placed in service at midyear and that they will have *zero* salvage value. As a result, only a half year of depreciation is allowed for the first year that property is placed in service. With half of one year's depreciation being taken in the first

TABLE 9.3 MACRS Depreciation Schedules for Personal Properties with Half-Year Convention, Declining-Balance Method

	Class	3	5	7	10	15	20
Year	Depreciation Rate	200%	200%	200%	200%	150%	150%
1		33.33	20.00	14.29	10.00	5.00	3.750
2		44.45	32.00	24.49	18.00	9.50	7.219
3		14.81*	19.20	17.49	14.40	8.55	6.677
4		7.41	11.52*	12.49	11.52	7.70	6.177
5			11.52	8.93*	9.22	6.93	5.713
6			5.76	8.92	7.37	6.23	5.285
7				8.93	6.55*	5.90*	4.888
8				4.46	6.55	5.90	4.522
9					6.56	5.91	4.462*
10					6.55	5.90	4.461
11					3.28	5.91	4.462
12						5.90	4.461
13						5.91	4.462
14						5.90	4.461
15						5.91	4.462
16						2.95	4.461
17							4.462
18							4.461
19							4.462
20							4.461
21							2.231

* Year to switch from declining balance to straight line. (*Source*: IRS Publication 534. *Depreciation*. Washington, DC: U.S. Government Printing Office, December 2005.)

year, a full year's depreciation is allowed in each of the remaining years of the asset's recovery period, and the remaining half-year's depreciation is taken in the year following the end of the recovery period. A half year of depreciation is also allowed for the year in which property is disposed of, or is otherwise retired from service, anytime before the end of the recovery period.

Switching from DB to the Straight-Line Method

The MACRS asset is depreciated initially by the DB method and then by SL depreciation. Consequently, the MACRS adopts the switching convention illustrated in Example 9.5. To demonstrate how the MACRS depreciation percentages were calculated by the IRS with the use of the half-year convention, consider Example 9.8.

EXAMPLE 9.8 MACRS Depreciation: Personal Property

A taxpayer wants to place in service a $10,000 asset that is assigned to the five-year class. Compute the MACRS percentages and the depreciation amounts for the asset.

SOLUTION

Given: Five-year asset, half-year convention, $\alpha = 40\%$, and $S = 0$.
Find: MACRS depreciation percentages D_n for $10,000 asset.

For this problem, we use the following equations:

$$\text{Straight-line rate} = \tfrac{1}{5} = 0.20,$$

$$200\% \text{ declining balance rate} = 2(0.20) = 40\%,$$

$$\text{Under MACRS, salvage } (S) = 0.$$

Then, beginning with the first taxable year and ending with the sixth year, MACRS deduction percentages are computed as follows:

Year	Calculation (%)		MACRS Percentage
1	$\tfrac{1}{2}$-year DDB depreciation $= 0.5(0.40)(100\%)$	$=$	20%
2	DDB depreciation $= (0.40)(100\% - 20\%)$	$=$	32%
	SL depreciation $= (1/4.5)(100\% - 20\%)$	$=$	17.78%
3	DDB depreciation $= (0.40)(100\% - 52\%)$	$=$	19.20%
	SL depreciation $= (1/3.5)(100\% - 52\%)$	$=$	13.71%
4	DDB depreciation $= (0.40)(100\% - 71.20\%)$	$=$	11.52%
	SL depreciation $= (1/2.5)(100\% - 71.20\%)$	$=$	11.52%
5	SL depreciation $= (1/1.5)(100\% - 82.72\%)$	$=$	11.52%
6	$\tfrac{1}{2}$-year SL depreciation $= (0.5)(11.52\%)$	$=$	5.76%

In year 2, we check to see what the SL depreciation would be. Since 4.5 years are left to depreciate, SL depreciation $= (1/4.5)(100\% - 20\%) = 17.78\%$. The DDB depreciation is greater than the SL depreciation, so DDB still applies. Note that SL depreciation $\geq$ DDB depreciation in year 4, so we switch to SL then.

We can calculate the depreciation amounts from the percentages we just determined. In practice, the percentages are taken directly from Table 9.3, supplied by the IRS. The results are as follows and are also shown in Figure 9.5.

Year n	MACRS Percentage (%)		Depreciation Basis		Depreciation Amount (D_n)
1	20	$\times$	$10,000	=	$2,000
2	32	$\times$	10,000	=	3,200
3	19.20	$\times$	10,000	=	1,920
4	11.52	$\times$	10,000	=	1,152
5	11.52	$\times$	10,000	=	1,152
6	5.76	$\times$	10,000	=	576

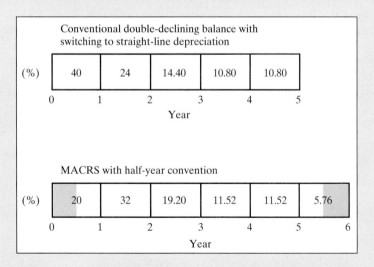

Figure 9.5 MACRS with a five-year recovery period (Example 9.8).

Note that when an asset is disposed of before the end of a recovery period, only half of the normal depreciation is allowed. If, for example, the $10,000 asset were to be disposed of in year 2, the MACRS deduction for that year would be $1,600.

Midmonth Convention for Real Property

Table 9.4 shows the useful lives and depreciation percentages for real property. In depreciating such property, the straight-line method and the midmonth convention are used. For example, a property placed in service in March would be allowed $9\frac{1}{2}$ months' depreciation for year 1. If it is disposed of before the end of the recovery period, the depreciation percentage must take into account the number of months the property was in service during the year of its disposal.

EXAMPLE 9.9 MACRS for Real Property

On May 1, Jack Costanza paid \$100,000 for a residential rental property. This purchase price represents \$80,000 for the cost of the building and \$20,000 for the cost of the land. Three years and five months later, on October 1, he sold the property for \$130,000. Compute the MACRS depreciation for each of the four calendar years during which Jack owned the property.

SOLUTION

Given: Residential real property, cost basis $= \$80,000$; the building was put into service on May 1.

Find: The depreciation in each of four tax years the property was in service.

In this example, the midmonth convention assumes that the property is placed in service on May 15, which gives $7\frac{1}{2}$ months of depreciation in the first year. Remembering that only the building (not the land) may be depreciated, we use the SL method to compute the depreciation over a $27\frac{1}{2}$-year recovery period:

Year	Calculation		D_n	Recovery Percentages
1	$\left(\dfrac{7.5}{12}\right)\dfrac{80,000 - 0}{27.5}$	$=$	\$1,818	2.273%
2	$\dfrac{80,000 - 0}{27.5}$	$=$	\$2,909	3.636%
3	$\dfrac{80,000 - 0}{27.5}$	$=$	\$2,909	3.636%
4	$\left(\dfrac{9.5}{12}\right)\dfrac{80,000 - 0}{27.5}$	$=$	\$2,303	2.879%

Notice that the midmonth convention also applies to the year of disposal. Now compare the percentages with those in Table 9.4. As with personal property, calculations for real property generally use the precalculated percentages in the table.

TABLE 9.4 MACRS Percentages for Real Property

Year	Month Property Is Placed in Service											
	1	2	3	4	5	6	7	8	9	10	11	12
(a) Residential Rental Property: Straight Line over 27½ Years with Midmonth Convention												
1	3.485%	3.182%	2.879%	2.576%	2.273%	1.970%	1.667%	1.364%	1.061%	0.758%	0.455%	0.152%
2–9	3.636	3.636	3.636	3.636	3.636	3.636	3.636	3.636	3.636	3.636	3.636	3.636
10	3.637	3.637	3.637	3.637	3.637	3.637	3.636	3.636	3.636	3.636	3.636	3.636
11	3.636	3.636	3.636	3.637	3.636	3.636	3.637	3.637	3.636	3.636	3.637	3.637
12	3.637	3.637	3.637	3.636	3.637	3.637	3.636	3.636	3.637	3.637	3.636	3.636
13	3.636	3.636	3.636	3.637	3.636	3.636	3.637	3.637	3.636	3.636	3.637	3.637
14	3.637	3.637	3.637	3.636	3.637	3.637	3.636	3.636	3.637	3.637	3.636	3.636
15	3.636	3.636	3.636	3.637	3.636	3.636	3.637	3.637	3.636	3.636	3.637	3.637
16	3.637	3.637	3.637	3.636	3.637	3.637	3.636	3.636	3.637	3.637	3.636	3.636
17	3.636	3.636	3.636	3.637	3.636	3.636	3.637	3.637	3.636	3.636	3.637	3.637
18	3.637	3.637	3.637	3.636	3.637	3.637	3.636	3.636	3.637	3.637	3.636	3.636
19	3.636	3.636	3.636	3.637	3.636	3.636	3.637	3.637	3.636	3.636	3.637	3.637
20	3.637	3.637	3.637	3.636	3.637	3.637	3.636	3.636	3.637	3.637	3.636	3.636
21	3.636	3.636	3.636	3.637	3.636	3.636	3.637	3.637	3.636	3.636	3.637	3.637
22	3.637	3.637	3.637	3.636	3.637	3.637	3.636	3.636	3.637	3.637	3.636	3.636
23	3.636	3.636	3.636	3.637	3.636	3.636	3.637	3.637	3.636	3.636	3.637	3.637
24	3.637	3.637	3.637	3.636	3.637	3.637	3.636	3.636	3.637	3.637	3.636	3.636
25	3.636	3.636	3.636	3.637	3.636	3.636	3.637	3.637	3.636	3.636	3.637	3.637
26	3.637	3.637	3.637	3.636	3.637	3.637	3.636	3.636	3.637	3.637	3.636	3.636
27	3.636	3.636	3.636	3.637	3.636	3.636	3.637	3.637	3.636	3.636	3.637	3.637
28	1.97	2.273	2.576	2.879	3.182	3.485	3.636	3.636	3.636	3.636	3.636	3.636
29							0.152	0.455	0.758	1.061	1.364	1.667
(b) Nonresidential Real Property: Straight Line over 39 Years with Midmonth Convention												
1	2.4573	2.2436	2.0299	1.8162	1.6026	1.3889	1.1752	0.9615	0.7479	0.5342	0.3205	0.1068
2–39	2.5641	2.5641	2.5641	2.5641	2.5641	2.5641	2.5641	2.5641	2.5641	2.5641	2.5641	2.5641
40	0.1068	0.3205	0.5342	0.7479	0.9615	1.1752	1.3889	1.6026	1.8162	2.0299	2.2436	2.4573

Source: IRS Publication No. 534. *Depreciation.* Washington, DC: U.S. Government Printing Office, 2000.

9.5 Depletion

If you own mineral property (distinguished from personal and real properties), such as oil, gas, a geothermal well, or standing timber, you may be able to claim a deduction as you deplete the resource. A capital investment in natural resources needs to be recovered as those resources are being removed and sold. The process of amortizing the cost of natural resources in the accounting periods is called depletion. The objective of depletion is the same as that of depreciation: to amortize the cost in a systematic manner over the asset's useful life.

There are two ways of figuring depletion: **cost depletion** and **percentage depletion**. These depletion methods are used for book as well as tax purposes. In most instances, depletion is calculated by both methods, and the larger value is taken as the depletion allowance for the year. For standing timber and for most oil and gas wells, only cost depletion is permissible.

Depletion: Unlike depreciation and amortization, which mainly describe the deduction of expenses due to the aging of equipment and property, depletion is the actual physical reduction of natural resources by companies.

9.5.1 Cost Depletion

The cost depletion method is based on the same concept as the units-of-production method. To determine the amount of cost depletion, the adjusted basis of the mineral property is divided by the total number of recoverable units in the deposit, and the resulting rate is multiplied by the number of units sold:

$$\text{Cost depletion} = \frac{(\text{Adjusted basis of mineral property})}{\text{Total number of recoverable units}}$$
$$\times \ (\text{Number of units sold}). \qquad (9.9)$$

The **adjusted basis** represents all the depletion allowed (or the allowable cost on the property). Estimating the number of recoverable units in a natural deposit is largely an engineering problem.

EXAMPLE 9.10 Cost Depletion for a Timber Tract

Suppose you bought a timber tract for $200,000 and the land was worth $80,000. The basis for the timber is therefore $120,000. The tract has an estimated 1.5 million board feet (1.5 MBF) of standing timber. If you cut 0.5 MBF of timber, determine your depletion allowance.

SOLUTION

Given: Basis = $120,000, total recoverable volume = 1.5 MBF, and amount sold this year = 0.5 MBF.

Find: The depletion allowance this year.

Timber depletion may be figured only by the cost method. Percentage depletion does not apply to timber, nor does your depletion basis include any part of the cost of the land. Because depletion takes place when standing timber is cut, you may figure your depletion deduction only after the timber is cut and the quantity is first accurately measured. We have

$$\text{Depletion allowance per MBF} = \$120{,}000/1.5 \text{ MBF}$$

$$= \$80{,}000/\text{MBF}$$

$$\text{Depletion allowance for the year} = 0.5 \text{ MBF} \times \$80{,}000/\text{MBF}$$

$$= \$40{,}000.$$

9.5.2 Percentage Depletion

Percentage depletion is an alternative method of calculating the depletion allowance for certain mineral properties. For a given mineral property, the depletion allowance calculation is based on a prescribed percentage of the gross income from the property during the tax year. Notice the distinction between depreciation and depletion: **Depreciation** is the allocation of cost over a useful life, whereas **percentage depletion** is an annual allowance of a percentage of the gross income from the property.

Since percentage depletion is computed on the basis of the income from, rather than the cost of, the property, the total depletion on a property may exceed the cost of the property. To prevent this from happening, the annual allowance under the percentage method cannot be more than 50% of the taxable income from the property (figured without the deduction for depletion). Table 9.5 shows the allowed percentages for selected mining properties.

TABLE 9.5 Percentage Depletion Allowances for Mineral Properties

Deposits	Percentage Allowed
Oil and gas wells (only for certain domestic and gas production)	15
Sulfur and uranium and, if from deposits in the United States, asbestos, lead, zinc, nickel, mica, and certain other ores and minerals	22
Gold, silver, copper, iron ore, and oil shale, if from deposits in the United States	15
Coal, lignite, and sodium chloride	10
Clay and shale to be used in making sewer pipe or bricks	7.5
Clay (used for roofing tile), gravel, sand, and stone	5
Most other minerals; includes carbon dioxide produced from a well and metallic ores	14

EXAMPLE 9.11 Percentage Depletion versus Cost Depletion

A gold mine with an estimated deposit of 300,000 ounces of gold has a basis of $30 million (cost minus land value). The mine has a gross income of $16,425,000 for the year from selling 45,000 ounces of gold (at a unit price of $365 per ounce). Mining expenses before depletion equal $12,250,000. Compute the percentage depletion allowance. Would it be advantageous to apply cost depletion rather than percentage depletion?

SOLUTION

Given: Basis = $30 million, total recoverable volume = 300,000 ounces of gold, amount sold this year = 45,000 ounces, gross income = $16,425,000, and this year's expenses before depletion = $12,250,000.

Find: Maximum depletion allowance (cost or percentage).

Percentage depletion: Table 9.5 indicates that gold has a 15% depletion allowance. The percentage depletion allowance is computed from the gross income:

Gross income from sale of 45,000 ounces	$16,425,000
Depletion percentage	× 15%
Computed percentage depletion	$ 2,463,750

Next, we need to compute the taxable income. The percentage depletion allowance is limited to the computed percentage depletion or 50% of the taxable income, whichever is smaller:

Gross income from sale of 45,000 ounces	$16,425,000
Less mining expenses	12,250,000
Taxable income from mine	4,175,000
Deduction limitation	× 50%
Maximum depletion deduction	$ 2,088,000

Since the maximum depletion deduction ($2,088,000) is less than the computed percentage depletion ($2,463,750), the allowable percentage deduction is $2,088,000.

Cost depletion: It is worth computing the depletion allowance with the cost depletion method:

$$\text{Cost depletion} = \left(\frac{\$30,000,000}{300,000} \right)(45,000)$$

$$= \$4,500,000.$$

Note that percentage depletion is less than the cost depletion. Since the law allows the taxpayer to take whichever deduction is larger in any one year, in this situation it

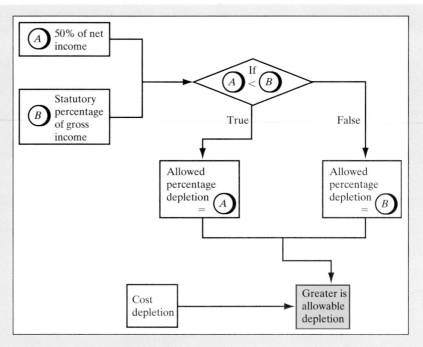

Figure 9.6 Calculating the allowable depletion deduction for federal tax purposes.

would be advantageous to apply the cost depletion method. Figure 9.6 illustrates the steps to be taken to apply that depletion method.

9.6 Repairs or Improvements Made to Depreciable Assets

If any major repairs (e.g., an engine overhaul) or improvements (say, an addition) are made during the life of the asset, we need to determine whether these actions will extend the life of the asset or will increase the originally estimated salvage value. When either of these situations arises, a revised estimate of the useful life of the asset should be made, and the periodic depreciation expense should be updated accordingly. We will examine how repairs or improvements affect both book and tax depreciations.

9.6.1 Revision of Book Depreciation

Recall that book depreciation rates are based on estimates of the useful lives of assets. Such estimates are seldom precise. Therefore, after a few years of use, you may find that the asset could last for a considerably longer or shorter period than was originally estimated. If this happens, the annual depreciation expense, based on the estimated useful life, may be either excessive or inadequate. (If repairs or improvements do not extend the life or increase the salvage value of the asset, these costs may be treated as maintenance expenses during the year in which they were incurred.) The procedure for correcting the book depreciation schedule is to revise the current book value and to allocate this cost over the remaining years of the asset's useful life.

9.6.2 Revision of Tax Depreciation

For tax depreciation purposes, repairs or improvements made to any property are treated as *separate* property items. The recovery period for a repair or improvement to the initial property normally begins on the date the repaired or improved property is placed in service. The recovery class of the repair or improvement is the recovery class that would apply to the property if it were placed in service at the same time as the repair or improvement. Example 9.12 illustrates the procedure for correcting the depreciation schedule for an asset that had repairs or improvements made to it during its depreciable life.

EXAMPLE 9.12 Depreciation Adjustment for an Overhauled Asset

In January 2001, Kendall Manufacturing Company purchased a new numerical control machine at a cost of $60,000. The machine had an expected life of 10 years at the time of purchase and a zero expected salvage value at the end of the 10 years.

- For book depreciation purposes, no major overhauls had been planned over the 10-year period, and the machine was being depreciated toward a zero salvage value, or $6,000 per year, with the straight-line method.
- For tax purposes, the machine was classified as a 7-year MACRS property.

In December 2003, however, the machine was thoroughly overhauled and rebuilt at a cost of $15,000. It was estimated that the overhaul would extend the machine's useful life by 5 years.

(a) Calculate the book depreciation for the year 2006 on a straight-line basis.

(b) Calculate the tax depreciation for the year 2006 for this machine.

SOLUTION

Given: $I = \$60,000$, $S = \$0$, $N = 10$ years, machine overhaul $= \$15,000$, and extended life $= 15$ years from the original purchase.

Find: D_6 for book depreciation and D_6 for tax depreciation.

(a) Since an improvement was made at the end of the year 2003, the book value of the asset at that time consisted of the original book value plus the cost added to the asset. First, the original book value at the end of 2003 is calculated:

$$B_3 \text{ (before improvement)} = \$60,000 - 3(\$6,000) = \$42,000.$$

After the improvement cost of $15,000 is added, the revised book value is

$$B_3 \text{ (after improvement)} = \$42,000 + \$15,000 = \$57,000.$$

To calculate the book depreciation in the year 2006, which is 3 years after the improvement, we need to calculate the annual straight-line depreciation amount with the extended useful life. The remaining useful life before the improvement was made was 7 years. Therefore, the revised remaining useful life should be 12 years. The revised annual depreciation is then $57,000/12 = $4,750. Using the straight-line depreciation method, we compute the depreciation amount for 2006 as follows:

$$D_6 = \$4,750.$$

(b) For tax depreciation purposes, the improvement made is viewed as a separate property with the same recovery period as that of the initial asset. Thus, we need to calculate both the tax depreciation under the original asset and that of the new asset. For the 7-year MACRS property, the 6th-year depreciation allowance is 8.92% of $60,000, or $5,352. The 3rd-year depreciation for the improved asset is 17.49% of $15,000, or $2,623. Therefore, the total tax depreciation in 2006 is

$$D_6 = \$5,352 + \$2,623 = \$7,975.$$

Figure 9.7 illustrates how additions or improvements are treated in revising depreciation amounts for book and tax purposes.

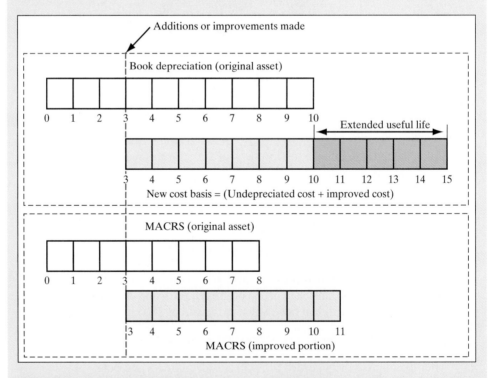

Figure 9.7 Revision of depreciation amount as additions or improvements are made as described in Example 9.12.

9.7 Corporate Taxes

Now that we have learned what elements constitute taxable income, we turn our attention to the process of computing income taxes. The corporate tax rate, is applied to the taxable income of a corporation. As we briefly discussed in Section 8.5, the allowable deductions include the cost of goods sold, salaries and wages, rent, interest, advertising, depreciation, amortization,[3] depletion, and various tax payments other than federal income tax. The following table is illustrative:

Item
Gross income
Expenses:
Cost of goods sold
Depreciation
Operating expenses
Taxable operating income
Income taxes
Net income

9.7.1 Income Taxes on Operating Income

The corporate tax rate structure for 2006 is relatively simple. As shown in Table 9.6, there are four basic rate brackets (15%, 25%, 34%, and 35%), plus two surtax rates (5% and 3%), based on taxable incomes. U.S. tax rates are progressive; that is, businesses with lower taxable incomes are taxed at lower rates than those with higher taxable incomes.

TABLE 9.6 Corporate Tax Schedule for 2006

Taxable Income (X)	Tax Rate	Tax Computation Formula
$0–$50,000	15%	$0 + 0.15X$
50,001–75,000	25%	$7,500 + 0.25(X - \$50,000)$
75,001–100,000	34%	$13,750 + 0.34(X - 75,000)$
100,001–335,000	34% + 5%	$22,250 + 0.39(X - 100,000)$
335,001–10,000,000	34%	$113,900 + 0.34(X - 335,000)$
10,000,001–15,000,000	35%	$3,400,000 + 0.35(X - 10,000,000)$
15,000,001–18,333,333	35% + 3%	$5,150,000 + 0.38(X - 15,000,000)$
18,333,334 and up	35%	$6,416,666 + 0.35(X - 18,333,333)$

[3] The **amortization expense** is a special form of depreciation for an intangible asset, such as patents, goodwill, and franchises. More precisely, the amortization expense is the systematic write-off to expenses of the cost of an intangible asset over the periods of its economic usefulness. Normally a straight-line method is used to calculate the amortization expense.

Marginal Tax Rate

Marginal tax rate: The amount of tax paid on an additional dollar of income.

The **marginal tax rate** is defined as the rate applied to the last dollar of income earned. Income of up to $50,000 is taxed at a 15% rate (meaning that if your taxable income is less than $50,000, your marginal tax rate is 15%); income between $50,000 and $75,000 is taxed at 25%; and income over $75,000 is taxed at a 34% rate.

An additional 5% surtax (resulting in 39%) is imposed on a corporation's taxable income in excess of $100,000, with the maximum additional tax limited to $11,750 (235,000 × 0.05). This surtax provision phases out the benefit of graduated rates for corporations with taxable incomes between $100,000 and $335,000. Another 3% surtax is imposed on corporate taxable income in the range from $15,000,001 to $18,333,333.

Corporations with incomes in excess of $18,333,333 in effect pay a flat tax of 35%. As shown in Table 9.6, the corporate tax is progressive up to $18,333,333 in taxable income, but essentially is constant thereafter.

Effective (Average) Tax Rate

Effective (average) tax rate: The rate a taxpayer would be taxed at if taxing was done at a constant rate, instead of progressively.

Effective tax rates can be calculated from the data in Table 9.6. For example, if your corporation had a taxable income of $16,000,000 in 2006, then the income tax owed by the corporation would be as follows:

Taxable Income	Tax Rate	Taxes	Cumulative Taxes
First $50,000	15%	$7,500	$7,500
Next $25,000	25%	6,250	13,750
Next $25,000	34%	8,500	22,250
Next $235,000	39%	91,650	113,900
Next $9,665,000	34%	3,286,100	3,400,000
Next $5,000,000	35%	1,750,000	5,150,000
Remaining $1,000,000	38%	380,000	$5,530,000

Alternatively, using the tax formulas in Table 9.6, we obtain

$$\$5,150,000 + 0.38(\$16,000,000 - \$15,000,000) = \$5,530,000.$$

The effective (average) tax rate would then be

$$\frac{\$5,530,000}{\$16,000,000} = 0.3456, \text{ or } 34.56,$$

as opposed to the marginal rate of 38%. In other words, on the average, the company paid 34.56 cents for each taxable dollar it generated during the accounting period.

EXAMPLE 9.13 Corporate Taxes

A mail-order computer company sells personal computers and peripherals. The company leased showroom space and a warehouse for $20,000 a year and installed $290,000 worth of inventory-checking and packaging equipment. The allowed depreciation expense for this capital expenditure ($290,000) amounted to $58,000 using the category of 5-year MACRS. The store was completed and operations began on January 1. The company had a gross income of $1,250,000 for the calendar year. Supplies and all operating expenses, other than the lease expense, were itemized as follows:

Merchandise sold in the year	$600,000
Employee salaries and benefits	150,000
Other supplies and expenses	90,000
	$840,000

Compute the taxable income for this company. How much will the company pay in federal income taxes for the year?

SOLUTION

Given: Income, preceding cost information and depreciation.
Find: Taxable income and federal income taxes.

First we compute the taxable income as follows:

Gross revenues	$1,250,000
Expenses	−840,000
Lease expense	−20,000
Depreciation	−58,000
Taxable income	$332,000

Note that capital expenditures are not deductible expenses. Since the company is in the 39% marginal tax bracket, its income tax can be calculated by using the formula given in Table 9.6, namely, $22,250 + 0.39 (X - 100,000)$:

$$\text{Income tax} = \$22,250 + 0.39(\$332,000 - \$100,000)$$
$$= \$112,730.$$

The firm's current marginal tax rate is 39%, but its average corporate tax rate is

$$\frac{\$112,730}{\$332,000} = 33.95\%.$$

COMMENTS: Instead of using the tax formula in Table 9.6, we can compute the federal income tax in the following fashion:

First $50,000 at 15%	$ 7,500
Next $25,000 at 25%	6,250
Next $25,000 at 34%	8,500
Next $232,000 at 39%	90,480
Income tax	$112,730

9.8 Tax Treatment of Gains or Losses on Depreciable Assets

As in the disposal of capital assets, gains or losses generally are associated with the sale (or exchange) of depreciable assets. To calculate a gain or loss, we first need to determine the book value of the depreciable asset at the time of its disposal.

9.8.1 Disposal of a MACRS Property

For a MACRS property, one important consideration at the time of disposal is whether the property is disposed of *during* or *before* its specified recovery period. Moreover, with the half-year convention, which is now mandated by all MACRS depreciation methods, the year of disposal is charged one-half of that year's annual depreciation amount, if it should occur during the recovery period.

EXAMPLE 9.14 Book Value in the Year of Disposal

Consider a five-year MACRS asset purchased for $10,000. Note that property belonging to the five-year MACRS class is depreciated over six years due to the half-year convention. The applicable depreciation percentages, shown in Table 9.3, are 20%, 32%, 19.20%, 11.52%, 11.52%, and 5.76%. Compute the allowed depreciation amounts and the book value when the asset is disposed of (a) in year 3, (b) in year 5, and (c) in year 6.

SOLUTION

Given: Five-year MACRS asset, cost basis = $10,000, and MACRS depreciation percentages as shown in Figure 9.8.

Find: Total depreciation and book value at disposal if the asset is sold in year 3, 5, or 6.

(a) If the asset is disposed of in year 3 (or at the end of year 3), the total accumulated depreciation amount and the book value would be

$$\text{Total depreciation} = \$10,000(0.20 + 0.32 + 0.192/2)$$

$$= \$6,160,$$

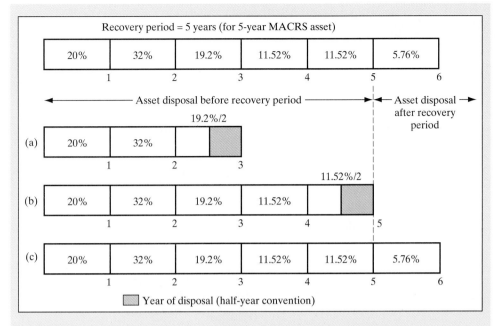

Figure 9.8 Disposal of a MACRS property and its effect on depreciation allowances (Example 9.14).

$$Book\ value = \$10,000 - \$6,160$$
$$= \$3,840.$$

(b) If the asset is disposed of during[4], year 5, the depreciation and book values will be

$$Total\ depreciation = \$10,000(0.20 + 0.32 + 0.192 + 0.1152 + 0.1152/2)$$
$$= \$8,848,$$
$$Book\ value = \$10,000 - \$8,848$$
$$= \$1,152.$$

(c) If the asset is disposed of after the recovery period, there will be no penalty due to early disposal. Since the asset is depreciated fully, we have

$$Total\ depreciation = \$10,000,$$
$$Book\ value = \$0.$$

---

[4] Note that even though you dispose of the asset at the *end* of the recovery period, the half-year convention still applies.

9.8.2 Calculations of Gains and Losses on MACRS Property

When a depreciable asset used in business is sold for an amount that differs from its book value, the gain or loss has an important effect on income taxes. The gain or loss is found as follows:

- **Case 1: Salvage value < Cost basis**

  If the salvage value of the asset is less than its cost basis, then

  $$\text{Gains (losses)} = \text{Salvage value} - \text{book value},$$

  where the salvage value represents the proceeds from the sale (i.e., the selling price) less any selling expense or removal cost. These gains, commonly known as **depreciation recapture**, are taxed as ordinary income under current tax law.

- **Case 2: Salvage value > Cost basis**

  In the unlikely event that an asset is sold for an amount greater than its cost basis, the gains (salvage value − book value) are divided into two parts for tax purposes:

  $$\text{Gains} = \text{Salvage value} - \text{book value}$$
  $$= \underbrace{(\text{Salvage value} - \text{cost basis})}_{\text{Capital gains}}$$
  $$+ \underbrace{(\text{Cost basis} - \text{book value})}_{\text{Ordinary gains}}. \tag{9.10}$$

  Recall from Section 9.2.2 that the cost basis is the cost of purchasing an asset, plus any incidental costs, such as freight and installation costs. As illustrated in Figure 9.9,

  $$\text{Capital gains} = \text{Salvage value} - \text{cost basis},$$
  $$\text{Ordinary gains} = \text{Cost basis} - \text{book value}.$$

Capital gain: An increase in the value of a capital asset that gives it a higher worth than the purchase price, when the asset is sold.

This distinction is necessary only when capital gains are taxed at the capital gains tax rate and ordinary gains (or depreciation recapture) at the ordinary income tax rate. Current tax

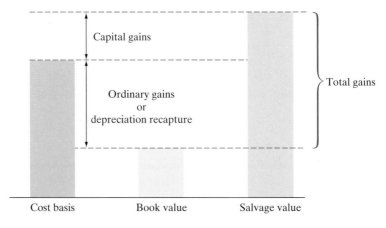

Figure 9.9   Capital gains and ordinary gains (or depreciation recapture) when the salvage value exceeds the cost basis.

law does not provide a special low rate of taxation for capital gains. Instead, capital gains are treated as ordinary income, but the maximum tax rate is set at the U.S. statutory rate of 35%. Nevertheless, the statutory structure for capital gains has been retained in the tax code. This provision could allow Congress to restore capital gains' preferential treatment at some future time.

EXAMPLE 9.15 Gains and Losses on Depreciable Assets

A company purchased a drill press costing $230,000 in year 0. The drill press, classified as seven-year recovery property, has been depreciated by the MACRS method. If it is sold at the end of three years, compute the gains (losses) for the following four salvage values: (a) $150,000, (b) $120,693, (c) $100,000, and (d) $250,000. Assume that both capital gains and ordinary income are taxed at 34%.

SOLUTION

Given: Seven-year MACRS asset, cost basis = $230,000, and the asset is sold three years after its purchase.

Find: Gains or losses, tax effects, and net proceeds from the sale if the asset is sold for $150,000, $120,693, $100,000, or $250,000.

In this example, we first compute the current book value of the machine. From the MACRS depreciation schedule in Table 9.3, the allowed annual depreciation percentages for the first three years of a seven-year MACRS property are 14.29%, 24.49%, and 17.49%. Since the asset is disposed of before the end of its recovery period, the depreciation amount in year 3 will be reduced by half. The total depreciation and the final book value will be

$$\text{Total allowed depreciation} = \$230,000\left(0.1429 + 0.2449 + \frac{0.1749}{2}\right)$$

$$= \$109,308,$$

$$\text{Book value} = \$230,000 - \$109,308$$

$$= \$120,693.$$

(a) Case 1: Book value < Salvage value < Cost basis

In this case, there are no capital gains to consider. All gains are ordinary gains. Thus, we have

$$\text{Ordinary gains} = \text{Salvage value} - \text{book value}$$

$$= \$150,000 - \$120,693 = \$29,308,$$

$$\text{Gains tax } (34\%) = 0.34(\$29,308) = \$9,965,$$

$$\text{Net proceeds from sale} = \text{Salvage value} - \text{gains tax}$$

$$= \$150,000 - \$9,965 = \$140,035.$$

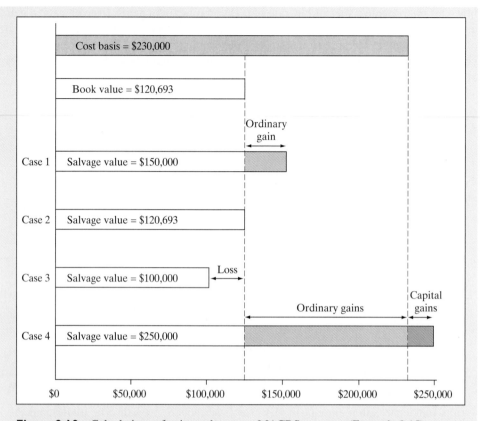

Figure 9.10 Calculations of gains or losses on MACRS property (Example 9.15).

This situation (in which the asset's salvage value exceeds its book value) is denoted as Case 1 in Figure 9.10.

(b) Case 2: Salvage value = Book value

In Case 2, the book value is again $120,693. Thus, if the drill press's salvage value equals $120,693—its book value—no taxes are levied on that salvage value. Therefore, the net proceeds equal the salvage value.

(c) Case 3: Salvage value < Book value

Case 3 illustrates a loss, when the salvage value (say, $100,000) is less than the book value. We compute the net salvage value after tax as follows:

$$\text{Gain (loss)} = \text{Salvage value} - \text{book value}$$

$$= \$100,000 - \$120,693$$

$$= (\$20,693),$$

$$\text{Tax savings} = 0.34(\$20,693)$$

$$= \$7,036,$$

$$\text{Net proceeds from sale} = \$100,000 + \$7,036$$
$$= \$107,036.$$

(d) Case 4: Salvage value $>$ Cost basis

This situation is not likely for most depreciable assets (except real property). But it is the only situation in which both capital gains and ordinary gains can be observed. Nevertheless, the tax treatment on this gain is as follows:

$$\text{Capital gains} = \text{Salvage value} - \text{cost basis}$$
$$= \$250,000 - \$230,000$$
$$= \$20,000,$$

$$\text{Capital gains tax} = \$20,000(0.34)$$
$$= \$6,800,$$

$$\text{Ordinary gains} = \$230,000 - \$120,693$$
$$= \$109,307,$$

$$\text{Gains tax} = \$109,307(0.34)$$
$$= \$37,164,$$

$$\text{Net proceeds from sale} = \$250,000 - (\$6,800 + \$37,164)$$
$$= \$206,036.$$

COMMENTS: Note that in (c) the reduction in tax, due to the loss, actually increases the net proceeds. This is realistic when the incremental tax rate (34% in this case) is positive, indicating the corporation is still paying tax, but less than if the asset had not been sold at a loss. The incremental tax rate will be discussed in Section 9.9.

9.9 Income Tax Rate to Be Used in Economic Analysis

As we have seen in earlier sections, average income tax rates for corporations vary with the level of taxable income from 0 to 35%. Suppose that a company now paying a tax rate of 25% on its current operating income is considering a profitable investment. What tax rate should be used in calculating the taxes on the investment's projected income?

9.9.1 Incremental Income Tax Rate

The choice of a corporation's rate depends on the incremental effect of an investment on the company's taxable income. In other words, the tax rate to use is the rate that applies to the additional taxable income projected in the economic analysis.

To illustrate, consider ABC Corporation, whose taxable income from operations is expected to be $70,000 for the current tax year. ABC management wishes to evaluate the incremental tax impact of undertaking a project during the same tax year. The revenues, expenses, and taxable incomes before and after the project are estimated as follows:

	Before	**After**	**Incremental**
Gross revenue	$200,000	$240,000	$40,000
Salaries	100,000	110,000	10,000
Wages	30,000	40,000	10,000
Taxable income	$ 70,000	$ 90,000	$20,000

Because the income tax rate is progressive, the tax effect of the project cannot be isolated from the company's overall tax obligations. The base operations of ABC without the project are expected to yield a taxable income of $70,000. With the new project, the taxable income increases to $90,000. From the tax computation formula in Table 9.6, the corporate income taxes with and without the project are as follows:

$$\text{Income tax without the project} = \$7,500 + 0.25(\$70,000 - \$50,000)$$
$$= \$12,500,$$

$$\text{Income tax with the project} = \$13,750 + 0.34(\$90,000 - \$75,000)$$
$$= \$18,850.$$

The additional income tax is then $18,850 - $12,500 = $6,350. This amount, on the additional $20,000 of taxable income, is based on a rate of 31.75%, which is an incremental rate. This is the rate we should use in evaluating the project in isolation from the rest of ABC's operations. As shown in Figure 9.11, the 31.75% is not an arbitrary figure, but a weighted average of two distinct marginal rates. Because the new project pushes ABC into a higher tax bracket, the first $5,000 it generates is taxed at 25%; the remaining $15,000 it generates is taxed in the higher bracket, at 34%. Thus, we could have calculated the incremental tax rate with the formula

$$0.25\left(\frac{\$5,000}{\$20,000}\right) + 0.34\left(\frac{\$15,000}{\$20,000}\right) = 31.75\%.$$

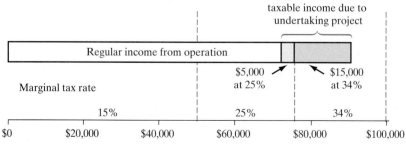

Figure 9.11 Illustration of incremental tax rate.

The average tax rates before and after the new project being considered is undertaken are as follows:

	Before	**After**	**Incremental**
Taxable income	$70,000	$90,000	$20,000
Income taxes	12,500	18,850	6,350
Average tax rate	17.86%	20.94%	
Incremental tax rate			31.75%

Note that neither 17.86% nor 20.94% is a correct tax rate to use in evaluating the new project.

A corporation with continuing base operations that place it consistently in the highest tax bracket will have both marginal and average federal tax rates of 35%. For such firms, the tax rate on an additional investment project is, naturally, 35%. But for corporations in lower tax brackets, and for those which fluctuate between losses and profits, the marginal and average tax rates are likely to vary. For such corporations, estimating a prospective incremental tax rate for a new investment project may be difficult. The only solution may be to perform scenario analysis, in which we examine how much the income tax fluctuates due to undertaking the project. (In other words, we calculate the total taxes and the incremental taxes for each scenario.) A typical scenario is presented in Example 9.16.

EXAMPLE 9.16 Scenario Analysis for a Small Company

EverGreen Grass Company expects to have an annual taxable income of $320,000 from its regular grass-sodding business over the next two years. EverGreen has just won a contract to sod grasses for a new golf complex for just those years. This two-year project requires a purchase of new equipment costing $50,000. The equipment falls into the MACRS five-year class, with depreciation allowances of 20%, 32%, 19.2%, 11.52%, 11.52%, and 5.76% in each of the six years, respectively, during which the equipment will be depreciated. After the contract is terminated, the equipment will be retained for future use (instead of being sold), indicating no salvage cash flow, gain, or loss on this asset. The project will bring in an additional annual revenue of $150,000, but it is expected to incur additional annual operating costs of $90,000. Compute the incremental (marginal) tax rates applicable to the project's operating profits for years 1 and 2.

SOLUTION

Given: Base taxable income = $320,000 per year, incremental income, expenses, and depreciation amounts as stated.

Find: Incremental tax rate for this new project in years 1 and 2.

First, we compute the additional taxable income from the golf course project over the next two years:

Year	1	2
Gross Revenue	$150,000	$150,000
Expenses	90,000	90,000
Depreciation	10,000	16,000
Taxable income	$ 50,000	$ 44,000

Next, we compute the income taxes. To do this, we need to determine the applicable marginal tax rate, but because of the progressive income tax rate on corporations, the project cannot be isolated from the other operations of EverGreen.

We can solve this problem much more efficiently by using the incremental tax rate concept discussed at the beginning of this section. Because the golf course project pushes EverGreen into the 34% tax bracket from the 39% bracket, we want to know what proportion of the incremental taxable income of $50,000 in year 1 is taxed at 39% and what proportion is taxable at 34%. Without the project, the firm's taxable income is $320,000 and its marginal tax rate is 39%. With the additional taxable income of $50,000, EverGreen's tax bracket reverts to 34%, as its combined taxable income changes from $320,000 to $370,000. Since the rate changes at $335,000, the first $15,000 of the $50,000 taxable income will still be in the 39% bracket, and the remaining $35,000 will be in the 34% bracket. In year 2, we can divide the additional taxable income of $44,000 in a similar fashion. Then we can calculate the incremental tax rates for the first two years as follows:

$$0.39\left(\frac{\$15,000}{\$50,000}\right) + 0.34\left(\frac{\$35,000}{\$50,000}\right) = 0.3550,$$

$$0.39\left(\frac{\$15,000}{\$44,000}\right) + 0.34\left(\frac{\$29,000}{\$44,000}\right) = 0.3570.$$

Note that these incremental tax rates vary slightly from year to year. Much larger changes could occur if a company's taxable income fluctuates drastically from its continuing base operation.

9.9.2 Consideration of State Income Taxes

For large corporations, the top federal marginal tax rate is 35%. In addition to federal income taxes, state income taxes are levied on corporations in most states. State income taxes are an allowable deduction in computing federal taxable income, and two ways are available to consider explicitly the effects of state income taxes in an economic analysis.

The first approach is to estimate explicitly the amount of state income taxes before calculating the federal taxable income. We then reduce the federal taxable income by the amount of the state taxes and apply the marginal tax rate to the resulting federal taxes. The total taxes would be the sum of the state taxes and the federal taxes.

The second approach is to calculate a single tax rate that reflects both state and federal income taxes. This single rate is then applied to the federal taxable income, without subtracting state income taxes. Taxes computed in this fashion represent total taxes. If state income taxes are considered, the combined state and federal marginal tax rate may be higher than 35%. Since state income taxes are deductible as expenses in determining federal taxes, the marginal rate for combined federal and state taxes can be calculated with the expression

$$t_m = t_f + t_s - (t_f)(t_s),$$ (9.11)

where

t_m = combined marginal tax rate,

t_f = federal marginal tax rate.

t_s = state marginal tax rate.

This second approach provides a more convenient and efficient way to handle taxes in an economic analysis in which the incremental tax rates are known. Therefore, incremental tax rates will be stated as combined marginal tax rates, unless indicated otherwise. (For large corporations, these would be about 40%, but they vary from state to state.)

EXAMPLE 9.17 Combined State and Federal Income Taxes

Consider a corporation whose revenues and expenses before income taxes are as follows:

Gross revenue	$1,000,000
All expenses	400,000

If the marginal federal tax rate is 35% and the marginal state rate is 7%, compute the combined state and federal taxes, using the two methods just described.

SOLUTION

Given: Gross income = $1,000,000, deductible expenses = $400,000, t_f = 35%, and t_s = 7%.

Find: Combined income taxes t_m.

(a) Explicit calculation of state income taxes:

Let's define FT as federal taxes and ST as state taxes. Then

State taxable income = $1,000,000 − $400,000

and

$$ST = (0.07)(\$600,000)$$
$$= \$42,000.$$

Also,

Federal taxable income = $1,000,000 − $400,000 − ST
$$= (\$558,000),$$

so that

$$FT = (0.35)(\$558,000)$$
$$= \$195,300.$$

Thus,

$$\text{Combined taxes} = FT + ST$$
$$= \$237,300.$$

(b) Tax calculation based on the combined tax rate:

Compute the combined tax rate directly from the formula:

$$\text{Combined tax rate } (t_m) = 0.35 + 0.07 - (0.35)(0.07)$$
$$= 39.55\%.$$

Hence,

$$\text{Combined taxes} = \$600,000(0.3955)$$
$$= \$237,300.$$

As expected, these two methods always produce exactly the same results.

9.10 The Need for Cash Flow in Engineering Economic Analysis

Traditional accounting stresses net income as a means of measuring a firm's profitability, but it is desirable to discuss why cash flows are relevant in project evaluation. As seen in Section 8.2, net income is an accounting measure based, in part, on the **matching concept**. Costs become expenses as they are matched against revenue. The actual timing of cash inflows and outflows is ignored.

9.10.1 Net Income versus Net Cash Flow

Over the life of a firm, net incomes and net cash inflows will usually be the same. However, the timing of incomes and cash inflows can differ substantially. Given the time value of money, it is better to receive cash now rather than later, because cash can be invested to earn more cash. (You cannot invest net income.) For example, consider two firms and their income and cash flow schedules over two years:

		Company A	Company B
Year 1	Net income	$1,000,000	$1,000,000
	Cash flow	1,000,000	0
Year 2	Net income	1,000,000	1,000,000
	Cash flow	1,000,000	2,000,000

Both companies have the same amount of net income and cash over two years, but Company A returns $1 million cash yearly, while Company B returns $2 million at the end of the second year. If you received $1 million at the end of the first year from Company A, you could, for example, invest it at 10%. While you would receive only $2 million in total from Company B at the end of the second year, you would receive $2.1 million in total from Company A.

9.10.2 Treatment of Noncash Expenses

Apart from the concept of the time value of money, certain expenses do not even require a cash outflow. Depreciation and amortization are the best examples of this type of expense. Even though depreciation (or amortization expense) is deducted from revenue on a daily basis, no cash is paid to anyone.

In Example 9.13, we learned that the annual depreciation allowance has an important impact on both taxable and net income. However, although depreciation has a direct impact on net income, it is *not* a cash outlay; hence, it is important to distinguish between annual income in the presence of depreciation and annual cash flow.

The situation described in Example 9.13 serves as a good vehicle to demonstrate the difference between depreciation costs as expenses and the cash flow generated by the purchase of a fixed asset. In that example, cash in the amount of $290,000 was expended in year 0, but the $58,000 depreciation charged against the income in year 1 was not a cash outlay. Figure 9.12 summarizes the difference.

Net income (**accounting profit**) is important for accounting purposes, but **cash flows** are more important for project evaluation purposes. However, as we will now demonstrate, net income can provide us with a starting point to estimate the cash flow of a project.

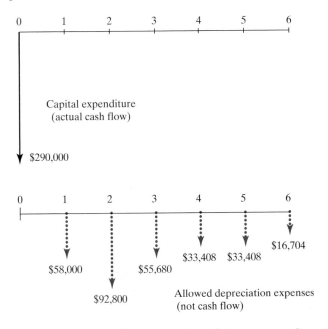

Figure 9.12 Cash flows versus depreciation expenses for an asset with a cost basis of $290,000, which was placed in service in year 0.

The procedure for calculating net cash flow (after tax) is identical to that used to obtain net income from operations, with the exception of depreciation, which is excluded from the net cash flow computation. (It is needed only for computing income taxes.) Assuming that revenues are received and expenses are paid in cash, we can obtain the net cash flow by adding the **noncash expense** (depreciation) to net income, which cancels the operation of subtracting it from revenues:

Item
Gross income
Expenses:
Cost of goods sold
Depreciation
Operating expenses
Taxable operating income
Income taxes
Net income
Net cash flow = Net income + Depreciation

Example 9.18 illustrates this relationship.

EXAMPLE 9.18 Cash Flow versus Net Income

A company buys a numerically controlled (NC) machine for $28,000 (year 0) and uses it for five years, after which it is scrapped. The allowed depreciation deduction during the first year is $4,000, as the equipment falls into the category of seven-year MACRS property. (The first-year depreciation rate is 14.29%.) The cost of the goods produced by this NC machine should include a charge for the depreciation of the machine. Suppose the company estimates the following revenues and expenses, including depreciation, for the first operating year:

$$\text{Gross income} = \$50,000,$$
$$\text{Cost of goods sold} = \$20,000,$$
$$\text{Depreciation on NC machine} = \$4,000,$$
$$\text{Operating expenses} = \$6,000.$$

(a) If the company pays taxes at the rate of 40% on its taxable income, what is its net income from the project during the first year?

(b) Assume that (1) all sales are cash sales and (2) all expenses except depreciation were paid during year 1. How much cash would be generated from operations?

SOLUTION

Given: Net-income components.
Find: Cash flow.

We can generate a cash flow statement simply by examining each item in the income statement and determining which items actually represent receipts or disbursements (some of the assumptions listed in the statement of the problem make this process simpler):

Item	Income	Cash Flow
Gross income (revenues)	$50,000	$50,000
Expenses:		
Cost of goods sold	20,000	−20,000
Depreciation	4,000	
Operating expenses	6,000	−6,000
Taxable income	20,000	
Taxes (40%)	8,000	−8,000
Net income	$12,000	
Net cash flow		$16,000

Column 2 shows the income statement, while Column 3 shows the statement on a cash flow basis. The sales of $50,000 are all cash sales. Costs other than depreciation were $26,000; these were paid in cash, leaving $24,000. Depreciation is not a cash flow: The firm did not pay out $4,000 in depreciation expenses. Taxes, however, are paid in cash, so the $8,000 for taxes must be deducted from the $24,000, leaving a net cash flow from operations of $16,000. As shown in Figure 9.13, this $16,000 is exactly equal to net income plus depreciation: $12,000 + $4,000 = $16,000.

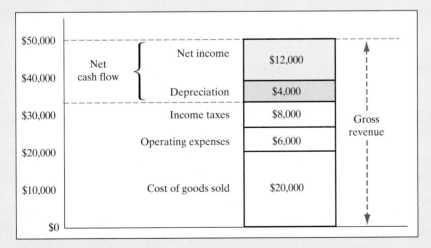

Figure 9.13 Net income versus net cash flow (Example 9.18). Cash flow versus depreciation expenses for an asset with a cost basis of $28,000, which was placed in service in year 0.

As we've just seen, depreciation has an important impact on annual cash flow in its role as an accounting expense that reduces taxable income and thus taxes. (Although depreciation expenses are not actual cash flows, depreciation has a positive impact on the after-tax cash flow of the firm.) Of course, during the year in which an asset is actually acquired, the cash disbursed to purchase it creates a significant negative cash flow, and during the depreciable life of the asset, the depreciation charges will affect the taxes paid and, therefore, cash flows.

As shown in Example 9.18, through its influence on taxes, depreciation plays a critical role in project cash flow analysis, which we will explore further in Chapter 10.

SUMMARY

- Machine tools and other manufacturing equipment, and even the factory buildings themselves, are subject to wear over time. However, it is not always obvious how to account for the cost of their replacement. Clearly, the choice of estimated service life for a machine, and the method used to calculate the cost of operating it, can have significant effects on an asset's management.

- The entire cost of replacing a machine cannot be properly charged to any one year's production; rather, the cost should be spread (or capitalized) over the years in which the machine is in service. The cost charged to operations during a particular year is called **depreciation**. Several different meanings and applications of depreciation have been presented in this chapter. From an engineering economics point of view, our primary concern is with **accounting depreciation**: the systematic allocation of an asset's value over its depreciable life.

- Accounting depreciation can be broken into two categories:

  1. Book depreciation—the method of depreciation used in financial reports and for pricing products;

  2. Tax depreciation—the method of depreciation used for calculating taxable income and income taxes; this method is governed by tax legislation.

- The four components of information required to calculate depreciation are as follows:

  1. The cost basis of the asset.

  2. The salvage value of the asset.

  3. The depreciable life of the asset.

  4. The method of its depreciation.

  Table 9.7 summarizes the differences in the way these components are treated for purposes of book and tax depreciation.

- Because it employs accelerated methods of depreciation and shorter-than-actual depreciable lives, the **Modified Accelerated Cost Recovery System (MACRS)** gives taxpayers a break, allowing them to take earlier and faster advantage of the tax-deferring benefits of depreciation.

- Many firms select straight-line depreciation for book depreciation because of its relative ease of calculation.

TABLE 9.7 Summary of Book versus Tax Depreciation

Component of Depreciation	Book Depreciation	Tax Depreciation (MACRS)
Cost basis	Based on the actual cost of the asset, plus all incidental costs, such as the cost of freight, site preparation, installation, etc.	Same as for book depreciation.
Salvage value	Estimated at the outset of depreciation analysis. If the final book value does not equal the estimated salvage value, we may need to make adjustments in our depreciation calculations.	Salvage value is zero for all depreciable assets.
Depreciable life	Firms may select their own estimated useful lives or follow government guidelines for asset depreciation ranges (ADRs).	Eight recovery periods—3, 5, 7, 10, 15, 20, 27.5, and 39 years—have been established; all depreciable assets fall into one of these eight categories.
Method of depreciation	Firms may select from the following: • straight line • accelerated methods (declining balance, double-declining balance, and sum-of-years'-digits) • units of production	Exact depreciation percentages are mandated by tax legislation, but are based largely on DDB and straight-line methods. The sum-of-years'-digits method is rarely used in the United States, except for some cost analysis in engineering valuation.

■ Depletion is a cost allocation method used particularly for natural resources. **Cost depletion** is based on the units-of-production method of depreciation. **Percentage depletion** is based on a prescribed percentage of the gross income of a property during a tax year.

■ Given the frequently changing nature of depreciation and tax law, we must use whatever percentages, depreciable lives, and salvage values are in effect *at the time an asset is acquired*.

■ Explicit consideration of taxes is a necessary aspect of any complete economic study of an investment project.

■ For corporations, the U.S. tax system has the following characteristics:

1. Tax rates are progressive: The more you earn, the more you pay.

2. Tax rates increase in stair-step fashion: four brackets for corporations and two additional surtax brackets, giving a total of six brackets.

3. Allowable exemptions and deductions may reduce the overall tax assessment.

■ Three distinct terms to describe taxes were used in this chapter: **marginal tax rate**, which is the rate applied to the last dollar of income earned; **effective (average) tax rate**, which is the ratio of income tax paid to net income; and **incremental tax rate**, which is the average rate applied to the incremental income generated by a new investment project.

■ **Capital gains** are currently taxed as ordinary income, and the maximum rate is capped at 35%. **Capital losses** are deducted from capital gains; net remaining losses may be carried backward and forward for consideration in years other than the current tax year.

■ Since we are interested primarily in the measurable financial aspects of depreciation, we consider its effects on two important measures of an organization's financial position: **net income** and **cash flow**. Once we understand that depreciation has a significant influence on the income and cash position of a firm, we will be able to appreciate fully the importance of utilizing depreciation as a means of maximizing the value both of engineering projects and of the organization as a whole.

PROBLEMS

Note: *Unless otherwise specified, use current tax rates for corporate taxes. Check the website (described in the preface) for the most current tax rates for corporations.*

Economic Depreciation

9.1   A machine now in use was purchased four years ago at a cost of $20,000. It has a book value of $6,246. It can be sold for $7,000, but could be used for three more years, at the end of which time it would have no salvage value. What is the current amount of economic depreciation for this asset?

Cost Basis

9.2   General Service Contractor Company paid $200,000 for a house and lot. The value of the land was appraised at $65,000 and the value of the house at $135,000. The house was then torn down at an additional cost of $5,000 so that a warehouse could be built on the lot at a cost of $250,000. What is the total value of the property with the warehouse? For depreciation purposes, what is the cost basis for the warehouse?

9.3   To automate one of its production processes, Milwaukee Corporation bought three flexible manufacturing cells at a price of $500,000 each. When they were delivered, Milwaukee paid freight charges of $25,000 and handling fees of $12,000. Site preparation for these cells cost $35,000. Six foremen, each earning $15 an hour, worked five 40-hour weeks to set up and test the manufacturing cells. Special wiring and other materials applicable to the new manufacturing cells cost $1,500. Determine the cost basis (amount to be capitalized) for these cells.

9.4   A new drill press was purchased for $126,000 by trading in a similar machine that had a book value of $39,000. Assuming that the trade-in allowance is $40,000 and that $86,000 cash is to be paid for the new asset, what is the cost basis of the new asset for depreciation purposes?

9.5   A lift truck priced at $35,000 is acquired by trading in a similar lift truck and paying cash for the remaining balance. Assuming that the trade-in allowance is $10,000 and the book value of the asset traded in is $6,000, what is the cost basis of the new asset for the computation of depreciation for tax purposes?

Book Depreciation Methods

9.6 Consider the following data on an asset:

Cost of the asset, I	$132,000
Useful life, N	5 years
Salvage value, S	$ 20,000

Compute the annual depreciation allowances and the resulting book values, using

(a) The straight-line depreciation method.

(b) The double-declining-balance method.

9.7 A firm is trying to decide whether to keep an item of construction equipment for another year. The firm is using DDB for book purposes, and this is the fourth year of ownership of the equipment, which cost $150,000 new. What is the depreciation in year 3?

9.8 Consider the following data on an asset:

Cost of the asset, I	$50,000
Useful life, N	7 years
Salvage value, S	$0

Compute the annual depreciation allowances and the resulting book values, using the DDB and switching to SL.

9.9 The double-declining-balance method is to be used for an asset with a cost of $68,000, an estimated salvage value of $12,000, and an estimated useful life of six years.

(a) What is the depreciation for the first three fiscal years, assuming that the asset was placed in service at the beginning of the year?

(b) If switching to the straight-line method is allowed, when is the optimal time to switch?

9.10 Compute the double-declining-balance (DDB) depreciation schedule for the following asset:

Cost of the asset, I	$76,000
Useful life, N	8 years
Salvage value, S	$ 6,000

9.11 Compute the DDB depreciation schedule for the following asset:

Cost of the asset, I	$46,000
Useful life, N	5 years
Salvage value, S	$10,000

(a) What is the value of α?

(b) What is the amount of depreciation for the second full year of use of the asset?

(c) What is the book value of the asset at the end of the fourth year?

9.12 Upjohn Company purchased new packaging equipment with an estimated useful life of five years. The cost of the equipment was $35,000, and the salvage value was estimated to be $5,000 at the end of year 5. Compute the annual depreciation expenses over the five-year life of the equipment under each of the following methods of book depreciation:

(a) Straight-line method.

(b) Double-declining-balance method. (Limit the depreciation expense in the fifth year to an amount that will cause the book value of the equipment at year-end to equal the $5,000 estimated salvage value.)

(c) Sum-of-years'-digits method.

9.13 A secondhand bulldozer acquired at the beginning of the fiscal year at a cost of $68,000 has an estimated salvage value of $9,500 and an estimated useful life of 12 years. Determine

(a) The amount of annual depreciation by the straight-line method.

(b) The amount of depreciation for the third year, computed by the double-declining-balance method.

(c) The amount of depreciation for the second year, computed by the sum-of-years'-digits method.

Units-of-Production Method

9.14 If a truck for hauling coal has an estimated net cost of $100,000 and is expected to give service for 250,000 miles, resulting in a salvage value of $5,000, depreciation would be charged at a rate of 38 cents per mile. Compute the allowed depreciation amount for the same truck's usage amounting to 55,000 miles.

9.15 A diesel-powered generator with a cost of $65,000 is expected to have a useful operating life of 50,000 hours. The expected salvage value of this generator is $7,500. In its first operating year, the generator was operated for 5,000 hours. Determine the depreciation for the year.

Tax Depreciation

9.16 Zerex Paving Company purchased a hauling truck on January 1, 2005, at a cost of $32,000. The truck has a useful life of eight years with an estimated salvage value of $5,000. The straight-line method is used for book purposes. For tax purposes, the truck would be depreciated with the MACRS method over its five-year class life. Determine the annual depreciation amount to be taken over the useful life of the hauling truck for both book and tax purposes.

9.17 The Harris Foundry Company purchased new casting equipment in 2006 at a cost of $220,000. Harris also paid $35,000 to have the equipment delivered and installed. The casting machine has an estimated useful life of 12 years, but it will be depreciated with MACRS over its seven-year class life.

(a) What is the cost basis of the casting equipment?

(b) What will be the depreciation allowance in each year of the seven-year class life of the casting equipment?

9.18 A machine is classified as seven-year MACRS property. Compute the book value for tax purposes at the end of three years. The cost basis is $145,000.

9.19 A piece of machinery purchased at a cost of $86,000 has an estimated salvage value of $12,000 and an estimated useful life of five years. It was placed in service on May 1 of the current fiscal year, which ends on December 31. The asset falls into a seven-year MACRS property category. Determine the depreciation amounts over the useful life.

9.20 Suppose that a taxpayer places in service a $20,000 asset that is assigned to the six-year class (say, a new property class) with a half-year convention. Develop the MACRS deductions, assuming a 200% declining-balance rate followed by switching to straight line.

9.21 On April 1, Leo Smith paid $250,000 for a residential rental property. This purchase price represents $200,000 for the building and $50,000 for the land. Five years later, on November 1, he sold the property for $300,000. Compute the MACRS depreciation for each of the five calendar years during which he had the property.

9.22 In 2006, you purchased a spindle machine (seven-year MACRS property) for $26,000, which you placed in service in January. Use the calendar year as your tax year. Compute the depreciation allowances.

9.23 On October 1, you purchased a residential home in which to locate your professional office for $250,000. The appraisal is divided into $80,000 for the land and $170,000 for the building.

(a) In your first year of ownership, how much can you deduct for depreciation for tax purposes?

(b) Suppose that the property was sold at $325,000 at the end of fourth year of ownership. What is the book value of the property?

9.24 For each of four assets in the following table, determine the missing amounts (for asset type III, the annual usage is 15,000 miles):

Types of Asset	I	II	III	IV
Depreciating Methods	SL	DDB	UP	MACRS
End of year	7	4	3	4
Initial cost ($)	10,000	18,000	30,000	8,000
Salvage value ($)	2,000	2,000	0	1,000
Book value ($)	3,000	2,320	☐	1,382
Depreciable life	8 yr	5 yr	90,000 mi	☐
Depreciable Amount ($)	☐	☐	☐	☐
Accumulated Depreciable ($)	☐	15,680	☐	☐

9.25 Flint Metal Shop purchased a stamping machine for $147,000 on March 1, 2006. The machine is expected to have a useful life of 10 years, a salvage value of $27,000, a production of 250,000 units, and working hours of 30,000. During 2006, Flint used the stamping machine for 2,450 hours to produce 23,450 units. From the information given, compute the book depreciation expense for 2006 under each of the following methods:

(a) Straight line.
(b) Units-of-production method.
(c) Working hours.
(d) Double-declining balance (without conversion to straight line).
(e) Double-declining balance (with conversion to straight line).

Depletion

9.26 Early in 2006, Atlantic Mining Company began operation at its West Virginia mine. The mine had been acquired at a cost of $6,900,000 in 2004 and is expected to contain 3 million tons of silver and to have a residual value of $1,500,000 (once the silver is depleted). Before beginning mining operations, the company installed equipment at a cost of $2,700,000. This equipment will have no economic useful-ness once the silver is depleted. Therefore, depreciation of the equipment is based upon the estimated number of tons of ore produced each year. Ore removed from the West Virginia mine amounted to 500,000 tons in 2006 and 682,000 tons in 2007.

(a) Compute the per ton depletion rate of the mine and the per ton depreciation rate of mining equipment.
(b) Determine the depletion expense for the mine and the depreciation expense for the mining equipment.

9.27 Suppose you bought a timber tract for $550,000, and the land was worth as much as $150,000. An estimated 4.8 million board feet (4.8 MBF) of standing timber was in the tract. If you cut 1.5 MBF of timber, determine your depletion allowance.

9.28 A gold mine with an estimated deposit of 500,000 ounces of gold is purchased for $40 million. The mine has a gross income of $22,623,000 for the year, obtained from selling 52,000 ounces of gold. Mining expenses before depletion equal $12,250,000. Compute the percentage depletion allowance. Would it be advanta-geous to apply cost depletion rather than percentage depletion?

9.29 Oklahoma Oil Company incurred acquisition, exploration, and development costs during 2006 as follows:

Items*	Site		
	Parcel A	**Parcel B**	**Total**
Acquisition costs	6	4	10
Exploration costs	13	9	22
Development costs	20	11	31
Recoverable oil (millions of barrels)	9	5	14

* Units are in millions of dollars, except recoverable oil.

The market price of oil during 2006 was $16 per barrel.

(a) Determine the cost basis for depletion on each parcel.

(b) During 2006, 1,200,000 barrels were extracted from parcel A at a production cost of $3,600,000. Determine the depletion charge allowed for parcel A.

(c) In (b), if Oklahoma Oil Company sold 1,000,000 of the 1,200,000 barrels extracted during 2006 at a price of $55 per barrel, the sales revenue would be $55,000,000. If it qualified for the use of percentage depletion (15%), what would be the allowed depletion amount for 2006?

(d) During 2006, 800,000 barrels were extracted from parcel B at a production cost of $3,000,000. Assume that during 2007 it is ascertained that the remaining proven reserves on parcel B total only 4,000,000 barrels, instead of the originally estimated 5,000,000. This revision in proven reserves is considered a change in an accounting estimate that must be corrected during the current and future years. (A correction of previous years' depletion amounts is not permitted.) If 1,000,000 barrels are extracted during 2007, what is the total depletion charge allowed, according to the unit cost method?

9.30 A coal mine expected to contain 6.5 million tons of coal was purchased at a cost of $30 million. One million tons of coal are produced this year. The gross income for this coal is $600,000, and operating costs (excluding depletion expenses) are $450,000. If you know that coal has a 10% depletion allowance, what will be the depletion allowance for

(a) Cost depletion?

(b) Percentage depletion?

Revision of Depreciation Rates

9.31 Perkins Construction Company bought a building for $800,000 to be used as a warehouse. A number of major structural repairs, completed at the beginning of the current year at a cost of $125,000, are expected to extend the life of the building 10 years beyond the original estimate. The building has been depreciated by the straight-line method for 25 years. The salvage value is expected to be negligible and has been ignored. The book value of the building before the structural repairs is $400,000.

(a) What has the amount of annual depreciation been in past years?

(b) What is the book value of the building after the repairs have been recorded?

(c) What is the amount of depreciation for the current year, according to the straight-line method? (Assume that the repairs were completed at the very beginning of the year.)

9.32 The Dow Ceramic Company purchased a glass-molding machine in 2001 for $140,000. The company has been depreciating the machine over an estimated useful life of 10 years, assuming no salvage value, by the straight-line method of depreciation. For tax purposes, the machine has been depreciated under 7-year MACRS property. At the beginning of 2004, Dow overhauled the machine at a cost of $25,000. As a result of the overhaul, Dow estimated that the useful life of the machine would extend 5 years beyond the original estimate.

(a) Calculate the book depreciation for year 2006.

(b) Calculate the tax depreciation for year 2006.

9.33 On January 2, 2004, Hines Food Processing Company purchased a machine that dispenses a premeasured amount of tomato juice into a can. The machine cost $75,000, and its useful life was estimated at 12 years, with a salvage value of $4,500. At the time it purchased the machine, Hines incurred the following additional expenses:

Freight-in	$800
Installation cost	2,500
Testing costs prior to regular operation	1,200

Book depreciation was calculated by the straight-line method, but for tax purposes, the machine was classified as a 7-year MACRS property. In January 2006, accessories costing $5,000 were added to the machine to reduce its operating costs. These accessories neither prolonged the machine's life nor provided any additional salvage value.

(a) Calculate the book depreciation expense for 2007.

(b) Calculate the tax depreciation expense for 2007.

Corporate Tax Systems

9.34 In tax year 1, an electronics-packaging firm had a gross income of $25,000,000, 5,000,000 in salaries, $4,000,000 in wages, $800,000 in depreciation expenses, a loan principal payment of $200,000, and a loan interest payment of $210,000. Determine the net income of the company in tax year 1.

9.35 A consumer electronics company was formed to develop cellphones that run on or are recharged by fuel cells. The company purchased a warehouse and converted it into a manufacturing plant for $6,000,000. It completed installation of assembly equipment worth $1,500,000 on December 31. The plant began operation on January 1. The company had a gross income of $8,500,000 for the calendar year. Manufacturing costs and all operating expenses, excluding the capital expenditures, were $2,280,000. The depreciation expenses for capital expenditures amounted to $456,000.

(a) Compute the taxable income of this company.

(b) How much will the company pay in federal income taxes for the year?

9.36 Huron Roofing Company had gross revenues of $1,200,000 from operations. Financial transactions as shown in Table P9.36 were posted during the year. The old equipment had a book value of $75,000 at the time of its sale.

(a) What is Huron's income tax liability?

(b) What is Huron's operating income?

Gains or Losses

9.37 Consider a five-year MACRS asset purchased at $60,000. (Note that a five-year MACRS property class is depreciated over six years, due to the half-year convention.

TABLE P9.36

Manufacturing expenses (including depreciation)	$450,000
Operating expenses (excluding interest expenses)	120,000
A new short-term loan from a bank	50,000
Interest expenses on borrowed funds (old and new)	40,000
Dividends paid to common stockholders	80,000
Old equipment sold	60,000

The applicable salvage values would be $20,000 in year 3, $10,000 in year 5, and $5,000 in year 6.) Compute the gain or loss amounts when the asset is disposed of

(a) In year 3.

(b) In year 5.

(c) In year 6.

9.38 In year 0, an electrical appliance company purchased an industrial robot costing $300,000. The robot, to be used for welding operations and classified as seven-year recovery property, has been depreciated by the MACRS method. If the robot is to be sold after five years, compute the amounts of gains (losses) for the following three salvage values (assume that both capital gains and ordinary incomes are taxed at 34%):

(a) $10,000.

(b) $125,460.

(c) $200,000.

9.39 AmSouth, Inc., bought a machine for $50,000 on January 2, 2004. Management expects to use the machine for 10 years, at the end of which time it will have a $1,000 salvage value. Consider the following questions independently:

(a) If AmSouth uses straight-line depreciation, what will be the book value of the machine on December 31, 2007?

(b) If AmSouth uses double-declining-balance depreciation, what will be the depreciation expense for 2007?

(c) If AmSouth uses double-declining-balance depreciation, followed by switching to straight-line depreciation, when will be the optimal time to switch?

(d) If Amsouth uses 7-year MACRS and sells the machine on April 1, 2007, at a price of $30,000, what will be the taxable gains?

Marginal Tax Rate in Project Evaluation

9.40 Boston Machine Shop expects to have an annual taxable income of $325,000 from its regular business over the next six years. The company is considering acquiring a new milling machine during year 0. The machine's price is $200,000, installed. The machine falls into the MACRS five-year class, and it will have an estimated salvage value of $30,000 at the end of six years. The machine is expected to generate additional before-tax revenue of $80,000 per year.

(a) What is the total amount of economic depreciation for the milling machine if the asset is sold at $30,000 at the end of six years?

(b) Determine the company's marginal tax rates over the next six years with the machine.

(c) Determine the company's average tax rates over the next six years with the machine.

9.41 Major Electrical Company expects to have an annual taxable income of $550,000 from its residential accounts over the next two years. The company is bidding on a two-year wiring service for a large apartment complex. This commercial service requires the purchase of a new truck equipped with wire-pulling tools at a cost of $50,000. The equipment falls into the MACRS five-year class and will be retained for future use (instead of being sold) after two years, indicating no gain or loss on the property. The project will bring in an additional annual revenue of $200,000, but it is expected to incur additional annual operating costs of $100,000. Compute the marginal tax rates applicable to the project's operating profits for the next two years.

9.42 Florida Citrus Corporation estimates its taxable income for next year at $2,000,000. The company is considering expanding its product line by introducing pineapple–orange juice for the next year. The market responses could be (1) good, (2) fair, or (3) poor. Depending on the market response, the expected additional taxable incomes are (1) $2,000,000 for a good response, (2) $500,000 for a fair response, and (3) a $100,000 loss for a poor response.

(a) Determine the marginal tax rate applicable to each situation.

(b) Determine the average tax rate that results from each situation.

9.43 Simon Machine Tools Company is considering purchasing a new set of machine tools to process special orders. The following financial information is available:

- Without the project, the company expects to have a taxable income of $300,000 each year from its regular business over the next three years.
- With the three-year project, the purchase of a new set of machine tools at a cost of $50,000 is required. The equipment falls into the MACRS three-year class. The tools will be sold for $10,000 at the end of project life. The project will be bringing in additional annual revenue of $80,000, but it is expected to incur additional annual operating costs of $20,000.

(a) What are the additional taxable incomes (due to undertaking the project) during each of years 1 through 3?

(b) What are the additional income taxes (due to undertaking the new orders) during each of years 1 through 3?

(c) Compute the gain taxes when the asset is disposed of at the end of year 3.

Combined Marginal Income Tax Rate

9.44 Consider a corporation whose taxable income without state income tax is as follows:

Gross revenue	$2,000,000
All expenses	1,200,000

If the marginal federal tax rate is 34% and the marginal state rate is 6%, compute the combined state and federal taxes using the two methods described in the text.

9.45 A corporation has the following financial information for a typical operating year:

Gross revenue	$4,500,000
Cost of goods sold	2,450,000
Operating costs	630,000
Federal taxes	352,000
State taxes	193,120

(a) On the basis of this financial information, determine both federal and state marginal tax rates.

(b) Determine the combined marginal tax rate for this corporation.

9.46 Van-Line Company, a small electronics repair firm, expects an annual income of $70,000 from its regular business. The company is considering expanding its repair business to include personal computers. The expansion would bring in an additional annual income of $30,000, but will require an additional expense of $10,000 each year over the next three years. Using applicable current tax rates, answer the following:

(a) What is the marginal tax rate in tax year 1?

(b) What is the average tax rate in tax year 1?

(c) Suppose that the new business expansion requires a capital investment of $20,000 (a three-year MACRS property). At $i = 10\%$, what is the PW of the total income taxes to be paid over the project life?

9.47 A company purchased a new forging machine to manufacture disks for airplane turbine engines. The new press cost $3,500,000, and it falls into a seven-year MACRS property class. The company has to pay property taxes to the local township for ownership of this forging machine at a rate of 1.2% on the beginning book value of each year.

(a) Determine the book value of the asset at the beginning of each tax year.

(b) Determine the amount of property taxes over the machine's depreciable life.

Short Case Studies

ST9.1 On January 2, 2000, Allen Flour Company purchased a new machine at a cost of $63,000. Installation costs for the machine were $2,000. The machine was expected to have a useful life of 10 years, with a salvage value of $4,000. The company uses straight-line depreciation for financial reporting. On January 3, 2003, the machine broke down, and an extraordinary repair had to be made to the machine at a cost of $6,000. The repair extended the machine's life to 13 years, but left the salvage value unchanged. On January 2, 2006, an improvement was made to the machine in the amount of $3,000 that increased the machine's productivity and increased the salvage value (to $6,000), but did not affect the remaining useful life. Determine depreciation expenses every December 31 for the years 2000, 2003, and 2006.

ST9.2 On March 17, 2003, Wildcat Oil Company began operations at its Louisiana oil field. The oil field had been acquired several years earlier at a cost of $11.6 million. The field is estimated to contain 4 million barrels of oil and to have a salvage value of $2 million both before and after all of the oil is pumped out. Equipment costing $480,000 was purchased for use at the oil field. The equipment will have no economic usefulness once the Louisiana field is depleted; therefore, it is depreciated on a units-of-production method. In addition, Wildcat Oil built a pipeline at a cost of $2,880,000 to serve the Louisiana field. Although this pipeline is physically capable of being used for many years, its economic usefulness is limited to the productive life of the Louisiana field; therefore, the pipeline has no salvage value. Depreciation of the pipeline is based on the estimated number of barrels of oil to be produced. Production at the Louisiana oil field amounted to 420,000 barrels in 2006 and 510,000 barrels in 2007.

(a) Compute the per barrel depletion rate of the oil field during the years 2006 and 2007.

(b) Compute the per barrel depreciation rates of the equipment and the pipeline during the years 2006 and 2007.

ST9.3 At the beginning of the fiscal year, Borland Company acquired new equipment at a cost of $65,000. The equipment has an estimated life of five years and an estimated salvage value of $5,000.

(a) Determine the annual depreciation (for financial reporting) for each of the five years of the estimated useful life of the equipment, the accumulated depreciation at the end of each year, and the book value of the equipment at the end of each year, all by (1) the straight-line method, and (2) the double-declining-balance method.

(b) Determine the annual depreciation for tax purposes, assuming that the equipment falls into a seven-year MACRS property class.

(c) Assume that the equipment was depreciated under seven-year MACRS. In the first month of the fourth year, the equipment was traded in for similar equipment priced at $82,000. The trade-in allowance on the old equipment was $10,000, and cash was paid for the balance. What is the cost basis of the new equipment for computing the amount of depreciation for income tax purposes?

ST9.4 Electronic Measurement and Control Company (EMCC) has developed a laser speed detector that emits infrared light, which is invisible to humans and radar detectors alike. For full-scale commercial marketing, EMCC needs to invest $5 million in new manufacturing facilities. The system is priced at $3,000 per unit. The company expects to sell 5,000 units annually over the next five years. The new manufacturing facilities will be depreciated according to a seven-year MACRS property class. The expected salvage value of the manufacturing facilities at the end of five years is $1.6 million. The manufacturing cost for the detector is $1,200 per unit, excluding depreciation expenses. The operating and maintenance costs are expected to run to $1.2 million per year. EMCC has a combined federal and state income tax rate of 35%, and undertaking this project will not change this current marginal tax rate.

(a) Determine, for the next five years, the incremental taxable income, income taxes, and net income due to undertaking this new product.

(b) Determine the gains or losses associated with the disposal of the manufacturing facilities at the end of five years.

ST9.5 Diamonid is a start-up diamond-coating company that is planning to manufacture a microwave plasma reactor which synthesizes diamonds. Diamonid anticipates that the industry demand for diamonds will skyrocket over the next decade, for use in industrial drills, high-performance microchips, and artificial human joints, among other things. Diamonid has decided to raise $50 million through issuing common stocks for investment in plant ($10 million) and equipment ($40 million). Each reactor can be sold at a price of $100,000 per unit. Diamonid can expect to sell 300 units per year during the next 8 years. The unit manufacturing cost is estimated at $30,000, excluding depreciation. The operating and maintenance cost for the plant is estimated at $12 million per year. Diamonid expects to phase out the operation at the end of 8 years, revamp the plant and equipment, and adopt a new diamond-manufacturing technology. At that time, Diamonid estimates that the salvage values for the plant and equipment will be about 60% and 10% of the original investments, respectively. The plant and equipment will be depreciated according to 39-year real property (placed in service in January) and 7-year MACRS, respectively. Diamonid pays 5% of state and local income taxes on its taxable income.

(a) If the 2006 corporate tax system continues over the project life, determine the combined state and federal income tax rate each year.

(b) Determine the gains or losses at the time the plant is revamped.

(c) Determine the net income each year over the plant life.

CHAPTER TEN

Developing Project Cash Flows

New Incentives for Being Green[1] Appliance makers are gearing up to push new energy-efficient systems in the wake of the energy bill passed in July 2005 that offers tax credits to homeowners who upgrade to electricity-saving appliances. In air-conditioning, industry giant **Carrier** Corporation says it has invested $250 million in developing new heat exchangers—a major component in air-conditioners—that use less energy and are about 20% smaller and 30% lighter than current energy-saving versions. Carrier's current energy-efficient models are almost double the size of its regular central air-conditioning units. The company speculates [that] this model's bulk may have been a deterrent for homeowners. The new air-conditioners were expected to hit the market in the first quarter of 2006.

Cooling efficiency is measured by a standard called SEER: Seasonal Energy Efficiency Ratio. It's kind of like the gas mileage system used on cars—the higher the number, the more money you save. Older systems had SEER numbers as low as 8; the new Carrier systems are rated at 18! (Federal standards required a minimum SEER of 13 from January 2006.) Translated into operating costs, this means that for every $100 you used to spend on electricity for cooling, you now can spend just $39.

There are several issues involved in Carrier's pushing more efficient air-conditioning units. First, the market demand is difficult to estimate. Second, it is even more difficult to predict the useful life of the product, as market competition is ever increasing. Carrier's gross margin is about 25%, its operating margin is 8.3%, and its net margin is 7.45%. The expected retail price of the Delux Puron unit is about $4,236.

---

[1] "New Incentives for Being Green," Cheryl Lu-Lien Tan, *The Wall Street Journal*, Thursday, August 4, 2005, p. D1.

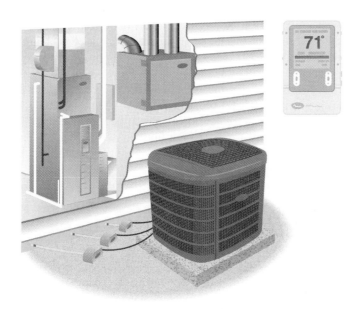

Of course, the main issue is how Carrier justified the capital investment of $250 million when it first decided to launch the product.

Clearly, Carrier's first step would be to estimate the magnitude of the revenue stream over the life of the product. In doing so, the final number of units sold would be the basis for possible revenue projections. The number of potential units sold is correlated with two important factors: housing market trends and how well their competitors are doing in terms of bringing more energy efficient products to market. Although the product is priced at $4,236 initially, there is no way of knowing how long Carrier will be able to sustain their desired profit margin. Under this circumstance, any project's justification depends upon the ability to estimate potential cash flows.

Projecting cash flows is the most important—and the most difficult—step in the analysis of a capital project. Typically, a capital project initially requires investment outlays and only later produces annual net cash inflows. A great many variables are involved in forecasting cash flows, and many individuals, ranging from engineers to cost accountants and marketing people, participate in the process. This chapter provides the general principles on which the determination of a project's cash flows are based.

To help us imagine the range of activities that are typically initiated by project proposals, we begin the chapter with an overview in Section 10.1 of how firms classify projects. A variety of projects exists, each having its own characteristic set of economic concerns. Section 10.2 then defines the typical cash flow elements of engineering projects. Once we have defined these elements, we will examine how to develop cash flow statements that are used to analyze the economic value of projects. In Section 10.3, we use several examples to demonstrate the development of after-tax project cash flow statements. Then, Section 10.4 presents some alternative techniques for developing a cash flow statement based on a generalized cash flow approach. By the time you have finished this chapter, not only will you understand the format and significance of after-tax cash flow statements, but also you will know how to develop them yourself.

CHAPTER LEARNING OBJECTIVES

After completing this chapter, you should understand the following concepts:

- What constitutes project cash flow elements.
- The use of the income statement approach in developing a project cash flow.
- How to treat the gains and losses related to disposal of an asset in the project cash flow statement.
- How to determine the working capital requirement and its impact on project cash flows.
- How to incorporate the cost associated with financing a project in developing the project's cash flow statement.
- How to develop a generalized cash flow model.
- The analysis of a lease-or-buy decision on an after-tax basis.

10.1 Cost–Benefit Estimation for Engineering Projects

Before the economics of an engineering project can be evaluated, it is necessary to estimate the various cost and revenue components that describe the project. Engineering projects may range from something as simple as the purchase of a new milling machine to the design and construction of a multibillion-dollar process or resource recovery complex.

The engineering projects appearing in this book as examples and problems already include the necessary cost and revenue estimates. Developing adequate estimates for these quantities is extremely important and can be a time-consuming activity. Although cost-estimating techniques are not the focus of this book, it is worthwhile to mention some of the approaches that are taken. Some are used in the context of simple projects that are straightforward and involve little or no engineering design; others are employed on complex projects that tend to be large and may involve many thousands of hours of engineering design. Obviously, there are also projects that fall between these extremes, and some combination of approaches may be appropriate in these cases.

10.1.1 Simple Projects

Projects in this category usually involve a single "off-the-shelf" component or a series of such components that are integrated together in a simple manner. The acquisition of a new milling machine is an example.

The installed cost is the price of the equipment as determined from catalogues or supplier quotations, shipping and handling charges, and the cost of building modifications and changes to utility requirements. The latter may call for some design effort to define the scope of the work, which is the basis for contractor quotations.

Project benefits are in the form of new revenue and/or cost reduction. Estimating new revenue requires agreement on the total number of units produced and the selling price per unit. These quantities are related through supply and demand considerations in the marketplace. In highly competitive product markets, sophisticated marketing studies are often required to establish price–volume relationships. These studies are undertaken as one of the first steps in an effort to define the appropriate scale for the project.

The ongoing costs required to operate and maintain equipment can be estimated at various levels of detail and accuracy. Familiarity with the cost–volume relationships for similar facilities allows the engineer to establish a "ballpark" cost. Some of this information may be used in conjunction with more detailed estimates in other areas. For example, maintenance costs may be estimated as a percentage of the installed cost. Such percentages are derived from historical data and are frequently available from equipment suppliers. Other costs, such as the cost of manpower and energy, may be estimated in detail to reflect specific local considerations. The most comprehensive and time-consuming type of estimate is a set of detailed estimates around each type of cost associated with the project.

10.1.2 Complex Projects

The estimates developed for complex projects involve the same general considerations as those which apply to simple projects. However, complex projects usually include specialized equipment that is not "off the shelf" and must be fabricated from detailed engineering drawings. For certain projects, such drawings are not even available until after a commitment has been made to proceed with the project. The typical phases of a project are as follows:

- Development.
- Conceptual design.
- Preliminary design.
- Detailed design.

Depending upon the project, some phases may be combined. Project economics is performed during each of these phases to confirm the attractiveness of the project and the incentive to continue working on it. The types of estimates and estimating techniques are a function of the stage of the project.

When the project involves the total or partial replacement of an existing facility, the benefits of the project are usually well known at the outset in terms of pricing–volume relationships or cost reduction potential. In the case of natural-resource projects, however, oil, gas, and mineral price forecasts are subject to considerable uncertainty.

At the development phase, work is undertaken to identify technologies that may be needed and to confirm their technical viability. Installed cost estimates and operating estimates are based on evaluations of the cost of similar existing facilities or their parts.

Conceptual design examines the issues of the scale of the project and alternative technologies. Again, estimates tend to be based on large pieces of, or processes within, the overall project that correspond to similar facilities already in operation elsewhere.

Preliminary design takes the most attractive alternative from the conceptual design phase to a level of detail that yields specific sizing and layout estimates for the actual equipment and associated infrastructure. Estimates at this stage tend to be based on individual pieces of equipment. Once more, the estimating basis is similar pieces of equipment in use elsewhere.

For very large projects, the cost of undertaking detailed engineering is prohibitive unless the project is going forward. At this stage, detailed fabrication and construction drawings that provide the basis for an actual vendor quotation become available.

The accuracy of the estimates improves with each phase as the project becomes defined in greater detail. Normally, the installed cost estimate from each phase includes a contingency that is some fraction of the actual estimate calculated. The contingency at the development phase can be 50 to 100%; after preliminary design, it decreases to something on the order of 10%.

More information on cost-estimating techniques can be found in reference books on project management and cost engineering. Industry-specific data books are also available for some sectors, for which costs are summarized on some normalized basis, such as dollars per square foot of building space or dollars per tonne of material moved in operating mines. In these cases, the data are categorized by type of building and type of mine. Engineering design companies maintain extensive databases of such cost information.

10.2 Incremental Cash Flows

When a company purchases a fixed asset such as equipment, it makes an investment. The company commits funds today in the expectation of earning a return on those funds in the future. Such investments are similar to those made by a bank when it lends money. In the case of a bank loan, the future cash flow consists of interest plus repayment of the principal. For a fixed asset, the future return is in the form of cash flows generated by the profitable use of the asset. In evaluating a capital investment, we are concerned only with those cash flows which result directly from the investment. These cash flows, called **differential** or **incremental cash flows**, represent the change in the firm's total cash flow that occurs as a direct result of the investment. In this section, we will look into some of the cash flow elements that are common to most investments.

10.2.1 Elements of Cash Outflows

We first consider the potential uses of cash in undertaking an investment project:

- **Purchase of New Equipment.** A typical project usually involves a cash outflow in the form of an initial investment in equipment. The relevant investment costs are incremental ones, such as the cost of the asset, shipping and installation costs, and the cost of training employees to use the new asset. If the purchase of a new asset results in the sale of an existing asset, the net proceeds from this sale reduce the amount of the incremental investment. In other words, the incremental investment represents the total amount of additional funds that must be committed to the investment project. When existing equipment is sold, the transaction results in an accounting gain or loss, which is dependent on whether the amount realized from the sale is greater or less than the equipment's book value. In either case, when existing assets are disposed of, the relevant amount by which the new investment is reduced consists of the proceeds of the sale, adjusted for tax effects.

- **Investments in Working Capital.** Some projects require an investment in nondepreciable assets. If, for example, a project increases a firm's revenues, then more funds will be needed to support the higher level of operation. Investment in nondepreciable assets is often called **investment in working capital**. In accounting, working capital is the amount carried in cash, accounts receivable, and inventory that is available to meet day-to-day operating needs. For example, additional working capital may be needed to meet the greater volume of business that will be generated by a project. Part of this increase in current assets may be supplied from increased accounts payable, but the remainder must come from permanent capital. This additional working capital is as much a part of the initial investment as the equipment itself. (We explain the amount of working capital required for a typical investment project in Section 10.3.2.)

  > **Working capital** measures how much in liquid assets a company has available to build its business.

- **Manufacturing, Operating, and Maintenance Costs.** The costs associated with manufacturing a new product need to be determined. Typical manufacturing costs include labor, materials, and overhead costs, the last of which cover items such as power, water, and indirect labor. Investments in fixed assets normally require periodic outlays for repairs and maintenance and for additional operating costs, all of which must be considered in investment analysis.

- **Leasing Expenses.** When a piece of equipment or a building is leased (instead of purchased) for business use, leasing expenses become cash outflows. Many firms lease computers, automobiles, industrial equipment, and other items that are subject to technological obsolescence.

- **Interest and Repayment of Borrowed Funds.** When we borrow money to finance a project, we need to make interest payments as well as payments on the principal. Proceeds from both short-term borrowing (bank loans) and long-term borrowing (bonds) are treated as cash inflows, but repayments of debts (both principal and interest) are classified as cash outflows.

- **Income Taxes and Tax Credits.** Any income tax payments following profitable operations should be treated as cash outflows for capital budgeting purposes. As we learned in Chapter 9, when an investment is made in depreciable assets, depreciation is an expense that offsets part of what would otherwise be additional taxable income. This strategy is called a **tax shield**, or **tax savings**, and we must take it into account when calculating income taxes. If any investment **tax credit** is allowed, it will directly reduce income taxes, resulting in cash inflows.

10.2.2 Elements of Cash Inflows

The following are potential sources of cash inflow over the project life:

- **Borrowed Funds.** If you finance your investment by borrowing, the borrowed funds will appear as cash inflow to the project at the time they are borrowed. From these funds, the purchase of new equipment or any other investment will be paid out.

- **Operating Revenues.** If the primary purpose of undertaking the project is to increase production or service capacity to meet increased demand, the new project will be bringing in additional revenues.
- **Cost Savings (or Cost Reduction).** If the primary purpose of undertaking a new project is to reduce operating costs, the amount involved should be treated as a cash inflow for capital budgeting purposes. A reduction in costs is equivalent to an increase in revenues, even though the actual sales revenues may remain unchanged.
- **Salvage Value.** In many cases, the estimated salvage value of a proposed asset is so small and occurs so far in the future that it may have no significant effect on the decision to undertake a project. Furthermore, any salvage value that is realized may be offset by removal and dismantling costs. In situations where the estimated salvage value is significant, the net salvage value is viewed as a cash inflow at the time of disposal. The **net salvage value** of the existing asset is the selling price of the asset, minus any costs incurred in selling, dismantling, and removing it, and this value is subject to taxable gain or loss.
- **Working-Capital Release.** As a project approaches termination, inventories are sold off and receivables are collected. That is, at the end of the project, these items can be liquidated at their cost. As this occurs, the company experiences an end-of-project cash flow that is equal to the net working-capital investment that was made when the project began. This recovery of working capital is not taxable income, since it merely represents a return of investment funds to the company.

Figure 10.1 sums up the various types of cash flows that are common in engineering investment projects.

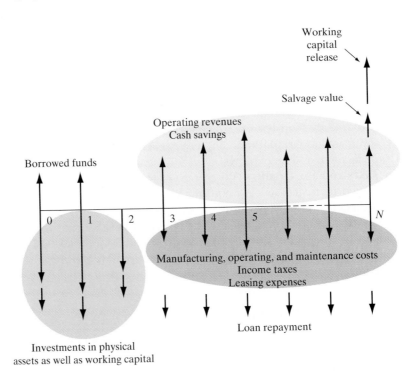

Figure 10.1 Types of cash flow elements used in project analysis.

10.2.3 Classification of Cash Flow Elements

Once the cash flow elements (both inflows and outflows) are determined, we may group them into three categories: (1) cash flow elements associated with operations, (2) cash flow elements associated with investment activities (capital expenditures), and (3) cash flow elements associated with project financing (such as borrowing). The main purpose of grouping cash flows this way is to provide information about the operating, investing, and financing activities of a project.

Operating Activities

In general, cash flows from operations include current sales revenues, the cost of goods sold, operating expenses, and income taxes. Cash flows from operations should generally reflect the cash effects of transactions entering into the determination of net income. The interest portion of a loan repayment is a deductible expense that is allowed in determining net income, and it is included in operating activities. Since we usually look only at yearly flows, it is logical to express all cash flows on a yearly basis.

As mentioned in Chapter 9, we can determine the net cash flow from operations on the basis of either (1) net income or (2) cash flow as determined by computing income taxes in a separate step. When we use net income as the starting point for cash flow determination, we should add any noncash expenses (mainly depreciation and amortization expenses) to net income to estimate the net cash flow from the operation. Recall that depreciation (or amortization) is not a cash flow, but is deducted, along with operating expenses and lease costs, from gross income to find taxable income and therefore taxes. Accountants calculate net income by subtracting taxes from taxable income, but depreciation—which is not a cash flow—was already subtracted to find taxable income, so it must be added back to taxable income if we wish to use the net-income figure as an intermediate step along the path to after-tax cash flow. Mathematically, it is easy to show that the two approaches are identical. Thus,

$$\text{Cash flow from operation} = \text{Net income} + (\text{Depreciation or amortization}).$$

Investing Activities

In general, three types of investment flows are associated with buying a piece of equipment: the original investment, the salvage value at the end of the useful life of the equipment, and the working-capital investment or recovery. We will assume that our outflow for both capital investment and working-capital investment take place in year 0. It is possible, however, that both investments will occur, not instantaneously, but rather, over a few months as the project gets into gear; we could then use year 1 as an investment year. (Capital expenditures may occur over several years before a large investment project becomes fully operational. In this case, we should enter all expenditures as they occur.) For a small project, either method of timing these flows is satisfactory, because the numerical differences are likely to be insignificant.

Financing Activities

Cash flows classified as financing activities include (1) the amount of borrowing and (2) the repayment of principal. Recall that interest payments are tax-deductible expenses, so they are classified as operating, not financing, activities.

The net cash flow for a given year is simply the sum of the net cash flows from operating, investing, and financing activities. Table 10.1 can be used as a checklist when you

TABLE 10.1 Classifying Cash Flow Elements and Their Equivalent Terms as Practiced in Business

Cash Flow Element	Other Terms Used in Business
Operating activities:	
Gross income	Gross revenue, sales revenue, gross profit, operating revenue
Cost savings	Cost reduction
Manufacturing expenses	Cost of goods sold, cost of revenue
Operations and maintenance cost	Operating expenses
Operating income	Operating profit, gross margin
Interest expenses	Interest payments, debt cost
Income taxes	Income taxes owed
Investing activities:	
Capital investment	Purchase of new equipment, capital expenditure
Salvage value	Net selling price, disposal value, resale value
Investment in working capital	Working-capital requirement
Release of working capital	Working-capital recovery
Gains taxes	Capital gains taxes, ordinary gains taxes
Financing activities:	
Borrowed funds	Borrowed amounts, loan amount
Repayment of principal	Loan repayment

set up a cash flow statement, because it classifies each type of cash flow element as an operating, investing, or financing activity.

10.3 Developing Cash Flow Statements

In this section, we will illustrate, through a series of numerical examples, how we actually prepare a project's cash flow statement; a generic version is shown in Figure 10.2, in which we first determine the net income from operations and then adjust the net income by adding any noncash expenses, mainly depreciation (or amortization). We will also consider a case in which a project generates a negative taxable income for an operating year.

10.3.1 When Projects Require Only Operating and Investing Activities

We will start with the simple case of generating after-tax cash flows for an investment project with only operating and investment activities. In the sections ahead, we will add complexities to this problem by including working-capital investments (Section 10.3.2) and borrowing activities (Section 10.3.3).

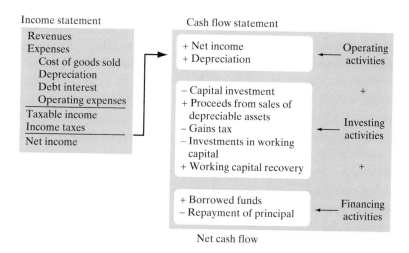

Figure 10.2 A popular format used for presenting a cash flow statement.

EXAMPLE 10.1 Cash Flow Statement: Operating and Investing Activities for an Expansion Project

A computerized machining center has been proposed for a small tool manufacturing company. If the new system, which costs $125,000, is installed, it will generate annual revenues of $100,000 and will require $20,000 in annual labor, $12,000 in annual material expenses, and another $8,000 in annual overhead (power and utility) expenses. The automation facility would be classified as a seven-year MACRS property.

The company expects to phase out the facility at the end of five years, at which time it will be sold for $50,000. Find the year-by-year after-tax net cash flow for the project at a 40% marginal tax rate based on the net income and determine the after-tax net present worth of the project at the company's MARR of 15%.

DISCUSSION: We can approach the problem in two steps by using the format shown in Figure 10.2 to generate an income statement and then a cash flow statement. We will follow this form in our subsequent listing of givens and unknowns. In year 0 (that is, at present) we have an investment cost of $125,000 for the equipment.[2] This cost will be depreciated in years 1 to 5. The revenues and costs are uniform annual flows in years 1 to 5. We can see that once we find depreciation allowances for each year, we can easily compute the results for years 1 to 4, which have fixed revenue and expense entries along with the variable depreciation charges. In year 5, we will need to incorporate the salvage value of the asset and any gains tax from its disposal.

We will use the business convention that no signs (positive or negative) be used in preparing the income statement, except in the situation where we have a negative taxable income or tax savings. In that situation, we will use parentheses to denote a

---

[2] We will assume that the asset is purchased and placed in service at the beginning of year 1 (or the end of year 0), and the first year's depreciation will be claimed at the end of year 1.

negative entry. However, in preparing the cash flow statement, we will explicitly observe the sign convention: A positive sign indicates a cash inflow; a negative sign or parentheses indicate a cash outflow.

SOLUTION

Given: Preceding cash flow information.

Find: After-tax cash flow.

Before presenting the cash flow table, we need to do some preliminary calculations. The following notes explain the essential items in Table 10.2:

- Calculation of depreciation

    1. If, contrary to expectations, the asset is held for eight years, we can depreciate a seven-year property in respective percentages of 14.29%, 24.49%, 17.49%, 12.49%, 8.93%, 8.92%, 8.93%, and 4.46%. (See Table 9.3.)

TABLE 10.2 Cash Flow Statement for the Automated Machining Center Project Using the Income Statement Approach (Example 10.1)

Year	0	1	2	3	4	5
Income Statement						
Revenues		$100,000	$100,000	$100,000	$100,000	$100,000
Expenses						
Labor		20,000	20,000	20,000	20,000	20,000
Material		12,000	12,000	12,000	12,000	12,000
Overhead		8,000	8,000	8,000	8,000	8,000
Depreciation		17,863	30,613	21,863	15,613	5,581
Taxable income		$ 42,137	$ 29,387	$ 38,137	$ 44,387	$ 54,419
Income taxes (40%)		16,855	11,755	15,255	17,755	21,768
Net income		$ 25,282	$ 17,632	$ 22,882	$ 26,632	$ 32,651
Cash Flow Statement						
Operating activities						
Net income		25,282	17,632	22,882	26,632	32,651
Depreciation		17,863	30,613	21,863	15,613	5,581
Investment activities						
Investment	(125,000)					
Salvage						50,000
Gains tax						(6,613)
Net cash flow	$(125,000)	$ 43,145	$ 48,245	$ 44,745	$ 42,245	$ 81,619

2. If the asset is sold at the end of the fifth tax year (during the recovery period), the applicable depreciation amounts would be $17,863, $30,613, $21,863, $15,613, and $5,581. Since the asset is disposed of in the fifth tax year, the last year's depreciation, which would ordinarily be $11,163, is halved due to the half-year convention.

- Salvage value and gain taxes

  In year 5, we must deal with two aspects of the asset's disposal: salvage value and gains (both ordinary as well as capital). We list the estimated salvage value as a positive cash flow. Taxable gains are calculated as follows:

  1. The total depreciation in years 1 to 5 is $17,863 + $30,613 + $21,863 + $15,613 + $5,581 = $91,533.
  2. The book value at the end of period 5 is the cost basis minus the total depreciation, or $125,000 − $91,533 = $33,467.
  3. The gains on the sale are the salvage value minus the book value, or $50,000 − $33,467 = $16,533. (The salvage value is less than the cost basis, so the gains are ordinary gains.)
  4. The tax on the ordinary gains is $16,533 × 40% = $6,613. This is the amount placed in the table under "gains tax."

  Table 10.2 presents a summary of the cash flow profile.[3]

- Investment analysis

  Once we obtain the project's after-tax net cash flows, we can determine their equivalent present worth at the firm's interest rate. The after-tax cash flow series from the cash flow statement is shown in Figure 10.3. Since this series

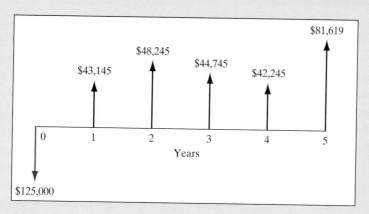

Figure 10.3 Cash flow diagram (Example 10.1).

---

[3] Even though gains from equipment disposal have an effect on income tax calculations, they should not be viewed as ordinary operating income. Therefore, in preparing the income statement, capital expenditures and related items such as gains tax and salvage value are not included. Nevertheless, these items represent actual cash flows in the year they occur and must be shown in the cash flow statement.

does not contain any patterns to simplify our calculations, we must find the net present worth of each payment. Using $i = 15\%$, we have

$$PW(15\%) = -\$125,000 + \$43,145(P/F, 15\%, 1)$$
$$+ \$48,245(P/F, 15\%, 2) + \$44,745(P/F, 15\%, 3)$$
$$+ \$42,245(P/F, 15\%, 4) + \$81,619(P/F, 15\%, 5)$$
$$= \$43,152.$$

This means that investing $125,000 in this automated facility would bring in enough revenue to recover the initial investment and the cost of funds, with a surplus of $43,152.

10.3.2 When Projects Require Working-Capital Investments

In many cases, changing a production process by replacing old equipment or by adding a new product line will have an impact on cash balances, accounts receivable, inventory, and accounts payable. For example, if a company is going to market a new product, inventories of the product and larger inventories of raw materials will be needed. Accounts receivable from sales will increase, and management might also decide to carry more cash because of the higher volume of activities. These investments in working capital are investments just like those in depreciable assets, except that they have no tax effects: The flows always sum to zero over the life of a project, but the inflows and outflows are shifted in time, so they do affect net present worth.

Consider the case of a firm that is planning a new product line. The new product will require a two-month's supply of raw materials at a cost of $40,000. The firm could provide $40,000 in cash on hand to pay them. Alternatively, the firm could finance these raw materials via a $30,000 increase in accounts payable (60-day purchases) by buying on credit. The balance of $10,000 represents the amount of net working capital that must be invested.

Working-capital requirements differ according to the nature of the investment project. For example, larger projects may require greater average investments in inventories and accounts receivable than would smaller ones. Projects involving the acquisition of improved equipment entail different considerations. If the new equipment produces more rapidly than the old equipment, the firm may be able to decrease its average inventory holdings because new orders can be filled faster as a result of using the new equipment. (One of the main advantages cited in installing advanced manufacturing systems, such as flexible manufacturing systems, is the reduction in inventory made possible by the ability to respond to market demand more quickly.) Therefore, it is also possible for working-capital needs to decrease because of an investment. If inventory levels were to decrease at the start of a project, the decrease would be considered a cash inflow, since the cash freed up from inventory could be put to use in other places. (See Example 10.5.)

Two examples illustrate the effects of working capital on a project's cash flows. Example 10.2 shows how the net working-capital requirement is computed, and Example 10.3 examines the effects of working capital on the automated machining center project discussed in Example 10.1.

EXAMPLE 10.2 Working-Capital Requirements

Suppose that in Example 10.1 the tool-manufacturing company's annual revenue projection of $100,000 is based on an annual volume of 10,000 units (or 833 units per month). Assume the following accounting information:

Price (revenue) per unit	$10
Unit variable manufacturing costs:	
Labor	$2
Material	$1.20
Overhead	$0.80
Monthly volume	833 units
Finished-goods inventory to maintain	2-month supply
Raw-materials inventory to maintain	1-month supply
Accounts payable	30 days
Accounts receivable	60 days

The accounts receivable period of 60 days means that revenues from the current month's sales will be collected two months later. Similarly, accounts payable of 30 days indicates that payment for materials will be made approximately one month after the materials are received. Determine the working-capital requirement for this operation.

SOLUTION

Given: Preceding information.
Find: Working-capital requirement.

Figure 10.4 illustrates the working-capital requirements for the first 12-month period.

During year 1	Income/Expense reported	Actual cash received/paid	Difference
Sales	$100,000 (10,000 units)	$83,333	–$16,666
Expenses	$40,000 (10,000 units)	$46,665 (11,667 units)	+$6,665
Income taxes	$16,855	$16,855	0
Net amount	$43,145	$19,814	–$23,333

This differential amount must be invested at the beginning of the year

Figure 10.4 Illustration of working-capital requirements (Example 10.2).

Accounts receivable of $16,666 (2 months' sales) means that in year 1 the company will have cash inflows of $83,333, less than the projected sales of $100,000 ($8,333 × 12). In years 2 to 5, collections will be $100,000, equal to sales, because the beginning and ending accounts receivable will be $16,666, with sales of $100,000. Collection of the final accounts receivable of $16,666 would occur in the first 2 months of year 6, but can be added to the year-5 revenue to simplify the calculations. The important point is that cash inflow lags sales by $16,666 in the first year.

Assuming that the company wishes to build up 2 months' inventory during the first year, it must produce 833 × 2 = 1,666 more units than are sold the first year. The extra cost of these goods in the first year is 1,666 units ($4 variable cost per unit), or $6,665. The finished-goods inventory of $6,665 represents the variable cost incurred to produce 1,666 more units than are sold in the first year. In years 2 to 4, the company will produce and sell 10,000 units per year, while maintaining its 1,666-unit supply of finished goods. In the final year of operations, the company will produce only 8,334 units (for 10 months) and will use up the finished-goods inventory. As 1,666 units of the finished-goods inventory get liquidated during the last year (exactly, in the first 2 months of year 6), a working capital in the amount of $6,665 will be released. Along with the collection of the final accounts receivable of $16,666, a total working-capital release of $23,331 will remain when the project terminates. Now we can calculate the working-capital requirements as follows:

Accounts receivable	
(833 units/month × 2 months × $10)	$16,666
Finished-goods inventory	
(833 units/month × 2 months × $4)	6,665
Raw-materials inventory	
(833 units/month × 1 month × $1.20)	1,000
Accounts payable (purchase of raw materials)	
(833 units/month × 1 month × $1.20)	(1,000)
Net change in working capital	$23,331

COMMENTS: During the first year, the company produces 11,666 units to maintain two months' finished-goods inventory, but it sells only 10,000 units. On what basis should the company calculate the net income during the first year (i.e., use 10,000 or 11,666 units)? *Any* increases in inventory expenses will reduce the taxable income; therefore, this calculation is based on 10,000 units. The reason is that the accounting measure of net income is based on the **matching concept**. If we report revenue when it is *earned* (whether it is actually received or not), and we report expenses when they are *incurred* (whether they are paid or not), we are using the *accrual method* of accounting. By tax law, this method *must* be used for purchases and sales whenever business transactions involve an inventory. Therefore, most manufacturing and merchandising businesses use the accrual basis in recording revenues and expenses. Any cash inventory expenses not accounted for in the net-income calculation will be reflected in changes in working capital.

EXAMPLE 10.3 Cash Flow Statement, Including Working Capital

Update the after-tax cash flows for the automated machining center project of Example 10.1 by including a working-capital requirement of $23,331 in year 0 and full recovery of the working capital at the end of year 5.

SOLUTION

Given: Flows as in Example 10.1, with the addition of a $23,331 working-capital requirement.

Find: Net after-tax cash flows with working capital and present worth.

Using the procedure just outlined, we group the net after-tax cash flows for this machining center project as shown in Table 10.3. As the table indicates, investments in working capital are cash outflows when they are expected to occur, and recoveries

TABLE 10.3 Cash Flow Statement for Automated Machining Center Project with Working-Capital Requirement (Example 10.3)

Year	0	1	2	3	4	5
Income Statement						
Revenues		$100,000	$100,000	$100,000	$100,000	$100,000
Expenses						
Labor		20,000	20,000	20,000	20,000	20,000
Material		12,000	12,000	12,000	12,000	12,000
Overhead		8,000	8,000	8,000	8,000	8,000
Depreciation		17,863	30,613	21,863	15,613	5,581
Taxable income		$42,137	$29,387	$38,137	$44,387	$54,419
Income taxes (40%)		16,855	11,755	15,255	17,755	21,768
Net income		$25,282	$17,632	$22,882	$26,632	$32,651
Cash Flow Statement						
Operating activities						
Net income		25,282	17,632	22,882	26,632	32,651
Depreciation		17,863	30,613	21,863	15,613	5,581
Investment activities						
Investment	(125,000)					
Salvage						50,000
Gains tax						(6,613)
Working capital	(23,331)					23,331
Net cash flow	$(148,331)	$43,145	$48,245	$44,745	$42,245	$104,950

are treated as cash inflows at the times they are expected to materialize. In this example, we assume that the investment in working capital made at period 0 will be recovered at the end of the project's life. (See Figure 10.5.)[4] Moreover, we also assume a full recovery of the initial working capital. However, in many situations, the investment in working capital may not be fully recovered (e.g., inventories may deteriorate in value or become obsolete). The equivalent net present worth of the after-tax cash flows, including the effects of working capital, is calculated as

$$PW(15\%) = -\$148,331 + \$43,145(P/F, 15\%, 1) + \cdots$$
$$+ \$104,950(P/F, 15\%, 5)$$
$$= \$31,420.$$

This present-worth value is $11,732 less than that in the situation with no working-capital requirement (Example 10.1), demonstrating that working-capital requirements play a critical role in assessing a project's worth.

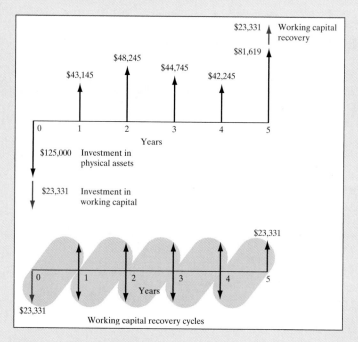

Figure 10.5   Cash flow diagram (Example 10.3).

---

[4] In fact, we could assume that the investment in working capital would be recovered at the end of the first operating cycle (say, year 1). However, the same amount of investment in working capital has to be made again at the beginning of year 2 for the second operating cycle, and the process repeats until the project terminates. Therefore, the net cash flow transaction looks as though the initial working capital will be recovered at the end of the project's life:

Period	0	1	2	3	4	5
Investment	−$23,331	−$23,331	−$23,331	−$23,331	−$23,331	0
Recovery	0	23,331	23,331	23,331	23,331	23,331
Net flow	−$23,331	0	0	0	0	$23,331

COMMENT: The $11,732 reduction in present worth is just the present worth of an annual series of 15% interest payments on the working capital, which is borrowed by the project at time 0 ar.d repaid at the end of year 5:

$$\$23,331(15\%)(P/A, 15\%, 5) = \$11,732.$$

The investment tied up in working capital results in lost earnings.

10.3.3 When Projects Are Financed with Borrowed Funds

Many companies use a mixture of debt and equity to finance their physical plant and equipment. The ratio of total debt to total investment, generally called the **debt ratio**, represents the percentage of the total initial investment provided by borrowed funds. For example, a debt ratio of 0.3 indicates that 30% of the initial investment is borrowed and the rest is provided from the company's earnings (also known as **equity**). Since interest is a tax-deductible expense, companies in high tax brackets may incur lower after-tax financing costs by financing through debt. (Along with the effect of debt on taxes, the method of repaying the loan can have a significant impact. We will discuss the issue of project financing in Chapter 15.)

Debt ratio: Debt capital divided by total assets. This will tell you how much the company relies on debt to finance assets.

EXAMPLE 10.4 Cash Flow Statement with Financing (Borrowing)

Rework Example 10.3, assuming that $62,500 of the $125,000 paid for the investment is obtained through debt financing (debt ratio = 0.5). The loan is to be repaid in equal annual installments at 10% interest over five years. The remaining $62,500 will be provided by equity (e.g., from retained earnings).

SOLUTION

Given: Same as in Example 10.3, but $62,500 is borrowed and then repaid in equal installments over five years at 10% interest.

Find: Net after-tax cash flows in each year.

We first need to compute the size of the annual loan repayment installments:

$$\$62,500(A/P, 10\%, 5) = \$16,487.$$

Next, we determine the repayment schedule of the loan by itemizing both the interest and principal represented in each annual repayment:

Year	Beginning Balance	Interest Payment	Principal Payment	Ending Balance
1	$62,500	$6,250	$10,237	$52,263
2	52,263	5,226	11,261	41,002
3	41,002	4,100	12,387	28,615
4	28,615	2,861	13,626	14,989
5	14,989	1,499	14,988	0

The resulting after-tax cash flow is detailed in Table 10.4. The present-value equivalent of the after-tax cash flow series is

$$PW(15\%) = -\$85,351 + \$29,158(P/F, 15\%, 1) + \cdots$$
$$+ \$89,063(P/F, 15\%, 5)$$
$$= \$44,439.$$

When this amount is compared with the amount found in the case when there was no borrowing ($31,420), we see that debt financing actually increases the present worth

TABLE 10.4 Cash Flow Statement for Automated Machining Center Project with Debt Financing (Example 10.4)

Year	0	1	2	3	4	5
Income Statement						
Revenues		$100,000	$100,000	$100,000	$100,000	$100,000
Expenses						
Labor		20,000	20,000	20,000	20,000	20,000
Material		12,000	12,000	12,000	12,000	12,000
Overhead		8,000	8,000	8,000	8,000	8,000
Depreciation		17,863	30,613	21,863	15,613	5,581
Debt interest		6,250	5,226	4,100	2,861	1,499
Taxable income		$ 35,887	$ 24,161	$ 34,037	$ 41,526	$ 52,920
Income taxes (40%)		14,355	12,664	13,615	16,610	21,168
Net income		$ 21,532	$ 14,497	$ 20,422	$ 24,916	$ 31,752
Cash Flow Statement						
Operating activities						
Net income		21,532	14,497	20,422	24,916	31,752
Depreciation		17,863	30,613	21,863	15,613	5,581
Investment activities						
Investment	(125,000)					
Salvage						50,000
Gains tax						(6,613)
Working capital	(23,331)					23,331
Financing activities						
Borrowed funds	62,500					
Repayment of principal		(10,237)	(11,261)	(12,387)	(13,626)	(14,988)
Net cash flow	$ (85,831)	$ 29,158	$ 33,849	$ 29,898	$ 26,903	$ 89,063

by $13,019. This surprising result is largely caused by the firm's being able to borrow the funds at a cheaper rate (10%) than its MARR (opportunity cost rate) of 15%. We should be careful in interpreting the result, however: It is true, to some extent, that firms can usually borrow money at lower rates than their MARR, but if the firm can borrow money at a significantly lower rate, that also affects its MARR, because the borrowing rate is one of the elements determining the MARR. Therefore, a significant difference in present values between "with borrowings" and "without borrowings" is not expected in practice. We will also address this important issue in Chapter 15.

10.3.4 When Projects Result in Negative Taxable Income

In a typical project year, revenues may not be large enough to offset expenses, thereby resulting in a negative taxable income. Now, a negative taxable income does *not* mean that a firm does not need to pay income tax; rather, the negative figure can be used to reduce the taxable incomes generated by other business operations.[5] Therefore, a negative taxable income usually results in a **tax savings**. When we evaluate an investment project with the use of an incremental tax rate, we also assume that the firm has sufficient taxable income from other activities, so that changes due to the project under consideration will not change the incremental tax rate.

Tex savings (shield): The reduction in income taxes that results from taking an allowable deduction from taxable income.

When we compare **cost-only** mutually exclusive projects (service projects), we have no revenues to consider in their cash flow analysis. In this situation, we typically assume no revenue (zero), but proceed as before in constructing the after-tax cash flow statement. With no revenue to match expenses, we have a negative taxable income, again resulting in tax savings. Example 10.5 illustrates how we may develop an after-tax cash flow statement for this type of project.

EXAMPLE 10.5 Project Cash Flows for a Cost-Only Project

Alcoa Aluminum's McCook plant produces aluminum coils, sheets, and plates. Its annual production runs at 400 million pounds. In an effort to improve McCook's current production system, an engineering team, led by the divisional vice-president, went on a fact-finding tour of Japanese aluminum and steel companies to observe their production systems and methods. Cited among the team's observations were large fans which the Japanese companies used to reduce the time that coils need to cool down after various processing operations. Cooling the hot process coils with the fans was estimated to significantly reduce the queue or work-in-process (WIP) inventory buildup allowed for cooling. The approach also reduced production lead time and improved delivery performance. The possibility of reducing production time and, as a consequence, the WIP inventory, excited the team members, particularly the vice-president. After the trip, Neal Donaldson, the plant engineer, was asked to

---

[5] Even if the firm does not have any other taxable income to offset in the current tax year, the operating loss can be carried back to each of the preceding 3 years and forward for the next 15 years to offset taxable income in those years.

investigate the economic feasibility of installing cooling fans at the McCook plant. Neal's job is to justify the purchase of cooling fans for his plant. He was given one week to prove that the idea was a good one. Essentially, all he knew was the number of fans, their locations, and the project cost. Everything else was left to Neal's devices. Suppose that he compiled the following financial data:

- The project will require an investment of $536,000 in cooling fans now.
- The cooling fans would provide 16 years of service with no appreciable salvage values, considering the removal costs.
- It is expected that the amount of time required between hot rolling and the next operation would be reduced from five days to two days. Cold-rolling queue time also would be reduced, from two days to one day for each cold-roll pass. The net effect of these changes would be a reduction in the WIP inventory at a value of $2,121,000. Because of the lead time involved in installing the fans, as well as the consumption of stockpiled WIP inventory, this working-capital release will be realized one year after the fans are installed.
- The cooling fans will be depreciated according to seven-year MACRS.
- Annual electricity costs are estimated to rise by $86,000.
- The firm's after-tax required rate of return is known to be 20% for this type of cost reduction project.

Develop the project cash flows over the service period of the fans, and determine whether the investment is a wise one, based on 20% interest.

SOLUTION

Given: Required investment = $536,000; service period = 16 years; salvage value = $0; depreciation method for cooling fans = 7-year MACRS; working-capital release = $2,121,000 1 year later; annual operating cost (electricity) = $86,000.

Find: (a) annual after-tax cash flows; (b) make the investment decision on the basis of the NPW; and (c) make the investment decision on the basis of the IRR.

(a) Because we can assume that the annual revenues would stay the same as they were before and after the installation of the cooling fans, we can treat these unknown revenue figures as zero. Table 10.5 summarizes the cash flow statement for the cooling-fan project. With no revenue to offset the expenses, the taxable income will be negative, resulting in tax savings. Note that the working-capital recovery (as opposed to working-capital investment for a typical investment project) is shown in year 1. Note also that there is no gains tax, because the cooling fans are fully depreciated with a zero salvage value.

(b) At $i = 20\%$, the NPW of this investment would be

$$PW(20\%) = -\$536,000 + \$2,100,038(P/F, 20\%, 1)$$
$$+ \$906(P/F, 20\%, 2) + \dots$$
$$- \$2,172,600(P/F, 20\%, 16)$$
$$= \$991,008.$$

TABLE 10.5 Cash Flow Statement for the Cooling Fan Project without Revenue (Example 10.5)

Year	0	1	2	3	4	5	6	7	8	9–15	16
Income Statement											
Revenues											
Expenses											
Depreciation		$76,594	$131,266	$93,746	$66,946	$47,865	$47,811	$47,865	$23,906		
Electricity cost		86,000	86,000	86,000	86,000	86,000	86,000	86,000	86,000	86,000	86,000
Taxable income		(162,594)	(217,266)	(179,746)	(152,946)	(133,865)	(133,811)	(133,865)	(109,906)	(86,000)	(86,000)
Income taxes		(65,038)	(86,906)	(71,898)	(61,178)	(53,546)	(53,524)	(53,546)	(43,962)	(34,400)	(34,400)
Net income		$ (97,556)	$(130,360)	$(107,848)	$ (91,768)	$ (80,319)	$ (80,287)	$ (80,319)	$ (65,944)	$ (51,600)	$ (51,600)
Cash Flow Statement											
Operating activities											
Net income		(97,556)	(130,360)	(107,848)	(91,768)	(80,319)	(80,287)	(80,319)	(65,944)	(51,600)	(51,600)
Depreciation		76,594	131,266	93,746	66,946	47,865	47,811	47,865	23,906	0	0
Investment activities											
Cooling fans	(536,000)										
Salvage value											
Gains tax											
Working capital		2,121,000									(2,121,000)
Net cash flow	$ (536,000)	$2,100,038	$ 906	$ (14,102)	$ (24,822)	$ (32,454)	$ (32,476)	$ (32,454)	$ (42,038)	$ (51,600)	$ (2,172,600)

Note: The working-capital release attributable to a reduction in work-in-process inventories will be realized at the end of year 1.

Even with only one time savings in WIP, this cooling-fan project is economically justifiable.

(c) If we are to justify the investment on the basis of the internal rate of return, we first need to determine whether the investment is a simple or nonsimple one. Since there are more than one sign changes in the cash flow series, this is not a simple investment, indicating the possibility of multiple rates of return. In fact, as shown in Figure 10.6, the project has two rates of return, one at 4.24% and the other at 291.56%. Neither is a true rate of return, as we learned in Section 7.3.2, so we may proceed to abandon the rate-of-return approach and use the NPW criterion. If you desire to find the true IRR for this project, you need to follow the procedures outlined in Section 7.3.4. At a MARR of 20%, the true IRR (or return on invested capital) is 241.87%, which is significantly larger than the MARR and indicates acceptance of the investment. Note that this is the same conclusion that we reached in (b).

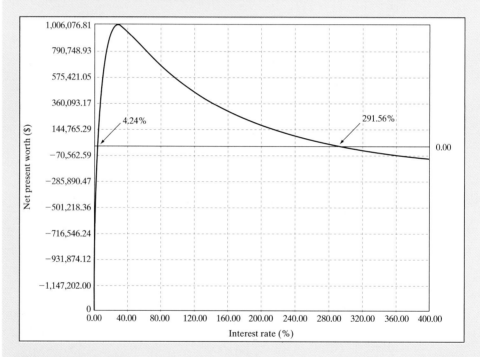

Figure 10.6 NPW plot for the cooling-fan project (Example 10.5).

COMMENTS: As the cooling fans reach the end of their service lives, we need to add the working-capital investment ($2,120,000) at the end of year 16, working under the assumption that the plant will return to the former system without the cooling fans and thus will require the additional investment in working capital.

If the new system has proven to be effective, and the plant will remain in service, we need to make another investment to purchase a new set of cooling fans at the end of year 16. As a consequence of this new investment, there will be a working-capital release in the amount of $2,120,000. However, this investment should bring benefits to the second cycle of the operation, so that it should be charged against the cash flows for the second cycle, not the first cycle.

10.3.5 When Projects Require Multiple Assets

Up to this point, our examples have been limited to situations in which only one asset was employed in a project. In many situations, however, a project may require the purchase of multiple assets with different property classes. For example, a typical engineering project may involve more than just the purchase of equipment: It may need a building in which to carry out manufacturing operations. The various assets may even be placed in service at different points in time. What we then have to do is itemize the timing of the investment requirement and the depreciation allowances according to the placement of the assets. Example 10.6 illustrates the development of project cash flows that require multiple assets.

EXAMPLE 10.6 A Project Requiring Multiple Assets

Langley Manufacturing Company (LMC), a manufacturer of fabricated metal products, is considering purchasing a new computer-controlled milling machine to produce a custom-ordered metal product. The following summarizes the relevant financial data related to the project:

- The machine costs $90,000. The costs for its installation, site preparation, and wiring are expected to be $10,000. The machine also needs special jigs and dies, which will cost $12,000. The milling machine is expected to last 10 years, the jigs and dies 5 years. The machine will have a $10,000 salvage value at the end of its life. The special jigs and dies are worth only $1,000 as scrap metal at any time in their lives. The milling machine is classified as a 7-year MACRS property and the jigs and dies as a 3-year MACRS property.

- LMC needs to either purchase or build an 8,000-ft^2 warehouse in which to store the product before it is shipped to the customer. LMC has decided to purchase a building near the plant at a cost of $160,000. For depreciation purposes, the warehouse cost of $160,000 is divided into $120,000 for the building (39-year real property) and $40,000 for land. At the end of 10 years, the building will have a salvage value of $80,000, but the value of the land will have appreciated to $110,000.

- The revenue from increased production is expected to be $150,000 per year. The additional annual production costs are estimated as follows: materials, $22,000; labor, $32,000; energy $3,500; and other miscellaneous costs, $2,500.

- For the analysis, a 10-year life will be used. LMC has a marginal tax rate of 40% and a MARR of 18%. No money is borrowed to finance the project. Capital gains will also be taxed at 40%.[6]

---

[6] Capital gains for corporations are taxed at a maximum rate of 35%. However, capital gains are also subject to state taxes, so the combined tax rate will be approximately 40%.

DISCUSSION: Three types of assets are to be considered in this problem: the milling machine, the jigs and dies, and the warehouse. The first two assets are personal properties and the last is a real property. The cost basis for each asset has to be determined separately. For the milling machine, we need to add the site-preparation expense to the cost basis, whereas we need to subtract the land cost from the warehouse cost to establish the correct cost basis for the real property. The various cost bases are as follows:

- The milling machine: $90,000 + $10,000 = $100,000.
- Jigs and dies: $12,000.
- Warehouse (building): $120,000.
- Warehouse (land): $40,000.

Since the jigs and dies last only five years, we need to make a specific assumption regarding the replacement cost at the end of that time. In this problem, we will assume that the replacement cost would be approximately equal to the cost of the initial purchase. We will also assume that the warehouse property will be placed in service in January, which indicates that the first year's depreciation percentage will be 2.4573%. (See Table 9.4.)

SOLUTION

Given: Preceding cash flow elements, $t_m = 40\%$, and MARR $= 18\%$.
Find: Net after-tax cash flow and NPW.

Table 10.6 and Figure 10.7 summarize the net after-tax cash flows associated with the multiple-asset investment. In the table, we see that the milling machine and the jigs and dies are fully depreciated during the project life, whereas the building is not. We need to calculate the book value of the building at the end of the project life. We assume that the building will be disposed of December 31 in the 10th year, so that the midmonth convention also applies to the book-value calculation:

$$B_{10} = \$120,000 - (\$2,949 + \$3,077 \times 8 + \$2,949) = \$89,486.$$

Then the gains (losses) are

$$\text{Salvage value} - \text{book value} = \$80,000 - \$89,486 = (\$9,486).$$

We can now calculate the gains or losses associated with the disposal of each asset as follows:

Property (Asset)	Cost Basis	Salvage Value	Book Value	Gains (Losses)	Gains Taxes
Land	$40,000	$110,000	$40,000	$70,000	$28,000
Building	120,000	80,000	89,486	(9,486)	(3,794)
Milling machine	100,000	10,000	0	10,000	4,000
Jigs and dies	12,000	1,000	0	1,000	400

TABLE 10.6 Cash Flow Statement for LMC's Machining Center Project with Multiple Assets (Example 10.6)

Year	0	1	2	3	4	5	6	7	8	9	10
Income Statement											
Revenues		$150,000	$150,000	$150,000	$150,000	$150,000	$150,000	$150,000	$150,000	$150,000	$150,000
Expenses											
Materials		22,000	22,000	22,000	22,000	22,000	22,000	22,000	22,000	22,000	22,000
Labor		32,000	32,000	32,000	32,000	32,000	32,000	32,000	32,000	32,000	32,000
Energy		3,500	3,500	3,500	3,500	3,500	3,500	3,500	3,500	3,500	3,500
Other		2,500	2,500	2,500	2,500	2,500	2,500	2,500	2,500	2,500	2,500
Depreciation											
Building		2,949	3,077	3,077	3,077	3,077	3,077	3,077	3,077	3,077	2,949
Machines		14,290	24,490	17,490	12,490	8,930	8,920	8,930	4,460		
Tools		4,000	5,333	1,778	889		4,000	5,333	1,778	889	
Taxable income		68,761	57,100	67,655	73,544	77,993	74,003	72,660	80,685	86,034	87,051
Income taxes		27,504	22,840	27,062	29,418	31,197	29,601	29,064	32,274	34,414	34,820
Net income		$ 41,257	$ 34,260	$ 40,593	$ 44,126	$ 46,796	$ 44,402	$ 43,596	$ 48,411	$ 51,620	$ 52,231
Cash Flow Statement											
Operating activities:											
Net income		41,257	34,260	40,593	44,126	46,796	44,402	43,596	48,411	51,620	52,231
Depreciation		21,239	32,900	22,345	16,456	12,007	15,997	17,340	9,315	3,966	2,949
Investment activities:											
Land	(40,000)										110,000
Building	(120,000)										80,000
Machines	(100,000)										10,000
Tools (first cycle)	(12,000)					1,000					
Tools (second cycle)						(12,000)					
Gains tax:											
Land									(28,000)		1,000
Building											3,794
Machines											(4,000)
Tools						(400)					(400)
Net cash flow	$ (272,000)	$ 62,496	$ 67,160	$ 62,938	$ 60,582	$ 47,403	$ 60,399	$ 60,936	$ 57,726	$ 55,586	$ 227,574

Note: Investment in tools (jigs and dies) will be repeated at the end of year 5, at the cost of the initial purchase.

The NPW of the project is

$$PW(18\%) = -\$272,000 + \$62,496(P/F, 18\%, 1) + \$67,160(P/F, 18\%, 2)$$

$$+ \ldots + \$227,574(P/F, 18\%, 10)$$

$$= \$32,343 > 0.$$

and the IRR for this investment is about 21%, which exceeds the MARR. Therefore, the project is acceptable.

COMMENT: Note that the gains (losses) posted in the preceding table can be classified into two types: ordinary gains (losses) and capital gains. Only the $70,000 for land represents true capital gains, whereas the other figures represent ordinary gains (losses).

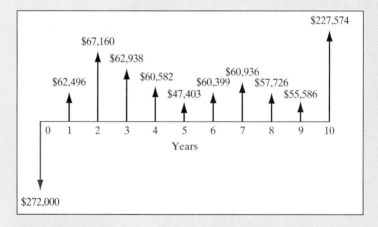

Figure 10.7 Cash flow diagram (Example 10.6).

10.4 Generalized Cash Flow Approach

If we analyze project cash flows for a corporation that consistently operates in the highest tax bracket, we can assume that the firm's marginal tax rate will remain the same, whether the project is accepted or rejected. In this situation, we may apply the top marginal tax rate (currently 35%) to each taxable item in the cash profile, thus obtaining the after-tax cash flows. By aggregating individual items, we obtain the project's net cash flows. This approach is referred to as the **generalized cash flow approach**. As we shall see in later chapters, it affords several analytical advantages. In particular, when we compare service projects, the generalized cash flow method is computationally efficient. (Examples are given in Chapters 12 and 14.)

10.4.1 Setting up Net Cash Flow Equations

To produce a generalized cash flow table, we first examine each cash flow element. We can do this as follows, using the scheme for classifying cash flows that we have just developed:

$$
A_n = \left.
\begin{array}{l}
+\text{Revenues at time } n,\ (R_n) \\
-\text{Expenses (excluding depreciation and} \\
\quad \text{debt interest) at time } n,\ (E_n) \\
-\text{Debt interest at time } n,\ (\text{IP}_n) \\
-\text{Income taxes at time } n,\ (T_n)
\end{array}
\right\} \text{Operating activities}
$$

$$
\left.
\begin{array}{l}
-\text{Investment at time } n,\ (I_n) \\
+\text{Net proceeds from sale at time } n,\ (S_n - G_n) \\
-\text{Working capital investment at time } n,\ (W_n)
\end{array}
\right\} \text{Investing activities}
$$

$$
\left.
\begin{array}{l}
+\text{Proceeds from loan } n,\ (B_n) \\
-\text{Repayment of principal } n,\ (\text{PP}_n),
\end{array}
\right\} \text{Borrowing activities.}
$$

Here, A_n is the net after-tax cash flow at the end of period n.

Depreciation (D_n) is *not* a cash flow and is therefore excluded from E_n (although it must be considered in calculating income taxes). Note that $(S_n - G_n)$ represents the net salvage value after adjustments for gains tax or loss credits (G_n). Not all terms are relevant in calculating a cash flow in every year; for example, the term $(S_n - G_n)$ appears only when the asset is disposed of.

In terms of symbols, we can express A_n as

$$
A_n = \begin{array}{ll}
R_n - E_n - \text{IP}_n - T_n & \leftarrow \text{Operating activities} \\
-I_n + (S_n - G_n) - W_n & \leftarrow \text{Investing activities} \\
+B_n - \text{PP}_n & \leftarrow \text{Financing activities.}
\end{array}
\tag{10.1}
$$

If we designate T_n as the total income taxes paid at time n and t_m as the marginal tax rate, income taxes on this project are

$$
\begin{aligned}
T_n &= (\text{Taxable income})(\text{marginal tax rate}) \\
&= (R_n - E_n - \text{IP}_n - D_n)t_m \\
&= (R_n - E_n - \text{IP}_n)t_m - D_n t_m \\
&= (R_n - E_n)t_m - (\text{IP}_n + D_n)t_m.
\end{aligned}
\tag{10.2}
$$

The term $(IP_n + D_n)t_m$ is known as the tax shield (or tax savings) from financing and asset depreciation. Now substituting the result of Eq. (10.2) into Eq. (10.1), we obtain

$$A_n = (R_n - E_n - IP_n)(1 - t_m) + t_mD_n$$
$$- I_n + (S_n - G_n) - W_n$$
$$+ B_n - PP_n. \tag{10.3}$$

10.4.2 Presenting Cash Flows in Compact Tabular Formats

After-tax cash flow components over the project life can be grouped by type of activity in a compact tabular format as follows:

	End of Period			
Cash Flow Elements	**0**	**1**	**2**	**...N**
Investment activities:				
$-I_n$				
$+S_n - G_n$				
$-W_n$				
Operating activities:				
$+(1 - t_m)(R_n)$				
$-(1 - t_m)(E_n)$				
$-(1 - t_m)(IP_n)$				
$+t_mD_n$				
Financing activities:				
$+B_n$				
$-PP_n$				
Net cash flow				
A_n				

However, in preparing their after-tax cash flow, most business firms adopt the income statement approach presented in previous sections, because they want to know the accounting income along with the cash flow statement.

EXAMPLE 10.7 Generalized Cash Flow Approach

Consider again Example 10.4. Use the generalized cash flow approach to obtain the after-tax cash flows:

SOLUTION

Given:

Investment cost $(I_0) = \$125,000,$

Investment in working capital (W_n) = $23,331,

Annual revenues (R_n) = $100,000,

Annual expenses other than depreciation and debt interest (E_n) = $40,000,

Debt interest (IP_n), years 1 to 5 = $6,250, $5,226, $4,100, $2,861, $1,499, respectively,

Principal repayment (PP_n), years 1 to 5 = $10,237, $11,261, $12,387, $13,626, $14,988, respectively,

Depreciation (D_n), years 1 to 5 = $17,863, $30,613, $21,863, $15,613, $5,581, respectively,

Marginal tax rate (t_m) = 40%, and

Salvage value (S_n) = $50,000.

Find: Annual after-tax cash flows (A_n).

Step 1: Find the cash flow at year 0:

1. Investment in depreciable asset (I_0) = $-$125,000.
2. Investment in working capital (W_0) = $-$23,331.
3. Borrowed funds (PP_0) = $62,500.
4. Net cash flow (A_0) = $-$125,000 $-$ $23,331 + $62,500 = $-$85,831.

Step 2: Find the cash flow in years 1 to 4:

$$A_n = (R_n - E_n - IP_n)(1 - t_m) + D_n t_m - PP_n.$$

n	Net Operating Cash Flow ($)
1	$(100,000 - 40,000 - 6,250)(0.60) + 17,863(0.40) - 10,237 = \$29,158$
2	$(100,000 - 40,000 - 5,226)(0.60) + 30,613(0.40) - 11,261 = \$33,849$
3	$(100,000 - 40,000 - 4,100)(0.60) + 21,863(0.40) - 12,387 = \$29,898$
4	$(100,000 - 40,000 - 2,861)(0.60) + 15,613(0.40) - 13,626 = \$26,903$

Step 3: Find the cash flow for year 5:

1. Operating cash flow:
   $(100,000 - 40,000 - 1,499)(0.60) + 5,581(0.40) - 14,988 = \$22,345.$

2. Net salvage value
   (as calculated in Example 10.1) = $50,000 $-$ $6,613 = $43,387.

3. Recovery of working capital, W_5 = $23,331.

4. Net cash flow in year 5, A_5 = $22,345 + $43,387 + $23,331 = $89,063.

TABLE 10.7	Cash Flow Statement for Example 10.4 Using the Generalized Cash Flow Approach (Example 10.8)					
	0	**1**	**2**	**3**	**4**	**5**
Investment	$(125,000)					
Net proceeds from sale						$43,387
Investment in working capital	(23,331)					
Recovery of working capital						23,331
(1 − 0.40) (Revenue)		$ 60,000	$ 60,000	$ 60,000	$ 60,000	$ 60,000
−(1 − 0.40) (Expenses)		(24,000)	(24,000)	(24,000)	(24,000)	(24,000)
−(1 − 0.40) (Debt interest)		(3,750)	(3,136)	(2,460)	(1,717)	(899)
+(0.40) (Depreciation)		7,145	12,245	8,745	6,245	2,232
Borrowed funds	62,500					
Repayment of principal		(10,237)	(11,261)	(12,387)	(13,626)	(14,988)
Net cash flow	$ (85,831)	$ 29,158	$ 33,849	$ 29,898	$ 26,903	$ 89,063

Our results and overall calculations are summarized in Table 10.7. Checking them against the results we obtained in Table 10.4 confirms them.

10.4.3 Lease-or-Buy Decision

A lease-or-buy decision begins only after a company has decided that the acquisition of a piece of equipment is necessary to carry out an investment project. Having made this decision, the company may be faced with several alternative methods of financing the acquisition: cash purchase, debt purchase, or acquisition via a lease.

With a debt purchase, the present-worth expression is similar to that for a cash purchase, except that it has additional items: loan repayments and a tax shield on interest payments. The only way that the lessee can evaluate the cost of a lease is to compare it against the best available estimate of what the cost would be if the lessee owned the equipment.

To lay the groundwork for a more general analysis, we shall first consider how to analyze the lease-or-buy decision for a project with a single fixed asset for which the company expects a service life of N periods. Since the net after-tax revenue is the same for both alternatives, we need only consider the incremental cost of owning the asset and the incremental cost of leasing. Using the generalized cash flow approach, we may express the incremental cost of owning the asset by borrowing as

$$PW(i)_{Buy} = + \text{PW of loan repayment}$$
$$+ \text{PW of after-tax O\&M costs}$$
$$- \text{PW of tax credit on depreciation and interest}$$
$$- \text{PW of net proceeds from sale.} \tag{10.4}$$

Note that the acquisition (investment) cost of the asset is offset by the same amount of borrowing at time 0, so that we need to consider only the loan repayment series.

Suppose that the firm can lease the asset at a constant amount per period. Then the project's incremental cost of leasing becomes

$$PW(i)_{Lease} = \text{PW of after-tax lease expenses.}$$

If the lease does not provide for the maintenance of the equipment leased, then the lessee must assume this responsibility. In that situation, the maintenance term in Eq. (10.4) can be dropped in the calculation of the incremental cost of owning the asset.

The criterion for the decision to lease as opposed to purchasing the asset thus reduces to a comparison between $PW(i)_{Buy}$ and $PW(i)_{Lease}$. Purchasing is preferred if the combined present value of the loan repayment series and the after-tax O&M expense, reduced by the present values of the depreciation tax shield and the net proceeds from disposal of the asset, is less than the present value of the net lease costs.

EXAMPLE 10.8 Lease-or-Buy Decision

The Dallas Electronics Company is considering replacing an old, 1,000-pound-capacity industrial forklift truck. The truck has been used primarily to move goods from production machines into storage. The company is working nearly at capacity and is operating on a two-shift basis six days per week. Dallas management is considering owning or leasing the new truck. The plant engineer has compiled the following data for management:

- The initial cost of a gas-powered truck is $20,000. The new truck would use about 8 gallons of gasoline (in a single shift of 8 hours per day) at a cost of $2.35 per gallon. If the truck is operated 16 hours per day, its expected life will be four years, and an engine overhaul at a cost of $1,500 will be required at the end of two years.

- The Austin Industrial Truck Company was servicing the old forklift truck, and Dallas would buy the new truck through Austin, which offers a service agreement to users of its trucks. The agreement costs $120 per month and provides for a monthly visit by an experienced service representative to lubricate and tune the truck. Insurance and property taxes for the truck are $650 per year.

- The truck is classified as five-year property under MACRS and will be depreciated accordingly. Dallas is in the 40% marginal tax bracket; the estimated resale value of the truck at the end of four years will be 15% of the original cost.

- Austin also has offered to lease a truck to Dallas. Austin will maintain the equipment and will guarantee to keep the truck in serviceable condition at all times. In the event of a major breakdown, Austin will provide a replacement truck at its expense. The cost of the operating lease plan is $850 per month. The contract term is three years at a minimum, with the option to cancel on 30 days' notice thereafter.

- On the basis of recent experience, the company expects that funds committed to new investments should earn at least a 12% rate of return after taxes.

Compare the cost of owning versus leasing the truck.

DISCUSSION: We may calculate the fuel costs for two-shift operation as

$$(8 \text{ gallons/shift})(2 \text{ shifts/day})(\$2.35/\text{gallon}) = \$37.60 \text{ per day.}$$

The truck will operate 300 days per year, so the annual fuel cost will be $11,280. However, both alternatives require the company to supply its own fuel, so the fuel cost is not relevant to our decision making. Therefore, we may drop this common cost item from our calculations.

SOLUTION

Given: Preceding cost information, MARR = 12%, and marginal tax rate = 40%.
Find: Incremental cost of owning versus leasing the truck.

(a) Incremental cost of owning the truck.

To compare the incremental cost of ownership with the incremental cost of leasing, we make the following additional estimates and assumptions:

- **Step 1.** The preventive-maintenance contract, which costs $120 per month (or $1,440 per year), will be adopted. With annual insurance and taxes of $650, the equivalent present worth of the after-tax O&M is

$$P_1 = (\$1,440 + \$650)(1 - 0.40)(P/A, 12\%, 4) = \$3,809.$$

- **Step 2.** The engine overhaul is not expected to increase either the salvage value or the service life of the truck. Therefore, the overhaul cost ($1,500) will be expensed all at once rather than capitalized. The equivalent present worth of this after-tax overhaul expense is

$$P_2 = \$1,500(1 - 0.40)(P/F, 12\%, 2) = \$717.$$

- **Step 3.** If Dallas decided to purchase the truck through debt financing, the first step in determining financing costs would be to compute the annual installments of the debt repayment schedule. Assuming that the entire investment of $20,000 is financed at a 10% interest rate, the annual payment would be

$$A = \$20,000(A/P, 10\%, 4) = \$6,309.$$

The equivalent present worth of this loan payment series is

$$P_3 = \$6,309 (P/A, 12\%, 4) = \$19,163.$$

- **Step 4.** The interest payment each year (10% of the beginning balance) is calculated as follows:

Year	Beginning Balance	Interest Charged	Annual Payment	Ending Balance
1	$20,000	$2,000	−$6,309	$15,691
2	15,691	1,569	−6,309	10,951
3	10,951	1,095	−6,309	5,737
4	5,737	573	−6,309	0

With the five-year MACRS depreciation schedule, the combined tax savings due to depreciation expenses and interest payments can be calculated as follows:

n	D_n	I_n	Combined Tax Savings
1	$4,000	$2,000	$6,000(0.40) = $2,400
2	6,400	1,569	7,969(0.40) = 3,188
3	3,840	1,095	4,935(0.40) = 1,974
4	1,152	573	1,725(0.40) = 690

Therefore, the equivalent present worth of the combined tax credit is

$$P_4 = \$2,400(P/F, 12\%, 1) + \$3,188(P/F, 12\%, 2)$$
$$+ \$1.974(P/F, 12\%, 3) + \$690(P/F, 12\%, 4)$$
$$= \$6,527.$$

- **Step 5.** With the estimated salvage value of 15% of the initial investment ($3,000) and with the five-year MACRS depreciation schedule given, we compute the net proceeds from the sale of the truck at the end of four years as follows:

Book value $= \$20,000 - \$15,392 = \$4,608,$

Gains (losses) $= \$3,000 - \$4,608 = (\$1,608),$

Tax savings $= 0.40(\$1,608) = \$643,$

Net proceeds from sale $= \$3,000 + \$643 = \$3,643.$

The present equivalent amount of the net salvage value is

$$P_5 = \$3,643(P/F, 12\%, 4) = \$2,315.$$

- **Step 6.** Therefore, the net incremental cost of owning the truck through 100% debt financing is

$$PW(12\%)_{Buy} = P_1 + P_2 + P_3 - P_4 - P_5$$
$$= \$14,847.$$

(b) Incremental cost of leasing the truck.

How does the cost of acquiring a forklift truck under the lease compare with the cost of owning the truck?

- **Step 1.** The lease payment is also a tax-deductible expense; however, the net cost of leasing has to be computed explicitly on an after-tax basis. The calculation of the annual incremental leasing costs is as follows:

Annual lease payments (12 months) $= \$10,200,$

Less 40% taxes $= 4,080,$

Annual net costs after taxes $= \$6,120.$

- **Step 2.** The total net present equivalent incremental cost of leasing is

$$PW(12\%)_{Lease} = \$6,120(P/A, 12\%, 4)$$
$$= \$18,589.$$

Purchasing the truck with debt financing would save Dallas $3,742 in NPW.

- **Step 3.** Here, we have assumed that the lease payments occur at the end of the period; however, many leasing contracts require the payments to be made at the beginning of each period. In the latter situation, we can easily modify the present-worth expression of the lease expense to reflect this cash-flow timing:

$$PW(12\%)_{Lease} = \$6,120 + \$6,120(P/A, 12\%, 3)$$
$$= \$20,819.$$

COMMENTS: In our example, leasing the truck appears to be more expensive than purchasing the truck with debt financing. You should not conclude, however, that leasing is *always* more expensive than owning. In many cases, analysis favors a lease option. The interest rate, salvage value, lease payment schedule, and debt financing all have an important effect on decision making.

SUMMARY

■ Identifying and estimating relevant project cash flows is perhaps the most challenging aspect of engineering economic analysis. All cash flows can be organized into one of the following three categories:

1. Operating activities.
2. Investing activities.
3. Financing activities.

■ The following types of cash flow are the most common flows a project may generate:

1. New investment and disposal of existing assets.
2. Salvage value (or net selling price).
3. Working capital.
4. Working-capital release.
5. Cash revenues or savings.
6. Manufacturing, operating, and maintenance costs.
7. Interest and loan payments.
8. Taxes and tax credits.

In addition, although not cash flows, the following elements may exist in a project analysis, and they must be accounted for:

1. Depreciation expenses.
2. Amortization expenses.

Table 10.1 summarizes these elements and organizes them as investment, financing, or operating elements.

■ The **income statement approach** is typically used in organizing project cash flows. This approach groups cash flows according to whether they are operating, investing, or financing functions.

■ The **generalized cash flow approach** (shown in Table 10.7) to organizing cash flows can be used when a project does not change a company's marginal tax rate. The cash flows can be generated more quickly and the formatting of the results is less elaborate than with the income statement approach. There are also analytical advantages, which we will discover in later chapters. However, the generalized approach is less intuitive and not commonly understood by businesspeople.

PROBLEMS

Generating Net Cash Flows

10.1 You are considering a luxury apartment building project that requires an investment of $12,500,000. The building has 50 units. You expect the maintenance cost for the apartment building to be $250,000 the first year and $300,000 the second year, after which it will continue to increase by $50,000 in subsequent years. The cost to hire a manager for the building is estimated to be $80,000 per year. After five years of operation, the apartment building can be sold for $14,000,000. What is the annual rent per apartment unit that will provide a return on investment of 15% after tax? Assume that the building will remain fully occupied during the five years. Assume also that your tax rate is 35%. The building will be depreciated according to 39-year MACRS and will be placed in service in January during the first year of ownership and sold in December during the fifth year of ownership.

10.2 An automobile-manufacturing company is considering purchasing an industrial robot to do spot welding, which is currently done by skilled labor. The initial cost of the robot is $185,000, and the annual labor savings are projected to be $120,000. If purchased, the robot will be depreciated under MACRS as a five-year recovery property. The robot will be used for seven years, at the end of which time the firm expects to sell it for $40,000. The company's marginal tax rate is 35% over the project period. Determine the net after-tax cash flows for each period over the project life.

10.3 A firm is considering purchasing a machine that costs $55,000. It will be used for six years, and the salvage value at that time is expected to be zero. The machine will save $25,000 per year in labor, but it will incur $7,000 operating and maintenance costs each year. The machine will be depreciated according to five-year MACRS. The firm's tax rate is 40% and its after-tax MARR is 15%. Should the machine be bought?

10.4 A Los Angeles company is planning to market an answering device for people working alone who want the prestige that comes with having a secretary, but who cannot afford one. The device, called Tele-Receptionist, is similar to a voice-mail system. It uses digital recording technology to create the illusion that a person is operating the switchboard at a busy office. The company purchased a 40,000-ft^2

building and converted it to an assembly plant for $600,000 ($100,000 worth of land and $500,000 worth of building). Installation of the assembly equipment worth $500,000 was completed on December 31. The plant will begin operation on January 1. The company expects to have a gross annual income of $2,500,000 over the next 5 years. Annual manufacturing costs and all other operating expenses (excluding depreciation) are projected to be $1,280,000. For depreciation purposes, the assembly plant building will be classified as 39-year real property and the assembly equipment as a 7-year MACRS property. The property value of the land and the building at the end of year 5 would appreciate as much as 15% over the initial purchase cost. The residual value of the assembly equipment is estimated to be about $50,000 at the end of year 5. The firm's marginal tax rate is expected to be about 40% over the project period. Determine the project's after-tax cash flows over the period of 5 years.

10.5 A highway contractor is considering buying a new trench excavator that costs $200,000 and can dig a 3-foot-wide trench at the rate of 16 feet per hour. With the machine adequately maintained, its production rate will remain constant for the first 1,200 hours of operation and then decrease by 2 feet per hour for each additional 400 hours thereafter. The expected average annual use is 400 hours, and maintenance and operating costs will be $15 per hour. The contractor will depreciate the equipment in accordance with a five-year MACRS. At the end of five years, the excavator will be sold for $40,000. Assuming that the contractor's marginal tax rate is 34% per year, determine the annual after-tax cash flow.

10.6 A small children's clothing manufacturer is considering an investment to computerize its management information system for material requirement planning, piece-goods coupon printing, and invoice and payroll. An outside consultant has been retained to estimate the initial hardware requirement and installation costs. He suggests the following:

PC systems	
(15 PCs, 4 printers)	$85,000
Local area networking system	15,000
System installation and testing	4,000

The expected life of the computer system is five years, with no expected salvage value. The proposed system is classified as a five-year property under the MACRS depreciation system. A group of computer consultants needs to be hired to develop various customized software packages to run on the system. Software development costs will be $20,000 and can be expensed during the first tax year. The new system will eliminate two clerks, whose combined annual payroll expenses are $52,000. Additional annual expenses to run this computerized system are expected to be $12,000. Borrowing is not considered an option for this investment, nor is a tax credit available for the system. The firm's expected marginal tax rate over the next six years will be 35%. The firm's interest rate is 13%. Compute the after-tax cash flows over the life of the investment.

10.7 A firm has been paying a print shop $18,000 annually to print the company's monthly newsletter. The agreement with this print shop has now expired, but it could be renewed for a further five years. The new subcontracting charges are expected to be 12% higher than they were under the previous contract. The company is also considering the purchase of a desktop publishing system with a high-quality laser printer driven by a microcomputer. With appropriate text and graphics software, the newsletter can be composed and printed in near-typeset quality. A special device is also required to print photos in the newsletter. The following estimates have been quoted by a computer vendor:

Personal computer	$4,500
Color laser printer	6,500
Photo device/scanner	5,000
Software	2,500
Total cost basis	$18,500
Annual O&M costs	10,000

The salvage value of each piece of equipment at the end of five years is expected to be only 10% of the original cost. The company's marginal tax rate is 40%, and the desktop publishing system will be depreciated by MACRS under its five-year property class.

(a) Determine the projected net after-tax cash flows for the investment.

(b) Compute the IRR for this project.

(c) Is the project justifiable at MARR $= 12\%$?

10.8 An asset in the five-year MACRS property class costs $120,000 and has a zero estimated salvage value after six years of use. The asset will generate annual revenues of $300,000 and will require $80,000 in annual labor and $50,000 in annual material expenses. There are no other revenues and expenses. Assume a tax rate of 40%.

(a) Compute the after-tax cash flows over the project life.

(b) Compute the NPW at MARR $= 12\%$. Is the investment acceptable?

10.9 An automaker is considering installing a three-dimensional (3-D) computerized car-styling system at a cost of $200,000 (including hardware and software). With the 3-D computer modeling system, designers will have the ability to view their design from many angles and to fully account for the space required for the engine and passengers. The digital information used to create the computer model can be revised in consultation with engineers, and the data can be used to run milling machines that make physical models quickly and precisely. The automaker expects to decrease the turnaround time for designing a new automobile model (from configuration to final design) by 22%. The expected savings in dollars is $250,000 per year. The training and operating maintenance cost for the new system is expected to be $50,000 per year. The system has a five-year useful life and can be depreciated according to five-year MACRS class. The system will have an estimated salvage value of $5,000. The automaker's marginal tax rate is 40%. Determine the annual cash flows for this investment. What is the return on investment for the project?

10.10 A facilities engineer is considering a $50,000 investment in an energy management system (EMS). The system is expected to save $10,000 annually in utility bills for N years. After N years, the EMS will have a zero salvage value. In an after-tax analysis, what would N need to be in order for the investment to earn a 10% return? Assume MACRS depreciation with a three-year class life and a 35% tax rate.

10.11 A corporation is considering purchasing a machine that will save $130,000 per year before taxes. The cost of operating the machine, including maintenance, is $20,000 per year. The machine will be needed for five years, after which it will have a zero salvage value. MACRS depreciation will be used, assuming a three-year class life. The marginal income tax rate is 40%. If the firm wants 12% IRR after taxes, how much can it afford to pay for this machine?

Investment in Working Capital

10.12 The Doraville Machinery Company is planning to expand its current spindle product line. The required machinery would cost $500,000. The building that will house the new production facility would cost $1.5 million. The land would cost $250,000, and $150,000 working capital would be required. The product is expected to result in additional sales of $675,000 per year for 10 years, at which time the land can be sold for $500,000, the building for $700,000, and the equipment for $50,000. All of the working capital will be recovered. The annual disbursements for labor, materials, and all other expenses are estimated to be $425,000. The firm's income tax rate is 40%, and any capital gains will be taxed at 35%. The building will be depreciated according to a 39-year property class. The manufacturing facility will be classified as a 7-year MACRS. The firm's MARR is known to be 15% after taxes.

(a) Determine the projected net after-tax cash flows from this investment. Is the expansion justified?

(b) Compare the IRR of this project with that of a situation with no working capital.

10.13 An industrial engineer proposed the purchase of scanning equipment for the company's warehouse and weave rooms. The engineer felt that the purchase would provide a better system of locating cartons in the warehouse by recording the locations of the cartons and storing the data in the computer.

The estimated investment, annual operating and maintenance costs, and expected annual savings are as follows:

- Cost of equipment and installation: $65,500.
- Project life: 6 years.
- Expected salvage value: $3,000.
- Investment in working capital (fully recoverable at the end of the project life): $10,000.
- Expected annual savings on labor and materials: $55,800.
- Expected annual expenses: $8,120.
- Depreciation method: 5-year MACRS.

The firm's marginal tax rate is 35%.

(a) Determine the net after-tax cash flows over the project life.

(b) Compute the IRR for this investment.

(c) At MARR $=$ 18%, is the project acceptable?

10.14 Delaware Chemical Corporation is considering investing in a new composite material. R&D engineers are investigating exotic metal–ceramic and ceramic–ceramic composites to develop materials that will withstand high temperatures, such as those to be encountered in the next generation of jet fighter engines. The company expects a 3-year R&D period before these new materials can be applied to commercial products.

The following financial information is presented for management review:

- **R&D cost.** $5 million over a 3-year period. Annual R&D expenditure of $0.5 million at the beginning of year 1, $2.5 million at the beginning of year 2, and $2 million at the beginning of year 3. For tax purposes, these R&D expenditures will be expensed rather than amortized.

- **Capital investment.** $5 million at the beginning of year 4. This investment consists of $2 million in a building and $3 million in plant equipment. The company already owns a piece of land as the building site.

- **Depreciation method.** The building (39-year real property class with the asset placed in service in January) and plant equipment (7-year MACRS recovery class).

- **Project life.** 10 years after a 3-year R&D period.

- **Salvage value.** 10% of the initial capital investment for the equipment and 50% for the building (at the end of the project life).

- **Total sales.** $50 million (at the end of year 4), with an annual sales growth rate of 10% per year (compound growth) during the next 5 years (year 5 through year 9) and -10% (negative compound growth) per year for the remaining project life.

- **Out-of-pocket expenditures.** 80% of annual sales.

- **Working capital.** 10% of annual sales (considered as an investment at the beginning of each production year and investments fully recovered at the end of the project life).

- **Marginal tax rate.** 40%.

   (a) Determine the net after-tax cash flows over the project life.

   (b) Determine the IRR for this investment.

   (c) Determine the equivalent annual worth for the investment at MARR $=$ 20%.

Effects of Borrowing

10.15 Refer to the data in Problem 10.1. If the firm expects to borrow the initial investment ($12,500,000) at 10% over five years (paying back the loan in equal annual payments of $3,297,469), determine the project's net cash flows.

10.16 In Problem 10.2, to finance the industrial robot, the company will borrow the entire amount from a local bank, and the loan will be paid off at the rate of $37,000 per year, plus 10% on the unpaid balance. Determine the net after-tax cash flows over the project life.

10.17 Refer to the financial data in Problem 10.9. Suppose that 50% of the initial investment of $200,000 will be borrowed from a local bank at an interest rate of 11% over five years (to be paid off in five equal annual payments). Recompute the after-tax cash flow.

10.18 A special-purpose machine tool set would cost $20,000. The tool set will be financed by a $10,000 bank loan repayable in two equal annual installments at 10% compounded annually. The tool is expected to provide annual (material) savings of $30,000 for two years and is to be depreciated by the MACRS three-year recovery period. The tool will require annual O&M costs in the amount of $5,000. The salvage value at the end of the two years is expected to be $8,000. Assuming a marginal tax rate of 40% and MARR of 15%, what is the net present worth of this project? You may use the following worksheet in your calculation:

Cash Flow Statement	0	1	2
Operating activities			
Net income		10,400	12,019
Depreciation		6,666	4,445
Investment activities			
Investment	−20,000		
Salvage			8,000
Gains tax (40%)			
Financial activities			
Borrowed funds	10,000		
Principal repayment	0		
Net cash flow	−$10,000		

10.19 A.M.I. Company is considering installing a new process machine for the firm's manufacturing facility. The machine costs $200,000 installed, will generate additional revenues of $80,000 per year, and will save $55,000 per year in labor and material costs. The machine will be financed by a $150,000 bank loan repayable in three equal annual principal installments, plus 9% interest on the outstanding balance. The machine will be depreciated using 7-year MACRS. The useful life of the machine is 10 years, after which it will be sold for $20,000. The combined marginal tax rate is 40%.

(a) Find the year-by-year after-tax cash flow for the project.

(b) Compute the IRR for this investment.

(c) At MARR = 18%, is the project economically justifiable?

10.20 Consider the following financial information about a retooling project at a computer manufacturing company:

• The project costs $2 million and has a five-year service life.

• The retooling project can be classified as seven-year property under the MACRS rule.

- At the end of the fifth year, any assets held for the project will be sold. The expected salvage value will be about 10% of the initial project cost.
- The firm will finance 40% of the project money from an outside financial institution at an interest rate of 10%. The firm is required to repay the loan with five equal annual payments.
- The firm's incremental (marginal) tax rate on the investment is 35%.
- The firm's MARR is 18%.
- With the preceding financial information,

    (a) Determine the after-tax cash flows.

    (b) Compute the annual equivalent worth for this project.

10.21 A fully automatic chucker and bar machine is to be purchased for $35,000, to be borrowed with the stipulation that it be repaid with six equal end-of-year payments at 12% compounded annually. The machine is expected to provide an annual revenue of $10,000 for six years and is to be depreciated by the MACRS seven-year recovery period. The salvage value at the end of six years is expected to be $3,000. Assume a marginal tax rate of 36% and a MARR of 15%.

(a) Determine the after-tax cash flow for this asset over six years.

(b) Determine whether the project is acceptable on the basis of the IRR criterion.

10.22 A manufacturing company is considering acquiring a new injection molding machine at a cost of $100,000. Because of a rapid change in product mix, the need for this particular machine is expected to last only eight years, after which time the machine is expected to have a salvage value of $10,000. The annual operating cost is estimated to be $5,000. The addition of the machine to the current production facility is expected to generate an annual revenue of $40,000. The firm has only $60,000 available from its equity funds, so it must borrow the additional $40,000 required at an interest rate of 10% per year, with repayment of principal and interest in eight equal annual amounts. The applicable marginal income tax rate for the firm is 40%. Assume that the asset qualifies for a seven-year MACRS property class.

(a) Determine the after-tax cash flows.

(b) Determine the NPW of this project at MARR $= 14\%$.

Generalized Cash Flow Method

10.23 Suppose an asset has a first cost of $6,000, a life of five years, a salvage value of $2,000 at the end of five years, and a net annual before-tax revenue of $1,500. The firm's marginal tax rate is 35%. The asset will be depreciated by three-year MACRS.

(a) Using the generalized cash flow approach, determine the cash flow after taxes.

(b) Rework part (a), assuming that the entire investment would be financed by a bank loan at an interest rate of 9%.

(c) Given a choice between the financing methods of parts (a) and (b), show calculations to justify your choice of which is the better one at an interest rate of 9%.

10.24 A construction company is considering acquiring a new earthmover. The purchase price is $100,000, and an additional $25,000 is required to modify the equipment for special use by the company. The equipment falls into the MACRS seven-year classification (the tax life), and it will be sold after five years (the project life) for $50,000. The purchase of the earthmover will have no effect on revenues, but the machine is expected to save the firm $60,000 per year in before-tax operating costs, mainly labor. The firm's marginal tax rate is 40%. Assume that the initial investment is to be financed by a bank loan at an interest rate of 10%, payable annually. Determine the after-tax cash flows by using the generalized cash flow approach and the worth of the investment for this project if the firm's MARR is known to be 12%.

10.25 Air South, a leading regional airline that is now carrying 54% of all the passengers that pass through the Southeast, is considering adding a new long-range aircraft to its fleet. The aircraft being considered for purchase is the McDonnell Douglas DC-9-532 "Funjet," which is quoted at $60 million per unit. McDonnell Douglas requires a 10% down payment at the time of delivery, and the balance is to be paid over a 10-year period at an interest rate of 12% compounded annually. The actual payment schedule calls for only interest payments over the 10-year period, with the original principal amount to be paid off at the end of the 10th year. Air South expects to generate $35 million per year by adding this aircraft to its current fleet, but also estimates an operating and maintenance cost of $20 million per year. The aircraft is expected to have a 15-year service life with a salvage value of 15% of the original purchase price. If the aircraft is bought, it will be depreciated by the 7-year MACRS property classifications. The firm's combined federal and state marginal tax rate is 38%, and its required minimum attractive rate of return is 18%.

(a) Use the generalized cash flow approach to determine the cash flow associated with the debt financing.
(b) Is this project acceptable?

Comparing Mutually Exclusive Alternatives

10.26 The Pittsburgh Division of Vermont Machinery, Inc., manufactures drill bits. One of the production processes of a drill bit requires tipping, whereby carbide tips are inserted into the bit to make it stronger and more durable. The tipping process usually requires four or five operators, depending on the weekly workload. The same operators are assigned to the stamping operation, in which the size of the drill bit and the company's logo are imprinted into the bit. Vermont is considering acquiring three automatic tipping machines to replace the manual tipping and stamping operations. If the tipping process is automated, Vermont engineers will have to redesign the shapes of the carbide tips to be used in the machines. The new design requires less carbide, resulting in a savings of material. The following financial data have been compiled:

- Project life: six years.
- Expected annual savings: reduced labor, $56,000; reduced material, $75,000; other benefits (reduction in carpal tunnel syndrome and related problems), $28,000; reduced overhead, $15,000.
- Expected annual O&M costs: $22,000.
- Tipping machines and site preparation: equipment costs (three machines), including delivery, $180,000; site preparation, $20,000.
- Salvage value: $30,000 (three machines) at the end of six years.
- Depreciation method: seven-year MACRS.

- Investment in working capital: $25,000 at the beginning of the project year, and that same amount will be fully recovered at the end of the project year.
- Other accounting data: marginal tax rate of 39% and MARR of 18%.

To raise $200,000, Vermont is considering the following financing options:

- **Option 1.** Use the retained earnings of the tipping machines to finance them.
- **Option 2.** Secure a 12% term loan over six years (to be paid off in six equal annual installments).
- **Option 3.** Lease the tipping machines. Vermont can obtain a six-year financial lease on the equipment (with, however, no maintenance service) for payments of $55,000 at the *beginning* of each year.

  (a) Determine the net after-tax cash flows for each financing option.
  (b) What is Vermont's present-value cost of owning the equipment by borrowing?
  (c) What is Vermont's present-value cost of leasing the equipment?
  (d) Recommend the best course of action for Vermont.

10.27 The headquarters building owned by a rapidly growing company is not large enough for the company's current needs. A search for larger quarters revealed two new alternatives that would provide sufficient room, enough parking, and the desired appearance and location. The company now has three options:

- **Option 1.** Lease the new quarters for $144,000 per year.
- **Option 2.** Purchase the new quarters for $800,000, including a $150,000 cost for land.
- **Option 3.** Remodel the current headquarters building.

It is believed that land values will not decrease over the ownership period, but the value of all structures will decline to 10% of the purchase price in 30 years. Annual property tax payments are expected to be 5% of the purchase price. The present headquarters building is already paid for and is now valued at $300,000. The land it is on is appraised at $60,000. The structure can be remodeled at a cost of $300,000 to make it comparable to other alternatives. However, the remodeling will occupy part of the existing parking lot. An adjacent, privately owned parking lot can be leased for 30 years under an agreement that the first year's rental of $9,000 will increase by $500 each year. The annual property taxes on the remodeled property will again be 5% of the present valuation, plus the cost of remodeling. The new quarters are expected to have a service life of 30 years, and the desired rate of return on investments is 12%. Assume that the firm's marginal tax rate is 40% and that the new building and remodeled structure will be depreciated under MACRS, using a real-property recovery period of 39 years. If the annual upkeep costs are the same for all three alternatives, which one is preferable?

10.28 An international manufacturer of prepared food items needs 50,000,000 kWh of electrical energy a year, with a maximum demand of 10,000 kW. The local utility currently charges $0.085 per kWh, a rate considered high throughout the industry. Because the firm's power consumption is so large, its engineers are considering installing a 10,000-kW steam-turbine plant. Three types of plant have been proposed (units in thousands of dollars):

	Plant A	Plant B	Plant C
Average station heat rate (BTU/kWh)	16,500	14,500	13,000
Total investment (boiler/turbine/electrical/structures)	$8,530	$9,498	$10,546
Annual operating cost:			
Fuel	1,128	930	828
Labor	616	616	616
O&M	150	126	114
Supplies	60	60	60
Insurance and property taxes	10	12	14

The service life of each plant is expected to be 20 years. The plant investment will be subject to a 20-year MACRS property classification. The expected salvage value of the plant at the end of its useful life is about 10% of its original investment. The firm's MARR is known to be 12%. The firm's marginal income tax rate is 39%.

(a) Determine the unit power cost ($/kWh) for each plant.

(b) Which plant would provide the most economical power?

Lease-versus-Buy Decisions

10.29 The Jacob Company needs to acquire a new lift truck for transporting its final product to the warehouse. One alternative is to purchase the truck for $40,000, which will be financed by the bank at an interest rate of 12%. The loan must be repaid in four equal installments, payable at the end of each year. Under the borrow-to-purchase arrangement, Jacob would have to maintain the truck at a cost of $1,200, payable at year-end. Alternatively, Jacob could lease the truck under a four-year contract for a lease payment of $11,000 per year. Each annual lease payment must be made at the beginning of each year. The truck would be maintained by the lessor. The truck falls into the five-year MACRS classification, and it has a salvage value of $10,000, which is the expected market value after four years, at which time Jacob plans to replace the truck irrespective of whether it leases or buys. Jacob has a marginal tax rate of 40% and a MARR of 15%.

(a) What is Jacob's cost of leasing, in present worth?

(b) What is Jacob's cost of owning, in present worth?

(c) Should the truck be leased or purchased?

This is an operating lease, so the truck would be maintained by the lessor.

10.30 Janet Wigandt, an electrical engineer for Instrument Control, Inc. (ICI), has been asked to perform a lease–buy analysis of a new pin-inserting machine for ICI's PC-board manufacturing.

- **Buy Option.** The equipment costs $120,000. To purchase it, ICI could obtain a term loan for the full amount at 10% interest, payable in four equal end-of-year

annual installments. The machine falls into a five-year MACRS property classification. Annual revenues of $200,000 and operating costs of $40,000 are
anticipated. The machine requires annual maintenance at a cost of $10,000.
Because technology is changing rapidly in pin-inserting machinery, the salvage value of the machine is expected to be only $20,000.

- **Lease Option.** Business Leasing, Inc. (BLI), is willing to write a four-year operating lease on the equipment for payments of $44,000 at the beginning of each
year. Under this arrangement, BLI will maintain the asset, so that the annual
maintenance cost of $10,000 will be saved. ICI's marginal tax rate is 40%, and
its MARR is 15% during the analysis period.

(a) What is ICI's present-value (incremental) cost of owning the equipment?
(b) What is ICI's present-value (incremental) cost of leasing the equipment?
(c) Should ICI buy or lease the equipment?

10.31 Consider the following lease-versus-borrow-and-purchase problem:

- **Borrow-and-purchase option:**

  1. Jensen Manufacturing Company plans to acquire sets of special industrial
  tools with a four-year life and a cost of $200,000, delivered and installed.
  The tools will be depreciated by the MACRS three-year classification.

  2. Jensen can borrow the required $200,000 at a rate of 10% over four years.
  Four equal end-of-year annual payments would be made in the amount of
  $63,094 = $200,000(A/P, 10%, 4). The annual interest and principal payment schedule, along with the equivalent present worth of these payments, is
  as follows:

End of Year	Interest	Principal
1	$20,000	$43,094
2	15,961	47,403
3	10,950	52,144
4	5,736	57,358

  3. The estimated salvage value for the tool sets at the end of four years is $20,000.

  4. If Jensen borrows and buys, it will have to bear the cost of maintenance,
  which will be performed by the tool manufacturer at a fixed contract rate of
  $10,000 per year.

- **Lease option:**

  1. Jensen can lease the tools for four years at an annual rental charge of $70,000,
  payable at the end of each year.

  2. The lease contract specifies that the lessor will maintain the tools at no additional charge to Jensen.

Jensen's tax rate is 40%. Any gains will also be taxed at 40%.

(a) What is Jensen's PW of after-tax cash flow of leasing at $i = 15\%$?
(b) What is Jensen's PW of after-tax cash flow of owning at $i = 15\%$?

10.32 Tom Hagstrom has decided to acquire a new car for his business. One alternative is to purchase the car outright for $16,170, financing the car with a bank loan for the net purchase price. The bank loan calls for 36 equal monthly payments of $541.72 at an interest rate of 12.6% compounded monthly. Payments must be made at the end of each month. The terms of each alternative are as follows:

Buy	Lease
$16,170	$425 per month
	36-month open-end lease
	Annual mileage allowed = 15,000 miles

If Tom takes the lease option, he is required to pay $500 for a security deposit, refundable at the end of the lease, and $425 a month at the beginning of each month for 36 months. If the car is purchased, it will be depreciated according to a five-year MACRS property classification. The car has a salvage value of $5,800, which is the expected market value after three years, at which time Tom plans to replace the car irrespective of whether he leases or buys. Tom's marginal tax rate is 35%. His MARR is known to be 13% per year.

(a) Determine the annual cash flows for each option.

(b) Which option is better?

10.33 The Boggs Machine Tool Company has decided to acquire a pressing machine. One alternative is to lease the machine under a three-year contract for a lease payment of $15,000 per year, with payments to be made at the beginning of each year. The lease would include maintenance. The second alternative is to purchase the machine outright for $100,000, financing the machine with a bank loan for the net purchase price and amortizing the loan over a three-year period at an interest rate of 12% per year (annual payment = $41,635).

Under the borrow-to-purchase arrangement, the company would have to maintain the machine at an annual cost of $5,000, payable at year-end. The machine falls into a five-year MACRS classification and has a salvage value of $50,000, which is the expected market value at the end of year 3, at which time the company plans to replace the machine irrespective of whether it leases or buys. Boggs has a tax rate of 40% and a MARR of 15%.

(a) What is Boggs' PW cost of leasing?

(b) What is Boggs' PW cost of owning?

(c) From the financing analysis in (a) and (b), what are the advantages and disadvantages of leasing and owning?

10.34 An asset is to be purchased for $25,000. The asset is expected to provide revenue of $10,000 a year and have operating costs of $2,500 a year. The asset is considered to be a seven-year MACRS property. The company is planning to sell the asset at the end of year 5 for $5,000. Given that the company's marginal tax rate is 30% and that it has a MARR of 10% for any project undertaken, answer the following questions:

(a) What is the net cash flow for each year, given that the asset is purchased with borrowed funds at an interest rate of 12%, with repayment in five equal end-of-year payments?

(b) What is the net cash flow for each year, given that the asset is leased at a rate of $3,500 a year (a financial lease)?

(c) Which method (if either) should be used to obtain the new asset?

10.35 Enterprise Capital Leasing Company is in the business of leasing tractors to construction companies. The firm wants to set a three-year lease payment schedule for a tractor purchased at $53,000 from the equipment manufacturer. The asset is classified as a five-year MACRS property. The tractor is expected to have a salvage value of $22,000 at the end of three years' rental. Enterprise will require a lessee to make a security deposit in the amount of $1,500 that is refundable at the end of the lease term. Enterprise's marginal tax rate is 35%. If Enterprise wants an after-tax return of 10%, what lease payment schedule should be set?

Short Case Studies

ST10.1 American Aluminum Company is considering making a major investment of $150 million ($5 million for land, $45 million for buildings, and $100 million for manufacturing equipment and facilities) to develop a stronger, lighter material, called aluminum lithium, that will make aircraft sturdier and more fuel efficient. Aluminum lithium, which has been sold commercially for only a few years as an alternative to composite materials, will likely be the material of choice for the next generation of commercial and military aircraft, because it is so much lighter than conventional aluminum alloys, which use a combination of copper, nickel, and magnesium to harden aluminum. Another advantage of aluminum lithium is that it is cheaper than composites. The firm predicts that aluminum lithium will account for about 5% of the structural weight of the average commercial aircraft within 5 years and 10% within 10 years. The proposed plant, which has an estimated service life of 12 years, would have a capacity of about 10 million pounds of aluminum lithium, although domestic consumption of the material is expected to be only 3 million pounds during the first 4 years, 5 million for the next 3 years, and 8 million for the remaining life of the plant. Aluminum lithium costs $12 a pound to produce, and the firm would expect to sell it at $17 a pound. The building will be depreciated according to the 39-year MACRS real property class, with the building placed in service July 1 of the first year. All manufacturing equipment and facilities will be classified as 7-year MACRS property. At the end of the project life, the land will be worth $8 million, the building $30 million, and the equipment $10 million. Assuming that the firm's marginal tax rate is 40% and its capital gains tax rate is 35%,

(a) Determine the net after-tax cash flows.

(b) Determine the IRR for this investment.

(c) Determine whether the project is acceptable if the firm's MARR is 15%.

ST10.2 Morgantown Mining Company is considering a new mining method at its Blacksville mine. The method, called longwall mining, is carried out by a robot. Coal is removed by the robot, not by tunneling like a worm through an apple, which leaves more of the target coal than is removed, but rather by methodically shuttling back and forth across the width of the deposit and devouring nearly everything. The method can extract about 75% of the available coal, compared with 50% for conventional mining, which is done largely with machines that dig tunnels. Moreover, the coal can be recovered far more inexpensively. Currently, at Blacksville alone, the company mines 5 million tons a year with 2,200 workers. By installing two longwall robot machines, the company can mine 5 million tons with only 860 workers. (A robot miner can dig more than 6 tons a minute.) Despite the loss of employment, the United Mine Workers union generally favors longwall mines, for two reasons: The union officials are quoted as saying, (1) "It would be far better to have highly productive operations that were able to pay our folks good wages and benefits than to have 2,200 shovelers living in poverty," and (2) "Longwall mines are inherently safer in their design." The company projects the following financial data upon installation of the longwall mining:

Robot installation (2 units)	$19.3 million
Total amount of coal deposit	50 million tons
Annual mining capacity	5 million tons
Project life	10 years
Estimated salvage value	$0.5 million
Working capital requirement	$2.5 million
Expected additional revenues:	
Labor savings	$6.5 million
Accident prevention	$0.5 million
Productivity gain	$2.5 million
Expected additional expenses:	
O&M costs	$2.4 million

(a) Estimate the firm's net after-tax cash flows over the project life if the firm uses the unit-production method to depreciate assets. The firm's marginal tax rate is 40%.

(b) Estimate the firm's net after-tax cash flows if the firm chooses to depreciate the robots on the basis of MACRS (seven-year property classification).

ST10.3 National Parts, Inc., an auto-parts manufacturer, is considering purchasing a rapid prototyping system to reduce prototyping time for form, fit, and function applications in automobile-parts manufacturing. An outside consultant has been called in to estimate the initial hardware requirement and installation costs. He suggests the following:

- Prototyping equipment: $187,000.
- Posturing apparatus: $10,000.

- Software: $15,000.
- Maintenance: $36,000 per year by the equipment manufacturer.
- Resin: Annual liquid polymer consumption of 400 gallons at $350 per gallon.
- Site preparation: Some facility changes are required for the installation of the rapid prototyping system (e.g., certain liquid resins contain a toxic substance, so the work area must be well vented).

The expected life of the system is six years, with an estimated salvage value of $30,000. The proposed system is classified as a five-year MACRS property. A group of computer consultants must be hired to develop customized software to run on the system. Software development costs will be $20,000 and can be expensed during the first tax year. The new system will reduce prototype development time by 75% and material waste (resin) by 25%. This reduction in development time and material waste will save the firm $114,000 and $35,000 annually, respectively. The firm's expected marginal tax rate over the next six years will be 40%. The firm's interest rate is 20%.

(a) Assuming that the entire initial investment will be financed from the firm's retained earnings (equity financing), determine the after-tax cash flows over the life of the investment. Compute the NPW of this investment.

(b) Assuming that the entire initial investment will be financed through a local bank at an interest rate of 13% compounded annually, determine the net after-tax cash flows for the project. Compute the NPW of the investment.

(c) Suppose that a financial lease is available for the prototype system at $62,560 per year, payable at the beginning of each year. Compute the NPW of the investment with lease financing.

(d) Select the best financing option, based on the rate of return on incremental investment.

ST10.4 National Office Automation, Inc. (NOAI), is a leading developer of imaging systems, controllers, and related accessories. The company's product line consists of systems for desktop publishing, automatic identification, advanced imaging, and office information markets. The firm's manufacturing plant in Ann Arbor, Michigan, consists of eight different functions: cable assembly, board assembly, mechanical assembly, controller integration, printer integration, production repair, customer repair, and shipping. The process to be considered is the transportation of pallets loaded with eight packaged desktop printers from printer integration to the shipping department. Several alternatives for minimizing operating and maintenance costs have been examined. The two most feasible alternatives are the following:

- **Option 1.** Use gas-powered lift trucks to transport pallets of packaged printers from printer integration to shipping. The truck also can be used to return printers that must be reworked. The trucks can be leased at a cost of $5,465 per year. With a maintenance contract costing $6,317 per year, the dealer will maintain the trucks. A fuel cost of $1,660 per year is expected. The truck requires a driver for each of the three shifts, at a total cost of $58,653 per year for labor. It is estimated that transportation by truck would cause damages to material and equipment totaling $10,000 per year.
- **Option 2.** Install an automatic guided vehicle system (AGVS) to transport pallets of packaged printers from printer integration to shipping and to return

products that require rework. The AGVS, using an electrical powered cart and embedded wire-guidance system, would do the same job that the truck currently does, but without drivers. The total investment costs, including installation, are itemized as follows:

Vehicle and system installation	$97,255
Staging conveyor	24,000
Power supply lines	5,000
Transformers	2,500
Floor surface repair	6,000
Batteries and charger	10,775
Shipping	6,500
Sales tax	6,970
Total AGVS system cost	$159,000

NOAI could obtain a term loan for the full investment amount ($159,000) at a 10% interest rate. The loan would be amortized over 5 years, with payments made at the end of each year. The AGVS falls into the 7-year MACRS classification, and it has an estimated service life of 10 years and no salvage value. If the AGVS is installed, a maintenance contract would be obtained at a cost of $20,000, payable at the beginning of each year. The firm's marginal tax rate is 35% and its MARR is 15%.

(a) Determine the net cash flows for each alternative over 10 years.
(b) Compute the incremental cash flows (Option 2 − Option 1), and determine the rate of return on this incremental investment.
(c) Determine the best course of action, based on the rate-of-return criterion.

Note: Assume a zero salvage value for the AGVS.

4

Handling Risk and Uncertainty

Inflation and Its Impact on Project Cash Flows

How Much Will It Cost to Send Your Child to College in the Year 2015? You may have heard your parents fondly remember the "good old days" of penny candy or 35-cents-per-gallon gasoline. But even a college student in his or her early twenties can relate to the phenomenon of escalating costs. Do you remember when a postage stamp cost a dime? When the price of a first-run movie was under $2? The accompanying table[1] demonstrates price differences between 2003 and 2005 for some commonly bought items. For example, unleaded gasoline cost $1.61 a gallon in 2003, whereas it cost $2.32 in 2005. Thus, in 2005, the same $1.61 bought only 0.69 gallon of the gasoline it would have bought in 2003. From 2003 to 2005, the $1.61 sum had lost 31% of its purchasing power!

On a brighter note, we open this chapter with some statistics about engineers' pay rates. The average starting salary of a first-year engineer was $7,500 in 1967. According to a recent survey by the National Society of Professional Engineers,[2] the average starting salary of a first-year engineer in 2005 was $51,509, which is equivalent to an annual increase of 5.20% over 38 years. During the same period, the general inflation rate remained at 4.73%, indicating that engineers' pay rates at least have kept up with the rate of inflation.

---

[1] *Source*: *The Wall Street Journal*, Tuesday, January 3, 2006 (Section R12, "Year-End Review of Markets and Finance, Consumer Purchases").

[2] *Source*: 2000 Income and Salary Survey by National Society of Professional Engineers.

Consumer Purchases	2003	2004	2005
Single-Family Home Median resale price	$170,000	$184,100	$215,900
Toyota Camry Sticker price, plus destination charge, for LE 5-speed	$19,560	$19,835	$20,125
Unleaded Gasoline Average national price per gallon for all grades combined, including all taxes, self-service	$1.61	$1.90	$2.32
Pair of Jeans Gap's Easy Fit, stonewashed, national price	$39.50	$39.50	$39.50
Internet Service Average monthly subscription for use of cable service from Comcast, standard tier	$42.95	$42.95	$42.95
Tax Preparation Average cost of federal, state, and local tax-return preparation by H&R Block	$130	$140	$151
McDonald's Big Mac Average recommended price	$2.22	$2.29	$2.39
Clearing Clogged Sink Roto-Rooter sewer and drain service, residential (nat'l avg.)	$189	$201	$212
Movie Ticket Adult ticket; first-run theater; evening, national average	$8.08	$8.39	$8.52
Airline Ticket Domestic roundtrip, based on a 2,000-mile trip, excluding aviation taxes	$246	$241	$235
Vacation One week for an adult at Club Med's Punta Cana resort, including airfare from New York	$1,162	$1,375	$1,315
Hospital Stay Average cost of one day in a semiprivate room, including ancillary services except private physician's fee (Cleveland)	$3,889	$4,416	$4,848
Birth Average hospital cost for mother and child, excluding private physician's fee (Cleveland)	$6,696	$7,187	$7,907
A Year in College In-state, including room and board and fees, undergraduate student at Penn State	$15,441	$16,862	$17,799
Funeral National average, excluding cemetery costs	$6,366	$6,530	$6,725

Up to this point, we have demonstrated how to develop cash flows in a variety of ways and how to compare them under constant conditions in the general economy. In other words, we have assumed that prices remain relatively unchanged over long periods. As you know from personal experience, that is not a realistic assumption. In this chapter, we define and quantify **inflation** and then go on to apply it in several economic analyses. We will demonstrate inflation's effect on depreciation, borrowed funds, the rate of return of a project, and working capital within the bigger picture of developing projects.

CHAPTER LEARNING OBJECTIVES

After completing this chapter, you should understand the following concepts:

- How to measure inflation.
- Conversion from actual dollars to constant dollars or from constant to actual dollars.
- How to compare the amount of dollars received at different points in time.
- Which inflation rate to use in economic analysis.
- Which interest rate to use in economic analysis (market interest rate versus inflation-free interest rate).
- How to handle multiple inflation rates in project analysis.
- The cost of borrowing and changes in working-capital requirements under inflation.
- How to conduct rate-of-return analysis under inflation.

11.1 Meaning and Measure of Inflation

Inflation is the rate at which the general level of prices and goods and services is rising, and, subsequently, purchasing power is falling.

Historically, the general economy has usually fluctuated in such a way as to exhibit **inflation,** a loss in the purchasing power of money over time. Inflation means that the cost of an item tends to increase over time, or, to put it another way, the same dollar amount buys less of an item over time. **Deflation** is the opposite of inflation, in that prices usually decrease over time; hence, a specified dollar amount gains in purchasing power. Inflation is far more common than deflation in the real world, so our consideration in this chapter will be restricted to accounting for inflation in economic analyses.

11.1.1 Measuring Inflation

CPI is an inflationary indicator that measures the change in the cost of a fixed basket of products and services.

Before we can introduce inflation into an engineering economic problem, we need a means of isolating and measuring its effect. Consumers usually have a relative, if not a precise, sense of how their purchasing power is declining. This sense is based on their experience of shopping for food, clothing, transportation, and housing over the years. Economists have developed a measure called the **Consumer Price Index** (**CPI**), which is based on a typical **market basket** of goods and services required by the average consumer. This market basket normally consists of items from eight major groups: (1) food and alcoholic beverages, (2) housing, (3) apparel, (4) transportation, (5) medical care, (6) entertainment, (7) personal care, and (8) other goods and services.

The CPI compares the cost of the typical market basket of goods and services in a current month with its cost 1 month ago, 1 year ago, or 10 years ago. The point in the past with which current prices are compared is called the **base period**. The index value for the base period is set at $100. The original base period used by the Bureau of Labor Statistics (BLS), of the U.S. Department of Labor, for the CPI index was 1967. For example, let us say that, in 1967, the prescribed market basket could have been purchased for $100. Now suppose the same combination of goods and services costs $578 in 2005. We can then compute the CPI for 2005 by multiplying the ratio of the current price to the base-period price by 100. In our example, the price index is ($578/$100)100 = 578, which means that the 2005 price of the contents of the market basket is 578% of its base-period price.

The revised CPI introduced by BLS in 1987 includes indexes for two populations: urban wage earners and clerical workers (CPI-W), and all urban consumers (CPI-U). As a result of the revision, both the CPI-W and the CPI-U utilize updated expenditure weights based upon data tabulated from the years 1982, 1983, and 1984 of the Consumer Expenditure Survey and incorporate a number of technical improvements. This method of assessing inflation does not imply, however, that consumers actually purchase the same goods and services year after year. Consumers tend to adjust their shopping practices to changes in relative prices, and they tend to substitute other items for those whose prices have increased greatly in relative terms. We must understand that the CPI does not take into account this sort of consumer behavior, because it is predicated on the purchase of a fixed market basket of the same goods and services, in the same proportions, month after month. For this reason, the CPI is called a **price index** rather than a **cost-of-living index**, although the general public often refers to it as a cost-of-living index.

The consumer price index is a good measure of the general increase in prices of consumer products. However, it is not a good measure of industrial price increases. In performing engineering economic analysis, the appropriate price indexes must be selected to estimate the price increases of raw materials, finished products, and operating costs. The *Survey of Current Business*, a monthly publication prepared by the BLS, provides the **Producer Price Index** (PPI) for various industrial goods.[3]

Table 11.1 lists the overall CPI, together with several price indexes for certain commodities over a number of years. From the table, we can easily calculate the inflation rate of gasoline from 2004 to 2005 as follows:

> **PPI** An inflationary indicator published by the U.S. Bureau of Labor Statistics to evaluate wholesale price levels in the economy.

$$\frac{162.5 - 126.1}{126.1} = 0.2887 = 28.87\%.$$

Since the inflation rate calculated is positive, the price of gasoline increased at a rate of 28.87% over the year 2004, one of the largest jumps in the price of gasoline in recent years.

Average Inflation Rate (*f*)

To account for the effect of varying yearly inflation rates over a period of several years, we can compute a single rate that represents an **average inflation rate**. Since each individual year's inflation rate is based on the previous year's rate, all these rates have a compounding

---

[3] Both CPI and PPI data are available on the Internet at the BLS home page on the Web: http://stats.bls.gov.

TABLE 11.1 Selected Price Indexes (Index for Base Year = 100, Calendar Month = April)

Year	New CPI, 1982–84	Old CPI, 1967	Gasoline, 1982	Semiconductor, 1982	Passenger Car, 1982
1995	152.2	455.0	67.7	133.1	133.7
1996	156.6	468.2	76.4	123.5	134.9
1997	160.2	479.7	72.7	114.0	134.8
1998	162.5	487.1	54.0	102.9	131.8
1999	166.2	497.8	64.4	98.5	130.9
2000	171.2	512.9	92.6	92.0	132.8
2001	176.9	529.9	104.8	87.7	133.2
2002	179.8	538.6	89.0	84.1	129.8
2003	183.8	550.5	100.1	79.4	129.0
2004	188.0	563.2	126.1	71.8	131.1
2005	194.6	582.9	162.5	70.0	133.2

Note: Years listed are base periods.

effect. As an example, suppose we want to calculate the average inflation rate for a two-year period: The first year's inflation rate is 4%, and the second year's rate is 8%, on a base price of $100.

Step 1. To find the price at the end of the second year, we use the process of compounding:

First year

$$\$100(1 + 0.04)(1 + 0.08) = \$112.32.$$

Second year

Step 2. To find the average inflation rate f, we establish the following equivalence equation:

$$\$100(1 + f)^2 = \$112.32 \text{ or } \$100(F/P, f, 2) = \$112.32.$$

Solving for f yields

$$f = 5.98\%.$$

We can say that the price increases in the last two years are equivalent to an average annual percentage rate of 5.98% per year. Note that the average is a geometric, not an arithmetic, average over a several-year period. Our computations are simplified by using a single average rate such as this, rather than a different rate for each year's cash flows.

EXAMPLE 11.1 Average Inflation Rate

Consider again the price increases for the 15 items listed in the table at the beginning of this chapter. Determine the average inflation rate for each item over the two-year period shown.

SOLUTION

Let's take the first item, the single-family home, for a sample calculation. Since we know the prices for 2003 and 2005, we can use the appropriate equivalence formula (single-payment compound amount factor, or growth formula). We state the problem as follows:

Given: $P = \$170{,}000$, $F = \$215{,}900$, and $N = 2005 - 2003 = 2$.

Find: f.

The equation we desire is $F = P(1 + f)^N$, so we have

$$\$215{,}900 = \$170{,}000(1 + f)^2,$$
$$f = \sqrt{1.27} - 1$$
$$= 12.69\%.$$

In a similar fashion, we can obtain the average inflation rates for the remaining items. The answers are as follows:

Item	2005 Price ($)	2003 Price ($)	Average Inflation Rate
Consumer Price Index (CPI)	194.60	183.80	2.90%
Single-family home	215,900.00	170,000.00	12.69%
Toyota Camry	20,125.00	19,560.00	1.43%
Unleaded gasoline	2.32	1.61	20.04%
Pair of jeans	39.50	39.50	0.00%
Internet service	42.95	42.95	0.00%
Tax preparation	151.00	130.00	7.77%
McDonald's Big Mac	2.39	2.22	3.76%
Clearing clogged sink	212.00	189.00	5.91%
Movie ticket	8.52	8.08	2.69%
Airline ticket	235.00	246.00	−2.26%
Vacation	1,315.00	1,162.00	6.38%
Hospital stay	4,848.00	3,889.00	11.65%
Birth	7,907.00	6,696.00	8.67%
A year in college	17,798.00	15,441.00	7.36%
Funeral	6,725.00	6,366.00	2.78%

The cost of unleaded gasoline increased most among the items listed in the table, whereas the price of an airline ticket exhibited "deflation" during the same period.

General Inflation Rate $\bar{f}$ Versus Specific Inflation Rate f_j

When we use the CPI as a base to determine the average inflation rate, we obtain the **general inflation rate**. We need to distinguish carefully between the general inflation rate and the average inflation rate for specific goods:

- **General inflation rate $\bar{f}$.** This average inflation rate is calculated on the basis of the CPI for all items in the market basket. The market interest rate is expected to respond to this general inflation rate.
- **Specific inflation rate f_j.** This rate is based on an index (or the CPI) specific to segment j of the economy. For example, we must often estimate the future cost of a particular item (e.g., labor, material, housing, or gasoline). When we refer to the average inflation rate for just one item, we will drop the subscript j for simplicity.

In terms of the CPI, we define the general inflation rate as

$$\text{CPI}_n = \text{CPI}_0 (1 + \bar{f})^n, \tag{11.1}$$

or

$$\bar{f} = \left[\frac{\text{CPI}_n}{\text{CPI}_0} \right]^{1/n} - 1, \tag{11.2}$$

where $\bar{f}$ = The general inflation rate,

CPI_n = The consumer price index at the end period n, and

CPI_0 = The consumer price index for the base period.

Knowing the CPI values for two consecutive years, we can calculate the annual general inflation rate as

$$\bar{f}_n = \frac{\text{CPI}_n - \text{CPI}_{n-1}}{\text{CPI}_{n-1}}, \tag{11.3}$$

where $\bar{f}_n$ = the general inflation rate for period n.

As an example, let us calculate the general inflation rate for the year 2005, where $\text{CPI}_{2004} = 188.0$ and $\text{CPI}_{2005} = 194.6$:

$$\frac{194.6 - 188.0}{188.0} = 0.0351 = 3.51\%.$$

This was an unusually good rate for the U.S. economy, compared with the average general inflation rate of 4.78%[4] over the last 38 years.

---

[4] To calculate the average general inflation rate from the base period (1967) to 2005, we evaluate

$$\bar{f} = \left[\frac{582.9}{100} \right]^{1/38} - 1 = 4.78\%.$$

EXAMPLE 11.2 Yearly and Average Inflation Rates

The following table shows a utility company's cost to supply a fixed amount of power to a new housing development:

Year	Cost
0	$504,000
1	538,400
2	577,000
3	629,500

The indices are specific to the utilities industry. Assume that year 0 is the base period, and determine the inflation rate for each period. Then calculate the average inflation rate over the three years listed in the table.

SOLUTION

Given: Annual cost to supply a fixed amount of power.
Find: Annual inflation rate (f_n) and average inflation rate $(\overline{f})$.

The inflation rate during year 1 (f_1) is

$$\frac{(\$538,400 - \$504,000)}{\$504,000} = 6.83\%.$$

The inflation rate during year 2 (f_2) is

$$\frac{(\$577,000 - \$538,400)}{\$538,400} = 7.17\%.$$

The inflation rate during year 3 (f_3) is

$$\frac{(\$629,500 - \$577,000)}{\$577,000} = 9.10\%.$$

The average inflation rate over the three years is

$$f = \left(\frac{\$629,500}{\$504,000}\right)^{1/3} - 1 = 0.0769 = 7.69\%.$$

Note that, although the average inflation rate[5] is 7.69% for the period taken as a whole, none of the years within the period had that particular rate.

---

[5] Since we obtained this average rate on the basis of costs that are specific to the utility industry, that rate is not the general inflation rate. Rather, it is a specific inflation rate for the utility in question.

11.1.2 Actual versus Constant Dollars

To introduce the effect of inflation into our economic analysis, we need to define several inflation-related terms:[6]

Actual (current) dollars: Out-of-pocket dollars paid at the time of purchasing goods and services.

Constant (nominal/real) dollars: Dollars as if in some base year, used to adjust for the effects of inflation.

- **Actual (current) dollars (A_n).** Actual dollars are estimates of future cash flows for year n that take into account any anticipated changes in amounts caused by inflationary or deflationary effects. Usually, these amounts are determined by applying an inflation rate to base-year dollar estimates.
- **Constant (real) dollars (A'_n).** Constant dollars represent constant purchasing power that is independent of the passage of time. In situations where inflationary effects were assumed when cash flows were estimated, the estimates obtained can be converted to constant dollars (base-year dollars) by adjustment with some readily accepted **general inflation rate**. We will assume that the base year is always time 0, unless we specify otherwise.

Conversion from Constant to Actual Dollars

Since constant dollars represent dollar amounts expressed in terms of the purchasing power of the base year, we may find the equivalent dollars in year n by using the general inflation rate $\bar{f}$ in the formula

$$A_n = A'_n(1 + \bar{f})^n = A'_n(F/P, \bar{f}, n), \tag{11.4}$$

where A'_n = Constant-dollar expression for the cash flow at the end of year n, and A_n = Actual-dollar expression for the cash flow at the end of year n.

If the future price of a specific cash flow element (j) is not expected to follow the general inflation rate, we will need to use the appropriate average inflation rate f_j applicable to that cash flow element, instead of $\bar{f}$. The conversion process is illustrated in Figure 11.1.

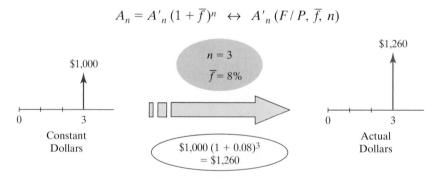

$$A_n = A'_n(1 + \bar{f})^n \leftrightarrow A'_n(F/P, \bar{f}, n)$$

Figure 11.1 Converting a $1,000 cash flow at year 3 estimated in constant dollars to its equivalent actual dollars at the same period, assuming an 8% inflation rate.

---

[6] The definitions presented are based on the ANSI Z94 Standards Committee on Industrial Engineering Terminology, *The Engineering Economist*. 1988; 33(2): 145–171.

EXAMPLE 11.3 Conversion from Constant to Actual Dollars

Jack Nicklaus won his first Masters Golf Championship in 1963. His prize money was $20,000. Phil Mickelson won his first Masters in 2004. The prize money was $1.17 million. It seems that Phil's prize money is much bigger. Compare these two prizes in terms of common purchasing power: Whose buying power is greater?

SOLUTION

Given: Two prize monies received in 1963 and 2004, respectively.

Find: Compare these two prizes in terms of purchasing power of 2004.

Both prize monies are in actual dollars at the time of winning. So, if we want to compare the prizes on the basis of the purchasing power of the year 2004, then we need to know the consumer price indexes for 1963 as well as 2004.

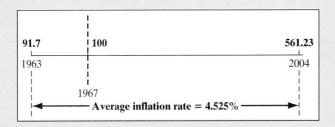

Using the old CPI, we find that the CPI for 1963 is 91.7 and the CPI for 2004 is 561.23. The average inflation rate between these two periods is 4.525%. To find the purchasing power of this $20,000 in 2004, we simply calculate the expression

$$20,000(1 + 0.04525)^{41} = \$122,760.$$

Given inflation and the changing value of money, $20,000 in 1963 is equivalent to $122,760 in terms of 2004 purchasing power. This is clearly much lower than Phil's actual 2004 winning prize, but Jack had 41 years to invest the amount to match Phil's winning prize.

Conversion from Actual to Constant Dollars

This conversion, shown in Figure 11.2, is the reverse process of converting from constant to actual dollars. Instead of using the compounding formula, we use a discounting formula (single-payment present-worth factor):

$$A'_n = \frac{A_n}{(1 + \overline{f})^n} = A_n(P/F, \overline{f}, n). \tag{11.5}$$

Once again, we may substitute f_j for $\overline{f}$ if future prices are not expected to follow the general inflation rate.

$$A'_n = A_n (1 + \overline{f})^{-n} \quad \leftrightarrow \quad A'_n (P/F, \overline{f}, n)$$

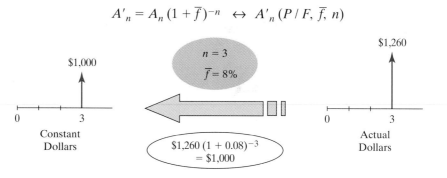

Figure 11.2 Conversion from actual to constant dollars. What it means is that $1,260 three years from now will have a purchasing power of $1,000 in terms of base dollars (year 0).

EXAMPLE 11.4 Conversion from Actual to Constant Dollars

Jagura Creek Fish Company, an aquacultural production firm, has negotiated a five-year lease on 20 acres of land, which will be used for fishponds. The annual cost stated in the lease is $20,000, to be paid at the beginning of each of the five years. The general inflation rate $\overline{f} = 5\%$. Find the equivalent cost in constant dollars during each period.

DISCUSSION: Although the $20,000 annual payments are *uniform*, they are not expressed in constant dollars. Unless an inflation clause is built into a contract, any stated amounts refer to *actual dollars*.

SOLUTION

Given: Actual dollars, $\overline{f} = 5\%$.
Find: Constant dollars during each period.

Using Eq. (11.5), we can determine the equivalent lease payments in constant dollars as follows:

End of Period	Cash Flow in Actual $	Conversion at f	Cash Flow in Constant $	Loss of Purchasing Power
0	$20,000	$(1 + 0.05)^0$	$20,000	0%
1	20,000	$(1 + 0.05)^{-1}$	19,048	4.76
2	20,000	$(1 + 0.05)^{-2}$	18,141	9.30
3	20,000	$(1 + 0.05)^{-3}$	17,277	13.62
4	20,000	$(1 + 0.05)^{-4}$	16,454	17.73

Note that, under the inflationary environment, the lender's receipt of the lease payment in year 5 is worth only 82.27% of the first lease payment.

11.2 Equivalence Calculations under Inflation

In previous chapters, our equivalence analyses took into consideration changes in the **earning power** of money (i.e., interest effects). To factor in changes in **purchasing power** as well—that is, inflation—we may use either (1) constant-dollar analysis or (2) actual-dollar analysis. Either method produces the same solution; however, each method requires the use of a different interest rate and procedure. Before presenting the two procedures for integrating interest and inflation, we will give a precise definition of the two interest rates.

11.2.1 Market and Inflation-Free Interest Rates

Two types of interest rates are used in equivalence calculations: the market interest rate and the inflation-free interest rate. The rate to apply depends on the assumptions used in estimating the cash flow:

- **Market interest rate (i).** This rate takes into account the combined effects of the earning value of capital (earning power) and any anticipated inflation or deflation (purchasing power). Virtually all interest rates stated by financial institutions for loans and savings accounts are market interest rates. Most firms use a market interest rate (also known as an **inflation-adjusted MARR**) in evaluating their investment projects.

  > **Market interest rate:** The interest rate quoted by financial institutions that accounts for both earning and purchasing power.

- **Inflation-free interest rate (i').** This rate is an estimate of the true earning power of money when the inflation effects have been removed. Commonly known as the **real interest rate**, it can be computed if the market interest rate and the inflation rate are known. As you will see later in this chapter, in the absence of inflation, the market interest rate is the same as the inflation-free interest rate.

In calculating any cash flow equivalence, we need to identify the nature of project cash flows. The three common cases are as follows:

Case 1. All cash flow elements are estimated in constant dollars.

Case 2. All cash flow elements are estimated in actual dollars.

Case 3. Some of the cash flow elements are estimated in constant dollars, and others are estimated in actual dollars.

For case 3, we simply convert all cash flow elements into one type—either constant or actual dollars. Then we proceed with either constant-dollar analysis as for case 1 or actual-dollar analysis as for case 2.

11.2.2 Constant-Dollar Analysis

Suppose that all cash flow elements are already given in constant dollars and that you want to compute the equivalent present worth of the constant dollars (A_n') occurring in year n. In the absence of an inflationary effect, we should use i' to account for only the earning power of the money. To find the present-worth equivalent of this constant-dollar amount at i', we use

$$P_n = \frac{A_n'}{(1 + i')^n}. \tag{11.6}$$

Constant-dollar analysis is common in the evaluation of many long-term public projects, because governments do not pay income taxes. Typically, income taxes are levied on the basis of taxable incomes in actual dollars.

EXAMPLE 11.5 Equivalence Calculation When Flows Are Stated in Constant Dollars

Transco Company is considering making and supplying computer-controlled traffic-signal switching boxes to be used throughout Arizona. Transco has estimated the market for its boxes by examining data on new road construction and on the deterioration and replacement of existing units. The current price per unit is $550; the before-tax manufacturing cost is $450. The start-up investment cost is $250,000. The projected sales and net before-tax cash flows in constant dollars are as follows:

Period	Unit Sales	Net Cash Flow in Constant $
0		−$250,000
1	1,000	100,000
2	1,100	110,000
3	1,200	120,000
4	1,300	130,000
5	1,200	120,000

If Transco managers want the company to earn a 12% inflation-free rate of return (i') before tax on any investment, what would be the present worth of this project?

SOLUTION

Given: Cash flow series stated in constant dollars, $i' = 12\%$.
Find: Present worth of the cash flow series.

Since all values are in constant dollars, we can use the inflation-free interest rate. We simply discount the dollar inflows at 12% to obtain the following:

$$PW(12\%) = -\$250,000 + \$100,000(P/A, 12\%, 5)$$
$$+ \$10,000(P/G, 12\%, 4) + \$20,000(P/F, 12\%, 5)$$
$$= \$163,099 \text{ (in year-0 dollars)}.$$

Since the equivalent net receipts exceed the investment, the project can be justified even before considering any tax effects.

11.2.3 Actual-Dollar Analysis

Now let us assume that all cash flow elements are estimated in actual dollars. To find the equivalent present worth of the actual dollar amount (A_n) in year n, we may use either the **deflation method** or the **adjusted-discount method**.

Deflation Method

The deflation method requires two steps to convert actual dollars into equivalent present-worth dollars. First we convert actual dollars into equivalent constant dollars by discounting by the general inflation rate, a step that removes the inflationary effect. Now we can use i' to find the equivalent present worth.

EXAMPLE 11.6 Equivalence Calculation When Cash Flows Are in Actual Dollars: Deflation Method

Applied Instrumentation, a small manufacturer of custom electronics, is contemplating an investment to produce sensors and control systems that have been requested by a fruit-drying company. The work would be done under a proprietary contract that would terminate in five years. The project is expected to generate the following cash flows in actual dollars:

n	Net Cash Flow in Actual Dollars
0	−$75,000
1	32,000
2	35,700
3	32,800
4	29,000
5	58,000

(a) What are the equivalent year-0 dollars (constant dollars) if the general inflation rate $(\bar{f})$ is 5% per year?

(b) Compute the present worth of these cash flows in constant dollars at $i' = 10\%$.

SOLUTION

Given: Cash flow series stated in actual dollars, $\bar{f} = 5\%$, and $i' = 10\%$.

Find: (a) Cash flow series converted into constant dollars and (b) the present worth of the cash flow series.

The net cash flows in actual dollars can be converted to constant dollars by deflating them, again assuming a 5% yearly deflation factor. The deflated, or constant-dollar, cash flows can then be used to determine the NPW at i'.

(a) We convert the actual dollars into constant dollars as follows:

n	Cash Flows in Actual Dollars	×	Deflation Factor	=	Cash Flows in Constant Dollars
0	−$75,000		1		−$75,000
1	32,000		$(1 + 0.05)^{-1}$		30,476
2	35,700		$(1 + 0.05)^{-2}$		32,381
3	32,800		$(1 + 0.05)^{-3}$		28,334
4	29,000		$(1 + 0.05)^{-4}$		23,858
5	58,000		$(1 + 0.05)^{-5}$		45,445

(b) We compute the equivalent present worth of the constant dollars by using $i' = 10\%$:

n	Cash Flows in Constant Dollars	×	Discounting	=	Equivalent Present Worth
0	−$75,000		1		−$75,000
1	30,476		$(1 + 0.10)^{-1}$		27,706
2	32,381		$(1 + 0.10)^{-2}$		26,761
3	28,334		$(1 + 0.10)^{-3}$		21,288
4	23,858		$(1 + 0.10)^{-4}$		16,295
5	45,445		$(1 + 0.10)^{-5}$		28,218
					$45,268

Adjusted-Discount Method

The two-step process shown in Example 11.6 can be greatly streamlined by the efficiency of the **adjusted-discount method**, which performs deflation and discounting in one step. Mathematically, we can combine this two-step procedure into one with the formula

$$P_n = \frac{\dfrac{A_n}{(1 + \overline{f})^n}}{(1 + i')^n}$$

$$= \frac{A_n}{(1 + \overline{f})^n(1 + i')^n}$$

$$= \frac{A_n}{[(1 + \overline{f})(1 + i')]^n}. \tag{11.7}$$

Since the market interest rate (i) reflects both the earning power and the purchasing power, we have

$$P_n = \frac{A_n}{(1 + i)^n}. \tag{11.8}$$

The equivalent present-worth values in Eqs. (11.7) and (11.8) must be equal at year 0. Therefore,

$$\frac{A_n}{(1 + i)^n} = \frac{A_n}{[(1 + \overline{f})(1 + i')]^n}.$$

This leads to the following relationship among $\overline{f}$, i', and i:

$$(1 + i) = (1 + \overline{f})(1 + i').$$

Simplifying the terms yields

$$i = i' + \overline{f} + i'\overline{f}. \tag{11.9}$$

Equation (11.9) implies that the market interest rate is a function of two terms: i' and $\overline{f}$. Note that without an inflationary effect, the two interest rates are the same ($\overline{f} = 0 \rightarrow i = i'$). As either i' or $\overline{f}$ increases, i also increases. For example, we can easily observe that, when prices increase due to inflation, bond rates climb, because lenders (i.e., anyone who invests in a money-market fund, a bond, or a certificate of deposit) demand higher rates to protect themselves against erosion in the value of their dollars. If inflation were to remain at 3%, you might be satisfied with an interest rate of 7% on a bond because your return would more than beat inflation. If inflation were running at 10%, however, you would not buy a 7% bond; you might insist instead on a return of at least 14%. By contrast, when prices are coming down, or at least are stable, lenders do not fear any loss of purchasing power with the loans they make, so they are satisfied to lend at lower interest rates.

In practice, we often approximate the market interest rate (i) simply by adding the inflation rate ($\overline{f}$) to the real interest rate (i') and ignoring the product ($i'\overline{f}$). This practice is fine as long as either i' or $\overline{f}$ is relatively small. With continuous compounding, however, the relationship among i, i', and $\overline{f}$ becomes precisely

$$i' = i - \overline{f}. \tag{11.10}$$

Real interest rate: The current interest rate minus the current inflation rate.

So, if we assume a nominal APR (market interest rate) of 6% per year compounded continuously and an inflation rate of 4% per year compounded continuously, the inflation-free interest rate is exactly 2% per year compounded continuously.

EXAMPLE 11.7 Equivalence Calculation When Flows Are in Actual Dollars: Adjusted-Discounted Method

Consider the cash flows in actual dollars in Example 11.6. Use the adjusted-discount method to compute the equivalent present worth of these cash flows.

SOLUTION

Given: Cash flow series in actual dollars, $\overline{f} = 5\%$, and $i' = 10\%$.
Find: Equivalent present worth.

First, we need to determine the market interest rate i. With $\overline{f} = 5\%$ and $i' = 10\%$, we obtain

$$i = i' + \overline{f} + i'\overline{f}$$
$$= 0.10 + 0.5 + (0.10)(0.05)$$
$$= 15.5\%.$$

Note that the equivalent present worth that we obtain with the adjusted-discount method ($i = 15.5\%$) is exactly the same as the result we obtained in Example 11.6:

n	Cash Flows in Actual Dollars	Multiplied by	=	Equivalent Present Worth
0	$-\$75,000$	1		$-\$75,000$
1	32,000	$(1 + 0.155)^{-1}$		27,706
2	35,700	$(1 + 0.155)^{-2}$		26,761
3	32,800	$(1 + 0.155)^{-3}$		21,288
4	29,000	$(1 + 0.155)^{-4}$		16,296
5	58,000	$(1 + 0.155)^{-5}$		28,217
				$\$45,268$

11.2.4 Mixed-Dollar Analysis

Let us consider another situation in which some cash flow elements are expressed in constant (or today's) dollars and the other elements in actual dollars. In this situation, we can convert all cash flow elements into the same dollar units (either constant or actual). If the cash flow is converted into actual dollars, the market interest rate (i) should be used in calculating the equivalence value. If the cash flow is converted into constant dollars, the inflation-free interest rate (i') should be used.

11.3 Effects of Inflation on Project Cash Flows

We now introduce inflation into some investment projects. We are especially interested in two elements of project cash flows: depreciation expenses and interest expenses. These two elements are essentially immune to the effects of inflation, as they are always given in actual dollars. We will also consider the complication of how to proceed when multiple price indexes have been used to generate various project cash flows.

Because depreciation expenses are calculated on some base-year purchase amount, they do not increase over time to keep pace with inflation. Thus, they lose some of their value to defer taxes, because inflation drives up the general price level and hence taxable income. Similarly, the selling prices of depreciable assets can increase with the general inflation rate, and because any gains on salvage values are taxable, they can result in increased taxes. Example 11.8 illustrates how a project's profitability changes under an inflationary economy.

EXAMPLE 11.8 Effects of Inflation on Projects with Depreciable Assets

Consider again the automated machining center investment project described in Example 10.1. The summary of the financial facts in the absence of inflation is as follows:

Item	Description or Data
Project	Automated machining center
Required investment	$125,000
Project life	5 years
Salvage value	$50,000
Depreciation method	7-year MACRS
Annual revenues	$100,000 per year
Annual expenses	
Labor	$20,000 per year
Material	$12,000 per year
Overhead	$8,000 per year
Marginal tax rate	40%
Inflation-free interest rate (i')	15%

The after-tax cash flow for the automated machining center project was given in Table 10.2, and the net present worth of the project in the absence of inflation was calculated to be $43,152.

What will happen to this investment project if the general inflation rate during the next five years is expected to increase by 5% annually? Sales and operating costs are assumed to increase accordingly. Depreciation will remain unchanged, but taxes, profits, and thus cash flow, will be higher. The firm's inflation-free interest rate (i') is known to be 15%.

(a) Use the deflation method to determine the NPW of the project.

(b) Compare the NPW with that in the inflation-free situation.

DISCUSSION: All cash flow elements, except depreciation expenses, are assumed to be in constant dollars. Since income taxes are levied on actual taxable income, we will use actual-dollar analysis, which requires that all cash flow elements be expressed in actual dollars.

- For the purposes of this illustration, all inflationary calculations are made as of year-end.

- Cash flow elements such as sales, labor, material, overhead, and the selling price of the asset will be inflated at the same rate as the general inflation rate.[7] For example,

---

[7] This is a simplistic assumption. In practice, these elements may have price indexes other than the CPI. Differential price indexes will be treated in Example 11.9.

whereas annual sales had been estimated at $100,000, under conditions of inflation they become 5% greater in year 1, or $105,000; 10.25% greater in year 2; and so forth.

The following table gives the sales conversion from constant to actual dollars for each of the five years of the project life:

Period	Sales in Constant $	Conversion $\bar{f}$	Sales in Actual $
1	$100,000	$(1 + 0.05)^1$	$105,000
2	100,000	$(1 + 0.05)^2$	110,250
3	100,000	$(1 + 0.05)^3$	115,763
4	100,000	$(1 + 0.05)^4$	121,551
5	100,000	$(1 + 0.05)^5$	127,628

Future cash flows in actual dollars for other elements can be obtained in a similar way. Note that

- No change occurs in the investment in year 0 or in depreciation expenses, since these items are unaffected by expected future inflation.
- The selling price of the asset is expected to increase at the general inflation rate. Therefore, the salvage value of the machine, in actual dollars, will be

$$\$50,000(1 + 0.05)^5 = \$63,814.$$

This increase in salvage value will also increase the taxable gains, because the book value remains unchanged. The calculations for both the book value and the gains tax are shown in Table 11.2.

SOLUTION

Given: Financial data in constant dollars, $\bar{f} = 5\%$, and $i' = 15\%$.
Find: The NPW.

Table 11.2 shows after-tax cash flows in actual dollars. Using the deflation method, we convert the cash flows to constant dollars with the same purchasing power as the dollars used to make the initial investment (year 0), assuming a general inflation rate of 5%. Then we discount these constant-dollar cash flows at i' to determine the NPW. Since NPW = $38,899 > 0$, the investment is still economically attractive.

COMMENTS: Note that the NPW in the absence of inflation was $43,152 in Example 10.1. The $4,253 decline just illustrated (known as inflation loss) in the NPW under inflation is due entirely to income tax considerations. The depreciation expense is a

TABLE 11.2 **Cash Flow Statement for the Automated Machining Center Project under Inflation**

	Inflation Rate	0	1	2	3	4	5
Income statement:							
Revenues	5%		$105,000	$110,250	$115,763	$121,551	$127,628
Expenses							
Labor	5%		21,000	22,050	23,153	24,310	25,526
Material	5%		12,600	13,230	13,892	14,586	15,315
Overhead	5%		8,400	8,820	9,261	9,724	10,210
Depreciation			17,863	30,613	21,863	15,613	5,581
Taxable income			$ 45,137	$ 35,537	$ 47,595	$ 57,317	$ 70,996
Income taxes (40%)			18,055	14,215	19,038	22,927	28,398
Net income			$ 27,082	$ 21,322	$ 28,557	$ 34,390	$ 42,598
Cash flow statement:							
Operating activities							
Net income			27,082	21,322	28,557	34,390	42,598
Depreciation			17,863	30,613	21,863	15,613	5,581
Investment activities							
Investment		(125,000)					
Salvage	5%						63,814
Gains tax							(12,139)
Net cash flow							
(in actual dollars)		$(125,000)	$ 44,945	$ 51,935	$ 50,420	$ 50,003	$ 99,854

charge against taxable income, which reduces the amount of taxes paid and, as a result, increases the cash flow attributable to an investment by the amount of taxes saved. But the depreciation expense under existing tax laws is based on historic cost. As time goes by, the depreciation expense is charged to taxable income in dollars of declining purchasing power; as a result, the "real" cost of the asset is not totally reflected in the depreciation expense. Depreciation costs are thereby understated, and the taxable income is overstated, resulting in higher taxes. In "real" terms, the amount of this additional income tax is $4,253, which is also known as the **inflation tax**. In general, any investment that, for tax purposes, is expensed over time, rather than immediately, is subject to the inflation tax. The following table gives the net cash flow conversion from actual to constant dollars for each of the five years of the project's life:

Year	Net Cash Flow in Actual $	Conversion $\bar{f}$	Net Cash Flow in Constant $	NPW at 15%
0	−$125,000	$(1 + 0.05)^0$	−$125,000	−$125,000
1	44,945	$(1 + 0.05)^{-1}$	42,805	37,222
2	51,935	$(1 + 0.05)^{-2}$	47,107	35,620
3	50,420	$(1 + 0.05)^{-3}$	43,555	28,638
4	50,003	$(1 + 0.05)^{-4}$	41,138	23,521
5	99,854	$(1 + 0.05)^{-5}$	78,238	38,898
				$38,899

11.3.1 Multiple Inflation Rates

As we noted previously, the inflation rate f_j represents a rate applicable to a specific segment j of the economy. For example, if we were estimating the future cost of a piece of machinery, we should use the inflation rate appropriate for that item. Furthermore, we may need to use several rates to accommodate the different costs and revenues in our analysis. The next example introduces the complexity of multiple inflation rates.

EXAMPLE 11.9 Applying Specific Inflation Rates

Let us rework Example 11.8, using different annual indexes (differential inflation rates) in the prices of cash flow components. Suppose that we expect the general rate of inflation, $\bar{f}$, to average 6% per year during the next five years. We also expect that the selling price of the equipment will increase 3% per year, that wages (labor) and overhead will increase 5% per year, and that the cost of materials will increase 4% per year. We expect sales revenue to climb at the general inflation rate. Table 11.3 shows the relevant calculations based on the income statement format. For simplicity, all cash flows and inflation effects are assumed to occur at year's end. Determine the net present worth of this investment, using the adjusted-discount method.

SOLUTION

Given: Financial data given in Example 11.8, multiple inflation rates.

Find: The NPW, using the adjusted-discount method.

Table 11.3 summarizes the after-tax cash flows in actual dollars. To evaluate the present worth using actual dollars, we must adjust the original discount rate of 15%, which is an inflation-free interest rate i'. The appropriate interest rate to use is the market interest rate:[8]

$$i = i + \bar{f} + i'\bar{f}$$
$$= 0.15 + 0.06 + (0.15)(0.06)$$
$$= 21.90\%.$$

---

[8] In practice, the market interest rate is usually given and the inflation-free interest rate can be calculated when the general inflation rate is known for years in the past or is estimated for time in the future. In our example, we are considering the opposite situation.

TABLE 11.3 Cash Flow Statement for the Automated Machining Center Project under Inflation, with Multiple Price Indexes

	Inflation Rate	0	1	2	3	4	5
Income statement:							
Revenues	6%		$106,000	$112,360	$119,102	$126,248	$133,823
Expenses							
Labor	5%		21,000	22,050	23,153	24,310	25,526
Material	4%		12,480	12,979	13,498	14,038	14,600
Overhead	5%		8,400	8,820	9,261	9,724	10,210
Depreciation			17,863	30,613	21,863	15,613	5,581
Taxable income			$ 46,257	$ 37,898	$ 51,327	$ 62,562	$ 77,906
Income taxes (40%)			18,503	15,159	20,531	25,025	31,162
Net income			$ 27,754	$ 22,739	$ 30,796	$ 37,537	$ 46,744
Cash flow statement:							
Operating activities							
Net income			27,754	22,739	30,796	37,537	46,744
Depreciation			17,863	30,613	21,863	15,613	5,581
Investment activities							
Investment		(125,000)					
Salvage	3%						57,964
Gains tax							(9,799)
Net cash flow							
(in actual dollars)		$(125,000)	$ 45,617	$ 53,352	$ 52,659	$ 53,150	$100,490

The equivalent present worth is obtained as follows:

$$PW(21.90\%) = -\$125,000 + \$45,617(P/F, 21.90\%, 1)$$
$$+ \$53,352(P/F, 21.90\%, 2) + \dots$$
$$+ \$100,490(P/F, 21.90\%, 5)$$
$$= \$38,801.$$

11.3.2 Effects of Borrowed Funds under Inflation

The repayment of a loan is based on the historical contract amount; the payment size does not change with inflation. Yet inflation greatly affects the value of these future payments, which are computed in year-0 dollars. First, we shall look at how the values of loan payments change under inflation. Interest expenses are usually already stated in the loan contract in actual dollars and need not be adjusted. Under the effect of inflation, the constant-dollar costs of both interest and principal repayments on a debt are reduced. Example 11.10 illustrates the effects of inflation on payments with project financing.

EXAMPLE 11.10 Effects of Inflation on Payments with Financing (Borrowing)

Let us rework Example 11.8 with a debt-to-equity ratio of 0.50, where the debt portion of the initial investment is borrowed at 15.5% annual interest. Assume, for simplicity, that the general inflation rate $\bar{f}$ of 5% during the project period will affect all revenues and expenses (except depreciation and loan payments) and the salvage value of the asset. Determine the NPW of this investment. (Note that the borrowing rate of 15.5% reflects the higher cost of debt financing under an inflationary environment.)

SOLUTION

Given: Cash flow data in Example 11.8, debt-to-equity ratio = 0.50, borrowing interest rate = 15.5%, $\bar{f}$ = 5%, and i' = 15%.
Find: The NPW.

For equal future payments, the actual-dollar cash flows for the financing activity are represented by the circles in Figure 11.3. If inflation were to occur, the cash flow, measured in year-0 dollars, would be represented by the shaded circles in the figure. Table 11.4 summarizes the after-tax cash flows in this situation. For simplicity, assume that all cash flows and inflation effects occur at year's end. To evaluate the present worth with the use of actual dollars, we must adjust the original discount rate (MARR) of 15%, which is an inflation-free interest rate i'. The appropriate interest rate to use is thus the market interest rate:

$$i = i' + \bar{f} + i'\bar{f}$$
$$= 0.15 + 0.05 + (0.15)(0.05)$$
$$= 20.75\%.$$

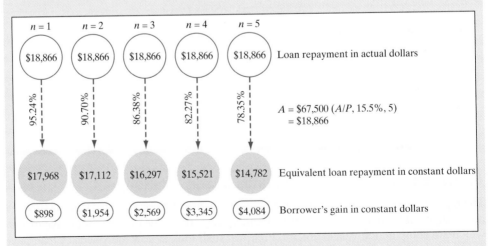

Figure 11.3 Equivalent loan repayment cash flows measured in year-0 dollars and borrower's gain over the life of the loan (Example 11.10).

TABLE 11.4 Cash Flow Statement for the Automated Machining Center Project under Inflation, with Borrowed Funds

	Inflation Rate	0	1	2	3	4	5
Income statement:							
Revenues	5%		$105,000	$110,250	$115,763	$121,551	$127,628
Expenses							
Labor	5%		21,000	22,050	23,153	24,310	25,526
Material	5%		12,600	13,230	13,892	14,586	15,315
Overhead	5%		8,400	8,820	9,261	9,724	10,210
Depreciation			17,863	30,613	21,863	15,613	5,581
Debt interest			9,688	8,265	6,622	4,724	2,532
Taxable income			$ 35,449	$ 27,272	$ 40,973	$ 52,593	$ 68,464
Income taxes (40%)			14,180	10,909	16,389	21,037	27,386
Net income			$ 21,269	$ 16,363	$ 24,584	$ 31,556	$ 41,078
Cash flow statement:							
Operating activities							
Net income			21,269	16,363	24,584	31,556	41,078
Depreciation			17,863	30,613	21,863	15,613	5,581
Investment activities							
Investment		$ (125,000)					
Salvage	5%						63,814
Gains tax							(12,139)
Financing activities							
Borrowed funds		62,500					
Principal repayment			(9,179)	(10,601)	(12,244)	(14,142)	(16,334)
Net cash flow (in actual dollars)		$ (62,500)	$ 29,953	$ 36,375	$ 34,203	$ 33,027	$ 82,000
Net cash flow (in constant dollars)	5%	$ (62,500)	$ 28,527	$ 32,993	$ 29,545	$ 27,171	$ 64,249
Equivalent present worth	15%	$ (62,500)	$ 24,806	$ 24,948	$ 19,427	$ 15,535	$ 31,943
Net present worth	$54,159						

Then, from Table 11.4, we compute the equivalent present worth of the after-tax cash flow as follows:

$$\text{PW}(20.75\%) = -\$62,500 + \$29,953(P/F, 20.75\%, 1) + \cdots$$
$$+ \$82,000(P/F, 20.75\%, 5)$$
$$= \$54,159.$$

COMMENTS: In the absence of debt financing, the project would have a net present worth of $38,899, as shown in Example 11.8. Compared with the preceding result of $54,159, the gain in present worth due to debt financing is $15,260. This increase in NPW is due primarily to the debt financing. An inflationary trend decreases the purchasing power of future dollars, which helps long-term borrowers because they can repay a loan with dollars with reduced buying power. That is, the debt-financing cost is reduced in an inflationary environment. In this case, the benefits of financing under inflation have more than offset the *inflation tax* effect on depreciation and the salvage value. The amount of the gain may vary with the interest rate on the loan: The interest rate for borrowing is also generally higher during periods of inflation because it is a market-driven rate.

11.4 Rate-of-Return Analysis under Inflation

In addition to affecting individual aspects of a project's income statement, inflation can have a profound effect on the overall return—that is, the very acceptability—of an investment project. In this section, we will explore the effects of inflation on the rate of return on an investment and present several examples.

11.4.1 Effects of Inflation on Return on Investment

The effect of inflation on the rate of return on an investment depends on how future revenues respond to the inflation. Under inflation, a company is usually able to compensate for increasing material and labor prices by raising its selling prices. However, even if future revenues increase to match the inflation rate, the allowable depreciation schedule, as we have seen, does not increase. The result is increased taxable income and higher income-tax payments. This increase reduces the available constant-dollar after-tax benefits and, therefore, the inflation-free after-tax rate of return (IRR'). The next example will help us to understand this situation.

EXAMPLE 11.11 IRR Analysis with Inflation

Hartsfield Company, a manufacturer of auto parts, is considering purchasing a set of machine tools at a cost of $30,000. The purchase is expected to generate increased sales of $24,500 per year and increased operating costs of $10,000 per year in each of the next four years. Additional profits will be taxed at a rate of 40%. The asset falls into the three-year MACRS property class for tax purposes. The project has a four-year life with zero salvage value. (All dollar figures represent constant dollars.)

(a) What is the expected internal rate of return?

(b) What is the expected IRR' if the general inflation is 10% during each of the next four years? (Here also, assume that $f_j = \overline{f} = 10\%$.)

(c) If the company requires an inflation-free MARR (or MARR') of 20%, should the company invest in the equipment?

SOLUTION

Given: Financial data in constant dollars, $t_m = 40\%$, and $N = 4$ years.

Find: (a) IRR, (b) IRR' when $\bar{f} = 10\%$, and (c) whether the project can be justified at MARR $= 20\%$.

(a) Rate-of-return analysis without inflation:

We find the rate of return by first computing the after-tax cash flow by the income statement approach, as illustrated in Table 11.5. The first part of the table shows the calculation of additional sales, operating costs, depreciation, and taxes. The asset will be depreciated fully over four years, with no expected salvage value. As we emphasized in Chapter 10, depreciation is not a cash expense, although it affects taxable income, and thus cash flow, indirectly. Therefore, we must add depreciation to net income to determine the net cash flow.

Thus, if the investment is made, we expect to receive additional annual cash flows of $12,700, $14,033, $10,478, and $9,589. This is a simple investment, so we can calculate the IRR for the project as follows:

$$PW(i') = -\$30,000 + \$12,700(P/F, i', 1) + \$14,033(P/F, i', 2)$$
$$+ \$10,478(P/F, i', 3) + \$9,589(P/F, i', 4)$$
$$= 0.$$

Solving for i' yields

$$IRR' = i'^* = 21.88\%.$$

TABLE 11.5 **Rate-of-Return Calculation without Inflation**

	0	1	2	3	4
Income statement:					
Revenues		$ 24,500	$ 24,500	$ 24,500	$ 24,500
Expenses					
O&M		10,000	10,000	10,000	10,000
Depreciation		10,000	13,333	4,445	2,222
Taxable income		4,500	1,167	10,055	12,278
Income taxes (40%)		1,800	467	4,022	4,911
Net income		$ 2,700	$ 700	$ 6,033	$ 7,367
Cash flow statement:					
Operating activities					
Net income		2,700	700	6,033	7,367
Depreciation		10,000	13,333	4,445	2,222
Investment activities					
Machine center	$ (30,000)				
Salvage					0
Gains tax					0
Net cash flow					
(in actual dollars)	$ (30,000)	$ 12,700	$ 14,033	$ 10,478	$ 9,589

The project has an inflation-free rate of return of 21.88%; that is, the company will recover its original investment ($30,000) plus interest at 21.88% each year for each dollar still invested in the project. Since IRR' > MARR' (20%), the company should buy the equipment.

(b) Rate-of-return analysis under inflation:

With inflation, we assume that sales, operating costs, and the future selling price of the asset will increase. Depreciation will be unchanged, but taxes, profits, and cash flow will be higher. We might think that higher cash flows will mean an increased rate of return. Unfortunately, this is not the case. We must recognize that cash flows for each year are stated in dollars of declining purchasing power. When the net after-tax cash flows are converted to dollars with the same purchasing power as those used to make the original investment, the resulting rate of return decreases. These calculations, assuming an inflation rate of 10% in sales and operating expenses and a 10% annual decline in the purchasing power of the dollar, are shown in Table 11.6. For example, whereas additional sales

TABLE 11.6 Rate-of-Return Calculation under Inflation

	0	1	2	3	4
Income statement:					
Revenues		$26,950	$29,645	$32,610	$35,870
Expenses					
O&M		11,000	12,100	13,310	14,641
Depreciation		10,000	13,333	4,445	2,222
Taxable income		5,950	4,212	14,855	19,007
Income taxes (40%)		2,380	1,685	5,942	7,603
Net income		$ 3,570	$ 2,527	$ 8,913	$11,404
Cash flow statement:					
Operating activities					
Net income		3,570	2,527	8,913	11,404
Depreciation		10,000	13,333	4,445	2,222
Investment activities					
Machine center	$(30,000)				
Salvage					0
Gains tax					0
Net cash flow (in actual dollars)	$(30,000)	$13,570	$15,860	$13,358	$13,626
Net cash flow (in constant dollars)	$(30,000)	$12,336	$13,108	$10,036	$ 9,307

Note: PW(20%) = $(321); IRR (actual dollars) = 31.34%; IRR' (constant dollars) = 19.40%.

had been \$24,500 yearly, under conditions of inflation they would be 10% greater in year 1, or \$26,950; 21% greater in year 2; and so forth. No change in investment or depreciation expenses will occur, since these items are unaffected by expected future inflation. We have restated the after-tax cash flows (actual dollars) in dollars of a common purchasing power (constant dollars) by deflating them, again assuming an annual deflation factor of 10%. The constant-dollar cash flows are then used to determine the real rate of return. First,

$$PW(i') = -\$30,000 + \$12,336(P/F, i', 1) + \$13,108(P/F, i', 2)$$
$$+ \$10,036(P/F, i', 3) + \$9,307(P/F, i', 4)$$
$$= 0.$$

Then, solving for i' yields

$$i'^* = 19.40\%.$$

The rate of return for the project's cash flows in constant dollars (year-0 dollars) is 19.40%, which is less than the 21.88% return in the inflation-free case. Since IRR$'$ < MARR$'$, the investment is no longer acceptable.

COMMENTS: We could also calculate the rate of return by setting the PW of the actual dollar cash flows to 0. This would give a value of IRR = 31.34%, but that is an inflation-adjusted IRR. We could then convert to an IRR$'$ by deducting the amount caused by inflation:

$$i' = \frac{(1 + i)}{(1 + \bar{f})} - 1$$
$$= \frac{(1 + 0.3134)}{(1 + 0.10)} - 1$$
$$= 19.40\%.$$

This approach gives the same final result of IRR$'$ = 19.40%.

11.4.2 Effects of Inflation on Working Capital

The loss of tax savings in depreciation is not the only way that inflation may distort an investment's rate of return. Another source is working-capital drain. Projects requiring increased levels of working capital suffer from inflation because additional cash must be invested to maintain new price levels. For example, if the cost of inventory increases, additional outflows of cash are required to maintain appropriate inventory levels over time. A similar phenomenon occurs with funds committed to accounts receivable. These additional working-capital requirements can significantly reduce a project's rate of return, as the next example illustrates.

EXAMPLE 11.12 Effect of Inflation on Profits with Working Capital

Suppose that, in Example 11.11, a $1,000 investment in working capital is expected and that all the working capital will be recovered at the end of the project's four-year life. Determine the rate of return on this investment.

SOLUTION

Given: Financial data in Example 11.11 and working-capital requirement = $1,000. Find: IRR.

Using the data in the upper part of Table 11.7, we can calculate IRR′ = IRR of 20.88% in the absence of inflation. Also, PW(20%) = $499. The lower part of the table includes the effect of inflation on the proposed investment. As illustrated in Figure 11.4, working-capital levels can be maintained only by additional infusions of cash; the working-capital drain also appears in the lower part of Table 11.7. For example,

TABLE 11.7 Effects of Inflation on Working Capital and After-Tax Rate of Return

Case 1: Without Inflation	0	1	2	3	4
Income statement:					
Revenue		$ 24,500	$ 24,500	$ 24,500	$ 24,500
Expenses					
O&M		10,000	10,000	10,000	10,000
Depreciation		10,000	13,333	4,445	2,222
Taxable income		4,500	1,167	10,055	12,278
Income taxes (40%)		1,800	467	4,022	4,911
Net income		$ 2,700	$ 700	$ 6,033	$ 7,367
Cash flow statement:					
Operating activities					
Net income		2,700	700	6,033	7,367
Depreciation		10,000	13,333	4,445	2,222
Investment activities					
Machine center	$ (30,000)				
Working capital	(1,000)				1,000
Salvage					0
Gains tax					0
Net cash flow					
(in actual dollars)	$ (31,000)	$ 12,700	$ 14,033	$ 10,478	$ 10,589

Note: PW(20%) = $499; IRR′ = 20.88%.

Case 2: With Inflation (10%)	0	1	2	3	4
Income statement:					
Revenue		$ 26,950	$ 29,645	$ 32,610	$ 35,870
Expenses					
O&M		11,000	12,100	13,310	14,641
Depreciation		10,000	13,333	4,445	2,222
Taxable income		5,950	4,212	14,855	19,007
Income taxes (40%)		2,380	1,685	5,942	7,603
Net income		$ 3,570	$ 2,527	$ 8,913	$ 11,404
Cash flow statement:					
Operating activities					
Net income		3,570	2,527	8,913	11,404
Depreciation		10,000	13,333	4,445	2,222
Investment activities					
Machine center	$ (30,000)				
Working capital	(1,000)	(100)	(110)	(121)	$ 1,331
Salvage					0
Gains tax					0
Net cash flow (in actual dollars)	$ (31,000)	$ 13,470	$ 15,750	$ 13,237	$ 14,957
Net cash flow (in constant dollars)	$ (31,000)	$ 12,245	$ 13,017	$ 9,945	$ 10,216

Note: PW(20%) = $(1,074); IRR (actual dollars) = 29.89%; IRR′ (constant dollars) = 18.09%.

the $1,000 investment in working capital made in year 0 will be recovered at the end of the first year, assuming a one-year recovery cycle. However, because of 10% inflation, the required working capital for the second year increases to $1,100. In addition to reinvesting the $1,000 revenues, an additional investment of $100 must be made. This $1,100 will be recovered at the end of the second year. However, the project will need a 10% increase, or $1,210, for the third year, and so forth.

As Table 11.7 illustrates, the combined effect of depreciation and working capital is significant. Given an inflationary economy and investment in working capital, the project's IRR′ drops to 18.09%, or PW(20%) = −$1,074. By using either IRR analysis or NPW analysis, we end up with the same result (as we must): Alternatives that are attractive when inflation does not exist may not be acceptable when inflation does exist.

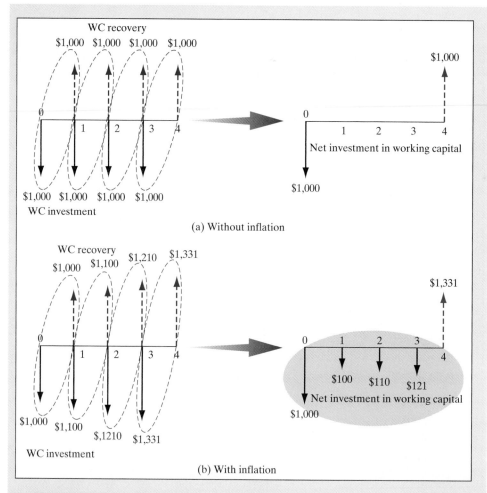

Figure 11.4 Working-capital requirements under inflation: (a) Requirements without inflation and (b) requirements with inflation, assuming a one-year recovery cycle.

SUMMARY

- The **Consumer Price Index (CPI)** is a statistical measure of change, over time, of the prices of goods and services in major expenditure groups—such as food, housing, apparel, transportation, and medical care—typically purchased by urban consumers. Essentially, the CPI compares the cost of a sample "market basket" of goods and services in a specific period with the cost of the same "market basket" in an earlier reference period, designated the **base period**.

- **Inflation** is the term used to describe a **decline in purchasing power** evidenced in an economic environment of rising prices.

- **Deflation** is the opposite: an increase in purchasing power evidenced by falling prices.

- The **general inflation rate** $\bar{f}$ is an average inflation rate based on the CPI. An annual general inflation rate $\bar{f}_j$ can be calculated with the following equation:

$$\overline{f}_n = \frac{\text{CPI}_n - \text{CPI}_{n-1}}{\text{CPI}_{n-1}}.$$

■ Specific, individual commodities do not always reflect the general inflation rate in their price changes. We can calculate an **average inflation rate** $\overline{f}_j$ for a specific commodity (j) if we have an index (i.e., a record of historical costs) for that commodity.

■ Project cash flows may be stated in one of two forms:

Actual dollars (A_n): Dollars that reflect the inflation or deflation rate.

Constant dollars (A'_n): Year-0 dollars.

■ Interest rates used in evaluating a project may be stated in one of two forms:

Market interest rate (i): A rate that combines the effects of interest and inflation; used with **actual-dollar** analysis.

Inflation-free interest rate (i'): A rate from which the effects of inflation have been removed; this rate is used with constant-dollar analysis.

■ To calculate the present worth of actual dollars, we can use a two-step or a one-step process:

Deflation method—two steps:

1. Convert actual dollars by deflating with the general inflation rate $\overline{f}$.
2. Calculate the PW of constant dollars by discounting at i'.

Adjusted-discount method—one step (use the market interest rate):

$$P_n = \frac{A_n}{((1 + \overline{f})(1 + i'))^n}$$

$$= \frac{A_n}{(1 + i)^n}.$$

■ A number of individual elements of project evaluations can be distorted by inflation. These elements are summarized in Table 11.8.

TABLE 11.8 Effects of Inflation on Project Cash Flows and Return

Item	Effects of Inflation
Depreciation expense	Depreciation expense is charged to taxable income in dollars of declining values; taxable income is overstated, resulting in higher taxes.
Salvage values	Inflated salvage values combined with book values based on historical costs result in higher taxable gains.
Loan repayments	Borrowers repay historical loan amounts with dollars of decreased purchasing power, reducing the cost of financing debt.
Working capital requirement	Known as a *working capital drain*, the cost of working capital increases in an inflationary economy.
Rate of return and NPW	Unless revenues are sufficiently increased to keep pace with inflation, tax effects and a working-capital drain result in a lower rate of return or a lower NPW.

PROBLEMS

Note: *In the problems that follow, the term "market interest rate" represents the "inflation-adjusted MARR" for project evaluation or the "interest rate" quoted by a financial institution for commercial loans.*

Measure of Inflation

11.1 The following data indicate the median price of unleaded gasoline during the last 10 years for California residents:

Period	Price ($)
1996	$1.10
2005	$2.62

Assuming that the base period (price index = 100) is 1996, compute the average price index for the unleaded gasoline price for the year 2005.

11.2 The following data indicate the price indexes of lumber (base period 1982 = 100) during the last five years:

Period	Price (cents)
1996	150.6
1997	155.1
1998	158.3
1999	161.8
2000	165.8
2005	?

(a) Assuming that the base period (price index = 100) is reset to the year 1996, compute the average price index for lumber between 1996 and 2000.

(b) If the past trend is expected to continue, how would you estimate the lumber product in 2005?

11.3 For prices that are increasing at an annual rate of 5% the first year and 8% the second year, determine the average inflation rate $(\overline{f})$ over the two years.

11.4 Because of general price inflation in our economy, the purchasing power of the dollar shrinks with the passage of time. If the average general inflation rate is expected to be 7% per year for the foreseeable future, how many years will it take for the dollar's purchasing power to be one-half of what it is now?

Actual versus Constant Dollars

11.5 An annuity provides for 10 consecutive end-of-year payments of $4,500. The average general inflation rate is estimated to be 5% annually, and the market interest rate is 12% annually. What is the annuity worth in terms of a single equivalent amount of today's dollars?

11.6 A company is considering buying workstation computers to support its engineering staff. In today's dollars, it is estimated that the maintenance costs for the computers (paid at the end of each year) will be $25,000, $30,000, $32,000, $35,000, and $40,000 for years 1 to 5, respectively. The general inflation rate $(\bar{f})$ is estimated to be 8% per year, and the company will receive 15% per year on its invested funds during the inflationary period. The company wants to pay for maintenance expenses in equivalent equal payments (in actual dollars) at the end of each of the five years. Find the amount of the company's payments.

11.7 The following cash flows are in actual dollars:

n	Cash Flow (in actual $)
0	$1,500
4	2,500
5	3,500
7	4,500

Convert to an equivalent cash flow in constant dollars if the base year is time 0. Keep cash flows at the same point in time—that is, years 0, 4, 5, and 7. Assume that the market interest rate is 16% and that the general inflation rate $(\bar{f})$ is 4% per year.

11.8 The purchase of a car requires a $25,000 loan to be repaid in monthly installments for four years at 12% interest compounded monthly. If the general inflation rate is 6% compounded monthly, find the actual- and constant-dollar value of the 20th payment.

11.9 A series of four annual constant-dollar payments beginning with $7,000 at the end of the first year is growing at the rate of 8% per year. (Assume that the base year is the current year $(n = 0)$.) If the market interest rate is 13% per year and the general inflation rate $(\bar{f})$ is 7% per year, find the present worth of this series of payments, based on

(a) constant-dollar analysis.

(b) actual-dollar analysis.

11.10 Consider the accompanying cash flow diagrams, where the equal payment cash flow in constant dollars (a) is converted from the equal payment cash flow in actual dollars (b) at an annual general inflation rate of $\bar{f} = 3.8\%$ and $i = 9\%$. What is the amount A in actual dollars equivalent to $A' = \$1,000$ in constant dollars?

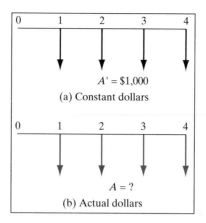

(a) Constant dollars

(b) Actual dollars

11.11 A 10-year $1,000 bond pays a nominal rate of 9% compounded semiannually. If the market interest rate is 12% compounded annually and the general inflation rate is 6% per year, find the actual- and constant-dollar amount (in time-0 dollars) of the 16th interest payment on the bond.

Equivalence Calculation under Inflation

11.12 Suppose that you borrow $20,000 at 12% compounded monthly over five years. Knowing that the 12% represents the market interest rate, you realize that the monthly payment in actual dollars will be $444.90. If the average monthly general inflation rate is expected to be 0.5%, determine the equivalent equal monthly payment series in constant dollars.

11.13 The annual fuel costs required to operate a small solid-waste treatment plant are projected to be $1.5 million, without considering any future inflation. The best estimates indicate that the annual inflation-free interest rate (i') will be 6% and the general inflation rate $(\bar{f})$ will be 5%. If the plant has a remaining useful life of five years, what is the present equivalent of its fuel costs? Use actual-dollar analysis.

11.14 Suppose that you just purchased a used car worth $6,000 in today's dollars. Suppose also that you borrowed $5,000 from a local bank at 9% compounded monthly over two years. The bank calculated your monthly payment at $228. Assuming that average general inflation will run at 0.5% per month over the next two years,

(a) Determine the annual inflation-free interest rate (i') for the bank.

(b) What equal monthly payments, in terms of constant dollars over the next two years, are equivalent to the series of actual payments to be made over the life of the loan?

11.15 A man is planning to retire in 20 years. Money can be deposited at 6% compounded monthly, and it is also estimated that the future general inflation $(\bar{f})$ rate will be 5% compounded annually. What monthly deposit must be made each month until the man retires so that he can make annual withdrawals of $40,000 in terms of today's dollars over the 15 years following his retirement? (Assume that his first withdrawal occurs at the end of the first six months after his retirement.)

11.16 On her 23rd birthday, a young woman engineer decides to start saving toward building up a retirement fund that pays 8% interest compounded quarterly (the market interest rate). She feels that $600,000 worth of purchasing power in today's dollars will be adequate to see her through her sunset years after her 63rd birthday. Assume a general inflation rate of 6% per year.

(a) If she plans to save by making 160 equal quarterly deposits, what should be the amount of her quarterly deposit in actual dollars?

(b) If she plans to save by making end-of-the-year deposits, increasing by $1,000 over each subsequent year, how much would her first deposit be in actual dollars?

11.17 A father wants to save for his 8-year-old son's college expenses. The son will enter college 10 years from now. An annual amount of $40,000 in constant dollars will be required to support the son's college expenses for 4 years. Assume that these college payments will be made at the beginning of the school year. The future general inflation rate is estimated to be 6% per year, and the market interest rate on the savings account will average 8% compounded annually. Given this information,

(a) What is the amount of the son's freshman-year expense in terms of actual dollars?

(b) What is the equivalent single-sum amount at the present time for these college expenses?

(c) What is the equal amount, in actual dollars, the father must save each year until his son goes to college?

Effects of Inflation on Project Cash Flows

11.18 Consider the following project's after-tax cash flow and the expected annual general inflation rate during the project period:

End of Year	Expected Cash Flow (in actual $)	General Inflation Rate
0	−$45,000	
1	26,000	6.5%
2	26,000	7.7
3	26,000	8.1

(a) Determine the average annual general inflation rate over the project period.

(b) Convert the cash flows in actual dollars into equivalent constant dollars with the base year 0.

(c) If the annual inflation-free interest rate is 5%, what is the present worth of the cash flow? Is this project acceptable?

11.19 Gentry Machines, Inc., has just received a special job order from one of its clients. The following financial data have been collected:

• This two-year project requires the purchase of special-purpose equipment for $55,000. The equipment falls into the MACRS five-year class.

• The machine will be sold for $27,000 (today's dollars) at the end of two years.

- The project will bring in additional annual revenue of $114,000 (actual dollars), but it is expected to incur an additional annual operating cost of $53,800 (today's dollars).
- The project requires an investment in working capital in the amount of $12,000 at $n = 0$. In each subsequent year, additional working capital needs to be provided at the general inflation rate. Any investment in working capital will be recovered after the project is terminated.
- To purchase the equipment, the firm expects to borrow $50,000 at 10% over a two-year period. The remaining $5,000 will be taken from the firm's retained earnings. The firm will make equal annual payments of $28,810 (actual dollars) to pay off the loan.
- The firm expects a general inflation of 5% per year during the project period. The firm's marginal tax rate is 40%, and its market interest rate is 18%.

  (a) Compute the after-tax cash flows in actual dollars.

  (b) What is the equivalent present value of this amount at time 0?

11.20 Hugh Health Product Corporation is considering purchasing a computer to control plant packaging for a spectrum of health products. The following data have been collected:

- First cost = $120,000, to be borrowed at 9% interest, with only interest paid each year and the principal due in a lump sum at end of year 2.
- Economic service life (project life) = 6 years.
- Estimated selling price in year-0 dollars = $15,000.
- Depreciation = Five-year MACRS property.
- Marginal income tax rate = 40% (remains constant).
- Annual revenue = $145,000 (today's dollars).
- Annual expense (not including depreciation and interest) = $82,000 (today's dollars).
- Market interest rate = 18%.

  (a) With an average general inflation rate of 5% expected during the project period (which will affect all revenues, expenses, and the salvage value of the computer), determine the cash flows in actual dollars.

  (b) Compute the net present value of the project under inflation.

  (c) Compute the net present-value loss (gain) due to inflation.

  (d) In (c), how much is the present-value loss (or gain) due to borrowing?

11.21 Norcross Textile Company is considering automating its piece-goods screen printing system at a cost of $20,000. The firm expects to phase out the new printing system at the end of five years due to changes in style. At that time, the firm could scrap the system for $2,000 in today's dollars. The expected net savings due to automation also are in today's dollars (constant dollars) as follows:

End of Year	Cash Flow (in constant $)
1	$15,000
2	17,000
3–5	14,000

The system qualifies as a five-year MACRS property and will be depreciated accordingly. The expected average general inflation rate over the next five years is approximately 5% per year. The firm will finance the entire project by borrowing at 10%. The scheduled repayment of the loan will be as follows:

End of Year	Principal Payment	Interest Payment
1	$6,042	$2,000
2	6,647	3,396
3	7,311	731

The firm's market interest rate for project evaluation during this inflation-ridden time is 20%. Assume that the net savings and the selling price will be responsive to this average inflation rate. The firm's marginal tax rate is known to be 40%.

(a) Determine the after-tax cash flows of this project, in actual dollars.

(b) Determine the net present-value reduction (or gains) in profitability due to inflation.

11.22 The J. F. Manning Metal Co. is considering the purchase of a new milling machine during year 0. The machine's base price is $135,000, and it will cost another $15,000 to modify it for special use. This results in a $150,000 cost base for depreciation. The machine falls into the MACRS seven-year property class. The machine will be sold after three years for $80,000 (actual dollars). Use of the machine will require an increase in net working capital (inventory) of $10,000 at the beginning of the project year. The machine will have no effect on revenues, but it is expected to save the firm $80,000 (today's dollars) per year in before-tax operating costs, mainly for labor. The firm's marginal tax rate is 40%, and this rate is expected to remain unchanged over the duration of the project. However, the company expects that the labor cost will increase at an annual rate of 5%, but that the working-capital requirement will grow at an annual rate of 8%, caused by inflation. The selling price of the milling machine is not affected by inflation. The general inflation rate is estimated to be 6% per year over the project period. The firm's market interest rate is 20%.

(a) Determine the project cash flows in actual dollars.

(b) Determine the project cash flows in constant (time-0) dollars.

(c) Is this project acceptable?

Rate of Return Analysis under Inflation

11.23 Fuller Ford Company is considering purchasing a vertical drill machine. The machine will cost $50,000 and will have an eight-year service life. The selling price of the machine at the end of eight years is expected to be $5,000 in today's dollars. The machine will generate annual revenues of $20,000 (today's dollars), but the company expects to have an annual expense (excluding depreciation) of $8,000 (today's dollars). The asset is classified as a seven-year MACRS property. The project requires a working-capital investment of $10,000 at year 0. The marginal income tax rate for the firm is averaging 35%. The firm's market interest rate is 18%.

(a) Determine the internal rate of return of this investment.

(b) Assume that the firm expects a general inflation rate of 5%, but that it also expects an 8% annual increase in revenue and working capital and a 6% annual increase in expense caused by inflation. Compute the real (inflation-free) internal rate of return. Is this project acceptable?

11.24 Sonja Jensen is considering the purchase of a fast-food franchise. Sonja will be operating on a lot that is to be converted into a parking lot in six years, but that may be rented in the interim for $800 per month. The franchise and necessary equipment will have a total initial cost of $55,000 and a salvage value of $10,000 (in today's dollars) after six years. Sonja is told that the future annual general inflation rate will be 5%. The projected operating revenues and expenses, in actual dollars, other than rent and depreciation for the business, are as follows:

End of Year	Revenue	Expenses
1	$30,000	$15,000
2	35,000	21,000
3	55,000	25,000
4	70,000	30,000
5	70,000	30,000
6	60,000	30,000

Assume that the initial investment will be depreciated under the five-year MACRS and that Sonja's tax rate will be 30%. Sonja can invest her money at a rate of least 10% in other investment activities during this inflation-ridden period.

(a) Determine the cash flows associated with the investment over its life.

(b) Compute the projected after-tax rate of return (real) for this investment opportunity.

11.25 Suppose you have $10,000 cash that you want to invest. Normally, you would deposit the money in a savings account that pays an annual interest rate of 6%. However, you are now considering the possibility of investing in a bond. Your alternatives are either a nontaxable municipal bond paying 9% or a taxable corporate bond paying 12%. Your marginal tax rate is 30% for both ordinary income and capital gains. You expect the general inflation to be 3% during the investment period. You can buy a high-grade municipal bond costing $10,000 and that pays interest of 9% ($900) per year. This interest is not taxable. A comparable high-grade corporate bond is also available that is just as safe as the municipal bond, but that pays an interest rate of 12% ($1,200) per year. This interest is taxable as ordinary income. Both bonds mature at the end of year 5.

(a) Determine the real (inflation-free) rate of return for each bond.

(b) Without knowing your MARR, can you make a choice between these two bonds?

11.26 Air Florida is considering two types of engines for use in its planes. Each engine has the same life, the same maintenance, and the same repair record.

- Engine A costs $100,000 and uses 50,000 gallons of fuel per 1,000 hours of operation at the average service load encountered in passenger service.
- Engine B costs $200,000 and uses 32,000 gallons of fuel per 1,000 hours of operation at the same service load.

Both engines are estimated to have 10,000 service hours before any major overhaul of the engines is required. If fuel currently costs $1.25 per gallon, and its price is expected to increase at the rate of 8% because of inflation, which engine should the firm install for an expected 2,000 hours of operation per year? The firm's marginal income tax rate is 40%, and the engine will be depreciated on the basis of the unit-of-production method. Assume that the firm's market interest rate is 20%. It is estimated that both engines will retain a market value of 40% of their initial cost (actual dollars) if they are sold on the market after 10,000 hours of operation.

(a) Using the present-worth criterion, which project would you select?

(b) Using the annual-equivalent criterion, which project would you select?

(c) Using the future-worth criterion, which project would you select?

11.27 Johnson Chemical Company has just received a special subcontracting job from one of its clients. The two-year project requires the purchase of a special-purpose painting sprayer of $60,000. This equipment falls into the MACRS five-year class. After the subcontracting work is completed, the painting sprayer will be sold at the end of two years for $40,000 (actual dollars). The painting system will require an increase of $5,000 in net working capital (for spare-parts inventory, such as spray nozzles). This investment in working capital will be fully recovered after the project is terminated. The project will bring in an additional annual revenue of $120,000 (today's dollars), but it is expected to incur an additional annual operating cost of $60,000 (today's dollars). It is projected that, due to inflation, sales prices will increase at an annual rate of 5%. (This implies that annual revenues will increase at an annual rate of 5%.) An annual increase of 4% for expenses and working-capital requirement is expected. The company has a marginal tax rate of 30%, and it uses a market interest rate of 15% for project evaluation during the inflationary period. If the firm expects a general inflation of 8% during the project period,

(a) Compute the after-tax cash flows in actual dollars.

(b) What is the rate of return on this investment (real earnings)?

(c) Is the special subcontracting project profitable?

11.28 Land Development Corporation is considering purchasing a bulldozer. The bulldozer will cost $100,000 and will have an estimated salvage value of $30,000 at the end of six years. The asset will generate annual before-tax revenues of $80,000 over the next six years. The asset is classified as a five-year MACRS property. The marginal tax rate is 40%, and the firm's market interest rate is known to be 18%. All dollar figures represent constant dollars at time 0 and are responsive to the general inflation rate $\bar{f}$.

(a) With $\bar{f} = 6\%$, compute the after-tax cash flows in actual dollars.

(b) Determine the real rate of return of this project on an after-tax basis.

(c) Suppose that the initial cost of the project will be financed through a local bank at an interest rate of 12%, with an annual payment of $24,323 over six years. With this additional condition, answer part (a) again.

(d) In part (a), determine the present-value loss due to inflation.

(e) In part (c), determine how much the project has to generate in additional before-tax annual revenue in actual dollars (equal amount) to make up the loss due to inflation.

Short Case Studies

ST11.1 Wilson Machine Tools, Inc., a manufacturer of fabricated metal products, is considering purchasing a high-tech computer-controlled milling machine at a cost of $95,000. The cost of installing the machine, preparing the site, wiring, and rearranging other equipment is expected to be $15,000. This installation cost will be added to the cost of the machine to determine the total cost basis for depreciation. Special jigs and tool dies for the particular product will also be required at a cost of $10,000. The milling machine is expected to last 10 years, the jigs and dies only 5 years. Therefore, another set of jigs and dies has to be purchased at the end of 5 years. The milling machine will have a $10,000 salvage value at the end of its life, and the special jigs and dies are worth only $300 as scrap metal at any time in their lives. The machine is classified as a 7-year MACRS property, and the special jigs and dies are classified as a 3-year MACRS property. With the new milling machine, Wilson expects an additional annual revenue of $80,000 due to increased production. The additional annual production costs are estimated as follows: materials, $9,000; labor, $15,000; energy, $4,500; and miscellaneous O&M costs, $3,000. Wilson's marginal income tax rate is expected to remain at 35% over the project life of 10 years. All dollar figures represent today's dollars. The firm's market interest rate is 18%, and the expected general inflation rate during the project period is estimated at 6%.

(a) Determine the project cash flows in the absence of inflation.

(b) Determine the internal rate of return for the project in (a).

(c) Suppose that Wilson expects price increases during the project period: material at 4% per year, labor at 5% per year, and energy and other O&M costs at 3% per year. To compensate for these increases in prices, Wilson is planning to increase annual revenue at the rate of 7% per year by charging its customers a higher price. No changes in salvage value are expected for the machine or the jigs and dies. Determine the project cash flows in actual dollars.

(d) In (c), determine the real (inflation-free) rate of return of the project.

(e) Determine the economic loss (or gain) in present worth caused by inflation.

ST11.2 Recent biotechnological research has made possible the development of a sensing device that implants living cells on a silicon chip. The chip is capable of detecting physical and chemical changes in cell processes. Proposed uses of the device include researching the mechanisms of disease on a cellular level, developing new therapeutic drugs, and substituting for animals in cosmetic and drug testing. Biotech Device Corporation (BDC) has just perfected a process for mass-producing the chip. The following information has been compiled for the board of directors:

- BDC's marketing department plans to target sales of the device to the larger chemical and drug manufacturers. BDC estimates that annual sales would be 2,000 units if the device were priced at $95,000 per unit (in dollars of the first operating year).

- To support this level of sales volume, BDC would need a new manufacturing plant. Once the "go" decision is made, this plant could be built and made ready for production within 1 year. BDC would need a 30-acre tract of land that would cost $1.5 million. If the decision were to be made, the land could be purchased on December 31, 2006. The building would cost $5 million and would be depreciated according to the MACRS 39-year class. The first payment of $1 million would be due to the contractor on December 31, 2007, and the remaining $4 million on December 31, 2008.
- The required manufacturing equipment would be installed late in 2008 and would be paid for on December 31, 2008. BDC would have to purchase the equipment at an estimated cost of $8 million, including transportation, plus a further $500,000 for installation. The equipment would fall into the MACRS 7-year class.
- The project would require an initial investment of $1 million in working capital. This investment would be made on December 31, 2008. Then, on December 31 of each subsequent year, net working capital would be increased by an amount equal to 15% of any sales increase expected during the coming year. The investments in working capital would be fully recovered at the end of the project year.
- The project's estimated economic life is 6 years (excluding the 2-year construction period). At that time, the land is expected to have a market value of $2 million, the building a value of $3 million, and the equipment a value of $1.5 million. The estimated variable manufacturing costs would total 60% of the dollar sales. Fixed costs, excluding depreciation, would be $5 million for the first year of operations. Since the plant would begin operations on January 1, 2009, the first operating cash flows would occur on December 31, 2009.
- Sales prices and fixed overhead costs, other than depreciation, are projected to increase with general inflation, which is expected to average 5 percent per year over the 6-year life of the project.
- To date, BDC has spent $5.5 million on research and development (R&D) associated with cell implantation. The company has already expensed $4 million R&D costs. The remaining $1.5 million will be amortized over 6 years (i.e., the annual amortization expense will be $250,000). If BDC decides not to proceed with the project, the $1.5 million R&D cost could be written off on December 31, 2006.
- BDC's marginal tax rate is 40%, and its market interest rate is 20%. Any capital gains will also be taxed at 40%.
  - (a) Determine the after-tax cash flows of the project, in actual dollars.
  - (b) Determine the inflation-free (real) IRR of the investment.
  - (c) Would you recommend that the firm accept the project?

Project Risk
and Uncertainty

Oil Forecasts Are a Roll of the Dice[1] You may know as much as the oil experts. That is, you know that a barrel of oil is pricey and getting pricier. Beyond that, nobody—not even those who get paid to prognosticate—has a real handle on the push and pull that goes into figuring how much oil people need, how much can be pumped, and how much can be refined.

The unreliable data and forecasts have plagued the industry for decades. They became more of a problem once demand and prices

[1] "Oil Forecasts Are a Roll of the Dice," Bhusan Bahree, *The Wall Street Journal*, Tuesday, August 2, 2005, Section C1.

starting climbing two years ago, because the substantial margins of error in these numbers are even larger than the oil industry's shrinking margin of spare pumping capacity.

Put another way, even a small error in predicting oil consumption can cause energy markets to gyrate if demand turns out higher than the market assumed, because the industry lacks the ability it had in the 1990s to gin up extra oil on the fly to meet a surge in buying. Yet traders, companies, and consumers have no choice but to rely on the numbers that are out there. One problem anyone who factors oil into an investing decision faces is that accurate oil data come only with a time lag. Oil data on actual demand and supply bounce around because of bad weather, accidents such as pipeline ruptures, or political shocks such as terrorist strikes. Amplifying the fuzziness, meanwhile, is that projections of economic growth, a critical factor in assessing energy needs, are forever changing.

Crude-Oil Price Forecast

West Texas Intermediate, Spot Price. USD/bbl. Average of Month

	Dec 2005	Jan 2006	Feb 2006	Mar 2006	Apr 2006	May 2006
Forecast Value	**59.3**	**58.7**	**61.8**	**61.6**	**62.1**	**63.8**
Standard Deviation	0.9	1.2	1.5	1.8	2.2	2.6
Correlation Coefficient	0.9925	0.9912	0.9899	0.9886	0.9873	0.9860

Suppose that your business depends on the price of oil. For example, the airline industry, UPS, FedEx, and rental companies are all heavily affected by the price of fuel. Now, if your proposed project also depends on the price of crude oil, how would you factor the fluctuation and uncertainty into the analysis?

In previous chapters, cash flows from projects were assumed to be known with complete certainty; our analysis was concerned with measuring the economic worth of projects and selecting the best ones to invest in. Although that type of analysis can provide a reasonable basis for decision making in many investment situations, we should certainly consider the more usual *un*certainty. In this type of situation, management rarely has precise expectations about the future cash flows to be derived from a particular project. In fact, the best that a firm can reasonably expect to do is to estimate the range of possible future costs and benefits and the relative chances of achieving a reasonable return on the investment. We use the term **risk** to describe an investment project whose cash flow is not known in advance with absolute certainty, but for which an array of alternative outcomes and their probabilities (odds) are known. We will also use the term **project risk** to refer to variability in a project's NPW. A greater project risk usually means a greater variability in a project's NPW, or simply that the *risk is the potential for loss*. This chapter begins by exploring the origins of project risk.

Risk: The chance that an investment's actual return will be different than expected.

CHAPTER LEARNING OBJECTIVES

After completing this chapter, you should understand the following concepts:

- How to describe the nature of project risk.
- How to conduct a sensitivity analysis of key input variables.
- How to conduct a break-even analysis.
- How to develop a net-present-worth probability distribution.
- How to compare mutually exclusive risky alternatives.
- How to develop a risk simulation model.
- How to make a sequential investment decision with a decision tree.

12.1 Origins of Project Risk

The decision to make a major capital investment such as introducing a new product requires information about cash flow over the life of a project. The profitability estimate of an investment depends on cash flow estimations, which are generally uncertain. The factors to be estimated include the total market for the product; the market share that the firm can attain; the growth in the market; the cost of producing the product, including labor and materials; the selling price; the life of the product; the cost and life of the equipment needed; and the effective tax rates. Many of these factors are subject to substantial uncertainty. A common approach is to make single-number "best estimates" for each of the uncertain factors and then to calculate measures of profitability, such as the NPW or rate of return for the project. This approach, however, has two drawbacks:

1. No guarantee can ever ensure that the "best estimates" will match actual values.
2. No provision is made to measure the risk associated with an investment, or the project risk. In particular, managers have no way of determining either the probability that a project will lose money or the probability that it will generate large profits.

Because cash flows can be so difficult to estimate accurately, project managers frequently consider a range of possible values for cash flow elements. If a range of values for individual cash flows is possible, it follows that a range of values for the NPW of a given project is also possible. Clearly, the analyst will want to gauge the probability and reliability of individual cash flows and, consequently, the level of certainty about the overall project worth.

12.2 Methods of Describing Project Risk

We may begin analyzing project risk by first determining the uncertainty inherent in a project's cash flows. We can do this analysis in a number of ways, which range from making informal judgments to calculating complex economic and statistical quantities. In this section, we will introduce three methods of describing project risk: (1) sensitivity analysis, (2) break-even analysis, and (3) scenario analysis. Each method will be explained with reference to a single example, involving the Boston Metal Company.

12.2.1 Sensitivity Analysis

One way to glean a sense of the possible outcomes of an investment is to perform a sensitivity analysis. This kind of analysis determines the effect on the NPW of variations in the input variables (such as revenues, operating cost, and salvage value) used to estimate after-tax cash flows. A **sensitivity analysis** reveals how much the NPW will change in response to a given change in an input variable. In calculating cash flows, some items have a greater influence on the final result than others. In some problems, the most significant item may be easily identified. For example, the estimate of sales volume is often a major factor in a problem in which the quantity sold varies with the alternatives. In other problems, we may want to locate the items that have an important influence on the final results so that they can be subjected to special scrutiny.

Sensitivity analysis is sometimes called "what-if" analysis, because it answers questions such as "What if incremental sales are only 1,000 units, rather than 2,000 units? Then what will the NPW be?" Sensitivity analysis begins with a base-case situation, which is developed by using the most likely values for each input. We then change the specific variable of interest by several specified percentage points above and below the most likely value, while holding other variables constant. Next, we calculate a new NPW for each of the values we obtained. A convenient and useful way to present the results of a sensitivity analysis is to plot **sensitivity graphs**. The slopes of the lines show how sensitive the NPW is to changes in each of the inputs: The steeper the slope, the more sensitive the NPW is to a change in a particular variable. Sensitivity graphs identify the crucial variables that affect the final outcome most. Example 12.1 illustrates the concept of sensitivity analysis.

EXAMPLE 12.1 Sensitivity Analysis

Boston Metal Company (BMC), a small manufacturer of fabricated metal parts, must decide whether to enter the competition to become the supplier of transmission housings for Gulf Electric, a company that produces the housings in its own in-house manufacturing facility, but that has almost reached its maximum production capacity. Therefore, Gulf is looking for an outside supplier. To compete, BMC must design a new fixture for the production process and purchase a new forge. The available details for this purchase are as follows:

- The new forge would cost $125,000. This total includes retooling costs for the transmission housings.
- If BMC gets the order, it may be able to sell as many as 2,000 units per year to Gulf Electric for $50 each, in which case variable production costs,[2] such as direct labor and direct material costs, will be $15 per unit. The increase in fixed costs,[3] other than depreciation, will amount to $10,000 per year.
- The firm expects that the proposed transmission-housings project will have about a five-year product life. The firm also estimates that the amount ordered by Gulf Electric in the first year will be ordered in each of the subsequent four years. (Due to the nature of contracted production, the annual demand and unit price would remain the same over the project after the contract is signed.)
- The initial investment can be depreciated on a MACRS basis over a seven-year period, and the marginal income tax rate is expected to remain at 40%. At the end of five years, the forge is expected to retain a market value of about 32% of the original investment.
- On the basis of this information, the engineering and marketing staffs of BMC have prepared the cash flow forecasts shown in Table 12.1. Since the NPW is positive ($40,168) at the 15% opportunity cost of capital (MARR), the project appears to be worth undertaking.

What Makes BMC Managers Worry: BMC's managers are uneasy about this project, because too many uncertain elements have not been considered in the analysis:

- If it decided to take on the project, BMC would have to invest in the forging machine to provide Gulf Electric with some samples as a part of the bidding process. If Gulf Electric were not to like BMC's sample, BMC would stand to lose its entire investment in the forging machine.
- If Gulf were to like BMC's sample, then if it was overpriced, BMC would be under pressure to bring the price in line with competing firms. Even the possibility that BMC would get a smaller order must be considered, as Gulf may utilize its overtime capacity to produce some extra units. Finally, BMC is not certain about its projections of variable and fixed costs.

Recognizing these uncertainties, the managers want to assess the various possible future outcomes before making a final decision. Put yourself in BMC's management position, and describe how you may resolve the uncertainty associated with the project. In doing so, perform a sensitivity analysis for each variable and develop a sensitivity graph.

DISCUSSION: Table 12.1 shows BMC's expected cash flows—but that they will indeed materialize cannot be assumed. In particular, BMC is not very confident in its revenue forecasts. The managers think that if competing firms enter the market,

---

[2] These are expenses that change in direct proportion to the change in volume of sales or production, as defined in Section 8.3.
[3] These are expenses that do not vary with the volume of sales or production. For example, property taxes, insurance, depreciation, and rent are usually fixed expenses.

TABLE 12.1 After-Tax Cash Flow for BMC's Transmission-Housings Project (Example 12.1)

	0	1	2	3	4	5
Revenues:						
Unit price		$ 50	$ 50	$ 50	$ 50	$ 50
Demand (units)		2,000	2,000	2,000	2,000	2,000
Sales revenue		$100,000	$100,000	$100,000	$100,000	$100,000
Expenses:						
Unit variable cost		$ 15	$ 15	$ 15	$ 15	$ 15
Variable cost		30,000	30,000	30,000	30,000	30,000
Fixed cost		10,000	10,000	10,000	10,000	10,000
Depreciation		17,863	30,613	21,863	15,613	5,575
Taxable income		$ 42,137	$ 29,387	$ 38,137	$ 44,387	$ 54,425
Income taxes (40%)		16,855	11,755	15,255	17,755	21,770
Net income		$ 25,282	$ 17,632	$ 22,882	$ 26,632	$ 32,655
Cash flow statement:						
Operating activities:						
Net income		25,282	17,632	22,882	26,632	32,655
Depreciation		17,863	30,613	21,863	15,613	5,575
Investment activities:						
Investment	(125,000)					
Salvage						40,000
Gains tax						(2,611)
Net cash flow	$(125,500)	$ 43,145	$ 48,245	$ 44,745	$ 42,245	$ 75,619

BMC will lose a substantial portion of the projected revenues by not being able to increase its bidding price. Before undertaking the project, the company needs to identify the key variables that will determine whether it will succeed or fail. The marketing department has estimated revenue as follows:

$$\text{Annual revenue} = (\text{Product demand})(\text{unit price})$$
$$= (2,000)(\$50) = \$100,000.$$

The engineering department has estimated variable costs, such as those of labor and materials, at $15 per unit. Since the projected sales volume is 2,000 units per year, the total variable cost is $30,000.

After first defining the unit sales, unit price, unit variable cost, fixed cost, and salvage value, we conduct a sensitivity analysis with respect to these key input variables. This is done by varying each of the estimates by a given percentage and determining what effect the variation in that item will have on the final results. If the effect is large, the result is sensitive to that item. Our objective is to locate the most sensitive item(s).

SOLUTION

Sensitivity analysis: We begin the sensitivity analysis with a consideration of the base-case situation, which reflects the best estimate (expected value) for each input variable. In developing Table 12.2, we changed a given variable by 20% in 5% increments, above and below the base-case value, and calculated new NPWs, while other variables were held constant. The values for both sales and operating costs were the expected, or base-case, values, and the resulting $40,169 is the base-case NPW. Now we ask a series of "what-if" questions: What if sales are 20% below the expected level? What if operating costs rise? What if the unit price drops from $50 to $45? Table 12.2 summarizes the results of varying the values of the key input variables.

Sensitivity graph: Next, we construct a sensitivity graph for five of the transmission project's key input variables. (See Figure 12.1.) We plot the base-case NPW on the

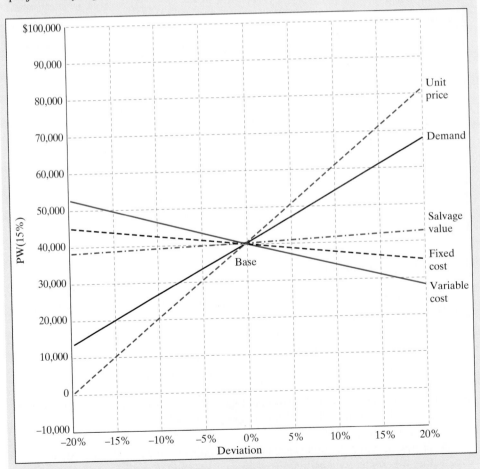

Figure 12.1 Sensitivity graph for BMC's transmission-housings project (Example 12.1).

TABLE 12.2 Sensitivity Analysis for Five Key Input Variables (Example 12.1)

Deviation	–20%	–15%	–10%	–5%	0%	5%	10%	15%	20%
Unit price	$ (57)	$ 9,999	$ 20,055	$ 30,111	$ 40,169	$ 50,225	$ 60,281	$ 70,337	$ 80,393
Demand	12,010	19,049	26,088	33,130	40,169	47,208	54,247	61,286	68,325
Variable cost	52,236	49,219	46,202	43,186	40,169	37,152	34,135	31,118	28,101
Fixed cost	44,191	43,185	42,179	41,175	40,169	39,163	38,157	37,151	36,145
Salvage value	37,782	38,378	38,974	39,573	40,169	40,765	41,361	41,957	42,553

ordinate of the graph at the value 1.0 on the abscissa (or 0% deviation). Then we reduce the value of product demand to 0.95 of its base-case value and recompute the NPW with all other variables held at their base-case value. We repeat the process by either decreasing or increasing the relative deviation from the base case. The lines for the variable unit price, variable unit cost, fixed cost, and salvage value are obtained in the same manner. In Figure 12.1, we see that the project's NPW is (1) very sensitive to changes in product demand and unit price, (2) fairly sensitive to changes in variable costs, and (3) relatively insensitive to changes in the fixed cost and the salvage value.

Graphic displays such as the one in Figure 12.1 provide a useful means to communicate the relative sensitivities of the different variables to the corresponding NPW value. However, sensitivity graphs do not explain any interactions among the variables or the likelihood of realizing any specific deviation from the base case. Certainly, it is conceivable that an answer might not be very sensitive to changes in either of two items, but very sensitive to combined changes in them.

12.2.2 Break-Even Analysis

When we perform a sensitivity analysis of a project, we are asking how serious the effect of lower revenues or higher costs will be on the project's profitability. Managers sometimes prefer to ask instead how much sales can decrease below forecasts before the project begins to lose money. This type of analysis is known as **break-even analysis**. In other words, break-even analysis is a technique for studying the effect of variations in output on a firm's NPW (or other measures). We will present an approach to break-even analysis based on the project's cash flows.

To illustrate the procedure of break-even analysis based on NPW, we use the generalized cash flow approach we discussed in Section 10.4. We compute the PW of cash inflows as a function of an unknown variable (say, x), perhaps annual sales. For example,

$$\text{PW of cash inflows} = f(x)_1.$$

Next, we compute the PW of cash outflows as a function of x:

$$\text{PW of cash outflows} = f(x)_2.$$

NPW is, of course, the difference between these two numbers. Accordingly, we look for the break-even value of x that makes

$$f(x)_1 = f(x)_2.$$

Note that this break-even value is similar to that used to calculate the internal rate of return when we want to find the interest rate that makes the NPW equal zero. The break-even value is also used to calculate many other similar "cutoff values" at which a choice changes.

EXAMPLE 12.2 Break-Even Analysis

Through the sensitivity analysis in Example 12.1, BMC's managers become convinced that the NPW is most sensitive to changes in annual sales volumes. Determine the break-even NPW value as a function of that variable.

SOLUTION

The analysis is shown in Table 12.3, in which the revenues and costs of the BMC transmission-housings project are set out in terms of an unknown amount of annual sales X.

We calculate the PWs of cash inflows and outflows as follows:

- PW of cash inflows:

$$
\begin{aligned}
\text{PW}(15\%)_{\text{Inflow}} &= (\text{PW of after-tax net revenue}) \\
&\quad + (\text{PW of net salvage value}) \\
&\quad + (\text{PW of tax savings from depreciation}) \\
&= 30X(P/A, 15\%, 5) + \$37{,}389(P/F, 15\%, 5)
\end{aligned}
$$

TABLE 12.3 Break-Even Analysis with Unknown Annual Sales (Example 12.2)

	0	1	2	3	4	5
Cash inflow:						
Net salvage						37,389
Revenue:						
$X(1 - 0.4)(\$50)$		30X	30X	30X	30X	30X
Depreciation credit						
0.4 (depreciation)		7,145	12,245	8,745	6,245	2,230
Cash outflow:						
Investment	−125,000					
Variable cost:						
$-X(1 - 0.4)(\$15)$		−9X	−9X	−9X	−9X	−9X
Fixed cost:						
$-(1 - 0.4)(\$10{,}000)$		−6,000	−6,000	−6,000	−6,000	−6,000
Net cash flow	−125,000	21X+1,145	21X+6,245	21X+2,745	21X+245	21X+33,617

$$+ \ \$7{,}145(P/F, 15\%, 1) \ + \ \$12{,}245(P/F, 15\%, 2)$$
$$+ \ \$8{,}745(P/F, 15\%, 3) \ + \ \$6{,}245(P/F, 15\%, 4)$$
$$+ \ \$2{,}230(P/F, 15\%, 5)$$
$$= \ 30X(P/A, 15\%, 5) \ + \ \$44{,}490$$
$$= \ 100.5650X \ + \ \$44{,}490.$$

- PW of cash outflows:

$$\text{PW}(15\%)_{\text{Outflow}} = (\text{PW of capital expenditure})$$
$$+ \ (\text{PW of after-tax expenses})$$
$$= \ \$125{,}000 \ + \ (9X \ + \ \$6{,}000)(P/A, 15\%, 5)$$
$$= \ 30.1694X \ + \ \$145{,}113.$$

The NPW of cash flows for the BMC is thus

$$\text{PW}(15\%) = 100.5650X \ + \ \$44{,}490$$
$$= \ -(30.1694X \ + \ \$145{,}113)$$
$$= \ 70.3956X \ - \ \$100{,}623.$$

In Table 12.4, we compute the PW of the inflows and the PW of the outflows as a function of demand (X).

The NPW will be just slightly positive if the company sells 1,430 units. Precisely calculated, the zero-NPW point (break-even volume) is 1,429.43 units:

$$\text{PW}(15\%) = 70.3956X \ - \ \$100{,}623$$
$$= \ 0,$$
$$X_b = 1{,}430 \text{ units.}$$

TABLE 12.4 **Determination of Break-Even Volume Based on Project's NPW (Example 12.3)**

Demand (X)	PW of Inflow (100.5650X + $44,490)	PW of Outflow (30.1694X + $145,113)	NPW (70.3956X − $100,623)
0	$ 44,490	$ 145,113	(100,623)
500	94,773	160,198	(65,425)
1,000	145,055	175,282	(30,227)
1,429	188,197	188,225	(28)
1,430	188,298	188,255	43
1,500	195,338	190,367	4,970
2,000	245,620	205,452	40,168
2,500	295,903	220,537	75,366

Break-even volume = 1,430 units.

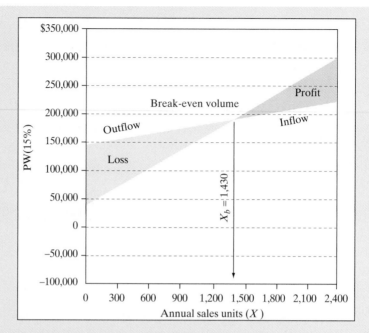

Figure 12.2 Break-even analysis based on net cash flow (Example 12.2).

In Figure 12.2, we have plotted the PWs of the inflows and outflows under various assumptions about annual sales. The two lines cross when sales are 1,430 units, the point at which the project has a zero NPW. Again we see that, as long as sales are greater or equal to 1,430, the project has a positive NPW.

12.2.3 Scenario Analysis

Although both sensitivity and break-even analyses are useful, they have limitations. Often, it is difficult to specify precisely the relationship between a particular variable and the NPW. The relationship is further complicated by interdependencies among the variables. Holding operating costs constant while varying unit sales may ease the analysis, but in reality, operating costs do not behave in this manner. Yet, it may complicate the analysis too much to permit movement in more than one variable at a time.

Scenario analysis is a technique that considers the sensitivity of NPW both to changes in key variables and to the range of likely values of those variables. For example, the decision maker may examine two extreme cases: a "worst-case" scenario (low unit sales, low unit price, high variable cost per unit, high fixed cost, and so on) and a "best-case" scenario. The NPWs under the worst and the best conditions are then calculated and compared with the expected, or base-case, NPW. Example 12.3 illustrates a plausible scenario analysis for BMC's transmission-housings project.

EXAMPLE 12.3 Scenario Analysis

Consider again BMC's transmission-housings project first presented in Example 12.1. Assume that the company's managers are fairly confident of their estimates of all the project's cash flow variables, except the estimates of unit sales. Assume further that they regard a drop in unit sales below 1,600 or a rise above 2,400 as extremely unlikely. Thus, a decremental annual sale of 400 units defines the lower bound, or the worst-case scenario, whereas an incremental annual sale of 400 units defines the upper bound, or the best-case scenario. (Remember that the most likely value was 2,000 in annual unit sales.) Discuss the worst- and best-case scenarios, assuming that the unit sales for all five years are equal.

DISCUSSION: To carry out the scenario analysis, we ask the marketing and engineering staffs to give optimistic (best-case) and pessimistic (worst-case) estimates for the key variables. Then we use the worst-case variable values to obtain the worst-case NPW and the best-case variable values to obtain the best-case NPW.

SOLUTION

The results of our analysis are summarized as follows:

Variable Considered	Worst-Case Scenario	Most-Likely-Case Scenario	Best-Case Scenario
Unit demand	1,600	2,000	2,400
Unit price ($)	48	50	53
Variable cost ($)	17	15	12
Fixed cost ($)	11,000	10,000	8,000
Salvage value ($)	30,000	40,000	50,000
PW(15%)	−$5,856	$40,169	$104,295

We see that the base case produces a positive NPW, the worst case produces a negative NPW, and the best case produces a large positive NPW. Still, by just looking at the results in the table, it is not easy to interpret the scenario analysis or to make a decision based on it. For example, we could say that there is a chance of losing money on the project, but we do not yet have a specific probability for that possibility. Clearly, we need estimates of the probabilities of occurrence of the worst case, the best case, the base (most likely) case, and all the other possibilities.

The need to estimate probabilities leads us directly to our next step: developing a probability distribution (or, put another way, the probability that the variable in question takes on a certain value). If we can predict the effects on the NPW of variations in the parameters, why should we not assign a probability distribution to the possible outcomes of each parameter and combine these distributions in some way to produce a probability distribution for the possible outcomes of the NPW? We shall consider this issue in the next two sections.

12.3 Probability Concepts for Investment Decisions

In this section, we shall assume that the analyst has available the probabilities (likelihoods) of future events from either previous experience with a similar project or a market survey. The use of probability information can provide management with a range of possible outcomes and the likelihood of achieving different goals under each investment alternative.

12.3.1 Assessment of Probabilities

We begin by defining terms related to probability, such as *random variable*, *probability distribution*, and *cumulative probability distribution*. Quantitative statements about risk are given as numerical probabilities or as likelihoods (odds) of occurrence. Probabilities are given as decimal fractions in the interval from 0.0 to 1.0. An event or outcome that is certain to occur has a probability of 1.0. As the probability of an event approaches 0, the event becomes increasingly less likely to occur. The assignment of probabilities to the various outcomes of an investment project is generally called **risk analysis**.

Risk analysis is a technique to identify and assess factors that may jeopardize the success of a project.

Random Variables

A **random variable** is a parameter or variable that can have more than one possible value (though not simultaneously). The value of a random variable at any one time is unknown until the event occurs, but the probability that the random variable will have a specific value is known in advance. In other words, associated with each possible value of the random variable is a likelihood, or probability, of occurrence. For example, when your college team plays a football game, only two events regarding the outcome of the game are possible: win or lose. The outcome is a random variable, dictated largely by the strength of your opponent.

To indicate random variables, we will adopt the convention of a capital italic letter (for example, X). To denote the situation in which the random variable takes a specific value, we will use a lowercase italic letter (for example, x). Random variables are classified as either discrete or continuous:

- Any random variables that take on only isolated (countable) values are **discrete random variables**.
- **Continuous random variables** may have any value within a certain interval.

For example, the outcome of a game should be a discrete random variable. By contrast, suppose you are interested in the amount of beverage sold on a given day that the game is played. The quantity (or volume) of beverage sold will depend on the weather conditions, the number of people attending the game, and other factors. In this case, the quantity is a continuous random variable—a random variable that takes a value from a continuous set of values.

Probability Distributions

For a discrete random variable, the probability of each random event needs to be assessed. For a continuous random variable, the probability function needs to be assessed, as the event takes place over a continuous domain. In either case, a range of probabilities for each feasible outcome exists. Together, these probabilities make up a **probability distribution**.

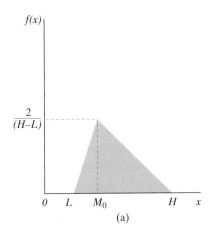
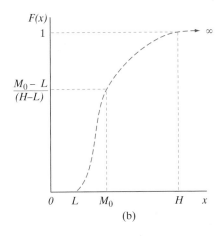

Figure 12.3 A triangular probability distribution: (a) Probability function and (b) cumulative probability distribution.

Probability assessments may be based on past observations or historical data if the trends that were characteristic of the past are expected to prevail in the future. Forecasting weather or predicting the outcome of a game in many professional sports is done on the basis of compiled statistical data. Any probability assessments based on objective data are called **objective probabilities**. However, in many real investment situations, no objective data are available to consider. In these situations, we assign **subjective probabilities** that we think are appropriate to the possible states of nature. As long as we act consistently with our beliefs about the possible events, we may reasonably account for the economic consequences of those events in our profitability analysis.

For a continuous random variable, we usually try to establish a range of values; that is, we try to determine a **minimum value** (L) and a **maximum value** (H). Next, we determine whether any value within these limits might be *more likely* to occur than the other values; in other words, does the distribution have a **mode** (M_O), or a most frequently occurring value?

- If the distribution does have a mode, we can represent the variable by a **triangular distribution**, such as that shown in Figure 12.3.
- If we have no reason to assume that one value is any more likely to occur than any other, perhaps the best we can do is to describe the variable as a **uniform distribution**, as shown in Figure 12.4.

These two distributions are frequently used to represent the variability of a random variable when the only information we have is its minimum, its maximum, and whether the distribution has a mode. For example, suppose the best judgment of the analyst was that the sales revenue could vary anywhere from $2,000 to $5,000 per day, and any value within the range is equally likely. This judgment about the variability of sales revenue could be represented by a uniform distribution.

For BMC's transmission-housings project, we can think of the discrete random variables (X and Y) as variables whose values cannot be predicted with certainty at the time of decision making. Let us assume the probability distributions in Table 12.5. We

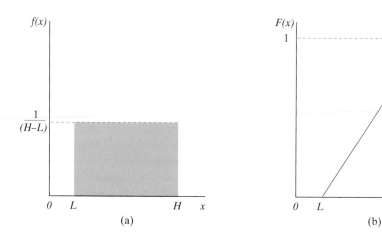

Figure 12.4 A uniform probability distribution: (a) Probability function and (b) cumulative probability distribution.

TABLE 12.5 Probability Distributions for Unit Demand (*X*) and Unit Price (*Y*) for BMC's Project

Product Demand (*X*)		Unit Sale Price (*Y*)	
Units (*x*)	$P(X = x)$	Unit Price (*y*)	$P(Y = y)$
1,600	0.20	$48	0.30
2,000	0.60	50	0.50
2,400	0.20	53	0.20

X and *Y* are independent random variables.

see that the product demand with the highest probability is 2,000 units, whereas the unit sale price with the highest probability is $50. These, therefore, are the most likely values. We also see a substantial probability that a unit demand other than 2,000 units will be realized. When we use only the most likely values in an economic analysis, we are in fact ignoring these other outcomes.

Cumulative Distribution

As we have observed in the previous section, the probability distribution provides information regarding the probability that a random variable will assume some value *x*. We can use this information, in turn, to define the cumulative distribution function. The **cumulative distribution** function gives the probability that the random variable will attain a value smaller than or equal to some value *x*. A common notation for the cumulative distribution is

$$F(x) = P(X \le x) = \begin{cases} \displaystyle\sum_{j=1}^{j} p_j & \text{(for a discrete random variable)} \\ \displaystyle\int_{L}^{x} f(x)\, dx & \text{(for a continuous random variable)} \end{cases},$$

where p_j is the probability of occurrence of the x_jth value of the discrete random variable and $f(x)$ is a probability function for a continuous variable. With respect to a continuous random variable, the cumulative distribution rises continuously in a smooth (rather than stairwise) fashion.

Example 12.4 reveals the method by which probabilistic information can be incorporated into our analysis. Again, BMC's transmission-housings project will be used. In the next section, we will show you how to compute some composite statistics with the use of all the data.

EXAMPLE 12.4 Cumulative Probability Distributions

Suppose that the only parameters subject to risk are the number of unit sales (X) to Gulf Electric each year and the unit sales price (Y). From experience in the market, BMC assesses the probabilities of outcomes for each variable as shown in Table 12.5. Determine the cumulative probability distributions for these random variables.

SOLUTION

Consider the demand probability distribution (X) given in Table 12.5 for BMC's transmission-housings project:

Unit Demand (X)	Probability, $P(X = x)$
1,600	0.2
2,000	0.6
2,400	0.2

If we want to know the probability that demand will be less than or equal to any particular value, we can use the following cumulative probability function:

$$F(x) = P(X \le x) = \begin{cases} 0.2 & x \le 1,600 \\ 0.8 & x \le 2,000. \\ 1.0 & x \le 2,400 \end{cases}$$

For example, if we want to know the probability that the demand will be less than or equal to 2,000, we can examine the appropriate interval $(x \le 2,000)$, and we shall find that the probability is 80%.

We can find the cumulative distribution for Y in a similar fashion. Figures 12.5 and 12.6 illustrate probability and cumulative probability distributions for X and Y, respectively.

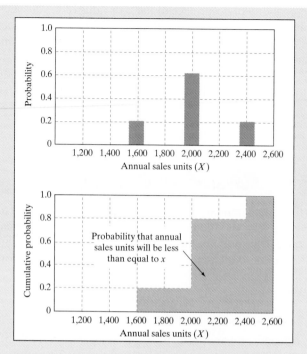

Figure 12.5 Probability and cumulative probability distributions for random variable X (annual sales).

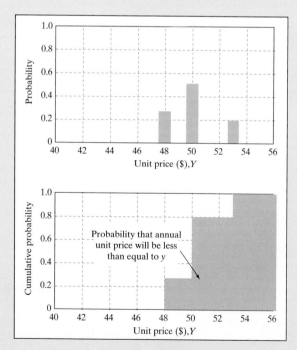

Figure 12.6 Probability and cumulative probability distributions for random variable Y from Table 12.5.

12.3.2 Summary of Probabilistic Information

Although knowledge of the probability distribution of a random variable allows us to make a specific probability statement, a single value that may characterize the random variable and its probability distribution is often desirable. Such a quantity is the **expected value** of the random variable. We also want to know something about how the values of the random variable are dispersed about the expected value (i.e., the **variance**). In investment analysis, this dispersion information is interpreted as the degree of project risk. The expected value indicates the weighted average of the random variable, and the variance captures the variability of the random variable.

Measure of Expectation

The **expected value** (also called the **mean**) is a weighted average value of the random variable, where the weighting factors are the probabilities of occurrence. All distributions (discrete and continuous) have an expected value. We will use $E[X]$ (or μ_x) to denote the expected value of random variable X. For a random variable X that has either discrete or continuous values, we compute the expected value with the formula

> **Expected value** represents the average amount one "expects" to win per bet if bets with identical odds are repeated many times.

$$E[X] = \mu_x = \begin{cases} \displaystyle\sum_{j=1}^{J} (p_j)x_j & \text{(discrete case)} \\ \displaystyle\int_{L}^{H} xf(x)\,dx & \text{(continuous case)} \end{cases}, \qquad (12.1)$$

where J is the number of discrete events and L and H are the lower and upper bounds of the continuous probability distribution.

The expected value of a distribution gives us important information about the "average" value of a random variable, such as the NPW, but it does not tell us anything about the variability on either side of the expected value. Will the range of possible values of the random variable be very small, and will all the values be located at or near the expected value? For example, the following represents the temperatures recorded on a typical winter day for two cities in the United States:

Location	Low	High	Average
Atlanta	15°F	67°F	41°F
Seattle	36°F	48°F	42°F

Even though both cities had almost identical mean (average) temperatures on that particular day, they had different variations in extreme temperatures. We shall examine this variability issue next.

Measure of Variation

Another measure needed when we are analyzing probabilistic situations is a measure of the risk due to the variability of the outcomes. Among the various measures of the variation of a set of numbers that are used in statistical analysis are the **range**, the **variance**, and the **standard deviation**. The variance and the standard deviation are used most commonly in the analysis of risk. We will use $\text{Var}[X]$ or σ_x^2 to denote the variance, and σ_x to denote the standard deviation, of random variable X. (If there is only one random variable in an analysis, we normally omit the subscript.)

The variance tells us the degree of spread, or dispersion, of the distribution on either side of the mean value. As the variance increases, the spread of the distribution increases; the smaller the variance, the narrower is the spread about the expected value.

To determine the variance, we first calculate the deviation of each possible outcome x_j from the expected value $(x_j - \mu)$. Then we raise each result to the second power and multiply it by the probability of x_j occurring (i.e., p_j). The summation of all these products serves as a measure of the distribution's variability. For a random variable that has only discrete values, the equation for computing the variance is[4]

Variance of a random variable is a measure of its statistical dispersion, indicating how far from the expected value its values typically are.

$$\text{Var}[X] = \sigma_x^2 = \sum_{j=1}^{J}(x_j - \mu)^2(p_j), \tag{12.2}$$

where p_j is the probability of occurrence of the jth value of the random variable (x_j), and μ is as defined by Eq. (12.1). To be most useful, any measure of risk should have a definite value (unit). One such measure is the **standard deviation**, which we may calculate by taking the positive square root of Var[X], measured in the same units as X:

$$\sigma_x = \sqrt{\text{Var}[X]}. \tag{12.3}$$

The standard deviation is a probability-weighted deviation (more precisely, the square root of the sum of squared deviations) from the expected value. Thus, it gives us an idea of how far above or below the expected value the actual value is likely to be. For most probability distributions, the actual value will be observed within the $\pm 3\sigma$ range.

In practice, the calculation of the variance is somewhat easier if we use the formula

$$\text{Var}[X] = \sum p_j x_j^2 - \left(\sum p_j x_j\right)^2$$
$$= E[X^2] - (E[X])^2. \tag{12.4}$$

The term $E[X^2]$ in Eq. (12.4) is interpreted as the mean value of the squares of the random variable (i.e., the actual values squared). The second term is simply the mean value squared. Example 12.5 illustrates how we compute measures of variation.

EXAMPLE 12.5 Calculation of Mean and Variance

Consider once more BMC's transmission-housings project. Unit sales (X) and unit price (Y) are estimated as in Table 12.5. Compute the means, variances, and standard deviations for the random variables X and Y.

SOLUTION

For the product demand variable (X), we have

x_j	p_j	$x_j(p_j)$	$(x_j - E[X])^2$	$(x_j - E[X])^2(p_j)$
1,600	0.20	320	$(-400)^2$	32,000
2,000	0.60	1,200	0	0
2,400	0.20	480	$(400)^2$	32,000
		$E[X] = 2,000$		Var[X] = 64,000
				$\sigma_x = 252.98$

[4] For a continuous random variable, we compute the variance as follows: $\text{Var}[X] = \int_{L}^{H}(x - \mu)^2 f(x)\, dx.$

For the variable unit price (Y), we have

y_j	p_j	$y_j(p_j)$	$(y_j - E[Y])^2$	$(y_j - E[Y])^2(p_j)$
$48	0.30	$14.40	$(-2)^2$	1.20
50	0.50	25.00	$(0)^2$	0
53	0.20	10.60	$(3)^2$	1.80
		$E[Y] = 50.00$		$\text{Var}[Y] = 3.00$
				$\sigma_y = 1.73$

12.3.3 Joint and Conditional Probabilities

Thus far, we have not looked at how the values of some variables can influence the values of others. It is, however, entirely possible—indeed, it is likely—that the values of some parameters will be dependent on the values of others. We commonly express these dependencies in terms of conditional probabilities. An example is product demand, which will probably be influenced by unit price.

We define a **joint probability** as

$$P(x, y) = P(X = x|Y = y)P(Y = y), \tag{12.5}$$

where $P(X = x|Y = y)$ is the **conditional probability** of observing x, given $Y = y$, and $P(Y = y)$ is the **marginal probability** of observing $Y = y$. Certainly, important cases exist in which a knowledge of the occurrence of event X does *not* change the probability of an event Y. That is, if X and Y are **independent**, then the joint probability is simply

$$P(x, y) = P(x)P(y). \tag{12.6}$$

The concepts of joint, marginal, and conditional distributions are best illustrated by numerical examples.

Suppose that BMC's marketing staff estimates that, for a given unit price of $48, the conditional probability that the company can sell 1,600 units is 0.10. The probability of this joint event (unit sales = 1,600 and unit sales price = $48) is

$$P(x, y) = P(1,600, \$48)$$
$$= P(x = 1,600|y = \$48)P(y = \$48)$$
$$= (0.10)(0.30)$$
$$= 0.03.$$

We can obtain the probabilities of other joint events in a similar fashion; several are presented in Table 12.6.

TABLE 12.6 Assessments of Conditional and Joint Probabilities

Unit Price Y	Probability	Conditional Unit Sales X	Joint Probability	Probability
		1,600	0.10	0.03
$48	0.30	2,000	0.40	0.12
		2,400	0.50	0.15
		1,600	0.10	0.05
50	0.50	2,000	0.64	0.32
		2,400	0.26	0.13
		1,600	0.50	0.10
53	0.20	2,000	0.40	0.08
		2,400	0.10	0.02

From Table 12.6, we can see that the unit demand (X) ranges from 1,600 to 2,400 units, the unit price (Y) ranges from \$48 to \$53, and nine joint events are possible. These joint probabilities must sum to unity:

Joint Event (xy)	P(x, y)
(1,600, \$48)	0.03
(2,000, \$48)	0.12
(2,400, \$48)	0.15
(1,600, \$50)	0.05
(2,000, \$50)	0.32
(2,400, \$50)	0.13
(1,600, \$53)	0.10
(2,000, \$53)	0.08
(2,400, \$53)	0.02
	Sum = 1.00

The marginal distribution for x can be developed from the joint event by fixing x and summing over y:

x_j	$P(x_j) = \sum_y P(x, y)$
1,600	$P(1{,}600, \$48) + P(1{,}600, \$50) + P(1{,}600, \$53) = 0.18$
2,000	$P(2{,}000, \$48) + P(2{,}000, \$50) + P(2{,}000, \$53) = 0.52$
2,400	$P(2{,}400, \$48) + P(2{,}400, \$50) + P(2{,}400, \$53) = 0.30$

This marginal distribution tells us that 52% of the time we can expect to have a demand of 2,000 units and 18% and 30% of the time we can expect to have a demand of 1,600 and 2,400, respectively.

12.3.4 Covariance and Coefficient of Correlation

When two random variables are not independent, we need some measure of their dependence on each other. The parameter that tells the degree to which two variables (X, Y) are related is the covariance $Cov(X, Y)$, denoted by σ_{xy}. Mathematically, we define

$$Cov(X, Y) = \sigma_{xy}$$
$$= E\{(X - E[X])(Y - E[Y])\}$$
$$= E(XY) - E(X)E(Y)$$
$$= \rho_{xy}\sigma_x\sigma_y, \tag{12.7}$$

where ρ_{xy} is the coefficient of correlation between X and Y. It is clear that if X tends to exceed its mean whenever Y exceeds its mean, $Cov(X, Y)$ will be positive. If X tends to fall below its mean whenever Y exceeds its mean, then $Cov(X, Y)$ will be negative. The sign of $Cov(X, Y)$, therefore, reveals whether X and Y vary directly or inversely with one another. We can rewrite Eq. (12.7) in terms of ρ_{xy}:

$$\rho_{xy} = \frac{Cov(X, Y)}{\sigma_x\sigma_y}. \tag{12.8}$$

The value of ρ_{xy} can vary within the range of -1 and $+1$, with $\rho_{xy} = 0$ indicating no correlation between the two random variables. As shown in Table 12.7, the coefficient of correlation between the product demand (X) and the unit price (Y) is negatively correlated, $\rho_{xy} = -0.439$, indicating that as the firm lowers the unit price, it tends to generate a higher demand.

12.4 Probability Distribution of NPW

After we have identified the random variables in a project and assessed the probabilities of the possible events, the next step is to develop the project's NPW distribution.

12.4.1 Procedure for Developing an NPW Distribution

We will consider the situation in which all the random variables used in calculating the NPW are independent. To develop the NPW distribution, we may follow these steps:

- Express the NPW as functions of unknown random variables.
- Determine the probability distribution for each random variable.
- Determine the joint events and their probabilities.
- Evaluate the NPW equation at these joint events.
- Rank the NPW values in increasing order of NPW.

These steps are illustrated in Example 12.6.

TABLE 12.7 Calculating the Correlation Coefficient between Two Random Variables X and Y

(x, y)	$(x - E[X])$	$(y - E[Y])$	$p(x, y)$	$(x - E[X])(y - E[Y])$	$p(x, y) \times (x - E[X])(y - E[Y])$
(1,600, 48)	(1,600 − 2,000)	(48 − 50)	0.03	800	24
(2,000, 48)	(2,000 − 2,000)	(48 − 50)	0.12	0	0
(2,400, 48)	(2,400 − 2,000)	(48 − 50)	0.15	−800	−120
(1,600, 50)	(1,600 − 2,000)	(50 − 50)	0.05	0	0
(2,000, 50)	(2,000 − 2,000)	(50 − 50)	0.32	0	0
(2,400, 50)	(2,400 − 2,000)	(50 − 50)	0.13	0	0
(1,600, 53)	(1,600 − 2,000)	(53 − 50)	0.10	−1,200	−120
(2,000, 53)	(2,000 − 2,000)	(53 − 50)	0.08	0	0
(2,400, 53)	(2,400 − 2,000)	(53 − 50)	0.02	1,200	24

$$\text{Cov}(X, Y) = -192$$

$$\rho_{xy} = \frac{\text{Cov}(X, Y)}{\sigma_x \sigma_y}$$

$$= \frac{-192}{\left(\sqrt{64,000} \times \sqrt{3.00} \right)}$$

$$= -0.439$$

EXAMPLE 12.6 Procedure for Developing an NPW Distribution

Consider again BMC's transmission-housings project first set forth in Example 12.1. Use the unit demand (X) and price (Y) given in Table 12.5, and develop the NPW distribution for the BMC project. Then calculate the mean and variance of the NPW distribution.

SOLUTION

Table 12.8 summarizes the after-tax cash flow for the BMC's transmission-housings project as functions of random variables X and Y. From the table, we can compute the PW of cash inflows as follows:

$$PW(15\%) = 0.6XY(P/A, 15\%, 5) + \$44,490$$
$$= 2.0113XY + \$44,490.$$

The PW of cash outflows is

$$PW(15\%) = \$125,000 + (9X + \$6,000)(P/A, 15\%, 5)$$
$$= 30.1694X + \$145,113.$$

Thus, the NPW is

$$PW(15\%) = 2.0113X(Y - \$15) - \$100,623.$$

If the product demand X and the unit price Y are independent random variables, then PW(15%) will also be a random variable. To determine the NPW distribution, we need to consider all the combinations of possible outcomes.[5] The first possibility is the event in which $x = 1,600$ and $y = \$48$. Since X and Y are considered to be independent random variables, the probability of this joint event is

$$P(x = 1,600, y = \$48) = P(x = 1,600)P(y = \$48)$$
$$= (0.20)(0.30)$$
$$= 0.06.$$

With these values as input, we compute the possible NPW outcome as follows:

$$PW(15\%) = 2.0113X(Y - \$15) - \$100,623$$
$$= 2.0113(1,600)(\$48 - \$15) - \$100,623$$
$$= \$5,574.$$

Eight other outcomes are possible; they are summarized with their joint probabilities in Table 12.9 and depicted in Figure 12.7.

The NPW probability distribution in Table 12.9 indicates that the project's NPW varies between $5,574 and $82,808, but that no loss occurs under any of the circumstances examined. On the one hand, from the cumulative distribution, we further observe that there is a 0.38 probability that the project would realize an NPW less than that forecast for the base case ($40,168). On the other hand, there is a 0.32 probability that the NPW will be greater than this value. Certainly, the probability distribution provides much more information on the likelihood of each possible event than does the scenario analysis presented in Section 12.2.3.

---

[5] If X and Y are dependent random variables, the joint probabilities developed in Table 12.6 should be used.

TABLE 12.8 After-Tax Cash Flow as a Function of Unknown Unit Demand (X) and Unit Price (Y)

Item	0	1	2	3	4	5
Cash inflow:						
Net salvage						37,389
Revenue:						
$X(1 - 0.4)Y$		0.6XY	0.6XY	0.6XY	0.6XY	0.6XY
Depreciation credit:						
0.4 (depreciation)		7,145	12,245	8,745	6,245	2,230
Cash outflow:						
Investment	−125,000					
Variable cost:						
$-X(1 - 0.4)(\$15)$		−9X	−9X	−9X	−9X	−9X
Fixed cost:						
$-(1 - 0.4)(\$10,000)$		−6,000	−6,000	−6,000	−6,000	−6,000
Net cash flow	−125,000	$0.6X(Y - 15)$ $+1,145$	$0.6X(Y - 15)$ $+6,245$	$0.6X(Y - 15)$ $+2,745$	$0.6X(Y - 15)$ $+245$	$0.6X(Y - 15)$ $+33,617$

TABLE 12.9 The NPW Probability Distribution with Independent Random Variables

Event No.	x	y	P(x,y)	Cumulative Joint Probability	NPW
1	1,600	$48.00	0.06	0.06	$ 5,574
2	1,600	50.00	0.10	0.16	12,010
3	1,600	53.00	0.04	0.20	21,664
4	2,000	48.00	0.18	0.38	32,123
5	2,000	50.00	0.30	0.68	40,168
6	2,000	53.00	0.12	0.80	52,236
7	2,400	48.00	0.06	0.86	58,672
8	2,400	50.00	0.10	0.96	68,326
9	2,400	53.00	0.04	1.00	82,808

We have developed a probability distribution for the NPW by considering random cash flows. As we observed, a probability distribution helps us to see what the data imply in terms of project risk. Now we can learn how to summarize the probabilistic information—the mean and the variance:

- For BMC's transmission-housings project, we compute the expected value of the NPW distribution as shown in Table 12.10. Note that this expected value is the same as the most likely value of the NPW distribution. This equality was expected because both X and Y have a symmetrical probability distribution.
- We obtain the variance of the NPW distribution, assuming independence between X and Y and using Eq. (12.2), as shown in Table 12.11. We could obtain the same result more easily by using Eq. (12.4).

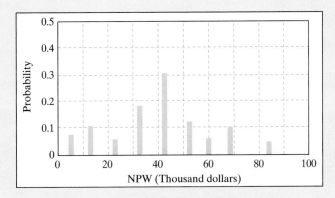

Figure 12.7 The NPW probability distribution when X and Y are independent (Example 12.6).

TABLE 12.10 Calculation of the Mean of the NPW Distribution

Event No.	x	y	P(x,y)	Cumulative Joint Probability	NPW	Weighted NPW
1	1,600	$ 48.00	0.06	0.06	$ 5,574	$ 334
2	1,600	50.00	0.10	0.16	12,010	1,201
3	1,600	53.00	0.04	0.20	21,664	867
4	2,000	48.00	0.18	0.38	32,123	5,782
5	2,000	50.00	0.30	0.68	40,168	12,050
6	2,000	53.00	0.12	0.80	52,236	6,268
7	2,400	48.00	0.06	0.86	58,672	3,520
8	2,400	50.00	0.10	0.96	68,326	6,833
9	2,400	53.00	0.04	1.00	82,808	3,312

$$E[\text{NPW}] = \$40,168$$

TABLE 12.11 Calculation of the Variance of the NPW Distribution

Event No.	x	y	P(x,y)	NPW	$(\text{NPW} - E[\text{NPW}])^2$	Weighted $(\text{NPW} - E[\text{NPW}])^2$
1	1,600	$48.00	0.06	$ 5,574	1,196,769,744	$71,806,185
2	1,600	50.00	0.10	12,010	792,884,227	79,288,423
3	1,600	53.00	0.04	21,664	342,396,536	13,695,861
4	2,000	48.00	0.18	32,123	64,725,243	11,650,544
5	2,000	50.00	0.30	40,168	0	0
6	2,000	53.00	0.12	52,236	145,631,797	17,475,816
7	2,400	48.00	0.06	58,672	342,396,536	20,543,792
8	2,400	50.00	0.10	68,326	792,884,227	79,288,423
9	2,400	53.00	0.04	82,808	1,818,132,077	72,725,283

$$\text{Var}[\text{NPW}] = 366,474,326$$
$$\sigma = \$19,144$$

COMMENTS: We can obtain the mean and variance of the NPW analytically by using the properties of the product of random variables. Let $W = XY$, where X and Y are random variables with known means and variances (μ_x, σ_x^2) and (μ_y, σ_y^2), respectively. If X and Y are independent of each other, then

$$E(W) = E(XY) = \mu_x \mu_y \qquad (12.9)$$

and

$$\text{Var}(W) = \text{Var}(XY) = \mu_x^2 \sigma_y^2 + \mu_y^2 \sigma_x^2 + \sigma_x^2 \sigma_y^2. \qquad (12.10)$$

The expected net present worth is then

$$E[PW(15\%)] = 2.0113E(XY) - (2.0113)(15)E(X) - 100,623$$
$$= 2.0113(2,000)(50) - (2.0113)(15)(2,000) - 100,623$$
$$= \$40,168.$$

Now let $Z = Y - 15$. Then $E(Z) = E(Y) - 15 = \mu_y - 15 = 50 - 15 = 35$ and $Var[Z] = Var[Y] = \sigma_y^2 = 3$. So

$$Var[PW(15\%)] = Var[2.0113X(Y - 15) - 100,623]$$
$$= Var[2.0113XZ]$$
$$= (2.0113)^2 Var[XY]$$
$$= (2.0113)^2(\mu_x^2\sigma_y^2 + \mu_y^2\sigma_x^2 + \sigma_x^2\sigma_y^2)$$
$$= (2.0113)^2[(2,000^2)(3) + (50^2)(64,000) + (64,000)(3)]$$
$$= 366,474,326.$$

Note that the mean and variance thus calculated are exactly the same as in Tables 12.10 and 12.11, respectively. Note also that if X and Y are correlated with each other, then Eq. (12.10) cannot be used.[6]

12.4.2 Aggregating Risk over Time

In the previous section, we developed an NPW equation by aggregating all cash flow components over time. Another approach to estimating the amount of risk present in a particular investment opportunity is to determine the mean and variance of cash flows in each period; then we may be able to aggregate the risk over the project life in terms of net present worth. We have two cases:

- **Independent random variables.** In this case,

$$E[PW(i)] = \sum_{n=0}^{N} \frac{E(A_n)}{(1 + i)^n} \tag{12.11}$$

and

$$Var[PW(i)] = \sum_{n=0}^{N} \frac{Var(A_n)}{(1 + i)^{2n}} \tag{12.12}$$

where

$$A_n = \text{cash flow in period } n,$$
$$E(A_n) = \text{expected cash flow in period } n,$$
$$Var(A_n) = \text{variance of the cash flow in period } n.$$

---

[6] Analytical treatment for the products of random variables including the correlated case is given by Chan S. Park and Gunter Sharp-Bette, *Advanced Engineering Economics*, New York, John Wiley, 1990 (Chapter 10).

In defining Eq. (12.12), we are also assuming the independence of cash flows, meaning that knowledge of what will happen to one particular period's cash flow will not allow us to predict what will happen to cash flows in other periods.

- **Dependent random variables.** In case we cannot assume a statistical independence among cash flows, we need to consider the degree of dependence among them. The expected-value calculation will not be affected by this dependence, but the project variance will be calculated as

$$\text{Var}[\text{PW}(i)] = \sum_{n=0}^{N} \frac{\text{Var}(A_n)}{(1+i)^{2n}} + 2 \sum_{n=0}^{N-1} \sum_{s=n+1}^{N} \frac{\rho_{ns} \sigma_n \sigma_s}{(1+i)^{n+s}}, \qquad (12.13)$$

where ρ_{ns} = correlation coefficient (degree of dependence) between A_n and A_s. The value of ρ_{ns} can vary within the range from -1 to $+1$, with $\rho_{ns} = -1$ indicating perfect negative correlation and $\rho_{ns} = +1$ indicating perfect positive correlation. The result $\rho_{ns} = 0$ implies that no correlation exists between A_n and A_s. If $\rho_{ns} > 0$, we can say that A_n and A_s are positively correlated, meaning that if the actual realization of cash flow for A_n is higher than its expected value, it is likely that you will also observe a higher cash flow than its expected value for A_s. If $\rho_{ns} < 0$, the opposite relation exists.

Example 12.7 illustrates the mechanics involved in calculating the mean and variance of a project's net present worth.

EXAMPLE 12.7 Aggregation of Risk over Time

Consider the following financial data for an investment project, where only random components are the operating expenses in each period:

- Investment required = $10,000
- Project life = 3 years
- Expected salvage value = $0
- Annual operating revenue = $20,000
- Annual operating expenses are random variables with the following means and variances:

n	X_1	X_2	X_3
Mean	$9,000	$13,000	$15,000
Variance	250,000	490,000	1,000,000

- Depreciation method is alternative MACRS (3 years):

n	1	2	3
D_n	$1,667	$3,333	$1,667

- Tax rate = 40%
- Discount rate (or MARR) = 12%

Determine the expected net present worth and the variance of the NPW assuming the following two situations: (a) X_i are independent random variables and (b) X_i are dependent random variables.

SOLUTION

Step 1: Calculate the net proceeds from disposal of the equipment at the end of the project life:

Salvage value = 0,

$$\text{Book value} = 10,000 - \sum_{n=1}^{3} D_n = \$3,333,$$

Taxable gain or (loss) = $0 - \$3,333 = (\$3,333)$,

Loss credit = $0.4(\$3,333) = \$1,333$,

Net proceeds from sale = Salvage value + Loss credit = $\$1,333$.

Step 2: Construct a generalized cash flow table as a function of the random variable X_n:

Description	Cash Flow			
	0	1	2	3
Investment	−10,000			
Revenue × (0.6)		12,000	12,000	12,000
−O&M × (0.6)		−0.6X_1	−0.6X_2	−0.6X_3
D_n × (0.4)		667	1,334	667
Net proceeds from sale				1,333
Net cash flow	−10,000	12,667	13,334	14,000
		−0.6X_1	−0.6X_2	−0.6X_3

$E[X_1] = 9,000,$ $\quad E[X_2] = 13,000,$ $\quad E[X_3] = 15,000,$
$\text{Var}[X_1] = 250,000,$ $\text{Var}[X_2] = 490,000,$ $\text{Var}[X_3] = 1,000,000.$

The net present worth is then

$$PW(12\%) = \left[-10,000 + \frac{12,667}{1.12} + \frac{13,334}{1.12^2} + \frac{14,000}{1.12^3} \right]$$
$$- 0.6 \left[\frac{X_1}{1.12} + \frac{X_2}{1.12^2} + \frac{X_3}{1.12^3} \right]$$
$$= 21,905 - 0.6 \left[\frac{X_1}{1.12} + \frac{X_2}{1.12^2} + \frac{X_3}{1.12^3} \right].$$

- **Case 1.** Independent Cash Flows: Using Eqs. (12.10) and (12.12), we obtain

$$E[PW(12\%)] = 21{,}905 - 0.6\left[\frac{E[X_1]}{1.12} + \frac{E[X_2]}{1.12^2} + \frac{E[X_3]}{1.12^3}\right]$$

$$= 21{,}905 - 17{,}446$$

$$= \$4{,}459,$$

$$\text{Var}[PW(12\%)] = \left(\frac{-0.6}{1.12}\right)^2 \text{Var}[X_1] + \left(\frac{-0.6}{1.12^2}\right)^2 \text{Var}[X_2]$$

$$+ \left(\frac{-0.6}{1.12^3}\right)^2 \text{Var}[X_3]$$

$$= 71{,}747.45 + 112{,}103.39 + 182{,}387.20$$

$$= 366{,}240,$$

$$\sigma[PW(12\%)] = \$605.18.$$

- **Case 2.** Dependent Cash Flows: If the random variables X_1, X_2, and X_3 are partially correlated with the correlation coefficients $\rho_{12} = 0.3$, $\rho_{13} = 0.5$, and $\rho_{23} = 0.4$, respectively (see accompanying diagram), then the expected value and the variance of the NPW can be calculated with Eq. (12.13):

$$E[PW(12\%)] = 21{,}905 - 0.6\left[\frac{E[X_1]}{1.12} + \frac{E[X_2]}{1.12^2} + \frac{E[X_3]}{1.12^3}\right]$$

$$= 21{,}905 - 17{,}446$$

$$= \$4{,}459,$$

$$\text{Var}[PW(12\%)] = (\text{Original variance}) + (\text{Covariance terms})$$

$$= 366{,}240$$

$$+ 2\left(\frac{-0.6}{1.12}\right)\left(\frac{-0.6}{1.12^2}\right)\rho_{12}\sigma_1\sigma_2$$

$$+ 2\left(\frac{-0.6}{1.12}\right)\left(\frac{-0.6}{1.12^3}\right)\rho_{13}\sigma_1\sigma_3$$

$$+ 2\left(\frac{-0.6}{1.12^2}\right)\left(\frac{-0.6}{1.12^3}\right)\rho_{23}\sigma_2\sigma_3$$

$$= 366{,}240$$

$$+ 2\left(\frac{0.6^2}{1.12^3}\right)(0.3)(500)(700)$$

$$+ 2\left(\frac{0.6^2}{1.12^4}\right)(0.5)(500)(1{,}000)$$

$$+ 2\left(\frac{0.6^2}{1.12^5}\right)(0.4)(700)(1{,}000)$$

$$= 366{,}240 + 53{,}810.59 + 114{,}393.25 + 114{,}393.25$$

$$= 648{,}837,$$

$$\sigma[\text{PW}(12\%)] = \$805.50.$$

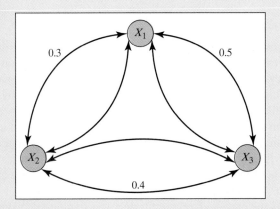

Note that the variance increases significantly when at least some of the random variables are positively correlated.

COMMENTS: How is information such as the preceding used in decision making? Most probability distributions are completely described by six standard deviations— three above, and three below, the mean. As shown in Figure 12.8, the NPW of this project would almost certainly fall between $2,643 and $6,275 for the independent case and between $2,043 and $6,876 for the dependent case. In either situation, the NPW below 3σ from the mean is still positive, so we may say that the project is quite safe. If that figure were negative, it would then be up to the decision maker to determine whether it is worth investing in the project, given its mean and standard deviation.

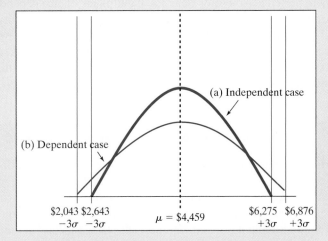

Figure 12.8 NPW distributions with $\pm 3\sigma$: (a) Independent case and (b) dependent case.

12.4.3 Decision Rules for Comparing Mutually Exclusive Risky Alternatives

Once the expected value has been located from the NPW distribution, it can be used to make an accept–reject decision in much the same way that a single NPW is used when a single possible outcome for an investment project is considered.

Expected-Value Criterion

The decision rule is called the **expected-value criterion**, and using it, we may accept a single project if its expected NPW value is positive. In the case of mutually exclusive alternatives, we select the one with the highest expected NPW. The use of the expected NPW has an advantage over the use of a point estimate, such as the likely value, because it includes all possible cash flow events and their probabilities.

> **Laws of large numbers** imply that the average of a random sample from a large population is likely to be close to the mean of the whole population.

The justification for the use of the expected-value criterion is based on the **law of large numbers**, which states that if many repetitions of an experiment are performed, the average outcome will tend toward the expected value. This justification may seem to negate the usefulness of the criterion, since, in project evaluation, we are most often concerned with a single, nonrepeatable "experiment" (i.e., an investment alternative). However, if a firm adopts the expected-value criterion as a standard decision rule for *all* of its investment alternatives, then, over the long term, the law of large numbers predicts that accepted projects tend to meet their expected values. Individual projects may succeed or fail, but the average project tends to meet the firm's standard for economic success.

Mean-and-Variance Criterion

The expected-value criterion is simple and straightforward to use, but it fails to reflect the variability of the outcome of an investment. Certainly, we can enrich our decision by incorporating information on variability along with the expected value. Since the variance represents the dispersion of the distribution, it is desirable to minimize it. In other words, the smaller the variance, the less the variability (the potential for loss) associated with the NPW. Therefore, when we compare mutually exclusive projects, we may select the alternative with the smaller variance if its expected value is the same as or larger than those of other alternatives.

In cases where preferences are not clear cut, the ultimate choice depends on the decision maker's trade-offs—how far he or she is willing to take the variability to achieve a higher expected value. In other words, the challenge is to decide what level of risk you are willing to accept and then, having decided on your tolerance for risk, to understand the implications of that choice. Example 12.8 illustrates some of the critical issues that need to be considered in evaluating mutually exclusive risky projects.

EXAMPLE 12.8 Comparing Risky Mutually Exclusive Projects

With ever-growing concerns about air pollution, the greenhouse effect, and increased dependence on oil imports in the United States, Green Engineering has developed a prototype conversion unit that allows a motorist to switch from gasoline to compressed natural gas (CNG) or vice versa. Driving a car equipped with Green's conversion kit is not much different from driving a conventional model. A small dial switch on the dashboard controls which fuel is to be used. Four different configurations are available, according to the type of vehicle: compact size, mid-size, large size, and trucks. In the past, Green has built a few prototype vehicles powered by alternative fuels other than gasoline, but has been reluctant to go into higher volume production without more evidence of public demand.

As a result, Green Engineering initially would like to target one market segment (one configuration model) in offering the conversion kit. Green Engineering's marketing group has compiled the potential NPW distribution for each different configuration when marketed independently.

Event (NPW) (unit: thousands)	Probabilities			
	Model 1	Model 2	Model 3	Model 4
$1,000	0.35	0.10	0.40	0.20
1,500	0	0.45	0	0.40
2,000	0.40	0	0.25	0
2,500	0	0.35	0	0.30
3,000	0.20	0	0.20	0
3,500	0	0	0	0
4,000	0.05	0	0.15	0
4,500	0	0.10	0	0.10

Evaluate the expected return and risk for each model configuration, and recommend which one, if any, should be selected.

SOLUTION

For model 1, we calculate the mean and variance of the NPW distribution as follows:

$$E[NPW]_1 = \$1,000(0.35) + \$2,000(0.40)$$
$$+ \$3,000(0.20) + \$4,000(0.05)$$
$$= \$1,950;$$
$$Var[NPW]_1 = 1,000^2(0.35) + 2,000^2(0.40)$$
$$+ 3,000^2(0.20) + 4,000^2(0.05) - (1,950)^2$$
$$= 747,500.$$

In a similar manner, we compute the other values as follows:

Configuration	E[NPW]	Var[NPW]
Model 1	$1,950	747,500
Model 2	2,100	915,000
Model 3	2,100	1,190,000
Model 4	2,000	1,000,000

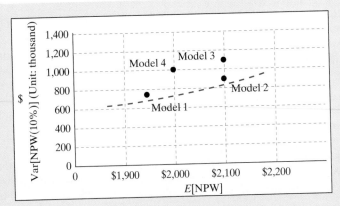

Figure 12.9 Mean–variance chart showing project dominance. Both model 3 and model 4 are dominated by model 2.

The results are plotted in Figure 12.9. If we make a choice solely on the basis of the expected value, we may select either model 2 or model 3, because they have the highest expected NPW.

If we consider the variability along with the expected NPW, however, the correct choice is not obvious. We will first eliminate those alternatives which are clearly inferior to others:

- *Model 2 versus Model 3.*   We see that models 2 and 3 have the same mean of $2,100, but model 2 has a much lower variance. In other words, model 2 dominates model 3, so we eliminate model 3 from further consideration.

- *Model 2 versus Model 4.*   Similarly, we see that model 2 is preferred over model 4, because model 2 has a higher expected value and a smaller variance; model 2 again dominates model 4, so we eliminate model 4. In other words, model 2 dominates both models 3 and 4 as we consider the variability of NPW.

- *Model 1 versus Model 2.*   Even though the mean-and-variance rule has enabled us to narrow down our choices to only two models (models 1 and 2), it does not indicate what the ultimate choice should be. In other words, comparing models 1 and 2, we see that E[NPW] increases from $1,950 to $2,100 at the price of a higher Var[NPW], which increases from 747,500 to 915,000. The choice, then, will depend on the decision maker's trade-offs between the incremental expected return ($150) and the incremental risk (167,500). We cannot choose between the two simply on the basis of mean and variance, so we must resort to other probabilistic information.[7]

Risk-Return Tradeoffs: The greater the amount of risk that an investor is willing to take on, the greater the potential return.

12.5 Risk Simulation

In the previous sections, we examined analytical methods of determining the NPW distributions and computing their means and variances. As we saw in Section 12.4.1, the NPW distribution offers numerous options for graphically presenting to the decision maker probabilistic information, such as the range and likelihoods of occurrence of possible levels

[7] As we seek further refinement in our decision under risk, we may consider the expected-utility theory, or stochastic dominance rules, which is beyond the scope of our text. (See Park, C. S., and Sharp-Bette, G. P., *Advanced Engineering Economics* (New York: John Wiley, 1990), Chapters 10 and 11.)

of NPW. Whenever we can adequately evaluate the risky investment problem by analytical methods, it is generally preferable to do so. However, many investment situations are such that we cannot solve them easily by analytical methods, especially when many random variables are involved. In these situations, we may develop the NPW distribution through computer simulation.

12.5.1 Computer Simulation

Before we examine the details of risk simulation, let us consider a situation in which we wish to train a new astronaut for a future space mission. Several approaches exist for training this astronaut. One (somewhat unlikely) possibility is to place the trainee in an actual space shuttle and to launch him or her into space. This approach certainly would be expensive; it also would be extremely risky, because any human error made by the trainee would have tragic consequences. As an alternative, we can place the trainee in a flight simulator designed to mimic the behavior of the actual space shuttle in flight. The advantage of this approach is that the astronaut trainee learns all the essential functions of space operation in a simulated space environment. The flight simulator generates test conditions approximating operational conditions, and any human errors made during training cause no harm to the astronaut or to the equipment being used.

The use of computer simulation is not restricted to simulating a physical phenomenon such as the flight simulator. In recent years, techniques for testing the results of some investment decisions before they are actually executed have been developed. As a result, many phases of business investment decisions have been simulated with considerable success. In fact, we can analyze BMC's transmission-housings project by building a simulation model. The general approach is to assign a subjective (or objective) probability distribution to each unknown factor and to combine all these distributions into a probability distribution for the project's profitability as a whole. The essential idea is that if we can simulate the actual state of nature for unknown investment variables on a computer, we may be able to obtain the resulting NPW distribution.

The unit demand (X) in our BMC's transmission-housing project was one of the random variables in the problem. We can know the exact value for this random variable only after the project is implemented. Is there any way to predict the actual value before we make any decision about the project?

The following logical steps are often suggested for a computer program that simulates investment scenarios:

Step 1. Identify all the variables that affect the measure of investment worth (e.g., NPW after taxes).

Step 2. Identify the relationships among all the variables. The relationships of interest here are expressed by the equations or the series of numerical computations by which we compute the NPW of an investment project. These equations make up the model we are trying to analyze.

Step 3. Classify the variables into two groups: the parameters whose values are known with certainty and the random variables for which exact values cannot be specified at the time of decision making.

Step 4. Define distributions for all the random variables.

Step 5. Perform Monte Carlo sampling (see Section 12.5.3) and describe the resulting NPW distribution.

Step 6. Compute the distribution parameters and prepare graphic displays of the results of the simulation.

Monte Carlo methods are a class of computational algorithms for simulating the behavior of various physical and mathematical system.

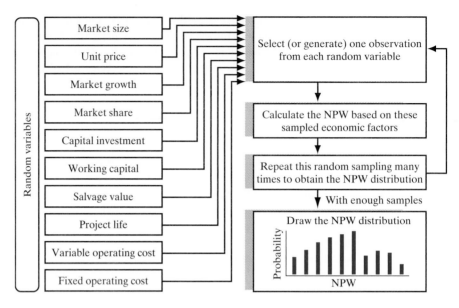

Figure 12.10 Logical steps involved in simulating a risky investment.

Figure 12.10 illustrates the logical steps involved in simulating a risky investment project. The risk simulation process has two important advantages compared with the analytical approach discussed in Section 12.4:

1. The number of variables that can be considered is practically unlimited, and the distributions used to define the possible values of each random variable can be of any type and any shape. The distributions can be based on statistical data if they are available or, more commonly, on subjective judgment.

2. The method lends itself to sensitivity analyses. By defining some factors that have the most significant effect on the resulting NPW values and using different distributions (in terms of either shape or range) for each variable, we can observe the extent to which the NPW distribution is changed.

12.5.2 Model Building

In this section, we shall present some of the procedural details related to the first three steps (model building) outlined in Section 12.5.1. To illustrate the typical procedure involved, we shall work with the investment setting for BMC's transmission-housings project described in Example 12.6.

The initial step is to define the measure of investment worth and the factors which affect that measure. For our presentation, we choose the measure of investment worth as an after-tax NPW computed at a given interest rate i. In fact, we are free to choose any measure of worth, such as annual worth or future worth. In the second step, we must divide into two groups all the variables that we listed in Step 1 as affecting the NPW. One group consists of all the parameters for which values are known. The second group includes all remaining parameters for which we do not know exact values at the time of analysis. The third step is to define the relationships that tie together all the variables. These relationships may take the form of a single equation or several equations.

EXAMPLE 12.9 Developing a Simulation Model

Consider again BMC's transmission-housings project, as set forth in Example 12.6. Identify the input factors related to the project and develop the simulation model for the NPW distribution.

DISCUSSION: For the BMC project, the variables that affect the NPW are the investment required, unit price, demand, variable production cost, fixed production cost, tax rate, and depreciation expenses, as well as the firm's interest rate. Some of the parameters that might be included in the known group are the investment cost and interest rate (MARR). If we have already purchased the equipment or have received a price quote, then we also know the depreciation amount. In addition, assuming that we are operating in a stable economy, we would probably know the tax rates for computing income taxes due.

The group of parameters with unknown values would usually include all the variables relating to costs and future operating expense and to future demand and sales prices. These are the random variables for which we must assess the probability distributions.

For simplicity, we classify the input parameters or variables for BMC's transmission-housings project as follows:

Assumed to Be Known Parameters	Assumed to Be Unknown Parameters
MARR	Unit price
Tax rate	Demand
Depreciation amount	Salvage value
Investment amount	
Project life	
Fixed production cost	
Variable production cost	

Note that, unlike the situation in Example 12.6, here we treat the salvage value as a random variable. With these assumptions, we are now ready to build the NPW equation for the BMC project.

SOLUTION

Recall that the basic investment parameters assumed for BMC's five-year project in Example 12.6 were as follows:

- Investment = $125,000
- Marginal tax rate = 0.40
- Annual fixed cost = $10,000
- Variable unit production cost = $15/unit
- MARR(i) = 15%

- Annual depreciation amounts:

n	D_n
1	$17,863
2	30,613
3	21,863
4	15,613
5	5,575

The after-tax annual revenue is expressed in terms of functions of product demand (X) and unit price (Y):

$$R_n = XY(1 - t_m) = 0.6XY.$$

The after-tax annual expenses excluding depreciation are also expressed as a function of product demand (X):

$$E_n = (\text{Fixed cost} + \text{variable cost})(1 - t_m)$$
$$= (\$10,000 + 15X)(0.60)$$
$$= \$6,000 + 9X.$$

Then the net after-tax cash revenue is

$$V_n = R_n - E_n$$
$$= 0.6XY - 9X - \$6,000.$$

The present worth of the net after-tax cash inflow from revenue is

$$\sum_{n=1}^{5} V_n(P/F, 15\%, n) = [0.6X(Y - 15) - \$6,000](P/A, 15\%, 5)$$
$$= 0.6X(Y - 15)(3.3522) - \$20,113.$$

The present worth of the total depreciation credits is

$$\sum_{n=1}^{5} D_n t_m(P/F, i, n) = 0.40[\$17,863(P/F, 15\%, 1) + \$30,613(P/F, 15\%, 2)$$
$$+ \$21,863(P/F, 15\%, 3) + \$15,613(P/F, 15\%, 4)$$
$$+ \$5,575(P/F, 15\%, 5)]$$
$$= \$25,901.$$

Since the total depreciation amount is $91,527, the book value at the end of year 5 is $33,473 ($125,000 − $91,527). Any salvage value greater than this book value is treated as a gain taxed at the rate t_m. In our example, the salvage value is considered to be a random variable. Thus, the amount of taxable gains (losses) also becomes a random variable. Therefore, the net salvage value after tax adjustment is

$$S - (S - \$33,473)t_m = S(1 - t_m) + 33,473t_m$$
$$= 0.6S + \$13,389.$$

Then the equivalent present worth of this amount is

$$(0.6S + \$13,389)(P/F, 15\%, 5) = (0.6S + \$13,389)(0.4972).$$

Now the NPW equation can be summarized as

$$PW(15\%) = -\$125,000 + 0.6X(Y - 15)(3.3522) - \$20,113 + \$25,901$$
$$+ (0.6S + \$13,389)(0.4972)$$
$$= -\$112,555 + 2.0113X(Y - 15) + 0.2983S.$$

Note that the NPW function is now expressed in terms of the three random variables X, Y, and S.

12.5.3 Monte Carlo Sampling

With some variables, we may base the probability distribution on objective evidence gleaned from the past if the decision maker feels that the same trend will continue to operate in the future. If not, we may use subjective probabilities as discussed in Section 12.3.1. Once we specify a distribution for a random variable, we need to determine how to generate samples from that distribution. **Monte Carlo sampling** is a simulation method in which a random sample of outcomes is generated for a specified probability distribution. In this section, we shall discuss the Monte Carlo sampling procedure for an *independent* random variable.

Random Numbers

The sampling process is the key part of the analysis. It must be done such that the sequence of values sampled will be distributed in the same way as the original distribution. To accomplish this objective, we need a source of independent, identically distributed uniform random numbers between 0 and 1. Toward that end, we can use a table of random numbers, but most digital computers have programs available to generate "equally likely (uniform)" random decimals between 0 and 1. We will use $U(0,1)$ to denote such a statistically reliable uniform random-number generator, and we will use $U_1, U_2, U_3, \ldots$ to represent uniform random numbers generated by this routine. (In Microsoft Excel, the RAND function can be used to generate such a sequence of random numbers.)

Sampling Procedure

For any given random numbers, the question is, How are they used to sample a distribution in a simulation analysis? The first task is to convert the distribution into its corresponding cumulative frequency distribution. Then the random number generated is set equal to its numerically equivalent percentile and is used as the entry point on the $F(x)$-axis of the cumulative-frequency graph. The sampled value of the random variable is the x value corresponding to this cumulative percentile entry point.

This method of generating random values works because choosing a random decimal between 0 and 1 is equivalent to choosing a random percentile of the distribution. Then the random value is used to convert the random percentile to a particular value. The method is general and can be used for any cumulative probability distribution, either continuous or discrete.

EXAMPLE 12.10 **Monte Carlo Sampling**

In Example 12.9, we developed an NPW equation for BMC's transmission-housings project as a function of three random variables—demand (X), unit price (Y), and salvage value (S):

$$PW(15\%) = -\$112{,}555 + 2.0113X(Y - 15) + 0.2983S.$$

- For random variable X, we will assume the same discrete distribution as defined in Table 12.5.
- For random variable Y, we will assume a triangular distribution with $L = \$48$, $H = \$53$, and $M_O = \$50$.
- For random variable S, we will assume a uniform distribution with $L = \$30{,}000$ and $H = \$50{,}000$.

With the random variables (X, Y, and S) distributed as just set forth, and assuming that these random variables are *mutually independent* of each other, we need three uniform random numbers to sample one realization from each random variable. We determine the NPW distribution on the basis of 200 iterations.

DISCUSSION: As outlined in Figure 12.11, a simulation analysis consists of a series of repetitive computations of the NPW. To perform the sequence of repeated simulation trials, we generate a sample observation for each random variable in the model and

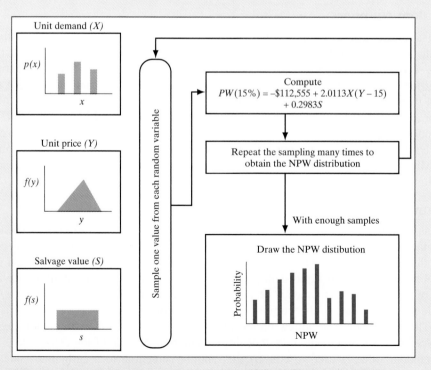

Figure 12.11 A logical sequence of a Monte Carlo simulation to obtain the NPW distribution for BMC's transmission-housings project.

substitute all the values into the NPW equation. Each trial requires that we use a different random number in the sequence to sample each distribution. Thus, if three random
variables affect the NPW, we need three random numbers for each trial. After each trial,
the computed NPW is stored in the computer. Each value of the NPW computed in this
manner represents one state of nature. The trials are continued until a sufficient number
of NPW values is available to define the NPW distribution.

SOLUTION

Suppose the following three uniform random numbers are generated for the first iteration: $U_1 = 0.12135$ for X, $U_2 = 0.82592$ for Y, and $U_3 = 0.86886$ for S.

- **Demand (X).** The cumulative distribution for X is already given in Example
  12.4. To generate one sample (observation) from this discrete distribution, we
  first find the cumulative probability function, as depicted in Figure 12.12(a).

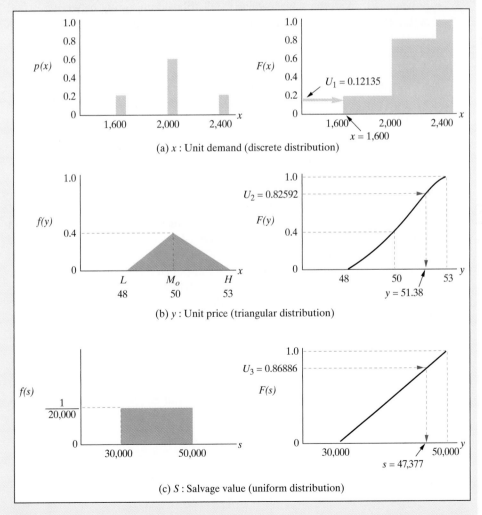

(a) x : Unit demand (discrete distribution)

(b) y : Unit price (triangular distribution)

(c) S : Salvage value (uniform distribution)

Figure 12.12 Illustration of a sampling scheme for discrete and continuous random variables.

In a given trial, suppose the computer gives the random number 0.12135. We then enter the vertical axis at the 12.135th percentile (the percentile numerically equivalent to the random number), read across to the cumulative curve, and then read down to the x-axis to find the corresponding value of the random variable X; this value is 1,600, and it is the value of x that we use in the NPW equation. In the next trial, we sample another value of x by obtaining another random number, entering the ordinate at the numerically equivalent percentile and reading the corresponding value of x from the x-axis.

- **Price (Y).** Assuming that the unit-price random variable can be estimated by the three parameters, its probability distribution is shown in Figure 12.12(b). Note that Y takes a continuous value (unlike the assumption of discreteness in Table 12.5). The sampling procedure is again similar to that in the discrete situation. Using the random number $U = 0.82592$, we can approximate $y = \$51.38$ by performing a linear interpolation.

- **Salvage (S).** With the salvage value (S) distributed uniformly between \$30,000 and \$50,000, and a random number $U = 0.86886$, the sample value is $s = \$30,000 + (50,000 - 30,000)0.86886$, or $s = \$47,377$. The sampling scheme is shown in Figure 12.12(c).

Now we can compute the NPW equation with these sample values, yielding

$$PW(15\%) = -\$112,555 + 2.0113(1,600)(\$51.3841 - \$15)$$
$$+ 0.2983(\$47,377)$$
$$= \$18,665.$$

This result completes the first iteration of NPW_1 computation.

For the second iteration, we need to generate another set of three uniform random numbers (suppose they are 0.72976, 0.79885, and 0.41879), to generate the respective sample from each distribution, and to compute $NPW_2 = \$44,752$. If we repeat this process for 200 iterations, we obtain the NPW values listed in Table 12.12.

By ordering the observed data by increasing NPW value and tabulating the ordered NPW values, we obtain the frequency distribution shown in Table 12.13. Such a tabulation results from dividing the entire range of computed NPWs into a series of subranges (20 in this case) and then counting the number of computed values that fall in each of the 20 intervals. Note that the sum of all the observed frequencies (column 3) is the total number of trials.

Column 4 simply expresses the frequencies of column 3 as a fraction of the total number of trials. At this point, all we have done is arrange the 200 numerical values of NPW into a table of relative frequencies.

TABLE 12.12 Observed NPW Values ($) for BMC's Simulation Project

18,665	44,752	41,756	75,804	67,508
74,177	47,664	43,106	17,394	35,841
46,400	36,237	12,686	47,947	48,411
50,365	64,287	44,856	47,970	47,988
42,071	43,421	36,991	59,551	9,788
50,147	46,010	11,777	71,202	76,259
20,937	48,250	33,837	37,378	7,188
47,875	14,891	48,362	35,894	75,360
38,726	73,987	45,346	62,508	36,738
44,849	46,076	48,374	47,287	70,702
47,168	42,348	8,213	6,689	14,415

No.	U_1	U_2	U_3	X	Y	S	NPW
1	0.12135	0.82592	0.86886	1,600	$51.38	$47,377	$18,665
2	0.72976	0.79885	0.41879	2,000	$51.26	$38,376	$44,752
⋮	⋮	⋮	⋮	⋮	⋮	⋮	⋮
200	0.57345	0.75553	0.61251	2,000	$51.08	$42,250	$45,204

41,962	45,207	46,487	77,414	67,723
79,089	51,846	76,157	49,960	17,091
15,438	37,214	49,542	45,830	46,679
36,393	15,714	44,899	48,109	76,593
34,927	42,545	45,452	48,089	43,163
78,925	39,179	38,883	80,994	45,202
18,807	74,707	44,787	5,944	6,377
45,811	44,608	42,448	34,584	48,677
10,303	35,597	37,910	62,624	62,060
44,813	36,629	75,111	15,739	43,621
34,288	49,116	76,778	46,067	62,853
9,029	42,485	51,817	44,096	35,198
36,345	75,337	14,034	18,253	41,865
48,089	64,934	81,532	74,407	57,465
32,969	46,529	19,653	10,250	31,269
33,300	47,552	19,139	48,664	46,410
33,457	44,605	35,160	51,496	34,247
46,657	37,249	44,171	34,189	36,673
47,567	62,654	65,062	34,519	45,204

TABLE 12.13 Simulated NPW Frequency Distribution for BMC's Transmission-Housings Project

Cell No.	Cell Interval	Observed Frequency	Relative Frequency	Cumulative Frequency
1	$5{,}944 \leq \text{NPW} \leq \$9{,}723$	8	0.04	0.04
2	$9{,}723 < \text{NPW} \leq 13{,}502$	6	0.03	0.07
3	$13{,}502 < \text{NPW} \leq 17{,}281$	9	0.05	0.12
4	$17{,}281 < \text{NPW} \leq 21{,}061$	10	0.05	0.17
5	$21{,}061 < \text{NPW} \leq 24{,}840$	0	0.00	0.17
6	$24{,}840 < \text{NPW} \leq 28{,}620$	0	0.00	0.17
7	$28{,}620 < \text{NPW} \leq 32{,}399$	2	0.01	0.18
8	$32{,}399 < \text{NPW} \leq 36{,}179$	20	0.10	0.28
9	$36{,}179 < \text{NPW} \leq 39{,}958$	18	0.09	0.37
10	$39{,}958 < \text{NPW} \leq 43{,}738$	13	0.07	0.43
11	$43{,}378 < \text{NPW} \leq 47{,}517$	39	0.19	0.63
12	$47{,}517 < \text{NPW} \leq 51{,}297$	28	0.14	0.77
13	$51{,}297 < \text{NPW} \leq 55{,}076$	3	0.02	0.78
14	$55{,}076 < \text{NPW} \leq 58{,}855$	1	0.01	0.78
15	$58{,}855 < \text{NPW} \leq 62{,}635$	6	0.03	0.81
16	$62{,}635 < \text{NPW} \leq 66{,}414$	6	0.03	0.84
17	$66{,}414 < \text{NPW} \leq 70{,}194$	4	0.02	0.86
18	$70{,}194 < \text{NPW} \leq 73{,}973$	4	0.02	0.88
19	$73{,}973 < \text{NPW} \leq 77{,}752$	18	0.09	0.97
20	$77{,}752 < \text{NPW} \leq 81{,}532$	5	0.03	1.00

12.5.4 Simulation Output Analysis

After a sufficient number of repetitive simulation trials has been run, the analysis is essentially completed. The only remaining tasks are to tabulate the computed NPW values in order to determine the expected value and to make various graphic displays that will be useful to management.

Interpretation of Simulation Results

Once we obtain an NPW frequency distribution (such as that shown in Table 12.13), we need to make the assumption that the actual relative frequencies of column 4 in the table are representative of the probability of having an NPW in each range. That is, we assume that the relative frequencies we observed in the sampling are representative of the proportions we would have obtained had we examined all the possible combinations.

This sampling is analogous to polling the opinions of voters about a candidate for public office. We could speak to every registered voter if we had the time and resources, but a simpler procedure would be to interview a smaller group of persons selected with an unbiased sampling procedure. If 60% of this scientifically selected sample supports the candidate, it probably would be safe to assume that 60% of all registered voters support the candidate. Conceptually, we do the same thing with simulation. As long as we ensure that a sufficient number of representative trials has been made, we can rely on the simulation results.

Once we have obtained the probability distribution of the NPW, we face the crucial question, How do we use this distribution in decision making? Recall that the probability distribution provides information regarding the probability that a random variable will attain some value x. We can use this information, in turn, to define the cumulative distribution, which expresses the probability that the random variable will attain a value smaller than or equal to some x [i.e., $F(x) = P(X \le x)$]. Thus, if the NPW distribution is known, we can also compute the probability that the NPW of a project will be negative. We use this probabilistic information in judging the profitability of the project.

With the assurance that 200 trials was a sufficient number for the BMC project, we may interpret the relative frequencies in column 4 of Table 12.13 as probabilities. The NPW values range between \$5,944 and \$81,532, thereby indicating no loss for any situation. The NPW distribution has an expected value of \$44,245 and a standard deviation of \$18,585.

Creation of Graphic Displays

Using the output data (the relative frequency) in Table 12.13, we can create the distribution in Figure 12.13(a). A picture such as this can give the decision maker a feel for the

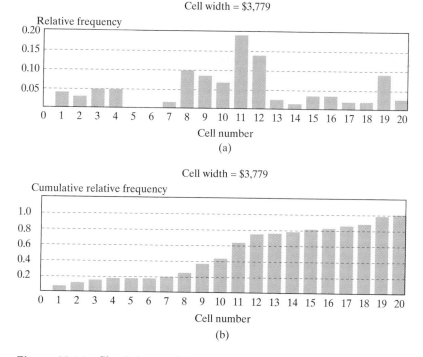

Figure 12.13 Simulation result for BMC's transmission-housings project based on 200 iterations.

ranges of possible NPWs, for the relative likelihoods of loss versus gain, for the range of NPWs that are most probable, and so on.

Another useful display is the conversion of the NPW distribution to the equivalent cumulative frequency, as shown in Figure 12.13(b). Usually, a decision maker is concerned with the likelihood of attaining at least a given level of NPW. Therefore, we construct the cumulative distribution by accumulating the areas under the distribution as the NPW decreases. The decision maker can use Figure 12.13(b) to answer many questions—for example, what is the likelihood of making at least a 15% return on investment (i.e., the likelihood that the NPW will be at least 0)? In our example, this probability is virtually 100%.

Dependent Random Variables

All our simulation examples have involved independent random variables. We must recognize that some of the random variables affecting the NPW may be related to one another. If they are, we need to sample from distributions of the random variables in a manner that accounts for any such dependencies. For example, in the BMC project, both the demand and the unit price are not known with certainty. Each of these parameters would be on our list of variables for which we need to describe distributions, but they could be related inversely. When we describe distributions for these two parameters, we have to account for that dependency. This issue can be critical, as the results obtained from a simulation analysis can be misleading if the analysis does not account for dependencies. The sampling techniques for these dependent random variables are beyond the scope of this text, but can be found in many textbooks on simulation.

12.5.5 Risk Simulation with @RISK[8]

One practical way to conduct a risk simulation is to use an Excel-based program such as @RISK. All we have to do is to develop a worksheet for a project's cash flow statement. Then we identify the random variables in the cash flow elements. With @RISK, we have a variety of probability distributions to choose from to describe our beliefs about the random variables of interest. We illustrate how to conduct a risk simulation on @RISK, using the same financial data described in Example 12.9. @RISK uses Monte Carlo simulation to show you all possible outcomes. Running an analysis with @RISK involves three simple steps.

1. Create a Cash Flow Statement with Excel

@RISK is an add-in to Microsoft Excel. As an add-in, @RISK becomes seamlessly integrated with your spreadsheet, adding risk analysis to your existing models. Table 12.14 is an Excel spreadsheet in which the cash flow entries are a function of the input variables. Here what we are looking for is the NPW of the project when we assigned specific values to the random variables. As in Example 12.9, we treat the unit price (Y), the demand (X), and the salvage value (S) as random variables.

---

[8] A risk simulation software developed by Palisade Corporation, Ithaca, NY 14850.

TABLE 12.14 An Excel Worksheet of the Transmission-Housing Project Prepared for @RISK

	A	B	C	D	E	F	G	H
1	@Risk Simulation for Example 12.9							
2								
3								
4	Input Data (Base):				Type of Distributions for the random variables:			
5								
6	**Unit Price ($)**		**50.33**		Category		Distribution	
7	**Demand**		**2000**					
8	Var. cost ($/unit)		15		Unit price (Y)		Triangular(48,50,53)	
9	Fixed cost ($)		10000		Demand (X)		Discrete (1600,2000,2400)	
10	**Salvage ($)**		**40000**		Salvage (S)		Uniform(30000,50000)	
11	Tax rate (%)		40%					
12	MARR (%)		15%					
13					Output (NPW)		$41,509	
14								
15			0	1	2	3	4	5
16	**Income Statement**							
17								
18	Revenues:							
19	Unit Price			$ 50.33	$ 50.33	$ 50.33	$ 50.33	$ 50.33
20	Demand (units)			2,000	2,000	2,000	2,000	2,000
21	Sales Revenue			$ 100,667	$ 100,667	$ 100,667	$ 100,667	$ 100,667
22	Expenses:							
23	Unit Variable Cost			$ 15	$ 15	$ 15	$ 15	$ 15
24	Variable Cost			30,000	30,000	30,000	30,000	30,000
25	Fixed Cost			10,000	10,000	10,000	10,000	10,000
26	Depreciation			17,863	30,613	21,863	15,613	5,581
27								
28	Taxable Income			$ 42,804	$ 30,054	$ 38,804	$ 45,054	$ 55,086
29	Income Taxes (40%)			17,121	12,021	15,521	18,021	22,034
30								
31	Net Income			$ 25,682	$ 18,032	$ 23,282	$ 27,032	$ 33,051
32								
33	**Cash Flow Statement**							
34								
35	Operating Activities:							
36	Net Income			25,682	18,032	23,282	27,032	33,051
37	Depreciation			17,863	30,613	21,863	15,613	5,581
38	Investment Activities:							
39	Investment		(125,000)					
40	Salvage							40,000
41	Gains Tax							(2,613)
42								
43	Net Cash Flow		$ (125,000)	$ 43,545	$ 48,645	$ 45,145	$ 42,645	$ 76,019

2. Define Uncertainty

We start by replacing uncertain values in our spreadsheet model with @RISK probability distribution functions. These @RISK functions simply represent a range of different possible values that cell could take instead of limiting it to just one value.

Random Variable	Probability Distribution		Probability Functions Provided by @RISK	Cell
Unit price	Triangular (48, 50, 53)		**=RiskTriang (48,50,53)**	C6
Demand	Event	Probability		
	1,600	0.2		
	2,000	0.6	**=RiskDiscrete({1600,2000,2400}, {0.2,0.6,0.2})**	C8
	2,400	0.2		
Salvage	Uniform (30,000, 50,000)		**=RiskUniform(30000,50000)**	C10

As shown in Figure 12.14, the probability functions are found from the "Define Distribution" menu in the tool bar.

3. Pick Your Bottom Line

We select our output cells, the "bottom line" cells whose values we are interested in. In our case, this is the net present value for the transmission-housing project, located at cell G13. To designate cell G13 as the output cell, we move the cursor to cell G13 and select the "Add Output" command in the tool bar. Then the cell formula in G13 will look like **RiskOutput("Output (NPW)")** + NPV(C12, D43:H43) + C43.

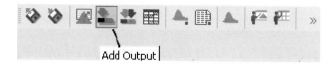

Add Output

4. Simulate

Now we are ready to simulate. Click the "Start Simulation" button and watch. @RISK recalculates our spreadsheet model hundreds or thousands of times. We can specify the number of iterations through the "Simulation Settings" command. There is no limit to the number of different scenarios we can look at in our simulations. Each time, @RISK samples random values from the @RISK functions we entered and records the resulting outcome. The result: a look at a whole range of possible outcomes, including the probabilities that they will occur! Almost instantly, you will see what critical situations to seek out—or avoid! With @RISK, we can answer questions like, "What is the probability of NPW exceeding $50,000?" or "What are the chances of losing money on this investment?"

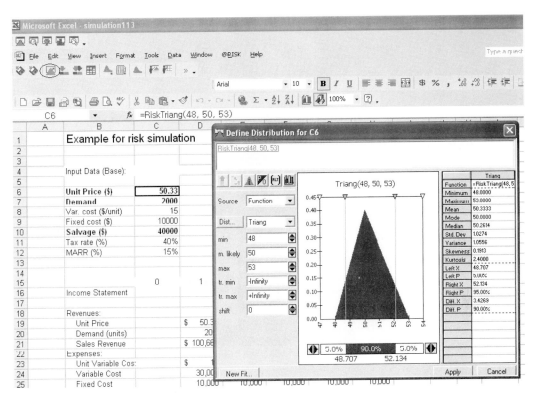

Figure 12.14 @RISK comes with RISKview, a built-in distribution viewer that lets you preview various distributions before selecting them. You can choose distributions from a gallery of thumbnail distribution pictures, and then watch as @RISK builds a graph of the distribution for you while you enter your parameters.

5. **Analyzing the Simulation Result Screen**

@RISK provides a wide range of graphing options for interpreting and presenting the simulation results. @RISK also gives us a full statistical report on our simulations, as well as access to all the data generated. Quick Reports include cumulative graphs, Tornado charts for sensitivity analysis, histograms, and summary statistics. As shown in Figure 12.15, based on 200 iterations, we find that the NPW ranges between $6,813 and $80,380, with the mean value of $41,468 and the standard deviation of $18,361. These results are comparable to those obtained in Example 12.9.

12.6 Decision Trees and Sequential Investment Decisions

Most investment problems that we have discussed so far involved only a single decision at the time of investment (i.e., an investment is accepted or rejected), or a single decision that entails a different decision option, such as "make a product in-house" or "farm out," and so on. Once the decision is made, there are no later contingencies or decision options to follow up on. However, certain classes of investment decisions cannot be analyzed in a single step. As an oil driller, for example, you have an opportunity to bid on an offshore lease. Your first decision is whether to bid. But if you decide to bid and eventually win the

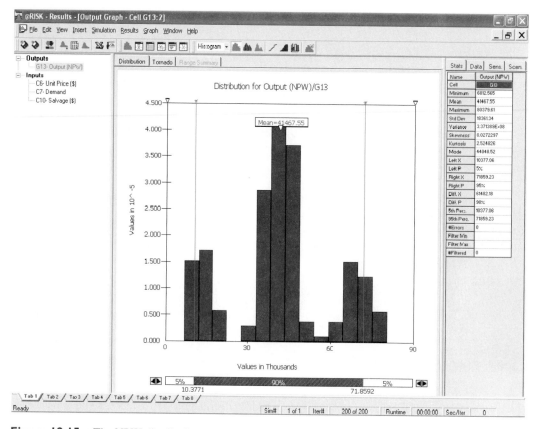

Figure 12.15 The NPW distribution created by @RISK based on 200 iterations. @RISK lets you use the output data to accurately describe what the NPW distribution will look like using the BestFit command, @RISK's built-in distribution fitting tool.

contract, you will have another decision regarding whether to drill immediately or run more seismic (drilling) tests. If you drill a test well that turns out to be dry, you have another decision whether to drill another well or drop the entire drilling project. In a sequential decision problem such as this, in which the actions taken at one stage depend on actions taken in earlier stages, the evaluation of investment alternatives can be quite complex. Certainly, all these future options must be considered when one is evaluating the feasibility of bidding at the initial stage. In this section, we describe a general approach for dealing with more complex decisions that is useful both for structuring the decision problems and for finding a solution to them. The approach utilizes a decision tree—a graphic tool for describing the actions available to the decision maker, the events that can occur, and the relationship between the actions and events.

12.6.1 Structuring a Decision-Tree Diagram

To illustrate the basic decision-tree concept, let's consider an investor, named Bill Henderson, who wants to invest some money in the financial market. He wants to choose between a highly speculative growth stock (d_1) and a very safe U.S. Treasury bond (d_2).

Figure 12.16 illustrates this situation, with the decision point represented by a square box (□) or decision node. The alternatives are represented as branches emanating from the node. Suppose Bill were to select some particular alternative, say, "invest in the stock." There are three chance events that can happen, each representing a potential return on investment. These events are shown in Figure 12.16 as branches emanating from a circular node (○). Note that the branches represent uncertain events over which Bill has no control. However, Bill can assign a probability to each chance event and enter it beside each branch in the decision tree. At the end of each branch is the conditional monetary transaction associated with the selected action and given event.

Relevant Net After-Tax Cash Flow

Once the structure of the decision tree is determined, the next step is to find the relevant cash flow (monetary value) associated with each of the alternatives and the possible chance outcomes. As we have emphasized throughout this book, the decision has to be made on an after-tax basis. Therefore, the relevant monetary value should be on that basis. Since the costs and revenues occur at different points in time over the study period, we also need to convert the various amounts on the tree's branches to their equivalent lump-sum amounts,

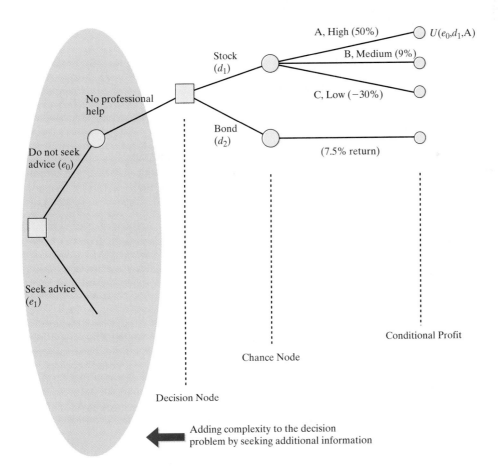

Figure 12.16 Structure of a typical decision tree (Bill's investment problem).

say, net present value. The conditional net present value thus represents the profit associated with the decisions and events along the path from the first part of the tree to the end.

Rollback Procedure

To analyze a decision tree, we begin at the end of the tree and work backwards—the **rollback** procedure. In other words, starting at the tips of the decision tree's branches and working toward the initial node, we use the following two rules:

1. For each chance node, we calculate the expected monetary value (EMV). This is done by multiplying probabilities by conditional profits associated with branches emanating from that node and summing these conditional profits. We then place the EMV in the node to indicate that it is the expected value calculated over all branches emanating from that node.

2. For each decision node, we select the one with the highest EMV (or minimum cost). Then those decision alternatives which are not selected are eliminated from further consideration. On the decision-tree diagram, we draw a mark across the nonoptimal decision branches, indicating that they are not to be followed.

Example 12.11 illustrates how Bill could transform his investment problem into a tree format using numerical values.

EXAMPLE 12.11 Bill's Investment Problem in a Decision-Tree Format

Suppose Bill has $50,000 to invest in the financial market for one year. His choices have been narrowed to two options:

- **Option 1.** Buy 1,000 shares of a technology stock at $50 per share that will be held for one year. Since this is a new initial public offering (IPO), there is not much research information available on the stock; hence, there will be a brokerage fee of $100 for this size of transaction (for either buying or selling stocks). For simplicity, assume that the stock is expected to provide a return at any one of three different levels: a high level (A) with a 50% return ($25,000), a medium level (B) with a 9% return ($4,500), or a low level (C) with a 30% loss ($-15,000$). Assume also that the probabilities of these occurrences are assessed at 0.25, 0.40, and 0.35, respectively. No stock dividend is anticipated for such a growth-oriented company.

- **Option 2.** Purchase a $50,000 U.S. Treasury bond, which pays interest at an effective annual rate of 7.5% ($3,750). The interest earned from the Treasury bond is nontaxable income. However, there is a $150 transaction fee for either buying or selling the bond.

Bill's dilemma is which alternative to choose to maximize his financial gain. At this point, Bill is not concerned about seeking some professional advice on the stock before making a decision. We will assume that any long-term capital gains will be taxed at 20%. Bill's minimum attractive rate of return is known to be 5% after taxes. Determine the payoff amount at the tip of each branch.

SOLUTION

Figure 12.17(a) shows the costs and revenues on the branches, transformed to their present equivalents.

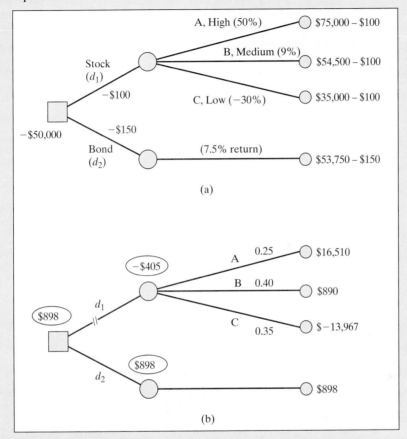

Figure 12.17   Decision tree for Bill's investment problem: (a) Relevant cash flows (before tax) and (b) net present worth for each decision path.

* **Option 1**

   1. With a 50% return (a real winner) over a one-year holding period,

      * The net cash flow associated with this event at Period 0 will include the amount of investment and the brokerage fee. Thus, we have

         Period 0: $(-\$50,000 - \$100) = -\$50,100.$

      * When you sell stock, you need to pay another brokerage fee. However, any investment expenses, such as brokerage commissions, must be included in determining the cost basis for investment. The taxable capital gains will be calculated as $(\$75,000 - \$50,000 - \$100 - \$100) = \$24,800.$ Therefore, the net cash flow at Period 1 would be

   Period 1: $(+\$75,000 - \$100) - 0.20(\$24,800) = \$69,940.$

- Then the conditional net present value of this stock transaction is

$$PW(5\%) = -\$50,100 + \$69,940(P/F, 5\%, 1) = \$16,510.$$

This amount of \$16,510 is entered at the tip of the corresponding branch. This procedure is repeated for each possible branch, and the resulting amounts are shown in Figure 12.17(b).

2. With a 9% return, we have
   - Period 0: $-\$50,100$
   - Period 1: \$53,540
   - $PW(5\%) = -50,100 + 53,540(P/F, 5\%, 1) = \$890$

3. With a 30% loss, we have
   - Period 0: $-\$50,100$
   - Period 1: \$37,940
   - $PW(5\%) = -\$50,100 + \$37,940(P/F, 5\%, 1) = -\$13,967$

- **Option 2.** Since the interest income on the U.S. government bond will not be taxed, there will be no capital-gains tax. Considering only the brokerage commission, we find that the relevant cash flows would be as follows:
  - Period 0: $(-\$50,000 - \$150) = \$50,150$
  - Period 1: $(+\$53,750 - \$150) = \$53,600$
  - $PW(5\%) = -\$50,150 + \$53,600(P/F, 5\%, 1) = \$898$

Figure 12.17(b) shows the complete decision tree for Bill's investment problem. Now Bill can calculate the expected monetary value (EMV) at each chance node. The EMV of Option 1 represents the sum of the products of the probabilities of high, medium, and low returns and the respective conditional profits (or losses):

$$EMV = \$16,510(0.25) + \$890(0.40) - \$13,967(0.35) = -\$405.$$

For Option 2, the EMV is simply

$$EMV = \$898.$$

In Figure 12.17, the expected monetary values are shown in the event nodes. Bill must choose which action to take, and this would be the one with the highest EMV, namely, Option 2, with EMV = \$898. This expected value is indicated in the tree by putting \$898 in the decision node at the beginning of the tree. Note that the decision tree uses the idea of maximizing expected monetary value developed in the previous section. In addition, the mark $\|$ is drawn across the nonoptimal decision branch (Option 1), indicating that it is not to be followed. In this simple example, the benefit of using a decision tree may not be evident. However, as the decision problem becomes more complex, the decision tree becomes more useful in organizing the information flow needed to make the decision. This is true in particular if Bill must make a sequence of decisions, rather than a single decision, as we next illustrate.

12.6.2 Worth of Obtaining Additional Information

In this section, we introduce a general method for evaluating whether it is worthwhile to seek more information. Most of the information that we can obtain is imperfect, in the sense that it will not tell us exactly which event will occur. Such imperfect information may have some value if it improves the chances of making a correct decision—that is, if it improves the expected monetary value. The problem is whether the reduced uncertainty is valuable enough to offset its cost. The gain is in the improved efficiency of decisions that may become possible with better information.

We use the term *experiment* in a broad sense in what follows. An experiment may be a market survey to predict sales volume for a typical consumer product, a statistical sampling of the quality of a product, or a seismic test to give a well-drilling firm some indications of the presence of oil.

The Value of Perfect Information

Let us take Bill's prior decision to "Purchase U.S. Treasury bonds" as a starting point. How do we determine whether further strategic steps would be profitable? We could do more to obtain additional information about the potential return on an investment in stock, but such steps cost money. Thus, we have to balance the monetary value of reducing uncertainty with the cost of obtaining additional information. In general, we can evaluate the worth of a given experiment only if we can estimate the reliability of the resulting information. In our stock investment problem, an expert's opinion may be helpful in deciding whether to abandon the idea of investing in a stock. This can be of value, however, only if Bill can say beforehand how closely the expert can estimate the performance of the stock.

The best place to start a decision improvement process is with a determination of how much we might improve our incremental profit by removing uncertainty. Although we probably couldn't ever really obtain such uncertainty, the value of perfect information is worth computing as an upper bound to the value of additional information.

We can easily calculate the value of perfect information. First we merely note the mathematical difference of the incremental profit from an optimal decision based on perfect information and the original decision to "Purchase Treasury bonds" made without foreknowledge of the actual return on the stock. This difference is called **opportunity loss** and must be computed for each potential level of return on the stock. Then we compute the expected opportunity loss by weighting each potential loss by the assigned probability associated with that event. What turns out in this expected opportunity loss is exactly the **expected value of perfect information** (EVPI). In other words, if we had perfect information about the performance of the stock, we should have made a correct decision about each situation, resulting in no regrets (i.e., no opportunity losses). Example 12.12 illustrates the concept of opportunity loss and how it is related to the value of perfect information.

> **EVPI** is the maximum price that one would be willing to pay in order to gain access to perfect information.

EXAMPLE 12.12 Expected Value of Perfect Information

Reluctant to give up a chance to make a larger profit with the stock option, Bill may wonder whether to obtain further information before acting. As with the decision to "Purchase U.S. Treasury bonds," Bill needs a single figure to represent the expected value of perfect information (EVPI) about investing in the stock. Calculate the EVPI on the basis of the financial data in Example 12.11 and Figure 12.17(b).

TABLE 12.15 Opportunity Loss Cost Associated with Investing in Bonds

Potential Return	Probability	Option 1: Invest in Stock	Option 2: Invest in Bonds	Optimal Choice with Perfect Information	Opportunity Loss Associated with Investing in Bonds
		Decision Option			
High (A)	0.25	$16,510	$898	Stock	$15,612
Medium (B)	0.40	890	898	Bond	0
Low (C)	0.35	−13,967	898	Bond	0
Expected Value		−$405	$898		$ 3,903
			↑		↑
			Prior optimal decision		EVPI

SOLUTION

As regards stock, the decision whether to purchase may hinge on the return on the stock during the next year—the only unknown, subject to a probability distribution. The opportunity loss associated with the decision to purchase bonds is shown in Table 12.15.

- For example, the conditional net present value of $16,510 is the net profit if the return is high. Also, recall that, without Bill's having received any new information, the indicated action (the prior optimal decision) was to select Option 2 ("Purchase Treasury bond"), with an expected net present value of $898.
- However, if we know that the stock will be a definite winner (yield a high return), then the prior decision to "Purchase Treasury bond" is inferior to "Invest in stock" to the extent of $16,510 − $898 = $15,612. With either a medium or low return, however, there is no opportunity loss, as a decision strategy with and without perfect information is the same.

Again, an average weighted by the assigned chances is used, but this time weights are applied to the set of opportunity losses (regrets) in Table 12.15. The result is

$$\text{EVPI} = (0.25)(\$15,612) + (0.40)(\$0) + (0.35)(\$0) = \$3,903.$$

This EVPI, representing the maximum expected amount that could be gained in incremental profit from having perfect information, places an upper limit on the sum Bill would be willing to pay for additional information.

Updating Conditional Profit (or Loss) after Paying a Fee to the Expert

With a relatively large EVPI in Example 12.12, there is a huge potential that Bill can benefit by obtaining some additional information on the stock. Bill can seek an expert's advice by receiving a detailed research report on the stock under consideration. This report, available at a nominal fee, say, $200, will suggest either of two possible outcomes: (1) The report may come up with a "buy" rating on the stock after concluding that business conditions were favorable, or (2) the report may not recommend any

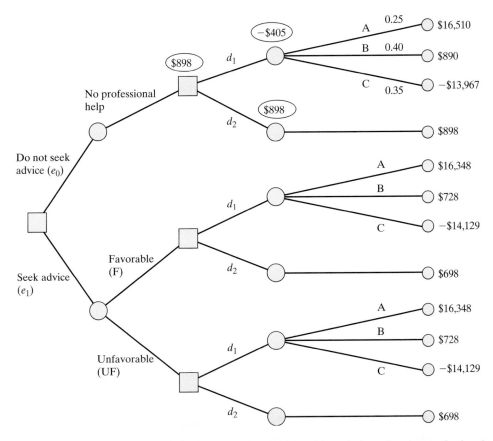

Figure 12.18 Decision tree for Bill's investment problem with an option of getting professional advice.

action or may take a neutral stance after concluding that business conditions were unfavorable or that the business model on which the decision is based was not well received in the financial marketplace. Bill's alternatives are either to pay for this service before making any further decision or simply to make his own decision without seeing the report.

If Bill seeks advice before acting, he can base his decision upon the stock report. This problem can be expressed in terms of a decision tree, as shown in Figure 12.18. The upper part of the tree shows the decision process if no advice is sought. This is the same as Figure 12.17, with probabilities of 0.25, 0.40, and 0.35 for high, medium, and low returns, respectively, and an expected loss of $405 for investing in the stock and an expected profit of $898 from purchasing Treasury bonds.

The lower part of the tree, following the branch "Seek advice," displays the results and the subsequent decision possibilities. Using the analyst's research report, Bill can expect two possible events: the expert's recommendation and the actual performance of the stock. After reading either of the two possible report outcomes, Bill must make a decision about whether to invest in the stock. Since Bill has to pay a nominal fee ($200) to get the report, the profit or loss figures at the tips of the branches must also be updated. For the stock investment option, if the actual return proves to be very high (50%), the resulting cash flows would be as follows:

Period 0: $(-\$50,000 - \$100 - \$200) = -\$50,300;$

Period 1: $(+\$75,000 - \$100) - (0.20)(\$25,000 - \$400) = \$69,980;$

$$PW(5\%) = -\$50,300 + \$69,980(P/F, 5\%, 1) = \$16,348.$$

Since the cost of obtaining the report ($200) will be a part of the total investment expenses, along with the brokerage commissions, the taxable capital gains would be adjusted accordingly. With a medium return, the equivalent net present value would be $728. With a low return, there would be a net loss in the amount of $14,129.

Similarly, Bill can compute the net profit for Option 2 as follows:

Period 0: $(-\$50,000 - \$150 - \$200) = -\$50,350;$

Period 1: $(+\$53,750 - \$150) = \$53,600;$

$$PW(5\%) = -\$50,350 + \$53,600(P/F, 5\%, 1) = \$698.$$

Once again, note that gains earned from the U.S. government bond will not be taxed.

12.6.3 Decision Making after Having Imperfect Information

In Figure 12.16, the analyst's recommendation precedes the actual investment decision. This allows Bill to change his prior probabilistic assessments regarding the performance of the stock. In other words, if the analyst's report has any bearing on Bill's reassessments of the stock's performance, then Bill must change his prior probabilistic assessments accordingly. But what types of information should be available and how should the prior probabilistic assessments be updated?

Conditional Probabilities of the Expert's Predictions, Given a Potential Return on the Stock

Suppose that Bill knows an expert (stock analyst) to whom he can provide his research report on the stock described under Option 1. The expert will charge a fee in the amount of $200 to provide Bill with a report on the stock that is either favorable or unfavorable. This research report is not infallible, but it is known to be pretty reliable. From past experience, Bill estimates that when the stock under consideration is relatively highly profitable (A), there is an 80% chance that the report will be favorable; when the stock return is medium (B), there is a 65% chance that the report will be favorable; and when the stock return is relatively unprofitable (C), there is a 20% chance that the report will be favorable. Bill's estimates are summarized in Table 12.16.

These conditional probabilities can be expressed in a tree format as shown in Figure 12.19. For example, if the stock indeed turned out to be a real winner, the probability

TABLE 12.16 Conditional Probabilities of the Expert's Predictions

	Given Level of Stock Performance		
What the Report Will Say	**High (A)**	**Medium (B)**	**Low (C)**
Favorable (F)	0.80	0.65	0.20
Unfavorable (UF)	0.20	0.35	0.80

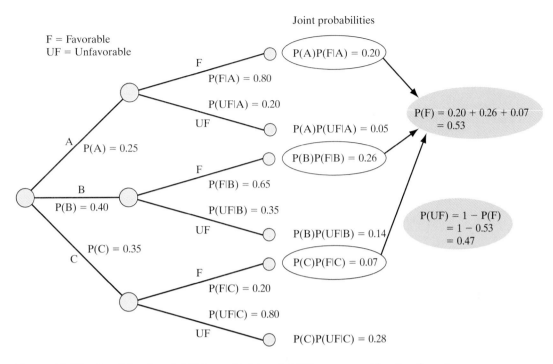

Figure 12.19 Conditional probabilities and joint probabilities expressed in a tree format.

that the report would have predicted the same (with a favorable recommendation) is 0.80, and the probability that the report would have issued an "unfavorable recommendation" is 0.20. Such probabilities would reflect past experience with the analyst's service of this type, modified perhaps by Bill's own judgment. In fact, these probabilities represent the reliability or accuracy of the expert opinion. It is a question of counting the number of times the expert's forecast was "on target" in the past and the number of times the expert "missed the mark," for each of the three actual levels of return on the stock. With this past history of the expert's forecasts, Bill can evaluate the economic worth of the service. Without such a history, no specific value can be attached to seeking the advice.

Joint and Marginal Probabilities

Suppose now that, before seeing the research report, Bill has some critical information on (1) all the probabilities for the three levels of stock performance (prior probabilities) and (2) with what probability the report will say that the return on the stock is favorable or unfavorable, given the three levels of the stock performance in the future. A glance at the sequence of events in Figure 12.18, however, shows that these probabilities are not the ones required to find "expected" values of various strategies in the decision path. What is really needed are the probabilities pertaining to what the report will say about the stock—favorable (a recommendation to buy) or unfavorable (no recommendation or a neutral stance)—and the conditional probabilities of each of the three potential returns after the report is seen. In other words, the probabilities of Table 12.16 are not directly useful in Figure 12.18. Rather, we need the *marginal* probabilities of the "favorable" and "unfavorable" recommendations.

Marginal probability is the probability of one event, regardless of the other event.

To remedy this situation, the probabilities must be put in a different form: We must construct a joint probabilities table. To start with, we have the original (prior) probabilities assessed by Bill before seeing the report: a 0.25 chance that the stock will result in a high return (A), a 0.40 chance of a medium return (B), and a 0.35 chance of a loss (C). From these and the conditional probabilities of Table 12.16, the joint probabilities of Figure 12.19 can be calculated. Thus, the joint probability of both a prediction of a favorable stock market (F) and a high return on the stock (A) is calculated by multiplying the conditional probability of a favorable prediction, given a high return on the stock (which is 0.80 from Table 12.16), by the probability of a high return (A):

$$P(A, F) = P(F|A)P(A) = (0.80)(0.25) = 0.20.$$

Similarly,

$$P(A, UF) = P(UF|A)P(A) = (0.20)(0.25) = 0.05,$$
$$P(B, F) = P(F|B)P(B) = (0.65)(0.40) = 0.26,$$
$$P(B, UF) = P(UF|B)P(B) = (0.35)(0.40) = 0.14,$$

and so on.

The marginal probabilities of return in Table 12.17 are obtained by summing the values across the columns. Note that these are precisely the original probabilities of a high, medium, and low return, and they are designated prior probabilities because they were assessed before any information from the report was obtained.

In understanding Table 12.17, it is useful to think of it as representing the results of 100 past situations identical to the one under consideration. The probabilities then represent the frequency with which the various outcomes occurred. For example, in 25 of the 100 cases, the actual return on the stock turned out to be high; and in these 25 high-return cases, the report predicted a favorable condition in 20 instances [that is $P(A, F) = 0.20$] and predicted an unfavorable one in 5 instances.

TABLE 12.17 Joint as Well as Marginal Probabilities

When Potential Level of Return Is Given	What Report Will Say — Joint Probabilities		Marginal Probabilities of Return
	Favorable (F)	Unfavorable (UF)	
High (A)	0.20	0.05	0.25
Medium (B)	0.26	0.14	0.40
Low (C)	0.07	0.28	0.35
Marginal probabilities	0.53	0.47	1.00

Similarly, the marginal probabilities of prediction in the bottom row of Table 12.17 can be interpreted as the relative frequency with which the report predicted favorable and unfavorable conditions. For example, the survey predicted favorable conditions 53 out of 100 times, 20 of these times when stock returns actually were high, 26 times when returns were medium, and 7 times when returns were low. These marginal probabilities of what the report will say are important to our analysis, for they give us the probabilities associated with the information received by Bill before the decision to invest in the stock is made. The probabilities are entered beside the appropriate branches in Figure 12.18.

Determining Revised Probabilities

We still need to calculate the probabilities for the branches that represent "high return," "medium return," and "low return" in the lower part of Figure 12.18. We cannot use the values 0.25, 0.40, and 0.35 for these events, as we did in the upper part of the tree, because those probabilities were calculated prior to seeking the advice. At this point on the decision tree, Bill will have received information from the analyst, and the probabilities should reflect that information. (If Bill's judgment has not changed even after seeing the report, then there will be no changes in the probabilities.) The required probabilities are the conditional probabilities for the various levels of return on the stock, given the expert's opinion. Thus, using the data from Table 12.17, we can compute the probability $P(A|F)$ of seeing a high return (A), given the expert's buy recommendation (F), directly from the definition of conditional probability:

$$P(A|F) = P(A, F)/P(F) = 0.20/0.53 = 0.38.$$

The probabilities of a medium and a low return, given a buy recommendation, are, respectively,

$$P(B|F) = P(B, F)/P(F) = 0.26/0.53 = 0.49$$

and

$$P(C|F) = P(C, F)/P(F) = 0.07/0.53 = 0.13.$$

These probabilities are called *revised* (or *posterior*) probabilities, since they come after the inclusion of the information received from the report. To understand the meaning of the preceding calculations, think again of Table 12.17 as representing 100 past identical situations. Then, in 53 cases [since $P(F) = 0.53$], 20 actually resulted in a high return. Hence, the posterior probability of a high return is $20/53 = 0.38$, as calculated.

The posterior probabilities of other outcomes after seeing an unfavorable recommendation can be calculated similarly:

$$P(A|UF) = P(A, UF)/P(UF) = 0.05/0.47 = 0.11,$$
$$P(B|UF) = P(B, UF)/P(UF) = 0.14/0.47 = 0.30,$$
$$P(C|UF) = P(C, UF)/P(UF) = 0.28/0.47 = 0.59.$$

These values are also shown in Figure 12.20 at the appropriate points in the decision tree.

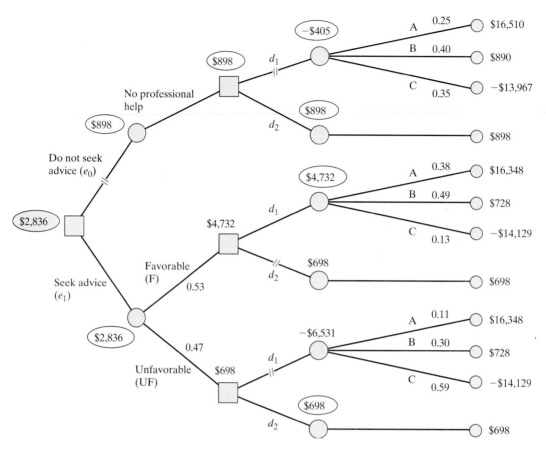

Figure 12.20   Decision tree for Bill's investment problem with an option of having professional advice.

EXAMPLE 12.13 Decision Making after Getting Some Professional Advice

On the basis of Figure 12.16, determine how much the research report is worth to Bill. Should Bill seek an expert's advice (research report)?

SOLUTION

All the necessary information is now available, and Figure 12.20 can be analyzed, starting from the right and working backward. The expected values are shown in the circles. For example, follow the branches "Seek Advice (e_1)," "Favorable (Buy recommendation)," and "Invest in the Stock (d_1)." The expected value of $4,732 shown in the circle at the end of these branches is calculated as

$$0.38(\$16,348) + 0.49(\$728) + 0.13(-\$14,129) = \$4,732.$$

Since Bill can expect a profit of $4,732 from investing in the stock after receiving a buy recommendation, the investment in Treasury bonds ($698) becomes a less profitable alternative. Thus, it can be marked out ($\neq$) of the "Purchase Treasury bonds" branch to indicate that it is not optimal.

There is an expected loss of $6,531 if the report gives an unfavorable recommendation, but Bill goes ahead with buying the stock. In this situation, the "Treasury bonds" option becomes a better choice, and the "Buy stock" branch is marked out. The part of the decision tree relating to seeking the advice is now reduced to two chance events (earning $4,732 and earning $698), and the expected value in the circular node is calculated as

$$0.53(\$4,732) + 0.47(\$698) = \$2,836.$$

Thus, if the expert's opinion is taken and Bill acts on the basis of the information received, the expected profit is $2,836. Since this amount is greater than the $898 profit that would be obtained in the absence of the expert's advice, we conclude that it is worth spending $200 to receive some additional information from the expert.

In sum, the optimal decision is as follows: (1) Bill should seek professional advice at the outset. (2) If the expert's report indicates a buy recommendation, go ahead and invest in the stock. (3) If the report says otherwise, invest in the Treasury bonds.

SUMMARY

■ Often, cash flow amounts and other aspects of an investment project analysis are uncertain. Whenever such uncertainty exists, we are faced with the difficulty of **project risk**: the possibility that an investment project will not meet our minimum requirements for acceptability and success.

■ Three of the most basic tools for assessing project risk are as follows:

   1. Sensitivity analysis—a means of identifying those project variables which, when varied, have the greatest effect on the acceptability of the project.

   2. Break-even analysis—a means of identifying the value of a particular project variable that causes the project to exactly break even.

   3. Scenario analysis—a means of comparing a "base-case" or expected project measurement (such as the NPW) with one or more additional scenarios, such as the best case and the worst case, to identify the extreme and most likely project outcomes.

■ Sensitivity, break-even, and scenario analyses are reasonably simple to apply, but also somewhat simplistic and imprecise in cases where we must deal with multifaceted project uncertainty. **Probability concepts** allow us to further refine the analysis of project risk by assigning numerical values to the likelihood that project variables will have certain values.

■ The end goal of a probabilistic analysis of project variables is to produce an NPW distribution from which we can extract such useful information as (1) the **expected NPW value,** (2) the extent to which other NPW values vary from, or are clustered around, the expected value (the **variance**), and (3) the best- and worst-case NPWs.

■ Our real task is not try to find "risk-free" projects; they don't exist in real life. The challenge is to decide what level of risk we are willing to assume and then, having decided on our tolerance for risk, to understand the implications of our choice.

■ **Risk simulation**, in general, is the process of modeling reality to observe and weigh the likelihood of the possible outcomes of a risky undertaking.

▪ **Monte Carlo sampling** is a specific type of randomized sampling method in which a random sample of outcomes is generated for a specified probability distribution. Because Monte Carlo sampling and other simulation techniques often rely on generating a significant number of outcomes, they can be more conveniently performed on the computer than manually.

▪ The **decision tree** is another technique that can facilitate investment decision making when uncertainty prevails, especially when the problem involves a sequence of decisions. Decision-tree analysis involves the choice of a decision criterion—say, to maximize expected profit. If possible and feasible, an experiment is conducted, and the prior probabilities of the states of nature are revised on the basis of the experimental results. The expected profit associated with each possible decision is computed, and the act with the highest expected profit is chosen as the optimum action.

PROBLEMS

Sensitivity Analysis

12.1 Ford Construction Company is considering acquiring a new earthmover. The mover's basic price is $90,000, and it will cost another $18,000 to modify it for special use by the company. This earthmover falls into the MACRS five-year class. It will be sold after four years for $30,000. The purchase of the earthmover will have no effect on revenues, but it is expected to save the firm $35,000 per year in before-tax operating costs, mainly labor. The firm's marginal tax rate (federal plus state) is 40%, and its MARR is 10%.

   (a) Is this project acceptable, based on the most likely estimates given?

   (b) Suppose that the project will require an increase in net working capital (spare-parts inventory) of $5,000, which will be recovered at the end of year 5. Taking this new requirement into account, would the project still be acceptable?

   (c) If the firm's MARR is increased to 18%, what would be the required savings in labor so that the project remains profitable?

12.2 Minnesota Metal Forming Company has just invested $500,000 of fixed capital in a manufacturing process that is estimated to generate an after-tax annual cash flow of $200,000 in each of the next five years. At the end of year 5, no further market for the product and no salvage value for the manufacturing process is expected. If a manufacturing problem delays the start-up of the plant for one year (leaving only four years of process life), what additional after-tax cash flow will be needed to maintain the same internal rate of return as would be experienced if no delay occurred?

12.3 A real-estate developer seeks to determine the most economical height for a new office building, which will be sold after five years. The relevant net annual revenues and salvage values are as follows:

| | Height | | | |
	2 Floors	3 Floors	4 Floors	5 Floors
First cost (net after tax)	$500,000	$750,000	$1,250,000	$2,000,000
Lease revenue	199,100	169,200	149,200	378,150
Net resale value (after tax)	600,000	900,000	2,000,000	3,000,000

(a) The developer is uncertain about the interest rate (i) to use, but is certain that it is in the range from 5% to 30%. For each building height, find the range of values of i for which that building height is the most economical.

(b) Suppose that the developer's interest rate is known to be 15%. What would be the cost, in terms of net present value, of a 10% overestimation of the resale value? (In other words, the true value was 10% lower than that of the original estimate.)

12.4 A special-purpose milling machine was purchased 4 years ago for $20,000. It was estimated at that time that this machine would have a life of 10 years and a salvage value of $1,000, with a cost of removal of $1,500. These estimates are still good. This machine has annual operating costs of $2,000, and its current book value is $13,000. If the machine is retained for its entire 10-year life, the remaining annual depreciation schedule would be $2,000 for years 5 through 10. A new machine that is more efficient will reduce operating costs to $1,000, but will require an investment of $12,000. The life of the new machine is estimated to be 6 years, with a salvage value of $2,000. The new machine would fall into the 5-year MACRS property class. An offer of $6,000 for the old machine has been made, and the purchaser would pay for removal of the machine. The firm's marginal tax rate is 40%, and its required minimum rate of return is 10%.

(a) What incremental cash flows will occur at the end of years 0 through 6 as a result of replacing the old machine? Should the old machine be replaced now?

(b) Suppose that the annual operating costs for the old milling machine would increase at an annual rate of 5% over the remaining service life of the machine. With this change in future operating costs for the old machine, would the answer in (a) change?

(c) What is the minimum trade-in value for the old machine so that both alternatives are economically equivalent?

12.5 A local telephone company is considering installing a new phone line for a new row of apartment complexes. Two types of cables are being examined: conventional copper wire and fiber optics. Transmission by copper wire cables, although cumbersome, involves much less complicated and less expensive support hardware than does fiber optics. The local company may use five different types of copper wire cables: 100 pairs, 200 pairs, 300 pairs, 600 pairs, and 900 pairs per cable. In calculating the cost per foot of cable, the following equation is used:

Cost per length = [Cost per foot +

cost per pair (number of pairs)](length),

where

22-gauge copper wire = $1.692 per foot

and

Cost per pair = $0.013 per pair.

The annual cost of the cable as a percentage of the initial cost is 18.4%. The life of the system is 30 years.

In fiber optics, a cable is referred to as a ribbon. One ribbon contains 12 fibers, grouped in fours; therefore, one ribbon contains three groups of 4 fibers. Each group can produce 672 lines (equivalent to 672 pairs of wires), and since each ribbon contains three groups, the total capacity of the ribbon is 2,016 lines. To transmit signals via fiber optics, many modulators, wave guides, and terminators are needed to convert the signals from electric currents to modulated light waves. Fiber-optic ribbon costs $15,000 per mile. At each end of the ribbon, three terminators are needed, one for each group of 4 fibers, at a cost of $30,000 per terminator. Twenty-one modulating systems are needed at each end of the ribbon, at a cost of $12,092 for a unit in the central office and $21,217 for a unit in the field. Every 22,000 feet, a repeater is required to keep the modulated light waves in the ribbon at an intensity that is intelligible for detection. The unit cost of this repeater is $15,000. The annual cost, including income taxes for the 21 modulating systems, is 12.5% of the initial cost of the units. The annual cost of the ribbon itself is 17.8% initially. The life of the whole system is 30 years. (All figures represent after-tax costs.)

(a) Suppose that the apartments are located 5 miles from the phone company's central switching system and that about 2,000 telephones will be required. This would require either 2,000 pairs of copper wire or one fiber-optic ribbon and related hardware. If the telephone company's interest rate is 15%, which option is more economical?

(b) In (a), suppose that the apartments are located 10 miles or 25 miles from the phone company's central switching system. Which option is more economically attractive under each scenario?

12.6 A small manufacturing firm is considering purchasing a new boring machine to modernize one of its production lines. Two types of boring machine are available on the market. The lives of machine A and machine B are 8 years and 10 years, respectively. The machines have the following receipts and disbursements:

Item	Machine A	Machine B
First cost	$6,000	$8,500
Service life	8 years	10 years
Salvage value	$500	$1,000
Annual O&M costs	$700	$520
Depreciation (MACRS)	7 years	7 years

Use a MARR (after tax) of 10% and a marginal tax rate of 30%, and answer the following questions:

(a) Which machine would be most economical to purchase under an infinite planning horizon? Explain any assumption that you need to make about future alternatives.

(b) Determine the break-even annual O&M costs for machine A so that the present worth of machine A is the same as that of machine B.

(c) Suppose that the required service life of the machine is only 5 years. The salvage values at the end of the required service period are estimated to be $3,000 for machine A and $3,500 for machine B. Which machine is more economical?

12.7 The management of Langdale Mill is considering replacing a number of old looms in the mill's weave room. The looms to be replaced are two 86-inch President looms, sixteen 54-inch President looms, and twenty-two 72-inch Draper X-P2 looms. The company may either replace the old looms with new ones of the same kind or buy 21 new shutterless Pignone looms. The first alternative requires purchasing 40 new President and Draper looms and scrapping the old looms. The second alternative involves scrapping the 40 old looms, relocating 12 Picanol looms, and constructing a concrete floor, plus purchasing the 21 Pignone looms and various related equipment.

Description	Alternative 1	Alternative 2
Machinery/related equipment	$2,119,170	$1,071,240
Removal cost of old looms/site preparation	26,866	49,002
Salvage value of old looms	62,000	62,000
Annual sales increase with new looms	7,915,748	7,455,084
Annual labor	261,040	422,080
Annual O&M	1,092,000	1,560,000
Depreciation (MACRS)	7 years	7 years
Project life	8 years	8 years
Salvage value	169,000	54,000

The firm's MARR is 18%, set by corporate executives, who feel that various investment opportunities available for the mills will guarantee a rate of return on investment of at least 18%. The mill's marginal tax rate is 40%.

(a) Perform a sensitivity analysis on the project's data, varying the operating revenue, labor cost, annual maintenance cost, and MARR. Assume that each of these variables can deviate from its base-case expected value by ±10%, by ±20%, and by ±30%.

(b) From the results of part (a), prepare sensitivity diagrams and interpret the results.

12.8 Mike Lazenby, an industrial engineer at Energy Conservation Service, has found that the anticipated profitability of a newly developed water-heater temperature control device can be measured by present worth with the formula

$$NPW = 4.028V(2X - \$11) - 77,860,$$

where V is the number of units produced and sold and X is the sales price per unit. Mike also has found that the value of the parameter V could occur anywhere over the range from 1,000 to 6,000 units and that of the parameter X anywhere between $20 and $45 per unit. Develop a sensitivity graph as a function of the number of units produced and the sales price per unit.

12.9 A local U.S. Postal Service office is considering purchasing a 4,000-pound fork-lift truck, which will be used primarily for processing incoming as well as outgoing postal packages. Forklift trucks traditionally have been fueled by either gasoline, liquid propane gas (LPG), or diesel fuel. Battery-powered electric fork-lifts, however, are increasingly popular in many industrial sectors due to the economic and environmental benefits that accrue from their use. Therefore, the postal service is interested in comparing forklifts that use the four different types of fuel. The purchase costs as well as annual operating and maintenance costs are provided by a local utility company and the Lead Industries Association. Annual fuel and maintenance costs are measured in terms of number of shifts per year, where one shift is equivalent to 8 hours of operation.

	Electrical Power	LPG	Gasoline	Diesel Fuel
Life expectancy	7 years	7 years	7 years	7 years
Initial cost	$29,739	$21,200	$20,107	$22,263
Salvage value	$3,000	$2,000	$2,000	$2,200
Maximum shifts per year	260	260	260	260
Fuel consumption/shift	31.25 kWh	11 gal	11.1 gal	7.2 gal
Fuel cost/unit	$0.10/kWh	$3.50/gal	$2.24/gal	$2.45/gal
Fuel cost per shift	$3.125	$38.50	$24.86	$17.64
Annual maintenance cost Fixed cost	$500	$1,000	$1,000	$1,000
Variable cost/shift	$4.5	$7	$7	$7

The postal service is unsure of the number of shifts per year, but it expects it should be somewhere between 200 and 260 shifts. Since the U.S. Postal Service does not pay income taxes, no depreciation or tax information is required. The U.S. government uses 10% as an interest rate for any project evaluation of this nature. Develop a sensitivity graph that shows how the choice of alternatives changes as a function of number of shifts per year.

Break-Even Analysis

12.10 Susan Campbell is thinking about going into the motel business near Disney World in Orlando. The cost to build a motel is $2,200,000. The lot costs $600,000. Furniture and furnishings cost $400,000 and should be recovered in 7 years (7-year MACRS property), while the motel building should be recovered in 39 years

(39-year MACRS real property placed in service on January 1). The land will appreciate at an annual rate of 5% over the project period, but the building will have a zero salvage value after 25 years. When the motel is full (100% capacity), it takes in (receipts) $4,000 per day, 365 days per year. Exclusive of depreciation, the motel has fixed operating expenses of $230,000 per year. The variable operating expenses are $170,000 at 100% capacity, and these vary directly with percent capacity down to zero at 0% capacity. If the interest is 10% compounded annually, at what percent capacity over 25 years must the motel operate in order for Susan to break even? (Assume that Susan's tax rate is 31%.)

12.11 A plant engineer wishes to know which of two types of lightbulbs should be used to light a warehouse. The bulbs that are currently used cost $45.90 per bulb and last 14,600 hours before burning out. The new bulb (at $60 per bulb) provides the same amount of light and consumes the same amount of energy, but lasts twice as long. The labor cost to change a bulb is $16.00. The lights are on 19 hours a day, 365 days a year. If the firm's MARR is 15%, what is the maximum price (per bulb) the engineer should be willing to pay to switch to the new bulb? (Assume that the firm's marginal tax rate is 40%.)

12.12 Robert Cooper is considering purchasing a piece of business rental property containing stores and offices at a cost of $250,000. Cooper estimates that annual receipts from rentals will be $35,000 and that annual disbursements, other than income taxes, will be about $12,000. The property is expected to appreciate at the annual rate of 5%. Cooper expects to retain the property for 20 years once it is acquired. Then it will be depreciated as a 39-year real-property class (MACRS), assuming that the property will be placed in service on January 1. Cooper's marginal tax rate is 30% and his MARR is 15%. What would be the minimum annual total of rental receipts that would make the investment break even?

12.13 Two different methods of solving a production problem are under consideration. Both methods are expected to be obsolete in six years. Method A would cost $80,000 initially and have annual operating costs of $22,000 a year. Method B would cost $52,000 and costs $17,000 a year to operate. The salvage value realized would be $20,000 with Method A and $15,000 with Method B. Method A would generate $16,000 revenue income a year more than Method B. Investments in both methods are subject to a five-year MACRS property class. The firm's marginal income tax rate is 40%. The firm's MARR is 20%. What would be the required additional annual revenue for Method A such that an engineer would be indifferent to choosing one method over the other?

12.14 Rocky Mountain Publishing Company is considering introducing a new morning newspaper in Denver. Its direct competitor charges $0.25 at retail, with $0.05 going to the retailer. For the level of news coverage the company desires, it determines the fixed cost of editors, reporters, rent, press-room expenses, and wire service charges to be $300,000 per month. The variable cost of ink and paper is $0.10 per copy, but advertising revenues of $0.05 per paper will be generated. To print the morning paper, the publisher has to purchase a new printing press, which will cost $600,000. The press machine will be depreciated according to a 7-year MACRS class. The press machine will be used for 10 years, at which time its salvage value would be about $100,000. Assume 20 weekdays in a month, a 40% tax rate, and a 13% MARR. How many copies per day must be sold to break even at a retail selling price of $0.25 per paper?

Probabilistic Analysis

12.15 A corporation is trying to decide whether to buy the patent for a product designed by another company. The decision to buy will mean an investment of $8 million, and the demand for the product is not known. If demand is light, the company expects a return of $1.3 million each year for three years. If demand is moderate, the return will be $2.5 million each year for four years, and high demand means a return of $4 million each year for four years. It is estimated the probability of a high demand is 0.4 and the probability of a light demand is 0.2. The firm's (risk-free) interest rate is 12%. Calculate the expected present worth of the patent. On this basis, should the company make the investment? (All figures represent after-tax values.)

12.16 Juan Carlos is considering two investment projects whose present values are described as follows:

- **Project 1.** $PW(10\%) = 20X + 8XY$, where X and Y are statistically independent discrete random variables with the following distributions:

X		Y	
Event	**Probability**	**Event**	**Probability**
$20	0.6	$10	0.4
40	0.4	20	0.6

- **Project 2.**

PW(10%)	Probability
$0	0.24
400	0.20
1,600	0.36
2,400	0.20

(*Note*: Cash flows between the two projects are also assumed to be statistically independent.)

(a) Develop the NPW distribution for Project 1.

(b) Compute the mean and variance of the NPW for Project 1.

(c) Compute the mean and variance of the NPW for Project 2.

(d) Suppose that Projects 1 and 2 are mutually exclusive. Which project would you select?

12.17 A financial investor has an investment portfolio worth $350,000. A bond in the portfolio will mature next month and provide him with $25,000 to reinvest. The choices have been narrowed down to the following two options:

- **Option 1.** Reinvest in a foreign bond that will mature in one year. This will entail a brokerage fee of $150. For simplicity, assume that the bond will provide interest of $2,450, $2,000, or $1,675 over the one-year period and that the probabilities of these occurrences are assessed to be 0.25, 0.45, and 0.30, respectively.

- **Option 2.** Reinvest in a $25,000 certificate with a savings-and-loan association. Assume that this certificate has an effective annual rate of 7.5%.

(a) Which form of reinvestment should the investor choose in order to maximize his expected financial gain?

(b) If the investor can obtain professional investment advice from Salomon Brothers, Inc., what would be the maximum amount the investor should pay for this service?

12.18 Kellogg Company is considering the following investment project and has estimated all cost and revenues in constant dollars. The project requires the purchase of a $9,000 asset, which will be used for only two years (the project life).

- The salvage value of this asset at the end of two years is expected to be $4,000.
- The project requires an investment of $2,000 in working capital, and this amount will be fully recovered at the end of the project year.
- The annual revenue, as well as general inflation, are discrete random variables, but can be described by the following probability distributions (both random variables are statistically independent):

Annual Revenue (X)	Probability	General Inflation Rate (Y)	Probability
$10,000	0.30	3%	0.25
20,000	0.40	5%	0.50
30,000	0.30	7%	0.25

- The investment will be classified as a three-year MACRS property (tax life).
- It is assumed that the revenues, salvage value, and working capital are responsive to the general inflation rate.
- The revenue and inflation rate dictated during the first year will prevail over the remainder of the project period.
- The marginal income tax rate for the firm is 40%. The firm's inflation-free interest rate (i') is 10%.

(a) Determine the NPW as a function of X.

(b) In (a), compute the expected NPW of this investment.

(c) In (a), compute the variance of the NPW of the investment.

Comparing Risky Projects

12.19 A manufacturing firm is considering two mutually exclusive projects, both of which have an economic service life of one year with no salvage value. The initial cost and the net year-end revenue for each project are given in Table P12.19.

We assume that both projects are statistically independent of each other.

(a) If you are an expected-value maximizer, which project would you select?

(b) If you also consider the variance of the project, which project would you select?

TABLE 12.19 Comparison of Mutually Exclusive Projects

First Cost	Project 1 ($1,000)		Project 2 ($800)	
	Probability	Revenue	Probability	Revenue
Net revenue, given in PW	0.2	$2,000	0.3	$1,000
	0.6	3,000	0.4	2,500
	0.2	3,500	0.3	4,500

12.20 A business executive is trying to decide whether to undertake one of two contracts or neither one. He has simplified the situation somewhat and feels that it is sufficient to imagine that the contracts provide alternatives as follows:

Contract A		Contract B	
NPW	Probability	NPW	Probability
$100,000	0.2	$40,000	0.3
50,000	0.4	10,000	0.4
0	0.4	−10,000	0.3

(a) Should the executive undertake either one of the contracts? If so, which one? What would he do if he made decisions with an eye toward maximizing his expected NPW?

(b) What would be the probability that Contract A would result in a larger profit than that of Contract B?

12.21 Two alternative machines are being considered for a cost-reduction project.

- Machine A has a first cost of $60,000 and a salvage value (after tax) of $22,000 at the end of 6 years of service life. The probabilities of annual after-tax operating costs of this machine are estimated as follows:

Annual O&M Costs	Probability
$5,000	0.20
8,000	0.30
10,000	0.30
12,000	0.20

- Machine B has an initial cost of $35,000, and its estimated salvage value (after tax) at the end of 4 years of service is negligible. The annual after-tax operating costs are estimated to be as follows:

Annual O&M Costs	Probability
$8,000	0.10
10,000	0.30
12,000	0.40
14,000	0.20

The MARR on this project is 10%. The required service period of these machines is estimated to be 12 years, and no technological advance in either machine is expected.

(a) Assuming independence, calculate the mean and variance for the equivalent annual cost of operating each machine.

(b) From the results of part (a), calculate the probability that the annual cost of operating Machine A will exceed the cost of operating Machine B.

12.22 Two mutually exclusive investment projects are under consideration. It is assumed that the cash flows are statistically independent random variables with means and variances estimated as follows:

End of Year	Project A Mean	Project A Variance	Project B Mean	Project B Variance
0	−$5,000	$1,000^2$	−$10,000	$2,000^2$
1	4,000	$1,000^2$	6,000	$1,500^2$
2	4,000	$1,500^2$	8,000	$2,000^2$

(a) For each project, determine the mean and standard deviation of the NPW, using an interest rate of 15%.

(b) On the basis of the results of part (a), which project would you recommend?

Decision-Tree Analysis

12.23 Delta College's campus police are quite concerned with ever-growing weekend parties taking place at the various dormitories on the campus, where alcohol is commonly served to underage college students. According to reliable information, on a given Saturday night, one may observe a party to take place 60% of the time. Police Lieutenant Shark usually receives a tip regarding student drinking that is to take place in one of the residence halls the next weekend. According to Officer Shark, this tipster has been correct 40% of the time when the party is planned. The tipster has been also correct 80% of the time when the party does not take place. (That is, the tipster says that no party is planned.) If Officer Shark does not raid the residence hall in question at the time of the supposed party, he loses 10 career progress points. (The police chief gets complete information on whether there was a party only after the weekend.) If he leads a raid and the tip is false, he loses 50 career progress points, whereas if the tip is correct, he earns 100 points.

(a) What is the probability that no party is actually planned even though the tipster says that there will be a party?

(b) If the lieutenant wishes to maximize his expected career progress points, what should he do?

(c) What is the EVPI (in terms of career points)?

12.24 As a plant manager of a firm, you are trying to decide whether to open a new factory outlet store, which would cost about $500,000. Success of the outlet store depends on demand in the new region. If demand is high, you expect to gain $1 million per year; if demand is average, the gain is $500,000; and if demand is low, you lose $80,000. From your knowledge of the region and your product, you feel that the chances are 0.4 that sales will be average and equally likely that they will be high or low (0.3, respectively). Assume that the firm's MARR is known to be 15%, and the marginal tax rate will be 40%. Also, assume that the salvage value of the store at the end of 15 years will be about $100,000. The store will be depreciated under a 39-year property class.

(a) If the new outlet store will be in business for 15 years, should you open it? How much would you be willing to pay to know the true state of nature?

(b) Suppose a market survey is available at $1,000, with the following reliability (the values shown were obtained from past experience, where actual demand was compared with predictions made by the survey):

Actual	Survey Prediction		
Demand	Low	Medium	High
Low	0.75	0.20	0.05
Medium	0.20	0.60	0.20
High	0.05	0.25	0.70

Determine the strategy that maximizes the expected payoff after taking the market survey. In doing so, compute the EVPI after taking the survey. What is the true worth of the sample information?

Short Case Studies

ST12.1 In Virginia's six-number lottery, or lotto, players pick six numbers from 1 to 44. The winning combination is determined by a machine that looks like a popcorn machine, except that it is filled with numbered table-tennis balls. On February 15, 1992, the Virginia lottery drawing offered the prizes shown in Table ST12.1, assuming that the first prize is not shared.

Common among regular lottery players is this dream: waiting until the jackpot reaches an astronomical sum and then buying every possible number, thereby guaranteeing a winner. Sure, it would cost millions of dollars, but the payoff would be much greater. Is it worth trying? How do the odds of winning the first prize change as you increase the number of tickets purchased?

TABLE ST12.1 Virginia Lottery Prizes

Number of Prizes	Prize Category	Total Amount
1	First prize	$27,007,364
228	Second prizes ($899 each)	204,972
10,552	Third prizes ($51 each)	538,152
168,073	Fourth prizes ($1 each)	168,073
	Total winnings	$27,918,561

ST12.2 Mount Manufacturing Company produces industrial and public safety shirts. As is done in most apparel-manufacturing factories, the cloth must be cut into shirt parts by marking sheets of paper in a particular way. At present, these sheet markings are done manually, and their annual labor cost is running around $103,718. Mount has the option of purchasing one of two automated marking systems: the Lectra System 305 and the Tex Corporation Marking System. The comparative characteristics of the two systems are as follows:

	Most Likely Estimates	
	Lectra System	Tex System
Annual labor cost	$ 51,609	$ 51,609
Annual material savings	$230,000	$274,000
Investment cost	$136,150	$195,500
Estimated life	6 years	6 years
Salvage value	$ 20,000	$ 15,000
Depreciation method (MACRS)	5 years	5 years

The firm's marginal tax rate is 40%, and the interest rate used for project evaluation is 12% after taxes.

(a) Based on the most likely estimates, which alternative is the best?

(b) Suppose that the company estimates the material savings during the first year for each system on the basis of the following probability distribution:

Lectra System	
Material Savings	Probability
$150,000	0.25
230,000	0.40
270,000	0.35

Tex System	
Material Savings	**Probability**
$200,000	0.30
274,000	0.50
312,000	0.20

Suppose further that the annual material savings for both Lectra and Tex are statistically independent. Compute the mean and variance for the equivalent annual value of operating each system.

(c) In part (b), calculate the probability that the annual benefit of operating Lectra will exceed the annual benefit of operating Tex. (Use @RISK or Excel to answer this question.)

ST12.3 The city of Opelika was having a problem locating land for a new sanitary landfill when the Alabama Energy Extension Service offered the solution of burning the solid waste to generate steam. At the same time, Uniroyal Tire Company seemed to be having a similar problem disposing of solid waste in the form of rubber tires. It was determined that there would be about 200 tons per day of waste to be burned, including municipal and industrial waste. The city is considering building a waste-fired steam plant, which would cost $6,688,800. To finance the construction cost, the city will issue resource recovery revenue bonds in the amount of $7,000,000 at an interest rate of 11.5%. Bond interest is payable annually. The differential amount between the actual construction costs and the amount of bond financing ($7,000,000 − $6,688,800 = $311,200) will be used to settle the bond discount and expenses associated with the bond financing. The expected life of the steam plant is 20 years. The expected salvage value is estimated to be about $300,000. The expected labor costs are $335,000 per year. The annual operating and maintenance costs (including fuel, electricity, maintenance, and water) are expected to be $175,000. The plant would generate 9,360 pounds of waste, along with 7,200 pounds of waste after incineration, which will have to be disposed of as landfill. At the present rate of $19.45 per pound, this will cost the city a total of $322,000 per year. The revenues for the steam plant will come from two sources: (1) sales of steam and (2) tipping fees for disposal. The city expects 20% downtime per year for the waste-fired steam plant. With an input of 200 tons per day and 3.01 pounds of steam per pound of refuse, a maximum of 1,327,453 pounds of steam can be produced per day. However, with 20% downtime, the actual output would be 1,061,962 pounds of steam per day. The initial steam charge will be approximately $4.00 per thousand pounds, which will bring in $1,550,520 in steam revenue the first year. The tipping fee is used in conjunction with the sale of steam to offset the total plant cost. It is the goal of the Opelika steam plant to phase out the tipping fee as soon as possible. The tipping fee will be $20.85 per ton in the first year of plant operation and will be phased out at the end of the eighth year. The scheduled tipping fee assessment is as follows:

Year	Tipping Fee
1	$976,114
2	895,723
3	800,275
4	687,153
5	553,301
6	395,161
7	208,585

(a) At an interest rate of 10%, would the steam plant generate sufficient revenue to recover the initial investment?

(b) At an interest rate of 10%, what would be the minimum charge (per thousand pounds) for steam sales to make the project break even?

(c) Perform a sensitivity analysis to determine the input variable of the plant's downtime.

ST12.4 Burlington Motor Carriers, a trucking company, is considering installing a two-way mobile satellite messaging service on its 2,000 trucks. On the basis of tests done last year on 120 trucks, the company found that satellite messaging could cut 60% from its $5 million bill for long-distance communication with truck drivers. More important, the drivers reduced the number of "deadhead" miles—those driven with nonpaying loads—by 0.5%. Applying that improvement to all 230 million miles covered by the Burlington fleet each year would produce an extra $1.25 million savings.

Equipping all 2,000 trucks with the satellite hookup will require an investment of $8 million and the construction of a message-relaying system costing $2 million. The equipment and onboard devices will have a service life of eight years and negligible salvage value; they will be depreciated under the five-year MACRS class. Burlington's marginal tax rate is about 38%, and its required minimum attractive rate of return is 18%.

(a) Determine the annual net cash flows from the project.

(b) Perform a sensitivity analysis on the project's data, varying savings in telephone bills and savings in deadhead miles. Assume that each of these variables can deviate from its base-case expected value by $\pm 10\%$, $\pm 20\%$, and $\pm 30\%$.

(c) Prepare sensitivity diagrams and interpret the results.

ST12.5 The following is a comparison of the cost structure of conventional manufacturing technology (CMT) with a flexible manufacturing system (FMS) at one U.S. firm.

	Most Likely Estimates	
	CMT	**FMS**
Number of part types	3,000	3,000
Number of pieces produced/year	544,000	544,000
Variable labor cost/part	$2.15	$1.30
Variable material cost/part	$1.53	$1.10
Total variable cost/part	$3.68	$2.40
Annual overhead	$3.15M	$1.95M
Annual tooling costs	$470,000	$300,000
Annual inventory costs	$141,000	$31,500
Total annual fixed operating costs	$3.76M	$2.28M
Investment	$3.5M	$10M
Salvage value	$0.5M	$1M
Service life	10 years	10 years
Depreciation method (MACRS)	7 years	7 years

(a) The firm's marginal tax rate and MARR are 40% and 15%, respectively. Determine the incremental cash flow (FMS − CMT), based on the most likely estimates.

(b) Management feels confident about all input estimates in CMT. How.ever, the firm does not have any previous experience operating an FMS. Therefore, many input estimates, except the investment and salvage value, are subject to variation. Perform a sensitivity analysis on the project's data, varying the elements of operating costs. Assume that each of these variables can deviate from its base-case expected value by ±10%, ±20%, and ±30%.

(c) Prepare sensitivity diagrams and interpret the results.

(d) Suppose that probabilities of the variable material cost and the annual inventory cost for the FMS are estimated as follows:

Material Cost	
Cost per Part	**Probability**
$1.00	0.25
1.10	0.30
1.20	0.20
1.30	0.20
1.40	0.05

Inventory Cost	
Annual Inventory Cost	**Probability**
$25,000	0.10
31,000	0.30
50,000	0.20
80,000	0.20
100,000	0.20

What are the best and the worst cases of incremental NPW?

(e) In part (d), assuming that the random variables of the cost per part and the annual inventory cost are statistically independent, find the mean and variance of the NPW for the incremental cash flows.

(f) In parts (d) and (e), what is the probability that the FMS will be the more expensive of the two investment options? (Use @RISK or Excel to answer this question.)

THIRTEEN

Real-Options Analysis

Wi-Fi Service on the Plane[1] U.S. airlines continue to flirt with high-speed Internet access on flights, but financial and technical obstacles may block it from becoming a routine travel feature. United Airlines is considering the possibility of offering in-flight Wi-Fi Internet service to its passengers.

There are two types of technology available in the marketplace. The first one is the "air-to-ground" (ATG) and the other is the satellite system. Using ground towers, ATG would be less expensive to install and provides less-spotty coverage, particularly on short-haul flights, than the satellite system. In the satellite system, currently offered only by Connexion by Boeing, signals are beamed between satellites and airplanes. It is a proven technology that works over land and water, making it ideal for international flights.

Unlike Connexion's satellite system, air-to-ground is not yet commercially available. It requires ground base stations that beam signals to a device on the aircraft. It works over and near land. Installing equipment for an air-to-ground system on a plane would be about one-third the cost of a satellite-based system. Satellites will be an integral part of future Wi-Fi development because of air-to-ground's limited use over water.

In-flight Wi-Fi, particularly on long-haul flights, could be a sizable business for airlines. About 38% of business travelers surveyed said they'd use the service even at a $25 flat fee per flight. Connexion customers now pay a flat fee of $29.95 per flight or $9.95 per half-hour. It is expected that up to 20% of all travelers will want in-flight wireless access. Whether it can contribute to an airline's bottom line is uncertain. Airlines are cutting back on many in-flight goodies, and Wi-Fi is an expensive perk requiring millions in outlay.

---

[1] "Airlines Get a Wi-Fi Service," *USA Today*, June 17, 2005.

How two wireless systems compare

In-flight Internet access can be provided by ground-based or satellite-based technology:

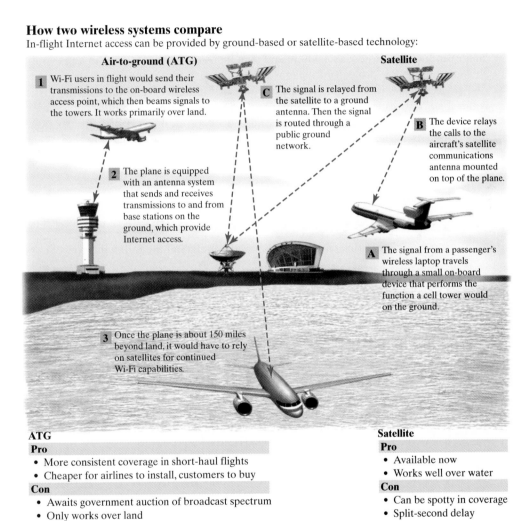

Air-to-ground (ATG)

1 Wi-Fi users in flight would send their transmissions to the on-board wireless access point, which then beams signals to the towers. It works primarily over land.

2 The plane is equipped with an antenna system that sends and receives transmissions to and from base stations on the ground, which provide Internet access.

3 Once the plane is about 150 miles beyond land, it would have to rely on satellites for continued Wi-Fi capabilities.

Satellite

C The signal is relayed from the satellite to a ground antenna. Then the signal is routed through a public ground network.

B The device relays the calls to the aircraft's satellite communications antenna mounted on top of the plane.

A The signal from a passenger's wireless laptop travels through a small on-board device that performs the function a cell tower would on the ground.

ATG

Pro
- More consistent coverage in short-haul flights
- Cheaper for airlines to install, customers to buy

Con
- Awaits government auction of broadcast spectrum
- Only works over land

Satellite

Pro
- Available now
- Works well over water

Con
- Can be spotty in coverage
- Split-second delay

Still, once the service is unleashed in the USA, it could prove difficult for airlines to ignore customer demand. Lufthansa boasts that its Internet service is helping to lure corporate customers. As the economy improves, more airlines will opt in. Business travelers are ideal customers because they like to stay productive, don't mind paying for convenience, usually travel with laptops, and often sit in premium seats with power ports.

Now United Airlines is wondering whether it is worth waiting until other airlines offer the Internet service or whether it should go ahead and provide the service on a limited basis.

A recent article in *BusinessWeek* magazine took a closer look at a new trend in corporate planning, known as "the real-options revolution in decision making." In a nutshell, real-options analysis simply says that companies benefit by keeping their options open. For example, suppose a company is deciding whether to fund a large R&D project that could either make or lose lots of money. A traditional calculation of net present worth (NPW), which discounts projected costs and revenues into today's dollars, examines the project as a whole and concludes that it is a no go. But a real-options analysis breaks the project into stages and concludes that it makes sense to fund the first stage at least. Real-options analysis rewards flexibility—and that is what makes it better than NPW, one of today's standard decision-making tools. In this chapter, we will examine the fundamentals of the real-options decision framework and discuss how we may apply its logic to approach solving capital-investment decision problems under uncertainty.

CHAPTER LEARNING OBJECTIVES

After completing this chapter, you should understand the following concepts:

■ How the financial option can be used to hedge the market risk.

■ How to price the financial options.

■ What features of financial options would be useful in managing risk for real assets.

■ How to structure various real-options models.

■ How to value the flexibility of various real options.

13.1 Risk Management: Financial Options

Let us start with an ordinary "option," which is just the *opportunity* to buy or sell stock at a specific price within a certain period. The two basic types of financial options are a *call option* and a *put option*. The call option gives the investor the right to buy a stock at a specified price within a specified period. The put option gives the investor the right to sell a stock at a specified price within a particular period. Financial options have several important features:

• **The contracting parties.** The party granting the right is referred to as the *option seller* (or option writer). The party purchasing the right is referred to as the *option buyer*. In the option world, the option buyer is said to have a *long* position in the option, while the option seller is said to have a *short* position in the option.

• **The right or obligation.** An option to buy a financial asset at a specified price is a call option on the asset. The call buyer has the *right* to purchase the asset; the call option seller has the *obligation* to sell. An option to sell an asset at a specified price is a put option on the asset. Consequently, the put seller has the *right* to sell and the put buyer has the *obligation* to buy.

- The **underlying asset** (S) dictates the value of the financial option. Typically, the underlying asset is a financial asset such as equities (stocks), equity indexes, and commodities.
- The **strike** or **exercise price** (K) gives the value at which the investor can purchase or sell the underlying security (financial asset).
- The **maturity** (T) of the option defines the period within which the investor can buy or sell the stock at the strike price.
- The **option premium** (C) is the price that an investor has to pay to own the option. If you buy a call option by paying a premium (C), you have the right to purchase a certain stock at a predetermined (strike) price (K) before or on a maturity date (or exercise date) T.
- An option that can be exercised earlier than its maturity date is called an ***American option***. An option that can be exercised only at the maturity date is a ***European option***.

American option: An option which can be exercised at any time between the purchase date and the expiration date.

- **Payoffs.** The payoff of an option depends on the price of the underlying asset (S_T) at the exercise date. Three terms—*at the money*, *in the money*, and *out of the money*—identify where the current stock price (S) is relative to the strike price (K). An option that is in the money is an option that, if exercised, would result in a positive value to the investor; an option that is out of the money is an option that, if exercised, would result in a negative value to the investor. You will exercise your option when the option is in the money. If not, you let the option expire. Your loss is limited to the option premium itself. Therefore, in every option contract, there are two sides: the one who holds the option (the long position) and the one who issues the option (the short position). Thus, there are four possible positions: a long position in a call option, a long position in a put option, a short position in a call option, and a short position in a put option. Figure 13.1 illustrates these four positions, together with their respective payoffs, for European options.

To explain the terminology further, we will use Figure 13.2, in which an investor could buy a call option with a strike price of $400 at $24 per share on Google stock (or $2,400 for 100 shares) or a put option at $42 per share with a strike price of $300. Either option has a maturity date of January 2007.

In general, the call and put options have the following characteristics:

- The greater the difference between the exercise price and the actual current price of the item, the cheaper the premium, because there is less of a chance that the option will be exercised.
- The closer the expiration date of an option that is *out of the money* (the market price is higher than the strike price), the cheaper the price.
- The more time there is until expiration, the larger the premium, because the chance of reaching the strike price is greater and the carrying costs are more.
- Call and put options move in opposition. Call options rise in value as the underlying market prices go up. Put options rise in value as market prices go down.

We will consider two types of investors: one who thinks that the share price of the stock will continue to go up (a call-option buyer) and the other who thinks that the share

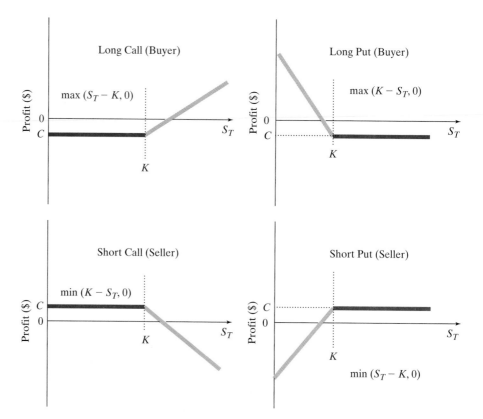

Call option: An option contract that gives the holder the right to buy a certain quantity of an underlying security from the writer of the option, at a specified price up to a specified date.

Figure 13.1 Payoffs from positions in financial options. If you long a call position, your loss is limited to your option premium (C) when $S_T < K$, but your potential for profit is unlimited when $S_T > K$.

Put option: An option contract that gives the holder the right to sell a certain quantity of an underlying security to the writer of the option, at a strike price up to an expiration date.

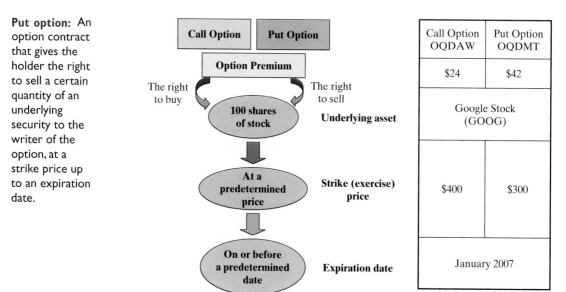

Figure 13.2 Buying a call or put option for Google stock on July 25, 2005. The stock was trading at $295.55.

price of the stock will go down (a put-option buyer). Let's examine how much profit each investor would make if his or her prediction proves to be true.

13.1.1 Buy Call Options when You Expect the Price to Go Up

An investor purchases a call option when he or she is bullish on a company and thinks that the stock price will rise within a specific amount of time—usually three, six, or nine months. For this right, you pay the call-option seller a fee (called a "premium"), which is forfeited if you do not exercise the option before the agreed-upon expiration date. For example, suppose you bought a January 2007 call option with a strike price of $400 on Google stock on July 25, 2005. The stock was trading at $295.55 and the option premium was quoted at $24 per share. As shown in Figure 13.3, you will exercise your option when the stock price is higher than $400. Otherwise your option is worthless and you let the option expire and take a $24 loss per share. Clearly, the investor is betting that the stock price will be much higher than $400 by January 2007. (Otherwise he or she would not own the call option.)

13.1.2 Buy Put Options when You Expect the Price to Go Down

One way to protect the value of financial assets from a significant downside risk is to buy put-option contracts. When an investor purchases a put, he or she is betting that the underlying investment is going to decrease in value within a certain amount of time. This means that if the value does go down, you will *exercise* the *put*. Buying a put-option contract is equivalent to purchasing a home insurance policy. You pay an annual premium for protection in case, say, your house burns to the ground. If that happens, you will exercise the put and get the money. If it never happens during the contract period, your loss is

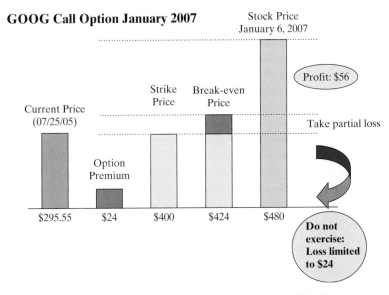

Figure 13.3 Buying a call option at $24 as of July 25, 2005. If the stock price increases to $480 on January 6, 2007, the investor's profit will be $56 per share. If the stock is trading below $400 on that date, the investor's loss will be limited to $24 per share.

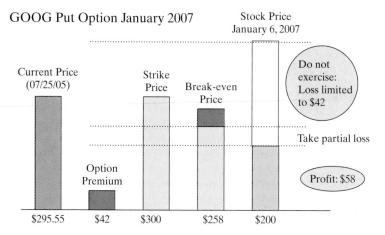

Figure 13.4 Buying a put option on Google stock. If the stock price decreases to $200, the investor will earn a $58 profit; otherwise the investor will lose money from holding the put option.

limited to the cost of the premium. Figure 13.4 illustrates how an investor might profit from buying a put option on Google stock. Once again, with a strike price of $300 and an option premium of $42, you would exercise your put only when the share price of the stock dipped below $300. At $258, you will break even; between $258 and $300, you will have a partial loss by recovering some of your premium. Once again, the reason you are buying a put option with a strike price of $300 is that the current price of the stock is too high to sustain that value in the future.

13.2 Option Strategies

Now that we have described how the basic financial options work, we will review two basic strategies on how to hedge the financial risk when we use these options. The first strategy is to reduce the capital that is at risk by buying call options, and the second strategy is to buy a put option to hedge the market risk.

13.2.1 Buying Calls to Reduce Capital That Is at Risk

Once again, buying a call gives the owner a right, but not an obligation. The risk for the call buyer is limited to the premium paid for the call (the price of the call), plus commissions. The value of the call tends to increase as the price of the underlying stock rises. This gain will increasingly reflect a rise in the value of the underlying stock when the market price moves above the option's strike price. As an investor, you could buy the underlying security, or you could buy call options on the underlying security. As a call buyer, you have three options: (1) Hold the option to maturity and trade at the strike price, (2) trade for profit before the option expires (known as exercising your option), and (3) let the option expire if doing so is advantageous to you. Example 13.1 illustrates how you might buy a call to participate in the upward movement of a stock while limiting your downside risk.

EXAMPLE 13.1 How to Buy a Call to Participate in the Upward Movement of a Stock while Limiting Your Downside Risk

Dell Computer (DELL) is trading at $44\frac{1}{4}$. Instead of spending $22,125 for 500 shares of DELL stock, an investor could purchase a six-month call with a 45-strike price for $3\frac{3}{8}$. By doing this, the investor is saying that he or she anticipates that DELL will rise above the strike of 45 (which is where DELL can be purchased no matter how high DELL has risen), plus another $3\frac{3}{8}$ (the option premium), or $48\frac{3}{8}$, by expiration. Without considering commissions and taxes, discuss the amount of risk presented in each of the following scenarios, as opposed to purchasing 500 shares:

- DELL is trading above $48\frac{3}{8}$ by expiration.
- DELL is trading between 45 and $48\frac{3}{8}$ at expiration.
- DELL is trading at or below 45 at expiration.

DISCUSSION: Each call represents 100 shares of stock, so 5 calls could be bought in place of 500 shares of stock. The cost of 5 calls at $3\frac{3}{8}$ is $1,687.50 (5 calls $\times$ $3\frac{3}{8} \times$ 100). Thus, instead of spending $22,125 on stock, the investor needs to spend only $1,687.50 for the purchase of the 5 calls. The balance of $20,437.50 could then be invested in short-term instruments. This investor has unlimited profit potential when DELL rises above $48\frac{3}{8}$. The risk for the option buyer is limited to the premium paid, which, in this example, is $1,687.50.

SOLUTION

Given: Buying 5 DELL six-month 45 calls at $3\frac{3}{8}$ versus buying 500 shares of DELL at $44\frac{1}{4}$.

Find: Profit or loss for three possible scenarios at expiration.

(a) DELL is above $48\frac{3}{8}$ by expiration:

- If DELL is at $51 at expiration, the option will be worth the difference between the strike and the current price of the stock:

$$\$51 \text{ (current price)} - 45 \text{ (strike price)} = \$6 \text{ (current option value)}$$

The option could be sold (exercised), and a 77% return would be earned on the initial investment. That is,

$$\$6 \text{ (current option value)} - \$3\frac{3}{8} \text{ (premium paid for option)}$$

$$= \$2\frac{5}{8} \text{ (profit if option is sold)}$$

- Had the stock been purchased at $44\frac{1}{4}$ (at a cost of $22,125), and had it risen to 51, it would now be worth $25,500. This would be a 15.3% increase in value over the original cost of $22,125. But the call buyer spent only $1,687.50 and earned 77% on his options. The following tables compare both scenarios:

Had Stock Been Purchased				
Stock Purchase Price	Initial Cost of Stock, 500 Shares	Stock Price at Expiration	Value of Stock at Expiration*	Change in Stock Value
44 1/4	$22,125	51	$25,500	$3,375
44 1/4	$22,125	47	$23,500	$1,375
44 1/4	**$22,125**	**40**	**$20,000**	**($2,125)**

Had Call Options Been Purchased				
Option Price per Contract	Initial Total Cost, 5 Options	Option Price Per Contract at Expiration	Total Value of Options	Change in Options Value
3 3/8	$1,687.50	6	$3,000	$1,312.50[†]
3 3/8	$1,687.50	2	$1,000	($687.50)[†]
3 3/8	**$1,687.50**	**0**	**0**	**($1,687.50)[†]**

*Plus dividends if any.

[†]Plus interest earned on cash not used, cost of stock less call premium.

(b) DELL is between 45 and $48\frac{3}{8}$ at expiration:

- The investor's option will still hold some value if DELL is between 45 and $48\frac{3}{8}$ but not enough to break even on the position. The option can still be sold to recoup some of the cost. For example, suppose DELL is at 47 on the last day of trading, usually the third Friday of the expiration month. Then the option can be sold to close out the position through the last trading day of the call. What happened is that DELL did rise in value, but not as much as anticipated. The option that cost $3\frac{3}{8}$ is now worth just 2 points. Instead of letting the option expire, you can sell the call and recoup some of your losses:

$$3\frac{3}{8} \, (\text{Cost}) - 2 \, (\text{Sale}) = \$1\frac{3}{8} \, (\text{Net loss excluding commissions})$$

- If just the stock had been bought and it rose to 47, $1,375 would have been earned, while the holder of calls would have lost $687.50. However, the holder of calls would have been earning interest on $20,437.50, which would offset some of the loss in the options.

(c) DELL is at or below 45 at expiration:

- Suppose DELL is now at 40 and the option has expired worthless. The premium that was paid for the calls has been lost. However, had the stock been bought, the investor would now be in a losing stock position, hoping to break even. By purchasing a call he or she had limited capital at risk. Now the investor still has most of the money that would have gone into buying the stock, plus interest. Thus, the investor can make another investment decision.

- If the investor had purchased DELL at $44\frac{1}{4}$, and the stock did not move as he or she anticipated, the investor would have had two choices: Sell the stock and, after commission costs, incur some losses; or hold onto it and hope that it rises over the long term. Had the investor bought the options and been wrong, the options would expire worthless and the loss would be limited to the premium paid.

13.2.2 Protective Puts as a Hedge[2]

People insure their valuable assets, but most investors have not realized that many of their stock positions also can be insured. That is exactly what a *protective put* does. Typically, by paying a relatively small premium (compared with the market value of the stock), an investor knows that no matter how far the stock drops, it can be sold at the strike price of the put anytime up until the put expires. Buying puts against an existing stock position or simultaneously purchasing stock and puts can supply the insurance needed to overcome the uncertainty of the marketplace. Although a protective put may not be suitable for all investors, it can provide the protection needed to invest in individual stocks in volatile markets, because it ensures limited downside risk and unlimited profit potential for the life of the option.

Financial hedging: An investment made in order to reduce the risk of adverse price movements in a security.

Buying a protective put involves the purchase of one put contract for every 100 shares of stock already owned or purchased. Purchasing a put against stock is similar to purchasing insurance. The investor pays a premium (the cost of the put) to insure against a loss in the stock position. No matter what happens to the price of the stock, the put owner can sell it at the strike price at any time prior to expiration.

EXAMPLE 13.2 How to Use a Protective Put as Insurance

Suppose you are investing in Qualcomm (QCOM) stock. QCOM is trading at $50 per share, and the QCOM six-month, 50-put contract can be purchased at $2\frac{1}{4}$. To take a look at what happens to a protective put position as the underlying stock (QCOM) moves up or down, compare the amount of protection anticipated under the following two possible scenarios at expiration (do not consider commissions and taxes for now):

- **Option 1.** Buy QCOM at $50, without owning a put for protection.
- **Option 2.** Buy QCOM at $50 and buy a QCOM 50-put contract.

SOLUTION

Given: (a) Option 1 and (b) Option 2.
Find: Compare buying QCOM stock with buying QCOM with a protective put.

---

[2] This section is based on the materials prepared by the Chicago Board Options Exchange (CBOE). These materials are available on the company's website at http://www.cboe.com.

(a) Option 1. Buy QCOM at $50:

If stock is bought at $50 per share, the investor begins to lose money as soon as the stock drops below the purchase price. The entire $50 purchase price is at risk.

- *Upside potential.* If the price increases, the investor benefits from the entire increase without incurring the cost of the put premium or insurance.
- *Downside risk.* When only the stock is bought, there is no protection or insurance. The investor is at risk of losing the entire investment.

(b) Option 2. Buy QCOM at $50 and buy a QCOM 50-put contract

A six-month put with a strike price of 50 can be bought for $2\frac{1}{4}$, or $225 per contract $\left(\$2\frac{1}{4} \times 100\right)$. This put can be considered insurance "without a deductible," because the stock is purchased at $50 and an "at-the-money" put with the same strike price, 50, is purchased. If the stock drops below $50, the put or insurance will begin to offset any loss in QCOM (less the cost of the put). If QCOM remains at $50 or above, the put will expire worthless and the premium would be lost. If just the stock had been bought, the investor would begin to profit as soon as the stock rose above $50. However, the investor would have no protection from the risk of the stock declining in value. Owning a put along with stock ensures limited risk while increasing the break-even on the stock by the cost of the put, but still allowing for unlimited profit potential above the break-even.

	Buy QCOM	Buy QCOM and Six-Month 50 Put
Stock Cost	$50	$50
Put Cost	0	$ $2\frac{1}{4}$
Total Cost	$50	52\frac{1}{4}$
Risk	$50	$ $2\frac{1}{4}$

- *Upside potential.* This strategy gives an investor the advantage of having downside protection without limiting upside potential above the total cost of the position, or 52\frac{1}{4}$. The only disadvantage is that the investor will not begin to profit until the stock rises above 52\frac{1}{4}$.
- *Downside risk.* No matter how low QCOM falls, buying the six-month put with a 50-strike price gives the investor the right to sell QCOM at $50 up until expiration. The downside risk is only $2\frac{1}{4}$: the total cost for this position, 52\frac{1}{4}$, less 50 (the strike price).

COMMENTS: If you buy QCOM 45 put instead of 50 put, your behavior is viewed as if you want some downside protection on a stock position, but are willing to have a deductible in exchange for a lower insurance cost. The potential volatility of the equity markets can be of great concern to investors. The purchase of a protective put can give the investor the comfort level needed to purchase individual securities. This strategy is actually more conservative than purchasing stock. As long as a put is held against a stock position, there is limited risk; you know where the stock can be sold. The only disadvantages are that money cannot be made until the stock moves above the combined cost of the stock and the put and that the put has a finite life. Once the stock rises above the total cost of the position, however, an investor has the potential for unlimited profit.

13.3 Option Pricing

In this section, we illustrate the conceptual foundation of how to determine the price of an option. We know exactly the value of a financial option at maturity. However, our interest is not in the value at maturity, but rather in the value *today*. Here, we are basically looking at the pricing of a European option on a share of stock that pays no dividend. In this simplified approach to option pricing, we are going to envision our continuous-process world as a series of snapshots. First we will define the symbols to use in pricing a financial option:

Δ = the number of shares of the underlying asset to purchase,

b = the amount of cash borrowed or bond purchased at the risk-free rate,

R = 1 plus the risk free rate $(1 + r)$,

S_0 = the value of the underlying asset today,

uS = the upward movement in the value of the underlying asset in the future at some time t,

dS = the downward movement in the value of the underlying asset in the future at some time t,

K = the strike (or exercise) price of the option,

C = the value of the call option,

C_u = the upward movement in the value of the call option,

C_d = the downward movement in the value of the call option.

> **The option premium** is primarily affected by the difference between the stock price and the strike price, the time remaining for the option to be exercised, and the volatility of the underlying stock.

In what follows, we will review three different approaches to valuing an option. The first approach is to use the replicating-portfolio concept, the second approach is to use the risk-free financing concept, and the third approach is to use the risk-neutral concept. All approaches yield the same result, so the choice depends on the preference of the analyst.

13.3.1 Replicating-Portfolio Approach with a Call Option

Let's suppose that today, day 0, the price of a particular share of stock, say, GOOG, is $300. Let's suppose further that tomorrow it could rise or fall by 5%. Now consider the value of a one-day call option on this share of stock. Suppose that the exercise price of this one-day call option is $300. Then the value of the call option at maturity—day 1—is as shown in Figure 13.5. However, if you were thinking about buying this call option, you would be interested in knowing its value today. The issue is how we calculate that value.

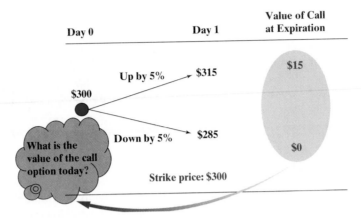

Figure 13.5 A simplified approach to a one-day call option.

We will value the option by creating an arbitrage portfolio that contains the option. An **arbitrage (replicating) portfolio** is a portfolio that earns a sure return. Let's form such a portfolio out of two risky assets: the share of stock and the call option on the share of the stock. We will create a portfolio such that the gains made on one of the assets will be exactly offset by losses on the other. As the following table indicates, creating a portfolio in which we are selling (writing) two call options against each share of stock could form such a portfolio:

Arbitrage: Attempting to profit by exploiting price differences of identical or similar financial instruments, on different markets or in different forms.

Share Price S	Option Value C	Hedge Ratio $2C$	Arbitrage Portfolio $S - 2C$
$315	$15	$30	$285
285	0	0	285

In option language, we are *long* one share of stock and *short* (selling) two call options. (Note that you (the writer) have already been paid for selling the option at day 0. So on day 1, you are concerned only about what the option buyer will do with his or her position.)

As illustrated in the table, if, at expiration, the share price is higher than the exercise price, then the owner of the call option benefits at the expense of the seller: The payoff to the option owner is $15 (or $315 − $300). By contrast, the payoff (expense) to the seller of the option is the reverse. If the share price at expiration is less than the exercise price, the call option is worthless: The payoff to both parties is zero. Therefore, this approach is equivalent to buying an insurance policy to protect any potential loss caused by changes in the price of a share.

- On day 1, the value of this arbitrage portfolio $(S - 2C)$ will be $285, regardless of the value of the share of stock. Hence, we have formed a portfolio that has no risk. Here,

$2C$ refers to the number of calls necessary to form the arbitrage portfolio. This number is easily found by dividing the spread between the share prices ($\$315 - \$285$) by the spread between the option values ($\$15 - 0$) at maturity. For example, the share price could change by $30 and the value of the call could change by $15, so it would take two calls to cover the possible change in the value of the share.

- On day 0, the value of the share is $300, but we do not know the option value. However, we know the value of the arbitrage portfolio: $S - 2C$, or $300 - 2C$. We also know that on day 1 the value of the portfolio is $285. Therefore, we conclude that

$$(300 - 2C)_{\text{day } 0} < (285)_{\text{day } 1}.$$

To equate these two option values, it is necessary to discount day-1 values to day-0 values. So we can express the present value of the $285 to be received one day earlier as

$$300 - 2C = \frac{285}{(1 + r)},$$

where r is a one-day interest rate. Since the arbitrage portfolio is riskless, the interest rate used is the risk-free interest rate—say, 6%. The interest rate for one day is $(1/365) \times 0.06 = 0.00016$, and the preceding equation becomes

$$300 - 2C = \frac{285}{(1 + 0.00016)}.$$

Solving for C, we obtain the value of the one-day call option on day 0, namely, $7.52.

Now we can summarize what we have demonstrated in this simple example:

- Assuming no arbitrage opportunities, the portfolio consisting of the stock and the option can be set up such that we know with certainty what the value of the portfolio will be at the expiration date.
- Because there is no uncertainty in the portfolio's value, the portfolio has no risk premium; hence, discounting can be done at the risk-free rate.
- As long as a portfolio consisting of shares of stock plus a short position in call options is set up, the value of this portfolio at expiration will be the same in both the up state and the down state.
- In essence, this portfolio mitigates all risk associated with the underlying asset's price movement.
- Because the investor has hedged against all risk, the appropriate discount rate to account for the time value of money is the risk-free rate!

13.3.2 Risk-Free Financing Approach

Another way of pricing financial options is to use the risk-free financing approach, which is still based on the replicating-portfolio concept illustrated in Section 13.3.1. As before, let's assume that the value of an asset today is $300 and it is known that in one day the price of the asset will go up by $u = 1.05$ with probability p or down by $d = 0.95$ with probability $(1 - p)$ (i.e., $uS = \$315$ and $dS = \$285$). Here, we are interested in knowing the value today of a European call option with an exercise price of $300 that expires on day 1.

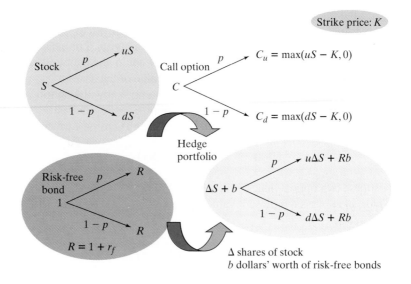

Figure 13.6 Creating a replicating portfolio by owning Δ shares of stock and b dollars worth of a risk-free asset such as a U.S. Treasury bond.

We want to create a replicating portfolio that consists of the underlying asset (a stock in this example), together with a risk-free bond instead of a call option. As shown in Figure 13.6, we are creating a portfolio that consists of Δ shares of stock and b dollars worth of risk-free bonds. On day 1, we know that the value of the up branch will be $u\Delta S + Rb$, and the down branch will be $d\Delta S + Rb$. This replicating portfolio should have the same value with the call option; otherwise investors will profit through an arbitrage opportunity. Therefore, we can set

$$u\Delta S + Rb = C_u,$$
$$d\Delta S + Rb = C_d.$$

Solving for ΔS and b respectively, we obtain

$$\Delta S = \frac{C_u - C_d}{u - d}, \tag{13.1}$$

$$b = \frac{C_u - u\Delta S}{R} = \frac{uC_d - dC_u}{R(u - d)}. \tag{13.2}$$

Using Eqs. (13.1) and (13.2) yields

$$\Delta S = \frac{C_u - C_d}{u - d} = \frac{15 - 0}{1.05 - 0.95} = \$150,$$

$$b = \frac{C_u - u\Delta S}{R} = \frac{uC_d - dC_u}{R(u - d)} = \frac{15 - 1.05(150)}{1.00016} = -\$142.48,$$

$$C = \Delta S + b = \$150.00 - \$142.48 = \$7.52.$$

These equations tell us that a replicating portfolio needs to be formed with $150 worth of stock financed in part by $142.48 at the risk-free rate of 6% interest and that the option value at day 0 should be $7.52. Note that this is exactly the same value we obtained earlier.

13.3.3 Risk-Neutral Probability Approach

Let us now assume a risk-neutral world in which to value the option. Recall from Section 13.3.2 that the replicating portfolio hedged against all risks associated with the underlying asset's movement. Because all risk has been mitigated, this approach is equivalent to valuing the option in a risk-neutral world.

As before, let's define p and $(1 - p)$ as the *objective* probabilities of obtaining the up state (uS) and down state (dS) volatiles, respectively, as shown in Figure 13.7. Then the value of a replicating portfolio will change according to these probabilities. Since no opportunity for arbitrage exists, the following equality for the asset, based on Eqs. (13.1) and (13.2), must hold:

$$\Delta S + b = \frac{C_u - C_d}{u - d} + \frac{uC_d - dC_u}{R(u - d)} = \frac{1}{R}\left(\frac{R - d}{u - d}C_u + \frac{u - R}{u - d}C_u\right). \quad (13.3)$$

Now let $q = (R - d)/(u - d)$ in Eq. (13.3). Then we can rewrite Eq. (13.3) as

$$C = \Delta S + b = \frac{1}{R}[qC_u + (1 - q)C_d]. \quad (13.4)$$

Notice that the option value is found by taking the expected value of the option with probability q and then discounting this value according to the risk-free rate. Here, the probability (q) is a **risk-neutral probability**, and the objective probability (p) never enters to the calculation. In other words, we do not need a risk-adjusted discount rate in valuing an option either.

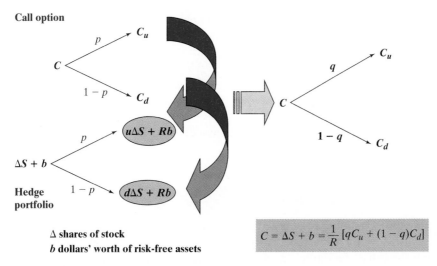

Figure 13.7 Risk-neutral probability approach to valuing an option.

If we assume continuous *discounting*, then the risk-neutral probability of an up movement q and the value of a call option can be calculated as follows:

$$q = \frac{R - d}{u - d} = \frac{(e^{rt} - d)}{u - d} \tag{13.5}$$

and

$$C = \Delta S + b = \frac{1}{e^{rt}}[qC_u + (1 - q)C_d]$$

$$= e^{-rt}[qC_u + (1 - q)C_d]. \tag{13.6}$$

Note that t represents the unit of time, expressed in years. If the duration is exactly one month, then $t = \frac{1}{12}$.

In a risk-neutral world, the following equality for the asset must hold:

$$q = \frac{R - d}{u - d} = \frac{1.00016 - 0.95}{1.05 - 0.95}$$

$$= 0.5016,$$

$$C = \Delta S + b = \frac{1}{R}[qC_u + (1 - q)C_d]$$

$$= \frac{1}{1.00016}[0.5016(\$15) + (1 - 0.5016)(\$0)]$$

$$= \$7.52.$$

Once again, we obtain the same result as the other approaches gave. The risk-neutral valuation approach is the most popular method, due to its ease of implementation and computational simplicity. Therefore, we will use that approach in most of our option valuations.

13.3.4 Put-Option Valuation

A put option is valued the same way as a call option, except with a different payoff function. Under a risk-neutral approach, the valuation formulas are

$$q = \frac{(e^{rt} - d)}{u - d}, \tag{13.7}$$

$$C = e^{-rt}[qP_u + (1 - q)P_d], \tag{13.8}$$

where P_u and P_d are the put-option values associated with up and down movements, respectively.

EXAMPLE 13.3 A Put-Option Valuation

Suppose the value of an asset today is $50 and it is known that in one year the asset price will go up by $u = 1.2$ or down by $d = 0.8$ (i.e., $uS = \$60$ and $dS = \$40$). What is the value of a European put option with an exercise price of $55 that expires in $T = 1$ year? Assume a risk-free rate of 6%.

SOLUTION

Given: $u = 1.2, d = 0.8, T = 1, r = 6\%, S_0 = \$50$, and $K = \$55$.
Find: The risk-neutral probabilities and the put-option value.

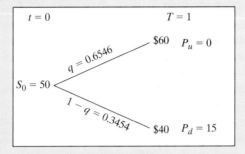

Figure 13.8 Binomial lattice model with risk-neutral probabilities.

From Figure 13.8,

$$q = \frac{e^{0.06} - 0.8}{1.2 - 0.8} = 0.6546,$$

$$C_{put} = e^{-0.06}[(0)(0.6546) + (15)(0.3454)]$$

$$= \$4.88.$$

13.3.5 Two-Period Binomial Lattice Option Valuation

The examples in the previous section assume a one-period movement of the underlying asset. Let's see what happens when we let the option run for two days. Continuing to assume that the share price can move up or down by 5% every day, we find that the distribution of share prices over the three days—and the resulting values of the call option—is as follows:

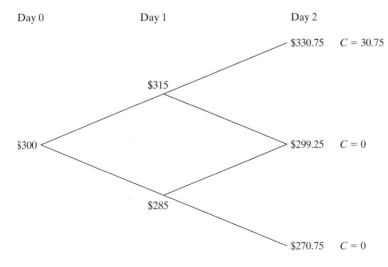

If we want to derive the value of the call option at day 0, we first must determine the values of the option for the share prices that could exist on day 1 (i.e., 315 and 285) and then use these values to determine what the option will be worth on day 0:

- On day 1, if the value of the share is $315, the arbitrage portfolio would be $S - (1.02439)C$. That is, the number of call options necessary to hedge one share of stock would be

$$N = \frac{330.75 - 299.25}{30.75 - 0} = 1.02439.$$

So $(315 - 1.02439C)_{\text{day } 1} < (299.25)_{\text{day } 2}$; therefore,

$$(315 - 1.02439C) = \frac{299.25}{1 + 0.00016}.$$

Solving for C, we obtain $15.42 for the value of the call.
- On day 1, if the value of the share is 285, the value of the call would be zero, since the value of the call will be zero regardless of whether the value of the share rises to $299.25 or falls to $270.75. (Note that our strike price is still $300 per share).
- On day 0, the relevant lattice has become

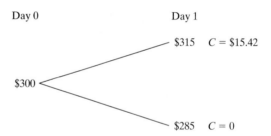

Hence, the number of call options necessary to hedge one share of stock is

$$N = \frac{315 - 285}{15.42 - 0} = 1.9455.$$

Thus, the arbitrage portfolio is $S - 1.9455C$, and we have

$$(300 - 1.9455C)_{\text{day } 0} < (285)_{\text{day } 1}.$$

Therefore, continuing to use 6% as the relevant annualized rate for a one-day interest, we obtain

$$(300 - 1.9455C) = \frac{285}{1 + 0.00016},$$

so the value of the call option would be $7.73. Note that the value of the option increases from $7.52 to $7.73 merely by extending the maturity of the option for one more period.

13.3.6 Multiperiod Binomial Lattice Model

If we can value a two-day option, we can calculate a three-day, or four-day, or in general, n-day option. The logic is exactly the same: We solve iteratively from expiration to period 0; the only thing that changes is the magnitude of the problem. The general form of n-period binomial lattice is shown in Figure 13.9. The stock price can be visualized as moving from node to node in a rightward direction. As before, the probability of an upward movement from any node is p and the probability of downward movement is $1 - p$.

The binomial lattice models described in Figure 13.9(a) or Figure 13.9(b) are still unrealistically simple, if we consider only a few time steps in a year. When binomial trees are used in practice, the life of the option is typically divided into 30 or more steps. In each time step, there is a binomial stock price movement, meaning that 2^{30} or about 1.073 billion possible stock price paths are possible in a year, resulting in a price distribution very close to a lognormal (Figure 13.9(c)). Because the binomial model is multiplicative in nature, the stock price will never become negative. Percentage changes in the stock price in a short period of time are also normally distributed. If we define

$$\mu = \text{Expected yearly growth rate}$$

$$\sigma = \text{Volatility yearly growth rate}$$

then the mean of the percentage change in time Δt is $\mu \Delta t$ and the standard deviation of the percentage change is $\sigma \sqrt{\Delta t}$. To capture the movement of the stock price, we need to select values for u and d and the probability p. If a period length of Δt is chosen as a time step, which is small compared to 1 (one year), the parameters of the binomial lattice can be selected as a function of the stock volatility σ.

Volatility is found by calculating the annualized standard deviation of daily change in price.

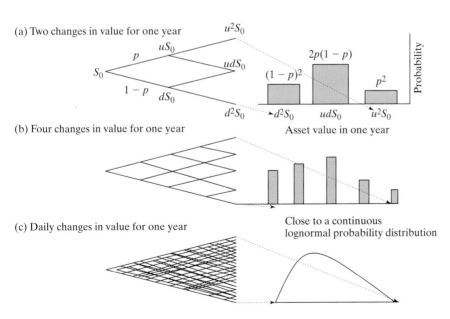

(a) Two changes in value for one year

(b) Four changes in value for one year

(c) Daily changes in value for one year

Close to a continuous lognormal probability distribution

Figure 13.9 The binominal representation of asset price over one year: (a) With only two changes in value, (b) with four changes in value, and (c) with daily changes in value.

$$p = \frac{1}{2} + \frac{1}{2}\left(\frac{\mu}{\sigma}\right)\sqrt{\Delta t},$$

$$u = e^{\sigma\sqrt{\Delta t}},$$

$$d = \frac{1}{u} = e^{-\sigma\sqrt{\Delta t}}. \qquad (13.9)$$

With this choice, the binomial model will closely match the values of the expected growth rate of the stock μ and its variance σ^2. The closeness of the match improves if Δt is made smaller, becoming exact as Δt goes to zero.

13.3.7 Black–Scholes Option Model

As shown in Figure 13.9(c), if we take a sufficiently small time interval and the expiration time becomes long, we can approximate the resulting share distribution by a lognormal function, and the option value at the current point can be calculated with the Black–Scholes option model. In 1973, Black and Scholes developed an option-pricing model based on *risk-free arbitrage,* which means that, over a short time interval, an investor is able to replicate the future payoff of the stock option by constructing a portfolio involving the stock and a risk-free asset. The model provides a trading strategy in which the investor is able to profit with a portfolio return equal to the risk-free rate. The Black–Scholes model is a continuous-time model and assumes that the resulting share distribution at expiration would be distributed lognormally.

- **Call Option.** A standard call option gives its holder the right, but not the obligation, to buy a fixed number of shares at the exercise price (K) on the maturity date. If the current price of the stock is S_0, the Black–Scholes formula for the price of the call is

$$C_{\text{call}} = S_0 N(d_1) - K e^{-r_f T} N(d_2), \qquad (13.10)$$

where

$$d_1 = \frac{\ln(S_0/K) + \left(r_f + \frac{\sigma^2}{2}\right)T}{\sigma\sqrt{T}},$$

$$d_2 = \frac{\ln(S_0/K) + \left(r_f - \frac{\sigma^2}{2}\right)T}{\sigma\sqrt{T}} = d_1 - \sigma\sqrt{T}.$$

$N(.)$ is the standard cumulative normal distribution function, T is the time to maturity, r_f is the risk-free rate of return, and σ^2 is the volatility of the stock return. The model is independent of the expected rate of return and the risk preference of investors. The advantage of the model is that all the input variables are observable except the variance of the return, which can easily be estimated from historical stock price data.

- **Put Option.** We can value a put option in a similar fashion. The new option formula is

$$C_{\text{put}} = K e^{-r_f T} N(-d_2) - S_0 N(-d_1), \qquad (13.11)$$

where

$$d_1 = \frac{\ln(S_0/K) + \left(r_f + \frac{\sigma^2}{2}\right)T}{\sigma\sqrt{T}},$$

$$d_2 = \frac{\ln(S_0/K) + \left(r_f - \frac{\sigma^2}{2}\right)T}{\sigma\sqrt{T}} = d_1 - \sigma\sqrt{T}.$$

Note that, in developing the preceding continuous-time model, Black and Scholes address only the valuation of a European option, which pays no dividend. For an American option, we still need to use the *discrete-time valuation* given in the binomial lattice approach. Of course, we can consider the effect of paying a cash dividend in the model, but we will not address this embellishment or others, as our focus is not the evaluation of financial options.

EXAMPLE 13.4 Option Valuation under a Continuous-Time Process

Consider a stock currently trading at $40. For a strike price of $44, you want to price both a call and a put option that mature two years from now. The volatility of the stock (σ^2) is 0.40^2 and the risk-free interest rate is 6%.

SOLUTION

Given: $S_0 = 40$, $K = \$44$, $r = 6\%$, $T = 2$ years, and $\sigma = 40\%$.
Find: C_{call} and C_{put}.

- The call-option value is calculated as follows:

$$d_1 = \frac{\ln(40/44) + \left(0.06 + \frac{0.4^2}{2}\right)2}{0.4\sqrt{2}} = 0.3265,$$

$$d_2 = 0.3265 - 0.4\sqrt{2} = -0.2392,$$

$$N(d_1) = N(0.3265) = 0.628,$$

$$N(d_2) = N(-0.2392) = 0.405,$$

$$C_{call} = 40(0.628) - 44e^{-0.06(2)}(0.405)$$

$$= \$9.32.$$

- The put-option value is calculated as follows:

$$N(-d_1) = N(-0.3265) = 0.372,$$

$$N(-d_2) = N(0.2392) = 0.595,$$

$$C_{put} = 44e^{-0.06(2)}(0.595) - 40(0.372)$$

$$= \$8.34.$$

> **COMMENTS:** Note that the put-option premium is smaller than the call-option premium, indicating that the upside potential is higher than the downside risk.

13.4 Real-Options Analysis

So far, we have discussed the conceptual foundation for pricing financial options. The idea is, "Can we apply the same logic to value the real assets?" To examine this possibility, we will explore a new way of thinking about corporate investment decisions. The idea is based on the premise that any corporate decision to invest or divest real assets is simply an option, giving the option holder a right to make an investment without any obligation to act. The decision maker therefore has more flexibility, and the value of this operating flexibility should be taken into consideration. Let's consider the following scenario as a starting point for our discussion:

Current Practices—the New Math in Action.[3] Let's say a company is deciding whether to fund a large Internet project that could make or lose lots of money—most likely, lose it. A traditional calculation of net present value, which discounts projected costs and revenues into today's dollars, examines the project as a whole and concludes that it's a no go. But a real-options analysis breaks the project into stages and concludes that it makes sense to fund at least the value of the first stage. Here is how it works:

Step 1: Evaluate each stage of the project separately. Say the first stage, setting up a website, has a net present value of −$50 million. The second stage, an e-commerce venture to be launched in one year, is tough to value, but let's say that the best guess of its net present value is −$300 million.

Step 2: Understand your options. Setting up the website gives you the opportunity—but not the obligation—to launch the e-commerce venture later. In a year, you will know better whether that opportunity is worth pursuing. If it's not, all you have lost is the investment in the website. However, the second stage could be immensely valuable.

Step 3: Reevaluate the project, using an options mind-set. In the stock market, formulas such as Black and Scholes calculate how much you should pay for an option to buy, say, IBM at $90 a share by June 30 if its current price is $82. Think of the first stage of your Internet project as buying such an option—risky and out of the money, but cheap.

Step 4: Go figure. Taking into account the limited downside of building a website and the huge, albeit iffy, opportunities it creates, we see that a real-options analysis could give the overall project a present value of, say, $70 million. So the no go changes to a go.

Now we will see how the logical procedure just outlined can be applied to address the investment risk inherent in strategic business decisions.

---

[3] From "Exploiting Uncertainty—The 'Real-Options' Revolution in Decision-Making," *BusinessWeek*, June 7, 1999, p. 119.

13.4.1 A Conceptual Framework for Real Options in Engineering Economics

In this section, we will conceptualize how the financial options approach can be used to value the flexibility associated with a real investment opportunity. A decision maker with an opportunity to invest in real assets can be viewed as having a *right, but not an obligation*, to invest. He or she therefore owns a real option similar to a simple call option on a stock:

- An investment opportunity can be compared to a call option on the present value of the cash flows arising from that investment (V).
- The investment outlay that is required to acquire the assets is the exercise price (I).
- The time to maturity is the time it takes to make the investment decision or the time until the opportunity disappears.

The analogy between a call option on stocks and a real option in capital budgeting is shown in Figure 13.10. The value of the real option at expiration depends on the value of the asset and would influence the decision as to whether to exercise the option. The decision maker would exercise the real option only if doing so were favorable.

- **Concept of Real Calls.** The NPW of an investment obtained with the options framework in order to capture strategic concerns is called the *strategic* NPW (SNPW) and is equivalent to the option value calculated by the Black–Scholes model in Eq. (13.10). The value of the right to undertake the investment now is ($R_{\min}$), which is the payoff if the option is exercised immediately. If $V > I$, the payoff is $V - I$, and if $V < I$, the payoff is 0. The true value of the option (R) that one needs to find is the SNPW of the real option. Since one would undertake the investment later only if the outcome is favorable, the SNPW is greater than the conventional NPW. The value of the flexibility associated with the option to postpone the investment is the difference

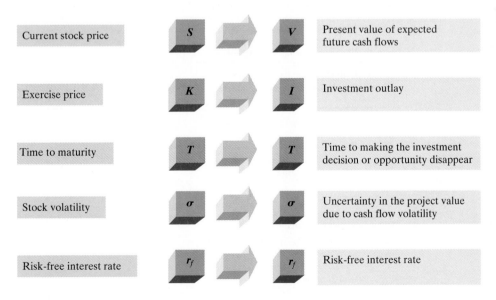

Current stock price	S	V	Present value of expected future cash flows
Exercise price	K	I	Investment outlay
Time to maturity	T	T	Time to making the investment decision or opportunity disappear
Stock volatility	σ	σ	Uncertainty in the project value due to cash flow volatility
Risk-free interest rate	r_f	r_f	Risk-free interest rate

Figure 13.10 The analogy between a call option on a stock and a real option.

between the SNPW and the conventional NPW. This is the real-option premium (ROP), or the value of flexibility, defined as

$$\text{Value of flexibility (ROP)} = \text{SNPW} - \text{Conventional NPW.} \qquad (13.12)$$

- **Concept of a Real Put Option.** The concept of a real put option is important from a strategic perspective. A put option gives its owner the right to dispose of an asset when it is favorable to do so. The put works like a guarantee or insurance when things go bad. An early-abandonment decision can be viewed as a simple put option. The option to abandon a project early may have value, as when an asset has a higher resale value in a secondary market than its use value. The put guarantees that the use value (V) of an asset does not fall below its market resale value (I). If it does, the option holder will exercise the put. In most instances, it is not possible to make an exact comparison between a standard put option on a stock and a real put option in capital budgeting.

13.4.2 Types of Real-Option Models

Most common types of real options can be classified into three categories as summarized in Table 13.1. Some of the unique features of each option will be examined with numerical examples.

TABLE 13.1 Types of Real Options

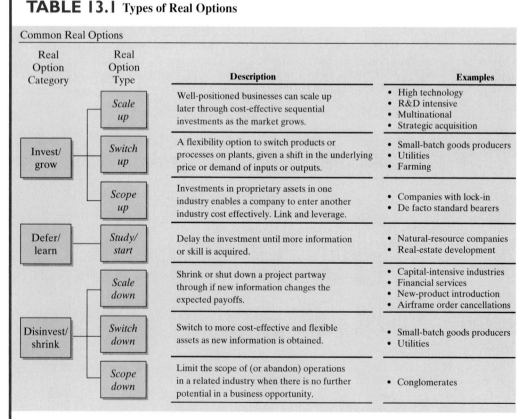

Common Real Options			
Real Option Category	**Real Option Type**	**Description**	**Examples**
Invest/ grow	Scale up	Well-positioned businesses can scale up later through cost-effective sequential investments as the market grows.	• High technology • R&D intensive • Multinational • Strategic acquisition
	Switch up	A flexibility option to switch products or processes on plants, given a shift in the underlying price or demand of inputs or outputs.	• Small-batch goods producers • Utilities • Farming
	Scope up	Investments in proprietary assets in one industry enables a company to enter another industry cost effectively. Link and leverage.	• Companies with lock-in • De facto standard bearers
Defer/ learn	Study/ start	Delay the investment until more information or skill is acquired.	• Natural-resource companies • Real-estate development
Disinvest/ shrink	Scale down	Shrink or shut down a project partway through if new information changes the expected payoffs.	• Capital-intensive industries • Financial services • New-product introduction • Airframe order cancellations
	Switch down	Switch to more cost-effective and flexible assets as new information is obtained.	• Small-batch goods producers • Utilities
	Scope down	Limit the scope of (or abandon) operations in a related industry when there is no further potential in a business opportunity.	• Conglomerates

Source: "Get Real—Using Real Options in Security Analysis," *Frontiers of Finance,* Volume 10, by Michael J. Mauboussin, Credit Suisse First Boston, June 23, 1999.

Option to Defer Investment

The option to defer an investment is similar to a call option on stock. Suppose that you have a new product that is currently selling well in the United States and your firm is considering expanding the market to China. Because of many uncertainties and risks in the Chinese market (pricing, competitive pressures, market size, and logistics), the firm is thinking about hiring a marketing firm in China who can conduct a test market for the product. Any new or credible information obtained through the test market will determine whether or not the firm will launch the product. If the market is ready for the product, the firm will execute the expansion. If the market is not ready for the product, then the firm may wait another year or walk away and abandon the expansion plan altogether. In this case, the firm will lose only the cost associated with testing the market. An American expansion option will provide the firm with an estimate of the value to spend in the market research phase.

EXAMPLE 13.5 Delaying Investment: Value of Waiting

A firm is preparing to manufacture and sell a new brand of digital phone. Consider the following financial information regarding the digital phone project:

- The investment costs are estimated at $50M today, and a "most likely" estimate for net cash inflows is $12M per year for the next five years.
- Due to the high uncertainty in the demand for this new type of digital phone, the volatility of cash inflows is estimated at 50%.
- It is assumed that a two-year "window of opportunity" exists for the investment decision.
- If the firm delays the investment decision, the investment costs are expected to increase 10% per year.
- The firm's risk-adjusted discount rate (MARR) is 12% and the risk-free rate is 6%.

Should the firm invest in this project today? If not, is there value associated with delaying the investment decision?

SOLUTION

- **Conventional Approach: Should the Firm Invest Today?** If the traditional NPW criterion is used, the decision is not to undertake the investment today, as it has a negative NPW of $6.74 million:

$$PW(12\%) = -\$50M + \$12M(P/A, 12\%, 5)$$
$$= -\$6.74M$$

- **Real-Options Approach.** Is there value to waiting two years? From an options approach, the investment will be treated as an opportunity to wait and then to undertake the investment two years later if events are favorable. As shown in Figure 13.11, this delay option can be viewed as a call option. If the value V of the project is greater than the investment cost I two years from now,

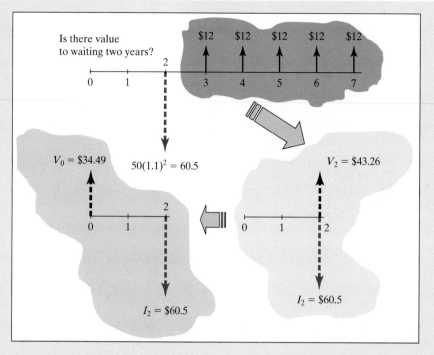

Figure 13.11 Transforming cash flow data to obtain input parameters V_0 and I_2 for option valuation.

then the option to undertake the project should be exercised. If not, it expires worthless. Therefore,

$$\text{Delay option} = \max[V_2 - I_2, 0],$$

where V_2 is a random variable that is dependent upon market demand.

Now, assume that one can use the Black–Scholes formula to value[4] the opportunity to undertake this investment in year 2. The opportunity to wait can be considered as a European call option on the present value of the future cash flows $V_0 = \$34.49$ million, with an exercise price of $I = \$60.5$ million, expiring two years from now ($T = 2$). Since $r_f = 6\%$ and $\sigma = 50\%$, using the Black–Scholes formula, we obtain $d_1 = -0.2715$, $d_2 = -0.976$, $N(d_1) = 0.3930$, and $N(d_2) = 0.1639$. The value of the real option is therefore

$$C = \text{SNPW} = 34.49(0.3930) - 60.5e^{-(0.06)2}(0.1639)$$

$$= 13.5546 - 8.7947$$

$$= \$4.76,$$

$$\text{ROP} = \$4.76 - (-\$6.74)$$

$$= \$11.50 \text{ M}.$$

---

[4] One can arrive at approximately the same value with the binomial model by dividing the one-year time to maturity into sufficiently small time intervals.

Thus, the value of retaining the flexibility of having the delay option is worth $11.50M. The ability to wait provides a decision maker a strategic NPW of $4.61M, instead of a negative NPW of $6.74M. On the one hand, if the market for the product turns out to be favorable, the company will exercise the real option in the money and undertake the investment in year 2. On the other hand, if the market two years later turns out to be unfavorable, the decision will be not to undertake the investment. The opportunity cost is only $4.61M, compared with the actual investment cost of $50 million.

COMMENTS: What exactly does the $4.76M delay value imply? Suppose the firm's project portfolio consists of 10 projects, including this digital phone project. Then if the other 9 projects have a net present value of $100M, the value of the firm using standard NPW would still be $100M, because the digital phone project would not be accepted, since it has a negative NPW. If the option to invest in the digital phones is included in the valuation, then the value of the firm's portfolio is $104.76M. Therefore, the delay option that the firm possesses is worth an additional $4.76M.

Patent and License Valuation

A patent or license provides a firm the right, but not the obligation, to develop a product (or land) over some prescribed time interval. The right to use the patent has value if and only if the expected benefits (V) exceed the projected development costs (I), or $\max\{V - I, 0\}$. This right is considered as a real option that can be used to value the worth of a license to a firm.

EXAMPLE 13.6 Option Valuation for Patent Licensing

A technology firm is contemplating purchasing a patent on a new type of digital phone. The patent would provide the firm with three years of exclusive rights to the digital phone technology. Estimates of market demand show net revenues for seven years of $50M per annum. The estimated cost of production is $200M. Assume that MARR $= 12\%, r = 6\%$, and the volatility due to market uncertainty related to product demand is 35%. Determine the value of the patent.

SOLUTION

The patent provides the right to use the digital phone technology anytime over the next three years. Therefore,

$$V_3 = \$50M(P/A, 12\%, 7) = \$228.19M,$$

$$V_0 = \$228.19M(P/F, 12\%, 3) = \$162.42M,$$

$$I_3 = \$200M,$$

$$T = 3,$$

$$r_f = 6\%,$$
$$\sigma = 35\%.$$

Substituting these values into Eq. (13.10), we obtain the Black–Scholes value of this call option: $36.95M. With the real-options analysis, the firm now has an upper limit on the value of the patent. At the most, the firm should pay $36.95M for it, and the firm can use this value as part of the negotiation process.

Growth Option

A growth option occurs when an initial investment is required to support follow-on investments, such as (1) an investment in phased expansion, (2) a Web-based technology investment, and (3) an investment in market positioning. The amount of loss from the initial investment represents the call-option premium. Making the initial investment provides the option to invest in any follow-on opportunity. For example, suppose a firm plans on investing in an initial small-scale project. If events are favorable, then the firm will invest in a large-scale project. The growth option values the flexibility to invest in the large-scale project if events are favorable. The loss on the initial project is viewed as the option premium.

EXAMPLE 13.7 Valuation of a Growth Option

A firm plans to market and sell its product in two markets: locally and regionally. The local market will require an initial investment, followed by some estimated cash inflows for three years. If events are favorable, the firm will invest and sell the product regionally, expecting benefits for four years. Figure 13.12 details the investment opportunities related to the growth options of this project.

Assume that the firm's MARR is 12% and the risk-free interest rate is 6%. Determine the value of this growth option.

SOLUTION

The NPW of each phase is as follows:

$$\text{PW}(12\%)_{\text{Small, 0}} = -\$30 + \$10(P/F, 12\%, 1) + \$12(P/F, 12\%, 2)$$
$$+ \$14(P/F, 12\%, 3)$$
$$= -\$1.54\text{M [in year 0]}.$$

$$\text{PW}(12\%)_{\text{Large, 3}} = -\$60 + \$16(P/F, 12\%, 1) + \$18(P/F, 12\%, 2)$$
$$+ \$20(P/F, 12\%, 3) + \$20(P/F, 12\%, 4)$$
$$= -\$4.42\text{M [in year 3]}.$$

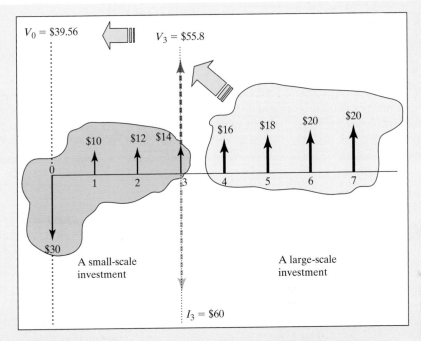

Figure 13.12 Cash flows associated with the two investment opportunities: a small-scale investment followed by a large-scale investment. The large-scale investment is contingent upon the small-scale investment.

Therefore, the total net present worth of the two-phased investment project is

$$PW(12\%)_{Total, 0} = -\$1.54M - \$4.42M(P/F, 12\%, 3)$$
$$= -\$1.54M - 3.14M$$
$$= -\$4.68M < 0$$

After investing in the small-scale project, the firm is not obligated to invest in the large-scale phase; hence, it is an option. The value of the two investment opportunities can be calculated in two steps:

Step 1: The option to expand can be valued as a European call option with the use of the Black–Scholes equation (Eq. 13.10). The option inputs are

$$V_{Large, 3} = \$55.8M,$$
$$V_{Large, 0} = \$55.8M(P/F, 12\%, 3) = \$39.56M,$$
$$I_3 = 60M,$$
$$T = 3,$$
$$r_f = 6\%,$$
$$\sigma = 40\%.$$

The value of the option is $7.54M.

Step 2: The total value of the two investment opportunities is

$$\text{Combined Option value} = -1.54\text{M} + 7.54\text{M}$$

$$= \$6\text{M} > 0.$$

Because the option premium for these combined investment opportunities is positive, the investments should be pursued. Even though the initial investment in the local market will lose money, the benefits of the large-scale investment later will offset the initial losses.

Scale-Up Option

A firm has the right to scale up an investment *at any time* if the initial project is favorable. In other words, a firm has the option to increase its investment in a project, in return for increased revenues. In a way, this is just a growth option, but it will be valued on a binomial basis. Example 13.8 illustrates how we value a scale-up option.

EXAMPLE 13.8 A Scale-Up Option Valuation Using Binomial Lattice Approach

A firm has undertaken a project in which the firm has the option to invest in additional manufacturing and distribution resources (or scale up). The project's current value is $V_0 = \$10\text{M}$. Anytime over the next three years, the firm can invest an additional $I = \$3\text{M}$ and receive an expected 30% increase in net cash flows and, therefore, a 30% increase in project value. The risk-free interest rate is 6% and the volatility of the project's value is 30%. Use a binomial lattice with a one-year time increment to value the scale-up option as an American option. (You can scale up anytime you see fit to do so.)

SOLUTION

We will solve the problem in two steps, the first of which is to determine how the value of the underlying asset changes over time and the second of which is to create the option valuation lattice:

Step 1: Binomial Approach: Lattice Evolution of the Underlying Asset. First, we determine the following parameters to build a binomial lattice that shows how the value of the project changes over time:

$$u = e^{0.30(1)} = 1.35,$$

$$d = \frac{1}{u} = 0.74,$$

$$q = \frac{1.06 - 0.74}{1.35 - 0.74} = 0.53.$$

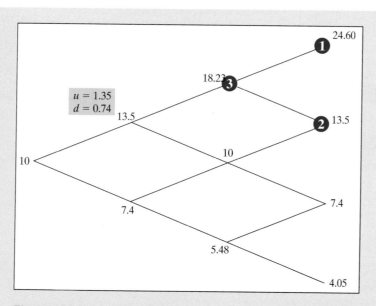

Figure 13.13   Event tree: Lattice evolution of the underlying value of a project.

The equity binomial lattice looks like the one in Figure 13.13; the values are created in a forward multiplication of up and down factors, from left to right.

Step 2: Binomial Approach: Option Valuation Lattice. Once you obtain the lattice evolution of the underlying project, we can create the option valuation lattice in two steps: the valuation of the terminal nodes and the valuation of the intermediate nodes. The calculation proceeds in a backward manner, starting from the terminal nodes: The nodes at the end of the lattice are valued first, going from right to left.

All the required calculations and steps are shown in Figure 13.14. To illustrate, we will take a sample circled node (❶) in the figure:

- Node ❶ reveals a value of $24.60, which is the value the firm can expect to achieve without any scale-up. Now, with the scale-up option, the project value can increase 30% at the expense of $3 million of investment, or $1.3(24.60) - 3 = \$28.98$ million. Since this figure is larger than $24.60, the firm will opt for scaling up the operation.

- Similarly, the firm will reach the same decision for node ❷, which yields a scale-up result of $14.55.

- Moving on to node ❸, we see that the firm again has one of two options: to scale up the operation at that point rather than in the future, or to keep the option open for the future in the hope that when the market is up, the firm will have the ability to execute the option. Mathematically, we can state the options as follows:

Scale up now ($n = 2$): $1.3(\$18.23) - \$3 = \$20.70$;

Keep the option open: $C = \dfrac{(0.53)(\$28.98) + (0.47)(\$14.55)}{(1 + 0.06)} = \$20.87.$

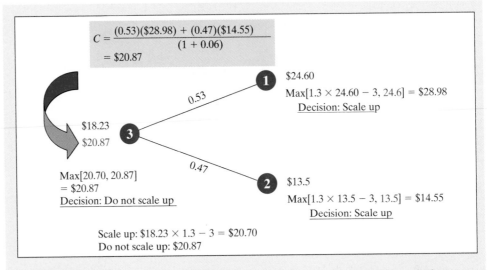

Figure 13.14 Decision tree diagram illustrating the process of reaching a decision at node ❸.

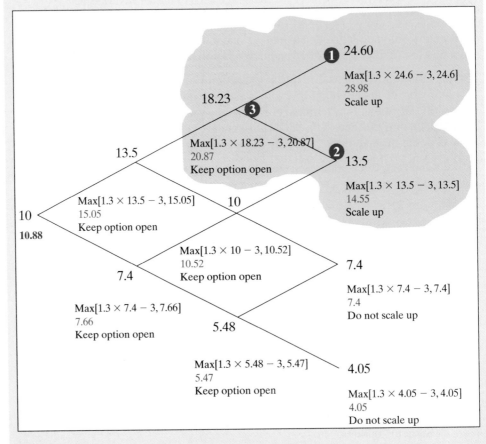

Figure 13.15 Decision tree for a scale-up option (valuation lattice).

Thus, the decision is to keep the option open. Finally at year 0, we compute

$$C = \frac{(0.53)(\$15.05) + (0.47)(\$7.66)}{(1 + 0.06)}$$

$$= \$10.88$$

The end-result option value on the binomial tree shown in Figure 13.15 represents the value of the project with the scale-up option. The value of the scale-up option itself is

$$\text{Option value} = \$10.88 - \$10 = \$0.88\text{M}.$$

Therefore, the option to expand the project in the future is worth $880,000.

13.5 Estimating Volatility at the Project Level

Up to this point, we assumed that the volatility of project cash flows is known to us or can be estimated somehow. Unfortunately, the volatility of cash flows is probably one of the most difficult input parameters to estimate in a real-options analysis. One approach, suggested by Tom Copeland,[5] is to develop an NPW distribution. The volatility that we need for the binomial tree is the volatility of the project's rate of return. So we convert values generated by a Monte Carlo simulation by using the relationship (with continuous compounding)

$$PW_t = PW_0 e^{rt},$$

or

$$\ln\left(\frac{PW_t}{PW_0}\right) = rt. \tag{13.13}$$

For $t = 1$, this is a simple transformation that helps to convert between consecutive random draws of present-value estimates in a Monte Carlo simulation, and the standard deviation of the rate of return is the project's volatility σ. We use a simple deferral option example to illustrate the conceptual process without resorting to a Monte Carlo simulation.

13.5.1 Estimating a Project's Volatility through a Simple Deferral Option

Consider a project that requires a $200 million investment and expects to last three years with the following cash flows:

Demand	Probability	Annual Cash Flow
Good	0.25	$250 million
Moderate	0.30	$100 million
Poor	0.45	$ 35 million

[5] Tom Copeland and Vladimir Antikarov, *Real Options—A Practitioner's Guide*. New York: Texere, 2001 (see especially Chapter 9). For a complete technical discussion, see Hemantha Herath and Chan S. Park, "Multi-Stage Investment Opportunities as Compound Real Options," The Engineering Economist, Vol. 47, No. 1, 2002.

The cost of capital that accounts for the market risk, known as the risk-adjusted discount rate, is 10%, and a risk-free interest rate is known to be 6%.

Conventional Approach: We will determine the acceptability of the project on the basis of the expected NPW as follows:

$$E[A_n] = 0.25(\$250) + (0.30)(\$100) + 0.45(\$35)$$

$$= \$108.25;$$

$$E[PW(10\%)]_{\text{Benefits}} = \$108.25(P/A, 10\%, 3)$$

$$= \$269.20;$$

$$E[PW(10\%)] = \$269.20 - \$200$$

$$= \$69.20 > 0.$$

Since the expected NPW is positive, the project would be considered for immediate action. Note that we need to use the market interest rate, which reflects the risk inherent in estimating the project cash flows. If we proceed immediately with the project, its expected NPW is $69.20 million. However, the project is very risky. If demand is good, NPW = $421.72 million; if demand is moderate, NPW = $48.69 million; if demand is poor, NPW = −$112.96 million. (See Figure 13.16.) Certainly, we are uneasy about the prospect of realizing a poor demand, even though its expected NPW is positive.

Investment Timing Option: Can we defer the project by one year and then implement it only if demand is either moderate or good? Suppose we have an option to defer the investment decision by *one year*, and then we will gain additional information regarding demand. If demand is low, we will not implement the project. If we wait, the up-front cost and cash flows will stay the same, except they will be shifted ahead by a year. The following table depicts the demand scenarios:

Demand Scenarios	0	Future Cash Flows				NPW	Prob.
		1	2	3	4		
Good	▶	−$200	$250	$250	$250	$376.51	25%
Moderate	▶	−$200	$100	$100	$100	$37.40	30%
Poor	▶	0	0	0	0	$0	45%

In finding the NPW, *we will use two different discount rates*. We will use 6% to discount the cost of the project at the risk-free rate, since the cost is known. Then we will discount the operating cash flows at the cost of capital. Under the various scenarios, we have

$$PW_{\text{Good}} = -\$200(P/F, 6\%, 1) + \$250(P/A, 10\%, 3)(P/F, 10\%, 1)$$

$$= \$376.51,$$

$$PW_{\text{Moderate}} = -\$200(P/F, 6\%, 1) + \$100(P/A, 10\%, 3)(P/F, 10\%, 1)$$

$$= \$37.40,$$

$$PW_{\text{Poor}} = \$0.$$

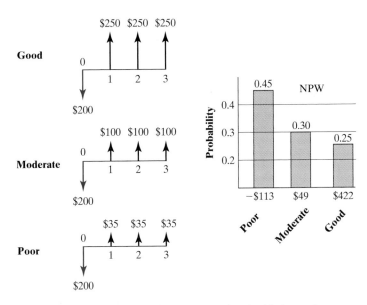

Figure 13.16 Cash flow diagrams associated with demand scenarios in Example 13.9.

Next, we calculate the project's expected NPW if we wait, by using the preceding three scenarios, with their given probabilities:

$$E[\text{NPW}]_{\text{wait}} = (0.25)(\$376.51) + 0.30(\$37.40) + 0.45(\$0)$$

$$= \$105.35.$$

Now we can compare the option to wait against the conventional NPW analysis. Note that the option to wait is higher ($105.35 million versus $69.20). In other words, the waiting option is worth $36.15 million. If we implement the project today, we gain $69.20 million, but lose the option worth $36.15 million. Therefore, we should wait and decide next year whether to implement the project, based on the observed demand. Note also that the cash flows are less risky under the option to wait, since we can avoid the low cash flows. Also, the cost of implementing the project may not be risk free. Given the change in risk, perhaps we should use different rates to discount the cash flows.

13.5.2 Use the Existing Model of a Financial Option to Estimate σ^2

The option to wait resembles a financial call option: We get to "buy" the project for $200 million in one year if value of the project is greater than $200 million after one year. This is like a call option with an exercise price of $200 million and an expiration date of one year. If we use the Black–Scholes model to value the option to wait, the required inputs are as follows:

- $K = I$, exercise price = cost to implement the project = $200 million.
- r_f = risk-free rate = 6%.
- T = time to maturity = 1 year.
- $S_0 = V$ = current stock price = estimated current value of the project.
- σ^2 = variance of stock return = estimated variance on project return.

Here, we need to estimate two unknown parameters: V and σ^2. Estimating V is equivalent to estimating the current stock price in a financial option. In a real option, V is the value of the project today, which is found by computing the present value of all of the project's future expected cash flows at its cost of capital (or risk-adjusted discount rate). We will demonstrate how we compute this value in a step-by-step procedure:

Step 1: Estimate the project's current value. We find the PW of future cash flows in two substeps: First we compute the value at the option's exercise year; then we bring it back to the current time, year 0.

| Demand | Future Cash Flows | | | | | PW at | |
Scenarios	0	1	2	3	4	Year 1	Probability
Good			$250	$250	$250	$621.71	25%
Moderate			$100	$100	$100	$248.69	30%
Poor			$ 35	$ 35	$ 35	$ 87.03	45%

$$PW_{Good} = \$250(P/A, 10\%, 3) = \$621.71,$$

$$PW_{Moderate} = \$100(P/A, 10\%, 3) = \$248.69,$$

$$PW_{Poor} = \$35(P/A, 10\%, 3) = \$87.03,$$

$$V = E[NPW]_{n=0}$$

$$= [0.25(\$621.71) + 0.30(\$248.69) + 0.45(\$87.03)](P/F, 10\%, 1)$$

$$= \$244.73.$$

Thus, the present value of the project's expected future cash flows is $V = \$244.73$. Note that we used a cost of capital in discounting the estimated future project cash flows, as these are risky cash flows.

Step 2: Estimating σ^2. Note that, for a financial option, σ^2 is the variance of the stock's rate of return. For a real option, σ^2 is the variance of the project's rate of return. There are several ways to estimate σ^2. One direct approach is to use the previous scenario analysis to estimate the return from the present until the option must be exercised. We do this for each scenario. Then we find the variance of these returns, given the probability of each scenario. In other words, we find the returns from the present until the option expires. Here is what we have from the previous scenario and calculation of the present value at year 0:

Demand Scenario	PW at Year 0	PW at Year 1	Return
Good		$621.71	154.03%
Moderate	$244.73	$248.69	1.62%
Poor		$ 87.03	−64.44%

For example, to find the return for the "good" demand case, we use

$$154.03\% = \frac{\$621.71 - \$244.73}{\$244.73}.$$

Now we use these scenarios, with their given probabilities, to find the expected return and the variance of the return:

$$E(R) = 0.25(1.5403) + 0.30(0.0162) + 0.45(-0.6444)$$
$$= 0.10 = 10\%,$$
$$\sigma^2 = 0.25(1.5403 - 0.10)^2 + 0.30(0.0162 - 0.10)^2$$
$$+ 0.45(-0.6444 - 0.10)^2$$
$$= 0.77,$$
$$\sigma = 0.8775 = 87.75\%.$$

Step 3: Model uncertainty with the Black–Scholes model. Substituting the estimated parameters into Eq. (13.10), we find that the option value is $104.03:

$$V = \$244.73; K = \$200, r_f = 0.06; t = 1; \sigma^2 = 0.77,$$

$$C_{call} = 244.73N(d_1) - 200e^{-0.06(1)}N(d_2),$$

$$d_1 = \frac{\ln(244.73/200) + \left(0.06 + {0.77}/{2}\right)(1)}{(0.77)^{1/2}\sqrt{1}}$$

$$= 0.73714,$$

$$d_2 = 0.73714 - (0.77)^{1/2}\sqrt{1}$$

$$= -0.14036,$$

$$N(d_1) = 0.7695,$$

$$N(d_2) = 0.4443,$$

$$C_{call} = 244.73(0.7695) - 200e^{-0.06(1)}(0.4443)$$

$$= \$104.03.$$

Step 4: Model uncertainty with the binomial lattice. If we use the discrete version of the binomial lattice model, we need to create an event tree that models the potential values of the underlying risky asset. This tree contains no decision nodes and simply models the evolution of the underlying asset. Defining T as the number of years per upward movement and σ^2 as the annualized volatility of the underlying project value, we can determine the up and down movements with Eq. (13.10):

$$\text{Up movement} = u = e^{\sigma\sqrt{T}},$$

$$\text{Down movement} = d = \frac{1}{u}.$$

Substituting numerical values into these formulas yields

$$u = e^{\sigma\sqrt{T}} = e^{0.8775\sqrt{1}} = 2.4049,$$

$$d = \frac{1}{u} = \frac{1}{2.4049} = 0.4158.$$

- **Model flexibility by means of a decision tree.** When we add decision points to an event tree, it becomes a decision tree.

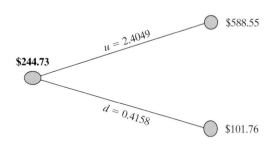

- **Estimate the value of flexibility.** To determine the value of the project with the flexibility to defer it, we work backward through the decision tree, using the risk-neutral probability approach at each node:

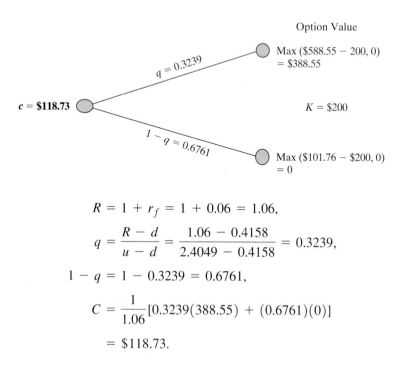

$$R = 1 + r_f = 1 + 0.06 = 1.06,$$

$$q = \frac{R - d}{u - d} = \frac{1.06 - 0.4158}{2.4049 - 0.4158} = 0.3239,$$

$$1 - q = 1 - 0.3239 = 0.6761,$$

$$C = \frac{1}{1.06}[0.3239(388.55) + (0.6761)(0)]$$

$$= \$118.73.$$

Note that this amount ($118.73) calculated under the assumption discreteness is different from the value ($104.03) obtained under the Black–Scholes model earlier. This is not surprising, as the discrete model with $T = 1$ is a rough approximate of the continuous version. Since the option itself is worth $118.73 million, the ROP is

$$\text{ROP} = \$118.73 - \$69.20 = \$49.53.$$

Clearly, it makes sense to defer the investment project for one year in this case.

13.6 Compound Options

In a compound-option analysis, the value of the option depends on the value of another option. A sequential compound option exists when a project has multiple phases and later phases depend on the success of earlier phases. The logical steps to use in handling this sequential compound-option problem are as follows:

Step 1: Calculate the initial underlying asset lattice by first calculating the up and down factors and evolving the present value of the future cash flow over the planning horizon.

Step 2: Calculate the longer term option first and then the shorter term option, because the value of a sequential compound option is based on earlier options.

Step 3: Calculate the combined-option value.

Example 13.9 illustrates how we value the compound option.

EXAMPLE 13.9 A Real-World Way to Manage Real Options[6]

Suppose a firm is considering building a new chemical plant. The plant costs $60 million, including permits and preparation. At the end of year 1, the firm has the right to invest $400 million on design phase. Upon completion of the design, the firm has another right to invest $800 million in building the plant over the next two years. The firm's risk-adjusted discount rate is 10.83%. If the plant existed today, its market value would be $1 billion. The volatility of the plant's value (σ) is 18.23%. The risk-free interest rate is known to be 8%. Determine the NPW of the project, the option values at each stage of the decision point, and the combined-option value of the project.

STRATEGY: Note that there are three investment phases: the permit-and-preparation phase (Phase 0), which provides an option to invest in the design phase; (2) the design phase (Phase 1), which provides the option to invest in plant construction; and (3) the construction phase (Phase 2). Each phase depends upon a decision made during the previous phase. Therefore, investing in Phase 2 is contingent upon investing in Phase 1, which in turn is contingent upon the results of Phase 0. The first task at hand is to develop an event tree or a binomial lattice tree to see how the project's value changes over time. We need two parameters: upward movement (u) and downward

---

[6] This example is based on the article "A Real World Way to Manage Real Options," by Tom Copeland, *Harvard Business Review*, March 2004.

movement (d). Since the time unit is one year, $\Delta t = 1$; and since $\sigma = 0.1823$, we compute u and d as follows:

$$u = e^{\sigma\sqrt{\Delta t}} = e^{0.1823\sqrt{1}} = 1.2,$$

$$d = \frac{1}{u} = 0.833.$$

The lattice evolution of the underlying project value will look like the event tree shown in Figure 13.17.

SOLUTION

(a) Traditional NPW calculation:

Using a traditional discounted cash flow model, we may calculate the present value of the expected future cash flows, discounted at 10.83%, as follows:

$$PW(10.83\%)_{\text{Investment}} = 60 + \frac{400}{1.1083} + \frac{800}{1.1083^3}$$

$$= 1,008.56 \text{ million},$$

$$PW(10.83\%)_{\text{Value of investment}} = \$1,000 \text{ million},$$

$$NPW = \$1,000 - \$1,008.56$$

$$= -\$8.56 < 0 \text{ (Reject)}.$$

Since the NPW is negative, the project would be a no-go one. Note that we used a cost of capital (k) of 10.83% in discounting the expected cash flows. This cost of capital represents a market risk-adjusted discount rate.

(b) Real-options analysis:

Using a process known as backward induction, we may proceed to create the option valuation lattice in two steps: the valuation of the terminal nodes and the valuation of the intermediate nodes. In our example, we start with the nodes at year 3.

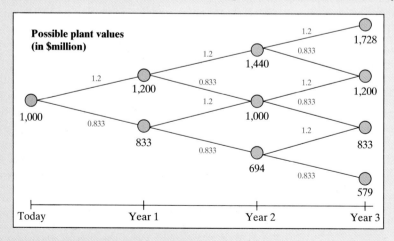

Figure 13.17 The event tree that illustrates how the project's value changes over time.

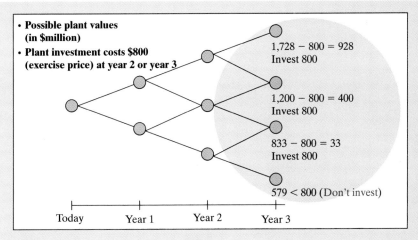

Figure 13.18 Valuing Phase 2 options.

Step 1: Valuing Phase 2 Options. We pretend that we are at the end of year 3, at which time the project's value would be one of the four possible values ($1,728, $1,200, $833, and $579), as shown in Figure 13.18. Clearly, the project values at the first three nodes exceed the investment cost of $800 million, so our decision should be to invest in Phase 2. Only if we reach the last node ($579) will we walk away from the project.

Step 2: What to Do in Year 2. In order to determine the option value at each node at year 2, we first need to calculate the risk-neutral probabilities. For node B, the option value comes out to be $699 million:

$$u = e^{\sigma \sqrt{\Delta t}} = e^{0.1823\sqrt{1}} = 1.2,$$

$$d = \frac{1}{u} = 0.833,$$

$$R = 1 + r_f = 1 + 0.08 = 1.08,$$

$$q = \frac{R - d}{u - d} = \frac{1.08 - 0.833}{1.2 - 0.833} = 0.673,$$

$$1 - q = 1 - 0.673 = 0.327,$$

$$C = \frac{1}{1.08}[0.673(928) + 0.327(400)]$$

$$= 699.$$

Thus, keeping the option open is more desirable than exercising it (committing $800 million), as the net investment value is smaller than the option value ($1,440 − $800 = $640).

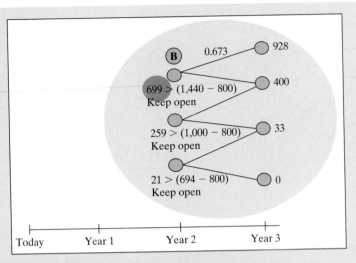

Figure 13.19 Sequential compound options (valuation lattice).

The option values for the rest of the nodes at the end of year 2 are calculated in a similar fashion and are shown in Figure 13.19. Note that in year 2, regardless of which node we arrive at, we should not exercise the option.

Step 3: **What to Do in Year 1.** Moving on to the nodes at year 1, we see that, at node C in Figure 13.20, the value of executing the option is $514 − $400 = $114 million, and

$$\text{Max}[\$514 - \$400, 0] = \$114.$$

Keep in mind that the value $514 million comes from the option valuation lattice from year 2:

$$\frac{1}{1 + 0.08}[(0.673)(699) + (0.327)(259)] = \$514.$$

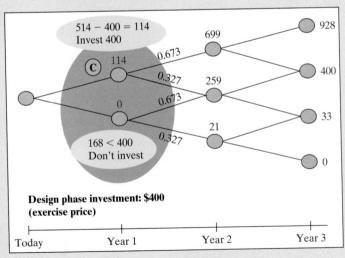

Figure 13.20 Valuing Phase 1 options.

To realize the option value of $514 million at node C, you need to invest $400 million. Therefore, the value of continuing the project is $514 − $400 = $114 million. Note that at the end of first period, the second option expires. Therefore, it must be exercised at a cost of $400 million, or left un-exercised (at no cost). If it is exercised, the payouts are directly dependent not on the value of the underlying project ($1,200 million), but on the value provided by the option to invest at the next stage.

Step 4: Standing at Year 0. We can estimate the present value of the compound op-tion by recognizing that we can either keep the first option open or exercise it at a cost of $60 million. As shown in Figure 13.21, the value of that option is determined by

$$C = \frac{1}{1.08}[(0.673)(\$114) + (0.327)(0)]$$

$$= \$71.039;$$

$$\text{Max}[\$71 - \$60, 0] = \$11.$$

We can interpret $71.039 million as the net present value of a project that has a present value of $1,000 million today, has a standard deviation of 18.23% per year, and requires completing three investments: $60 million for the first stage, $400 million for a design stage, and $800 million for a con-struction phase that must start by the end of the third year. If the start-up cost is greater than $71.039 million, the project would be rejected; otherwise it would be accepted. Figure 13.22 illustrates how we go about making various investment decisions along the paths of the event tree.

Now we can summarize what we have done:

- The initial project value is $71 million. Since the required initial investment is only $60 million, the option value is $11 million. Because this number is posi-tive, it is worth keeping the option by initiating the permit and preparation work.
- Note that, under the NPW approach, the project could be rejected altogether, as its NPW is negative.

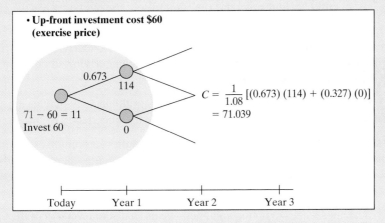

Figure 13.21 The value of Phase 0 options.

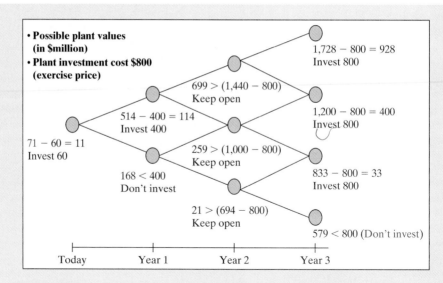

Figure 13.22 Sequential compound options (combined lattice).

SUMMARY

In this chapter, we have introduced a new tool to evaluate a risk associated with strategic investment decision problems. The tool is known as "real options" and is one of the most recent developments in corporate finance. Since the use of real options is based on financial options on stock, we reviewed some fundamentals of option valuation techniques and demonstrated how the financial option theory can be extended to evaluate the risk associated with real assets.

- An option to buy a financial asset at a specified price is a **call option** on the asset. The call buyer has the *right* to purchase the asset, whereas the call-option seller has the *obligation* to sell. An option to sell an asset at a specified price is a **put option** on the asset. Consequently, the put seller has the *right* to sell; the put buyer has the *obligation* to buy.

- The value of the call tends to increase as the price of the underlying stock rises. This gain will increasingly reflect a rise in the value of the underlying stock when the market price moves above the option's strike price.

- Buying puts against an existing stock position or simultaneously purchasing stock and puts can supply the insurance needed to overcome the uncertainty of the marketplace.

- There are three conceptual approaches to valuing option premiums: the replicating-portfolio approach with a call option, the risk-free financing approach, and the risk-neutral probabilistic approach. All three approaches lead to the same option valuation, so the choice is dependent on the preference of the analyst.

- As analytical approaches to option valuation, we use a discrete version of the binominal lattice model and the continuous version of the Black–Scholes model. As the number of time steps gets larger, the value calculated with the binomial lattice model approaches the closed-form solution to the Black–Scholes model.

■ Real-options analysis is the process of valuing managerial strategic and operating flexibilities. Real options exist when managers can influence the size and risk of a project's cash flows by taking different actions during the project's life in response to changing market dynamics.

■ The single most important characteristic of an option is that it does not obligate its owner to take any action. It merely gives the owner the right to buy or sell an asset.

■ Financial options have an underlying asset that is traded—usually a security such as a stock.

■ Real options have an underlying asset that is not a security—for example, a project or a growth opportunity—and it is not traded.

■ The payoffs for financial options are specified in the contract, whereas real options are found or created inside of projects. Their payoffs vary.

■ Among the different types of real-option models are the option to defer investment, an abandonment option, follow-on (compound) options, and the option to adjust production.

■ One of the most critical parameters in valuing real options is the volatility of the return on the project.

■ The fundamental difference between the traditional NPW approach and real-options analysis is in how they treat managing the project risk: The traditional NPW approach is to avoid risk whenever possible, whereas the real-options approach is to manage the risk.

PROBLEMS

Financial Options

13.1 Use a binomial lattice with the following attributes to value a European call option:
   (a) Current underlying asset value of 60.
   (b) Exercise price of 60.
   (c) Volatility of 30%.
   (d) Risk-free rate of 5%.
   (e) Time to expiration equal to 18 months.
   (f) A two-period lattice.

13.2 Use a binomial lattice with the following attributes to value an American put option:
   (a) Current underlying asset value of 40.
   (b) Exercise price of 45.
   (c) Volatility of 40%.
   (d) Risk-free rate of 5%.
   (e) Dividend yield of 3%.
   (f) Time to expiration equal to three years.
   (g) A three-period lattice.

13.3 An investor has a portfolio consisting of the following assets and instruments:
- A long call option with $X = \$40$ with a call premium of $3 at the time of purchase.
- A short put option with $X = \$45$ with a put premium of $4 at the time of purchase.
- Two short call options with $X = \$35$ with a call premium of $5 at the time of purchase.
- Two short stock positions that cost $40 per share at the time of purchase.

Assume that each of these contracts has the same expiration date, and ignore the time value of money. If the stock price at expiration is $S_T = \$60$, what is the net profit of this portfolio?

13.4 A put option premium is currently $4, with $S_0 = \$30$, $X = \$32$, and $T = 6$ months. Calculate the intrinsic value and time premium for this put option. In addition, explain *why* the time to contract maturity and the underlying asset volatility affect a put option's time premium.

Real-Options Analysis

13.5 A company is planning to undertake an investment of $2 million to upgrade one of its products for an emerging market. The market is highly volatile, but the company owns a product patent that will protect it from competitive entry until the next year. Because of the uncertainty of the demand for the upgraded product, there is a chance that the market will be in favor of the company. The present value of expected future cash flows is estimated to be $1.9 million. Assume a risk-free interest rate of 8% and a standard deviation of 40% per annum for the PV of future cash flows. What is the value of delaying the investment?

13.6 A mining firm has the opportunity to purchase a license on a plot of land to mine for gold. Consider the following financial information:
- The investment cost to mine is $40M.
- It costs $320 per ounce to mine gold.
- The spot price of gold is $340 per ounce.
- The licensing agreement provides exclusive rights for three years.
- The historical volatility of gold prices is 20%.
- The estimated gold reserve on the plot of land is 1.5M ounces.
- The firm's MARR is 12% and $r = 6\%$.

Determine the maximum amount the firm should pay for a license to mine for gold on the property in question, or, equivalently, what is the value of the gold mine today?

Switching Options

13.7 A firm has invested in, and is currently receiving benefits from, Project A. The current value of Project A is $4M. Over the next five years, the firm has the option to use most of the same equipment from Project A and switch to Project B. Switching over would entail a $2M investment cost. The expected net cash inflow of Project B is $1M per annum for 10 years. What is the total value of this investment scenario? Assume that MARR $= 12\%$, $r = 6\%$, and $\sigma = 50\%$.

R&D Options

13.8 A pharmaceutical company needs to estimate the maximum amount to spend on R&D for a new type of diet drug. It is estimated that three years of R&D spending will be required to develop and test market the drug. After the initial three years, an investment in manufacturing and production will be required in year 4. It is estimated today that net cash inflows for six years will be received from sales of the drug. The following cash flow diagram summarizes the firm's estimates:

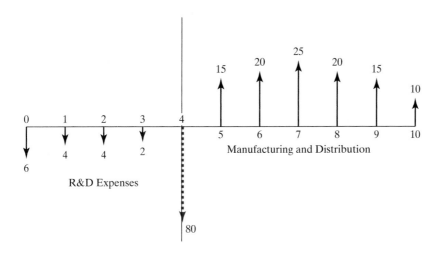

Assuming a MARR of 12%, $\sigma = 50\%$, and a risk-free interest rate of 6%, determine the maximum amount the firm should spend on R&D for this diet drug.

Abandonment Option

13.9 A firm is considering purchasing equipment to manufacture a new product. The equipment will cost $3M, and expected net cash inflows are $0.35M indefinitely. If market demand for the product is low, then over the next five years the firm will have the option of discarding the equipment on a secondary market for $2.2M. Assume that MARR $= 12\%$, $\sigma = 50\%$, and $r = 6\%$. What is the value of this investment opportunity for the firm?

Scale-Down Options

13.10 A firm that has undertaken a project has the option to sell some of its equipment and facilities and sublet out some of the project workload. The project's current value is $V_0 = \$10M$. Anytime over the next three years, the firm can sell off $4M in resources, but receive an expected 20% decrease in net cash flows (and, therefore, a 20% decrease in project value). Let $r = 6\%$ and $\sigma = 30\%$. A binomial lattice will be used with a one-year time increment. Determine the value of this scale-down investment opportunity.

Expansion–Contraction Options

13.11 Suppose a large manufacturing firm decides to hedge against risk through the use of strategic options. Specifically, it has the option to choose between two strategies:

- Expanding its current manufacturing operations.
- Contracting its current manufacturing operations at any time within the next two two years.

Suppose that the firm has a current operating structure whose static valuation of future profitability based on a discounted cash flow model is found to be $100 million. Suppose also that the firm estimates the implied volatility of the logarithmic returns on the projected future cash flows to be 15%. The risk-free interest rate is 5%. Finally, suppose that the firm has the option to contract 10% of its current operations at any time over the next two years, creating an additional $25 million in savings after the contraction. The expansion option will increase the firm's operations by 30% with a $20 million implementation cost.[7]

(a) Show the binomial lattice of the underlying asset over two years.

(b) What is the value of retaining the option to choose between both alternatives (i.e., consider both the expanding option and the contracting option together)?

Compound Options

13.12 A firm relies on R&D to maintain profitability. The firm needs to determine the maximum amount to invest today (or invest in Phase I) for its three-phased project:

- Phase I: Research. (Invest R_0 today.)
- Phase II: Development. (Invest I_1 one year from now.)
- Phase III: Implementation. (Invest I_2 in facilities, manpower, etc., three years from today.)

The three-phased investment cash flows are as follows:

Using a MARR of 12% $\sigma = 50\%$, and $r = 6\%$, determine the best investment strategy for the firm.

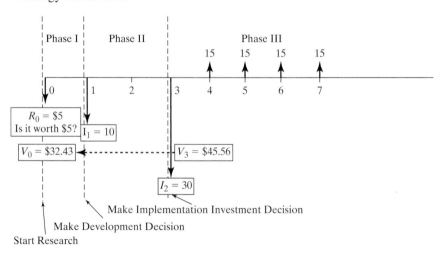

---

[7] This problem is adapted from Johnathan Mun, *Real Options Analysis*, New York: John Wiley, 2002, p. 182.

Short Case Studies

ST13.1 A drug company is considering developing a new drug. Due to the uncertain nature of the drug's progress in development, market demand, success in human and animal testing, and FDA approval, management has decided that it will create a strategic abandonment option. That is, anytime within the next five years, management can review the progress of the R&D effort and decide whether to terminate the development of the drug. If the program is terminated, the firm can sell off its intellectual property rights to the drug to another pharmaceutical firm.

- The present value of the expected future cash flows, discounted at an appropriate market-adjusted rate, is $150 million.
- Monte Carlo simulation indicates that the volatility of the logarithmic returns on future cash flows is 30%.
- The risk-free rate on a riskless asset for the same time frame is 5%.
- The drug's patent is worth $100 million if sold within the next five years.

Given that $S = \$150$, $\sigma = 30\%$, $T = 5$, and the risk-free interest rate $(r) = 5\%$, determine the value of the abandonment option if it is

(a) an American option using a binomial lattice approach.

(b) a European option using a Black–Scholes approach.

ST13.2 The pharmaceutical industry is composed of both large and small firms competing for new research, the introduction of new products, and the sales of existing products. Performing research and development for new drugs can be financially risky, as few researched drugs actually make it to the market. Sometimes, large firms will enter into agreements with smaller firms to conduct research. There are benefits to both parties in these situations. The biggest advantage is that the large firm does not need to invest in-house in required expertise and facilities to develop a new type of drug. In essence, large firms contract out their research work to smaller firms for an investment amount smaller than what the large firms would have to spend if they conducted the new research internally. On their part, small firms receive much-needed funds to carry out their research.

Merck Co. and Genetics, Inc., are publicly traded large and small pharmaceutical firms, respectively. Merck Co. is considering entering into an agreement with Genetics. The agreement would require Merck to give $4M to Genetics to develop a new drug over the next two years. By doing so, Merck would obtain the right to acquire Genetics for $60 per share three years from today. If the new drug is successful, Genetics' share price is expected to double or even triple. However, if the new drug is unsuccessful, Genetics' share price is expected to decrease in value significantly. Genetics has a current stock price of $30 per share, has an expected share price volatility of 50% (including the riskiness surrounding the new-drug research), and has 1.2 million shares outstanding.

(a) Is the initial $4M that Merck is agreeing to pay Genetics justified? (Quantify your answer, and assume the risk-free rate is 6%.)

(b) Assume that Merck could also buy Genetics today for $40 per share. Compare buying Genetics today versus entering into the aforesaid agreement. Briefly discuss the advantages and disadvantages of both alternatives. If you utilize the binomial lattice approach, then only use a *one-period lattice* for your computations.

Acknowledgement: Many of the end-of-chapter problems in this chapter were provided by Dr. Luke Miller at Fort Lewis College.

Special Topics in Engineering Economics

FOURTEEN

Replacement Decisions

Options for Replacing Alaskan Viaduct[1] The Alaskan Way Viaduct in Seattle, Washington, will be either rebuilt or replaced with a tunnel, the Washington State Department of Transportation (WSDOT) said. The urgency to replace the bridge, which carries 110,000 cars a day, came after the February 2001 Nisqually quake, which caused major damage to the viaduct. The road was built in 1953 to carry 64,000 cars a day.

Rebuilding the viaduct and replacing it with a tunnel are the most likely options for replacing the bridge and seem to be the most popular among those who responded to an environmental-impact statement produced for the project.

Option 1: Build a Tunnel This plan, supported by arts organizations and the Greater Seattle Chamber of Commerce, would replace the viaduct with a tunnel along the central waterfront carrying three lanes in each direction. This is the most expensive option, with cost estimates of $3.6 billion to $4.1 billion. It would take from seven to nine years to build.

Option 2: Rebuild the Viaduct This alternative, backed by a group of Magnolia residents, would replace the viaduct in its existing location with a structure similar to what is there now, including ramps into downtown at Seneca and Columbia Streets. Unlike the existing structure, this new viaduct would be designed to current earthquake standards. It would cost $2.7–$3.1 billion and take six to seven years.

[1]*SR 99—Alaskan Way Viaduct and Seawall Replacement*, Washington State Department of Transportation, http://www.wsdot.wa.gov/Projects/Viaduct/default.htm, and "Options for replacing viaduct down to two," *The Seattle Times*, September 8, 2004.

Option 1

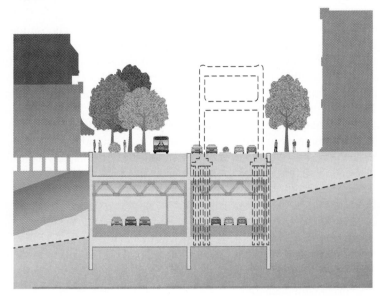

Option 2

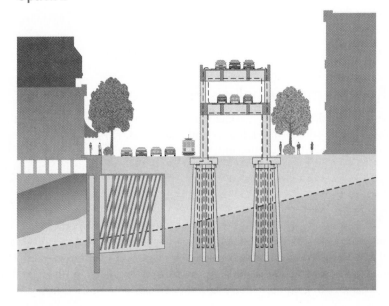

Currently, there is no money to replace the viaduct, other than $177 million as part of the nickel-a-gallon gas-tax increase approved by the legislature. The state is envisioning that it would contribute $2 billion to the project, and the rest would come from other sources, such as the Army Corps of Engineers, possible federal funds dedicated to megaprojects, the city, the Port of Seattle, and a possible voter-approved regional-transportation package. However, the WSDOT has to decide soon whether the state should replace the damaged viaduct with a tunnel or rebuild the viaduct in its current existing structure.

In Chapters 7 through 9, we presented methods that helped us choose the best of a number of investment alternatives. The problems we examined in those chapters concerned primarily profit-adding projects. However, economic analysis is also frequently performed on projects with existing facilities or profit-maintaining projects—those projects whose primary purpose is not to increase sales, but rather, simply to maintain ongoing operations. In practice, profit-maintaining projects less frequently involve the comparison of new machines; instead, the problem often facing management is whether to buy new and more efficient equipment or to continue to use existing equipment. This class of decision problems is known as the **replacement problem**. In this chapter, we examine the basic concepts and techniques related to replacement analysis.

CHAPTER LEARNING OBJECTIVES

After completing this chapter, you should understand the following concepts:

- What makes the replacement decision problems differ from the other capital investment decisions.
- What types of financial information should be collected to conduct a typical replacement decision problem.
- How to compare a defender with a challenger on the basis of opportunity cost concept.
- How to determine the economic service life for any given asset.
- How to determine the optimal time to replace a defender.
- How to consider the tax effects in replacement analysis.

14.1 Replacement Analysis Fundamentals

In this section and the next two, we examine three aspects of the replacement problem: (1) approaches to comparing defender and challenger, (2) the determination of economic service life, and (3) replacement analysis when the required service period is long. The impact of income tax regulations will be ignored; in Section 14.4, we revisit these replacement problems, taking income taxes into account.

14.1.1 Basic Concepts and Terminology

Replacement projects are decision problems involving the replacement of existing obsolete or worn-out assets. The continuation of operations is dependent on these assets. The failure to make an appropriate decision results in a slowdown or shutdown of the operations. The question is when existing equipment should be replaced with more efficient equipment.

This situation has given rise to the use of the terms **defender** and **challenger**, terms commonly used in the boxing world. In every boxing class, the current defending champion is constantly faced with a new challenger. In replacement analysis, the defender is the existing machine (or system), and the challenger is the best available replacement equipment.

An existing piece of equipment will be removed at some future time, either when the task it performs is no longer necessary or when the task can be performed more efficiently by newer and better equipment. The question is not *whether* the existing piece of equipment will be removed, but *when* it will be removed. A variation of this question is why we should replace existing equipment at the current time, rather than postponing replacement of the equipment by repairing or overhauling it. Another aspect of the defender–challenger comparison concerns deciding exactly which equipment is the best challenger. If the defender is to be replaced by the challenger, we would generally want to install the very best of the possible alternatives.

Current Market Value

The most common problem encountered in considering the replacement of existing equipment is the determination of what financial information is actually relevant to the analysis. Often, a tendency to include irrelevant information in the analysis is apparent. To illustrate this type of decision problem, let us consider Example 14.1.

EXAMPLE 14.1 Information Relevant to Replacement Analysis

Macintosh Printing, Inc., purchased a $20,000 printing machine two years ago. The company expected this machine to have a five-year life and a salvage value of $5,000. The company spent $5,000 last year on repairs, and current operating costs are running at the rate of $8,000 per year. Furthermore, the anticipated salvage value of the machine has been reduced to $2,500 at the end of the its remaining useful life. In addition, the company has found that the current machine has a market value of $10,000 today. The equipment vendor will allow the company this full amount as a trade-in on a new machine. What values for the defender are relevant to our analysis?

SOLUTION

In this example, three different dollar amounts relating to the defender are presented:

1. **Original cost.** The printing machine was purchased for $20,000.
2. **Market value.** The company estimates the old machine's market value at $10,000.
3. **Trade-in allowance.** This is the same as the market value. (In other problems, however, it could be different from the market value.)

COMMENTS: In this example and in all defender analyses, the relevant cost is the **current market value** of the equipment. The original cost, repair cost, and trade-in value are irrelevant. A common misconception is that the trade-in value is the same as the current market value of the equipment and thus could be used to assign a suitable current value to the equipment. This is not always true, however. For example, a car dealer typically offers a trade-in value on a customer's old car to reduce the price of a new car. Would the dealer offer the same value on the old car if he or she were not also selling the new one? The answer is, Not generally. In many instances, the trade-in allowance is inflated to make the deal look good, and the price of the new car is also inflated to compensate for the dealer's trade-in cost. In this type of situation, the trade-in value does not represent the true value of the item, so we should not use it in economic analysis.[2]

Sunk Costs

Sunk costs are costs that have already been incurred and which cannot be recovered to any significant degree.

As mentioned in Section 3.4.3, a **sunk cost** is any past cost that is unaffected by any future investment decision. In Example 14.1, the company spent $20,000 to buy the machine two years ago. Last year, $5,000 more was spent on the machine. The total accumulated expenditure is $25,000. If the machine were sold today, the company could get only $10,000 back (Figure 14.1). It is tempting to think that the company would lose $15,000 in addition to the cost of the new machine if the old machine were to be sold and replaced with a new one. This is an incorrect way of doing economic analysis, however. In a proper engineering economic analysis, only future costs should be considered; past or sunk costs should be ignored. Thus, the value of the defender that should be used in a replacement analysis should be its current market value, not what it cost when it was originally purchased and not the cost of repairs that have already been made to the machine.

Sunk costs are money that is gone, and no present action can recover them. They represent past actions—the results of decisions made in the past. In making economic decisions at the present time, one should consider only the possible outcomes of the various decisions and pick the one with the best possible future results. Using sunk costs in arguing one option over the other would only lead to more bad decisions.

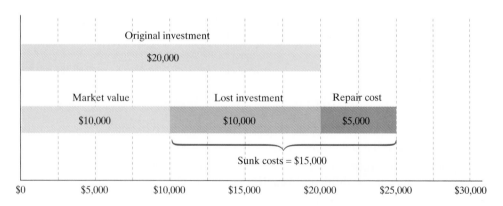

Figure 14.1 Sunk cost associated with an asset's disposal as described in Example 14.1.

---

[2] If we do make the trade, however, the actual net cash flow at the time of the trade, properly used, is certainly relevant.

Operating Costs

The driving force for replacing existing equipment is that it becomes more expensive to operate with time. The total cost of operating a piece of equipment may include repair and maintenance costs, wages for the operators, energy consumption costs, and costs of materials. Increases in any one or a combination of these cost items over a period of time may impel us to find a replacement for the existing asset. The challenger is usually newer than the defender and often incorporates improvements in design and newer technology. As a result, some or all of the cost items for the challenger are likely to be less expensive than those for the defender.

We will call the sum of the various cost items related to the operation of an asset the **operating costs**. As is illustrated in the sections that follow, keeping the defender involves a lower initial cost than purchasing the challenger, but higher annual operating costs. Usually, operating costs increase over time for both the defender and the challenger. In many instances, the labor costs, material costs, and energy costs are the same for the defender and the challenger and do not change with time. It is the repair and maintenance costs that increase and cause the operating costs to increase each year as an asset ages.

When repair and maintenance costs are the only cost items that differ between the defender and the challenger on a year-by-year basis, we need to include only those costs in the operating costs used in the analysis. Regardless of which cost items we choose to include in the operating costs, it is essential that the same items be included for both the defender and the challenger. For example, if energy costs are included in the operating costs of the defender, they should also be included in the operating costs of the challenger. A more comprehensive discussion of the various types of costs incurred in a complex manufacturing facility was provided in Chapter 8.

14.1.2 Opportunity Cost Approach to Comparing Defender and Challenger

Although replacement projects are a subcategory of the mutually exclusive categories of project decisions we studied in Chapter 5, they do possess unique characteristics that allow us to use specialized concepts and analysis techniques in their evaluation. We consider a basic approach to analyzing replacement problems commonly known as the **opportunity cost approach**.

The basic issue is how to treat the proceeds from the sale of the old equipment. In fact, if you decide to keep the old machine, this potential sales receipt is forgone. The opportunity cost approach views the net proceeds from sale as the opportunity cost of keeping the defender. In other words, we consider the salvage value as a cash outflow for the defender (or an investment required in order to keep the defender).

EXAMPLE 14.2 Replacement Analysis Using the Opportunity Cost Approach

Consider again Example 14.1. The company has been offered a chance to purchase another printing machine for $15,000. Over its three-year useful life, the machine will reduce the usage of labor and raw materials sufficiently to cut operating costs from $8,000 to $6,000. It is estimated that the new machine can be sold for $5,500 at the end of year 3. If the new machine were purchased, the old machine would be sold to another company, rather than traded in for the new machine.

Suppose that the firm will need either machine (old or new) for only three years and that it does not expect a new, superior machine to become available on the market during this required service period. Assuming that the firm's interest rate is 12%, decide whether replacement is justified now.

SOLUTION

- **Option 1: Keep the defender.**
  If the decision is to keep the defender, the opportunity cost approach treats the $10,000 current salvage value of the defender as an incurred cost. The annual operating cost for the next three years will be $8,000 per year, and the defender's salvage value three years from today will be $2,500. The cash flow diagram for the defender is shown in Figure 14.2(a).

Opportunity cost approach views the net proceeds from the sale of the old machine as an investment required to keep the old asset.

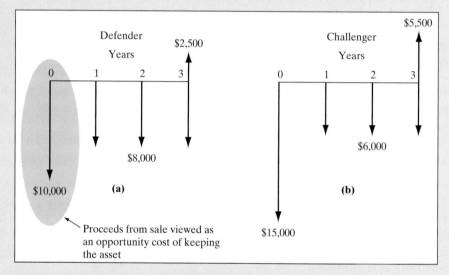

Figure 14.2 Comparison of defender and challenger based on the opportunity cost approach (Example 14.2).

- **Option 2: Replace the defender with the challenger.**
  The new machine costs $15,000. The annual operating cost of the challenger is $6,000. The salvage value of the challenger three years later will be $5,500. The actual cash flow diagram for this option is shown in Figure 14.2(b). We calculate the net present worth and annual equivalent cost for each of the two options as follows:

$$
\begin{aligned}
\text{PW}(12\%)_D &= \$10,000 + \$8,000(P/A, 12\%, 3) \\
&\quad - \$2,500(P/F, 12\%, 3) \\
&= \$27,435,
\end{aligned}
$$

$$AEC(12\%)_D = PW(12\%)_D \ (A/P, 12\%, 3)$$
$$= \$11,423,$$

$$PW(12\%)_C = \$15,000 + \$6,000(P/A, 12\%, 3) - \$5,500(P/F, 12\%, 3)$$
$$= \$25,496,$$

$$AEC(12\%)_C = PW(12\%)_C \ (A/P, 12\%, 3)$$
$$= \$10,615.$$

Because of the annual difference of $808 in favor of the challenger, the replacement should be made now.

COMMENTS: If our analysis showed instead that the defender should not be replaced now, we still need to address the question of whether the defender should be kept for one or two years and then replaced with the challenger. This is a valid question that requires more data on market values over time. We address the situation later, in Section 14.3. Recall that we assumed the same service life for both the defender and the challenger in Examples 14.1 and 14.2. In general, however, old equipment has a relatively short remaining life compared with new equipment, so this assumption is overly simplistic. In the next section, we discuss how to find the economic service life of equipment.

14.2 Economic Service Life

Perhaps you have seen a 50-year-old automobile that is still in service. Provided that it receives the proper repair and maintenance, almost anything can be kept operating for an extended period of time. If it's possible to keep a car operating for an almost indefinite period, why aren't more old cars spotted on the streets? Two reasons are that some people may get tired of driving the same old car, and other people may want to keep a car as long as it will last, but they realize that repair and maintenance costs will become excessive.

> **Economic service life** is the remaining useful life of an asset that results in the minimum annual equivalent cost.

In general, we need to consider economically how long an asset should be held once it is placed in service. For instance, a truck-rental firm that frequently purchases fleets of identical trucks may wish to arrive at a policy decision on how long to keep each vehicle before replacing it. If an appropriate life span is computed, a firm could stagger a schedule of truck purchases and replacements to smooth out annual capital expenditures for its overall truck purchases.

The costs of owning and operating an asset can be divided into two categories: **capital costs** and **operating costs**. Capital costs have two components: the initial investment and the salvage value at the time of disposal of the asset. The initial investment for the challenger is simply its purchase price. For the defender, we should treat the opportunity cost as its initial investment. We will use N to represent the length of time in years the asset will be kept, I to denote the initial investment, and S_N to designate the salvage value at the end of the ownership period of N years.

The annual equivalent of capital costs, which is called the capital recovery cost (see Section 8.2), over the period of N years can be calculated with the following equation:

$$CR(i) = I(A/P, i, N) - S_N(A/F, i, N).$$ (14.1)

Generally speaking, as an asset becomes older, its salvage value becomes smaller. As long as the salvage value is less than the initial cost, the capital recovery cost is a decreasing function of N. In other words, the longer we keep an asset, the lower the capital recovery cost becomes. If the salvage value is equal to the initial cost no matter how long the asset is kept, the capital recovery cost is constant.

As described earlier, the operating costs of an asset include operating and maintenance (O&M) costs, labor costs, material costs, and energy consumption costs. Labor costs, material costs, and energy costs are often constant for the same equipment from year to year if the usage of the equipment remains constant. However, O&M costs tend to increase as a function of the age of the asset. Because of the increasing trend of the O&M costs, the total operating costs of an asset usually increase as well as the asset ages. We use OC_n to represent the total operating costs in year n of the ownership period and $OC(i)$ to represent the annual equivalent of the operating costs over a life span of N years. Then $OC(i)$ can be expressed as

$$OC(i) = \left(\sum_{n=1}^{N} OC_n (P/F, i, n) \right)(A/P, i, N).$$ (14.2)

As long as the annual operating costs increase with the age of the equipment, $OC(i)$ is an increasing function of the life of the asset. If the annual operating costs are the same from year to year, $OC(i)$ is constant and equal to the annual operating costs, no matter how long the asset is kept.

The total annual equivalent costs of owning and operating an asset ($AEC(i)$) are a summation of the capital recovery costs and the annual equivalent of operating costs of the asset:

$$AEC(i) = CR(i) + OC(i).$$ (14.3)

The economic service life of an asset is defined to be the period of useful life that minimizes the annual equivalent costs of owning and operating the asset. On the basis of the foregoing discussions, we need to find the value of N that minimizes AE as expressed in Eq. (14.3). If $CR(i)$ is a decreasing function of N and $OC(i)$ is an increasing function of N, as is often the case, AE will be a convex function of N with a unique minimum point. (See Figure 14.3.) In this book, we assume that AE has a unique minimum point. If the salvage value of the asset is constant and equal to the initial cost, and if the annual operating cost increases with time, then AE is an increasing function of N and attains its minimum at $N = 1$. In this case, we should try to replace the asset as soon as possible. If, however, the annual operating cost is constant and the salvage value is less than the initial cost and decreases with time, then AE is a decreasing function of N. In this case, we would try to delay the replacement of the asset as much as possible. Finally, if the salvage value is

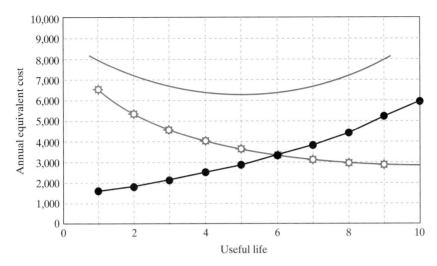

Figure 14.3   A schematic illustrating the trends of capital recovery cost (ownership cost), annual operating cost, and total annual equivalent cost.

constant and equal to the initial cost and the annual operating costs are constant, then AE will also be constant. In this case, when to replace the asset does not make any economic difference.

If a new asset is purchased and operated for the length of its economic life, the annual equivalent cost is minimized. If we further assume that a new asset of identical price and features can be purchased repeatedly over an indefinite period, we would always replace this kind of asset at the end of its economic life. By replacing perpetually according to an asset's economic life, we obtain the minimum AE cost stream over an indefinite period. However, if the identical-replacement assumption cannot be made, we will have to use the methods to be covered in Section 14.3 to carry out a replacement analysis. The next example explains the computational procedure for determining an asset's economic service life.

EXAMPLE 14.3 Economic Service Life of a Lift Truck

Suppose a company has a forklift, but is considering purchasing a new electric-lift truck that would cost $18,000, have operating costs of $1,000 in the first year, and have a salvage value of $10,000 at the end of the first year. For the remaining years, operating costs increase each year by 15% over the previous year's operating costs. Similarly, the salvage value declines each year by 25% from the previous year's salvage value. The lift truck has a maximum life of seven years. An overhaul costing $3,000 and $4,500 will be required during the fifth and seventh years of service, respectively. The firm's required rate of return is 15%. Find the economic service life of this new machine.

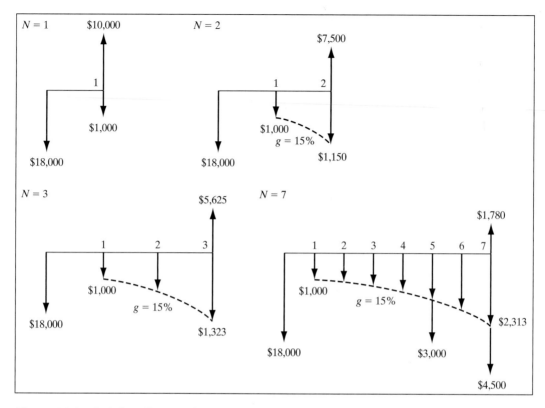

Figure 14.4 Cash flow diagrams for the options of keeping the asset for one year, two years, three years, and seven years (Example 14.3).

DISCUSSION: For an asset whose revenues are either unknown or irrelevant, we compute its economic life on the basis of the costs for the asset and its year-by-year salvage values. To determine an asset's economic service life, we need to compare the options of keeping the asset for one year, two years, three years, and so forth. The option that results in the lowest annual equivalent cost (AEC) gives the economic service life of the asset.

- **$N = 1$: One-year replacement cycle.** In this case, the machine is bought, used for one year, and sold at the end of year 1. The cash flow diagram for this option is shown in Figure 14.4. The annual equivalent cost for this option is

$$\text{AEC}(15\%) = \$18,000(A/P, 15\%, 1) + \$1,000 - \$10,000$$
$$= \$11,700.$$

Note that $(A/P, 15\%, 1) = (F/P, 15\%, 1)$ and the annual equivalent cost is the equivalent cost at the end of year 1, since $N = 1$. Because we are calculating the annual equivalent cost in the computation of AEC(15%), we have treated cost items with a positive sign, while the salvage value has a negative sign.

- $N = 2$: **Two-year replacement cycle.** In this case, the truck will be used for two years and disposed of at the end of year 2. The operating cost in year 2 is 15% higher than that in year 1, and the salvage value at the end of year 2 is 25% lower than that at the end of year 1. The cash flow diagram for this option is also shown in Figure 14.4. The annual equivalent cost over the two-year period is

$$\text{AEC}(15\%) = [\$18,000 + \$1,000(P/A_1\ 15\%, 15\%, 2)](A/P, 15\%, 2)$$

$$-\$7,500(A/F, 15\%, 2)$$

$$= \$8,653.$$

- $N = 3$: **Three-year replacement cycle.** In this case, the truck will be used for three years and sold at the end of year 3. The salvage value at the end of year 3 is 25% lower than that at the end of year 2; that is, $\$7,500(1 - 25\%) = \$5,625$. The operating cost per year increases at a rate of 15%. The cash flow diagram for this option is also shown in Figure 14.4. The annual equivalent cost over the three-year period is

$$\text{AEC}(15\%) = [\$18,000 + \$1,000(P/A_1\ 15\%, 15\%, 3)](A/P, 15\%, 3)$$

$$- \$5,625(A/F, 15\%, 3)$$

$$= \$7,406.$$

- Similarly, we can find the annual equivalent costs for the options of keeping the asset for four, five, six, and seven years. One has to note that there is an additional cost of overhaul in year 5. The cash flow diagram when $N = 7$ is shown in Figure 14.4. The computed annual equivalent costs for each of these options are

$$N = 4, \text{AEC}(15\%) = \$6,678,$$

$$N = 5, \text{AEC}(15\%) = \$6,642,$$

$$N = 6, \text{AEC}(15\%) = \$6,258,$$

$$N = 7, \text{AEC}(15\%) = \$6,394.$$

From the preceding calculated AEC values for $N = 1, \ldots, 7$, we find that AEC(15%) is smallest when $N = 6$. If the truck were to be sold after six years, it would have an annual cost of \$6,258 per year. If it were to be used for a period other than six years, the annual equivalent costs would be higher than \$6,258. Thus, a life span of six years for this truck results in the lowest annual cost. We conclude that the economic service life of the truck is six years. By replacing the assets perpetually according to

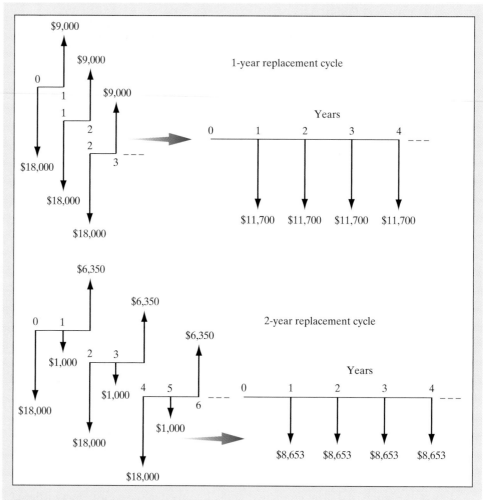

Figure 14.5 Conversion of an infinite number of replacement cycles to infinite AE cost streams (Example 14.3).

an economic life of six years, we obtain the minimum annual equivalent cost stream. Figure 14.5 illustrates this concept. Of course, we should envision a long period of required service for this kind of asset.

14.3 Replacement Analysis when the Required Service Is Long

Now that we understand how the economic service life of an asset is determined, the next question is how to use these pieces of information to decide whether now is the time to replace the defender. If now is not the right time, when *is* the optimal time to replace the defender? Before presenting an analytical approach to answer this question, we consider several important assumptions.

14.3.1 Required Assumptions and Decision Frameworks

In deciding whether now is the time to replace the defender, we need to consider the following three factors:

- Planning horizon (study period).
- Technology.
- Relevant cash flow information.

Planning Horizon (Study Period)

By the planning horizon, we simply mean the service period required by the defender and a sequence of future challengers. The infinite planning horizon is used when we are simply unable to predict when the activity under consideration will be terminated. In other situations, it may be clear that the project will have a definite and predictable duration. In these cases, replacement policy should be formulated more realistically on the basis of a finite planning horizon.

Technology

Predictions of technological patterns over the planning horizon refer to the development of types of challengers that may replace those under study. A number of possibilities exist in predicting purchase cost, salvage value, and operating cost that are dictated by the efficiency of a new machine over the life of an existing asset. If we assume that all future machines will be the same as those now in service, we are implicitly saying that no technological progress in the area will occur. In other cases, we may explicitly recognize the possibility of machines becoming available in the future that will be significantly more efficient, reliable, or productive than those currently on the market. (Personal computers are a good example.) This situation leads to the recognition of technological change and obsolescence. Clearly, if the best available machine gets better and better over time, we should certainly investigate the possibility of delaying an asset's replacement for a couple of years—a viewpoint that contrasts with the situation in which technological change is unlikely.

Revenue and Cost Patterns over the Life of an Asset

Many varieties of predictions can be used to estimate the patterns of revenue, cost, and salvage value over the life of an asset. Sometimes revenue is constant, but costs increase, while salvage value decreases, over the life of a machine. In other situations, a decline in revenue over the life of a piece of equipment can be expected. The specific situation will determine whether replacement analysis is directed toward cost minimization (with constant revenue) or profit maximization (with varying revenue). We formulate a replacement policy for an asset whose salvage value does not increase with age.

Decision Frameworks

To illustrate how a decision framework is developed, we indicate a replacement sequence of assets by the notation $(j_0, n_0), (j_1, n_1), (j_2, n_2), \ldots, (j_K, n_K)$. Each pair of numbers (j, n) indicates a type of asset and the lifetime over which that asset will be retained. The defender, asset 0, is listed first; if the defender is replaced now, $n_0 = 0$. A sequence of pairs may cover a finite period or an infinite period. For example, the sequence $(j_0, 2), (j_1, 5), (j_2, 3)$ indicates retaining the defender for two years, then replacing the defender with an asset of type j_1 and using it for five years, and then replacing j_1 with an asset of type j_2 and using it for three years. In this situation, the total planning horizon

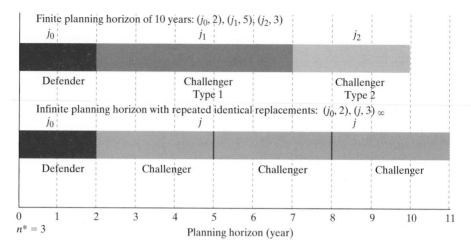

Figure 14.6 Types of typical replacement decision frameworks.

covers 10 years $(2 + 5 + 3)$. The special case of keeping the defender for n_0 periods, followed by infinitely repeated purchases and the use of an asset of type j for n^* years, is represented by $(j_0, n_0), (j, n^*)_\infty$. This sequence covers an infinite period, and the relationship is illustrated in Figure 14.6.

Decision Criterion

Although the economic life of the defender is defined as the additional number of years of service which minimizes the annual equivalent cost (or maximizes the annual equivalent revenue), that is *not* necessarily the *optimal* time to replace the defender. The correct replacement time depends on data on the challenger, as well as on data on the defender.

As a decision criterion, the AE method provides a more direct solution when the planning horizon is infinite. When the planning horizon is finite, the PW method is more convenient to use. We will develop the replacement decision procedure for both situations. We begin by analyzing an infinite planning horizon without technological change. Even though a simplified situation such as this is not likely to occur in real life, the analysis of this replacement situation introduces methods that will be useful in analyzing infinite-horizon replacement problems with technological change.

14.3.2 Replacement Strategies under the Infinite Planning Horizon

Consider a situation in which a firm has a machine that is in use in a process which is expected to continue for an indefinite period. Presently, a new machine will be on the market that is, in some ways, more effective for the application than the defender is. The problem is when, if at all, the defender should be replaced with the challenger.

Under the infinite planning horizon, the service is required for a very long time. Either we continue to use the defender to provide the service, or we replace the defender with the best available challenger for the same service requirement. In this case, we may apply the following procedure in replacement analysis:

1. Compute the economic lives of both the defender and the challenger. Let's use N_D^* and N_C^* to indicate the economic lives of the defender and the challenger, respectively. The annual equivalent costs for the defender and the challenger at their respective economic lives are indicated by AEC_D^* and AEC_C^*.

2. Compare AEC_D^* and AEC_C^*. If AEC_D^* is bigger than AEC_C^*, it is more costly to keep the defender than to replace it with the challenger. Thus, the challenger should replace the defender now. If AEC_D^* is smaller than AEC_C^*, it costs less to keep the defender than to replace it with the challenger. Thus, the defender should *not* be replaced now. The defender should continue to be used at least for the duration of its economic life if there are no technological changes over that life.

3. If the defender should not be replaced now, when should it be replaced? First we need to continue to use it until its economic life is over. Then we should calculate the cost of running the defender for one more year after its economic life. If this cost is greater than AEC_C^*, the defender should be replaced at the end of its economic life. Otherwise, we should calculate the cost of running the defender for the second year after its economic life. If this cost is bigger than AEC_C^*, the defender should be replaced one year after its economic life. The process should be continued until we find the optimal replacement time. This approach is called **marginal analysis**; that is, we calculate the incremental cost of operating the defender for just one more year. In other words, we want to see whether the cost of extending the use of the defender for an additional year exceeds the savings resulting from delaying the purchase of the challenger. Here, we have assumed that the best available challenger does not change.

Note that this procedure might be applied dynamically. For example, it may be performed annually for replacement analysis. Whenever there are updated data on the costs of the defender or new challengers available on the market, the new data should be used in the procedure. Example 14.4 illustrates the procedure.

EXAMPLE 14.4 Replacement Analysis under an Infinite Planning Horizon

Advanced Electrical Insulator Company is considering replacing a broken inspection machine, which has been used to test the mechanical strength of electrical insulators, with a newer and more efficient one.

- If repaired, the old machine can be used for another five years, although the firm does not expect to realize any salvage value from scrapping it at that time. However, the firm can sell it now to another firm in the industry for $5,000. If the machine is kept, it will require an immediate $1,200 overhaul to restore it to operable condition. The overhaul will neither extend the service life originally estimated nor increase the value of the inspection machine. The operating costs are estimated at $2,000 during the first year, and these are expected to increase by $1,500 per year thereafter. Future market values are expected to decline by $1,000 per year.

- The new machine costs $10,000 and will have operating costs of $2,000 in the first year, increasing by $800 per year thereafter. The expected salvage value is $6,000 after one year and will decline 15% each year. The company requires a rate of return of 15%. Find the economic life for each option, *and* determine when the defender should be replaced.

SOLUTION

1. **Economic service life:**
   - **Defender.** If the company retains the inspection machine, it is in effect deciding to overhaul the machine and invest the machine's current market value in that alternative. The opportunity cost of the machine is $5,000. Because an overhaul costing $1,200 is also needed to make the machine operational, the total initial investment in the machine is $5,000 + $1,200 = $6,200. Other data for the defender are summarized as follows:

n	Overhaul	Forecasted Operating Cost	Market Value if Disposed of
0	$1,200		$5,000
1	0	$2,000	$4,000
2	0	$3,500	$3,000
3	0	$5,000	$2,000
4	0	$6,500	$1,000
5	0	$8,000	0

We can calculate the annual equivalent costs if the defender is to be kept for one year, two years, three years, and so forth. For example, the cash flow diagram for $N = 4$ years is shown in Figure 14.7. The annual equivalent costs for four years are as follows:

$$N = 4 \text{ years: AEC}(15\%) = \$6,200(A/P, 15\%, 4) + \$2,000$$
$$+ \$1,500(A/G, 15\%, 4) - \$1,000(A/F, 15\%, 4)$$
$$= \$5,961.$$

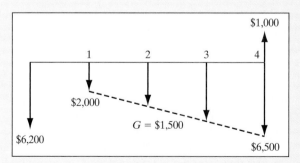

Figure 14.7 Cash flow diagram for defender when $N = 4$ years (Example 14.4).

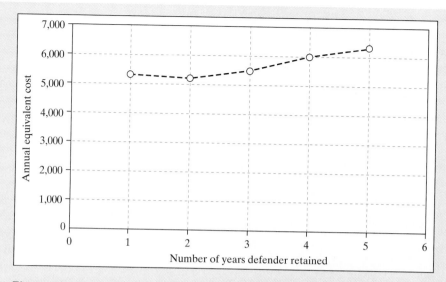

Figure 14.8 AEC as a function of the life of the defender (Example 14.4).

The other AE cost figures can be calculated with the following equation:

$$\text{AEC}(15\%)_N = \$6{,}200(A/P, 15\%, N) + \$2{,}000 + \$1{,}500$$
$$(A/G, 15\%, N)$$
$$- \$1{,}000(5 - N)(A/F, 15\%, N) \text{ for } N = 1, 2, 3, 4, 5;$$
$$N = 1: \text{AEC}(15\%) = \$5{,}130,$$
$$N = 2: \text{AEC}(15\%) = \$5{,}116,$$
$$N = 3: \text{AEC}(15\%) = \$5{,}500,$$
$$N = 4: \text{AEC}(15\%) = \$5{,}961,$$
$$N = 5: \text{AEC}(15\%) = \$6{,}434.$$

When $N = 2$ years, we get the lowest AEC value. Thus, the defender's economic life is two years. Using the notation we defined in the procedure, we have

$$N_D{}^* = 2 \text{ years,}$$
$$\text{AEC}_D{}^* = \$5{,}116.$$

The AEC values as a function of N are plotted in Figure 14.8. Actually, after computing AEC for $N = 1, 2,$ and 3, we can stop right there. There is no need to compute AEC for $N = 4$ and $N = 5$, because AEC is increasing when $N > 2$ and we have assumed that AEC has a unique minimum point.

- **Challenger.** The economic life of the challenger can be determined with the same procedure we used in this example for the defender and in Example 14.3. A summary of the general equation for calculating AEC for the challenger follows. You don't have to summarize such an equation when you need to determine the economic life of an asset, as long as you follow the procedure illustrated in Example 14.3. The equation is

$$AEC(15\%)_N = \$10,000(A/P, 15\%, N) + \$2,000$$
$$+ \$800(A/G, 15\%, N)$$
$$- \$6,000(1 - 15\%)^{N-1}(A/F, 15\%, N).$$

The results of "plugging in" the values and solving are as follows:

$$N = 1 \text{ year: } AEC(15\%) = \$7,500,$$
$$N = 2 \text{ years: } AEC(15\%) = \$6,151,$$
$$N = 3 \text{ years: } AEC(15\%) = \$5,857,$$
$$N = 4 \text{ years: } AEC(15\%) = \$5,826,$$
$$N = 5 \text{ years: } AEC(15\%) = \$5,897.$$

The economic life of the challenger is four years; that is,

$$N_C^* = 4 \text{ years.}$$

Thus,

$$AEC_C^* = \$5,826.$$

2. **Should the defender be replaced now?**
   Since $AEC_D^* = \$5,116 < AEC_C^* = \$5,826$, the defender should not be replaced now. If there are no technological advances in the next few years, the defender should be used for at least $N_D^* = 2$ more years. However, it is not necessarily best to replace the defender right at the high point of its economic life.

3. **When should the defender be replaced?**
   If we need to find the answer to this question today, we have to calculate the cost of keeping and using the defender for the third year from today. That is, what is the cost of not selling the defender at the end of year 2, using it for the third year, and replacing it at the end of year 3? The following cash flows are related to this question:

   (a) Opportunity cost at the end of year 2: equal to the market value then, or $3,000.

   (b) Operating cost for the third year: $5,000.

   (c) Salvage value of the defender at the end of year 3: $2,000.

   The following diagram represents these cash flows:

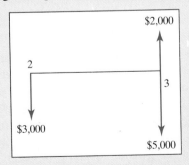

The cost of using the defender for one more year from the end of its economic life is

$$\$3,000 \times 1.15 + \$5,000 - \$2,000 = \$6,450.$$

Now compare this cost with the $AEC_C^* = \$5,826$ of the challenger. It is greater than AEC_C^*. Thus, it is more expensive to keep the defender for the third year than to replace it with the challenger. Accordingly, we conclude that we should replace the defender at the end of year 2. If this one-year cost is still smaller than AEC_C^*, we need to calculate the cost of using the defender for the fourth year and then compare that cost with the AEC_C^* of the challenger.

In replacement analysis, it is common for a defender and its challenger to have different economic service lives. The annual-equivalent approach is frequently used, but it is important to know that we use the AEC method in replacement analysis, not because we have to deal with the problem of unequal service lives, but rather because the AEC approach provides some computational advantage for a special class of replacement problems.

In Chapter 5, we discussed the general principle for comparing alternatives with unequal service lives. In particular, we pointed out that use of the AEC method relies on the concept of repeatability of projects and one of two assumptions: an infinite planning horizon or a common service period. In defender–challenger situations, however, repeatability of the defender cannot be assumed. In fact, by virtue of our definition of the problem, we are not repeating the defender, but replacing it with its challenger, an asset that in some way constitutes an improvement over the current equipment. Thus, the assumptions we made for using an annual cash flow analysis with unequal service life alternatives are not valid in the usual defender–challenger situation.

The complication—the unequal-life problem—can be resolved if we recall that the replacement problem at hand is not *whether* to replace the defender, but *when* to do so. When the defender is replaced, it will always be by the challenger—the best available equipment. An identical challenger can then replace the challenger repeatedly. In fact, we really are comparing the following two options in replacement analysis:

1. **Replace the defender now.** The cash flows of the challenger will be used from today and will be repeated because an identical challenger will be used if replacement becomes necessary again in the future. This stream of cash flows is equivalent to a cash flow of AEC_C^* each year for an infinite number of years.
2. **Replace the defender, say, x years later.** The cash flows of the defender will be used in the first x years. Starting in year $x + 1$, the cash flows of the challenger will be used indefinitely.

The annual-equivalent cash flows for the years beyond year x are the same for these two options. We need only to compare the annual-equivalent cash flows for the first x years to determine which option is better. This is why we can compare AEC_D^* with AEC_C^* to determine whether now is the time to replace the defender.

14.3.3 Replacement Strategies under the Finite Planning Horizon

If the planning period is finite (for example, eight years), a comparison based on the AE method over a defender's economic service life does not generally apply. The procedure for solving such a problem with a finite planning horizon is to establish all "reasonable" replacement patterns and then use the PW value for the planning period to select the most economical pattern. To illustrate this procedure, consider Example 14.5.

EXAMPLE 14.5 Replacement Analysis under the Finite Planning Horizon (PW Approach)

Consider again the defender and the challenger in Example 14.4. Suppose that the firm has a contract to perform a given service, using the current defender or the challenger for the next eight years. After the contract work, neither the defender nor the challenger will be retained. What is the best replacement strategy?

SOLUTION

Recall again the annual equivalent costs for the defender and challenger under the assumed holding periods (a boxed number denotes the minimum AEC value at $N_D^* = 2$ and $N_C^* = 4$, respectively):

	Annual Equivalent Cost ($)	
n	Defender	Challenger
1	5,130	7,500
2	5,116	6,151
3	5,500	5,857
4	5,961	5,826
5	6,434	5,897

Many ownership options would fulfill an eight-year planning horizon, as shown in Figure 14.9. Of these options, six appear to be the most likely by inspection. These options are listed, and the present equivalent cost for each option is calculated, as follows:

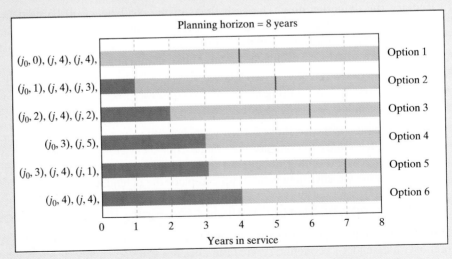

Figure 14.9 Some likely replacement patterns under a finite planning horizon of eight years (Example 14.5).

- **Option 1:** $(j_0, 0), (j, 4), (j, 4)$

$$\text{PW}(15\%)_1 = \$5,826(P/A, 15\%, 8)$$
$$= \$26,143.$$

- **Option 2:** $(j_0, 1), (j, 4), (j, 3)$

$$\text{PW}(15\%)_2 = \$5,130(P/F, 15\%, 1)$$
$$+ \$5,826(P/A, 15\%, 4)(P/F, 15\%, 1)$$
$$+ \$5,857(P/A, 15\%, 3)(P/F, 15\%, 5)$$
$$= \$25,573.$$

- **Option 3:** $(j_0, 2), (j, 4), (j, 2)$

$$\text{PW}(15\%)_3 = \$5,116(P/A, 15\%, 2)$$
$$+ \$5,826(P/A, 15\%, 4)(P/F, 15\%, 2)$$
$$+ \$6,151(P/A, 15\%, 2)(P/F, 15\%, 6)$$
$$= \$25,217 \leftarrow \text{minimum cost.}$$

- **Option 4:** $(j_0, 3), (j, 5)$

$$\text{PW}(15\%)_4 = \$5,500(P/A, 15\%, 3)$$
$$+ \$5,897(P/A, 15\%, 5)(P/F, 15\%, 3)$$
$$= \$25,555.$$

- **Option 5:** $(j_0, 3), (j, 4), (j, 1)$

$$\text{PW}(15\%)_5 = \$5,500(P/A, 15\%, 3)$$
$$+ \$5,826(P/A, 15\%, 4)(P/F, 15\%, 3)$$
$$+ \$7,500(P/F, 15\%, 8)$$
$$= \$25,946.$$

- **Option 6:** $(j_0, 4), (j, 4)$

$$\text{PW}(15\%)_6 = \$5,961(P/A, 15\%, 4)$$
$$+ \$5,826(P/A, 15\%, 4)(P/F, 15\%, 4)$$
$$= \$26,529.$$

An examination of the present equivalent cost of a planning horizon of eight years indicates that the least-cost solution appears to be Option 3: Retain the defender for two years, purchase the challenger and keep it for four years, and purchase another challenger and keep it for two years.

COMMENTS: In this example, we examined only six decision options that were likely to lead to the best solution, but it is important to note that several other possibilities

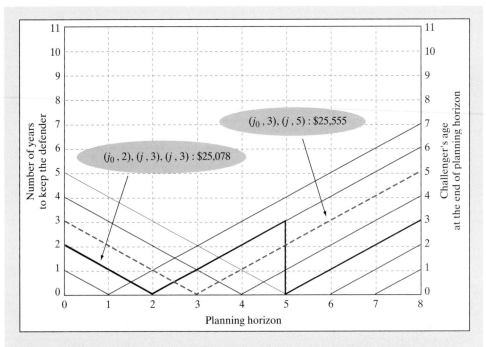

Figure 14.10 Graphical representations of replacement strategies under a finite planning horizon (Example 14.5).

have not been looked at. To explain, consider Figure 14.10, which shows a graphical representation of various replacement strategies under a finite planning horizon. For example, the replacement strategy $[(j_0, 2), (j, 3), (j, 3)]$ (shown as a solid line in the figure) is certainly feasible, but we did not include it in the previous computation. Naturally, as we extend the planning horizon, the number of possible decision options can easily multiply. To make sure that we indeed find the optimal solution for such a problem, an optimization technique such as dynamic programming can be used.[3]

14.3.4 Consideration of Technological Change

Thus far, we have defined the challenger simply as the best available replacement for the defender. It is more realistic to recognize that the replacement decision often involves an asset now in use versus a candidate for replacement—that is, in some way, an improvement on the current asset. This, of course, reflects technological progress that is ongoing continually. Future models of a machine are likely to be more effective than a current model. In most areas, technological change appears as a combination of gradual advances in effectiveness; the occasional technological breakthrough, however, can revolutionize the character of a machine.

The prospect of improved future challengers makes a current challenger a less desirable alternative. By retaining the defender, we may have an opportunity to acquire an improved challenger later. If this is the case, the prospect of improved future challengers may affect a current decision between a defender and its challenger. It is difficult to forecast future technological trends in any precise fashion. However, in developing a long-term replacement policy, we need to take technological change into consideration.

[3] F. S. Hillier and G. S. Lieberman, *Introduction to Operations Research*, 8th ed. (New York: McGraw-Hill, 2005).

14.4 Replacement Analysis with Tax Considerations

Up to this point, we have covered various concepts and techniques that are useful in replacement analysis in general. In this section, we illustrate how to use those concepts and techniques to conduct replacement analysis on an after-tax basis.

To apply the concepts and methods covered in Sections 14.1 through 14.3 in an after-tax comparison of defender and challenger, we have to incorporate the tax effects (gains or losses) whenever an asset is disposed of. Whether the defender is kept or the challenger is purchased, we also need to incorporate the tax effects of depreciation allowances into our analysis.

Replacement studies require a knowledge of the depreciation schedule and of taxable gains or losses at disposal of the asset. Note that the depreciation schedule is determined at the time the asset is acquired, whereas the relevant tax law determines the gains tax effects at the time of disposal. In this section, we will use the same examples (Example 14.1 through 14.4) to illustrate how to do the following analyses on an after-tax basis:

1. Calculate the net proceeds due to disposal of the defender (Example 14.6).
2. Use the opportunity cost approach in comparing defender and challenger (Example 14.7).
3. Calculate the economic life of the defender or the challenger (Example 14.8).
4. Conduct replacement analysis under the infinite planning horizon (Example 14.9).

EXAMPLE 14.6 Net Proceeds from the Disposal of an Old Machine

Suppose that, in Example 14.1, the $20,000 capital expenditure was set up to be depreciated on a seven-year MACRS (allowed annual depreciation: $2,858, $4,898, $3,498, $2,498, $1,786, $1,784, $1,786, and $892). If the firm's marginal income tax rate is 40%, determine the taxable gains (or losses) and the net proceeds from disposal of the old printing machine.

SOLUTION

First we need to find the current book value of the old printing machine. The original cost minus the accumulated depreciation, calculated with the half-year convention (if the machine sold now), is

$$\$20,000 - (\$2,858 + \$4,898/2) = \$14,693,$$

so we compute the following:

Allowed book value	=	$14,693
Current market value	=	$10,000
Losses	=	$4,693
Tax savings = $4,693(0.40)	=	$1,877
Net proceeds from the sale	=	$10,000 + $1,877
	=	$11,877

This calculation is illustrated in Figure 14.11.

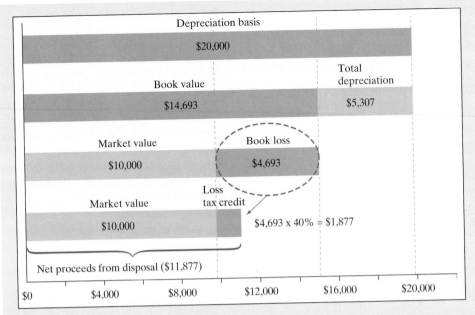

Figure 14.11 Net proceeds from the sale of the old printing machine—defender (Example 14.6).

EXAMPLE 14.7 Replacement Analysis Using the Opportunity Cost Approach

Suppose, in Example 14.2, that the new machine would fall into the same seven-year MACRS class as the old machine. The firm's after-tax interest rate (MARR) is 12%, and the marginal income tax rate is 40%. Use the opportunity cost approach to decide whether replacement of the old machine is justified now.

SOLUTION

Tables 14.1 and 14.2 show the worksheet formats the company uses to analyze a typical replacement project with the generalized cash flow approach. Each line is numbered, and a line-by-line description of the table follows:

- **Option 1: Keep the defender.**

  Lines 1–4: If the old machine is kept, the depreciation schedule would be ($3,498, $2,498, and $1,786). Following the half-year convention, it is assumed that the asset will be retired at the end of three years; thus, the depreciation for year 3 is $(0.5)($1,786) = 893. This results in total depreciation in the amount of $14,645 and a remaining book value of $20,000 - $14,645 = $5,355$.

  Lines 5–6: Repair costs in the amount of $5,000 were already incurred before the replacement decision. This is a sunk cost and should not be considered in the analysis. If a repair in the amount of $5,000 is required to keep the defender in serviceable condition, it will show as an expense in year 0. If the old machine is

TABLE 14.1 Replacement Worksheet: Option 1—Keep the Defender (Example 14.6)

n	−2	−1	0	1	2	3
Financial data						
(cost information):						
(1) Depreciation		$2,858	$ 4,898	$3,498	$2,498	$ 893
(2) Book value	$20,000	$17,142	$12,244	$8,746	$6,248	$5,355
(3) Salvage value						$2,500
(4) Loss from sale						−$2,855
(5) Repair cost		$5,000				
(6) O&M costs				$8,000	$8,000	$8,000
Cash flow statement:						
(7) Opportunity cost			−$11,877			
(8) Net proceeds from sale						$3,642
(9) −O&M cost × (0.6)				−$4,800	−$4,800	−$4,800
(10) +Depreciation × (0.4)				$1,399	$ 999	$ 357
(11) Net cash flow			−$11,877	−$3,401	−$3,801	−$ 801

Note: The highlighted data represent sunk costs.

TABLE 14.2 Replacement Worksheet: Option 2—Replace the Defender (Example 14.6)

n	−2	−1	0	1	2	3
Financial data						
(cost information):						
(1) Cost of new printer			$15,000			
(2) Depreciation				$ 2,144	$3,674	$1,312
(3) Book value				$12,856	$9,182	$7,870
(4) Salvage value						$5,500
(5) Loss from sale						−$2,370
(6) O&M costs				$6,000	$6,000	$6,000
Cash flow statement:						
(7) Investment cost			−$15,000			
(8) Net proceeds from sale						$6,448
(9) −O&M cost × (0.6)				−$3,600	−$3,600	−$3,600
(10) +Depreciation × (0.4)				$858	$1,470	$525
(11) Net cash flow			−$15,000	−$2,742	−$2,130	$3,373

retained for the next three years, the before-tax annual O&M costs are as shown in Line 6.

Lines 7–11: Recall that the depreciation allowances result in a tax reduction equal to the depreciation amount multiplied by the tax rate. The operating expenses are multiplied by the factor of (1 − the tax rate) to obtain the after-tax O&M. For a situation in which the asset is retained for one year, Table 14.1 summarizes the cash flows obtained by using the generalized cash flow approach.

- **Option 2: Replace the defender.**

  Line 1: The purchase price of the new machine, including installation and freight charges, is listed in Table 14.2.

  Line 2: The depreciation schedule, along with the book values for the new machine (seven-year MACRS), is shown. The depreciation amount of $1,312 in year 3 reflects the half-year convention.

  Line 5: With the salvage value estimated at $5,500, we expect a loss ($2,370 = $7,878 − $5,500) on the sale of the new machine at the end of year 3.

  Line 6: The O&M costs for the new machine are listed.

If the decision to keep the defender had been made, the opportunity cost approach would treat the $11,877 current salvage value of the defender as an incurred cost. Figure 14.12 illustrates the cash flows related to these decision options.

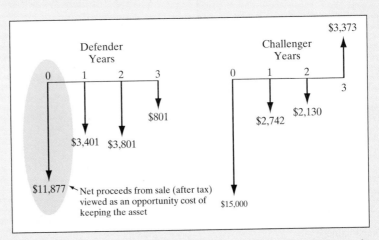

Figure 14.12 Comparison of defender and challenger on the basis of the opportunity cost approach (Example 14.7).

Since the lifetimes of the defender and challenger are the same, we can use either PW or AE analysis as follows:

$$PW(12\%)_{Old} = \$11,877 + \$3,401(P/F, 12\%, 1) + \$3,801(P/F, 12\%, 2)$$
$$+ \$801(P/F, 12\%, 3)$$
$$= \$18,514;$$
$$AEC(12\%)_{Old} = \$18,514(A/P, 12\%, 3)$$
$$= \$7,708;$$

$$\begin{aligned}
\text{PW}(12\%)_{\text{New}} &= \$15,000 + \$2,742(P/F, 12\%, 1) \\
&\quad + \$2,130(P/F, 12\%, 2) - \$3,373(P/F, 12\%, 3) \\
&= \$16,745; \\
\text{AEC}(12\%)_{\text{New}} &= \$16,745(A/P, 12\%, 3) \\
&= \$6,972.
\end{aligned}$$

Because of the annual difference of $736 in favor of the challenger, the replacement should be made now.

COMMENTS: Recall that we assumed the same service life for both the defender and the challenger. In general, however, old equipment has a relatively short remaining life compared with new equipment, so that assumption is too simplistic. When the defender and challenger have unequal lifetimes, we must make an assumption in order to obtain a common analysis period. A typical assumption is that, after the initial decision, we make **perpetual replacements** with assets similar to the challenger. Certainly, we can still use PW analysis with actual cash flows, but that would require evaluating *infinite* cash flow streams.

EXAMPLE 14.8 Economic Service Life of a Lift Truck

Consider again Example 14.3, but with the following additional data: The asset belongs to a five-year MACRS property class with the following annual depreciation allowances: 20%, 32%, 19.20%, 11.52%, 11.52%, and 5.76%. The firm's marginal tax rate is 40%, and its after-tax MARR is 15%. Find the economic service life of this new machine.

DISCUSSION: To determine an asset's economic service life, we first list the gains or losses that will be realized if the truck were to be disposed of at the end of each operating year. In doing so, we need to compute the book values at the end of each operating year, assuming that the asset would be disposed of at that time. Recall that, with the half-year convention, the book value (for the MACRS property) at the end of the year is based on its disposal during the year. As summarized in Table 14.3, these values provide a basis for identifying the relevant after-tax cash flows at the end of an assumed operating period.

SOLUTION

Two approaches may be used to find the economic life of an asset: (a) the generalized cash flow approach and (b) the tabular approach.

(a) Generalized cash flow approach:
  Since we have only a few cash flow elements (O&M, depreciation, and salvage value), an efficient way to obtain the after-tax cash flow is to use the generalized cash flow approach discussed in Section 9.4. Table 14.4 summarizes the cash flows for two-year ownership obtained by using the generalized cash flow approach.
  If we use the expected operating costs and the salvage values from Table 14.3, we can continue to generate yearly after-tax entries for the asset's remaining

TABLE 14.3 Forecasted Operating Costs and Net Proceeds from Sale as a Function of Holding Period (Example 14.8)

| Holding Period | O&M | Permitted Annual Depreciation Amount over the Holding Period | | | | | | | Total Depreciation | Book Value | Expected Market Value | Taxable Gains | Gains Tax | Net A/T Salvage Value |
		1	2	3	4	5	6	7						
1	$1,000	$3,600							$3,600	$14,400	$10,000	$(4,400)	$(1,760)	$11,760
2	1,150	3,600	$2,880						6,480	11,520	7,500	(4,020)	(1,608)	9,108
3	1,323	3,600	5,760	$1,728					11,088	6,912	5,625	(1,287)	(515)	6,140
4	1,521	3,600	5,760	3,456	$1,037				13,853	4,147	4,219	72	29	4,190
5	4,749	3,600	5,760	3,456	2,074	$1,037			15,927	2,073	3,164	1,091	436	2,728
6	2,011	3,600	5,760	3,456	2,074	2,074	$1,036	$0	18,000	0	2,373	2,373	949	1,424
7	6,813	3,600	5,760	3,456	2,074	2,074	1,036	$0	18,000	0	1,780	1,780	712	1,068

Note: Asset price of $18,000, depreciated under MACRS for five-year property with the half-year convention; in year 5, normal operating expense ($1,749) + overhaul ($3,000); in year 7, normal operating expense ($2,313) + another engine overhaul ($4,500).

TABLE 14.4 After-Tax Cash Flow Calculation for Owning and Operating the Asset for Two Years (Example 14.10)

n	0	1	2
Financial data (cost information):			
(1) Cost of new printer	$18,000		
(2) Depreciation		$3,600	$2,880
(3) Book value		$14,400	$11,520
(4) Salvage value			$7,500
(5) Gain (loss) from sale			−$4,020
(6) O&M costs		$1,000	$1,150
Cash flow statement:			
(7) Investment cost	−$18,000		
(8) Net proceeds from sale			$9,108
(9) −O&M cost × (0.6)		−$600	−$690
(10) +Depreciation × (0.4)		$1,440	$1,152
(11) Net cash flow	−$18,000	$840	$9,570

physical life. For the first two operating years, we compute the equivalent annual costs of owning and operating the asset as follows:

- $n = 1$, *one-year replacement cycle:*

$$\text{AEC}(15\%) = \left\{ \$18,000 + \left[\begin{array}{c} (0.6)(\$1,000) - (0.4)(\$3,600) \\ -\$11,760 \end{array} \right] \right.$$
$$\left. (P/F, 15\%, 1) \right\} (A/P, 15\%, 1)$$
$$= \$7,043(1.15)$$
$$= \$8,100.$$

- $n = 2$, *two-year replacement cycle:*

$$\text{AEC}(15\%) = \left\{ \begin{array}{l} \$18,000 + [0.6(\$1,000) - 0.4(\$3,600)(P/F, 15\%, 1)] \\ + [0.6(\$1,150) - 0.4(\$2,880) + \$9,108](P/F, 15\%, 2) \end{array} \right\}$$
$$(A/P, 15\%, 2)$$
$$= \$10,033(0.6151)$$
$$= \$6,171.$$

Similarly, the annual equivalent costs for the subsequent years can be computed as shown in Table 14.5 (column 12). If the truck were to be sold after six years, it would have a minimum annual cost of $4,344 per year, and this is the life that is most favorable for comparison purposes. That is, by replacing the asset perpetually according to an economic life of six years, we obtain the minimum infinite AE cost stream. Figure 14.13 illustrates this concept. Of course, we should envision a long period of required service for the asset, its life no

TABLE 14.5 Tabular Calculation of Economic Service Life (Example 14.8)

(1)	(2)	(3)	(4)	(5)	(6)	(7)	(8)	(9)	(10)	(11)	(12)
Holding Period		Before-Tax Operating Expenses				After-Tax Cash Flow if the Asset Is Kept for N More Years			Equivalent Annual Cost		
N	n	O&M	Depreciation	A/T O&M	Depreciation Credit	Net Operating Cost	Investment and Net Salvage	Net A/T Cash Flow	Capital Cost	Operating Cost	Total Cost
1	0	—	—	—	—	—	$(18,000)	(18,000)	$(8,940)	$ 840	$(8,100)
	1	$1,000	$3,600	$ (600)	$1,440	$ 840	11,760	12,600			
2	0	—		—		—	(18,000)	(18,000)	(6,835)	664	(6,171)
	1	1,000	3,600	(600)	1,440	840		840			
	2	1,150	2,880	(690)	1,152	462	9,108	9,570			
3	0	—		—		—	(18,000)	(18,000)	(6,115)	825	(5,291)
	1	1,000	3,600	(600)	1,440	840		840			
	2	1,150	5,760	(690)	2,304	1,614		1,614			
	3	1,323	1,728	(794)	691	(103)	6,140	6,037			
4	0	—		—		—	(18,000)	(18,000)	(5,466)	719	(4,746)
	1	1,000	3,600	(600)	1,440	840		840			
	2	1,150	5,760	(690)	2,304	1,614		1,614			
	3	1,323	3,456	(794)	1,382	589		589			
	4	1,521	1,037	(913)	415	(498)	4,190	3,692			
5	0	—		—		—	(18,000)	(18,000)	(4,965)	322	(4,643)
	1	1,000	3,600	(600)	1,440	840		840			
	2	1,150	5,760	(690)	2,304	1,614		1,614			
	3	1,323	3,456	(794)	1,382	589		589			
	4	1,521	2,074	(913)	830	(83)		(83)			
	5	4,749	1,037	(2,849)	415	(2,435)	2,728	293			
6	0	—		—		—	(18,000)	(18,000)	(4,594)	249	(4,344)
	1	1,000	3,600	(600)	1,440	840		840			
	2	1,150	5,760	(690)	2,304	1,614		1,614			
	3	1,323	3,456	(794)	1,382	589		589			
	4	1,521	2,074	(913)	830	(83)		(83)			
	5	4,749	2,074	(2,849)	830	(2,020)		(2,020)			
	6	2,011	1,037	(1,207)	415	(792)	1,424	632			
7	0	—		—		—	(18,000)	(18,000)	(4,230)	(143)	(4,372)
	1	1,000	3,600	(600)	1,440	840		840			
	2	1,150	5,760	(690)	2,304	1,614		1,614			
	3	1,323	3,456	(794)	1,382	589		589			
	4	1,521	2,074	(913)	830	(83)		(83)			
	5	4,749	2,074	(2,849)	830	(2,020)		(2,020)			
	6	2,011	1,037	(1,207)	415	(792)		(792)			
	7	6,813	—	(4,088)	—	(4,088)	1,068	(3,020)			

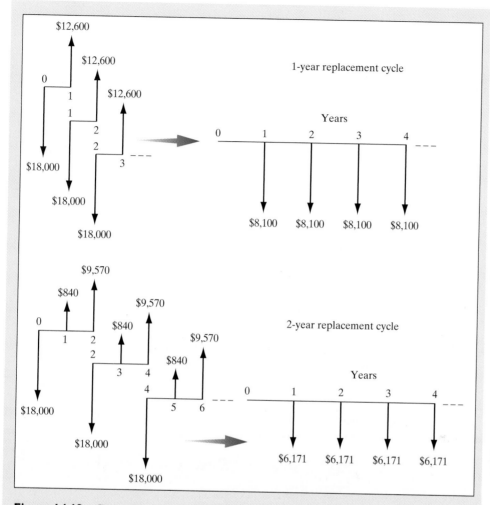

Figure 14.13 Conversion of an infinite number of replacement cycles to infinite AE cost streams (Example 14.8).

doubt being heavily influenced by market values, O&M costs, and depreciation credits.

(b) Tabular approach:

The tabular approach separates the annual cost elements into two parts, one associated with the capital recovery of the asset and the other associated with operating the asset. In computing the capital recovery cost, we need to determine the after-tax salvage values at the end of each holding period, as calculated previously in Table 14.3. Then, we compute the total annual equivalent cost of the asset for any given year's operation using Eq. (14.3).

If we examine the equivalent annual costs itemized in Table 14.5 (columns 10 and 11), we see that, as the asset ages, the equivalent annual O&M cost savings decrease. At the same time, capital recovery costs decrease with prolonged use of the asset. The combination of decreasing capital recovery costs and increasing annual O&M costs

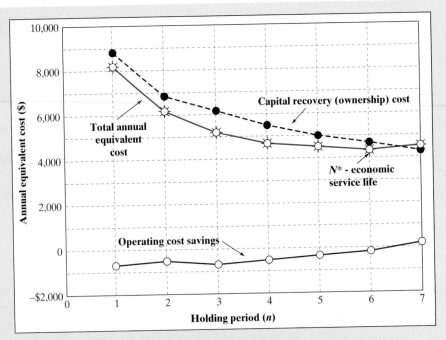

Figure 14.14 Economic service life obtained by finding the minimum AEC (Example 14.8). Note that we treat the cost items with a positive sign.

results in the total annual equivalent cost taking on a form similar to that depicted in Figure 14.14. Even though an expensive overhaul is required during the fifth year of service, it is more economical to keep the equipment over a six-year life.

EXAMPLE 14.9 Replacement Analysis under the Infinite Planning Horizon

Recall Example 14.4, in which Advanced Electrical Insulator Company is considering replacing a broken inspection machine. Let's assume the following additional data:

- The old machine has been fully depreciated, so it has zero book value. The machine could be used for another five years, but the firm does not expect to realize any salvage value from scrapping it in five years.
- The new machine falls into the five-year MACRS property class and will be depreciated accordingly.

The marginal income tax rate is 40%, and the after-tax MARR is 15%. Find the useful life for each option presented in Example 14.4, *and* decide whether the defender should be replaced now or later.

SOLUTION

1. Economic service life:

- **Defender.** The defender is fully depreciated, so that all salvage values can be treated as ordinary gains and taxed at 40%. The after-tax salvage values are thus as follows:

n	Current Market Value	After-Tax Salvage Value
0	$5,000	$5,000(1 − 0.40) = $3,000
1	4,000	4,000(1 − 0.40) = 2,400
2	3,000	3,000(1 − 0.40) = 1,800
3	2,000	2,000(1 − 0.40) = 1,200
4	1,000	1,000(1 − 0.40) = 600
5	0	0
6	0	0
⋮	⋮	⋮

If the company retains the inspection machine, it is in effect deciding to overhaul the machine and invest the machine's current market value (after taxes) in that alternative. Although the company will make no physical cash flow transaction, it is withholding the market value of the inspection machine (the opportunity cost) from the investment. The after-tax O&M costs are as follows:

n	Overhaul	Forecasted O&M Cost	After-Tax O&M Cost
0	$1,200		$1,200(1 − 0.40) = $720
1	0	$2,000	2,000(1 − 0.40) = 1,200
2	0	3,500	3,500(1 − 0.40) = 2,100
3	0	5,000	5,000(1 − 0.40) = 3,000
4	0	6,500	6,500(1 − 0.40) = 3,900
5	0	8,000	8,000(1 − 0.40) = 4,800

Using the current year's market value as the investment required to retain the defender, we obtain the data in Table 14.6, indicating that the remaining useful life of the defender is two years, *in the absence of future challengers*. The overhaul (repair) cost of $1,200 in year 0 can be treated as a deductible operating expense for tax purposes, as long as it does not add value to the property. (Any repair or improvement expenses that increase the value of the property must be capitalized by depreciating them over the estimated service life.)

- **Challenger.** Because the challenger will be depreciated over its tax life, we must determine the book value of the asset at the end of each period to compute the after-tax salvage value. This is done in Table 14.7. With the after-tax salvage values computed in that table, we are now ready to find the economic service life of the challenger by generating AEC value entries. These calculations are summarized in Table 14.8. The economic life of the challenger is four years, with an AEC(15%) value of $4,065.

2. **Optimal time to replace the defender:** Since the AEC value for the defender's remaining useful life (two years) is $3,070, which is less than $4,065, the decision will be to keep the defender for now. Of course, the defender's remaining

TABLE 14.6 Economics of Retaining the Defender for *N* More Years (Example 14.9)

(1)	(2)	(3)	(4)	(5)	(6)	(7)	(8)	(9)	(10)	(11)	(12)
Holding Period		Before-Tax Operating Expenses			After-Tax Cash Flow if the Asset Is Kept for *N* More Years				Equivalent Annual Cost		
N	*n*	O&M	Depreciation	A/T O&M	Depreciation Credit	Net Operating Cost	Investment and Net Salvage	Net A/T Cash Flow	Capital Cost	Operating Cost	Total Cost
0	0	$1,200		$ (720)		$ (720)	$(3,000)	$(3,720)			
1	1	2,000		(1,200)		(1,200)	2,400	1,200	$(1,050)	$(2,028)	$(3,078)
2	0	1,200		(720)		(720)	(3,000)	(3,720)			
	1	2,000		(1,200)		(1,200)		(1,200)			
	2	3,500		(2,100)		(2,100)	1,800	(300)	(1,008)	(2,061)	(3,070)
3	0	1,200		(720)		(720)	(3,000)	(3,720)			
	1	2,000		(1,200)		(1,200)		(1,200)			
	2	3,500		(2,100)		(2,100)		(2,100)			
	3	5,000		(3,000)		(3,000)	1,200	(1,800)	(968)	(2,332)	(3,300)
4	0	1,200		(720)		(720)	(3,000)	(3,720)			
	1	2,000		(1,200)		(1,200)		(1,200)			
	2	3,500		(2,100)		(2,100)		(2,100)			
	3	5,000		(3,000)		(3,000)		(3,000)			
	4	6,500		(3,900)		(3,900)	600	(3,300)	(931)	(2,646)	(3,576)
5	0	1,200		(720)		(720)	(3,000)	(3,720)			
	1	2,000		(1,200)		(1,200)		(1,200)			
	2	3,500		(2,100)		(2,100)		(2,100)			
	3	5,000		(3,000)		(3,000)		(3,000)			
	4	6,500		(3,900)		(3,900)		(3,900)			
	5	8,000		(4,800)		(4,800)	—	(4,800)	(895)	(2,965)	(3,860)

TABLE 14.7 Forecasted Operating Costs and Net Proceeds from Sale as a Function of Holding Period—Challenger (Example 14.9)

Holding Period	O&M	Permitted Annual Depreciation Amount over the Holding Period							Expected Total Depreciation	Book Value	Market Value	Net A/T Taxable Gains	Gains Tax	Salvage Value
		1	2	3	4	5	6	7						
1	$2,000	$2,000							$2,000	$8,000	$6,000	$(2,000)	$ (800)	$6,800
2	3,000	2,000	$1,600						3,600	6,400	5,100	(1,300)	(520)	5,620
3	4,000	2,000	3,200	$ 960					6,160	3,840	4,335	495	198	4,137
4	5,000	2,000	3,200	1,920	$ 576				7,696	2,304	3,685	1,381	552	3,133
5	6,000	2,000	3,200	1,920	1,152	$ 576			8,848	1,152	3,132	1,980	792	2,340
6	7,000	2,000	3,200	1,920	1,152	1,152	$ 576		10,000	0	2,662	2,662	1,065	1,597
7	8,000	2,000	3,200	1,920	1,152	1,152	576	$ 0	10,000	0	2,263	2,263	905	1,358

Note: Asset price of $10,000, depreciated under MACRS for five-year property with the half-year convention.

TABLE 14.8 Economics of Owning and Operating the Challenger for N More Years (Example 14.9)

(1)	(2)	(3)	(4)	(5)	(6)	(7)	(8)	(9)	(10)	(11)	(12)
Holding Period		Before-Tax Operating Expenses				After-Tax Cash Flow if the Asset Is Kept for N More Years			Equivalent Annual Cost		
N	n	O&M	Depreciation	A/T O&M	Depreciation Credit	Net Operating Cost	Investment and Net Salvage	Net A/T Cash Flow	Capital Cost	Operating Cost	Total
1	0						$(10,000)	$(10,000)			
	1	$2,000	$2,000	$(1,200)	$ 800	$(400)	6,800	6,400	$(4,700)	$(400)	$(5,100)
2	0						(10,000)	(10,000)			
	1	2,000	2,000	(1,200)	800	(400)	(400)				
	2	3,000	1,600	(1,800)	640	(1,160)	5,620	4,460	(3,536)	(753)	(4,290)
3	0						(10,000)	(10,000)			
	1	2,000	2,000	(1,200)	800	(400)		(400)			
	2	3,000	3,200	(1,800)	1,280	(520)		(520)			
	3	4,000	960	(2,400)	384	(2,016)	4,137	2,121	(3,188)	(905)	(4,094)
4	0						(10,000)	(10,000)			
	1	2,000	2,000	(1,200)	800	(400)		(400)			
	2	3,000	3,200	(1,800)	1,280	(520)		(520)			
	3	4,000	1,920	(2,400)	768	(1,632)		(1,632)			
	4	5,000	576	(3,000)	230	(2,770)	3,133	363	(2,875)	(1,190)	(4,065)
5	0						(10,000)	(10,000)			
	1	2,000	2,000	(1,200)	800	(400)		(400)			
	2	3,000	3,200	(1,800)	1,280	(520)		(520)			
	3	4,000	1,920	(2,400)	768	(1,632)		(1,632)			
	4	5,000	1,152	(3,000)	461	(2,539)		(2,539)			
	5	6,000	576	(3,600)	230	(3,370)	2,340	1,030	(2,636)	(1,474)	(4,110)

useful life of two years does not imply that the defender should actually be kept for two years before the company switches to the challenger. The reason for this is that the defender's remaining useful life of two years was calculated without considering what type of challenger would be available in the future. When a challenger's financial data are available, we need to enumerate all replacement timing possibilities. Since the defender can be used for another five years, six replacement strategies exist:

- Replace the defender with the challenger now.
- Replace the defender with the challenger in year 1.
- Replace the defender with the challenger in year 2.
- Replace the defender with the challenger in year 3.
- Replace the defender with the challenger in year 4.
- Replace the defender with the challenger in year 5.

The possible replacement cash patterns associated with each of these alternatives are shown in Figure 14.15, assuming that the costs and efficiency of the current challenger remain unchanged in future years. From the figure, we observe that, on an annual basis, the cash flows after the remaining physical life of the defender are the same.

Before we evaluate the economics of various replacement-decision options, recall the AEC values for the defender and the challenger under the assumed service lives (a boxed figure denotes the minimum AEC value at $n_0 = 2$ and $n^* = 4$):

	Annual Equivalent Cost	
n	**Defender**	**Challenger**
1	$3,078	$5,100
2	$\boxed{3,070}$	4,290
3	3,300	4,094
4	3,576	$\boxed{4,065}$
5	3,860	4,110
6		4,189
7		4,287

Instead of using the marginal analysis in Example 14.4, we will use PW analysis, which requires an evaluation of infinite cash flow streams. (The result is the same under both analyses.) Immediate replacement of the defender by the challenger is equivalent to computing the PW for an infinite cash flow stream of $-\$4,065$. If we use the capitalized equivalent-worth approach of Chapter 6 ($CE(i) = A/i$), we obtain

- $n = 0$:

$$PW(15\%)_{n_0=0} = (1/0.15)(\$4,065)$$
$$= \$27,100.$$

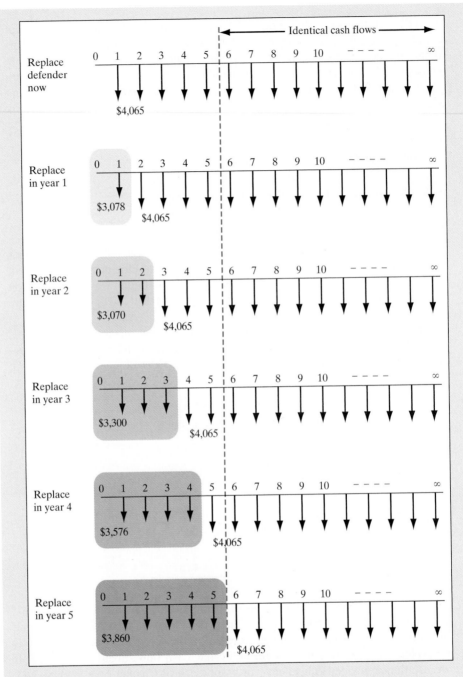

Figure 14.15 Equivalent annual cash flow streams when the defender is kept for *n* years followed by infinitely repeated purchases of the challenger every four years (Example 14.9).

Suppose we retain the old machine n more years and then replace it with the new one. Now we will compute $PW(i)_{n_0=n}$:

- $n = 1$:
$$PW(15\%)_{n_0=1} = \$3,078(P/A, 15\%, 1) + \$27,100(P/F, 15\%, 1)$$
$$= \$26,242.$$

- $n = 2$:
$$PW(15\%)_{n_0=2} = \$3,070(P/A, 15\%, 2) + \$27,100(P/F, 15\%, 2)$$
$$= \$25,482.$$

- $n = 3$:
$$PW(15\%)_{n_0=3} = \$3,300(P/A, 15\%, 3) + \$27,100(P/F, 15\%, 3)$$
$$= \boxed{\$25,353.}$$

- $n = 4$:
$$PW(15\%)_{n_0=4} = \$3,576(P/A, 15\%, 4) + \$27,100(P/F, 15\%, 4)$$
$$= \$25,704.$$

- $n = 5$:
$$PW(15\%)_{n_0=5} = \$3,860(P/A, 15\%, 5) + \$27,100(P/F, 15\%, 5)$$
$$= \$26,413.$$

This leads us to conclude that the defender should be kept for three more years. The present worth of $25,353 represents the net cost associated with retaining the defender for three years, replacing it with the challenger, and then replacing the challenger every four years for an indefinite period.

SUMMARY

- In replacement analysis, the **defender** is an existing asset; the **challenger** is the best available replacement candidate.

- The **current market value** is the value to use in preparing a defender's economic analysis. **Sunk costs**—past costs that cannot be changed by any future investment decision—should not be considered in a defender's economic analysis.

- The basic approach to analyzing replacement problems is the **opportunity cost approach**. The opportunity cost approach views the net proceeds from the sale of the defender as an opportunity cost of keeping the defender. That is, instead of deducting the salvage value from the purchase cost of the challenger, we consider the net proceeds as an investment required to keep the asset.

- The **economic service life** is the remaining useful life of a defender *or* a challenger that results in the minimum equivalent annual cost or maximum annual equivalent revenue. We should use the respective economic service lives of the defender and the challenger when conducting a replacement analysis.

■ Ultimately, in replacement analysis, the question is not *whether* to replace the defender, but *when* to do so. The AE method provides a marginal basis on which to make a year-by-year decision about the best time to replace the defender. As a general decision criterion, the PW method provides a more direct solution to a variety of replacement problems, with either an infinite or a finite planning horizon, or a technological change in a future challenger.

■ The role of **technological change** in improving assets should be evaluated in making long-term replacement plans: If a particular item is undergoing rapid, substantial technological improvements, it may be prudent to delay replacement (to the extent where the loss in production does not exceed any savings from improvements in future challengers) until a desired future model is available.

PROBLEMS

Sunk Costs, Opportunity Costs, and Cash Flows

14.1   Columbus Electronics Company is considering replacing a 1,000-pound-capacity forklift truck that was purchased three years ago at a cost of $15,000. The diesel-operated forklift was originally expected to have a useful life of eight years and a zero estimated salvage value at the end of that period. The truck has not been dependable and is frequently out of service while awaiting repairs. The maintenance expenses of the truck have been rising steadily and currently amount to about $3,000 per year. The truck could be sold for $6,000. If retained, the truck will require an immediate $1,500 overhaul to keep it in operating condition. This overhaul will neither extend the originally estimated service life nor increase the value of the truck. The updated annual operating costs, engine overhaul cost, and market values over the next five years are estimated as follows:

n	O&M	Depreciation	Engine Overhaul	Market Value
−3				
−2		$3,000		
−1		4,800		
0		2,880	$1,500	$6,000
1	$3,000	1,728		4,000
2	3,500	1,728		3,000
3	3,800	864		1,500
4	4,500	0		1,000
5	4,800	0	5,000	0

A drastic increase in O&M costs during the fifth year is expected due to another overhaul, which will again be required to keep the truck in operating condition. The firm's MARR is 15%.

(a) If the truck is to be sold now, what will be its sunk cost?

(b) What is the opportunity cost of not replacing the truck now?

(c) What is the equivalent annual cost of owning and operating the truck for two more years?

(d) What is the equivalent annual cost of owning and operating the truck for five years?

14.2 Komatsu Cutting Technologies is considering replacing one of its CNC machines with one that is newer and more efficient. The firm purchased the CNC machine 10 years ago at a cost of $135,000. The machine had an expected economic life of 12 years at the time of purchase and an expected salvage value of $12,000 at the end of the 12 years. The original salvage estimate is still good, and the machine has a remaining useful life of 2 years. The firm can sell this old machine now to another firm in the industry for $30,000. The new machine can be purchased for $165,000, including installation costs. It has an estimated useful (economic) life of 8 years. The new machine is expected to reduce cash operating expenses by $30,000 per year over its 8-year life, at the end of which the machine is estimated to be worth only $5,000. The company has a MARR of 12%.

(a) If you decided to retain the old machine, what is the opportunity (investment) cost of retaining the old asset?

(b) Compute the cash flows associated with retaining the old machine in years 1 to 2.

(c) Compute the cash flows associated with purchasing the new machine in years 1 to 8. (Use the opportunity cost concept.)

(d) If the firm needs the service of these machines for an indefinite period and no technology improvement is expected in future machines, what will be your decision?

14.3 Air Links, a commuter airline company, is considering replacing one of its baggage-handling machines with a newer and more efficient one. The current book value of the old machine is $50,000, and it has a remaining useful life of five years. The salvage value expected from scrapping the old machine at the end of five years is zero, but the company can sell the machine now to another firm in the industry for $10,000. The new baggage-handling machine has a purchase price of $120,000 and an estimated useful life of seven years. It has an estimated salvage value of $30,000 and is expected to realize economic savings on electric power usage, labor, and repair costs and also to reduce the amount of damaged luggage. In total, an annual savings of $50,000 will be realized if the new machine is installed. The firm uses a 15% MARR. Using the opportunity cost approach,

(a) What is the initial cash outlay required for the new machine?

(b) What are the cash flows for the defender in years 0 to 5?

(c) Should the airline purchase the new machine?

14.4 Duluth Medico purchased a digital image-processing machine three years ago at a cost of $50,000. The machine had an expected life of eight years at the time of purchase and an expected salvage value of $5,000 at the end of the eight years. The old machine has been slow at handling the increased business volume, so management is considering replacing the machine. A new machine can be purchased for $75,000, including installation costs. Over its five-year life, the machine will reduce cash operating expenses by $30,000 per year. Sales are not expected to change. At the end of its useful life, the machine is estimated to be worthless. The old machine can be sold today for $10,000. The firm's interest rate for project justification is known to be 15%. The firm does not expect a better machine (other than the current challenger) to be available for the next five years. Assuming that the economic service life of the new machine, as well as the remaining useful life of the old machine, is five years,

(a) Determine the cash flows associated with each option (keeping the defender versus purchasing the challenger).

(b) Should the company replace the defender now?

14.5 The Northwest Manufacturing Company is currently manufacturing one of its products on a hydraulic stamping press machine. The unit cost of the product is $12, and 3,000 units were produced and sold for $19 each during the past year. It is expected that both the future demand of the product and the unit price will remain steady at 3,000 units per year and $19 per unit. The old machine has a re-maining useful life of three years. The old machine could be sold on the open market now for $5,500. Three years from now, the old machine is expected to have a salvage value of $1,200. The new machine would cost $36,500, and the unit manufacturing cost on the new machine is projected to be $11. The new machine has an expected economic life of five years and an expected salvage value of $6,300. The appropriate MARR is 12%. The firm does not expect a sig-nificant improvement in technology, and it needs the service of either machine for an indefinite period.

(a) Compute the cash flows over the remaining useful life of the old machine if the firm decides to retain it.

(b) Compute the cash flows over the economic service life if the firm decides to purchase the machine.

(c) Should the machine be acquired now?

Economic Service Life

14.6 A firm is considering replacing a machine that has been used to make a certain kind of packaging material. The new, improved machine will cost $31,000 in-stalled and will have an estimated economic life of 10 years, with a salvage value of $2,500. Operating costs are expected to be $1,000 per year throughout the service life of the machine. The old machine (still in use) had an original cost of $25,000 four years ago, and at the time it was purchased, its service life (physical life) was estimated to be seven years, with a salvage value of $5,000. The old machine has a current market value of $7,700. If the firm retains the old machine, its updated market values and operating costs for the next four years will be as follows:

Year-End	Market Value	Book Value	Operating Costs
0	$7,700	$7,889	
1	4,300	5,578	$3,200
2	3,300	3,347	3,700
3	1,100	1,116	4,800
4	0	0	5,850

The firm's MARR is 12%.

(a) Working with the updated estimates of market values and operating costs over the next four years, determine the remaining useful life of the old machine.

(b) Determine whether it is economical to make the replacement now.

(c) If the firm's decision is to replace the old machine, when should it do so?

14.7 The University Resume Service has just invested $8,000 in a new desktop publishing system. From past experience, the owner of the company estimates its after-tax cash returns as

$$A_n = \$8,000 - \$4,000(1 + 0.15)^{n-1},$$

$$S_n = \$6,000(1 - 0.3)^n,$$

where A_n stands for the net after-tax cash flows from operation of the system during period n and S_n stands for the after-tax salvage value at the end of period n.

(a) If the company's MARR is 12%, compute the economic service life of the system.

(b) Explain how the economic service life varies with the interest rate.

14.8 A special-purpose machine is to be purchased at a cost of $15,000. The following table shows the expected annual operating and maintenance cost and the salvage values for each year of the machine's service:

Year of Service	O&M Costs	Market Value
1	$2,500	$12,800
2	3,200	8,100
3	5,300	5,200
4	6,500	3,500
5	7,800	0

(a) If the interest rate is 10%, what is the economic service life for this machine?

(b) Repeat (a), using $i = 15\%$.

Replacement Decisions with an Infinite Planning Horizon and No Technological Change

14.9 A special-purpose turnkey stamping machine was purchased four years ago for $20,000. It was estimated at that time that this machine would have a life of 10 years and a salvage value of $3,000, with a removal cost of $1,500. These estimates are still good. The machine has annual operating costs of $2,000. A new machine that is more efficient will reduce the operating costs to $1,000, but it will require an investment of $20,000, plus $1,000 for installation. The life of the new machine is estimated to be 12 years, with a salvage of $2,000 and a removal cost of $1,500. An offer of $6,000 has been made for the old machine, and the purchaser is willing to pay for removal of the machine. Find the economic advantage of replacing or of continuing with the present machine. State any assumptions that you make. (Assume that MARR = 8%).

14.10 A five-year-old defender has a current market value of $4,000 and expected O&M costs of $3,000 this year, increasing by $1,500 per year. Future market values are expected to decline by $1,000 per year. The machine can be used for another three years. The challenger costs $6,000 and has O&M costs of $2,000 per year, increasing by $1,000 per year. The machine will be needed for only three years, and the salvage value at the end of that time is expected to be $2,000. The MARR is 15%.

(a) Determine the annual cash flows for retaining the old machine for three years.

(b) Determine whether now is the time to replace the old machine. First show the annual cash flows for the challenger.

14.11 Greenleaf Company is considering purchasing a new set of air-electric quill units to replace an obsolete one. The machine currently being used for the operation has a market value of zero: however, it is in good working order, and it will last for at least an additional five years. The new quill units will perform the operation with so much more efficiency that the firm's engineers estimate that labor, material, and other direct costs will be reduced $3,000 a year if the units are installed. The new set of quill units costs $10,000, delivered and installed, and its economic life is estimated to be five years with zero salvage value. The firm's MARR is 10%.

(a) What investment is required to keep the old machine?

(b) Compute the cash flow to use in the analysis for each option.

(c) If the firm uses the internal-rate-of-return criterion, should the firm buy the new machine on that basis?

14.12 Wu Lighting Company is considering replacing an old, relatively inefficient vertical drill machine that was purchased 7 years ago at a cost of $10,000. The machine had an original expected life of 12 years and a zero estimated salvage value at the end of that period. The divisional manager reports that a new machine can be bought and installed for $12,000. Further, over its 5-year life, the machine will expand sales from $10,000 to $11,500 a year and will reduce the usage of labor and raw materials sufficiently to cut annual operating costs from $7,000 to $5,000. The new machine has an estimated salvage value of $2,000 at the end of its 5-year life. The old machine's current market value is $1,000; the firm's MARR is 15%.

(a) Should the new machine be purchased now?

(b) What current market value of the old machine would make the two options equal?

14.13 Advanced Robotics Company is faced with the prospect of replacing its old call-switching system, which has been used in the company's headquarters for 10 years. This particular system was installed at a cost of $100,000, and it was assumed that it would have a 15-year life with no appreciable salvage value. The current annual operating costs for the old system are $20,000, and these costs would be the same for the rest of its life. A sales representative from North Central Bell is trying to sell the company a computerized switching system that would require an investment of $200,000 for installation. The economic life of this computerized system is estimated to be 10 years, with a salvage value of $18,000, and the system will reduce annual operating costs to $5,000. No detailed agreement has been made with the sales representative about the disposal of the old system. Determine the ranges of resale value associated with the old system that would justify installation of the new system at a MARR of 14%.

14.14 A company is currently producing chemical compounds by a process that was installed 10 years ago at a cost of $100,000. It was assumed that the process would have a 20-year life with a zero salvage value. The current market value of the process however, is $60,000, and the initial estimate of its economic life is still good. The annual operating costs associated with the process are $18,000. A sales representative from U.S. Instrument Company is trying to sell a new chemical-compound-making process to the company. This new process will cost $200,000, have a service life of 10 years and a salvage value of $20,000, and reduce annual operating costs to $4,000. Assuming that the company desires a return of 12% on all investments, should it invest in the new process?

14.15 Eight years ago, a lathe was purchased for $45,000. Its operating expenses were $8,700 per year. An equipment vendor offers a new machine for $53,500 whose operating costs are $5,700 per year. An allowance of $8,500 would be made for the old machine when the new one is purchased. The old machine is expected to be scrapped at the end of five years. The new machine's economic service life is five years with a salvage value of $12,000. The new machine's O&M cost is estimated to be $4,200 for the first year, increasing at an annual rate of $500 thereafter. The firm's MARR is 12%. What option would you recommend?

14.16 The New York Taxi Cab Company has just purchased a new fleet of models for the year 2000. Each brand-new cab cost $20,000. From past experience, the company estimates after-tax cash returns for each cab as

$$A_n = \$65,800 - 30,250(1 + 0.15)^{n-1},$$
$$S_n = \$20,000(1 - 0.15)^n,$$

where, again, A_n stands for net after-tax cash flows from the cab's operation during period n and S_n stands for the after-tax salvage value of the cab at the end of period n. The management views the replacement process as a constant and infinite chain.
(a) If the firm's MARR is 10% and it expects no major technological and functional change in future models, what is the optimal period (constant replacement cycle) to replace its cabs? (Ignore inflation.)
(b) What is the internal rate of return for a cab if it is retired at the end of its economic service life? What is the internal rate of return for a sequence of identical cabs if each cab in the sequence is replaced at the optimal time?

14.17 Four years ago, an industrial batch oven was purchased for $23,000. It has been depreciated over a 10-year life and has a $1,000 salvage value. If sold now, the machine will bring $2,000. If sold at the end of the year, it will bring $1,500. Annual operating costs for subsequent years are $3,800. A new machine will cost $50,000 with a 12-year life and have a $3,000 salvage value. The operating cost will be $3,000 as of the end of each year, with the $6,000-per-year savings due to better quality control. If the firm's MARR is 10%, should the machine be purchased now?

14.18 Georgia Ceramic Company has an automatic glaze sprayer that has been used for the past 10 years. The sprayer can be used for another 10 years and will have a zero salvage value at that time. The annual operating and maintenance costs for the sprayer amount to $15,000 per year. Due to an increase in business, a new sprayer must be purchased. Georgia Ceramic is faced with two options:
- **Option 1.** If the old sprayer is retained, a new smaller capacity sprayer will be purchased at a cost of $48,000, and it will have a $5,000 salvage value in 10 years. This new sprayer will have annual operating and maintenance costs of $12,000. The old sprayer has a current market value of $6,000.

- **Option 2.** If the old sprayer is sold, a new sprayer of larger capacity will be purchased for $84,000. This sprayer will have a $9,000 salvage value in 10 years and will have annual operating and maintenance costs of $24,000.

  Which option should be selected at MARR = 12%?

Replacement Problem with a Finite Planning Horizon

14.19 The annual equivalent after-tax costs of retaining a defender machine over 4 years (physical life) or operating its challenger over 6 years (physical life) are as follows:

n	Defender	Challenger
1	$3,200	$5,800
2	2,500	4,230
3	2,650	3,200
4	3,300	3,500
5		4,000
6		5,500

If you need the service of either machine for only the next 10 years, what is the best replacement strategy? Assume a MARR of 12% and no improvements in technology in future challengers.

14.20 The after-tax annual equivalent worth of retaining a defender over four years (physical life) or operating its challenger over six years (physical life) are as follows:

n	Defender	Challenger
1	$13,400	$12,300
2	13,500	13,000
3	13,800	13,600
4	13,200	13,400
5		13,000
6		12,500

If you need the service of either machine for only the next eight years, what is the best replacement strategy? Assume a MARR of 12% and no improvements in technology in future challengers.

14.21 An existing asset that cost $16,000 two years ago has a market value of $12,000 today, an expected salvage value of $2,000 at the end of its remaining useful life of six more years, and annual operating costs of $4,000. A new asset under consideration as a replacement has an initial cost of $10,000, an expected salvage value of $4,000 at the end of its economic life of three years, and annual operating costs of $2,000. It is assumed that this new asset could be replaced by another one identical in every respect after three years at a salvage value of $4,000, if desired. Use a MARR of 11%, a six-year study period, and PW calculations to decide whether the existing asset should be replaced by the new one.

14.22 Repeat Problem 14.21, using the AE criterion.

Replacement Analysis with Tax Considerations

14.23 Redo Problem 14.1, but with the following additional information: The asset is classified as a five-year MACRS property and has a book value of $5,760 if disposed of now. The firm's marginal tax rate is 40% and its after-tax MARR is 15%.

14.24 Redo Problem 14.2, but with the following additional information: The asset is classified as a seven-year MACRS. The firm's marginal tax rate is 40%, and its after-tax MARR is 12%.

14.25 Redo Problem 14.3, but with the following additional information:

- The current book value of the old machine is $50,000. The old machine is being depreciated toward a zero salvage value by means of conventional straight-line methods, or by $10,000 per year.
- The new machine will be depreciated under a seven-year MACRS class.
- The company's marginal tax rate is 40%, and the firm uses a 15% after-tax MARR.

14.26 Redo Problem 14.4, but with the following additional information:

- The old machine has been depreciated under a five-year MACRS property class.
- The new machine will be depreciated under a five-year MACRS class.
- The marginal tax rate is 35%, and the firm's after-tax MARR is 15%.

14.27 Redo Problem 14.5, but with the following additional information:

- The old stamping machine has been fully depreciated.
- For tax purposes, the entire cost of $36,500 can be depreciated according to a five-year MACRS property class.
- The firm's marginal tax rate is 40%, and the after-tax MARR is 12%.

14.28 Redo Problem 14.6, but with the following additional information:

- The current book value of the old machine is $7,889. The anticipated book value for the next four years are as follows: Year 1: $5,578; Year 2: $3,347; Year 3: $1,116; and Year 4: $0. The new machine will be depreciated under a seven-year MACRS class.
- The company's marginal tax rate is 35%, and the firm uses a 12% after-tax MARR.

14.29 A machine has a first cost of $10,000. End-of-year book values, salvage values, and annual O&M costs are provided over its useful life as follows:

Year End	Book Value	Salvage Value	Operating Costs
1	$8,000	$5,300	$1,500
2	4,800	3,900	2,100
3	2,880	2,800	2,700
4	1,728	1,800	3,400
5	1,728	1,400	4,200
6	576	600	4,900

(a) Determine the economic life of the machine if the MARR is 15% and the marginal tax rate is 40%.

(b) Determine the economic life of the machine if the MARR is 10% and the marginal tax rate remains at 40%.

14.30 Given the data

$$I = \$20,000,$$

$$S_n = 12,000 - 2,000n,$$

$$B_n = 20,000 - 2,500n,$$

$$O\&M_n = 3,000 + 1,000(n - 1), \text{ and}$$

$$t_m = 0.40,$$

where I = Asset purchase price,

S_n = Market value at the end of year n,

B_n = Book value at the end of year n,

$O\&M_n$ = O&M cost during year n, and

t_m = Marginal tax rate,

(a) Determine the economic service life of the asset if $i = 10\%$.

(b) Determine the economic service life of the asset if $i = 25\%$.

(c) Assume that $i = 0$ and determine the economic service life of the asset mathematically (i.e., use the calculus technique for finding the minimum point, as described in Chapter 8).

14.31 Redo Problem 14.8, but with the following additional information:

- For tax purposes, the entire cost of $15,000 can be depreciated according to a five-year MACRS property class.

- The firm's marginal tax rate is 40%.

14.32 Quintana Electronic Company is considering purchasing new robot-welding equipment to perform operations currently being performed by less efficient equipment. The new machine's purchase price is $150,000, delivered and installed. A Quintana industrial engineer estimates that the new equipment will produce savings of $30,000 in labor and other direct costs annually, compared with the current equipment. He estimates the proposed equipment's economic life at 10 years, with a zero salvage value. The current equipment is in good working order and will last, physically, for at least 10 more years. Quintana Company expects to pay income taxes of 40%, and any gains will also be taxed at 40%. Quintana uses a 10% discount rate for analysis performed on an after-tax basis. Depreciation of the new equipment for tax purposes is computed on the basis of a seven-year MACRS property class.

(a) Assuming that the current equipment has zero book value and zero salvage value, should the company buy the proposed equipment?

(b) Assuming that the current equipment is being depreciated at a straight-line rate of 10%, has a book value of $72,000 (cost, $120,000; accumulated depreciation, $48,000), and zero net salvage value today, should the company buy the proposed equipment?

(c) Assuming that the current equipment has a book value of $72,000 and a salvage value today of $45,000 and that, if the current equipment is retained for 10 more years, its salvage value will be zero, should the company buy the proposed equipment?

(d) Assume that the new equipment will save only $15,000 a year, but that its economic life is expected to be 12 years. If other conditions are as described in part (a), should the company buy the proposed equipment?

14.33 Quintana Company decided to purchase the equipment described in Problem 14.32 (hereafter called "Model A" equipment). Two years later, even better equipment (called "Model B") came onto the market, making Model A obsolete, with no resale value. The Model B equipment costs $300,000 delivered and installed, but it is expected to result in annual savings of $75,000 over the cost of operating the Model A equipment. The economic life of Model B is estimated to be 10 years, with a zero salvage value. (Model B also is classified as a seven-year MACRS property.)

(a) What action should the company take?

(b) If the company decides to purchase the Model B equipment, a mistake must have been made, because good equipment, bought only two years previously, is being scrapped. How did this mistake come about?

14.34 Redo Problem 14.9, but with the following additional information:
- The current book value of the old machine is $6,248, and the old asset has been depreciated as a seven-year MACRS property.
- The new asset is also classified as a seven-year MACRS property.
- The company's marginal tax rate is 30%, and the firm uses an 8% after-tax MARR.

14.35 Redo Problem 14.10, but with the following additional information:
- The old machine has been fully depreciated.
- The new machine will be depreciated under a three-year MACRS class.
- The marginal tax rate is 40%, and the firm's after-tax MARR is 15%.

14.36 Redo Problem 14.11, but with the following additional information:
- The current book value of the old machine is $4,000, and the annual depreciation charge is $800 if the firm decides to keep the old machine for the additional five years.
- The new asset is classified as a seven-year MACRS property.
- The company's marginal tax rate is 40%, and the firm uses a 10% after-tax MARR.

14.37 Redo Problem 14.12, but with the following additional information:
- The old machine has been fully depreciated.
- The new machine will be depreciated under a seven-year MACRS class.
- The marginal tax rate is 40%, and the firm's after-tax MARR is 15%.

14.38 Redo Problem 14.13, with the following additional information:
- The old switching system has been fully depreciated.
- The new system falls into a five-year MACRS property class.
- The company's marginal tax rate is 40%, and the firm uses a 14% after-tax MARR.

14.39 Five years ago, a conveyor system was installed in a manufacturing plant at a cost of $35,000. It was estimated that the system, which is still in operating condition, would have a useful life of eight years, with a salvage value of $3,000. It was also estimated that if the firm continues to operate the system, its market values and operating costs for the next three years would be as follows:

Year-End	Market Value	Book Value	Operating Cost
0	$11,500	$15,000	
1	5,200	11,000	$4,500
2	3,500	7,000	5,300
3	1,200	3,000	6,100

A new system can be installed for $43,500; it would have an estimated economic life of 10 years, with a salvage value of $3,500. Operating costs are expected to be $1,500 per year throughout the service life of the system. The firm's MARR is 18%. The system belongs to the seven-year MACRS property class. The firm's marginal tax rate is 35%.

(a) Decide whether to replace the existing system now.

(b) If the decision is to replace the existing system, when should replacement occur?

14.40 Redo Problem 14.14, but with the following additional information:
- The old machine has been depreciated on a straight-line basis.
- The new machine will be depreciated under a seven-year MACRS class.
- The marginal tax rate is 40%, and the firm's after-tax MARR is 12%.

14.41 Redo Problem 14.15, but with the following additional information:
- The old machine has been fully depreciated.
- The new machine will be depreciated under a seven-year MACRS class.
- The marginal tax rate is 35%, and the firm's after-tax MARR is 12%.

14.42 Redo Problem 14.17, but with the following additional information:
- The old machine has been depreciated according to a seven-year MACRS.
- The new machine will also be depreciated under a seven-year MACRS class.
- The marginal tax rate is 40%, and the firm's after-tax MARR is 10%.

14.43 Redo Problem 14.18, but with the following additional information:
- **Option 1:** The old sprayer has been fully depreciated. The new sprayer is classified as having a seven-year MACRS recovery period.
- **Option 2:** The larger capacity sprayer is classified as a seven-year MACRS property.
- The company's marginal tax rate is 40%, and the firm uses a 12% after-tax MARR.

14.44 A six-year-old computer numerical control (CNC) machine that originally cost $8,000 has been fully depreciated, and its current market value is $1,500. If the machine is kept in service for the next five years, its O&M costs and salvage value are estimated as follows:

End of Year	O&M Costs Operation and Repairs	O&M Costs Delays Due to Breakdowns	Salvage Value
1	$1,300	$600	$1,200
2	1,500	800	1,000
3	1,700	1,000	500
4	1,900	1,200	0
5	2,000	1,400	0

It is suggested that the machine be replaced by a new CNC machine of improved design at a cost of $6,000. It is believed that this purchase will completely eliminate breakdowns and the resulting cost of delays and that operation and repair costs will be reduced $200 a year from what they would be with the old machine. Assume a five-year life for the challenger and a $1,000 terminal salvage value. The new machine falls into a five-year MACRS property class. The firm's MARR is 12%, and its marginal tax rate is 30%. Should the old machine be replaced now?

14.45 Redo Problem 14.21, but with the following additional information:
- The old asset has been depreciated according to a five-year MACRS.
- The new asset will also be depreciated under a five-year MACRS class.
- The marginal tax rate is 30%, and the firm's after-tax MARR is 11%.

Short Case Studies

ST14.1 Chevron Overseas Petroleum, Inc., entered into a 1993 joint venture with the Republic of Kazakhstan, a former republic of the old Soviet Union, to develop the huge Tengiz oil field.[4] Unfortunately, the climate in the region is harsh, making it difficult to keep oil flowing. The untreated oil comes out of the ground at 114°F. Even though the pipelines are insulated, as the oil gets further from the well on its way to be processed, hydrate salts begin to precipitate out of the liquid phase as the oil cools. These hydrate salts create a dangerous condition by forming plugs in the line.

The method for preventing this trap pressure condition is to inject methanol (MeOH) into the oil stream. This keeps the oil flowing and prevents hydrate salts from precipitating out of the liquid phase. The present methanol loading and storage facility is a completely manual controlled system, with no fire protection and with a rapidly deteriorating tank that causes leaks. The scope of repairs and upgrades is extensive. The storage tanks are rusting and are leaking at their riveted joints. The manual control system causes frequent tank overfills. There is no fire protection system, as water is not available at the site.

The present storage facility has been in service for 5 years. Permit requirements mandate upgrades to achieve minimum acceptable Kazakhstan standards. Upgrades in the amount of $104,000 will extend the life of the current facility to about 10 years. However, upgrades will not completely stop the leaks. The expected spill and leak losses will amount to $5,000 a year. The annual operating costs are expected to be $36,000.

As an alternative to the old facility, a new methanol storage facility can be designed on the basis of minimum acceptable international oil industry practices. The new facility, which would cost $325,000, would last about 12 years before a major upgrade would be required. However, it is believed that oil transfer technology will be such that methanol will not be necessary in 10 years. The pipeline heating and insulation systems will make methanol storage and use systems obsolete. With a lower risk of leaks, spills, and evaporation loss and a more closely monitored system, the expected annual operating cost would be $12,000.

(a) Assume that the storage tanks (the new ones as well as the upgraded ones) will have no salvage value at the end of their useful lives (after considering the removal costs) and that the tanks will be depreciated by the straight-line method according to the Kazakhstan's tax law. If Chevron's interest rate is 20% for foreign projects, which option is a better choice?

(b) How would the decision change as you consider the risk of spills (resulting in cleanup costs) and the evaporation of the product having an environmental impact?

---

[4] This example was provided by Mr. Joel M. Height of the Chevron Oil Company.

ST14.2 National Woodwork Company, a manufacturer of window frames, is considering replacing a conventional manufacturing system with a flexible manufacturing system (FMS). The company cannot produce rapidly enough to meet demand. One manufacturing problem that has been identified is that the present system is expected to be useful for another 5 years, but will require an estimated $105,000 per year in maintenance, which will increase $10,000 each year as parts become more scarce. The current market value of the existing system is $140,000, and the machine has been fully depreciated.

The proposed system will reduce or entirely eliminate setup times, and each window can be made as it is ordered by the customer, who phones the order into the head office, where details are fed into the company's main computer. These manufacturing details are then dispatched to computers on the manufacturing floor, which are, in turn, connected to a computer that controls the proposed FMS. This system eliminates the warehouse space and material-handling time that are needed when the conventional system is used.

Before the FMS is installed, the old equipment will be removed from the job shop floor at an estimated cost of $100,000. This cost includes needed electrical work on the new system. The proposed FMS will cost $1,200,000. The economic life of the machine is expected to be 10 years, and the salvage value is expected to be $120,000. The change in window styles has been minimal in the past few decades and is expected to continue to remain stable in the future. The proposed equipment falls into the 7-year MACRS category. The total annual savings will be $664,243: $12,000 attributed to a reduction in the number of defective windows, $511,043 from the elimination of 13 workers, $100,200 from the increase in productivity, and $41,000 from the near elimination of warehouse space and material handling. The O&M costs will be only $45,000, increasing by $2,000 per year. The National Woodwork's MARR is about 15%, and the expected marginal tax rate over the project years is 40%.

(a) What assumptions are required to compare the conventional system with the FMS?

(b) With the assumptions defined in (a), should the FMS be installed now?

ST14.3 In 2 × 4 and 2 × 6 lumber production, significant amounts of wood are present in sideboards produced after the initial cutting of logs. Instead of processing the sideboards into wood chips for the paper mill, Union Camp Company uses an "edger" to reclaim additional lumber, resulting in savings for the company. An edger is capable of reclaiming lumber by any of the following three methods: (1) removing rough edges, (2) splitting large sideboards, and (3) salvaging 2 × 4 lumber from low-quality 4 × 4 boards. Union Camp Company's engineers have discovered that a significant reduction in production costs could be achieved simply by replacing the original edger with a newer, laser-controlled model.

The old edger was placed in service 12 years ago and is fully depreciated. Any machine scrap value would offset the removal cost of the equipment. No market exists for this obsolete equipment. The old edger needs two operators. During the cutting operation, the operator makes edger settings, using his or her judgment. The operator has no means of determining exactly what dimension of lumber could be recovered from a given sideboard and must guess at the proper setting to recover the highest grade of lumber. Furthermore, the old edger is not capable of salvaging good-quality 2 × 4's from poor-quality 4 × 4's. The defender can continue in service for another 5 years with proper maintenance. Following are the financial data for the old edger:

Current market value	$0
Current book value	$0
Annual maintenance cost	$2,500 in year 1, increasing at a rate of 15% each year over the previous year's cost
Annual operating costs (labor and power)	$65,000

The new edger has numerous advantages over its defender, including laser beams that indicate where cuts should be made to obtain the maximum yield by the edger. The new edger requires a single operator, and labor savings will be reflected in lower operating and maintenance costs of $35,000 a year. The following gives the estimated costs and the depreciation methods associated with the new edger:

Estimated Cost	
Equipment	$ 35,700
Equipment installation	21,500
Building	47,200
Conveyor modification	14,500
Electrical (wiring)	16,500
Subtotal	$135,400
Engineering	7,000
Construction management	20,000
Contingency	16,200
Total	$178,600
Useful life of new edger	10 years
Salvage value	
Building (tear down)	$0
Equipment	10% of the original cost
Annual O&M costs	$35,000

Depreciation Methods	
Building	39-year MACRS
Equipment and installation	7-year MACRS

Twenty-five percent of the total mill volume passes through the edger. A 12% improvement in yield is expected to be realized with the new edger, resulting in an improvement of $(0.25)(0.12) = 3\%$ in the total mill volume. The annual savings due to the improvement in productivity is thus expected to be $57,895.

(a) Should the defender be replaced now if the mill's MARR and marginal tax rate are 16% and 40%, respectively?

(b) If the defender will eventually be replaced by the current challenger, when is the optimal time to perform the replacement?

ST14.4 Rivera Industries, a manufacturer of home heating appliances, is considering purchasing an Amada Turret Punch Press, a more advanced piece of machinery, to replace its present system that uses four old presses. Currently, the four smaller presses are used (in varying sequences, depending on the product) to produce one component of a product until a scheduled time when all machines must retool to set up for a different component. Because the setup cost is high, production runs of individual components take a long time and result in large inventory buildups of one component. These buildups are necessary to prevent extended backlogging while other products are being manufactured.

The four presses in use now were purchased six years ago at a price of $100,000. The manufacturing engineer expects that these machines can be used for eight more years, but they will have no market value after that. The presses have been depreciated by the MACRS method (seven-year property). Their current book value is $13,387, and their present market value is estimated to be $40,000. The average setup cost, which is determined by the number of labor hours required times the labor rate for the old presses, is $80 per hour, and the number of setups per year expected by the production control department is 200, yielding a yearly setup cost of $16,000. The expected operating and maintenance cost for each year in the remaining life of the old system is estimated as follows:

Year	Setup Costs	O&M Costs
1	$16,000	$15,986
2	16,000	16,785
3	16,000	17,663
4	16,000	18,630
5	16,000	19,692
6	16,000	20,861
7	16,000	22,147
8	16,000	23,562

These costs, which were estimated by the manufacturing engineer, with the aid of data provided by the vendor, represent a reduction in efficiency and an increase in needed service and repair over time.

The price of the two-year-old Amada Turret Punch Press is $135,000 and would be paid for with cash from the company's capital fund. In addition, the company would incur installation costs totaling $1,200. An expenditure of $12,000 would be required to recondition the press to its original condition. The reconditioning would extend the Amada's economic service life to eight years, after which time the machine would have no salvage value. The Amada would be depreciated under the MACRS as a seven-year property with the half-year convention. The cash savings of the Amada over the present system are due to the reduced setup time. The average setup cost of the Amada is $15, and the machine would incur 1,000 setups per year, yielding a yearly setup cost of $15,000. The savings due to the reduced setup time occur because of the reduction in carrying costs associated with that level of inventory at which the production run and ordering quantity are reduced. The Accounting Department has estimated that at least $26,000, and probably $36,000, per year could be saved by shortening production runs. The operating and maintenance costs of the Amada, as estimated by the manufacturing engineer, are similar to, but somewhat less, than the O&M costs of the present system.

Year	Setup Costs	O&M Costs
1	$15,000	$11,500
2	15,000	11,950
3	15,000	12,445
4	15,000	12,990
5	15,000	13,590
6	15,000	14,245
7	15,000	14,950
8	15,000	15,745

The reduction in the O&M costs is caused by the age difference of the machines and the reduced power requirements of the Amada.

If Rivera Industries delays the replacement of the current four presses for another year, the secondhand Amada machine will no longer be available, and the company will have to buy a brand-new machine at a price of $200,450, installed. The expected setup costs would be the same as those for the secondhand machine, but the annual operating and maintenance costs would be about 10% lower than the estimated O&M costs for the secondhand machine. The expected economic service life of the brand-new press would be eight years, with no salvage value. The brand-new press also falls into a seven-year MACRS.

Rivera's MARR is 12% after taxes, and the marginal income tax rate is expected to be 40% over the life of the project.

(a) Assuming that the company would need the service of either press for an indefinite period, what would you recommend?

(b) Assuming that the company would need the press for only five more years, what would you recommend?

ST14.5 Tiger Construction Company purchased its current bulldozer (a Caterpillar D8H) and placed it in service six years ago. Since the purchase of the Caterpillar, new technology has produced changes in machines resulting in an increase in productivity of approximately 20%. The Caterpillar worked in a system with a fixed (required) production level to maintain overall system productivity. As the Caterpillar aged and logged more downtime, more hours had to be scheduled to maintain the required production. Tiger is considering purchasing a new bulldozer (a Komatsu K80A) to replace the Caterpillar. The following data have been collected by Tiger's civil engineer:

	Defender (Caterpillar D8H)	Challenger (Komatsu K80A)
Useful life	Not known	Not known
Purchase price		$400,000
Salvage value if kept for		
0 year	$75,000	$400,000
1 year	60,000	300,000
2 years	50,000	240,000
3 years	30,000	190,000
4 years	30,000	150,000
5 years	10,000	115,000
Fuel use (gallon/hour)	11.30	16
Maintenance costs:		
1	$46,800	$35,000
2	46,800	38,400
3	46,800	43,700
4	46,800	48,300
5	46,800	58,000

	Defender (Caterpillar D8H) (continued)	Challenger (Komatsu K80A) (continued)
Operating hours (hours/year):		
1	1,800	2,500
2	1,800	2,400
3	1,700	2,300
4	1,700	2,100
5	1,600	2,000
Productivity index	1.00	1.20
Other relevant information:		
Fuel cost ($/gallon)		$1.20
Operator's wages ($/hour)		$23.40
Market interest rate (MARR)		15%
Marginal tax rate		40%
Depreciation methods	Fully Depreciated	MACRS (5-year)

(a) A civil engineer notices that both machines have different working hours and hourly production capacities. To compare the different units of capacity, the engineer needs to devise a combined index that reflects the machine's productivity as well as actual operating hours. Develop such a combined productivity index for each period.

(b) Adjust the operating and maintenance costs by the index you have come up with.

(c) Compare the two alternatives. Should the defender be replaced now?

(d) If the following price index were forecasted for the next five project years, should the defender be replaced now?

Year	Forecasted Price Index			
	General Inflation	Fuel	Wage	Maintenance
0	100	100	100	100
1	108	110	115	108
2	116	120	125	116
3	126	130	130	124
4	136	140	135	126
5	147	150	140	128

CHAPTER FIFTEEN

Capital-Budgeting Decisions

Hotels Go to the Mattresses[1] Marriott International, Inc., today will launch a major initiative to replace nearly every bed in seven of its chains. At the Marriott chain, which is slated for the most extensive upgrade, each king-size bed will be getting 300-thread-count 60% cotton sheets, seven pillows instead of five, a pillowy mattress cover, a white duvet, and a "bed scarf" that will be draped along the bottom of the bed. The yearlong project will cost an estimated $190 million, and the company is saying that the cost, along with the planned marketing efforts, make it its biggest initiative ever.

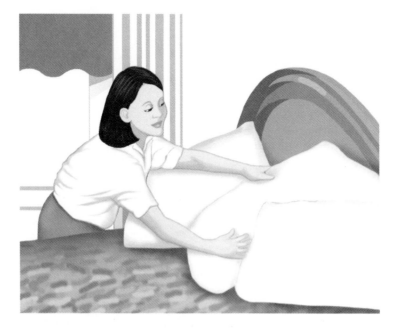

[1] "Hotels Go to the Mattresses," by Christina Binkley, *The Wall Street Journal*, January 25, 2005.

The bed replacements raise some awkward issues, such as what to do with all the old beds. Some hotels are trying to give them to charity, but "homeless shelters don't really have a use for king-size beds," says one Marriott executive. Housekeepers complain that stuffing down comforters into all the new duvets is more time consuming than making a traditional bed with bedspread.

All the pressure to make hotel beds better is finally forcing the industry to reveal—and start to abandon—its dirtiest little secret: the fact that those colorful bedspreads on hotel-room beds sometimes get washed only a few times a year at most. Marriott hopes it will top rivals with a pledge to wash its white duvets between each guest visit, an initiative the company will call "Clean for You." When it comes to bedspread sanitation, hotel chains are typically loath to reveal how often they wash the bedcovers. Regular laundering is costly, in terms of both housekeeping and laundering, as well as in wear and tear on expensive linens.

Hotels are doing all this because their research shows that people are willing to pay more for luxurious beds. Bill Marriott says he expects to be able to charge as much as $30 a night more in Marriott hotels once the new beds are installed.

Marriott plans to purchase 628,000 beds for hotels at seven chains. However, only the full-service Marriott and Renaissance chains will get the new white duvets, which the company is planning to launder regularly as part of the Clean for You campaign. Less expensive Marriott chains like Courtyard and SpringHill Suites, along with Residence Inn, will triple sheet their beds by putting the extra sheet on top of the (less frequently laundered) outermost bedspread.

All this comes as the hotel industry is experiencing an economic boom, with room rates and occupancies rising faster than any time since the dot-com boom, leaving hotels extra cash to spend on improvements. Much of the increase in travel is coming from business travelers, who tend to be pickier and less price sensitive than vacationers—meaning they want things like better beds and often don't mind paying for them since their expense accounts are picking up the tab.

Without budget limitations, the replacement problem would be more of a logistic issue: simply select the option with the most revenue-enhancing potential for each targeted hotel. However, to replace or upgrade all beds over a short period of time will cost in excess of $190 million, and the company needs to find a way to finance this large-scale project. Because of the size of the financing involved, the firm's cost of capital will tend to increase during the project period. In this circumstance, the choice of an appropriate interest rate (MARR) for use in the project evaluation becomes a critical issue. Given these budget and other restrictions, the company would certainly like to determine the least-cost replacement/ upgrade strategy.

Capital budgeting is the planning process used to determine a firm's long term investments.

In this chapter, we present the basic framework of **capital budgeting**, which involves investment decisions related to fixed assets. Here, the term **capital budget** includes planned expenditures on fixed assets; **capital budgeting** encompasses the entire process of analyzing projects and deciding whether they should be included in the capital budget. In previous chapters, we focused on how to evaluate and compare investment projects—the analysis aspect of capital budgeting. In this chapter, we focus on the budgeting aspect. Proper capital-budgeting decisions require a choice of the method of project financing, a schedule of investment opportunities, and an estimate of the minimum attractive rate of return (MARR).

CHAPTER LEARNING OBJECTIVES

After completing this chapter, you should understand the following concepts:

- How a corporation raises its capital to finance a project.
- How to determine the cost of debt.
- How to determine the cost of equity.
- How to determine the marginal cost of capital.
- How to determine the MARR in project evaluation.
- How to create an optimal project portfolio under capital rationing.

15.1 Methods of Financing

In previous chapters, we focused on problems relating to investment decisions. In reality, investment decisions are not always independent of the source of finance. For convenience, however, in economic analysis investment decisions are usually separated from finance decisions: First the investment project is selected, and then the source of financing is considered. After the source is chosen, appropriate modifications to the investment decision are made.

We have also assumed that the assets employed in an investment project are obtained with the firm's own capital (retained earnings) or from short-term borrowings. In practice, this arrangement is not always attractive or even possible. If the investment calls for a significant infusion of capital, the firm may raise the needed capital by issuing stock. Alternatively, the firm may borrow the funds by issuing bonds to finance such purchases. In this section, we will first discuss how a typical firm raises new capital from external sources. Then we will discuss how external financing affects after-tax cash flows and how the decision to borrow affects the investment decision.

The two broad choices a firm has for financing an investment project are **equity financing** and **debt financing**.[2] We will look briefly at these two options for obtaining external investment funds and also examine their effects on after-tax cash flows.

15.1.1 Equity Financing

Equity financing can take one of two forms: (1) the use of retained earnings otherwise paid to stockholders or (2) the issuance of stock. Both forms of equity financing use funds invested by the current or new owners of the company.

Until now, most of our economic analyses presumed that companies had cash on hand to make capital investments; implicitly, we were dealing with cases of financing by retained earnings. If a company had not reinvested these earnings, it might have paid them to the company's owners—the stockholders—in the form of a dividend, or it might have kept these earnings on hand for future needs.

If a company does not have sufficient cash on hand to make an investment and does not wish to borrow in order to fund the investment, financing can be arranged by selling common stock to raise the required funds. (Many small biotechnology and computer firms raise capital by going public and selling common stock.) To do this, the company has to decide how much money to raise, the type of securities to issue (common stock or preferred stock), and the basis for pricing the issue.

Once the company has decided to issue common stock, it must estimate **flotation costs**—the expenses it will incur in connection with the issue, such as investment bankers' fees, lawyers' fees, accountants' costs, and the cost of printing and engraving. Usually, an investment banker will buy the issue from the company at a discount, below the price at which the stock is to be offered to the public. (The discount usually represents the *flotation costs*.) If the company is already publicly owned, the offering price will commonly be based on the existing market price of the stock. If the company is going public for the first time, no established price will exist, so investment bankers have to estimate the expected market price at which the stock will sell after the stock issue. Example 15.1 illustrates how the flotation cost affects the cost of issuing common stock.

> **Flotation cost:** The costs associated with the issuance of new securities.

EXAMPLE 15.1 Issuing Common Stock

Scientific Sports, Inc. (SSI), a golf club manufacturer, has developed a new metal club (Driver). The club is made out of titanium alloy, an extremely light and durable metal with good vibration-damping characteristics (Figure 15.1). The company expects to acquire considerable market penetration with this new product. To produce it, the company needs a new manufacturing facility, which will cost $10 million. The company decided to raise this $10 million by selling common stock. The firm's current stock price is $30 per share. Investment bankers have informed management that the new public issue must be priced at $28 per share because of decreasing demand, which

---

[2] A hybrid financing method, known as *lease financing*, was discussed in Section 10.4.3.

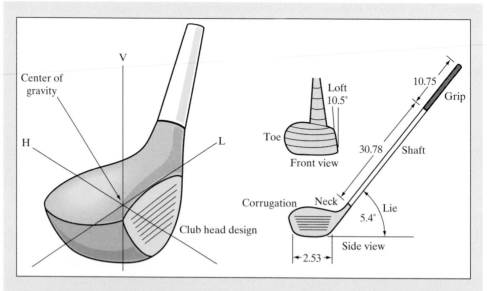

Figure 15.1 SSI's new golf club (Driver) design, developed with the use of advanced engineering materials (Example 15.1).

will occur as more shares become available on the market. The flotation costs will be 6% of the issue price, so SSI will net $26.32 per share. How many shares must SSI sell to net $10 million after flotation expenses?

SOLUTION

Let X be the number of shares to be sold. Then total flotation cost will be

$$(0.06)(\$28)(X) = 1.68X.$$

To net $10 million, we must have

$$\text{Sales proceeds} - \text{flotation cost} = \text{Net proceeds},$$
$$28X - 1.68X = \$10,000,000,$$
$$26.32X = \$10,000,000,$$
$$X = 379,940 \text{ shares}.$$

Now we can figure out the flotation cost for issuing the common stock. The cost is

$$1.68(379,940) = \$638,300.$$

15.1.2 Debt Financing

The second major type of financing a company can select is **debt financing**, which includes both short-term borrowing from financial institutions and the sale of long-term bonds, wherein money is borrowed from investors for a fixed period. With debt financing, the interest paid on the loans or bonds is treated as an expense for income-tax purposes.

Since interest is a tax-deductible expense, companies in high tax brackets may incur lower after-tax financing costs with a debt. In addition to influencing the borrowing interest rate and tax bracket, a loan-repayment method can affect financing costs.

When the debt-financing option is used, we need to separate the interest payments from the repayment of the loan for our analysis. The interest-payment schedule depends on the repayment schedule established at the time the money is borrowed. The two common debt-financing methods are as follows:

1. **Bond Financing.** This type of debt financing does not involve the partial payment of principal; only interest is paid each year (or semiannually). The principal is paid in a lump sum when the bond matures. (See Section 4.6.3 for bond terminologies and valuation.) Bond financing is similar to equity financing in that flotation costs are involved when bonds are issued.

2. **Term Loans.** Term loans involve an equal repayment arrangement according to which the sum of the interest payments and the principal payments is uniform; interest payments decrease, while principal payments increase, over the life of the loan. Term loans are usually negotiated directly between the borrowing company and a financial institution, generally a commercial bank, an insurance company, or a pension fund.

Example 15.2 illustrates how these different methods can affect the cost of issuing bonds or term loans.

EXAMPLE 15.2 Debt Financing

Consider again Example 15.1. Suppose SSI has instead decided to raise the $10 million by debt financing. SSI could issue a mortgage bond or secure a term loan. Conditions for each option are as follows:

- **Bond financing.** The flotation cost is 1.8% of the $10 million issue. The company's investment bankers have indicated that a five-year bond issue with a face value of $1,000 can be sold at $985 per share. The bond would require annual interest payments of 12%.

- **Term loan.** A $10 million bank loan can be secured at an annual interest rate of 11% for five years; it would require five equal annual installments.

(a) How many $1,000 par value bonds would SSI have to sell to raise the $10 million?

(b) What are the annual payments (interest and principal) on the bond?

(c) What are the annual payments (interest and principal) on the term loan?

SOLUTION

(a) To net $10 million, SSI would have to sell

$$\frac{\$10,000,000}{(1 - 0.018)} = \$10,183,300$$

TABLE 15.1 Two Common Methods of Debt Financing (Example 15.2)

	0	1	2	3	4	5
1. Bond financing: No principal repayments until end of life						
Beginning balance	$10,338,380	$10,338,380	$10,338,380	$10,338,380	$10,338,380	$10,338,380
Interest owed		1,240,606	1,240,606	1,240,606	1,240,606	1,240,606
Repayment						
Interest payment		(1,240,606)	(1,240,606)	(1,240,606)	(1,240,606)	(1,240,606)
Principal payment						(10,338,380)
Ending balance	$10,338,380	$10,338,380	$10,338,380	$10,338,380	$10,338,380	0
2. Term loan: Equal annual repayments [$10,000,000$(A/P, 11\%, 5) = $2,705,703$]						
Beginning balance	$10,000,000	$10,000,000	$ 8,394,297	$ 6,611,967	$ 4,633,580	$ 2,437,571
Interest owed		1,100,000	923,373	727,316	509,694	268,133
Repayment						
Interest payment		(1,100,000)	(923,373)	(727,316)	(509,694)	(268,133)
Principal payment		(1,605,703)	(1,782,330)	(1,978,387)	(2,196,009)	(2,437,570)
Ending balance	$10,000,000	$ 8,394,297	$ 6,611,967	$ 4,633,580	$ 2,437,571	0

worth of bonds and pay $183,300 in flotation costs. Since the $1,000 bond will be sold at a 1.5% discount, the total number of bonds to be sold would be

$$\frac{\$10,183,300}{\$985} = \$10,338.38.$$

(b) For the bond financing, the annual interest is equal to

$$\$10,338,380(0.12) = \$1,240,606.$$

Only the interest is paid each period; thus, the principal amount owed remains unchanged.

(c) For the term loan, the annual payments are

$$\$10,000,000(A/P, 11\%, 5) = \$2,705,703.$$

The principal and interest components of each annual payment are summarized in Table 15.1.

15.1.3 Capital Structure

The ratio of total debt to total capital, generally called the **debt ratio**, or **capital structure**, represents the percentage of the total capital provided by borrowed funds. For example, a debt ratio of 0.4 indicates that 40% of the capital is borrowed and the remaining funds are provided from the company's equity (retained earnings or stock offerings). This type of financing is called **mixed financing**.

Borrowing affects a firm's capital structure, and firms must determine the effects of a change in the debt ratio on their market value before making an ultimate financing decision. Even if debt financing is attractive, you should understand that companies do not simply borrow funds to finance projects. A firm usually establishes a **target capital structure**, or **target debt ratio**, after considering the effects of various financing methods. This target may change over time as business conditions vary, but a firm's management always strives to achieve the target whenever individual financing decisions are considered. On the one hand, the actual debt ratio is below the target level, any new capital will probably be raised by issuing debt. On the other hand, if the debt ratio is currently above the target, expansion capital will be raised by issuing stock.

How does a typical firm set the target capital structure? This is a rather difficult question to answer, but we can list several factors that affect the capital-structure policy. First, capital-structure policy involves a trade-off between risk and return. As you take on more debt for business expansion, the inherent business risk[3] also increases, but investors view business expansion as a healthy indicator for a corporation with higher expected earnings. When investors perceive higher business risk, the firm's stock price tends to be depressed. By contrast, when investors perceive higher expected earnings, the firm's stock price tends to increase. The optimal capital structure is thus the one that strikes a balance between business risk and expected future earnings. The greater the firm's business risk, the lower is its optimal debt ratio.

Second, a major reason for using debt is that interest is a deductible expense for business operations, which lowers the effective cost of borrowing. Dividends paid to common stockholders, however, are not deductible. If a company uses debt, it must pay interest on this debt, whereas if it uses equity, it pays dividends to its equity investors (shareholders). A company needs $1 in before-tax income to pay $1 of interest, but if the company is in the 34% tax bracket, it needs $1/(1 - 0.34) = 1.52 of before-tax income to pay a $1 dividend.

Third, financial flexibility—the ability to raise capital on reasonable terms from the financial market—is an important consideration. Firms need a steady supply of capital for stable operations. When money is tight in the economy, investors prefer to advance funds to companies with a healthy capital structure (lower debt ratio). These three elements (business risk, taxes, and financial flexibility) are major factors that determine the firm's optimal capital structure. Example 15.3 illustrates how a typical firm finances a large-scale engineering project by maintaining the predetermined capital structure.

> **Capital structure:** The means by which a firm is financed.

EXAMPLE 15.3 Project Financing Based on an Optimal Capital Structure

Consider again SSI's $10 million venture project in Example 15.1. Suppose that SSI's optimal capital structure calls for a debt ratio of 0.5. After reviewing SSI's capital structure, the investment banker convinced management that it would be better

---

[3] Unlike equity financing, in which dividends are optional, debt interest and principal (face value) must be repaid on time. Also, uncertainty is involved in making projections of future operating income as well as expenses. In bad times debt can be devastating, but in good times the tax deductibility of interest payments increases profits to owners.

off, in view of current market conditions, to limit the stock issue to $5 million and to raise the other $5 million as debt by issuing bonds. Because the amount of capital to be raised in each category is reduced by half, the flotation cost would also change. The flotation cost for common stock would be 8.1%, whereas the flotation cost for bonds would be 3.2%. As in Example 15.2, the 5-year, 12% bond will have a par value of $1,000 and will be sold for $985.

Assuming that the $10 million capital would be raised from the financial market, the engineering department has detailed the following financial information:

- The new venture will have a 5-year project life.
- The $10 million capital will be used to purchase land for $1 million, a building for $3 million, and equipment for $6 million. The plant site and building are already available, and production can begin during the first year. The building falls into a 39-year MACRS property class and the equipment into a 7-year MACRS class. At the end of year 5, the salvage value of each asset is as follows: the land $1.5 million, the building $2 million, and the equipment $3 million.
- For common stockholders, an annual cash dividend in the amount of $2 per share is planned over the project life. This steady cash dividend payment is deemed necessary to maintain the market value of the stock.
- The unit production cost is $50.31 (material, $22.70; labor and overhead (excluding depreciation), $10.57; and tooling, $17.04).
- The unit price is $250, and SSI expects an annual demand of 20,000 units.
- The operating and maintenance cost, including advertising expenses, would be $600,000 per year.
- An investment of $500,000 in working capital is required at the beginning of the project; the amount will be fully recovered when the project terminates.
- The firm's marginal tax rate is 40%, and this rate will remain constant throughout the project period.

(a) Determine the after-tax cash flows for this investment with external financing.

(b) Is this project justified at an interest rate of 20%?

DISCUSSION: As the amount of financing and flotation costs change, we need to recalculate the number of shares (or bonds) to be sold in each category. For a $5 million common stock issue, the flotation cost increases to 8.1%.[4] The number of shares to be sold to net $5 million is 5,000,000/(0.919)(28) = 194,311 shares (or $5,440,708). For a $5 million bond issue, the flotation cost is 3.2%. Therefore, to net $5 million, SSI has to sell 5,000,000/(0.968)(985) = 5,243.95 units of $1,000 par value. This implies that SSI is effectively borrowing $5,243,948, upon which figure the annual bond interest will be calculated. The annual bond interest payment is $5,243,948(0.12) = $629,274.

[4] Flotation costs are higher for small issues than for large ones due to the existence of fixed costs: Certain costs must be incurred regardless of the size of the issue, so the percentage of flotation costs increases as the size of the issue gets smaller.

SOLUTION

(a) *After-tax cash flows*. Table 15.2 summarizes the after-tax cash flows for the new venture. The following calculations and assumptions were used in developing the table:

- Revenue: $250 \times 20{,}000 = \$5{,}000{,}000$ per year.
- Costs of goods: $\$50.31 \times 20{,}000 = \$1{,}006{,}200$ per year.
- Bond interest: $\$5{,}243{,}948 \times 0.12 = \$629{,}274$ per year.

TABLE 15.2 Effects of Project Financing on After-Tax Cash Flows (Example 15.3)

	0	1	2	3	4	5
Income statement:						
Revenue		$ 5,000,000	$ 5,000,000	$ 5,000,000	$ 5,000,000	$ 5,000,000
Expenses:						
Cost of goods		1,006,200	1,006,200	1,006,200	1,006,200	1,006,200
O&M		600,000	600,000	600,000	600,000	600,000
Bond interest		629,274	629,274	629,274	629,274	629,274
Depreciation:						
Building		73,718	76,923	76,923	76,923	73,718
Equipment		857,400	1,469,400	1,049,400	749,400	267,900
Taxable income		1,833,408	1,218,203	1,638,203	1,938,203	2,422,908
Income taxes		733,363	487,281	655,281	775,281	969,163
Net income		$ 1,100,045	$ 730,922	$ 982,922	$ 1,162,922	$ 1,453,745
Cash flow statement:						
Operating activities:						
Net income		$ 1,100,045	$ 730,922	$ 982,922	$ 1,162,922	$ 1,453,745
Noncash expense		931,118	1,546,323	1,126,323	826,323	341,618
Investment activities:						
Land	(1,000,000)					1,500,000
Building	(3,000,000)					2,000,000
Equipment	(6,000,000)					2,500,000
Working capital	(500,000)					500,000
Gains tax						(308,682)
Financing activities:						
Common stock	5,000,000					(5,440,708)
Bond	5,000,000					(5,243,948)
Cash dividend		(388,622)	(388,622)	(388,622)	(388,622)	(388,622)
Net cash flow	$ (500,000)	$ 1,642,541	$ 1,888,623	$ 1,720,623	$ 1,600,623	$ (3,086,597)

- Depreciation: Assuming that the building is placed in service in January, the first year's depreciation percentage is 2.4573%. Therefore, the allowed depreciation amount is $3,000,000 × 0.024573 = $73,718. The percentages for the remaining years would be 2.5641% per year, or $76,923. Equipment is depreciated according to a 7-year MACRS.
- Gains tax:

Property	Salvage Value	Book Value	Gains (Losses)	Gains Tax
Land	$1,500,000	$1,000,000	$500,000	$200,000
Building	2,000,000	2,621,795	(621,795)	(248,718)
Equipment	2,500,000	1,606,500	893,500	357,400
				$308,682

- Cash dividend: 194,311 shares × $2 = $388,622.
- Common stock: When the project terminates and the bonds are retired, the debt ratio is no longer 0.5. If SSI wants to maintain the constant capital structure (0.5), SSI would have to repurchase the common stock in the amount of $5,440,708 at the prevailing market price. In developing Table 15.2, we assumed that this repurchase of common stock had taken place at the ends of project years. In practice, a firm may or may not repurchase the common stock. As an alternative means of maintaining the desired capital structure, the firm may use this extra debt capacity released to borrow for other projects.
- Bond: When the bonds mature at the end of year 5, the total face value in the amount of $5,243,948 must be paid to the bondholders.

(b) *Measure of project worth.* The NPW for this project is then

$$\text{PW}(20\%) = -\$500,000 + \$1,642,541(P/F, 20\%, 1) + \dots$$

$$= -\$3,086,597(P/F, 20\%, 5)$$

$$= \$2,707,530.$$

The investment is nonsimple, and it is also a mixed investment. The RIC at MARR of 20% is 327%. Even though the project requires a significant amount of cash expenditure at the end of its life, it still appears to be a very profitable one.

In Example 15.3, we neither discussed the cost of capital required to finance this project nor explained the relationship between the cost of capital and the MARR. In the remaining sections, these issues will be discussed. As we will see later, in Section 15.3, we can completely ignore the detailed cash flows related to project financing if we adjust our discount rate according to the capital structure, namely, by using the weighted cost of capital.

15.2 Cost of Capital

In most of the capital-budgeting examples in earlier chapters, we assumed that the firms under consideration were financed entirely with equity funds. In those cases, the cost of capital may have represented the firm's required return on equity. However, most firms finance a substantial portion of their capital budget with long-term debt (bonds), and many also use preferred stock as a source of capital. In these cases, a firm's cost of capital must reflect the average cost of the various sources of long-term funds that the firm uses, not only the cost of equity. In this section, we will discuss the ways in which the cost of each individual type of financing (retained earnings, common stock, preferred stock, and debt) can be estimated,[5] given a firm's target capital structure.

15.2.1 Cost of Equity

Whereas debt and preferred stocks are contractual obligations that have easily determined costs, it is not easy to measure the cost of equity. In principle, the cost of equity capital involves an **opportunity cost**. In fact, the firm's after-tax cash flows belong to the stockholders. Management may either pay out these earnings in the form of dividends, or retain the earnings and reinvest them in the business. If management decides to retain the earnings, an opportunity cost is involved: Stockholders could have received the earnings as dividends and invested the money in other financial assets. Therefore, the firm should earn on its retained earnings at least as much as the stockholders themselves could earn in alternative, but comparable, investments.

Cost of equity is the minimum rate of return a firm must offer shareholders to compensate for waiting for their returns, and for bearing some risk.

What rate of return can stockholders expect to earn on retained earnings? This question is difficult to answer, but the value sought is often regarded as the rate of return stockholders require on a firm's common stock. If a firm cannot invest retained earnings so as to earn at least the rate of return on equity, it should pay these funds to the stockholders and let them invest directly in other assets that do provide this return.

When investors are contemplating buying a firm's stock, they have two things in mind: (1) cash dividends and (2) gains (appreciation of shares) at the time of sale. From a conceptual standpoint, investors determine market values of stocks by discounting expected future dividends at a rate that takes into account any future growth. Since investors seek growth companies, a desired growth factor for future dividends is usually included in the calculation.

To illustrate, let's take a simple numerical example. Suppose investors in the common stock of ABC Corporation expect to receive a dividend of $5 by the end of the first year. The future annual dividends will grow at an annual rate of 10%. Investors will hold the stock for two more years and will expect the market price of the stock to rise to $120 by the end of the third year. Given these hypothetical expectations, ABC expects that investors would be willing to pay $100 for this stock in today's market. What is the required rate of return k_r on ABC's common stock? We may answer this question by solving the following equation for k_r:

$$\$100 = \frac{\$5}{(1 + k_r)} + \frac{\$5(1 + 0.1)}{(1 + k_r)^2} + \frac{\$5(1 + 0.1)^2 + \$120}{(1 + k_r)^3}.$$

[5] Estimating or calculating the cost of capital in any precise fashion is very difficult task.

In this case, $k_r = 11.44\%$. This implies that if ABC finances a project by retaining its earnings or by issuing additional common stock at the going market price of $100 per share, it must realize at least 11.44% on new investment just to provide the minimum rate of return required by the investors. Therefore, 11.44% is the specific cost of equity that should be used in calculating the weighted-average cost of capital. Because flotation costs are involved in issuing new stock, the cost of equity will increase. If investors view ABC's stock as risky and therefore are willing to buy the stock at a price lower than $100 (but with the same expectations), the cost of equity will also increase. Now we can generalize the preceding result.

Cost of Retained Earnings

Let's assume the same hypothetical situation for ABC. Recall that ABC's retained earnings belong to holders of its common stock. If ABC's current stock is traded for a market price of P_0, with a first-year dividend[6] of D_1, but growing at the annual rate of g thereafter, the specific cost of retained earnings for an infinite period of holding (stocks will change hands over the years, but it does not matter who holds the stock) can be calculated as

$$P_0 = \frac{D_1}{(1 + k_r)} + \frac{D_1(1 + g)}{(1 + k_r)^2} + \frac{D_1(1 + g)^2}{(1 + k_r)^3} + \cdots$$

$$= \frac{D_1}{1 + k_r} \sum_{n=0}^{\infty} \left[\frac{(1 + g)}{(1 + k_r)} \right]^n$$

$$= \frac{D_1}{1 + k_r} \left[\frac{1}{1 - \dfrac{1 + g}{1 + k_r}} \right], \text{ where } g < k_r.$$

Solving for k_r, we obtain

$$k_r = \frac{D_1}{P_0} + g. \tag{15.1}$$

If we use k_r as the discount rate for evaluating the new project, it will have a positive NPW only if the project's IRR exceeds k_r. Therefore, any project with a positive NPW, calculated at k_r, induces a rise in the market price of the stock. Hence, by definition, k_r is the rate of return required by shareholders and should be used as the cost of the equity component in calculating the weighted average cost of capital.

Issuing New Common Stock

Again, because flotation costs are involved in issuing new stock, we can modify the cost of retained earnings k_r by

$$k_e = \frac{D_1}{P_0(1 - f_c)} + g, \tag{15.2}$$

where k_e is the cost of common equity and f_c is the flotation cost as a percentage of the stock price.

---

[6] When we check the stock listings in the newspaper, we do not find the expected first-year dividend D_1. Instead, we find the dividend paid out most recently, D_0. So if we expect growth at a rate g, the dividend at the end of one year from now, D_1 may be estimated as $D_1 = D_0(1 + g)$.

Either calculation is deceptively simple, because, in fact, several ways are available to determine the cost of equity. In reality, the market price fluctuates constantly, as do a firm's future earnings. Thus, future dividends may not grow at a constant rate, as the model indicates. For a stable corporation with moderate growth, however, the cost of equity as calculated by evaluating either Eq. (15.1) or Eq. (15.2) serves as a good approximation.

Cost of Preferred Stock

A preferred stock is a hybrid security in the sense that it has some of the properties of bonds and other properties that are similar to common stock. Like bondholders, holders of preferred stock receive a fixed annual dividend. In fact, many firms view the payment of the preferred dividend as an obligation just like interest payments to bondholders. It is therefore relatively easy to determine the cost of preferred stock. For the purposes of calculating the weighted average cost of capital, the specific cost of a preferred stock will be defined as

$$k_p = \frac{D^*}{P^*(1 - f_c)},$$ (15.3)

where D^* is the fixed annual dividend, P^* is the issuing price, and f_c is as defined in Eq. (15.2).

Cost of Equity

Once we have determined the specific cost of each equity component, we can determine the weighted-average cost of equity (i_e) for a new project. We have

$$i_e = \left(\frac{c_r}{c_e}\right)k_r + \left(\frac{c_c}{c_e}\right)k_e + \left(\frac{c_p}{c_e}\right)k_p,$$ (15.4)

where c_r is the amount of equity financed from retained earnings, c_c is the amount of equity financed from issuing new stock, c_p is the amount of equity financed from issuing preferred stock, and $c_r + c_c + c_p = c_e$. Example 15.4 illustrates how we may determine the cost of equity.

EXAMPLE 15.4 Determining the Cost of Equity

Alpha Corporation needs to raise $10 million for plant modernization. Alpha's target capital structure calls for a debt ratio of 0.4, indicating that $6 million has to be financed from equity.

- Alpha is planning to raise $6 million from the following equity sources:

Source	Amount	Fraction of Total Equity
Retained earnings	$1 million	0.167
New common stock	4 million	0.666
Preferred stock	1 million	0.167

- Alpha's current common stock price is $40, the market price that reflects the firm's future plant modernization. Alpha is planning to pay an annual cash dividend of $5 at the end of the first year, and the annual cash dividend will grow at an annual rate of 8% thereafter.
- Additional common stock can be sold at the same price of $40, but there will be 12.4% flotation costs.
- Alpha can issue $100 par preferred stock with a 9% dividend. (This means that Alpha will calculate the dividend on the basis of the par value, which is $9 per share.) The stock can be sold on the market for $95, and Alpha must pay flotation costs of 6% of the market price.

Determine the cost of equity to finance the plant modernization.

SOLUTION

We will itemize the cost of each component of equity:

- Cost of retained earnings: With $D_1 = \$5$, $g = 8\%$, and $P_0 = \$40$,

$$k_r = \frac{5}{40} + 0.08 = 20.5\%.$$

- Cost of new common stock: With $D_1 = \$5$, $g = 8\%$, and $f_c = 12.4\%$,

$$k_e = \frac{5}{40(1 - 0.124)} + 0.08 = 22.27\%.$$

- Cost of preferred stock: With $D^* = \$9$, $P^* = \$95$, and $f_c = 0.06$,

$$k_p = \frac{9}{95(1 - 0.06)} = 10.08\%.$$

- Cost of equity: With $\dfrac{c_r}{c_e} = 0.167$, $\dfrac{c_c}{c_e} = 0.666$, and $\dfrac{c_p}{c_e} = 0.167$,

$$i_e = (0.167)(0.205) + (0.666)(0.2227) + (0.167)(0.1008)$$

$$= 19.96\%.$$

An Alternative Way of Determining the Cost of Equity

Whereas debt and preferred stocks are contractual obligations that have easily determined costs, it is not easy to measure the cost of equity. In principle, the cost of equity capital involves an **opportunity cost**. In fact, the firm's after-tax cash flows belong to the stockholders. Management may either pay out these earnings in the form of dividends or retain the earnings and reinvest them in the business. If management decides to retain the earnings, an opportunity cost is involved: Stockholders could have received the earnings as dividends and invested that money in other financial assets. Therefore, the firm should earn on its retained earnings at least as much as the stockholders themselves could earn in alternative, but comparable, investments.

What rate of return can stockholders expect to earn on retained earnings? This question is difficult to answer, but the value sought is often regarded as the rate of return stockholders require on a firm's common stock. If a firm cannot invest retained earnings so as to earn at least the rate of return on equity, it should pay these funds to its stockholders and let them invest directly in other assets that do provide that rate of return. In general, the expected return on any risky asset is composed of three factors:[7]

$$\begin{pmatrix} \text{Expected return} \\ \text{on risky asset} \end{pmatrix} = \begin{pmatrix} \text{Risk-free} \\ \text{interest rate} \end{pmatrix} + \begin{pmatrix} \text{Inflation} \\ \text{premium} \end{pmatrix} + \begin{pmatrix} \text{Risk} \\ \text{premium} \end{pmatrix}.$$

This equation says that the owner of a risky asset should expect to earn a return from three sources:

- Compensation from the opportunity cost incurred in holding the asset, known as the risk-free interest rate.
- Compensation for the declining purchasing power of the investment over time, known as the inflation premium.
- Compensation for bearing risk, known as the risk premium.

Fortunately, we do not need to treat the first two terms as separate factors because together they equal the expected return on a default-free bond such as a government bond. In other words, owners of government bonds expect a return from the first two sources, but not the third—a state of affairs we may express as

$$\begin{pmatrix} \text{Expected return} \\ \text{on risky asset} \end{pmatrix} = \begin{pmatrix} \text{Interest rate on} \\ \text{government bond} \end{pmatrix} + \begin{pmatrix} \text{Risk} \\ \text{premium} \end{pmatrix}.$$

When investors are contemplating buying a firm's stock, they have two primary things in mind: (1) cash dividends and (2) gains (appreciation of shares) at the time of sale. From a conceptual standpoint, investors determine market values of stocks by discounting expected future dividends at a rate that takes into account any future growth. Since investors seek growth companies, a desired growth factor for future dividends is usually included in the calculation.

The cost of equity is the risk-free cost of debt (e.g., 20-year U.S. Treasury bills around 6%), plus a premium for taking a risk as to whether a return will be received. The risk premium is the average return on the market, typically the mean of Standard & Poor's 500 large U.S. stocks, or S&P 500 (say, 12.5%), less the risk-free cost of debt. This premium is multiplied by *beta* (β), an approximate measure of stock price volatility that quantifies risk. β measures one firm's stock price relative to the market stock price as a whole. A number greater than unity means that the stock is *more* volatile than the market, on average; a number less than unity means that the stock is *less* volatile than the market, on average. Values for β for most publicly traded stocks are commonly found in various sources, such as Value-Line.[8] The following formula quantifies the cost of equity (i_e):

$$i_e = r_f + \beta[r_M - r_f]. \tag{15.5}$$

Inflation risk: The possibility that the value of assets or income will decrease as inflation shrinks the purchasing power of a currency.

Risk premium: The return in excess of the risk-free rate of return that an investment is expected to yield.

Beta: A measure of the volatility, or systematic risk, of a security or a portfolio in comparison to the market as a whole.

---

[7] An excellent discussion of this subject matter is found in Robert C. Higgins, *Analysis for Financial Management*, 5th ed. (Boston: Irwin/McGraw-Hill, 1998).

[8] Value Line reports are presently available for over 5,000 public companies, and the number is growing. The Value Line reports contain the following information: (1) total assets, (2) total liabilities, (3) total equity, (4) long-term debt as a percentage of capital, (5) equity as a percentage of capital, (6) financial strength (which is used to determine the interest rate), (7) β, and (8) return on invested capital.

Here, r_f is the risk-free interest rate (commonly referenced to the U.S. Treasury bond yield, adjusted for inflation) and r_M is the market rate of return (commonly referenced to the average return on S&P 500 stock index funds, adjusted for inflation).

Note that the cost of equity is almost always higher than the cost of debt. This is because the U.S. Tax Code allows the deduction of interest expense, but does not allow the deduction of the cost of equity, which could be considered more subjective and complicated. Example 15.5 illustrates how we may determine the cost of equity.

EXAMPLE 15.5 Determining the Cost of Equity by the Financial Market

Alpha Corporation needs to raise $10 million for plant modernization. Alpha's target capital structure calls for a debt ratio of 0.4, indicating that $6 million has to be financed from equity.

- Alpha is planning to raise $6 million from the financial market.
- Alpha's β is known to be 1.99, which is greater than unity, indicating that the firm is perceived as more risky than the market average.
- The risk-free interest rate is 6%, and the average market return is 13%. (All these interest rates are adjusted to reflect inflation in the economy.)

Determine the cost of equity to finance the plant modernization.

SOLUTION

Given: $r_M = 13\%$, $r_f = 6\%$, and $\beta = 1.99$.
Find: i_e.

$$i_e = 0.06 + 1.99(0.13 - 0.06)$$

$$= 19.93\%.$$

COMMENTS: In this example, we purposely selected the value of β to approximate the cost of equity derived from Example 15.4. What does this 19.93% represent? If Alpha finances the project entirely from its equity funds, the project must earn at least a 19.93% return on investment.

15.2.2 Cost of Debt

Now let us consider the calculation of the specific cost that is to be assigned to the debt component of the weighted-average cost of capital. The calculation is relatively straightforward and simple. As we said in Section 15.1.2, the two types of debt financing are term loans and bonds. Because the interest payments on both are tax deductible, the effective cost of debt will be reduced.

To determine the after-tax cost of debt (i_d), we evaluate the expression

$$i_d = \left(\frac{c_s}{c_d}\right)k_s(1 - t_m) + \left(\frac{c_b}{c_d}\right)k_b(1 - t_m), \tag{15.6}$$

where c_s is the amount of the short-term loan, k_s is the before-tax interest rate on the term loan, t_m is the firm's marginal tax rate, k_b is the before-tax interest rate on the bond, c_b is the amount of bond financing, and $c_s + c_b = c_d$.

As for bonds, a new issue of long-term bonds incurs flotation costs. These costs reduce the proceeds to the firm, thereby raising the specific cost of the capital raised. For example, when a firm issues a $1,000 par bond, but nets only $940, the flotation cost will be 6%. Therefore, the effective after-tax cost of the bond component will be higher than the nominal interest rate specified on the bond. We will examine this problem with a numerical example.

EXAMPLE 15.6 Determining the Cost of Debt

Consider again Example 15.4, and suppose that Alpha has decided to finance the remaining $4 million by securing a term loan and issuing 20-year $1,000 par bonds under the following conditions:

Cost of debt: The effective rate that a company pays on its current debt.

Source	Amount	Fraction	Interest Rate	Flotation Cost
Term loan	$1 million	0.333	12% per year	
Bonds	3 million	0.667	10% per year	6%

If the bond can be sold to net $940 (after deducting the 6% flotation cost), determine the cost of debt to raise $4 million for the plant modernization. Alpha's marginal tax rate is 38%, and it is expected to remain constant in the future.

SOLUTION

First, we need to find the effective after-tax cost of issuing the bond with a flotation cost of 6%. The before-tax specific cost is found by solving the equivalence formula

$$\$940 = \frac{\$100}{(1 + k_b)} + \frac{\$100}{(1 + k_b)^2} + \cdots + \frac{\$100 + \$1,000}{(1 + k_b)^{20}}$$

$$= \$100(P/A, k_b, 20) + \$1,000(P/F, k_b, 20).$$

Solving for k_b, we obtain $k_b = 10.74\%$. Note that the cost of the bond component increases from 10% to 10.74% after the 6% flotation cost is taken into account.

The after-tax cost of debt is the interest rate on debt, multiplied by $(1 - t_m)$. In effect, the government pays part of the cost of debt because interest is tax deductible. Now we are ready to compute the after-tax cost of debt as follows:

$$i_d = (0.333)(0.12)(1 - 0.38) + (0.667)(0.1074)(1 - 0.38)$$

$$= 6.92\%.$$

15.2.3 Calculating the Cost of Capital

With the specific cost of each financing component determined, we are ready to calculate the tax-adjusted weighted-average cost of capital based on total capital. Then we will define the marginal cost of capital that should be used in project evaluation.

Weighted-Average Cost of Capital

Cost of capital for a firm is a weighted sum of the cost of equity and the cost of debt.

Assuming that a firm raises capital on the basis of the target capital structure and that the target capital structure remains unchanged in the future, we can determine a **tax-adjusted weighted-average cost of capital** (or, simply stated, the **cost of capital**). This cost of capital represents a composite index reflecting the cost of raising funds from different sources. The cost of capital is defined as

$$k = \frac{i_d c_d}{V} + \frac{i_e c_e}{V}, \tag{15.7}$$

where c_d = Total debt capital (such as bonds) in dollars,

c_e = Total equity capital in dollars,

$V = c_d + c_e$,

i_e = Average equity interest rate per period, taking into account all equity sources,

i_d = After-tax average borrowing interest rate per period, taking into account all debt sources, and

k = Tax-adjusted weighted-average cost of capital.

Note that the cost of equity is already expressed in terms of after-tax cost, because any return to holders of either common stock or preferred stock is made after payment of income taxes. Figure 15.2 summarizes the process of determining the cost of capital.

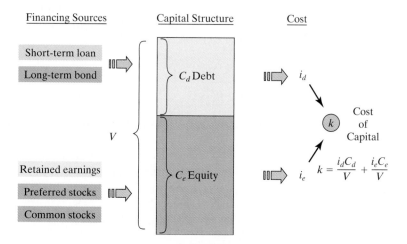

Figure 15.2 Process of calculating the cost of capital. Recall that there are two different ways of determining the cost of equity—one by the traditional approach and the other by the financial market, commonly known as "capital asset pricing model."

Marginal Cost of Capital

Now that we know how to calculate the cost of capital, we can ask, Could a typical firm raise unlimited new capital at the same cost? The answer is no. As a practical matter, as a firm tries to attract more new dollars, the cost of raising each additional dollar will at some point rise. As this occurs, the weighted-average cost of raising each additional new dollar also rises. Thus, the **marginal cost of capital** is defined as the cost of obtaining another dollar of new capital, and the marginal cost rises as more and more capital is raised during a given period. In evaluating an investment project, we are using the concept of the marginal cost of capital. The formula to find the marginal cost of capital is exactly the same as Eq. (15.6); however, the costs of debt and equity in that equation are the interest rates on new debt and equity, not outstanding (or combined) debt or equity. In other words, we are interested in the marginal cost of capital—specifically, to use it in evaluating a new investment project. The rate at which the firm has borrowed in the past is less important for this purpose. Example 15.7 works through the computations for finding the cost of capital (k).

EXAMPLE 15.7 Calculating the Marginal Cost of Capital

Consider again Examples 15.5 and 15.6. The marginal income tax rate (t_m) for Alpha is expected to remain at 38% in the future. Assuming that Alpha's capital structure (debt ratio) also remains unchanged, determine the cost of capital (k) for raising $10 million in addition to existing capital.

SOLUTION

With c_d = $4 million, c_e = $6 million, V = $10 million, i_d = 6.92%, i_e = 19.93%, and Eq. (15.7), we calculate

$$k = \frac{(0.0692)(4)}{10} + \frac{(0.1993)(6)}{10}$$

$$= 14.73\%.$$

This 14.73% would be the marginal cost of capital that a company with the given financial structure would expect to pay to raise $10 million.

15.3 Choice of Minimum Attractive Rate of Return

Thus far, we have said little about what interest rate, or minimum attractive rate of return (MARR), is suitable for use in a particular investment situation. Choosing the MARR is a difficult problem; no single rate is always appropriate. In this section, we will discuss briefly how to select a MARR for project evaluation. Then we will examine the relationship between capital budgeting and the cost of capital.

MARR: The required return necessary to make a capital budgeting project—such as building a new factory—worthwhile.

15.3.1 Choice of MARR when Project Financing Is Known

In Chapter 10, we focused on calculating after-tax cash flows, including situations involving debt financing. When cash flow computations reflect interest, taxes, and debt repayment, what is left is called **net equity flow**. If the goal of a firm is to maximize the wealth of its stockholders, why not focus only on the after-tax cash flow to equity, instead of on the flow to all suppliers of capital? Focusing on only the equity flows will permit us to use the cost of equity as the appropriate discount rate. In fact, we have implicitly assumed that all after-tax cash flow problems, in which financing flows are explicitly stated in earlier chapters, represent net equity flows, so the MARR used represents the **cost of equity** (i_e). Example 15.8 illustrates project evaluation by the net equity flow method.

EXAMPLE 15.8 Project Evaluation by Net Equity Flow

Suppose the Alpha Corporation, which has the capital structure described in Example 15.7, wishes to install a new set of machine tools, which is expected to increase revenues over the next five years. The tools require an investment of $150,000, to be financed with 60% equity and 40% debt. The equity interest rate (i_e), which combines the two sources of common and preferred stocks, is 19.96%. Alpha will use a 12% short-term loan to finance the debt portion of the capital ($60,000), with the loan to be repaid in equal annual installments over five years. Depreciation is according to a MACRS over a three-year property class life, and zero salvage value is expected. Additional revenues and operating costs are expected to be as follows:

n	Revenues ($)	Operating Cost
1	$68,000	$20,500
2	73,000	20,000
3	79,000	20,500
4	84,000	20,000
5	90,000	20,500

The marginal tax rate (combined federal, state, and city rate) is 38%. Evaluate this venture by using net equity flows at $i_e = 19.96\%$.

SOLUTION

The calculations are shown in Table 15.3. The NPW and IRR calculations are as follows:

$$PW(19.96\%) = -\$90,000 + \$34,541(P/F, 19.96\%, 1)$$
$$+ \$43,854(P/F, 19.96\%, 2) + \$29,893(P/F, 19.96\%, 3)$$
$$+ \$28,540(P/F, 19.96\%, 4) + \$27,124(P/F, 19.96\%, 5)$$
$$= \$11,285,$$
$$IRR = 25.91\% > 19.96\%.$$

The internal rate of return for this cash flow is 25.91%, which exceeds $i_e = 19.96\%$. Thus, the project would be profitable.

TABLE 15.3 After-Tax Cash Flow Analysis when Project Financing Is Known: Net Equity Cash Flow Method (Example 15.8)

End of Year	0	1	2	3	4	5
Income statement:						
Revenue		$ 68,000	$ 73,000	$ 79,000	$ 84,000	$ 90,000
Expenses:						
Operating cost		20,500	20,000	20,500	20,000	20,500
Interest payment		7,200	6,067	4,797	3,376	1,783
Depreciation		50,000	66,667	22,222	11,111	0
Taxable income		(9,700)	(19,734)	31,481	49,513	67,717
Income taxes (38%)		(3,686)	(7,499)	11,963	18,815	25,732
Net income		$ (6,014)	$ (12,235)	$ 19,518	$ 30,698	$ 41,985
Cash flow statement:						
Operating activities:						
Net income		$ (6,014)	$ (12,235)	$ 19,518	$ 30,698	$ 41,985
Depreciation		50,000	66,667	22,222	11,111	0
Investment activities:						
Investment	$ (150,000)					
Salvage value						0
Gains tax						0
Financing activities:						
Borrowed funds	60,000					
Principal repayment		(9,445)	(10,578)	(11,847)	(13,269)	(14,861)
Net equity flow	$ (90,000)	$ 34,541	$ 43,854	$ 29,893	$ 28,540	$ 27,124

COMMENT: In this problem, we assumed that the Alpha Corporation would be able to raise the additional equity funds at the same rate of 19.96%, so this 19.96% can be viewed as the marginal cost of capital.

15.3.2 Choice of MARR when Project Financing Is Unknown

You might well ask, Why, if we use the cost of equity (i_e) exclusively, of what use is the k? The answer to this question is that, by using the value of k, we may evaluate investments without explicitly treating the debt flows (both interest and principal). In this case, we make a tax adjustment to the discount rate by employing the effective after-tax cost of debt. This approach recognizes that the net interest cost is effectively transferred from the tax collector to the creditor in the sense that there is a dollar-for-dollar reduction in taxes up to the debt interest payments. Therefore, debt financing is treated implicitly. The method would be appropriate when debt financing is not identified with individual investments, but rather enables the company to engage in a set of investments. (Except where financing flows are explicitly stated, all previous examples in this book have implicitly assumed the more realistic and appropriate situation wherein debt financing is not identified with individual investment. Therefore, the MARRs represent the weighted cost of capital [k].) Example 15.9 illustrates this concept.

EXAMPLE 15.9 Project Evaluation by Marginal Cost of Capital

In Example 15.8, suppose that Alpha Corporation has not decided how the $150,000 will be financed. However, Alpha believes that the project should be financed according to its target capital structure, with a debt ratio of 40%. Using k, find the NPW and IRR.

SOLUTION

By not accounting for the cash flows related to debt financing, we calculate the after-tax cash flows as shown in Table 15.4. Notice that when we use this procedure, interest and the resulting tax shield are ignored in deriving the net incremental after-tax cash flow. In other words, no cash flow is related to any financing activity. Thus, taxable income is overstated, as are income taxes. To compensate for these overstatements, the discount rate is reduced accordingly. The implicit assumption is that the tax overpayment is exactly equal to the reduction in interest implied by i_d.

TABLE 15.4 After-Tax Cash Flow Analysis when Project Financing Is Unknown: Cost-of-Capital Approach (Example 15.9)

End of Year	0	1	2	3	4	5
Income statement:						
Revenue		$ 68,000	$ 73,000	$ 79,000	$ 84,000	$ 90,000
Expenses:						
Operating cost		20,500	20,000	20,500	20,000	20,500
Depreciation		50,000	66,667	22,222	11,111	0
Taxable income		$ (2,500)	$ (13,667)	$ 36,278	$ 52,889	$ 69,500
Income taxes (38%)		(950)	(5,193)	13,786	20,098	26,410
Net income		$ (1,550)	$ (8,474)	$ 22,492	$ 32,791	$ 43,090
Cash flow statement:						
Operating activities						
Net income		$ (1,550)	$ (8,474)	$ 22,492	$ 32,791	$ 43,090
Depreciation		50,000	66,667	22,222	11,111	0
Investment activities:						
Investment	$ (150,000)					
Salvage value						0
Gains tax						0
Net cash flow	$ (150,000)	$ 48,450	$ 58,193	$ 44,714	$ 43,902	$ 43,090

The flow at time 0 is simply the total investment, $150,000 in this example. Recall that Alpha's k was calculated to be 14.73%. The internal rate of return for the after-tax flow in the last line of Table 15.4 is calculated as follows:

$$\text{PW}(i) = -\$150,000 + \$48,450(P/F, i, 1)$$
$$+ \$58,193(P/F, i, 2) + \$44,714(P/F, i, 3)$$
$$+ \$43,902(P/F, i, 4) + \$43,090(P/F, i, 5)$$
$$= 0,$$

$$\text{IRR} = 18.47\% > 14.73\%.$$

Since the IRR exceeds the value of k, the investment would be profitable. Here, we evaluated the after-tax flow by using the value of k, and we reached the same conclusion about the desirability of the investment as we did in Example 15.8.

COMMENTS: The net equity flow and the cost-of-capital methods usually lead to the same accept/reject decision for independent projects (assuming the same amortization schedule for debt repayment, such as term loans) and usually rank projects identically for mutually exclusive alternatives. Some differences may be observed as special financing arrangements may increase (or even decrease) the attractiveness of a project by manipulating tax shields and the timing of financing inflows and payments.

In sum, in cases where the exact debt-financing and repayment schedules are known, we recommend the use of the net equity flow method. The appropriate MARR would be the cost of equity, i_e. If no specific assumption is made about the exact instruments that will be used to finance a particular project (but we do assume that the given capital-structure proportions will be maintained), we may determine the after-tax cash flows without incorporating any debt cash flows. Then we use the marginal cost of capital (k) as the appropriate MARR.

15.3.3 Choice of MARR under Capital Rationing

It is important to distinguish between the cost of capital (k), as calculated in Section 15.2.3, and the MARR (i) used in project evaluation under **capital rationing**—situations in which the funds available for capital investment are not sufficient to cover all potentially acceptable projects. When investment opportunities exceed the available money supply, we must decide which opportunities are preferable. Obviously, we want to ensure that all the selected projects are more profitable than the best rejected project (or the worst accepted project), which is the best opportunity forgone and whose value is called the **opportunity cost**. When a limit is placed on capital, the MARR is assumed to be equal to this opportunity cost, which is usually greater than the marginal cost of capital. In other words, the value of i represents the corporation's time-value trade-offs and partially reflects the available investment opportunities. Thus, there is nothing illogical about borrowing money at k and evaluating cash flows according to the different rate i. Presumably, the money will be invested to earn a rate i or greater. In the next example, we will provide guidelines for selecting a MARR for project evaluation under capital rationing.

A company may borrow funds to invest in profitable projects, or it may return to (invest in) its **investment pool** any unused funds until they are needed for other investment activities. Here, we may view the borrowing rate as a marginal cost of capital (k), as defined in Eq. (15.7). Suppose that all available funds can be placed in investments yielding a return equal to l, the **lending rate**. We view these funds as an investment pool. The firm may withdraw funds from this pool for other investment purposes, but if left in the pool, the funds will earn at the rate r (which is thus the opportunity cost). The MARR is thus related to either the borrowing interest rate or the lending interest rate. To illustrate the relationship among the borrowing interest rate, the lending interest rate, and the MARR, let us define the following variables:

$$k = \text{Borrowing rate (or the cost of capital)},$$

$$l = \text{Lending rate (or the opportunity cost)},$$

$$i = \text{MARR}.$$

Generally (but not always), we might expect k to be greater than or equal to l: We must pay more for the use of someone else's funds than we can receive for "renting out" our own funds (unless we are running a lending institution). Then we will find that the appropriate MARR would be between l and k. The concept for developing a discount rate (MARR) under capital rationing is best understood by a numerical example.

EXAMPLE 15.10 Determining an Appropriate MARR as a Function of the Budget

Sand Hill Corporation has identified six investment opportunities that will last one year. The firm drew up a list of all potentially acceptable projects. The list shows each project's required investment, projected annual net cash flows, life, and IRR. Then it ranks the projects according to their IRR, listing the highest IRR first:

Project	Cash Flow A0	A1	IRR
1	−$10,000	$12,000	20%
2	− 10,000	11,500	15
3	− 10,000	11,000	10
4	− 10,000	10,800	8
5	− 10,000	10,700	7
6	− 10,000	10,400	4

Suppose $k = 10\%$, which remains constant for the budget amount up to $60,000, and $l = 6\%$. Assuming that the firm has available (a) $40,000, (b) $60,000, and (c) $0 for investments, what is the reasonable choice for the MARR in each case?

SOLUTION

We will consider the following steps to determine the appropriate discount rate (MARR) under a capital-rationing environment:

Step 1: Develop the firm's cost-of-capital schedule as a function of the capital budget. For example, the cost of capital can increase as the amount of financing increases. Also, determine the firm's lending rate if any unspent money is lent out or remains invested in the company's investment pool.

Step 2: Plot this investment opportunity schedule by showing how much money the company could invest at different rates of return as shown in Figure 15.3.

(a) If the firm has $40,000 available for investing, it should invest in Projects 1, 2, 3, and 4. Clearly, it should not borrow at 10% to invest in either Project 5 or Project 6. In these cases, the best rejected project is Project 5. The worst accepted project is Project 4. If you view the opportunity cost as the cost associated with accepting the worst project, the MARR could be 8%.

(b) If the firm has $60,000 available, it should invest in Projects 1, 2, 3, 4, and 5. It could lend the remaining $10,000 rather than invest these funds in Project 6. For this new situation, we have MARR $= l = 6\%$.

(c) If the firm has no funds available, it probably would borrow to invest in Projects 1 and 2. The firm might also borrow to invest in Project 3, but it would be indifferent to this alternative, unless some other consideration was involved. In this case, MARR $= k = 10\%$; therefore, we can say that $l \leq$ MARR $\leq k$ when we have complete certainty about future investment opportunities. Figure 15.4 illustrates the concept of selecting a MARR under capital rationing.

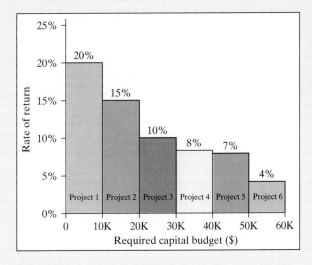

Figure 15.3 An investment opportunity schedule ranking alternatives by the rate of return (Example 15.10).

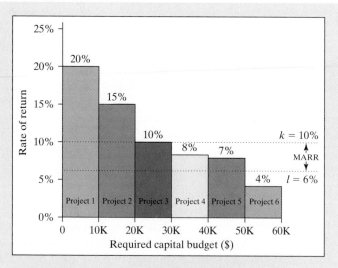

Figure 15.4 A range of MARRs as a function of a budget under capital rationing (Example 15.10): If you have an unlimited budget, MARR = lending rate; if you have no budget, but are allowed to borrow, MARR = 10%.

COMMENTS: In this example, for simplicity, we assumed that the timing of each investment is the same for all competing proposals—say, period 0. If each alternative requires investments over several periods, the analysis will be significantly complicated, as we have to consider both the investment amount and its timing in selecting the appropriate MARR. This is certainly beyond the scope of any introductory engineering economics text, but can be found in C. S. Park and G. P. Sharp-Bette, *Advanced Engineering Economics* (New York: John Wiley, 1990).

Now we can generalize what we have learned. If a firm finances investments through borrowed funds, it should use MARR = k; if the firm is a lender, it should use MARR = l. A firm may be a lender in one period and a borrower in another; consequently, the appropriate rate to use may vary from period to period. In fact, whether the firm is a borrower or a lender may well depend on its investment decisions.

In practice, most firms establish a single MARR for all investment projects. Note the assumption that we made in Example 15.10: **Complete certainty** about investment opportunities was assumed. Under highly uncertain economic environments, the MARR would generally be much greater than k, the firm's cost of capital. For example, if $k = 10\%$, a MARR of 15% would not be considered excessive. Few firms are willing to invest in projects earning only slightly more than their cost of capital, because of elements of *risk* in the project.

If the firm has a large number of current and future opportunities that will yield the desired return, we can view the MARR as the minimum rate at which the firm is willing to invest, and we can also assume that proceeds from current investments can be reinvested to earn at the MARR. Furthermore, *if we choose the "do-nothing" alternative, all available funds are invested at the MARR*. In engineering economics, we also normally separate the risk issue from the concept of MARR. As seen in Chapter 12, we treat the effects of risk explicitly when we must. Therefore, any reference to the MARR in this book refers strictly to the risk-free interest rate.

15.4 Capital Budgeting

In this section, we will examine the process of deciding whether projects should be included in the capital budget. In particular, we will consider decision procedures that should be applied when we have to evaluate a set of multiple investment alternatives for which we have a limited capital budget.

15.4.1 Evaluation of Multiple Investment Alternatives

In Chapters 5, 6, and 7, we learned how to compare two or more mutually exclusive projects. Now we shall extend the comparison techniques to a set of multiple decision alternatives that are not necessarily mutually exclusive. Here, we distinguish a **project** from an **investment alternative**, which is a decision option. For a single project, we have two investment alternatives: to accept or reject the project. For two independent projects, we can have four investment alternatives: (1) to accept both projects, (2) to reject both projects, (3) to accept only the first project, and (4) to accept only the second project. As we add interrelated projects, the number of investment alternatives to consider grows exponentially.

To perform a proper capital-budgeting analysis, a firm must group all projects under consideration into decision alternatives. This grouping requires the firm to distinguish between projects that are independent of one another and projects that are dependent on one another in order to formulate alternatives correctly.

Independent Projects

An **independent project** is a project that may be accepted or rejected without influencing the accept–reject decision of another independent project. For example, the purchase of a milling machine, office furniture, and a forklift truck constitutes three independent projects. Only projects that are economically independent of one another can be evaluated separately. (Budget constraints may prevent us from selecting one or more of several independent projects; this external constraint does not alter the fact that the projects are independent.)

Dependent Projects

In many decision problems, several investment projects are related to one another such that the acceptance or rejection of one project influences the acceptance or rejection of others. Two such types of dependencies are mutually exclusive projects and contingent projects. We say that two or more projects are **contingent** if the acceptance of one requires the acceptance of the other. For example, the purchase of a computer printer is dependent upon the purchase of a computer, but the computer may be purchased without purchasing the printer.

15.4.2 Formulation of Mutually Exclusive Alternatives

We can view the selection of investment projects as a problem of selecting a single decision alternative from a set of mutually exclusive alternatives. Note that each investment project is an investment alternative, but that a single investment alternative may entail a whole group of investment projects. The common method of handling various project relationships is to arrange the investment projects so that the selection decision involves only mutually exclusive alternatives. To obtain this set of mutually exclusive alternatives, we need to enumerate all of the feasible combinations of the projects under consideration.

Independent Projects

With a given number of independent investment projects, we can easily enumerate mutually exclusive alternatives. For example, in considering two projects, A and B, we have four decision alternatives, including a do-nothing alternative:

Alternative	Description	X_A	X_B
1	Reject A, Reject B	0	0
2	Accept A, Reject B	1	0
3	Reject A, Accept B	0	1
4	Accept A, Accept B	1	1

In our notation, X_j is a decision variable associated with investment project j. If $X_j = 1$, project j is accepted; if $X_j = 0$, project j is rejected. Since the acceptance of one of these alternatives will exclude any other, the alternatives are mutually exclusive.

Mutually Exclusive Projects

Suppose we are considering two independent sets of projects (A and B). Within each independent set are two mutually exclusive projects (A1, A2) and (B1, B2). The selection of either A1 or A2, however, is also independent of the selection of any project from the set (B1, B2). In other words, you can select A1 and B1 together, but you cannot select A1 and A2 together. For this set of investment projects, the mutually exclusive alternatives are as follows:

Alternative	(X_{A1}, X_{A2})	(X_{B1}, X_{B2})
1	(0, 0)	(0, 0)
2	(1, 0)	(0, 0)
3	(0, 1)	(0, 0)
4	(0, 0)	(1, 0)
5	(0, 0)	(0, 1)
6	(1, 0)	(1, 0)
7	(0, 1)	(1, 0)
8	(1, 0)	(0, 1)
9	(0, 1)	(0, 1)

Note that, with two independent sets of mutually exclusive projects, we have nine different decision alternatives.

Contingent Projects

Suppose the acceptance of C is contingent on the acceptance of both A and B, and the acceptance of B is contingent on the acceptance of A. Then the number of decision alternatives can be formulated as follows:

Alternative	X_A	X_B	X_C
1	0	0	0
2	1	0	0
3	1	1	0
4	1	1	1

Thus, we can easily formulate a set of mutually exclusive investment alternatives with a limited number of projects that are independent, mutually exclusive, or contingent merely by arranging the projects in a logical sequence.

One difficulty with the enumeration approach is that, as the number of projects increases, the number of mutually exclusive alternatives increases exponentially. For example, for 10 independent projects, the number of mutually exclusive alternatives is 2^{10}, or 1,024. For 20 independent projects, 2^{20}, or 1,048,576, alternatives exist. As the number of decision alternatives increases, we may have to resort to mathematical programming to find the solution to an investment problem. Fortunately, in real-world business, the number of engineering projects to consider at any one time is usually manageable, so the enumeration approach is a practical one.

15.4.3 Capital-Budgeting Decisions with Limited Budgets

Recall that capital rationing refers to situations in which the funds available for capital investment are not sufficient to cover all the projects. In such situations, we enumerate all investment alternatives as before, but eliminate from consideration any mutually exclusive alternatives that exceed the budget. The most efficient way to proceed in a capital-rationing situation is to select the group of projects that maximizes the total NPW of future cash flows over required investment outlays. Example 15.11 illustrates the concept of an optimal capital budget under a rationing situation.

EXAMPLE 15.11 Four Energy-Saving Projects under Budget Constraints

The facilities department at an electronic instrument firm has four energy-efficiency projects under consideration:

Project 1 (electrical). This project requires replacing the existing standard-efficiency motors in the air-conditioners and exhaust blowers of a particular building with high-efficiency motors.

Project 2 (building envelope). This project involves coating the inside surface of existing fenestration in a building with low-emissivity solar film.

Project 3 (air-conditioning). This project requires the installation of heat exchangers between a building's existing ventilation and relief air ducts.

Project 4 (lighting). This project requires the installation of specular reflectors and the delamping of a building's existing ceiling grid-lighting troffers.

These projects require capital outlays in the $50,000 to $140,000 range and have useful lives of about eight years. The facilities department's first task was to estimate the annual savings that could be realized by these energy-efficiency projects. Currently, the company pays 7.80 cents per kilowatt-hour (kWh) for electricity and $4.85 per thousand cubic feet (MCF). Assuming that the current energy prices would continue for the next eight years, the company has estimated the cash flow and the IRR for each project as follows:

Project	Investment	Annual O&M Cost	Annual Savings (Energy)	Annual Savings (Dollars)	IRR
1	$46,800	$1,200	151,000 kWh	$11,778	15.43%
2	104,850	1,050	513,077 kWh	40,020	33.48%
3	135,480	1,350	6,700,000 CF	32,493	15.95%
4	94,230	942	385,962 kWh	30,105	34.40%

Because each project could be adopted in isolation, at least as many alternatives as projects are possible. For simplicity, assume that all projects are independent, as opposed to being mutually exclusive, that they are equally risky, and that their risks are all equal to those of the firm's average existing assets.

(a) Determine the optimal capital budget for the energy-saving projects.
(b) With $250,000 approved for energy improvement funds during the current fiscal year, the department did not have sufficient capital on hand to undertake all four projects without any additional allocation from headquarters. Enumerate the total number of decison alternatives and select the best alternative.

DISCUSSION: The NPW calculation cannot be shown yet, as we do not know the marginal cost of capital. Therefore, our first task is to develop the **marginal-cost-of-capital (MCC)** schedule, a graph that shows how the cost of capital changes as more and more new capital is raised during a given year. The graph in Figure 15.5 is the company's marginal-cost-of-capital schedule. The first $100,000 would be raised at 14%, the next $100,000 at 14.5%, the next $100,000 at 15%, and any amount over $300,000 at 15.5%. We then plot the IRR data for each project as the **investment opportunity schedule (IOS)** shown in the graph. The IOS shows how much money the firm could invest at different rates of return.

SOLUTION

(a) Optimal capital budget if projects can be accepted in part:
Consider Project A4. Its IRR is 34.40%, and it can be financed with capital that costs only 14%. Consequently, it should be accepted. Projects A2 and A3 can be analyzed similarly; all are acceptable because the IRR exceeds the marginal cost of capital. Project A1, by contrast, should be rejected because its IRR is less than the marginal cost of capital. Therefore, the firm should accept the three projects A4, A2, and A3, which have rates of return in excess of the cost of capital

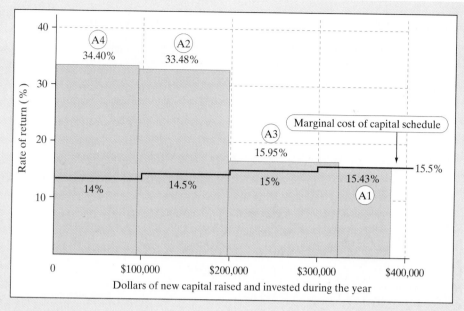

Figure 15.5 Combining the marginal-cost-of-capital schedule and investment opportunity schedule curves to determine a firm's optimal capital budget (Example 15.11).

that would be used to finance them if we end up with a capital budget of $334,560. This should be the amount of the *optimal capital budget*.

In Figure 15.5, even though two rate changes occur in the marginal cost of capital in funding Project A3 (the first change is from 14.5% to 15%, the second from 15% to 15.5%), the accept/reject decision for A3 remains unchanged, as its rate of return exceeds the marginal cost of capital. What would happen if the MCC cut through Project A3? For example, suppose that the marginal cost of capital for any project raised above $300,000 would cost 16% instead of 15.5%, thereby causing the MCC schedule to cut through Project A3. Should we then accept A3? If we can take A3 in part, we would take on only part of it up to 74.49%.

(b) Optimal capital budget if projects cannot be accepted in part:
If projects cannot be funded partially, we first need to enumerate the number of feasible investment decision alternatives within the budget limit. As shown in Table 15.5, the total number of mutually exclusive decision alternatives that can be obtained from four independent projects is 16, including the do-nothing alternative. However, decision alternatives 13, 14, 15, and 16 are not feasible, because of a $250,000 budget limit. So we need to consider only alternatives 1 through 12.

Now, how do we compare these alternatives as the marginal cost of capital changes for each one? Consider again Figure 15.5. If we take A1 first, it would be acceptable, because its 15.43% return would exceed the 14% cost of capital used to finance it. Why couldn't we do this? The answer is that we are seeking to maximize the excess of returns over costs, or the area above the MCC, but below the IOS. We accomplish that by accepting the most profitable projects

TABLE 15.5 Mutually Exclusive Decision Alternatives (Example 15.11)

j	Alternative	Required Budget	Combined Annual Savings	
1	0	0	0	
2	A1	$ (46,800)	$ 10,578	
3	A2	(104,850)	38,970	
4	A3	(135,480)	31,143	
5	A4	(94,230)	35,691	
6	A4,A1	(141,030)	46,269	
7	A2,A1	(151,650)	49,548	
8	A3,A1	(182,280)	41,721	
9	A4,A2	(199,080)	74,661	
10	A4,A3	(229,710)	66,834	
11	A2,A3	(240,330)	70,113	
12	A4,A2,A1	(245,880)	85,239	Best alternative
13	A4,A3,A1	(276,510)	77,412	
14	A2,A3,A1	(287,130)	80,691	
15	A4,A2,A3	(334,560)	105,804	Infeasible alternatives
16	A4,A2,A3,A1	(381,360)	116,382	

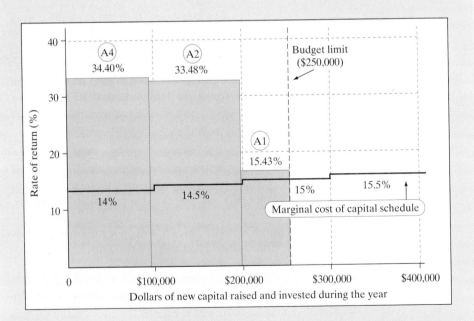

Figure 15.6 The appropriate cost of capital to be used in the capital-budgeting process for decision alternative 12, with a $250,000 budget limit (Example 15.11) is 15%.

first. This logic leads us to conclude that, as long as the budget permits, A4 should be selected first and A2 second. This will consume $199,080, which leaves us $50,920 in unspent funds. The question is, What are we going to do with this leftover money? Certainly, it is not enough to take A3 in full, but we can take A1 in full. Full funding for A1 will fit the budget, and the project's rate of return still exceeds the marginal cost of capital $(15.43\% > 15\%)$. Accordingly, unless the leftover funds earn more than 15.43% interest, alternative 12 becomes the best (Figure 15.6).

COMMENTS: In Example 15.10, the MARR was found by applying a capital limit to the investment opportunity schedule. The firm was then allowed to borrow or lend the money as the investment situation dictated. In this example, no such borrowing is explicitly assumed.

SUMMARY

■ Methods of financing fall into two broad categories:

1. **Equity financing** uses retained earnings or funds raised from an issuance of stock to finance a capital investment.

2. **Debt financing** uses money raised through loans or by an issuance of bonds to finance a capital investment.

■ Companies do not simply borrow funds to finance projects. Well-managed firms usually establish a **target capital structure** and strive to maintain the **debt ratio** when individual projects are financed.

■ The cost-of-capital formula is a composite index reflecting the cost of funds raised from different sources. The formula is

$$k = \frac{i_d c_d}{V} + \frac{i_e c_e}{V}, V = c_d + c_e.$$

■ The selection of an appropriate MARR depends generally upon the **cost of capital**—the rate the firm must pay to various sources for the use of capital:

1. The **cost of equity** (i_e) is used when debt-financing methods and repayment schedules are known explicitly.

2. The **cost of capital** (k) is used when exact financing methods are unknown, but a firm keeps its capital structure on target. In this situation, a project's after-tax cash flows contain no debt cash flows, such as principal and interest payment.

■ The **marginal cost of capital** is defined as the cost of obtaining another dollar of new capital. The marginal cost rises as more and more capital is raised during a given period.

▓ Without a capital limit, the choice of MARR is dictated by the availability of financing information:

1. In cases where the exact debt-financing and repayment schedules are known, we recommend the use of the net equity flow method. The proper MARR would be the cost of equity, i_e.

2. If no specific assumption is made about the exact instruments that will be used to finance a particular project (but we do assume that the given capital-structure proportions will be maintained), we may determine the after-tax cash flows without incorporating any debt cash flows. Then we use the marginal cost of capital (k) as the proper MARR.

Conditions	MARR
A firm borrows some capital from lending institutions at the borrowing rate k and some from its investment pool at the lending rate l.	$l < \text{MARR} < k$
A firm borrows all capital from lending institutions at the borrowing rate k.	$\text{MARR} = k$
A firm borrows all capital from its investment pool at the lending rate l.	$\text{MARR} = l$

▓ Under a highly uncertain economic environment, the MARR generally would be much greater than k, the firm's cost of capital, as the risk premium increases.

▓ Under conditions of capital rationing, the selection of the MARR is more difficult, but generally, the following possibilities exist:

• The cost of capital used in the capital-budgeting process is determined at the intersection of the **IOS** and **MCC** schedules. If the cost of capital at the intersection is used, then the firm will make correct accept/reject decisions, and its level of financing and investment will be optimal. This view assumes that the firm can invest and borrow at the rate where the two curves intersect.

• If a strict budget is placed in a capital-budgeting problem and no projects can be taken in part, all feasible investment decision scenarios need to be enumerated. Depending upon each such scenario, the cost of capital will also likely change. The task is then to find the best investment scenario in light of a changing-cost-of-capital environment. As the number of projects to consider increases, we may eventually resort to a more advanced technique, such as a mathematical programming procedure.

PROBLEMS

Methods of Financing

15.1 Optical World Corporation, a manufacturer of peripheral vision storage systems, needs $10 million to market its new robotics-based vision systems. The firm is considering two financing options: common stock and bonds. If the firm decides to raise the capital through issuing common stock, the flotation costs will be 6% and the share price will be $25. If the firm decides to use debt financing, it can sell a 10-year, 12% bond with a par value of $1,000. The bond flotation costs will be 1.9%.

(a) For equity financing, determine the flotation costs and the number of shares to be sold to net $10 million.

(b) For debt financing, determine the flotation costs and the number of $1,000 par value bonds to be sold to net $10 million. What is the required annual interest payment?

15.2 Consider a project whose initial investment is $300,000, financed at an interest rate of 12% per year. Assuming that the required repayment period is six years, determine the repayment schedule by identifying the principal as well as the interest payments for each of the following methods:

(a) Equal repayment of the principal.

(b) Equal repayment of the interest.

(c) Equal annual installments.

15.3 A chemical plant is considering purchasing a computerized control system. The initial cost is $200,000, and the system will produce net savings of $100,000 per year. If purchased, the system will be depreciated under MACRS as a five-year recovery property. The system will be used for four years, at the end of which time the firm expects to sell it for $30,000. The firm's marginal tax rate on this investment is 35%. Any capital gains will be taxed at the same income tax rate. The firm is considering purchasing the computer control system either through its retained earnings or through borrowing from a local bank. Two commercial banks are willing to lend the $200,000 at an interest rate of 10%, but each requires different repayment plans. Bank A requires four equal annual principal payments, with interest calculated on the basis of the unpaid balance:

Repayment Plan of Bank A		
End of Year	**Principal**	**Interest**
1	$50,000	$20,000
2	50,000	15,000
3	50,000	10,000
4	50,000	5,000

Bank B offers a payment plan that extends over five years, with five equal annual payments:

Repayment Plan of Bank B			
End of Year	**Principal**	**Interest**	**Total**
1	$32,759	$20,000	$52,759
2	36,035	16,724	52,759
3	39,638	13,121	52,759
4	43,602	9,157	52,759
5	47,998	4,796	52,759

(a) Determine the cash flows if the computer control system is to be bought through its retained earnings (equity financing).

(b) Determine the cash flows if the asset is financed through either bank A or bank B.

(c) Recommend the best course of financing the project. (Assume that the firm's MARR is known to be 10%.)

15.4 Edison Power Company currently owns and operates a coal-fired combustion turbine plant that was installed 20 years ago. Because of degradation of the system, 65 forced outages occurred during the last year alone and two boiler explosions during the last seven years. Edison is planning to scrap the current plant and install a new, improved gas turbine that produces more energy per unit of fuel than typical coal-fired boilers produce.

The 50-MW gas-turbine plant, which runs on gasified coal, wood, or agricultural wastes, will cost Edison $65 million. Edison wants to raise the capital from three financing sources: 45% common stock, 10% preferred stock (which carries a 6% cash dividend when declared), and 45% borrowed funds. Edison's investment banks quote the following flotation costs:

Financing Source	Flotation Costs	Selling Price	Par Value
Common stock	4.6%	$32/share	$10
Preferred stock	8.1	55/share	15
Bond	1.4	980	1,000

(a) What are the total flotation costs to raise $65 million?

(b) How many shares (both common and preferred) or bonds must be sold to raise $65 million?

(c) If Edison makes annual cash dividends of $2 per common share, and annual bond interest payments are at the rate of 12%, how much cash should Edison have available to meet both the equity and debt obligation? (Note that whenever a firm declares cash dividends to its common stockholders, the preferred stockholders are entitled to receive dividends of 6% of par value.)

Cost of Capital

15.5 Calculate the after-tax cost of debt under each of the following conditions:

(a) Interest rate, 12%; tax rate, 25%.

(b) Interest rate, 14%; tax rate, 34%.

(c) Interest rate, 15%; tax rate, 40%.

15.6 Sweeney Paper Company is planning to sell $10 million worth of long-term bonds with an 11% interest rate. The company believes that it can sell the $1,000 par value bonds at a price that will provide a yield to maturity of 13%. The flotation costs will be 1.9%. If Sweeney's marginal tax rate is 35%, what is its after-tax cost of debt?

15.7 Mobil Appliance Company's earnings, dividends, and stock price are expected to grow at an annual rate of 12%. Mobil's common stock is currently traded at $18 per share. Mobil's last cash dividend was $1.00, and its expected cash dividend for the end of this year is $1.12. Determine the cost of retained earnings (k_r).

15.8 Refer to Problem 15.7, and suppose that Mobil wants to raise capital to finance a new project by issuing new common stock. With the new project, the cash dividend is expected to be $1.10 at the end of the current year, and its growth rate is 10%. The stock now sells for $18, but new common stock can be sold to net Mobil $15 per share.

(a) What is Mobil's flotation cost, expressed as a percentage?

(b) What is Mobil's cost of new common stock (k_e)?

15.9 The Callaway Company's cost of equity is 22%. Its before-tax cost of debt is 13%, and its marginal tax rate is 40%. The firm's capital structure calls for a debt-to-equity ratio of 45%. Calculate Callaway's cost of capital.

15.10 Delta Chemical Corporation is expected to have the following capital structure for the foreseeable future:

Source of After-Tax Financing	Percent of Total Funds	Before-Tax Cost	Cost
Debt	30%		
Short term	10	14%	
Long term	20	12	
Equity	70%		
Common stock	55		30%

The flotation costs are already included in each cost component. The marginal income tax rate (t_m) for Delta is expected to remain at 40% in the future.

(a) Determine the cost of capital (k).

(b) If the risk-free rate is known to be 6% and the average return on S&P 500 is about 12%, determine the cost of equity with $\beta = 1.2$, based on the capital asset pricing principle.

(c) Determine the cost of capital on the basis of the cost of equity obtained in (b).

15.11 Charleston Textile Company is considering acquiring a new knitting machine at a cost of $200,000. Because of a rapid change in fashion styles, the need for this particular machine is expected to last only five years, after which the machine is expected to have a salvage value of $50,000. The annual operating cost is estimated at $10,000. The addition of this machine to the current production facility is expected to generate an additional revenue of $90,000 annually and will be depreciated in the seven-year MACRS property class. The income tax rate applicable to Charleston is 36%. The initial investment will be financed with 60% equity and 40% debt. The before-tax debt interest rate, which combines both short-term and long-term financing, is 12%, with the loan to be repaid in equal annual installments. The equity interest rate (i_e), which combines the two sources of common and preferred stocks, is 18%.

(a) Evaluate this investment project by using net equity flows.

(b) Evaluate this investment project by using k.

15.12 The Huron Development Company is considering buying an overhead pulley system. The new system has a purchase price of $100,000, an estimated useful life and MACRS class life of five years, and an estimated salvage value of $30,000. It is expected to allow the company to economize on electric power usage, labor, and repair costs, as well as to reduce the number of defective products. A total annual savings of $45,000 will be realized if the new pulley system is installed. The company is in the 30% marginal tax bracket. The initial investment will be financed with 40% equity and 60% debt. The before-tax debt interest rate, which combines both short-term and long-term financing, is 15%, with the loan to be repaid in equal annual installments over the project life. The equity interest rate (i_e), which combines the two sources of common and preferred stocks, is 20%.

(a) Evaluate this investment project by using net equity flows.

(b) Evaluate this investment project by using k.

15.13 Consider the following two mutually exclusive machines:

	Machine A	Machine B
Initial investment	$40,000	$60,000
Service life	6 years	6 years
Salvage value	$ 4,000	$ 8,000
Annual O&M cost	$ 8,000	$10,000
Annual revenues	$20,000	$28,000
MACRS property	5 year	5 year

The initial investment will be financed with 70% equity and 30% debt. The before-tax debt interest rate, which combines both short-term and long-term financing, is 10%, with the loan to be repaid in equal annual installments over the project life. The equity interest rate (i_e), which combines the two sources of common and preferred stock, is 15%. The firm's marginal income tax rate is 35%.

(a) Compare the alternatives, using $i_e = 15\%$. Which alternative should be selected?

(b) Compare the alternatives, using k. Which alternative should be selected?

(c) Compare the results obtained in (a) and (b).

Capital Budgeting

15.14 DNA Corporation, a biotech engineering firm, has identified seven R&D projects for funding. Each project is expected to be in the R&D stage for three to five years. The IRR figures shown in the following table represent the royalty income from selling the R&D results to pharmaceutical companies:

Project	Investment Type	Required IRR
1. Vaccines	$15M	22%
2. Carbohydrate chemistry	25M	40
3. Antisense	35M	80
4. Chemical synthesis	5M	15
5. Antibodies	60M	90
6. Peptide chemistry	23M	30
7. Cell transplant/gene therapy	19M	32

DNA Corporation can raise only $100 million. DNA's borrowing rate is 18% and its lending rate is 12%. Which R&D projects should be included in the budget?

15.15 Gene Fowler owns a house that contains 202 square feet of windows and 40 square feet of doors. Electricity usage totals 46,502 kWh: 7,960 kWh for lighting and appliances, 5,500 kWh for water heating, 30,181 kWh for space heating to 68°, and 2,861 kWh for space cooling to 78°F. Fowler can borrow money at 12% and lends money at 8%. The following 14 energy-savings alternatives have been suggested by the local power company for Fowler's 1,620-square-foot home:

Structural Improvement	Annual Savings	Estimated Costs	Payback Period (Years)
1. Add storm windows	$128–156	$455–556	3.5
2. Insulate ceilings to R-30	149–182	408–499	2.7
3. Insulate floors to R-11	158–193	327–399	2.1
4. Caulk windows and doors	25–31	100–122	4.0
5. Weather-strip windows and doors	31–38	224–274	7.2
6. Insulate ducts	184–225	1,677–2,049	9.1
7. Insulate space-heating water pipes	41–61	152–228	3.7
8. Install heat retardants on E, SE, SW, W windows	37–56	304–456	8.2
9. Install heat-reflecting film on E, SE, SW, W windows	21–31	204–306	9.9
10. Install heat-absorbing film on E, SE, SW, W windows	5–8	204–306	39.5
11. Upgrade 6.5-EER A/C to 9.5-EER unit	21–32	772–1,158	36.6
12. Install heat pump water-heating system	115–172	680–1,020	5.9
13. Install water heater jacket	26–39	32–48	1.2
14. Install clock thermostat to reduce heat from 70° to 60° for 8 hours each night	96–144	88–132	1.1

Note: EER = energy efficiency ratio, R-value indicates the degree of resistance to heat. The higher the number, the greater is the quality of the insulation.

(a) If Fowler stays in the house for the next 10 years, which alternatives would he select with no budget constraint? Assume that his interest rate is 8%. Assume also that all installations would last 10 years. Fowler will be conservative in calculating the net present worth of each alternative (using the minimum annual savings at the maximum cost). Ignore any tax credits available to energy-saving installations.

(b) If Fowler wants to limit his energy-savings investments to $1,800, which alternatives should he include in his budget?

Short Case Studies

ST15.1[9] Games, Inc., is a publicly traded company that makes computer software and accessories. Games's stock price over the last five years is plotted in Exhibit 15.1, Games's earnings per share for the last five years are shown in Exhibit 15.2, and Games's dividends per share for the last five years are shown in Exhibit 15.3. The company currently has 1,000,000 shares outstanding and a long-term debt of $12,000,000. The company also paid $1,200,000 in interest expenses last year, has other assets of $5,000,000, and had earnings before taxes of $3,500,000 last year.

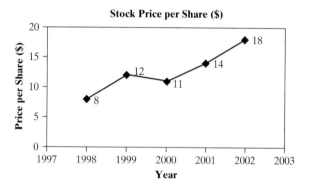

Exhibit 15.1 Stock price per share.

Exhibit 15.2 Earnings per share.

[9] Contributed by Dr. Luke Miller of Fort Lewis College.

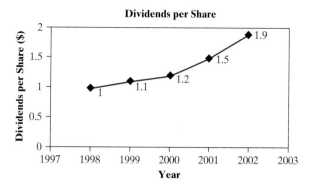

Exhibit 15.3 Dividends per share.

Games has decided to manufacture a new product. In order to make the new product, Games will need to invest in a new piece of equipment that costs $10,000,000. The equipment is classified as a seven-year MACRS property and is expected to depreciate 30% per year. Equipment installation will require 20 employees working for two weeks and charging $50 per hour. Once the equipment has been installed, the facility is expected to remain operational for two years.

Games intends to maintain its current debt-to-equity ratio and therefore plans on borrowing the appropriate amount today to cover the purchase of the equipment. The interest rate on the loan will be equal to its current cost of debt. The loan will require equal annual interest payments over its life (i.e., the loan rate times the principal borrowed for each year). The principal will be repaid in full at the end of year 2. Games plans on issuing new stock to cover the equity portion of the investment. The underwriter of the new stock issue charges an 11% flotation cost.

Games estimates that its new product will acquire 20% of all market share within the United States. Even though this product has never been produced before, Games has identified an older product that should have market attributes similar to those of the new product. The older product's unit market sales for the entire United States for the last five years are shown in Exhibit 15.4. It is estimated that the new product will sell for $20 per unit (in today's dollars). Costs are estimated at 80% of the selling price for the first year and 60% of the selling price for the second year. Inflation is estimated at 10% per annum for the new product.

Exhibit 15.4 Total unit sales in the United States for the older product.

Due to the large size of the investment required to manufacture the new product, Games's analysts predict that once the decision to accept the project is made, investors will revalue the company's stock price. Because you are the chief engineering economist for Games, you have been requested by upper management to determine how manufacturing this new product will change Games's stock price. Management would like answers to the following questions (*Note*: The first three questions are asked in a sequence that will assist you in answering the last two questions):

(a) Games's investors usually use the corporate value model (CVM) to determine the total market value of the company. In its most basic form, the CVM states that a firm's market value is nothing more than the present value of its expected future net cash flows plus the value of its assets. Using this logic, what must investors currently assess the present value of Games's future net cash flows to be (not including the investment in the new product)?

(b) Determine Games's tax rate.

(c) Determine an appropriate MARR to use in the analysis when the financing source is known. Now do the same when the financing source is unknown.

(d) Assuming that the new-product venture is accepted and the new piece of equipment is disposed of at the end of year 2, what is a most likely estimate for Games's stock price?

(e) Now assume that the new piece of equipment is not disposed of in year 2 and that Games has not decided how the $10,000,000 for the equipment will be financed. Instead, assume that the revenues generated from the new product will continue indefinitely. Estimate a pessimistic, a most likely, and an optimistic estimate for Games's stock price.

ST15.2 National Food Processing Company is considering investing in plant modernization and plant expansion. These proposed projects would be completed in two years, with varying requirements of money and plant engineering. Management is willing to use the following somewhat uncertain data in selecting the best set of proposals:

No.	Project	Investment Year 1	Investment Year 2	Engineering IRR	Engineering Hours
1	Modernize production line	$300,000	0	30%	4,000
2	Build new production line	100,000	$300,000	43%	7,000
3	Provide numerical control for new production line	0	200,000	18%	2,000
4	Modernize maintenance shops	50,000	100,000	25%	6,000
5	Build raw-material processing plant	50,000	300,000	35%	3,000
6	Buy present subcontractor's facilities for raw-material processing	200,000	0	20%	600
7	Buy a new fleet of delivery trucks	70,000	10,000	16%	0

The resource limitations are as follows:

- First-year expenditures: $450,000.
- Second-year expenditures: $420,000.
- Engineering hours: 11,000 hours.

The situation requires that a new or modernized production line be provided (Project 1 or Project 2). The numerical control (Project 3) is applicable only to the new line. The company obviously does not want to both buy (Project 6) and build (Project 5) raw-material processing facilities; it can, if desirable, rely on the present supplier as an independent firm. Neither the maintenance shop project (Project 4) nor the delivery-truck purchase (Project 7) is mandatory.

(a) Enumerate all possible mutually exclusive alternatives without considering the budget and engineering-hour constraints.

(b) Identify all feasible mutually exclusive alternatives.

(c) Suppose that the firm's marginal cost of capital will be 14% for raising the required capital up to $1 million. Which projects would be included in the firm's budget?

ST15.3 Consider the following investment projects:

	Project Cash Flows			
n	A	B	C	D
0	− $2,000	− $3,000	− $1,000	
1	1,000	4,000	1,400	−$1,000
2	1,000		−100	1,090
3	1,000			
i^*	23.38%	33.33%	32.45%	9%

Suppose that you have only $3,500 available at period 0. Neither additional budgets nor borrowing is allowed in any future budget period. However, you can lend out any remaining funds (or available funds) at 10% interest per period.

(a) If you want to maximize the future worth at period 3, which projects would you select? What is that future worth (the total amount available for lending at the end of period 3)? No partial projects are allowed.

(b) Suppose in (a) that, at period 0, you are allowed to borrow $500 at an interest rate of 13%. The loan has to be repaid at the end of year 1. Which project would you select to maximize your future worth at period 3?

(c) Considering a lending rate of 10% and a borrowing rate of 13%, what would be the most reasonable MARR for project evaluation?

ST15.4 American Chemical Corporation (ACC) is a multinational manufacturer of industrial chemical products. ACC has made great progress in reducing energy costs and has implemented several cogeneration projects in the United States and Puerto Rico, including the completion of a 35-megawatt (MW) unit in Chicago and a 29-MW unit in Baton Rouge. The division of ACC being considered for one of its more recent cogeneration projects is a chemical plant located in

Texas. The plant has a power usage of 80 million kilowatt hours (kWh) annually. However, on the average, it uses 85% of its 10-MW capacity, which would bring the average power usage to 68 million kWh annually. Texas Electric currently charges $0.09 per kWh of electric consumption for the ACC plant, a rate that is considered high throughout the industry. Because ACC's power consumption is so large, the purchase of a cogeneration unit would be desirable. Installation of the unit would allow ACC to generate its own power and to avoid the annual $6,120,000 expense to Texas Electric. The total initial investment cost would be $10,500,000, including $10,000,000 for the purchase of the power unit itself, a gas-fired 10-MW Allison 571, and engineering, design, and site preparation, and $500,000 for the purchase of interconnection equipment, such as poles and distribution lines, that will be used to interface the cogenerator with the existing utility facilities. ACC is considering two financing options:

- ACC could finance $2,000,000 through the manufacturer at 10% for 10 years and the remaining $8,500,000 through issuing common stock. The flotation cost for a common-stock offering is 8.1%, and the stock will be priced at $45 per share.

- Investment bankers have indicated that 10-year 9% bonds could be sold at a price of $900 for each $1,000 bond. The flotation costs would be 1.9% to raise $10.5 million.

(a) Determine the debt-repayment schedule for the term loan from the equipment manufacturer.

(b) Determine the flotation costs and the number of common stocks to sell to raise the $8,500,000.

(c) Determine the flotation costs and the number of $1,000 par value bonds to be sold to raise $10.5 million.

ST15.5 (Continuation of Problem ST15.4) After ACC management decided to raise the $10.5 million by selling bonds, the company's engineers estimated the operating costs of the cogeneration project. The annual cash flow is composed of many factors: maintenance, standby power, overhaul costs, and miscellaneous expenses. Maintenance costs are projected to be approximately $500,000 per year. The unit must be overhauled every 3 years at a cost of $1.5 million. Miscellaneous expenses, such as the cost of additional personnel and insurance, are expected to total $1 million. Another annual expense is that for standby power, which is the service provided by the utility in the event of a cogeneration unit trip or scheduled maintenance outage. Unscheduled outages are expected to occur four times annually, each averaging two hours in duration at an annual expense of $6,400. Overhauling the unit takes approximately 100 hours and occurs every 3 years, requiring another triennial power cost of $100,000. Fuel (spot gas) will be consumed at a rate of $8,000 BTU per kWh, including the heat recovery cycle. At $2.00 per million BTU, the annual fuel cost will reach $1,280,000. Due to obsolescence, the expected life of the cogeneration project will be 12 years, after which Allison will pay ACC $1 million for the salvage of all equipment.

Revenue will be incurred from the sale of excess electricity to the utility company at a negotiated rate. Since the chemical plant will consume, on average, 85% of the unit's 10-MW output, 15% of the output will be sold at $0.04

per kWh, bringing in an annual revenue of $480,000. ACC's marginal tax rate (combined federal and state) is 36%, and the company's minimum required rate of return for any cogeneration project is 27%. The anticipated costs and revenues are summarized as follows:

Initial investment:	
Cogeneration unit, engineering, design, and site preparation (15-year MACRS class)	$10,000,000
Interconnection equipment (5-year MACRS class)	500,000
Salvage after 12 years' use	1,000,000
Annual expenses:	
Maintenance	500,000
Misc. (additional personnel and insurance)	1,000,000
Standby power	6,400
Fuel	1,280,000
Other operating expenses:	
Overhaul every 3 years	1,500,000
Standby power during overhaul	100,000
Revenues	
Sale of excess power to Texas Electric	480,000

(a) If the cogeneration unit and other connecting equipment could be financed by issuing corporate bonds at an interest rate of 9% compounded annually, with the flotation expenses as indicated in Problem ST15.4, determine the net cash flow from the cogeneration project.

(b) If the cogeneration unit can be leased, what would be the maximum annual lease amount that ACC is willing to pay?

CHAPTER SIXTEEN

Economic Analysis in the Service Sector

***Keeping Cargo Safe from Terror**[1]* How do you keep a terrorist from smuggling a radiation-filled "dirty bomb" or other weapon in one of the seven-million-plus shipping containers that arrive at U.S. ports each year? Until now, U.S. Customs and Border Protection has sought to secure global shipping by relying on intelligence and scrutinizing suspicious cargo manifests—such as an unrefrigerated container full of "frozen fish"—to identify potentially dangerous shipments. Currently, less than 6% of the containers headed for American ports are deemed "high risk" by the U.S. Department of Homeland Security and pulled aside for examination by Custom inspectors.

Port officials in Hong Kong are testing a strategy that electronically scrutinizes every container full of sneakers, toys, gadgets, or other contents. Trucks haul each container passing through the port through two of the giant scanners. One checks for nuclear radiation, while the other uses gamma rays to seek out any dense, suspicious object made of steel or lead inside the containers that could shield a bomb from the nuclear detector. Proponents contend it better secures the global shipping system—without unacceptably slowing the flow of commerce. The Hong Kong project would cost shippers an additional $6.50 a container if its costs were passed on to them. That is a fraction of what it costs to transport a container: about $1,900 to send a 20-foot container from Hong Kong to Los Angeles.

Now, U.S. Customs and Border protection is examining the various options for inspecting the incoming cargo, including a 100% inspection. Clearly, one of the main issues to address is how to minimize the obvious congestion that would result at the ports and borders. This will

[1] "Keeping Cargo Safe from Terror," by Alex Ortolani and Robert Block, *The Wall Street Journal*, Friday, July 29, 2005, Section B1.

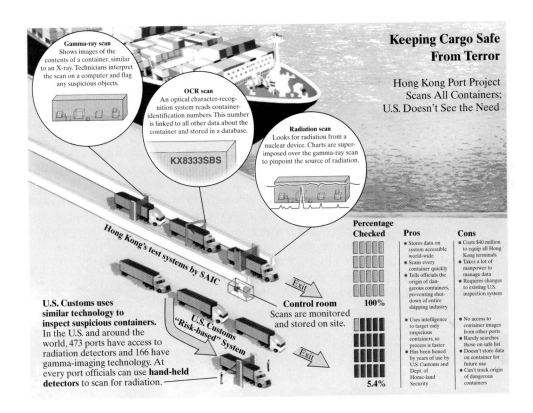

Keeping Cargo Safe From Terror

Hong Kong Port Project Scans All Containers; U.S. Doesn't See the Need

Gamma-ray scan
Shows images of the contents of a container, similar to an X-ray. Technicians interpret the scan on a computer and flag any suspicious objects.

OCR scan
An optical character-recognition system reads container-identification numbers. This number is linked to all other data about the container and stored in a database.

KX8333SBS

Radiation scan
Looks for radiation from a nuclear device. Charts are superimposed over the gamma-ray scan to pinpoint the source of radiation.

Hong Kong's test systems by SAIC

Exit

Control room
Scans are monitored and stored on site.

Exit

U.S. Customs uses **similar technology to inspect suspicious containers.** In the U.S. and around the world, 473 ports have access to radiation detectors and 166 have gamma-imaging technology. At every port officials can use **hand-held detectors** to scan for radiation.

U.S. Customs "Risk-based" System

Percentage Checked	Pros	Cons
100%	• Stores data on system accessible world-wide • Scans every container quickly • Tells officials the origin of dangerous containers, preventing shutdown of entire shipping industry	• Costs $40 million to equip all Hong Kong terminals • Takes a lot of manpower to manage data • Requires changes to existing U.S. inspection system
5.4%	• Uses intelligence to target only suspicious containers, so process is faster • Has been honed by years of use by U.S. Customs and Department of Homeland Security	• No access to container images from other ports • Rarely searches those on safe list • Doesn't store data on container for future use • Can't track origin of dangerous containers

undoubtedly add a huge burden to the U.S. economy, not to mention the cost of installing the scanners all over the border entry points.

Up to this point, we have focused our analysis on the economic issues related to the manufacturing sector of the U.S. economy. The main reason for doing this was that many engineers will be working in that sector. However, an increasing number of engineers are now seeking their careers in the service sector, such as health care, financial institutions, transportation and logistics, and government. In this chapter, we will present some unique features that must be considered in evaluating investment projects in the service sector.

CHAPTER LEARNING OBJECTIVES

After completing this chapter, you should understand the following concepts:

■ How to price service.

■ How to evaluate investment projects in the health care industry.

■ How to conduct cost benefit analysis and cost effectiveness analysis.

16.1 What Is the Service Sector?[2]

The service sector of the U.S. economy dominates both gross domestic product (GDP) and employment. It is also the fastest-growing part of the economy and the one offering the most fertile opportunities to engineers to improve their productivity. For example, service activities now approach 80% of U.S. employment, far outstripping sectors such as manufacturing (14%) and agriculture (2%), as shown in Figure 16.1.

The mere scale of the service sector makes it a critical element of the U.S. economy, employing, as it does, many millions of workers producing trillions of dollars in economic value. The service sector has expanded far beyond the traditional consumer or institutional service industries and currently includes distributive services, such as transportation, communications, and utilities; the rapidly expanding producer services, such as finance, insurance, real estate, and advertising; wholesale and retail trade and sales; the nonprofit sector, including health, philanthropy, and education; and government. The use of technology in the service sector promotes deskilling (i.e., automation). However, secure and reliable U.S. services provide much of the key infrastructure on which the whole nation, and indeed much of the world's commerce, depends. Tremendous new challenges have arisen as changes in processes and operations are developed to better assure the safety and reliability of critical services. These changes all lead to significant investment in infrastructures requiring detailed economic analysis.

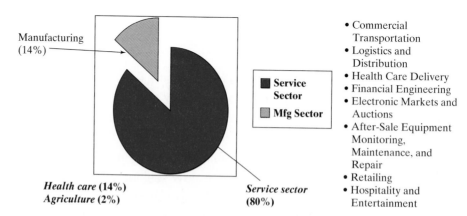

Figure 16.1 Contribution of the service sector to the U.S. gross domestic product (GDP).

[2] This section is taken from the National Science Foundation program "Exploratory Research on Engineering the Service Sector (ESS)," NSF 02-029.

16.1.1 Characteristics of the Service Sector

What makes the evaluation of a service sector project unique compared with one from the manufacturing sector? Some of the unique features are summarized as follows:

- Services are generally intangible. They have sometimes been defined as "anything of economic value that cannot be held or touched."
- It is usually impossible to build inventories of services. Either the demand for the service must be backlogged, or enough resources need to be provided to meet an acceptable fraction of the demand as it arises.
- Services are more dynamic and responsive to demand than are manufactured products. This means that variability and risk are more central issues in service industries. Indeed, the management of financial risk is an important service in itself.
- Many services (examples are medical treatment and equipment repair) require a diagnostic step to design the service as part of its delivery. Coproduction (i.e., active collaboration between the server and the customer) is also required in many settings.
- Service products are usually less standardized and less subject to design specifications than manufactured goods are, because the outputs are tailored to customer needs as they are delivered. This also makes it harder to distinguish service product design from product manufacture and delivery.
- The dimensions of service quality are more subtle and subjective than those of physical products. Not only are the parameters of services more difficult to express, but customers' perceptions play a much greater role in deciding what is satisfactory or valuable.
- Most service operations are more labor intensive than the production of goods.
- Compared with goods industries, the service economy has a much greater fraction of its operations performed by governments and institutions.
- Information technology is central to service industries. Often, it is the only significant means of multiplying human output.

Certainly, our objective is not to address all aspects of the service sector, but only some of the common economic analysis issues confronted by its engineers.

16.1.2 How to Price Service

Improving service can be many different things. For example, for a delivery business such as United Parcel Service (UPS) or FedEx, anything that reduces the total time taken from pickup to delivery is considered an improvement of service. For an airline, on-time departures and arrivals, a reduction in the number of mishandled pieces of checked-in baggage, and a speedy check-in at the gate could all be considered important service parameters, because they make airline passengers happy, which in turn will translate into more business volume. Accordingly, one of the critical questions related to improvement in service is "What do service providers gain by improving their own service to suppliers and customers?" If we can quantify improvements in service in terms of dollars, the economic analysis is rather straightforward: If the net increase in revenue due to improvements in service exceeds the investment required, the project can be justified. For this type of decision problem, we can simply use any of the measures of investment worth discussed in Chapter 5 through Chapter 7.

Improving service is one thing and putting a price on services is another thing. This is far more difficult than pricing products, because the benefits of services are less tangible and service companies often lack well-documented standard unit-production costs to go by. When

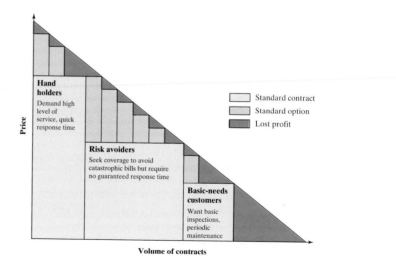

Figure 16.2 How to capture lost profits. By segmenting customers into three categories with appropriate service options, companies can capture much of the lost profit. Options may include guaranteed response time, remote equipment monitoring, or extended warranty.

a company designs an after-service contract to go with equipment sales, customers may be segmented according to their service needs rather than their size, industry, or type of equipment. Companies then develop the pricing, contracting, and monitoring capabilities to support the cost-effective delivery of the service.[3] For example, when customers are segmented according to the service level they need, they tend to fall into one of at least three common categories. The "basic-needs customers" want a standard level of service with basic inspections and periodic maintenance. The "risk avoiders" are looking for coverage to avoid big bills but care less about other elements, such as response times. And the "hand holders" need high levels of service, often with quick and reliable response times, and are willing to pay for the privilege. Therefore, to maximize return, companies need to capture tremendous value from their service businesses by taking a more careful, fact-based approach to designing and pricing services (Figure 16.2).

The types of service problems that we are interested in this chapter, however, are (1) those encountered by nonprofit organizations such as hospitals or public policy makers on health care and (2) those involving economic decisions by the public sector. In the next section, we will review some of the common decision tools adopted by the health-care service industry. The economic issues in the public sector will be discussed in Section 16.3.

16.2 Economic Analysis in Health-Care Service

The health-care service industry alone constitutes 14–15% of GDP when all its dimensions are included. Clearly, health care is one of the largest, most research-intensive service industries in the United States. Accordingly, its medical knowledge base is

[3] *Source*: Russell G. Bundschuh and Theodore M. Dezvane, "How to Make After-Sales Services Pay Off," *The McKinsey Quarterly*, 2003, no. 4, pp. 3–13.

expanding at a staggering pace, and many Americans enjoy unparalleled advancements in medical science and technology. At the same time, the nation's health-care system consistently fails to deliver high-quality care: Variation in both access to and delivery of care is considerable, errors are widespread, costs are spiraling, and few resources are devoted to optimizing system operations and improving the delivery of care.

16.2.1 Economic Evaluation Tools

Economic evaluation can be used to inform decision making and can provide information to assist in answering the following questions:

- What services do we provide (or improve), when, and at what level?
- How do we provide such services?
- Where do we provide the services?
- What are the costs associated with providing or improving the services?

The following three methods of economic evaluation are related to health service:

- **Cost-effectiveness analysis.** This technique is used in health economics to compare the financial costs of therapies whose outcomes can be measured purely in terms of their health effect (e.g., years of life saved or ulcers healed). Cost-effectiveness analysis (CEA) is the most commonly applied form of economic analysis in health economics. However, it does not allow comparisons to be made between courses of action that have different health effects.

- **Cost-utility analysis.** This technique is similar to CEA in that there is a defined outcome and the cost to achieve that outcome is measured in money. The outcome is measured in terms of survival and quality of life (for example, quality-adjusted life years, or QALYs). CEA can indicate which one of a number of alternative interventions represents the best value for money, but it is not as useful when comparisons need to be made across different areas of health care, since the outcome measures used may be very different. As long as the outcome measure is life years saved or gained, a comparison can still be made, but even in such situations CEA remains insensitive to the *quality*-of-life dimension. In order to know which areas of health care are likely to provide the greatest benefit in improving health status, a cost-utility analysis[4] needs to be undertaken using a common currency for measuring the outcomes across health-care areas.

- **Cost–benefit analysis.** If information is needed as to which interventions will result in overall resource savings, a cost–benefit analysis (CBA) has to be performed, although, like a cost-utility analysis, a cost-benefit analysis has its own drawbacks. In CBA, the benefit is measured as the associated economic benefit of an intervention; hence, both costs and benefits are expressed in money, and the CBA may ignore many intangible, but very important, benefits that are difficult to measure in monetary terms (e.g., relief of pain and anxiety). Even though the virtue of this analysis is that it enables comparisons to be made between schemes in very different areas of health care, the approach is not widely accepted for use in health economics.

Cost–benefit analysis is the process of weighing the total expected cost vs. the total expected benefits of one or more actions in order to choose the best or most profitable option.

---

[4] We will not discuss any technical details of the cost-utility approach in this chapter, but they can be found in a variety of health economics texts.

16.2.2 Cost-Effectiveness Analysis

In cost-effectiveness analysis (CEA), outcomes are reported in a single unit of measurement. CEA compares the costs and health effects of an intervention to assess whether it is worth doing from an economic perspective. First of all, CEA is a specific type of economic analysis in which all costs are related to a single, common effect. Decision makers can use it to compare different resource allocation options in like terms. A general misconception is that CEA is merely a means of finding the least expensive alternative or getting the "most bang for the buck." In reality, CEA is a comparison tool; it will not always indicate a clear choice, but it will evaluate options quantitatively and objectively on the basis of a defined model. CEA was designed to evaluate health-care interventions, but the methodology can be used for non-health-economic applications as well. CEA can compare any resource allocation with measurable outcomes.

What Constitutes a Cost?

In CEA, it is common to distinguish between the direct costs and the indirect costs associated with an intervention. Some interventions may also result in intangibles, which are difficult to quantify, but should be included in the cost profile. Examples of these different kinds of costs are as follows:

- **Direct cost.** Drugs, medical staff time, medical equipment, transport, and out-of-pocket expenses by the patients.
- **Indirect costs.** Loss of productive time by the patients during the intervention.
- **Intangibles.** Pain and suffering, and adverse effects from the intervention.

It is essential to specify which costs are included in a CEA and which are not, to ensure that the findings are not subject to misinterpretation.

Cost-Effectiveness Ratio

The cost-effectiveness ratio is simply the sum of all costs, divided by the sum of all health effects:

$$\text{Cost-effectiveness ratio} = \frac{\sum(\text{all costs})}{\sum(\text{all measured health effects})} \qquad (16.1)$$

The benefits are not measured in terms of just dollars, but in a ratio that incorporates both health outcomes and dollars.

Cost-effectiveness ratios should be related to the size of relevant budgets to determine the most cost-effective strategies. CEAs compare several program strategies and rank them by cost-effectiveness ratios. An analysis of two screening interventions might show you that one costs $10,000 per life year gained while the other costs $40,000 per life year gained. The first intervention requires monthly screening and the second requires biannual screening. Realizing that compliance is a greater problem with monthly screening, the decision maker would implement the most appropriate coverage strategy for the population in question. Sometimes, the analysis compares an option against a baseline option, such as "Do nothing" or "Give usual care." The last two are valid strategic options.

Discounting

There is often a significant time lag between the investment of health service resources and the arrival of the associated health gain. In general, we prefer to receive benefits now and pay costs in the future. In order to reflect this preference in economic evaluation, costs are discounted.

16.2.3 How to Use a CEA[5]

When we use a CEA, we need to distinguish between those interventions which are completely independent and those which are dependent. Two (or more) interventions are said to be **independent** if the costs and effects of one neither affect nor are affected by the costs and effects of the other. Two (or more) interventions are dependent if the implementation of one results in changes to the costs and effects of the other. The analysis proceeds as follows:

- **Independent interventions.** Using CEA with independent intervention programs requires that cost-effectiveness ratios be calculated for each program and placed in rank order. For example, in Table 16.1, there are three interventions for different patient groups, and each intervention has as an alternative "doing nothing." According to CEA, Program C should be given priority over Program A, since it has a lower cost-effectiveness ratio (CER), but in order to decide which program to implement, the extent of the resources available must be considered. (See Table 16.2.) Clearly, the choice of independent intervention is a function of the budget that is available to implement. For example, with $200,000, we will go with Program C, and the remaining $50,000 will be available for funding (up to 50%) of Program A.
- **Mutually exclusive interventions.** In reality, the likelihood is that choices will have to be made between different treatment regiments for the same condition and between

TABLE 16.1 Cost-Effectiveness of Three Independent Intervention Programs

Program	Cost ($)	Health Effect (Life Years Gained)	Cost-Effectiveness Ratio
A	100,000	1,200	83.33
B	120,000	1,350	88.89
C	150,000	1,850	81.08

TABLE 16.2 Choices of Program as a Function of Budget

Budget Available ($)	Programs to Be Implemented
Less than $150,000	As much of C as budget allows
$150,000	100% of C
$150,000–$250,000	C and as much of A as budget allows
$250,000	C and A
$250,000–$370,000	C and A, and as much of B as budget allows
$370,000	All three programs, A, B, C

[5] This section is based on the article "What Is Cost-effectiveness?" by Ceri Phillips and Guy Thompson, vol. 1, no. 3, Hayward Medical Communications, Copyright © 2001; on the Internet at www.evidence-based-medicine.co.uk.

TABLE 16.3 **Mutually Exclusive Intervention Programs**

Program	Cost	Effects (Life Years Gained)	Cost-Effectiveness (Life Years Gained)	Incremental Cost-Effectiveness Ratio
P1	$125,000	1,300	96.15	96.15
P2	$100,000	1,500	66.67	−125
P3	$160,000	2,000	80.00	120
P4	$140,000	2,200	63.63	−100
P5	$170,000	2,600	65.38	75

different dosages or treatments versus prophylaxis (i.e., mutually exclusive interventions). In this case, incremental cost-effectiveness ratios (ΔCERs) should be used:

$$\Delta CER_{2-1} = \frac{\text{Cost of P2} - \text{Cost of P1}}{\text{Effects of P2} - \text{Effects of P1}}$$

The alternative interventions are ranked according to their effectiveness—on the basis of securing the maximum effect rather than considering cost—and CERs are calculated as shown in Table 16.3. The analysis proceeds as follows:

- If money is no object, P5 is clearly the best alternative, as the number of life years gained is most significant.
- The least effective intervention (P1) has the same average CER as its incremental cost-effectiveness ratio (ICER), because it is compared with the alternative of "doing nothing":

$$\Delta CER_{1-0} = \frac{125,000 - 0}{1,300 - 0}$$
$$= 96.15.$$

- A comparison between P1 and P2 yields

$$\Delta CER_{2-1} = \frac{\text{Cost of P2} - \text{Cost of P1}}{\text{Effect of P2} - \text{Effect of P1}}$$
$$= \frac{100,000 - 125,000}{1,500 - 1,300}$$
$$= -125 < 96.15.$$

The negative ICER for P2 means that adopting P2 rather than P1 results in an improvement in life years gained and a reduction in costs. It also indicates that P2 dominates P1. Hence, we can eliminate P1 at this stage of the analysis.

- A comparison between P2 and P3 yields

$$\Delta CER_{3-2} = \frac{160,000 - 100,000}{2,000 - 1,500}$$
$$= 120 > 66.67.$$

The ICER for P3 works out to be 120, which means that it costs $120 to generate each additional life year gained compared with P2. Thus, there is no clear dominance between P2 and P3.

- A comparison between P4 and P3 yields

$$\Delta CER_{4-3} = \frac{140,000 - 160,000}{2,200 - 2,000}$$
$$= -100 < 80.$$

P4 is more effective than P3 as the incremental cost-effectiveness ratio becomes negative. Also, P4 dominates P3, so we can eliminate P3 from the analysis. Having excluded P1 and P3, we now recalculate for P2, P4, and P5, as shown in Table 16.4.

- A comparison between P2 and P4 yields

$$\Delta CER_{4-2} = \frac{140,000 - 100,000}{2,200 - 1,500}$$
$$= \$57.14 < 66.67.$$

Thus, P2 is dominated by P4, since the latter is more effective and costs less to produce an additional unit of effect ($57.14 compared with $66.67). The dominated alternative is then excluded and the ICERs are recalculated again (Table 16.5).

TABLE 16.4 Remaining Mutually Exclusive Alternatives after Eliminating More Costly and Less Effective Programs

Program	Cost	Effects (Life Years Gained)	Cost-Effectiveness (Life Years Gained)	Incremental Cost-Effectiveness Ratio
P2	$100,000	1,500	66.67	66.67
P4	$140,000	2,200	63.63	57.14
P5	$170,000	2,600	65.38	75.00

TABLE 16.5 Remaining Mutually Exclusive Alternatives after Eliminating All Dominated Programs

Program	Cost	Effects (Life Years Gained)	Cost-Effectiveness (Life Years Gained)	Incremental Cost-Effectiveness Ratio
P4	$140,000	2,200	63.63	63.63
P5	$170,000	2,600	65.38	75.00

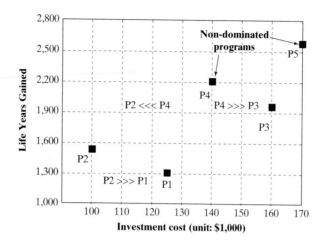

Figure 16.3 Cost-effectiveness diagram—Life years gained.

- Finally, a comparison between P4 and P5 yields

$$\Delta CER_{5-4} = \frac{170,000 - 140,000}{2,600 - 2,200}$$

$$= 75 > 63.63.$$

No clear dominance exists between P4 and P5. As shown in Figure 16.3, these two programs are therefore the ones that deserve funding consideration. In deciding between them, the size of the available budget must be brought to bear on the matter. If the available budget is $140,000, all patients should receive intervention P4, whereas if the available budget is $170,000, all patients should receive the more effective P5. However, if the budget is, say, $150,000, then, since the cost difference between P4 and P5 is $30,000 and the budget surplus is $10,000, it is possible to switch one-third of the patients to P5 and still remain within budget.

16.3 Economic Analysis in the Public Sector

In earlier chapters, we have focused attention on investment decisions in the private sector; the primary objective of these investments was to increase the wealth of corporations. In the public sector, federal, state, and local governments spend hundreds of billions of dollars annually on a wide variety of public activities, such as the port project described in the chapter opener. In addition, governments at all levels regulate the behavior of individuals and businesses by influencing the use of enormous quantities of productive resources. How can public decision makers determine whether their decisions, which affect the use of these productive resources, are, in fact, in the best public interest?

Many civil engineers work on public-works areas such as highway construction, airport construction, and water projects. In the port expansion scenario presented at the beginning of the chapter, each option requires a different level of investment and produces a different degree of benefits. One of the most important aspects of airport expansion is to quantify the cost of airport delays in dollar terms. In other words, planners ask, "What is the economic benefit of reducing airport delays?" From the airline's point of view, taxiing and arrival delays mean added fuel costs. For the airport, delays mean lost revenues in landing and departure fees. From the public's point of view, delays mean lost earnings, as people then have to spend more time on transportation. Comparing the investment costs with the potential benefits, an approach known as benefit–cost analysis, is an important feature of economic analysis.

16.3.1 What Is Benefit–Cost Analysis?

Benefit–cost analysis is a decision-making tool that is used to systematically develop useful information about the desirable and undesirable effects of public projects. In a sense, we may view benefit–cost analysis in the public sector as profitability analysis in the private sector. In other words, benefit–cost analysis attempts to determine whether the social benefits of a proposed public activity outweigh the social costs. Usually, public investment decisions involve a great deal of expenditure, and their benefits are expected to occur over an extended period of time. Examples of benefit–cost analysis include studies of (1) public transportation systems, (2) environmental regulations on noise and pollution, (3) public-safety programs, (4) education and training programs, (5) public-health programs, (6) flood control, (7) water resource development, and (8) national defense programs.

> **Benefit–cost analysis:** A technique designed to determine the feasibility of a project or plan by quantifying its costs and benefits.

The three typical aims of benefit–cost analyses are (1) to maximize the benefits for any given set of costs (or budgets), (2) to maximize the net benefits when both benefits and costs vary, and (3) to minimize costs to achieve any given level of benefits (often called "cost-effectiveness" analysis). Three types of decision problems, each having to do with one of these aims, will be considered in this chapter.

16.3.2 Framework of Benefit–Cost Analysis

To evaluate public projects designed to accomplish widely differing tasks, we need to measure the benefits or costs in the same units in all projects so that we have a common perspective by which to judge the different projects. In practice, this means expressing both benefits and costs in monetary units, a process that often must be performed without accurate data. In performing benefit–cost analysis, we define **users** as the public and **sponsors** as the government.

The general framework for benefit–cost analysis can be summarized as follows:

1. Identify all users' benefits expected to arise from the project.
2. Quantify these benefits in dollar terms as much as possible, so that different benefits may be compared against one another and against the costs of attaining them.
3. Identify sponsors' costs.
4. Quantify these costs in dollar terms as much as possible, to allow comparisons.
5. Determine the equivalent benefits and costs during the base period; use an interest rate appropriate for the project.
6. Accept the project if the equivalent users' benefits exceed the equivalent sponsor's costs.

We can use benefit–cost analysis to choose among such alternatives as allocating funds for the construction of a mass-transit system, a dam with irrigation, highways, or an air-traffic control system. If the projects are on the same scale with respect to cost, it is merely a question of choosing the project for which the benefits exceed the costs by the greatest amount. The steps just outlined are for a single (or independent) project evaluation. As in the case of the internal-rate-of-return criterion, in comparing mutually exclusive alternatives, an incremental benefit–cost ratio must be used. Section 16.3.3 illustrates this important issue in detail.

16.3.3 Valuation of Benefits and Costs

In the abstract, the framework we just developed for benefit–cost analysis is no different from the one we have used throughout this text to evaluate private investment projects. The complications, as we shall discover in practice, arise in trying to identify and assign values to all the benefits and costs of a public project.

Users' Benefits

To begin a benefit–cost analysis, we identify all project **benefits** (favorable outcomes) and **disbenefits** (unfavorable outcomes) to the user. We should also consider the indirect consequences resulting from the project—the so-called **secondary effects**. For example, the construction of a new highway will create new businesses such as gas stations, restaurants, and motels (benefits), but it will divert some traffic from the old road, and as a consequence, some businesses would be lost (disbenefits). Once the benefits and disbenefits are quantified, we define the users' benefits as follows:

$$\text{Users' benefits } (B) = \text{Benefits} - \text{Disbenefits}.$$

In identifying user's benefits, we should classify each one as a **primary benefit**—a benefit that is directly attributable to the project—or a **secondary benefit**—a benefit that is indirectly attributable to the project. As an example, at one time the U.S. government was considering building a superconductor research laboratory in Texas. If the project ever materializes, it could bring many scientists and engineers, along with other supporting population, to the region. Primary national benefits might include the long-term benefits that could accrue as a result of various applications of the research to U.S. businesses. Primary regional benefits might include economic benefits created by the research laboratory activities, which would generate many new supporting businesses. Secondary benefits might include the creation of new economic wealth as a consequence of a possible increase in international trade and an increase in the incomes of various regional producers attributable to a growing population.

The reason for making this distinction is that it may make our analysis more efficient: If primary benefits alone are sufficient to justify project costs, we can save time and effort by *not* quantifying the secondary benefits.

Sponsor's Costs

We determine the cost to the sponsor by identifying and classifying the expenditures required and any savings (or revenues) to be realized. The sponsor's costs should include both capital investment and annual operating costs. Any sales of products or services that take place upon completion of the project will generate some revenues—for example, toll revenues on highways. These revenues reduce the sponsor's costs. Therefore, we calculate the sponsor's costs by combining these cost elements:

Sponsor's cost = Capital cost + Operating and maintenance costs − Revenues.

Social Discount Rate

As we learned in Chapter 15, the selection of an appropriate MARR for evaluating an investment project is a critical issue in the private sector. In public-project analyses, we also need to select an interest rate, called the **social discount rate**, to determine equivalent benefits as well as the equivalent costs. The selection of a social discount rate in public project evaluation is as critical as the selection of a MARR in the private sector.

Ever since present-value calculations were initiated to evaluate public water resources and related land-use projects in the 1930s, a tendency to use relatively low rates of discount compared with those existing in markets for private assets has persisted. During the 1950s and into the 1960s, the rate for water resource projects was 2.63%, which, for much of that period, was even below the yield on long-term government securities. The persistent use of a lower interest rate for water resource projects is a political issue. In recent years, with the growing interest in performance budgeting and systems analysis in the 1960s, the tendency on the part of government agencies has been to examine the appropriateness of the discount rate in the public sector in relation to the efficient allocation of resources in the economy as a whole. Two views of the basis for determining the social discount rate prevail:

1. **Projects without private counterparts.** *The social discount rate should reflect only the prevailing government borrowing rate.* Projects such as dams designed purely for flood control, access roads for noncommercial uses, and reservoirs for community water supply may not have corresponding private counterparts. In those areas of government activity where benefit–cost analysis has been employed in evaluation, the rate of discount traditionally used has been the cost of government borrowing. In fact, water resource project evaluations follow this view exclusively.

2. **Projects with private counterparts.** *The social discount rate should represent the rate that could have been earned had the funds not been removed from the private sector.* For public projects that are financed by borrowing at the expense of private investment, we may focus on the opportunity cost of capital in alternative investments in the private sector to determine the social discount rate. In the case of public capital projects, similar to some in the private sector that produce a commodity or a service (such as electric power) to be sold on the market, the rate of discount employed would be the average cost of capital as discussed in Chapter 15. The reasons for using the private rate of return as the opportunity cost of capital in projects similar to those in the private sector are (1) to prevent the public sector from transferring capital from higher yielding to lower yielding investments and (2) to force public-project evaluators to employ market standards in justifying projects.

The Office of Management and Budget (OMB) holds the second view. Since 1972, OMB has required that a social discount rate of 10% be used to evaluate federal public projects. (Exceptions include water resource projects.)

16.3.4 Quantifying Benefits and Costs

Now that we have defined the general framework for benefit–cost analysis and discussed the appropriate discount rate, we will illustrate the process of quantifying the benefits and costs associated with a public project. Our context is a motor-vehicle inspection program initiated by the state of New Jersey.[6]

Many states in the United States employ inspection systems for motor vehicles. Critics often claim that these programs lack efficacy and have a poor benefit-to-cost ratio in terms of reducing fatalities, injuries, accidents, and pollution.

Elements of Benefits and Costs

The state of New Jersey identified the primary and secondary benefits of its new inspection program as follows:

- **Users' Benefits**

  Primary benefits. Deaths and injuries related to motor-vehicle accidents impose financial costs on individuals and society. Preventing such costs through the inspection program has the following primary benefits:

  1. Retention of contributions to society that might be lost due to an individual's death.

  2. Retention of productivity that might be lost while an individual recuperates from an accident.

  3. Savings of medical, legal, and insurance services.

  4. Savings on property replacement or repair costs.

  Secondary benefits. Some secondary benefits are not measurable (e.g., the avoidance of pain and suffering); others can be quantified. Both types of benefits should be considered. A list of secondary benefits is as follows:

  1. Savings of income of families and friends of accident victims who might otherwise be tending to the victims.

  2. Avoidance of air and noise pollution and savings on fuel costs.

  3. Savings on enforcement and administrative costs related to the investigation of accidents.

  4. Avoidance of pain and suffering.

- **Users' Disbenefits**

  1. Cost of spending time to have a vehicle inspected (including travel time), as opposed to devoting that time to an alternative endeavor (opportunity cost).

  2. Cost of inspection fees.

  3. Cost of repairs that would not have been made if the inspection had not been performed.

---

[6] Based on Loeb, P. D. and Gilad, B. "The Efficacy and Cost Effectiveness of Vehicle Inspection," Journal of Transport Economics and Policy, May 1984: 145--164. The original cost data, which were given in 1981 dollars, were converted to the equivalent cost data in 2000 by using the prevailing consumer price indices during the period.

4. Value of time expended in repairing the vehicle (including travel time).

5. Cost in time and direct payment for reinspection.

- **Sponsor's Costs**

    1. Capital investments in inspection facilities.

    2. Operating and maintenance costs associated with inspection facilities. (These include all direct and indirect labor, personnel, and administrative costs.)

- **Sponsor's Revenues or Savings**

    1. Inspection fee.

Valuation of Benefits and Costs

The aim of benefit–cost analysis is to maximize the equivalent value of all benefits minus that of all costs (expressed either in present values or in annual values). This objective is in line with promoting the economic welfare of citizens. In general, the benefits of public projects are difficult to measure, whereas the costs are more easily determined. For simplicity, we will attempt only to quantify the primary users' benefits and sponsor's costs on an annual basis.

- **Calculation of Primary Users' Benefits**

    1. **Benefits due to the reduction of deaths.** The equivalent value of the average income stream lost by victims of fatal accidents[7] was estimated at $571,106 per victim, in 2000 dollars. The state estimated that the inspection program would reduce the number of annual fatal accidents by 304, resulting in a potential savings of

    $$(304)(\$571,106) = \$173,616,200.$$

    2. **Benefits due to the reduction of damage to property.** The average cost of damage to property per accident was estimated at $1,845. This figure includes the cost of repairs for damages to the vehicle, the cost of insurance, the cost of legal and court administration, the cost of police accident investigation, and the cost of traffic delay due to accidents. The state estimated that accidents would be reduced by 37,910 per year and that about 63% of all accidents would result in damage to property only. Therefore, the estimated annual value of benefits due to a reduction in property damage was estimated at

    $$\$1,845(37,910)(0.63) = \$44,073,286.$$

    The overall annual primary benefits are estimated as the following sum:

Value of reduction in fatalities	$173,616,200
Value of reduction in property damage	44,073,286
Total	$217,689,486

---

[7] These estimates were based on the total average income that these victims could have generated had they lived. The average value on human life was calculated by considering several factors, such as age, sex, and income group.

- **Calculation of Primary Users' Disbenefits**

  1. **Opportunity cost associated with time spent bringing vehicles in for inspection.** This cost is estimated as

  $$C_1 = \text{(Number of cars inspected)}$$
  $$\times \text{(Average duration involved in travel)}$$
  $$\times \text{(Average wage rate)}.$$

  With an estimated average duration of 1.02 travel-time hours per car, an average wage rate of $14.02 per hour, and 5,136,224 inspected cars per year, we obtain

  $$C_1 = 5,136,224(1.02)(\$14.02)$$
  $$= \$73,450,058.$$

  2. **Cost of inspection fee.** This cost may be calculated as

  $$C_2 = \text{(Inspection fee)} \times \text{(number of cars inspected)}.$$

  Assuming that an inspection fee of $5 is to be paid for each car, the total annual inspection cost is estimated as

  $$C_2 = (\$5)(5,136,224)$$
  $$= \$25,681,120.$$

  3. **Opportunity cost associated with time spent waiting during the inspection process.** This cost may be calculated by the formula

  $$C_3 = \text{(Average waiting time in hours)}$$
  $$\times \text{(Average wage rate per hour)}$$
  $$\times \text{(Number of cars inspected)}.$$

  With an average waiting time of 9 minutes (or 0.15 hours),

  $$C_3 = 0.15(\$14.02)(5,136,224) = \$10,801,479.$$

  4. **Vehicle usage costs for the inspection process.** These costs are estimated as

  $$C_4 = \text{(Number of inspected cars)}$$
  $$\times \text{(Vehicle operating cost per mile)}$$
  $$\times \text{(Average round-trip miles to inspection station)}.$$

  Assuming a $0.35 operating cost per mile and 20 round-trip miles, we obtain

  $$C_4 = 5,136,224(\$0.35)(20) = \$35,953,568.$$

  The overall primary annual disbenefits are estimated as follows:

Item	Amount
C_1	$73,450,058
C_2	25,681,120
C_3	10,801,479
C_4	35,453,568
Total disbenefits	$145,886,225, or $28.40 per vehicle

- **Calculation of Primary Sponsor's Costs**

  New Jersey's Division of Motor Vehicles reported an expenditure of $46,376,703 for inspection facilities (this value represents the annualized capital expenditure) and another annual operating expenditure of $10,665,600 for inspection, adding up to $57,042,303.

- **Calculation of Primary Sponsor's Revenues**

  The sponsor's costs are offset to a large degree by annual inspection revenues, which must be subtracted to avoid double counting. Annual inspection revenues are the same as the direct cost of inspection incurred by the users (C_2), which was calculated as $20,339,447.

Reaching a Final Decision

Finally, a discount rate of 6% was deemed appropriate, because the state of New Jersey finances most state projects by issuing a 6% long-term tax-exempt bond. The streams of costs and benefits are already discounted so as to obtain their present and annual equivalent values.

From the state's estimates, the primary benefits of inspection are valued at $217,689,486, compared with the primary disbenefits of inspection, which total $145,886,225. Therefore, the user's net benefits are

$$\text{User's net benefits} = \$217,689,486 - \$145,886,225$$
$$= \$71,803,261.$$

The sponsor's net costs are

$$\text{Sponsor's net costs} = \$57,042,303 - \$20,339,447$$
$$= \$36,702,856.$$

Since all benefits and costs are expressed in annual equivalents, we can use these values directly to compute the degree of benefits that exceeds the sponsor's costs:

$$\$71,803,261 - \$36,702,856 = \$35,100,405 \text{ per year.}$$

This positive AE amount indicates that the New Jersey inspection system is economically justifiable. We can assume the AE amount would have been even greater had we also factored in secondary benefits. (For example, for simplicity, we have not explicitly considered vehicle growth in the state of New Jersey. For a complete analysis, this growth factor must be considered to account for all related benefits and costs in equivalence calculations.)

16.3.5 Difficulties Inherent in Public-Project Analysis

As we observed in the motor-vehicle inspection program in the previous section, public benefits are very difficult to quantify in a convincing manner. For example, consider the valuation of a saved human life in any category. Conceptually, the total benefit associated with saving a human life may include the avoidance of the insurance administration costs as well as legal and court costs. In addition, the average potential income lost because of premature death (taking into account age and sex) must be included. Obviously, the difficulties associated with any attempt to put precise numbers on human life are insurmountable.

Consider this example: A few years ago, a 50-year-old business executive was killed in a plane accident. The investigation indicated that the plane was not properly maintained according to federal guidelines. The executive's family sued the airline, and the court eventually ordered the airline to pay $5,250,000 to the victim's family. The judge calculated the value of the lost human life assuming that if the executive had lived and worked in the same capacity until his retirement, his remaining lifetime earnings would have been equivalent to $5,250,000 at the time of award. This is an example of how an individual human life was assigned a dollar value, but clearly any attempt to establish an average amount that represents the general population is controversial. We might even take exception to this individual case: Does the executive's salary adequately represent his worth to his family? Should we also assign a dollar value to their emotional attachment to him, and if so, how much?

Now consider a situation in which a local government is planning to widen a typical municipal highway to relieve chronic traffic congestion. Knowing that the project will be financed by local and state taxes, but that many out-of-state travelers also are expected to benefit, should the planner justify the project solely on the benefits to local residents? Which point of view should we take in measuring the benefits—the municipal level, the state level, or both? It is important that any benefit measure be performed from the appropriate *point of view*.

In addition to valuation and point-of-view issues, many possibilities for tampering with the results of benefit–cost analyses exist. Unlike private projects, many public projects are undertaken because of political pressure rather than on the basis of their economic benefits alone. In particular, whenever the benefit–cost ratio becomes marginal or less than unity, a potential to inflate the benefit figures to make the project look good exists.

16.4 Benefit–Cost Ratios

An alternative way of expressing the worthiness of a public project is to compare the user's benefits (*B*) with the sponsor's cost (*C*) by taking the ratio *B/C*. In this section, we shall define the benefit–cost (*B/C*) ratio and explain the relationship between it and the conventional NPW criterion.

16.4.1 Definition of Benefit–Cost Ratio

For a given benefit–cost profile, let *B* and *C* be the present values of benefits and costs defined respectively by

$$B = \sum_{n=0}^{N} b_n (1 + i)^{-n} \qquad (16.2)$$

and

$$C = \sum_{n=0}^{N} c_n(1 + i)^{-n}, \qquad (16.3)$$

where b_n = Benefit at the end of period n, $b_n \geq 0$,

c_n = Expense at the end of period n, $c_n \geq 0$,

$A_n = b_n - c_n$,

N = Project life, and

i = Sponsor's interest rate (discount rate).

The sponsor's costs (C) consist of the equivalent capital expenditure (I) and the equivalent annual operating costs (C') accrued in each successive period. (Note the sign convention we use in calculating a benefit–cost ratio. Since we are using a ratio, all benefits and cost flows are expressed in positive units. Recall that in previous equivalent-worth calculations our sign convention was to explicitly assign "+" for cash inflows and "−" for cash outflows.) Let's assume that a series of initial investments is required during the first K periods, while annual operating and maintenance costs accrue in each subsequent period. Then the equivalent present value for each component is

$$I = \sum_{n=0}^{K} c_n(1 + i)^{-n} \qquad (16.4)$$

and

$$C' = \sum_{n=K+1}^{N} c_n(1 + i)^{-n}, \qquad (16.5)$$

and $C = I + C'$.

The B/C ratio[8] is defined as

$$BC(i) = \frac{B}{C} = \frac{B}{I + C'}, I + C' > 0. \qquad (16.6)$$

If we are to accept a project, $BC(i)$ must be greater than unity.

---

[8] An alternative measure, called the **net B/C ratio**, $B'C(i)$, considers only the initial capital expenditure as a cash outlay, and annual net benefits are used:

$$B'C(i) = \frac{B - C'}{I} = \frac{B'}{I}, I > 0.$$

The decision rule has not changed — the ratio must still be greater than one. It can be easily shown that a project with $BC(i) > 1$ will always have $B'C(i) > 1$, as long as both C and I are > 0, as they must be for the inequalities in the decision rules to maintain the stated senses. The magnitude of $BC(i)$ will generally be different than that for $B'C(i)$, but the magnitudes are irrelevant for making decisions. All that matters is whether the ratio exceeds the threshold value of one. However, some analysts prefer to use $B'C(i)$ because it indicates the net benefit (B') expected per dollar invested. But why do they care if the choice of ratio does not affect the decision? They may be trying to increase or decrease the magnitude of the reported ratio in order to influence audiences who do not understand the proper decision rule. People unfamiliar with benefit/cost analysis often assume that a project with a higher B/C ratio is better. This is not generally true, as is shown in 16.4.3. An incremental approach must be used to properly compare mutually exclusive alternatives.

Note that we must express the values of B, C', and I in present-worth equivalents. Alternatively, we can compute these values in terms of annual equivalents and use them in calculating the B/C ratio. The resulting B/C ratio is not affected.

EXAMPLE 16.1 Benefit–Cost Ratio

A public project being considered by a local government has the following estimated benefit–cost profile (Figure 16.4):

n	b_n	c_n	A_n
0		$10	−$10
1		10	−10
2	$20	5	15
3	30	5	25
4	30	8	22
5	20	8	12

Assume that $i = 10\%$, $N = 5$, and $K = 1$. Compute B, C, I, C', and BC(10%).

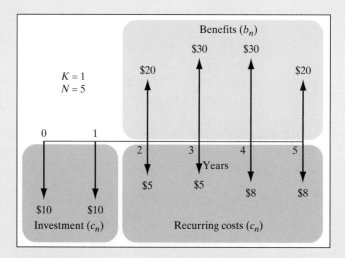

Figure 16.4 Classification of a project's cash flow elements (Example 16.1). Units are in millions of dollars.

SOLUTION

$$B = \$20(P/F, 10\%, 2) + \$30(P/F, 10\%, 3)$$
$$+ \$30(P/F, 10\%, 4) + \$20(P/F, 10\%, 5)$$
$$= \$71.98;$$
$$C = \$10 + \$10(P/F, 10\%, 1) + \$5(P/F, 10\%, 2)$$

$$+ \$5(P/F, 10\%, 3) + \$30(P/F, 10\%, 4) + \$20(P/F, 10\%, 5)$$

$$= \$37.41;$$

$$I = \$10 + \$10(P/F, 10\%, 1)$$

$$= \$19.09;$$

$$C' = C - I$$

$$= \$18.32.$$

Using Eq. (16.6), we can compute the B/C ratio as

$$BC(10\%) = \frac{71.98}{\$19.09 + \$18.32}$$

$$= 1.92 > 1.$$

The B/C ratio exceeds unity, so the users' benefits exceed the sponsor's costs.

16.4.2 Relationship between B/C Ratio and NPW

The B/C ratio yields the same decision for a project as does the NPW criterion. Recall that the BC(i) criterion for project acceptance can be stated as

$$\frac{B}{I + C'} > 1. \tag{16.7}$$

If we multiply both sides of the equation by the term $(I + C')$ and transpose $(I + C')$ to the left-hand side, we have

$$B > (I + C'),$$

$$B - (I + C') > 0, \tag{16.8}$$

$$PW(i) = B - C > 0, \tag{16.9}$$

which is the same decision rule[9] as that which accepts a project by the NPW criterion. This implies that we could use the benefit–cost ratio in evaluating private projects instead of using the NPW criterion, or we could use the NPW criterion in evaluating public projects. Either approach will signal consistent project selection. Recall that, in Example 16.1, $PW(10\%) = B - C = \$34.57 > 0$; the project would thus be acceptable under the NPW criterion.

16.4.3 Comparing Mutually Exclusive Alternatives: Incremental Analysis

Let us now consider how we choose among mutually exclusive public projects. As we explained in Chapter 7, we must use the incremental investment approach in comparing alternatives based on any relative measure such as IRR or B/C.

---

[9] We can easily verify a similar relationship between the net B/C ratio and the NPW criterion.

Incremental Analysis Based on BC(*i*)

To apply incremental analysis, we compute the incremental differences for each term (*B*, *I*, and *C'*) and take the *B/C* ratio on the basis of these differences. To use BC(*i*) on incremental investment, we may proceed as follows:

1. If one or more alternatives have *B/C* ratios greater than unity, eliminate any alternatives with a *B/C* ratio less than that.

2. Arrange the remaining alternatives in increasing order of the denominator $(I + C')$. Thus, the alternative with the smallest denominator should be the first (*j*), the alternative with the second smallest denominator should be second (*k*), and so forth.

3. Compute the incremental differences for each term (*B*, *I*, and *C'*) for the paired alternatives (*j*, *k*) in the list:

$$\Delta B = B_k - B_j,$$

$$\Delta I = I_k - I_j,$$

$$\Delta C' = C'_k - C'_j.$$

4. Compute the BC(*i*) on incremental investment by evaluating

$$BC(i)_{k-j} = \frac{\Delta B}{\Delta I + \Delta C'}.$$

If $BC(i)_{k-j} > 1$, select alternative *k*. Otherwise select alternative *j*.

5. Compare the alternative selected with the next one on the list by computing the incremental benefit–cost ratio.[10] Continue the process until you reach the bottom of the list. The alternative selected during the last pairing is the best one.

We may modify the foregoing decision procedure when we encounter the following situations:

- If $\Delta I + \Delta C' = 0$, we cannot use the benefit–cost ratio, because the equation implies that both alternatives require the same initial investment and operating expenditure. When this happens, we simply select the alternative with the largest *B* value.

- In situations where public projects with unequal service lives are to be compared, but the projects can be repeated, we may compute all component values (*B*, *C'*, and *I*) on an annual basis and use them in incremental analysis.

EXAMPLE 16.2 Incremental Benefit–Cost Ratios

Consider three investment projects: A1, A2, and A3. Each project has the same service life, and the present worth of each component value (*B*, *I*, and *C'*) is computed at 10% as follows:

---

[10] If we use the net *B/C* ratio as a basis, we need to order the alternatives in increasing order of *I* and compute the net *B/C* ratio on the incremental investment.

	Projects		
	A1	**A2**	**A3**
I	$5,000	$20,000	$14,000
B	12,000	35,000	21,000
C'	4,000	8,000	1,000
PW(*i*)	$3,000	$7,000	$6,000

(a) If all three projects are independent, which would be selected on the basis of BC(*i*)?

(b) If the three projects are mutually exclusive, which would be the best alternative? Show the sequence of calculations that would be required to produce the correct results. Use the *B*/*C* ratio on incremental investment.

SOLUTION

(a) Since PW$(i)_1$, PW$(i)_2$, and PW$(i)_3$ are positive, all of the projects are acceptable if they are independent. Also, the BC(*i*) values for each project are greater than unity, so the use of the benefit–cost ratio criterion leads to the same accept/reject conclusion as does the NPW criterion:

	A1	**A2**	**A3**
BC(*i*)	1.33	1.25	1.40

(b) If the projects are mutually exclusive, we must use the principle of incremental analysis. Obviously, if we attempt to rank the projects according to the size of the *B*/*C* ratio, we will observe a different project preference. For example, if we use the BC(*i*) on the total investment, we see that A3 appears to be the most desirable and A2 the least desirable, but selecting mutually exclusive projects on the basis of *B*/*C* ratios is incorrect. Certainly, with PW$(i)_2 >$ PW$(i)_3 >$ PW$(i)_1$, Project A2 would be selected under the NPW criterion. By computing the incremental *B*/*C* ratios, we will select a project that is consistent with that criterion.

We will first arrange the projects in increasing order of their denominators $(I + C')$ for the BC(*i*) criterion:[11]

Ranking Base	**A1**	**A3**	**A2**
I + *C'*	$9,000	$15,000	$28,000

---

[11] *I* is used as a ranking base for the $B'C(i)$ criterion. The order still remains unchanged: A1, A3, and A2.

- **A1 versus A3.** With the do-nothing alternative, we first drop from consideration any project that has a B/C ratio smaller than unity. In our example, the B/C
  ratios of all three projects exceed unity, so the first incremental comparison is
  between A1 and A3:

  $$BC(i)_{3-1} = \frac{\$21,000 - \$12,000}{(\$14,000 - \$5,000) + (\$1,000 - \$4,000)}$$

  $$= 1.51.$$

  Since the ratio is greater than unity, we prefer A3 to A1. Therefore, A3 becomes the "current best" alternative.
- **A3 versus A2.** Next, we must determine whether the incremental benefits to
  be realized from A2 would justify the additional expenditure. Therefore, we
  need to compare A2 and A3 as follows:

  $$BC(i)_{2-3} = \frac{\$35,000 - \$21,000}{(\$20,000 - \$14,000) + (\$8,000 - \$1,000)}$$

  $$= 1.081.$$

  The incremental B/C ratio again exceeds unity; therefore, we prefer A2 over
  A3. With no further projects to consider, A2 becomes our final choice.[12]

16.5 Analysis of Public Projects Based on Cost-Effectiveness

In evaluating public investment projects, we may encounter situations where competing
alternatives have the same goals, but the effectiveness with which those goals can be met
may or may not be measurable in dollars. In these situations, we compare decision alternatives directly on the basis of their **cost-effectiveness**. Here, we judge the effectiveness
of an alternative in dollars or some nonmonetary measure by the extent to which that alternative, if implemented, will attain the desired objective. The preferred alternative is
then either the one that produces the maximum effectiveness for a given level of cost or
the one that produces the minimum cost for a fixed level of effectiveness.

---

[12] Note that if we had to use the net B/C ratio on this incremental investment decision, we would obtain the
same conclusion. Since all $B'C(i)$ ratios exceed unity, no do-nothing alternative exists. By comparing the first
pair of projects on this list, we obtain

$$B'C(i)_{3-1} = \frac{(\$21,000 - \$12,000) - (\$1,000 - \$4,000)}{(\$14,000 - \$5,000)}$$

$$= 1.331.$$

Accordingly, Project A3 becomes the "current best." Next, a comparison of A2 and A3 yields

$$B'C(i)_{2-3} = \frac{(\$35,000 - \$21,000) - (\$8,000 - \$1,000)}{(\$20,000 - \$14,000)}$$

$$= 1.171.$$

Therefore, A2 becomes the best choice by the net B/C ratio criterion.

16.5.1 Cost-Effectiveness Studies in the Public Sector

A typical cost-effectiveness analysis procedure in the public sector involves the following steps:

Step 1. Establish the goals to be achieved by the analysis.

Step 2. Identify the restrictions imposed on achieving the goals, such as those having to do with the budget or with weight.

Step 3. Identify all the feasible alternatives for achieving the goals.

Step 4. Identify the social interest rate to use in the analysis.

Step 5. Determine the equivalent life-cycle cost of each alternative, including R&D costs, testing costs, capital investment, annual operating and maintenance costs, and the salvage value of the item under consideration.

Step 6. Determine the basis for developing the cost-effectiveness index. Two approaches may be used: (1) the fixed-cost approach and (2) the fixed-effectiveness approach. If the fixed-cost approach is used, determine the amount of effectiveness obtained at a given cost. If the fixed-effectiveness approach is used, determine the cost required to obtain the predetermined level of effectiveness.

Step 7. Compute the cost-effectiveness ratio for each alternative, based on the criterion selected in Step 6.

Step 8. Select the alternative with the maximum cost-effective index.

When either the cost or the level of effectiveness is clearly stated in achieving the declared program objective, most cost-effectiveness studies will be relatively straightforward. If that is not the case, however, the decision maker must come up with his or her own program objective by fixing either the cost required in the program or the level of effectiveness to be achieved. The next section shows how such a problem can easily evolve in many public projects or military applications.

16.5.2 A Cost-Effectiveness Case Study[13]

To illustrate the procedures involved in cost-effectiveness analysis, we shall present an example of how the U.S. Army selected the most cost-effective program for providing on-time delivery of time-sensitive, high-priority cargos and key personnel to the battle staging grounds.

Statement of the Problem

The U.S. Army has been engaged in recent conflicts that have transformed from the traditional set-piece battles into operations in distributed locations that require rapid sequences of activities. To support success in these new operations, the Army needs timely delivery of cargos and personnel in the distributed locations. Consequently, the dispersion of forces has created a condition of unsecured land lines of communication (LOC), and aerial delivery is considered as one of the practical solutions to the problem. The time-sensitive nature of these priority cargos makes it most preferable to have direct delivery from intermediate staging bases (ISB) to the forward brigade combat teams (BCT).

---

[13] This case study is provided by Dr. George C. Prueitt of CAS, Inc. All numbers used herein do not represent the actual values used by the U.S. Army.

The transportation network typically involves a long-haul move (a fixed-wing aerial movement), and a short-haul distribution (by helicopter or truck). The long-haul usually stops at the closest supporting airfield, but with the appropriate asset, a single, direct movement could be made. The long-haul portion has been the bigger problem—trucks take too long and are vulnerable to hostilities; the U.S. Air Force (USAF) assets are not always available or appropriate; and existing helicopters do not have the necessary range to fly. Further, the use of the larger USAF aircraft results in inefficient load, as the high-priority cargos use only a small fraction of the capacity of the aircraft. The Army is investigating how to best correct the problem. They are considering three options to handle the long haul: (1) Use the U.S. Air Force (C-130J) assets; (2) procure a commercially available Future Cargo Aircraft (FCA); or (3) use additional helicopters (CH-47).

Defining the Goals

The solution will be a "best value" selection, providing the best capability for the investment required. The selection process must address three operational perspectives.

- First, a "micro" examination: For a given specific battlefield arrangement (available transportation network, supported organizations and locations), the decision metric is the time required to delivery specific critical cargos and the solutions' comparative operating costs. This metric is a suitability measure, to identify the system that best, most quickly, meets the timely delivery requirement.
- Second, a "macro" examination: Given a broader, more prolonged theater support requirement, the decision metric becomes what percentage of critical cargos can be delivered, and what is the cost of procuring all the transportation assets needed to create the delivery network combinations (fixed-wing airplanes, helicopters, and trucks) to meet those delivery percentages. This metric is a feasibility measure to determine the number of assets needed to meet the quantities of cargos to be delivered.
- Third, given relatively comparable fixed-wing aircraft fleets (differing quantities for equal capability), what are the 15-year life-cycle costs associated with those alternatives? This is an affordability metric, to ensure that all life-cycle costs are properly identified and can be met within the Army budget.

Description of Alternatives

As mentioned earlier, the Army is considering three alternatives. They are:

- **Alternative 1—use the USAF C-130J assets.** The USAF C-130J aircraft has been proposed as an alternative for the long-haul task. The C-130J is a new acquisition, and will be considered in two employment concepts: first, in a direct movement pattern, and second, as part of a standard USAF Scheduled Tactical Air Re-supply (STAR) route (analogous to a bus route). Current operations almost exclusively use the STAR route method when C-130 aircraft are employed. It is assumed that transportation planning and theater priorities would be sufficiently high to permit the C-130J to be employed in the more direct delivery method, and not limited solely to STAR route operations.
- **Alternative 2—use future cargo aircraft.** Another long-haul alternative is to procure one of two commercially available products capable of delivering these cargos on a timely and efficient basis.

- **Alternative 3—use additional C-47 helicopters.** Looking first at the long-haul task, most of today's deliveries are provided by either the CH-47 helicopter or the C-130 aircraft in conjunction with another vehicle. The CH-47 is being used to perform this mission because it is the "best available" Army asset. Unfortunately, it is expensive to operate, and its range limitations make it an inefficient method of carrying out the mission. Further, the long distances between the ISB to the forward units are causing the helicopters to accumulate flight hours more rapidly than planned in their service-life projections, and this has generated a significant increase in their maintenance requirements.

These long-haul delivery alternatives are described in Table 16.6.

In addition to the long-haul task, the shorter delivery distribution legs will usually need a complementary ground or helicopter asset to make the final delivery when the fixed-wing assets cannot land close enough to the BCT, because of a lack of improved landing strips or runways of sufficient length. This creates a greater burden on the transportation network to ensure that adequate ground or rotary-wing assets are available to complete the mission. The combinations of long-haul alternatives and final delivery systems create the series of network transportation options described in Table 16.7.

TABLE 16.6 Descriptions of Alternatives for Transportation Network Long-Haul Requirement

Alternative	Description	Comment
Alternative 1—C-130J	Procure additional C-130Js and maximize their use to perform the long-haul task of the intra-theater delivery of critical, time-sensitive cargo and key personnel. Minimize requirement for additional CH-47s.	C-130Js will use both USAF Scheduled Tactical Air Re-supply (STAR) routes and more direct routes. The C-130J has longer run way requirements than smaller FCA aircraft does.
Alternative 2—Army FCA	Procure a variant of the FCA and maximize its use to perform the long-haul task of the intratheater delivery of mission-critical, time-sensitive cargo and key personnel, with minimal use of CH-47s.	In some situations, the FCA will be able to land either closer to or at the BCT locations. Two versions of the FCA (Alternative 2A and Alternative 2B) have been included in this examination.
Alternative 3—More CH-47s	Procure additional CH-47s to perform the long-haul task of the intratheater delivery of critical, time-sensitive cargo and key personnel.	This alternative will include those additional CH-47s needed to carry support systems for the CH-47s that are used to move cargo over extended distances.

TABLE 16.7 Descriptions of Transportation Network Options

Network Option (NO_i)	Description
NO_1	ALT 1 (C-130) direct from ISB to C-130 Supportable Forward Airfield,
NO_2	ALT 1 (C-130) direct from ISB to C-130 Supportable Forward Airfield, then CH-47 to BCT
NO_3	ALT 1 (C-130) STAR route from ISB to C-130 Supportable Forward Airfield (stop nearest to BCT), then truck to BCT
NO_4	ALT 1 (C-130) STAR route from ISB to C-130 Supportable Forward Airfield (stop nearest to BCT), then CH-47 to BCT
NO_5	ALT 2A direct from ISB to FCA Supportable Forward Airfield, then truck to BCT
NO_6	ALT 2A direct from ISB to FCA Supportable Forward Airfield, then CH-47 to BCT
NO_7	ALT 2A direct from ISB to BCT
NO_8	ALT 2B direct from ISB to FCA Supportable Forward Airfield, then truck to BCT
NO_9	ALT 2B direct from ISB to FCA Supportable Forward Airfield, then CH-47 to BCT
NO_{10}	ALT 2B direct from ISB to BCT
NO_{11}	ALT 3 (CH-47) direct from ISB to BCT, with multiple refueling stops en route

Figure 16.5 provides a descriptive portrayal of these transportation network options.

Micro Analysis

To perform a cost-effectiveness analysis, the micro comparisons were made by selecting the delivery time as the performance metric (vertical axis in Figure 16.6), and one-way operating cost as the other metric (horizontal axis). Lower cost and shorter delivery times cause the direction of preference to be in the lower left-hand corner of the figure. The respective operating costs for each system were: C-130J at $3,850 per flight hour, FCA 2A at $2,800, FCA 2B at $1,680, and CH-47 at $4,882. Transportation network option 10 (NO_{10}), based on Alternative 2B FCA, and NO_7, based on Alternative 2A FCA, with each flying directly from ISB to BCT, had the shortest times to accomplish the mission. Their respective operating costs pushed them into the most favored regions on the graph. Not surprisingly, the next quickest option was NO_2, based on Alternative 1 aircraft flying directly to a C-130 Supportable Forward Airfield, with a CH-47 delivering for the last leg; however, the higher operating costs of the C-130J (over the FCA alternatives) and the CH-47 made its operating cost more than double NO_{10}. For the remaining options, NO_3 and NO_4, using the C-130J on a

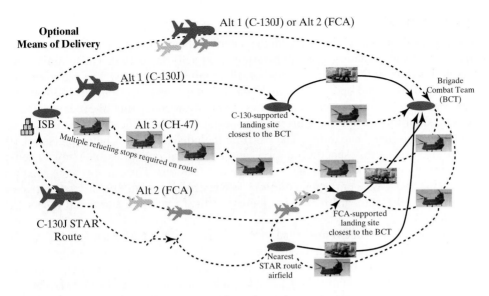

Figure 16.5 Transportation network option schematics.

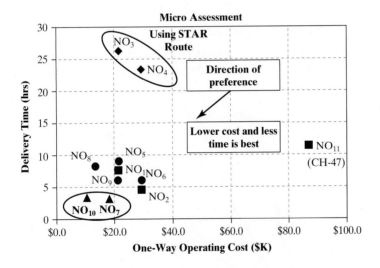

Figure 16.6 Assessment results from micro cost-effectiveness analysis.

STAR route, required the longest delivery time, although the operating costs were comparable to the most favored solutions. NO_{11}, the CH-47–based option, had the highest operating costs. Figure 16.6 summarizes the results for all 11 transportation network options.

Macro Analysis

The macro analysis was performed with the criterion of capability to deliver high-priority cargos in two simultaneous combat operations over a 30-day period. The minimum acceptable on-time delivery rate is 80% of all required shipments. The requirements were derived from one operation with 10 days of high-intensity combat, followed by 20 days of stability operations, and a second operation consisting of 30 days of low-intensity stability operations. Current transportation asset allocations were used to establish a "baseline" capability. The baseline included 28 CH-47 helicopters and 10 C-130J aircraft for this specific mission, and additional quantities of Army systems (other aircraft and trucks) to do the final cargo distributions. These assets were capable of satisfying about 12% of the total on-time delivery requirement. The shortfall between 12% and the 80% standard represents the capability "gap" that additional assets are required to fill.

To complete the macro analysis, additional quantities of Alternatives 1, 2A, 2B, and 3 were incrementally added, and the resultant increase in on-time deliveries was computed. The cost to procure the additional quantities was determined with the following unit prices: Alternative 1 at $75.6M each, Alternative 2A at $34.5M, Alternative 2B at $29.2M, and Alternative 3 at $26.1M. Figure 16.7 shows the changes in percent-on-time delivery versus the procurement cost of the respective alternatives. There is a significant procurement cost differentialy between alternatives, with Alternatives 2A and 2B being the most cost effective.

Life-Cycle Cost for Each Alternative

The third part of the case study is to perform the life-cycle cost (LCC) analysis for the various alternatives. Table 16.8 lists the cost items that are considered in this LCC analysis.

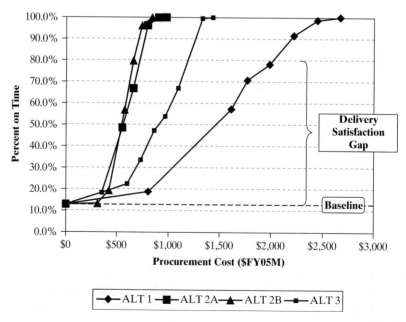

Figure 16.7 Assessment results from macro effectiveness analysis.

TABLE 16.8 Definitions of Life-Cycle Cost Variables

Life-Cycle Cost Element	Cost Variable	Description
1	RDT&E	Government and contractor costs associated with the research, development, test, and evaluation (RDT&E) of the material system.
2	Procurement	Costs resulting from the production and initial fielding of the material system in the Army's operational inventory.
3	Military construction	Cost of all military construction projects associated with a material system. No military construction money is anticipated to be needed for any of the FCA alternatives.
4	Military personnel	Cost of military personnel associated with the development and operation of the material system.
5	O&M	O&M funded cost associated with development, production, fielding, operation, and support of the material system for 20 years.

In fact, these cost items represent the common decision elements mandated by the U.S. Department of Defense, along with appropriate rules and assumptions, to ensure that each alternative be evaluated for comparable cost and affordability. First, an LCC estimate was developed for specific quantities of each primary long-haul aircraft in the alternatives. The number of aircraft was chosen to establish equal fleet capabilities (not necessarily fleet size). Second, to determine the number of aircraft required for each alternative, the U.S. Army used the equal-effectiveness standard of 80% on-time delivery of high-priority cargo and key personnel. Specifically,

- Alternatives 2A and 2B use a fleet size of 56, as derived from the macro analysis and with additional fleet support considerations.
- Alternative 1, the USAF C-130J, uses 40 aircraft, because it was assumed that all additional fleet support requirements would be addressed with the USAF larger program. Similarly,
- Alternative 3 uses 92 CH-47F as its mission requirement, with no additional fleet support requirements, as the 92 would be added to the existing fleet of about 400 CH-47 aircraft.

Cost estimates were developed for the respective alternatives. In a simple form, these estimates provide the future estimated costs for procurement and operations. The procurement costs include both RDT&E and procurement requirements. The operating cost includes operations and maintenance (O&M) and military personnel costs. There were some nominal military construction costs, but they were equal for all alternatives. A 15-year operating cycle (from first procurement through last operation) was assumed in developing the cost estimates. The projected costs for future years are provided in Table 16.9.

TABLE 16.9 Life-Cycle Cost Estimates for Each Alternative

Year	ALT 1 Proc	ALT 1 Operating	ALT 2A Proc	ALT 2A Operating	ALT 2B Proc	ALT 2B Operating	ALT 3 Proc	ALT 3 Operating
1	$702.80	$ 25.0	$257.5	$ 4.8	$221.1	$ 3.3	$620.2	$ 10.8
2	$860.27	$ 56.1	$529.9	$ 14.7	$450.3	$ 12.5	$654.4	$ 82.0
3	$877.47	$ 88.4	$800.8	$ 43.0	$687.2	$ 35.0	$451.4	$163.2
4	$447.51	$122.0	$431.3	$ 64.0	$376.1	$ 52.6	$145.2	$224.1
5	$ 0.00	$151.6	$206.1	$105.1	$189.2	$ 80.7	$351.0	$251.4
6	$ 0.00	$154.7	$ 0.0	$114.3	$ 0.0	$ 80.5	$262.6	$258.8
7	$ 0.00	$157.8	$ 0.0	$120.6	$ 0.0	$ 81.5	$ 0.0	$262.3
8	$ 0.00	$160.9	$ 0.0	$128.4	$ 0.0	$ 86.8	$ 0.0	$274.7
9	$ 0.00	$164.1	$ 0.0	$132.2	$ 0.0	$ 90.5	$ 0.0	$287.5
10	$ 0.00	$167.4	$ 0.0	$139.9	$ 0.0	$ 95.6	$ 0.0	$301.2
11	$ 0.00	$170.8	$ 0.0	$141.4	$ 0.0	$ 98.3	$ 0.0	$315.5
12	$ 0.00	$174.2	$ 0.0	$145.8	$ 0.0	$102.5	$ 0.0	$324.4
13	$ 0.00	$145.8	$ 0.0	$129.6	$ 0.0	$106.5	$ 0.0	$362.1
14	$ 0.00	$110.0	$ 0.0	$119.8	$ 0.0	$ 89.6	$ 0.0	$248.8
15	$ 0.00	$ 72.6	$ 0.0	$106.2	$ 0.0	$ 77.6	$ 0.0	$ 86.7
16	$ 0.00	$ 33.6	$ 0.0	$100.9	$ 0.0	$ 74.2	$ 0.0	$ 35.4

Notes: All costs are $M.

"Proc" means procurement cost estimate.

"Operating" includes operations, maintenance, and manpower costs.

For most of the procurements by the U.S. Department of Defense, future inflation factors must be specified in the analysis. For this case study, an annual inflation rate of 2% was used. In terms of interest rate, an inflation-adjusted discount rate of 8% was used to calculate the LCC for each alternative, which is summarized in Table 16.10.

Figure 16.8 illustrates the cost component of each alternative's LCC. The figure shows that Alternatives 2A and 2B, at 56 aircraft, have both lower procurement and operations and maintenance costs than either Alternative 1, the C-130J alternative, for its quantity of 40 aircraft, or Alternative 3, the CH-47F alternative, based on 92 helicopters. As anticipated earlier, the CH-47F alternative turns out to be the most expensive option due to its higher operating costs, even though its procurement of 92 helicopters is less expensive than 40 C-130J aircraft. The total operating costs differ in contribution by type of aircraft. The fixed-wing aircraft estimates are based on 600 hours of flying time per year, while the helicopters are scheduled for less than

TABLE 16.10 Present Value of Life-Cycle Cost for Each Alternative

Year	ALT 1 Proc	ALT 1 Operating	ALT 2A Proc	ALT 2A Operating	ALT 2B Proc	ALT 2B Operating	ALT 3 Proc	ALT 3 Operating
0	$ 702.8	$ 25.0	$ 257.5	$ 4.8	$ 221.1	$ 3.3	$ 620.2	$ 10.8
1	$ 796.5	$ 51.9	$ 490.6	$ 13.6	$ 417.0	$ 11.6	$ 606.0	$ 75.9
2	$ 752.3	$ 75.8	$ 686.6	$ 36.8	$ 589.1	$ 30.0	$ 387.0	$ 140.0
3	$ 355.2	$ 96.9	$ 342.4	$ 50.8	$ 298.6	$ 41.8	$ 115.2	$ 177.9
4	$ 0.0	$ 111.5	$ 151.5	$ 77.3	$ 139.1	$ 59.4	$ 258.0	$ 184.8
5	$ 0.0	$ 105.3	$ 0.0	$ 77.8	$ 0.0	$ 54.8	$ 178.7	$ 176.1
6	$ 0.0	$ 99.4	$ 0.0	$ 76.0	$ 0.0	$ 51.4	$ 0.0	$ 165.3
7	$ 0.0	$ 93.9	$ 0.0	$ 74.9	$ 0.0	$ 50.7	$ 0.0	$ 160.3
8	$ 0.0	$ 88.7	$ 0.0	$ 71.4	$ 0.0	$ 48.9	$ 0.0	$ 155.3
9	$ 0.0	$ 83.8	$ 0.0	$ 70.0	$ 0.0	$ 47.8	$ 0.0	$ 150.7
10	$ 0.0	$ 79.1	$ 0.0	$ 65.5	$ 0.0	$ 45.5	$ 0.0	$ 146.1
11	$ 0.0	$ 74.7	$ 0.0	$ 62.6	$ 0.0	$ 43.9	$ 0.0	$ 139.1
12	$ 0.0	$ 57.9	$ 0.0	$ 51.5	$ 0.0	$ 42.3	$ 0.0	$ 143.8
13	$ 0.0	$ 40.4	$ 0.0	$ 44.0	$ 0.0	$ 33.0	$ 0.0	$ 91.5
14	$ 0.0	$ 24.7	$ 0.0	$ 36.2	$ 0.0	$ 26.4	$ 0.0	$ 29.5
15	$ 0.0	$ 10.6	$ 0.0	$ 31.8	$ 0.0	$ 23.4	$ 0.0	$ 11.2
Subtotal	$2,606.9	$1,119.6	$1,928.5	$ 845.0	$1,664.9	$ 614.1	$2,165.1	$1,958.3
ALT Total		$3,726.5		$2,773.5		$2,278.9		$4,123.4

Notes: 1. All figures are in $M.

2. Sample calculation:

ALT 1 Present Value $(8\%) = (\$702.8 + \$25.0) + (\$796.5 + 51.9)(P/F, 8\%, 1)$
$$+ \ldots + \$10.6(P/F, 8\%, 15)$$
$$= \$3,726.5M.$$

200 hours of flying time. However, because the airspeed and flying range for the helicopter are much lower than for the fixed-wing aircraft, more CH-47F helicopters are required to provide the same level of capability. Alternative 1, using the C-130J as the primary aircraft, requires fewer total aircraft than Alternatives 2A and 2B, but with nearly twice the unit cost, it is a more expensive option. In conclusion, for the alternatives considered, the commercially available candidates for the FCA (Alternatives 2A and 2B) provide the best opportunity for the Army to satisfy its requirement to delivery of mission-critical, time-sensitive cargo and key personnel from an ISB directly to a BCT, with Alternative 2B having a slight edge over 2A, for the metrics used.

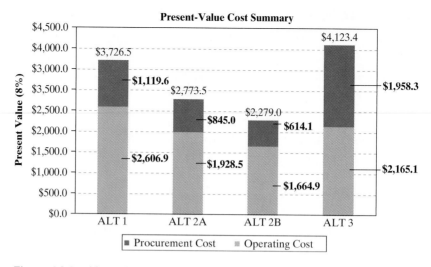

Figure 16.8 Alternatives' present value (8%) for procurement and operations.

SUMMARY

■ The service sector accounts for about 80% of U.S. GDP. Health care and government services account for almost 30% of the total GDP.

■ In health economics, cost-effectiveness analysis is the most widely used economic evaluation method. The cost-effectiveness ratio is defined as

$$\text{Cost-effectiveness ratio} = \frac{\sum(\text{all costs})}{\sum(\text{all measured health effects})}$$

If independent intervention programs are evaluated, we can rank the programs on the basis of their cost-effectiveness ratios and accept as many programs as the budget permits. When we compare mutually exclusive intervention programs, the incremental cost-effectiveness ratio should be used to determine whether an additional cost can be justified to increase health effects.

■ **Benefit–cost analysis** is commonly used to evaluate public projects; several facets unique to public-project analysis are neatly addressed by benefit–cost analysis:

1. Benefits of a nonmonetary nature can be quantified and factored into the analysis.

2. A broad range of project users distinct from the sponsor can and should be considered; benefits and disbenefits to *all* these users can and should be taken into account.

■ Difficulties involved in public-project analysis include the following:

1. Identifying all the users of the project.

2. Identifying all the benefits and disbenefits of the project.

3. Quantifying all the benefits and disbenefits in dollars or some other unit of measure.

4. Selecting an appropriate interest rate at which to discount benefits and costs to a present value.

■ The *B/C* ratio is defined as

$$\text{BC}(i) = \frac{B}{C} = \frac{B}{I + C'}, I + C' > 0.$$

The decision rule is that if $\text{BC}(i) \geq 1$, then the project is acceptable.

■ The net *B/C* ratio is defined as

$$\text{B'C}(i) = \frac{B - C'}{I} = \frac{B'}{I}, I > 0.$$

The net *B/C* ratio expresses the net benefit expected per dollar invested. The same decision rule applies as for the *B/C* ratio.

■ The **cost-effectiveness method** allows us to compare projects on the basis of cost and nonmonetary effectiveness measures. We may either maximize effectiveness for a given cost criterion or minimize cost for a given effectiveness criterion.

PROBLEMS

Cost-Effectiveness Analysis

16.1   The following table summarizes the costs of treatment of a disease, based on two different antibiotics and associated health benefits (effectiveness):

Which treatment option is the best?

Type of Treatment	Cost of Treating 100 Patients	Effectiveness (Percent Successful Treatment of Infections)
Antibiotic A	$12,000	75%
Antibiotic B	$13,500	80%
Antibiotic C	$14,800	82%

16.2   The Table P16.2 summarizes cervical cancer treatment options and their health effectiveness. Find the best strategy for treating cervical cancer.

TABLE P16.2 A CEA Examining Three Strategies

Strategy	Cost	Marginal Cost	Effectiveness	Marginal Effectiveness	CE Ratio
Nothing	$0	—	0 years	—	—
Simple	$5,000	$5,000	5 years	5 years	$1,000/yr
Complex	$50,000	$45,000	5.5 years	0.5 years	$90,000/yr

Valuation of Benefits and Costs

16.3 The state of Michigan is considering a bill that would ban the use of road salt on highways and bridges during icy conditions. Road salt is known to be toxic, costly, corrosive, and caustic. Chevron Chemical Company produces a calcium magnesium acetate (CMA) deicer and sells it for $600 a ton as Ice-B-Gon. Road salts, by contrast, sold for an average of $14 a ton in 1995. Michigan needs about 600,000 tons of road salt each year. (The state spent $9.2 million on road salt in 1995.) Chevron estimates that each ton of salt on the road costs $650 in highway corrosion, $525 in rust on vehicles, $150 in corrosion to utility lines, and $100 in damages to water supplies, for a total of $1,425. Unknown salt damage to vegetation and soil surrounding areas of highways has occurred. Michigan would ban road salt (at least on expensive steel bridges or near sensitive lakes) if state studies support Chevron's cost claims.

(a) What would be the users' benefits and sponsor's costs if a complete ban on road salt were imposed in Michigan?

(b) How would you go about determining the salt damages (in dollars) to vegetation and soil?

16.4 A public school system in Ohio is considering the adoption of a four-day school week as opposed to the current five-day school week in high schools. The community is hesitant about the plan, but the superintendent of the school system envisions many benefits associated with the four-day system, Wednesday being the "day off." The following pros and cons have been cited:

• Experiments with the four-day system indicate that the "day off" in the middle of the week will cut down on both teacher and pupil absences.

• The longer hours on school days will require increased attention spans, which is not an appropriate expectation for younger children.

• The reduction in costs to the federal government should be substantial, as the number of lunches served would be cut by approximately 20%.

• The state bases its expenditures on its local school systems largely on the average number of pupils attending school in the system. Since the number of absences will decrease, state expenditures on local systems should increase.

• Older students might want to work on Wednesdays. Unemployment is a problem in this region, however, and any influx of new job seekers could aggravate an existing problem. Community centers, libraries, and other public areas also may experience increased usage on Wednesdays.

• Parents who provide transportation for their children will see a savings in fuel costs. Primarily, only those parents whose children live less than 2 miles from the school would be involved. Children living more than 2 miles from school are eligible for free transportation provided by the local government.

• Decreases in both public and private transportation should result in fuel conservation, decreased pollution, and less wear on the roads. Traffic congestion should ease on Wednesdays in areas where congestion caused by school traffic is a problem.

• Working parents will be forced to make child-care arrangements (and possibly payments) for one weekday per week.

• Students will benefit from wasting less time driving to and from school; Wednesdays will be available for study, thus taking the heavy demand off most nights. Bused students will spend far less time per week waiting for buses.

- The local school board should see some ease in funding problems. The two areas most greatly affected are the transportation system and nutritional programs.
    - (a) For this type of public study, what do you identify as the users' benefits and disbenefits?
    - (b) What items would be considered as the sponsor's costs?
    - (c) Discuss any other benefits or costs associated with the four-day school week.

16.5 The Electric Department of the City of Tallahassee, Florida, operates generating and transmission facilities serving approximately 140,000 people in the city and surrounding Leon County. The city has proposed the construction of a $300 million, 235-MW circulating fluidized-bed combustor (CFBC) at Arvah B. Hopkins Station to power a turbine generator that is currently receiving steam from an existing boiler fueled by gas or oil. Among the advantages associated with the use of CFBC systems are the following:

- A variety of fuels can be burned, including inexpensive low-grade fuels with high ash and a high sulfur content.
- The relatively low combustion temperatures inhibit the formation of nitrogen oxides. Acid-gas emissions associated with CFBC units would be expected to be significantly lower than emissions from conventional coal-fueled units.
- The sulfur-removal method, low combustion temperatures, and high-combustion efficiency characteristic of CFBC units result in solid wastes, which are physically and chemically more amenable to land disposal than the solid wastes resulting from conventional coal-burning boilers equipped with flue-gas desulfurization equipment.

On the basis of the Department of Energy's (DOE's) projections of growth and expected market penetration, the demonstration of a successful 235-MW unit could lead to as much as 41,000 MW of CFBC generation being constructed by the year 2010. The proposed project would reduce the city's dependency on oil and gas fuels by converting its largest generating unit to coal-fuel capability. Consequently, substantial reductions in local acid-gas emissions could be realized in comparison to the permitted emissions associated with oil fuel. The city has requested a $50 million cost share from the DOE. Under the Clean Coal Technology Program, cost sharing is considered attractive because the DOE cost share would largely offset the risk of using such a new technology. To qualify for cost-sharing money, the city has to address the following questions for the DOE:

(a) What is the significance of the project at local and national levels?

(b) What items would constitute the users' benefits and disbenefits associated with the project?

(c) What items would constitute the sponsor's costs?

Put yourself in the city engineer's position, and respond to these questions.

Benefit–Cost Analyses

16.6 A city government is considering two types of town-dump sanitary systems. Design A requires an initial outlay of $400,000, with annual operating and maintenance costs of $50,000 for the next 15 years; design B calls for an investment of $300,000, with annual operating and maintenance costs of $80,000 per year for the next 15 years. Fee collections from the residents would be $85,000 per year. The interest rate is 8%, and no salvage value is associated with either system.

(a) Using the benefit–cost ratio BC(i), which system should be selected?

(b) If a new design (design C), which requires an initial outlay of $350,000 and annual operating and maintenance costs of $65,000, is proposed, would your answer in (a) change?

16.7 The U.S. government is considering building apartments for government employees working in a foreign country and living in locally owned housing. A comparison of two possible buildings indicates the following:

	Building X	Building Y
Original investment by government agencies	$8,000,000	$12,000,000
Estimated annual maintenance costs	240,000	180,000
Savings in annual rent now being paid to house employees	1,960,000	1,320,000

Assume the salvage or sale value of the apartments to be 60% of the first investment. Use 10% and a 20-year study period to compute the B/C ratio on incremental investment, and make a recommendation. (Assume no do-nothing alternative.)

16.8 Three public-investment alternatives with the same service life are available: A1, A2, and A3. Their respective total benefits, costs, and first costs are given in present worth as follows:

	Proposals		
Present worth	A1	A2	A3
I	100	300	200
B	400	700	500
C'	100	200	150

Assuming no do-nothing alternative, which project would you select on the basis of the benefit–cost ratio BC(i) on incremental investment?

16.9 A local city that operates automobile parking facilities is evaluating a proposal that it erect and operate a structure for parking in the city's downtown area. Three designs for a facility to be built on available sites have been identified (all dollar figures are in thousands):

	Design A	Design B	Design C
Cost of site	$240	$180	$200
Cost of building	2,200	700	1,400
Annual fee collection	830	750	600
Annual maintenance cost	410	360	310
Service life	30 years	30 years	30 years

At the end of the estimated service life, whichever facility had been constructed would be torn down, and the land would be sold. It is estimated that the proceeds from the resale of the land will be equal to the cost of clearing the site. If the city's interest rate is known to be 10%, which design alternative would be selected on the basis of the benefit–cost criterion?

16.10 The federal government is planning a hydroelectric project for a river basin. In addition to producing electric power, this project will provide flood control, irrigation, and recreation benefits. The estimated benefits and costs expected to be derived from the three alternatives under consideration are listed in the following table:

	Decision Alternatives		
	A	B	C
Initial cost	$8,000,000	$10,000,000	$15,000,000
Annual benefits or costs:			
Power sales	$1,000,000	$1,200,000	$1,800,000
Flood control savings	250,000	350,000	500,000
Irrigation benefits	350,000	450,000	600,000
Recreation benefits	100,000	200,000	350,000
O&M costs	200,000	250,000	350,000

The interest rate is 10%, and the life of each of the projects is estimated to be 50 years.

(a) Find the benefit–cost ratio for each alternative.

(b) Select the best alternative on the basis of BC(i).

16.11 Two different routes are under consideration for a new interstate highway:

	Length of Highway	First Cost	Annual Upkeep
The "long" route	22 miles	$21 million	$140,000
Transmountain shortcut	10 miles	$45 million	$165,000

For either route, the volume of traffic will be 400,000 cars per year. These cars are assumed to operate at $0.25 per mile. Assume a 40-year life for each road and an interest rate of 10%. Determine which route should be selected.

16.12 The government is considering undertaking four projects. These projects are mutually exclusive, and the estimated present worth of their costs and the present worth of their benefits are shown in millions of dollars in the following table:

Projects	PW of Benefits	PW of Costs
A1	$40	$85
A2	150	110
A3	70	25
A4	120	73

All of the projects have the same duration.

Assuming no do-nothing alternative, which alternative would you select? Justify your choice by using a benefit–cost (BC(i)) analysis on incremental investment.

Short Case Studies

ST16.1 Fast growth in the population of the city of Orlando and in surrounding counties—Orange County in particular—has resulted in insurmountable traffic congestion. The county has few places to turn for extra money for road improvements, except new taxes. County officials have said that the money they receive from current taxes is insufficient to widen overcrowded roads, improve roads that don't meet modern standards, and pave dirt roads. State residents now pay 12 cents in taxes on every gallon of gas. Four cents of that goes to the federal government, 4 cents to the state, 3 cents to the county in which the tax is collected, and 1 cent to the cities. The county commissioner has suggested that the county get the money by tacking an extra penny-a-gallon tax onto gasoline, bringing the total federal and state gas tax to 13 cents a gallon. This new tax would add about $2.6 million a year to the road-construction budget. The extra money would have a significant impact. With the additional revenue, the county could sell a $24 million bond issue. It would then have the option of spreading that amount among many smaller projects or concentrating on a major project. Assuming that voters would approve a higher gas tax, county engineers were asked to prepare a priority list outlining which roads would be improved with the extra money. The road engineers also computed the possible public benefits associated with each road-construction project; they accounted for a possible reduction in travel time, a reduction in the accident rate, land appreciation, and savings in the operating costs of vehicles. Following is a list of the projects and their characteristics:

District	Project	Type of Improvement	Construction Cost	Annual O&M	Annual Benefits
	27th Street	Four-lane	$ 980,000	$ 9,800	$313,600
	Holden Avenue	Four-lane	3,500,000	35,000	850,000
I	Forest City Road	Four-lane	2,800,000	28,000	672,000
	Fairbanks Avenue	Four-lane	1,400,000	14,000	490,000
	Oak Ridge Road	Realign	2,380,000	47,600	523,600
II	University Blvd.	Four-lane	5,040,000	100,800	1,310,400
	Hiawassee Road	Four-lane	2,520,000	50,400	831,600
	Lake Avenue	Four-lane	4,900,000	98,000	1,021,000
	Apopka-Ocoee Road	Realign	1,365,000	20,475	245,700
III	Kaley Avenue	Four-lane	2,100,000	31,500	567,000
	Apoka-Vineland Road	Two-lane	1,170,000	17,550	292,000
	Washington Street	Four-lane	1,120,000	16,800	358,400
	Mercy Drive	Four-lane	2,800,000	56,000	980,000
IV	Apopka Road	Reconstruct	1,690,000	33,800	507,000
	Old Dixie Highway	Widen	975,000	15,900	273,000
	Old Apopka Road	Widen	1,462,500	29,250	424,200

Assume a 20-year planning horizon and an interest rate of 10%. Which projects would be considered for funding in (a) and (b)?

(a) Due to political pressure, each district will have the same amount of funding, say, $6 million.

(b) The funding will be based on tourist traffic volumes. Districts I and II combined will get $15 million, and Districts III and IV combined will get $9 million. It is desirable to have at least one four-lane project from each district.

ST16.2 The City of Portland Sanitation Department is responsible for the collection and disposal of all solid waste within the city limits. The city must collect and dispose of an average of 300 tons of garbage each day. The city is considering ways to improve the current solid-waste collection and disposal system.

• The current system uses Dempster Dumpmaster Frontend Loaders for collection, and incineration or landfill for disposal. Each collecting vehicle has a load capacity of 10 tons, or 24 cubic yards, and dumping is automatic. The incinerator in use was manufactured in 1942 and was designed to incinerate 150 tons per 24 hours. A natural-gas afterburner has been added in an effort to reduce air pollution; however, the incinerator still does not meet state air-pollution requirements, and it is operating under a permit from the Oregon State Air and Water Pollution Control Board. Prison-farm labor is used for the operation of the incinerator. Because the capacity of the incinerator is relatively low, some trash is not incinerated, but is taken to the city landfill. The trash landfill is located approximately 11 miles, and the incinerator approximately 5 miles, from the center of the city. The mileage and costs in person-hours for delivery to the disposal sites are excessive; a high percentage of empty vehicle miles and person-hours is required because separate methods of disposal are used and the destination sites are remote from the collection areas. The operating cost for the present system is $905,400, including $624,635 to operate the prison-farm incinerator, $222,928 to operate the existing landfill, and $57,837 to maintain the current incinerator.

• The proposed system locates a number of portable incinerators, each with 100-ton-per-day capacity for the collection and disposal of refuse waste collected for three designated areas within the city. Collection vehicles will also be staged at these incineration-disposal sites, together with the plant and support facilities that are required for incineration, fueling and washing of the vehicles, a support building for stores, and shower and locker rooms for collection and site crew personnel. The pickup-and-collection procedure remains essentially the same as in the existing system. The disposal-staging sites, however, are located strategically in the city, on the basis of the volume and location of wastes collected, thus eliminating long hauls and reducing the number of miles the collection vehicles must retravel from pickup to disposal site.

Four variations of the proposed system are being considered, containing one, two, three, and four incinerator-staging areas, respectively. The type of incinerator is a modular prepackaged unit that can be installed at several sites in the city. Such units exceed all state and federal standards for exhaust emissions. The city of Portland needs 24 units, each with a rated capacity of 12.5 tons of garbage per 24 hours. The price per unit is $137,600, which means a capital investment of about $3,302,000. The plant facilities, such as housing and foundation, were estimated to cost $200,000 per facility, based on a plan incorporating four incinerator plants strategically located around the city. Each plant would house eight units and be capable of handling 100 tons of garbage per day. Additional plant features, such as landscaping, were estimated to cost $60,000 per plant.

The annual operating cost of the proposed system would vary according to the type of system configuration. It takes about 1.5 to 1.7 million cubic feet (MCF) of fuel to incinerate 1 ton of garbage. The conservative 1.7-MCF figure was used for total cost. This means that fuel cost $4.25 per ton of garbage at a cost of $2.50 per MCF. Electric requirements at each plant will be 230 kW per day, which means a $0.48-per-ton cost for electricity if the plant is operating at full capacity. Two men can easily operate one plant, but safety factors dictate three operators at a cost of $7.14 per hour. This translates to a cost of $1.72 per ton. The maintenance cost of each plant was estimated to be $1.19 per ton. Since three plants will require fewer transportation miles, it is necessary to consider the savings accruing from this operating advantage. Three plant locations will save 6.14 miles per truck per day, on the average. At an estimated cost of $0.30 per mile, this would mean that an annual savings of $6,750 is realized on minimum trips to the landfill disposer, for a total annual savings in transportation of $15,300. Savings in labor are also realized because of the shorter routes, which permit more pickups during the day. The annual savings from this source are $103,500. The following table summarizes all costs, in thousands of dollars, associated with the present and proposed systems:

Item	Present System	Costs for Proposed Systems Site Number			
		1	2	3	4
Capital costs:					
Incinerators		$3,302	$3,302	$3,302	$3,302
Plant facilities		600	900	1,260	1,920
Annex buildings		91	102	112	132
Additional features		60	80	90	100
Total		$4,053	$4,384	$4,764	$5,454
Annual O&M costs	$905.4	$342	$480	$414	$408
Annual savings:					
Pickup transportation		$13.2	$14.7	$15.3	$17.1
Labor		87.6	99.3	103.5	119.40

A bond will be issued to provide the necessary capital investment at an interest rate of 8% with a maturity date 20 years in the future. The proposed systems are expected to last 20 years, with negligible salvage values. If the current system is to be retained, the annual O&M costs would be expected to increase at an annual rate of 10%. The city will use the bond interest rate as the interest rate for any public-project evaluation.

(a) Determine the operating cost of the current system in terms of dollars per ton of solid waste.

(b) Determine the economics of each solid-waste disposal alternative in terms of dollars per ton of solid waste.

ST16.3 Because of a rapid growth in population, a small town in Pennsylvania is considering several options to establish a wastewater treatment facility that can handle a flow of 2 million gallons per day (MGD). The town has five treatment options available:

- **Option 1: No action.** This option will lead to continued deterioration of the environment. If growth continues and pollution results, fines imposed (as high as $10,000 per day) would soon exceed construction costs.

- **Option 2: Land-treatment facility.** This option will provide a system for land treatment of the wastewater to be generated over the next 20 years and will require the utilization of the most land for treatment of the wastewater. In addition to the need to find a suitable site, pumping of the wastewater for a considerable distance out of town will be required. The land cost in the area is $3,000 per acre. The system will use spray irrigation to distribute wastewater over the site. No more than 1 inch of wastewater can be applied in 1 week per acre.

- **Option 3: Activated sludge-treatment facility.** This option will provide an activated sludge-treatment facility at a site near the planning area. No pumping will be required, and only 7 acres of land will be needed for construction of the plant, at a cost of $7,000 per acre.

- **Option 4: Trickling filter-treatment facility.** Provide a trickling filter-treatment facility at the same site selected for the activated sludge plant of Option 3. The land required will be the same as that for Option 3. Both facilities will provide similar levels of treatment, but using different units.

- **Option 5: Lagoon-treatment system.** Utilize a three-cell lagoon system for treatment. The lagoon system requires substantially more land than Options 3 and 4 require, but less than Option 2. Due to the larger land requirement, this treatment system will have to be located some distance outside of the planning area and will require pumping of the wastewater to reach the site.

The following tables summarize, respectively, (1) the land cost and land value for each option, (2) the capital expenditures for each option, and (3) the O&M costs associated with each option:

Option Number	Land Required (acres)	Land Cost ($)	Land Value (in 20 years)
	Land Cost for Each Option		
2	800	$2,400,000	$4,334,600
3	7	49,000	88,500
4	7	49,000	88,500
5	80	400,000	722,400

Option Number	Equipment	Structure	Pumping	Total
	Capital Expenditures			
2	$500,000	$700,000	$100,000	$1,300,000
3	500,000	2,100,000	0	2,600,000
4	400,000	2,463,000	0	2,863,000
5	175,000	1,750,000	100,000	2,025,000

Option	Annual O&M costs			
Number	Energy	Labor	Repair	Total
2	$200,000	$95,000	$30,000	$325,000
3	125,000	65,000	20,000	210,000
4	100,000	53,000	15,000	168,000
5	50,000	37,000	5,000	92,000

The price of land is assumed to be appreciating at an annual rate of 3%, and the equipment installed will require a replacement cycle of 15 years. Its replacement cost will increase at an annual rate of 5% (over the initial cost), and its salvage value at the end of the planning horizon will be 50% of the replacement cost. The structure requires replacement after 40 years and will have a salvage value of 60% of the original cost. The cost of energy and repair will increase at an annual rate of 5% and 2%, respectively. The labor cost will increase at an annual rate of 4%.

With the following sets of assumptions, answer (a) and (b).

- Assume an analysis period of 120 years.
- Replacement costs for the equipment, as well as for the pumping facilities, will increase at an annual rate of 5%.
- Replacement cost for the structure will remain constant over the planning period. However, the salvage value of the structure will be 60% of the original cost. (Because it has a 40-year replacement cycle, any increase in the future replacement cost will have very little impact on the solution.)
- The equipment's salvage value at the end of its useful life will be 50% of the original replacement cost. For example, the equipment installed under Option 1 will cost $500,000. Its salvage value at the end of 15 years will be $250,000.
- All O&M cost figures are given in today's dollars. For example, the annual energy cost of $200,000 for Option 2 means that the actual energy cost during the first operating year will be $200,000(1.05) = $210,000.
- Option 1 is not considered a viable alternative, as its annual operating cost exceeds $3,650,000.

(a) If the interest rate (including inflation) is 10%, which option is the most cost effective?

(b) Suppose a household discharges about 400 gallons of wastewater per day through the facility selected in (a). What should be the monthly bill assessed for this household?

ST16.4 The Federal Highway Administration predicts that by the year 2005, Americans will be spending 8.1 billion hours per year in traffic jams. Most traffic experts believe that adding and enlarging highway systems will not alleviate the problem. As a result, current research on traffic management is focusing on three areas: (1) the development of computerized dashboard navigational systems, (2) the development of roadside sensors and signals that monitor and help manage the flow of traffic, and (3) the development of automated steering and speed controls that might allow cars to drive themselves on certain stretches of highway.

In Los Angeles, perhaps the most traffic-congested city in the United States, a Texas Transportation Institute study found that traffic delays cost motorists $8 billion per year. But Los Angeles has already implemented a system of computerized traffic-signal controls that, by some estimates, has reduced travel time by 13.2%, fuel consumption by 12.5%, and pollution by 10%. And between Santa Monica and downtown Los Angeles, testing of an electronic traffic and navigational system, including highway sensors and cars with computerized dashboard maps, is being sponsored by federal, state, and local governments and General Motors Corporation. This test program costs $40 million; to install it throughout Los Angeles could cost $2 billion.

On a national scale, the estimates for implementing "smart" roads and vehicles is even more staggering: It would cost $18 billion to build the highways, $4 billion per year to maintain and operate them, $1 billion for research and development of driver-information aids, and $2.5 billion for vehicle-control devices. Advocates say the rewards far outweigh the costs.

(a) On a national scale, how would you identify the users' benefits and disbenefits for this type of public project?

(b) On a national scale, what would be the sponsor's cost?

(c) Suppose that the users' net benefits grow at 3% per year and the sponsor's costs grow at 4% per year. Assuming a social discount rate of 10%, what would be the *B/C* ratio over a 20-year study period?

APPENDIX A

Interest Factors for Discrete Compounding

0.25%

	Single Payment		Equal Payment Series				Gradient Series		
	Compound Amount Factor	Present Worth Factor	Compound Amount Factor	Sinking Fund Factor	Present Worth Factor	Capital Recovery Factor	Gradient Uniform Series	Gradient Present Worth	
N	(F/P,i,N)	(P/F,i,N)	(F/A,i,N)	(A/F,i,N)	(P/A,i,N)	(A/P,i,N)	(A/G,i,N)	(P/G,i,N)	N
1	1.0025	0.9975	1.0000	1.0000	0.9975	1.0025	0.0000	0.0000	1
2	1.0050	0.9950	2.0025	0.4994	1.9925	0.5019	0.4994	0.9950	2
3	1.0075	0.9925	3.0075	0.3325	2.9851	0.3350	0.9983	2.9801	3
4	1.0100	0.9901	4.0150	0.2491	3.9751	0.2516	1.4969	5.9503	4
5	1.0126	0.9876	5.0251	0.1990	4.9627	0.2015	1.9950	9.9007	5
6	1.0151	0.9851	6.0376	0.1656	5.9478	0.1681	2.4927	14.8263	6
7	1.0176	0.9827	7.0527	0.1418	6.9305	0.1443	2.9900	20.7223	7
8	1.0202	0.9802	8.0704	0.1239	7.9107	0.1264	3.4869	27.5839	8
9	1.0227	0.9778	9.0905	0.1100	8.8885	0.1125	3.9834	35.4061	9
10	1.0253	0.9753	10.1133	0.0989	9.8639	0.1014	4.4794	44.1842	10
11	1.0278	0.9729	11.1385	0.0898	10.8368	0.0923	4.9750	53.9133	11
12	1.0304	0.9705	12.1664	0.0822	11.8073	0.0847	5.4702	64.5886	12
13	1.0330	0.9681	13.1968	0.0758	12.7753	0.0783	5.9650	76.2053	13
14	1.0356	0.9656	14.2298	0.0703	13.7410	0.0728	6.4594	88.7587	14
15	1.0382	0.9632	15.2654	0.0655	14.7042	0.0680	6.9534	102.2441	15
16	1.0408	0.9608	16.3035	0.0613	15.6650	0.0638	7.4469	116.6567	16
17	1.0434	0.9584	17.3443	0.0577	16.6235	0.0602	7.9401	131.9917	17
18	1.0460	0.9561	18.3876	0.0544	17.5795	0.0569	8.4328	148.2446	18
19	1.0486	0.9537	19.4336	0.0515	18.5332	0.0540	8.9251	165.4106	19
20	1.0512	0.9513	20.4822	0.0488	19.4845	0.0513	9.4170	183.4851	20
21	1.0538	0.9489	21.5334	0.0464	20.4334	0.0489	9.9085	202.4634	21
22	1.0565	0.9466	22.5872	0.0443	21.3800	0.0468	10.3995	222.3410	22
23	1.0591	0.9442	23.6437	0.0423	22.3241	0.0448	10.8901	243.1131	23
24	1.0618	0.9418	24.7028	0.0405	23.2660	0.0430	11.3804	264.7753	24
25	1.0644	0.9395	25.7646	0.0388	24.2055	0.0413	11.8702	287.3230	25
26	1.0671	0.9371	26.8290	0.0373	25.1426	0.0398	12.3596	310.7516	26
27	1.0697	0.9348	27.8961	0.0358	26.0774	0.0383	12.8485	335.0566	27
28	1.0724	0.9325	28.9658	0.0345	27.0099	0.0370	13.3371	360.2334	28
29	1.0751	0.9301	30.0382	0.0333	27.9400	0.0358	13.8252	386.2776	29
30	1.0778	0.9278	31.1133	0.0321	28.8679	0.0346	14.3130	413.1847	30
31	1.0805	0.9255	32.1911	0.0311	29.7934	0.0336	14.8003	440.9502	31
32	1.0832	0.9232	33.2716	0.0301	30.7166	0.0326	15.2872	469.5696	32
33	1.0859	0.9209	34.3547	0.0291	31.6375	0.0316	15.7736	499.0386	33
34	1.0886	0.9186	35.4406	0.0282	32.5561	0.0307	16.2597	529.3528	34
35	1.0913	0.9163	36.5292	0.0274	33.4724	0.0299	16.7454	560.5076	35
36	1.0941	0.9140	37.6206	0.0266	34.3865	0.0291	17.2306	592.4988	36
40	1.1050	0.9050	42.0132	0.0238	38.0199	0.0263	19.1673	728.7399	40
48	1.1273	0.8871	50.9312	0.0196	45.1787	0.0221	23.0209	1040.0552	48
50	1.1330	0.8826	53.1887	0.0188	46.9462	0.0213	23.9802	1125.7767	50
60	1.1616	0.8609	64.6467	0.0155	55.6524	0.0180	28.7514	1600.0845	60
72	1.1969	0.8355	78.7794	0.0127	65.8169	0.0152	34.4221	2265.5569	72
80	1.2211	0.8189	88.4392	0.0113	72.4260	0.0138	38.1694	2764.4568	80
84	1.2334	0.8108	93.3419	0.0107	75.6813	0.0132	40.0331	3029.7592	84
90	1.2520	0.7987	100.7885	0.0099	80.5038	0.0124	42.8162	3446.8700	90
96	1.2709	0.7869	108.3474	0.0092	85.2546	0.0117	45.5844	3886.2832	96
100	1.2836	0.7790	113.4500	0.0088	88.3825	0.0113	47.4216	4191.2417	100
108	1.3095	0.7636	123.8093	0.0081	94.5453	0.0106	51.0762	4829.0125	108
120	1.3494	0.7411	139.7414	0.0072	103.5618	0.0097	56.5084	5852.1116	120
240	1.8208	0.5492	328.3020	0.0030	180.3109	0.0055	107.5863	19398.9852	240
360	2.4568	0.4070	582.7369	0.0017	237.1894	0.0042	152.8902	36263.9299	360

0.50%

	Single Payment		Equal Payment Series				Gradient Series		
N	Compound Amount Factor (F/P,i,N)	Present Worth Factor (P/F,i,N)	Compound Amount Factor (F/A,i,N)	Sinking Fund Factor (A/F,i,N)	Present Worth Factor (P/A,i,N)	Capital Recovery Factor (A/P,i,N)	Gradient Uniform Series (A/G,i,N)	Gradient Present Worth (P/G,i,N)	N
1	1.0050	0.9950	1.0000	1.0000	0.9950	1.0050	0.0000	0.0000	1
2	1.0100	0.9901	2.0050	0.4988	1.9851	0.5038	0.4988	0.9901	2
3	1.0151	0.9851	3.0150	0.3317	2.9702	0.3367	0.9967	2.9604	3
4	1.0202	0.9802	4.0301	0.2481	3.9505	0.2531	1.4938	5.9011	4
5	1.0253	0.9754	5.0503	0.1980	4.9259	0.2030	1.9900	9.8026	5
6	1.0304	0.9705	6.0755	0.1646	5.8964	0.1696	2.4855	14.6552	6
7	1.0355	0.9657	7.1059	0.1407	6.8621	0.1457	2.9801	20.4493	7
8	1.0407	0.9609	8.1414	0.1228	7.8230	0.1278	3.4738	27.1755	8
9	1.0459	0.9561	9.1821	0.1089	8.7791	0.1139	3.9668	34.8244	9
10	1.0511	0.9513	10.2280	0.0978	9.7304	0.1028	4.4589	43.3865	10
11	1.0564	0.9466	11.2792	0.0887	10.6770	0.0937	4.9501	52.8526	11
12	1.0617	0.9419	12.3356	0.0811	11.6189	0.0861	5.4406	63.2136	12
13	1.0670	0.9372	13.3972	0.0746	12.5562	0.0796	5.9302	74.4602	13
14	1.0723	0.9326	14.4642	0.0691	13.4887	0.0741	6.4190	86.5835	14
15	1.0777	0.9279	15.5365	0.0644	14.4166	0.0694	6.9069	99.5743	15
16	1.0831	0.9233	16.6142	0.0602	15.3399	0.0652	7.3940	113.4238	16
17	1.0885	0.9187	17.6973	0.0565	16.2586	0.0615	7.8803	128.1231	17
18	1.0939	0.9141	18.7858	0.0532	17.1728	0.0582	8.3658	143.6634	18
19	1.0994	0.9096	19.8797	0.0503	18.0824	0.0553	8.8504	160.0360	19
20	1.1049	0.9051	20.9791	0.0477	18.9874	0.0527	9.3342	177.2322	20
21	1.1104	0.9006	22.0840	0.0453	19.8880	0.0503	9.8172	195.2434	21
22	1.1160	0.8961	23.1944	0.0431	20.7841	0.0481	10.2993	214.0611	22
23	1.1216	0.8916	24.3104	0.0411	21.6757	0.0461	10.7806	233.6768	23
24	1.1272	0.8872	25.4320	0.0393	22.5629	0.0443	11.2611	254.0820	24
25	1.1328	0.8828	26.5591	0.0377	23.4456	0.0427	11.7407	275.2686	25
26	1.1385	0.8784	27.6919	0.0361	24.3240	0.0411	12.2195	297.2281	26
27	1.1442	0.8740	28.8304	0.0347	25.1980	0.0397	12.6975	319.9523	27
28	1.1499	0.8697	29.9745	0.0334	26.0677	0.0384	13.1747	343.4332	28
29	1.1556	0.8653	31.1244	0.0321	26.9330	0.0371	13.6510	367.6625	29
30	1.1614	0.8610	32.2800	0.0310	27.7941	0.0360	14.1265	392.6324	30
31	1.1672	0.8567	33.4414	0.0299	28.6508	0.0349	14.6012	418.3348	31
32	1.1730	0.8525	34.6086	0.0289	29.5033	0.0339	15.0750	444.7618	32
33	1.1789	0.8482	35.7817	0.0279	30.3515	0.0329	15.5480	471.9055	33
34	1.1848	0.8440	36.9606	0.0271	31.1955	0.0321	16.0202	499.7583	34
35	1.1907	0.8398	38.1454	0.0262	32.0354	0.0312	16.4915	528.3123	35
36	1.1967	0.8356	39.3361	0.0254	32.8710	0.0304	16.9621	557.5598	36
40	1.2208	0.8191	44.1588	0.0226	36.1722	0.0276	18.8359	681.3347	40
48	1.2705	0.7871	54.0978	0.0185	42.5803	0.0235	22.5437	959.9188	48
50	1.2832	0.7793	56.6452	0.0177	44.1428	0.0227	23.4624	1035.6966	50
60	1.3489	0.7414	69.7700	0.0143	51.7256	0.0193	28.0064	1448.6458	60
72	1.4320	0.6983	86.4089	0.0116	60.3395	0.0166	33.3504	2012.3478	72
80	1.4903	0.6710	98.0677	0.0102	65.8023	0.0152	36.8474	2424.6455	80
84	1.5204	0.6577	104.0739	0.0096	68.4530	0.0146	38.5763	2640.6641	84
90	1.5666	0.6383	113.3109	0.0088	72.3313	0.0138	41.1451	2976.0769	90
96	1.6141	0.6195	122.8285	0.0081	76.0952	0.0131	43.6845	3324.1846	96
100	1.6467	0.6073	129.3337	0.0077	78.5426	0.0127	45.3613	3562.7934	100
108	1.7137	0.5835	142.7399	0.0070	83.2934	0.0120	48.6758	4054.3747	108
120	1.8194	0.5496	163.8793	0.0061	90.0735	0.0111	53.5508	4823.5051	120
240	3.3102	0.3021	462.0409	0.0022	139.5808	0.0072	96.1131	13415.5395	240
360	6.0226	0.1660	1004.5150	0.0010	166.7916	0.0060	128.3236	21403.3041	360

0.75%

	Single Payment		Equal Payment Series				Gradient Series		
	Compound Amount Factor	Present Worth Factor	Compound Amount Factor	Sinking Fund Factor	Present Worth Factor	Capital Recovery Factor	Gradient Uniform Series	Gradient Present Worth	
N	(F/P,i,N)	(P/F,i,N)	(F/A,i,N)	(A/F,i,N)	(P/A,i,N)	(A/P,i,N)	(A/G,i,N)	(P/G,i,N)	N
1	1.0075	0.9926	1.0000	1.0000	0.9926	1.0075	0.0000	0.0000	1
2	1.0151	0.9852	2.0075	0.4981	1.9777	0.5056	0.4981	0.9852	2
3	1.0227	0.9778	3.0226	0.3308	2.9556	0.3383	0.9950	2.9408	3
4	1.0303	0.9706	4.0452	0.2472	3.9261	0.2547	1.4907	5.8525	4
5	1.0381	0.9633	5.0756	0.1970	4.8894	0.2045	1.9851	9.7058	5
6	1.0459	0.9562	6.1136	0.1636	5.8456	0.1711	2.4782	14.4866	6
7	1.0537	0.9490	7.1595	0.1397	6.7946	0.1472	2.9701	20.1808	7
8	1.0616	0.9420	8.2132	0.1218	7.7366	0.1293	3.4608	26.7747	8
9	1.0696	0.9350	9.2748	0.1078	8.6716	0.1153	3.9502	34.2544	9
10	1.0776	0.9280	10.3443	0.0967	9.5996	0.1042	4.4384	42.6064	10
11	1.0857	0.9211	11.4219	0.0876	10.5207	0.0951	4.9253	51.8174	11
12	1.0938	0.9142	12.5076	0.0800	11.4349	0.0875	5.4110	61.8740	12
13	1.1020	0.9074	13.6014	0.0735	12.3423	0.0810	5.8954	72.7632	13
14	1.1103	0.9007	14.7034	0.0680	13.2430	0.0755	6.3786	84.4720	14
15	1.1186	0.8940	15.8137	0.0632	14.1370	0.0707	6.8606	96.9876	15
16	1.1270	0.8873	16.9323	0.0591	15.0243	0.0666	7.3413	110.2973	16
17	1.1354	0.8807	18.0593	0.0554	15.9050	0.0629	7.8207	124.3887	17
18	1.1440	0.8742	19.1947	0.0521	16.7792	0.0596	8.2989	139.2494	18
19	1.1525	0.8676	20.3387	0.0492	17.6468	0.0567	8.7759	154.8671	19
20	1.1612	0.8612	21.4912	0.0465	18.5080	0.0540	9.2516	171.2297	20
21	1.1699	0.8548	22.6524	0.0441	19.3628	0.0516	9.7261	188.3253	21
22	1.1787	0.8484	23.8223	0.0420	20.2112	0.0495	10.1994	206.1420	22
23	1.1875	0.8421	25.0010	0.0400	21.0533	0.0475	10.6714	224.6682	23
24	1.1964	0.8358	26.1885	0.0382	21.8891	0.0457	11.1422	243.8923	24
25	1.2054	0.8296	27.3849	0.0365	22.7188	0.0440	11.6117	263.8029	25
26	1.2144	0.8234	28.5903	0.0350	23.5422	0.0425	12.0800	284.3888	26
27	1.2235	0.8173	29.8047	0.0336	24.3595	0.0411	12.5470	305.6387	27
28	1.2327	0.8112	31.0282	0.0322	25.1707	0.0397	13.0128	327.5416	28
29	1.2420	0.8052	32.2609	0.0310	25.9759	0.0385	13.4774	350.0867	29
30	1.2513	0.7992	33.5029	0.0298	26.7751	0.0373	13.9407	373.2631	30
31	1.2607	0.7932	34.7542	0.0288	27.5683	0.0363	14.4028	397.0602	31
32	1.2701	0.7873	36.0148	0.0278	28.3557	0.0353	14.8636	421.4675	32
33	1.2796	0.7815	37.2849	0.0268	29.1371	0.0343	15.3232	446.4746	33
34	1.2892	0.7757	38.5646	0.0259	29.9128	0.0334	15.7816	472.0712	34
35	1.2989	0.7699	39.8538	0.0251	30.6827	0.0326	16.2387	498.2471	35
36	1.3086	0.7641	41.1527	0.0243	31.4468	0.0318	16.6946	524.9924	36
40	1.3483	0.7416	46.4465	0.0215	34.4469	0.0290	18.5058	637.4693	40
48	1.4314	0.6986	57.5207	0.0174	40.1848	0.0249	22.0691	886.8404	48
50	1.4530	0.6883	60.3943	0.0166	41.5664	0.0241	22.9476	953.8486	50
60	1.5657	0.6387	75.4241	0.0133	48.1734	0.0208	27.2665	1313.5189	60
72	1.7126	0.5839	95.0070	0.0105	55.4768	0.0180	32.2882	1791.2463	72
80	1.8180	0.5500	109.0725	0.0092	59.9944	0.0167	35.5391	2132.1472	80
84	1.8732	0.5338	116.4269	0.0086	62.1540	0.0161	37.1357	2308.1283	84
90	1.9591	0.5104	127.8790	0.0078	65.2746	0.0153	39.4946	2577.9961	90
96	2.0489	0.4881	139.8562	0.0072	68.2584	0.0147	41.8107	2853.9352	96
100	2.1111	0.4737	148.1445	0.0068	70.1746	0.0143	43.3311	3040.7453	100
108	2.2411	0.4462	165.4832	0.0060	73.8394	0.0135	46.3154	3419.9041	108
120	2.4514	0.4079	193.5143	0.0052	78.9417	0.0127	50.6521	3998.5621	120
240	6.0092	0.1664	667.8869	0.0015	111.1450	0.0090	85.4210	9494.1162	240
360	14.7306	0.0679	1830.7435	0.0005	124.2819	0.0080	107.1145	13312.3871	360

1.0%

	Single Payment		Equal Payment Series				Gradient Series		
	Compound Amount Factor	Present Worth Factor	Compound Amount Factor	Sinking Fund Factor	Present Worth Factor	Capital Recovery Factor	Gradient Uniform Series	Gradient Present Worth	
N	(F/P,i,N)	(P/F,i,N)	(F/A,i,N)	(A/F,i,N)	(P/A,i,N)	(A/P,i,N)	(A/G,i,N)	(P/G,i,N)	N
1	1.0100	0.9901	1.0000	1.0000	0.9901	1.0100	0.0000	0.0000	1
2	1.0201	0.9803	2.0100	0.4975	1.9704	0.5075	0.4975	0.9803	2
3	1.0303	0.9706	3.0301	0.3300	2.9410	0.3400	0.9934	2.9215	3
4	1.0406	0.9610	4.0604	0.2463	3.9020	0.2563	1.4876	5.8044	4
5	1.0510	0.9515	5.1010	0.1960	4.8534	0.2060	1.9801	9.6103	5
6	1.0615	0.9420	6.1520	0.1625	5.7955	0.1725	2.4710	14.3205	6
7	1.0721	0.9327	7.2135	0.1386	6.7282	0.1486	2.9602	19.9168	7
8	1.0829	0.9235	8.2857	0.1207	7.6517	0.1307	3.4478	26.3812	8
9	1.0937	0.9143	9.3685	0.1067	8.5660	0.1167	3.9337	33.6959	9
10	1.1046	0.9053	10.4622	0.0956	9.4713	0.1056	4.4179	41.8435	10
11	1.1157	0.8963	11.5668	0.0865	10.3676	0.0965	4.9005	50.8067	11
12	1.1268	0.8874	12.6825	0.0788	11.2551	0.0888	5.3815	60.5687	12
13	1.1381	0.8787	13.8093	0.0724	12.1337	0.0824	5.8607	71.1126	13
14	1.1495	0.8700	14.9474	0.0669	13.0037	0.0769	6.3384	82.4221	14
15	1.1610	0.8613	16.0969	0.0621	13.8651	0.0721	6.8143	94.4810	15
16	1.1726	0.8528	17.2579	0.0579	14.7179	0.0679	7.2886	107.2734	16
17	1.1843	0.8444	18.4304	0.0543	15.5623	0.0643	7.7613	120.7834	17
18	1.1961	0.8360	19.6147	0.0510	16.3983	0.0610	8.2323	134.9957	18
19	1.2081	0.8277	20.8109	0.0481	17.2260	0.0581	8.7017	149.8950	19
20	1.2202	0.8195	22.0190	0.0454	18.0456	0.0554	9.1694	165.4664	20
21	1.2324	0.8114	23.2392	0.0430	18.8570	0.0530	9.6354	181.6950	21
22	1.2447	0.8034	24.4716	0.0409	19.6604	0.0509	10.0998	198.5663	22
23	1.2572	0.7954	25.7163	0.0389	20.4558	0.0489	10.5626	216.0660	23
24	1.2697	0.7876	26.9735	0.0371	21.2434	0.0471	11.0237	234.1800	24
25	1.2824	0.7798	28.2432	0.0354	22.0232	0.0454	11.4831	252.8945	25
26	1.2953	0.7720	29.5256	0.0339	22.7952	0.0439	11.9409	272.1957	26
27	1.3082	0.7644	30.8209	0.0324	23.5596	0.0424	12.3971	292.0702	27
28	1.3213	0.7568	32.1291	0.0311	24.3164	0.0411	12.8516	312.5047	28
29	1.3345	0.7493	33.4504	0.0299	25.0658	0.0399	13.3044	333.4863	29
30	1.3478	0.7419	34.7849	0.0287	25.8077	0.0387	13.7557	355.0021	30
31	1.3613	0.7346	36.1327	0.0277	26.5423	0.0377	14.2052	377.0394	31
32	1.3749	0.7273	37.4941	0.0267	27.2696	0.0367	14.6532	399.5858	32
33	1.3887	0.7201	38.8690	0.0257	27.9897	0.0357	15.0995	422.6291	33
34	1.4026	0.7130	40.2577	0.0248	28.7027	0.0348	15.5441	446.1572	34
35	1.4166	0.7059	41.6603	0.0240	29.4086	0.0340	15.9871	470.1583	35
36	1.4308	0.6989	43.0769	0.0232	30.1075	0.0332	16.4285	494.6207	36
40	1.4889	0.6717	48.8864	0.0205	32.8347	0.0305	18.1776	596.8561	40
48	1.6122	0.6203	61.2226	0.0163	37.9740	0.0263	21.5976	820.1460	48
50	1.6446	0.6080	64.4632	0.0155	39.1961	0.0255	22.4363	879.4176	50
60	1.8167	0.5504	81.6697	0.0122	44.9550	0.0222	26.5333	1192.8061	60
72	2.0471	0.4885	104.7099	0.0096	51.1504	0.0196	31.2386	1597.8673	72
80	2.2167	0.4511	121.6715	0.0082	54.8882	0.0182	34.2492	1879.8771	80
84	2.3067	0.4335	130.6723	0.0077	56.6485	0.0177	35.7170	2023.3153	84
90	2.4486	0.4084	144.8633	0.0069	59.1609	0.0169	37.8724	2240.5675	90
96	2.5993	0.3847	159.9273	0.0063	61.5277	0.0163	39.9727	2459.4298	96
100	2.7048	0.3697	170.4814	0.0059	63.0289	0.0159	41.3426	2605.7758	100
108	2.9289	0.3414	192.8926	0.0052	65.8578	0.0152	44.0103	2898.4203	108
120	3.3004	0.3030	230.0387	0.0043	69.7005	0.0143	47.8349	3334.1148	120
240	10.8926	0.0918	989.2554	0.0010	90.8194	0.0110	75.7393	6878.6016	240
360	35.9496	0.0278	3494.9641	0.0003	97.2183	0.0103	89.6995	8720.4323	360

1.25%

	Single Payment		Equal Payment Series				Gradient Series		
	Compound Amount Factor	Present Worth Factor	Compound Amount Factor	Sinking Fund Factor	Present Worth Factor	Capital Recovery Factor	Gradient Uniform Series	Gradient Present Worth	
N	(F/P,i,N)	(P/F,i,N)	(F/A,i,N)	(A/F,i,N)	(P/A,i,N)	(A/P,i,N)	(A/G,i,N)	(P/G,i,N)	N
1	1.0125	0.9877	1.0000	1.0000	0.9877	1.0125	0.0000	0.0000	1
2	1.0252	0.9755	2.0125	0.4969	1.9631	0.5094	0.4969	0.9755	2
3	1.0380	0.9634	3.0377	0.3292	2.9265	0.3417	0.9917	2.9023	3
4	1.0509	0.9515	4.0756	0.2454	3.8781	0.2579	1.4845	5.7569	4
5	1.0641	0.9398	5.1266	0.1951	4.8178	0.2076	1.9752	9.5160	5
6	1.0774	0.9282	6.1907	0.1615	5.7460	0.1740	2.4638	14.1569	6
7	1.0909	0.9167	7.2680	0.1376	6.6627	0.1501	2.9503	19.6571	7
8	1.1045	0.9054	8.3589	0.1196	7.5681	0.1321	3.4348	25.9949	8
9	1.1183	0.8942	9.4634	0.1057	8.4623	0.1182	3.9172	33.1487	9
10	1.1323	0.8832	10.5817	0.0945	9.3455	0.1070	4.3975	41.0973	10
11	1.1464	0.8723	11.7139	0.0854	10.2178	0.0979	4.8758	49.8201	11
12	1.1608	0.8615	12.8604	0.0778	11.0793	0.0903	5.3520	59.2967	12
13	1.1753	0.8509	14.0211	0.0713	11.9302	0.0838	5.8262	69.5072	13
14	1.1900	0.8404	15.1964	0.0658	12.7706	0.0783	6.2982	80.4320	14
15	1.2048	0.8300	16.3863	0.0610	13.6005	0.0735	6.7682	92.0519	15
16	1.2199	0.8197	17.5912	0.0568	14.4203	0.0693	7.2362	104.3481	16
17	1.2351	0.8096	18.8111	0.0532	15.2299	0.0657	7.7021	117.3021	17
18	1.2506	0.7996	20.0462	0.0499	16.0295	0.0624	8.1659	130.8958	18
19	1.2662	0.7898	21.2968	0.0470	16.8193	0.0595	8.6277	145.1115	19
20	1.2820	0.7800	22.5630	0.0443	17.5993	0.0568	9.0874	159.9316	20
21	1.2981	0.7704	23.8450	0.0419	18.3697	0.0544	9.5450	175.3392	21
22	1.3143	0.7609	25.1431	0.0398	19.1306	0.0523	10.0006	191.3174	22
23	1.3307	0.7515	26.4574	0.0378	19.8820	0.0503	10.4542	207.8499	23
24	1.3474	0.7422	27.7881	0.0360	20.6242	0.0485	10.9056	224.9204	24
25	1.3642	0.7330	29.1354	0.0343	21.3573	0.0468	11.3551	242.5132	25
26	1.3812	0.7240	30.4996	0.0328	22.0813	0.0453	11.8024	260.6128	26
27	1.3985	0.7150	31.8809	0.0314	22.7963	0.0439	12.2478	279.2040	27
28	1.4160	0.7062	33.2794	0.0300	23.5025	0.0425	12.6911	298.2719	28
29	1.4337	0.6975	34.6954	0.0288	24.2000	0.0413	13.1323	317.8019	29
30	1.4516	0.6889	36.1291	0.0277	24.8889	0.0402	13.5715	337.7797	30
31	1.4698	0.6804	37.5807	0.0266	25.5693	0.0391	14.0086	358.1912	31
32	1.4881	0.6720	39.0504	0.0256	26.2413	0.0381	14.4438	379.0227	32
33	1.5067	0.6637	40.5386	0.0247	26.9050	0.0372	14.8768	400.2607	33
34	1.5256	0.6555	42.0453	0.0238	27.5605	0.0363	15.3079	421.8920	34
35	1.5446	0.6474	43.5709	0.0230	28.2079	0.0355	15.7369	443.9037	35
36	1.5639	0.6394	45.1155	0.0222	28.8473	0.0347	16.1639	466.2830	36
40	1.6436	0.6084	51.4896	0.0194	31.3269	0.0319	17.8515	559.2320	40
48	1.8154	0.5509	65.2284	0.0153	35.9315	0.0278	21.1299	759.2296	48
50	1.8610	0.5373	68.8818	0.0145	37.0129	0.0270	21.9295	811.6738	50
60	2.1072	0.4746	88.5745	0.0113	42.0346	0.0238	25.8083	1084.8429	60
72	2.4459	0.4088	115.6736	0.0086	47.2925	0.0211	30.2047	1428.4561	72
80	2.7015	0.3702	136.1188	0.0073	50.3867	0.0198	32.9822	1661.8651	80
84	2.8391	0.3522	147.1290	0.0068	51.8222	0.0193	34.3258	1778.8384	84
90	3.0588	0.3269	164.7050	0.0061	53.8461	0.0186	36.2855	1953.8303	90
96	3.2955	0.3034	183.6411	0.0054	55.7246	0.0179	38.1793	2127.5244	96
100	3.4634	0.2887	197.0723	0.0051	56.9013	0.0176	39.4058	2242.2411	100
108	3.8253	0.2614	226.0226	0.0044	59.0865	0.0169	41.7737	2468.2636	108
120	4.4402	0.2252	275.2171	0.0036	61.9828	0.0161	45.1184	2796.5694	120
240	19.7155	0.0507	1497.2395	0.0007	75.9423	0.0132	67.1764	5101.5288	240
360	87.5410	0.0114	6923.2796	0.0001	79.0861	0.0126	75.8401	5997.9027	360

1.5%

	Single Payment		Equal Payment Series				Gradient Series		
	Compound Amount Factor	Present Worth Factor	Compound Amount Factor	Sinking Fund Factor	Present Worth Factor	Capital Recovery Factor	Gradient Uniform Series	Gradient Present Worth	
N	(F/P,i,N)	(P/F,i,N)	(F/A,i,N)	(A/F,i,N)	(P/A,i,N)	(A/P,i,N)	(A/G,i,N)	(P/G,i,N)	N
1	1.0150	0.9852	1.0000	1.0000	0.9852	1.0150	0.0000	0.0000	1
2	1.0302	0.9707	2.0150	0.4963	1.9559	0.5113	0.4963	0.9707	2
3	1.0457	0.9563	3.0452	0.3284	2.9122	0.3434	0.9901	2.8833	3
4	1.0614	0.9422	4.0909	0.2444	3.8544	0.2594	1.4814	5.7098	4
5	1.0773	0.9283	5.1523	0.1941	4.7826	0.2091	1.9702	9.4229	5
6	1.0934	0.9145	6.2296	0.1605	5.6972	0.1755	2.4566	13.9956	6
7	1.1098	0.9010	7.3230	0.1366	6.5982	0.1516	2.9405	19.4018	7
8	1.1265	0.8877	8.4328	0.1186	7.4859	0.1336	3.4219	25.6157	8
9	1.1434	0.8746	9.5593	0.1046	8.3605	0.1196	3.9008	32.6125	9
10	1.1605	0.8617	10.7027	0.0934	9.2222	0.1084	4.3772	40.3675	10
11	1.1779	0.8489	11.8633	0.0843	10.0711	0.0993	4.8512	48.8568	11
12	1.1956	0.8364	13.0412	0.0767	10.9075	0.0917	5.3227	58.0571	12
13	1.2136	0.8240	14.2368	0.0702	11.7315	0.0852	5.7917	67.9454	13
14	1.2318	0.8118	15.4504	0.0647	12.5434	0.0797	6.2582	78.4994	14
15	1.2502	0.7999	16.6821	0.0599	13.3432	0.0749	6.7223	89.6974	15
16	1.2690	0.7880	17.9324	0.0558	14.1313	0.0708	7.1839	101.5178	16
17	1.2880	0.7764	19.2014	0.0521	14.9076	0.0671	7.6431	113.9400	17
18	1.3073	0.7649	20.4894	0.0488	15.6726	0.0638	8.0997	126.9435	18
19	1.3270	0.7536	21.7967	0.0459	16.4262	0.0609	8.5539	140.5084	19
20	1.3469	0.7425	23.1237	0.0432	17.1686	0.0582	9.0057	154.6154	20
21	1.3671	0.7315	24.4705	0.0409	17.9001	0.0559	9.4550	169.2453	21
22	1.3876	0.7207	25.8376	0.0387	18.6208	0.0537	9.9018	184.3798	22
23	1.4084	0.7100	27.2251	0.0367	19.3309	0.0517	10.3462	200.0006	23
24	1.4295	0.6995	28.6335	0.0349	20.0304	0.0499	10.7881	216.0901	24
25	1.4509	0.6892	30.0630	0.0333	20.7196	0.0483	11.2276	232.6310	25
26	1.4727	0.6790	31.5140	0.0317	21.3986	0.0467	11.6646	249.6065	26
27	1.4948	0.6690	32.9867	0.0303	22.0676	0.0453	12.0992	267.0002	27
28	1.5172	0.6591	34.4815	0.0290	22.7267	0.0440	12.5313	284.7958	28
29	1.5400	0.6494	35.9987	0.0278	23.3761	0.0428	12.9610	302.9779	29
30	1.5631	0.6398	37.5387	0.0266	24.0158	0.0416	13.3883	321.5310	30
31	1.5865	0.6303	39.1018	0.0256	24.6461	0.0406	13.8131	340.4402	31
32	1.6103	0.6210	40.6883	0.0246	25.2671	0.0396	14.2355	359.6910	32
33	1.6345	0.6118	42.2986	0.0236	25.8790	0.0386	14.6555	379.2691	33
34	1.6590	0.6028	43.9331	0.0228	26.4817	0.0378	15.0731	399.1607	34
35	1.6839	0.5939	45.5921	0.0219	27.0756	0.0369	15.4882	419.3521	35
36	1.7091	0.5851	47.2760	0.0212	27.6607	0.0362	15.9009	439.8303	36
40	1.8140	0.5513	54.2679	0.0184	29.9158	0.0334	17.5277	524.3568	40
48	2.0435	0.4894	69.5652	0.0144	34.0426	0.0294	20.6667	703.5462	48
50	2.1052	0.4750	73.6828	0.0136	34.9997	0.0286	21.4277	749.9636	50
60	2.4432	0.4093	96.2147	0.0104	39.3803	0.0254	25.0930	988.1674	60
72	2.9212	0.3423	128.0772	0.0078	43.8447	0.0228	29.1893	1279.7938	72
80	3.2907	0.3039	152.7109	0.0065	46.4073	0.0215	31.7423	1473.0741	80
84	3.4926	0.2863	166.1726	0.0060	47.5786	0.0210	32.9668	1568.5140	84
90	3.8189	0.2619	187.9299	0.0053	49.2099	0.0203	34.7399	1709.5439	90
96	4.1758	0.2395	211.7202	0.0047	50.7017	0.0197	36.4381	1847.4725	96
100	4.4320	0.2256	228.8030	0.0044	51.6247	0.0194	37.5295	1937.4506	100
108	4.9927	0.2003	266.1778	0.0038	53.3137	0.0188	39.6171	2112.1348	108
120	5.9693	0.1675	331.2882	0.0030	55.4985	0.0180	42.5185	2359.7114	120
240	35.6328	0.0281	2308.8544	0.0004	64.7957	0.0154	59.7368	3870.6912	240
360	212.7038	0.0047	14113.5854	0.0001	66.3532	0.0151	64.9662	4310.7165	360

1.75%

	Single Payment		Equal Payment Series				Gradient Series		
	Compound Amount Factor	Present Worth Factor	Compound Amount Factor	Sinking Fund Factor	Present Worth Factor	Capital Recovery Factor	Gradient Uniform Series	Gradient Present Worth	
N	(F/P,i,N)	(P/F,i,N)	(F/A,i,N)	(A/F,i,N)	(P/A,i,N)	(A/P,i,N)	(A/G,i,N)	(P/G,i,N)	N
1	1.0175	0.9828	1.0000	1.0000	0.9828	1.0175	0.0000	0.0000	1
2	1.0353	0.9659	2.0175	0.4957	1.9487	0.5132	0.4957	0.9659	2
3	1.0534	0.9493	3.0528	0.3276	2.8980	0.3451	0.9884	2.8645	3
4	1.0719	0.9330	4.1062	0.2435	3.8309	0.2610	1.4783	5.6633	4
5	1.0906	0.9169	5.1781	0.1931	4.7479	0.2106	1.9653	9.3310	5
6	1.1097	0.9011	6.2687	0.1595	5.6490	0.1770	2.4494	13.8367	6
7	1.1291	0.8856	7.3784	0.1355	6.5346	0.1530	2.9306	19.1506	7
8	1.1489	0.8704	8.5075	0.1175	7.4051	0.1350	3.4089	25.2435	8
9	1.1690	0.8554	9.6564	0.1036	8.2605	0.1211	3.8844	32.0870	9
10	1.1894	0.8407	10.8254	0.0924	9.1012	0.1099	4.3569	39.6535	10
11	1.2103	0.8263	12.0148	0.0832	9.9275	0.1007	4.8266	47.9162	11
12	1.2314	0.8121	13.2251	0.0756	10.7395	0.0931	5.2934	56.8489	12
13	1.2530	0.7981	14.4565	0.0692	11.5376	0.0867	5.7573	66.4260	13
14	1.2749	0.7844	15.7095	0.0637	12.3220	0.0812	6.2184	76.6227	14
15	1.2972	0.7709	16.9844	0.0589	13.0929	0.0764	6.6765	87.4149	15
16	1.3199	0.7576	18.2817	0.0547	13.8505	0.0722	7.1318	98.7792	16
17	1.3430	0.7446	19.6016	0.0510	14.5951	0.0685	7.5842	110.6926	17
18	1.3665	0.7318	20.9446	0.0477	15.3269	0.0652	8.0338	123.1328	18
19	1.3904	0.7192	22.3112	0.0448	16.0461	0.0623	8.4805	136.0783	19
20	1.4148	0.7068	23.7016	0.0422	16.7529	0.0597	8.9243	149.5080	20
21	1.4395	0.6947	25.1164	0.0398	17.4475	0.0573	9.3653	163.4013	21
22	1.4647	0.6827	26.5559	0.0377	18.1303	0.0552	9.8034	177.7385	22
23	1.4904	0.6710	28.0207	0.0357	18.8012	0.0532	10.2387	192.5000	23
24	1.5164	0.6594	29.5110	0.0339	19.4607	0.0514	10.6711	207.6671	24
25	1.5430	0.6481	31.0275	0.0322	20.1088	0.0497	11.1007	223.2214	25
26	1.5700	0.6369	32.5704	0.0307	20.7457	0.0482	11.5274	239.1451	26
27	1.5975	0.6260	34.1404	0.0293	21.3717	0.0468	11.9513	255.4210	27
28	1.6254	0.6152	35.7379	0.0280	21.9870	0.0455	12.3724	272.0321	28
29	1.6539	0.6046	37.3633	0.0268	22.5916	0.0443	12.7907	288.9623	29
30	1.6828	0.5942	39.0172	0.0256	23.1858	0.0431	13.2061	306.1954	30
31	1.7122	0.5840	40.7000	0.0246	23.7699	0.0421	13.6188	323.7163	31
32	1.7422	0.5740	42.4122	0.0236	24.3439	0.0411	14.0286	341.5097	32
33	1.7727	0.5641	44.1544	0.0226	24.9080	0.0401	14.4356	359.5613	33
34	1.8037	0.5544	45.9271	0.0218	25.4624	0.0393	14.8398	377.8567	34
35	1.8353	0.5449	47.7308	0.0210	26.0073	0.0385	15.2412	396.3824	35
36	1.8674	0.5355	49.5661	0.0202	26.5428	0.0377	15.6399	415.1250	36
40	2.0016	0.4996	57.2341	0.0175	28.5942	0.0350	17.2066	492.0109	40
48	2.2996	0.4349	74.2628	0.0135	32.2938	0.0310	20.2084	652.6054	48
50	2.3808	0.4200	78.9022	0.0127	33.1412	0.0302	20.9317	693.7010	50
60	2.8318	0.3531	104.6752	0.0096	36.9640	0.0271	24.3885	901.4954	60
72	3.4872	0.2868	142.1263	0.0070	40.7564	0.0245	28.1948	1149.1181	72
80	4.0064	0.2496	171.7938	0.0058	42.8799	0.0233	30.5329	1309.2482	80
84	4.2943	0.2329	188.2450	0.0053	43.8361	0.0228	31.6442	1387.1584	84
90	4.7654	0.2098	215.1646	0.0046	45.1516	0.0221	33.2409	1500.8798	90
96	5.2882	0.1891	245.0374	0.0041	46.3370	0.0216	34.7556	1610.4716	96
100	5.6682	0.1764	266.7518	0.0037	47.0615	0.0212	35.7211	1681.0886	100
108	6.5120	0.1536	314.9738	0.0032	48.3679	0.0207	37.5494	1816.1852	108
120	8.0192	0.1247	401.0962	0.0025	50.0171	0.0200	40.0469	2003.0269	120
240	64.3073	0.0156	3617.5602	0.0003	56.2543	0.0178	53.3518	3001.2678	240
360	515.6921	0.0019	29410.9747	0.0000	57.0320	0.0175	56.4434	3219.0833	360

2.0%

	Single Payment		Equal Payment Series				Gradient Series		
	Compound Amount Factor	Present Worth Factor	Compound Amount Factor	Sinking Fund Factor	Present Worth Factor	Capital Recovery Factor	Gradient Uniform Series	Gradient Present Worth	
N	(F/P,i,N)	(P/F,i,N)	(F/A,i,N)	(A/F,i,N)	(P/A,i,N)	(A/P,i,N)	(A/G,i,N)	(P/G,i,N)	N
1	1.0200	0.9804	1.0000	1.0000	0.9804	1.0200	0.0000	0.0000	1
2	1.0404	0.9612	2.0200	0.4950	1.9416	0.5150	0.4950	0.9612	2
3	1.0612	0.9423	3.0604	0.3268	2.8839	0.3468	0.9868	2.8458	3
4	1.0824	0.9238	4.1216	0.2426	3.8077	0.2626	1.4752	5.6173	4
5	1.1041	0.9057	5.2040	0.1922	4.7135	0.2122	1.9604	9.2403	5
6	1.1262	0.8880	6.3081	0.1585	5.6014	0.1785	2.4423	13.6801	6
7	1.1487	0.8706	7.4343	0.1345	6.4720	0.1545	2.9208	18.9035	7
8	1.1717	0.8535	8.5830	0.1165	7.3255	0.1365	3.3961	24.8779	8
9	1.1951	0.8368	9.7546	0.1025	8.1622	0.1225	3.8681	31.5720	9
10	1.2190	0.8203	10.9497	0.0913	8.9826	0.1113	4.3367	38.9551	10
11	1.2434	0.8043	12.1687	0.0822	9.7868	0.1022	4.8021	46.9977	11
12	1.2682	0.7885	13.4121	0.0746	10.5753	0.0946	5.2642	55.6712	12
13	1.2936	0.7730	14.6803	0.0681	11.3484	0.0881	5.7231	64.9475	13
14	1.3195	0.7579	15.9739	0.0626	12.1062	0.0826	6.1786	74.7999	14
15	1.3459	0.7430	17.2934	0.0578	12.8493	0.0778	6.6309	85.2021	15
16	1.3728	0.7284	18.6393	0.0537	13.5777	0.0737	7.0799	96.1288	16
17	1.4002	0.7142	20.0121	0.0500	14.2919	0.0700	7.5256	107.5554	17
18	1.4282	0.7002	21.4123	0.0467	14.9920	0.0667	7.9681	119.4581	18
19	1.4568	0.6864	22.8406	0.0438	15.6785	0.0638	8.4073	131.8139	19
20	1.4859	0.6730	24.2974	0.0412	16.3514	0.0612	8.8433	144.6003	20
21	1.5157	0.6598	25.7833	0.0388	17.0112	0.0588	9.2760	157.7959	21
22	1.5460	0.6468	27.2990	0.0366	17.6580	0.0566	9.7055	171.3795	22
23	1.5769	0.6342	28.8450	0.0347	18.2922	0.0547	10.1317	185.3309	23
24	1.6084	0.6217	30.4219	0.0329	18.9139	0.0529	10.5547	199.6305	24
25	1.6406	0.6095	32.0303	0.0312	19.5235	0.0512	10.9745	214.2592	25
26	1.6734	0.5976	33.6709	0.0297	20.1210	0.0497	11.3910	229.1987	26
27	1.7069	0.5859	35.3443	0.0283	20.7069	0.0483	11.8043	244.4311	27
28	1.7410	0.5744	37.0512	0.0270	21.2813	0.0470	12.2145	259.9392	28
29	1.7758	0.5631	38.7922	0.0258	21.8444	0.0458	12.6214	275.7064	29
30	1.8114	0.5521	40.5681	0.0246	22.3965	0.0446	13.0251	291.7164	30
31	1.8476	0.5412	42.3794	0.0236	22.9377	0.0436	13.4257	307.9538	31
32	1.8845	0.5306	44.2270	0.0226	23.4683	0.0426	13.8230	324.4035	32
33	1.9222	0.5202	46.1116	0.0217	23.9886	0.0417	14.2172	341.0508	33
34	1.9607	0.5100	48.0338	0.0208	24.4986	0.0408	14.6083	357.8817	34
35	1.9999	0.5000	49.9945	0.0200	24.9986	0.0400	14.9961	374.8826	35
36	2.0399	0.4902	51.9944	0.0192	25.4888	0.0392	15.3809	392.0405	36
40	2.2080	0.4529	60.4020	0.0166	27.3555	0.0366	16.8885	461.9931	40
48	2.5871	0.3865	79.3535	0.0126	30.6731	0.0326	19.7556	605.9657	48
50	2.6916	0.3715	84.5794	0.0118	31.4236	0.0318	20.4420	642.3606	50
60	3.2810	0.3048	114.0515	0.0088	34.7609	0.0288	23.6961	823.6975	60
72	4.1611	0.2403	158.0570	0.0063	37.9841	0.0263	27.2234	1034.0557	72
80	4.8754	0.2051	193.7720	0.0052	39.7445	0.0252	29.3572	1166.7868	80
84	5.2773	0.1895	213.8666	0.0047	40.5255	0.0247	30.3616	1230.4191	84
90	5.9431	0.1683	247.1567	0.0040	41.5869	0.0240	31.7929	1322.1701	90
96	6.6929	0.1494	284.6467	0.0035	42.5294	0.0235	33.1370	1409.2973	96
100	7.2446	0.1380	312.2323	0.0032	43.0984	0.0232	33.9863	1464.7527	100
108	8.4883	0.1178	374.4129	0.0027	44.1095	0.0227	35.5774	1569.3025	108
120	10.7652	0.0929	488.2582	0.0020	45.3554	0.0220	37.7114	1710.4160	120
240	115.8887	0.0086	5744.4368	0.0002	49.5686	0.0202	47.9110	2374.8800	240
360	1247.5611	0.0008	62328.0564	0.0000	49.9599	0.0200	49.7112	2483.5679	360

3.0%

	Single Payment		Equal Payment Series				Gradient Series		
N	Compound Amount Factor (F/P,i,N)	Present Worth Factor (P/F,i,N)	Compound Amount Factor (F/A,i,N)	Sinking Fund Factor (A/F,i,N)	Present Worth Factor (P/A,i,N)	Capital Recovery Factor (A/P,i,N)	Gradient Uniform Series (A/G,i,N)	Gradient Present Worth (P/G,i,N)	N
1	1.0300	0.9709	1.0000	1.0000	0.9709	1.0300	0.0000	0.0000	1
2	1.0609	0.9426	2.0300	0.4926	1.9135	0.5226	0.4926	0.9426	2
3	1.0927	0.9151	3.0909	0.3235	2.8286	0.3535	0.9803	2.7729	3
4	1.1255	0.8885	4.1836	0.2390	3.7171	0.2690	1.4631	5.4383	4
5	1.1593	0.8626	5.3091	0.1884	4.5797	0.2184	1.9409	8.8888	5
6	1.1941	0.8375	6.4684	0.1546	5.4172	0.1846	2.4138	13.0762	6
7	1.2299	0.8131	7.6625	0.1305	6.2303	0.1605	2.8819	17.9547	7
8	1.2668	0.7894	8.8923	0.1125	7.0197	0.1425	3.3450	23.4806	8
9	1.3048	0.7664	10.1591	0.0984	7.7861	0.1284	3.8032	29.6119	9
10	1.3439	0.7441	11.4639	0.0872	8.5302	0.1172	4.2565	36.3088	110
11	1.3842	0.7224	12.8078	0.0781	9.2526	0.1081	4.7049	43.5330	11
12	1.4258	0.7014	14.1920	0.0705	9.9540	0.1005	5.1485	51.2482	12
13	1.4685	0.6810	15.6178	0.0640	10.6350	0.0940	5.5872	59.4196	13
14	1.5126	0.6611	17.0863	0.0585	11.2961	0.0885	6.0210	68.0141	14
15	1.5580	0.6419	18.5989	0.0538	11.9379	0.0838	6.4500	77.0002	15
16	1.6047	0.6232	20.1569	0.0496	12.5611	0.0796	6.8742	86.3477	16
17	1.6528	0.6050	21.7616	0.0460	13.1661	0.0760	7.2936	96.0280	17
18	1.7024	0.5874	23.4144	0.0427	13.7535	0.0727	7.7081	106.0137	18
19	1.7535	0.5703	25.1169	0.0398	14.3238	0.0698	8.1179	116.2788	19
20	1.8061	0.5537	26.8704	0.0372	14.8775	0.0672	8.5229	126.7987	20
21	1.8603	0.5375	28.6765	0.0349	15.4150	0.0649	8.9231	137.5496	21
22	1.9161	0.5219	30.5368	0.0327	15.9396	0.0627	9.3186	148.5094	22
23	1.9736	0.5067	32.4529	0.0308	16.4436	0.0608	9.7093	159.6566	23
24	2.0328	0.4919	34.4265	0.0290	16.9355	0.0590	10.0954	170.9711	24
25	2.0938	0.4776	36.4593	0.0274	17.4131	0.0574	10.4768	182.4336	25
26	2.1566	0.4637	38.5530	0.0259	17.8768	0.0559	10.8535	194.0260	26
27	2.2213	0.4502	40.7096	0.0246	18.3270	0.0546	11.2255	205.7309	27
28	2.2879	0.4371	42.9309	0.0233	18.7641	0.0533	11.5930	217.5320	28
29	2.3566	0.4243	45.2189	0.0221	19.1885	0.0521	11.9558	229.4137	29
30	2.4273	0.4120	47.5754	0.0210	19.6004	0.0510	12.3141	241.3613	30
31	2.5001	0.4000	50.0027	0.0200	20.0004	0.0500	12.6678	253.3609	31
32	2.5751	0.3883	52.5028	0.0190	20.3888	0.0490	13.0169	265.3993	32
33	2.6523	0.3770	55.0778	0.0182	20.7658	0.0482	13.3616	277.4642	33
34	2.7319	0.3660	57.7302	0.0173	21.1318	0.0473	13.7018	289.5437	34
35	2.8139	0.3554	60.4621	0.0165	21.4872	0.0465	14.0375	301.6267	35
40	3.2620	0.3066	75.4013	0.0133	23.1148	0.0433	15.6502	361.7499	40
45	3.7816	0.2644	92.7199	0.0108	24.5187	0.0408	17.1556	420.6325	45
50	4.3839	0.2281	112.7969	0.0089	25.7298	0.0389	18.5575	477.4803	50
55	5.0821	0.1968	136.0716	0.0073	26.7744	0.0373	19.8600	531.7411	55
60	5.8916	0.1697	163.0534	0.0061	27.6756	0.0361	21.0674	583.0526	60
65	6.8300	0.1464	194.3328	0.0051	28.4529	0.0351	22.1841	631.2010	65
70	7.9178	0.1263	230.5941	0.0043	29.1234	0.0343	23.2145	676.0869	70
75	9.1789	0.1089	272.6309	0.0037	29.7018	0.0337	24.1634	717.6978	75
80	10.6409	0.0940	321.3630	0.0031	30.2008	0.0331	25.0353	756.0865	80
85	12.3357	0.0811	377.8570	0.0026	30.6312	0.0326	25.8349	791.3529	85
90	14.3005	0.0699	443.3489	0.0023	31.0024	0.0323	26.5667	823.6302	90
95	16.5782	0.0603	519.2720	0.0019	31.3227	0.0319	27.2351	853.0742	95
100	19.2186	0.0520	607.2877	0.0016	31.5989	0.0316	27.8444	879.8540	100

4.0%

	Single Payment		Equal Payment Series				Gradient Series		
	Compound Amount Factor	Present Worth Factor	Compound Amount Factor	Sinking Fund Factor	Present Worth Factor	Capital Recovery Factor	Gradient Uniform Series	Gradient Present Worth	
N	(F/P,i,N)	(P/F,i,N)	(F/A,i,N)	(A/F,i,N)	(P/A,i,N)	(A/P,i,N)	(A/G,i,N)	(P/G,i,N)	N
1	1.0400	0.9615	1.0000	1.0000	0.9615	1.0400	0.0000	0.0000	1
2	1.0816	0.9246	2.0400	0.4902	1.8861	0.5302	0.4902	0.9246	2
3	1.1249	0.8890	3.1216	0.3203	2.7751	0.3603	0.9739	2.7025	3
4	1.1699	0.8548	4.2465	0.2355	3.6299	0.2755	1.4510	5.2670	4
5	1.2167	0.8219	5.4163	0.1846	4.4518	0.2246	1.9216	8.5547	5
6	1.2653	0.7903	6.6330	0.1508	5.2421	0.1908	2.3857	12.5062	6
7	1.3159	0.7599	7.8983	0.1266	6.0021	0.1666	2.8433	17.0657	7
8	1.3686	0.7307	9.2142	0.1085	6.7327	0.1485	3.2944	22.1806	8
9	1.4233	0.7026	10.5828	0.0945	7.4353	0.1345	3.7391	27.8013	9
10	1.4802	0.6756	12.0061	0.0833	8.1109	0.1233	4.1773	33.8814	10
11	1.5395	0.6496	13.4864	0.0741	8.7605	0.1141	4.6090	40.3772	11
12	1.6010	0.6246	15.0258	0.0666	9.3851	0.1066	5.0343	47.2477	12
13	1.6651	0.6006	16.6268	0.0601	9.9856	0.1001	5.4533	54.4546	13
14	1.7317	0.5775	18.2919	0.0547	10.5631	0.0947	5.8659	61.9618	14
15	1.8009	0.5553	20.0236	0.0499	11.1184	0.0899	6.2721	69.7355	15
16	1.8730	0.5339	21.8245	0.0458	11.6523	0.0858	6.6720	77.7441	16
17	1.9479	0.5134	23.6975	0.0422	12.1657	0.0822	7.0656	85.9581	17
18	2.0258	0.4936	25.6454	0.0390	12.6593	0.0790	7.4530	94.3498	18
19	2.1068	0.4746	27.6712	0.0361	13.1339	0.0761	7.8342	102.8933	19
20	2.1911	0.4564	29.7781	0.0336	13.5903	0.0736	8.2091	111.5647	20
21	2.2788	0.4388	31.9692	0.0313	14.0292	0.0713	8.5779	120.3414	21
22	2.3699	0.4220	34.2480	0.0292	14.4511	0.0692	8.9407	129.2024	22
23	2.4647	0.4057	36.6179	0.0273	14.8568	0.0673	9.2973	138.1284	23
24	2.5633	0.3901	39.0826	0.0256	15.2470	0.0656	9.6479	147.1012	24
25	2.6658	0.3751	41.6459	0.0240	15.6221	0.0640	9.9925	156.1040	25
26	2.7725	0.3607	44.3117	0.0226	15.9828	0.0626	10.3312	165.1212	26
27	2.8834	0.3468	47.0842	0.0212	16.3296	0.0612	10.6640	174.1385	27
28	2.9987	0.3335	49.9676	0.0200	16.6631	0.0600	10.9909	183.1424	28
29	3.1187	0.3207	52.9663	0.0189	16.9837	0.0589	11.3120	192.1206	29
30	3.2434	0.3083	56.0849	0.0178	17.2920	0.0578	11.6274	201.0618	30
31	3.3731	0.2965	59.3283	0.0169	17.5885	0.0569	11.9371	209.9556	31
32	3.5081	0.2851	62.7015	0.0159	17.8736	0.0559	12.2411	218.7924	32
33	3.6484	0.2741	66.2095	0.0151	18.1476	0.0551	12.5396	227.5634	33
34	3.7943	0.2636	69.8579	0.0143	18.4112	0.0543	12.8324	236.2607	34
35	3.9461	0.2534	73.6522	0.0136	18.6646	0.0536	13.1198	244.8768	35
40	4.8010	0.2083	95.0255	0.0105	19.7928	0.0505	14.4765	286.5303	40
45	5.8412	0.1712	121.0294	0.0083	20.7200	0.0483	15.7047	325.4028	45
50	7.1067	0.1407	152.6671	0.0066	21.4822	0.0466	16.8122	361.1638	50
55	8.6464	0.1157	191.1592	0.0052	22.1086	0.0452	17.8070	393.6890	55
60	10.5196	0.0951	237.9907	0.0042	22.6235	0.0442	18.6972	422.9966	60
65	12.7987	0.0781	294.9684	0.0034	23.0467	0.0434	19.4909	449.2014	65
70	15.5716	0.0642	364.2905	0.0027	23.3945	0.0427	20.1961	472.4789	70
75	18.9453	0.0528	448.6314	0.0022	23.6804	0.0422	20.8206	493.0408	75
80	23.0498	0.0434	551.2450	0.0018	23.9154	0.0418	21.3718	511.1161	80
85	28.0436	0.0357	676.0901	0.0015	24.1085	0.0415	21.8569	526.9384	85
90	34.1193	0.0293	827.9833	0.0012	24.2673	0.0412	22.2826	540.7369	90
95	41.5114	0.0241	1012.7846	0.0010	24.3978	0.0410	22.6550	552.7307	95
100	50.5049	0.0198	1237.6237	0.0008	24.5050	0.0408	22.9800	563.1249	100

5.0%

	Single Payment		Equal Payment Series				Gradient Series		
	Compound Amount Factor	Present Worth Factor	Compound Amount Factor	Sinking Fund Factor	Present Worth Factor	Capital Recovery Factor	Gradient Uniform Series	Gradient Present Worth	
N	(F/P,i,N)	(P/F,i,N)	(F/A,i,N)	(A/F,i,N)	(P/A,i,N)	(A/P,i,N)	(A/G,i,N)	(P/G,i,N)	N
1	1.0500	0.9524	1.0000	1.0000	0.9524	1.0500	0.0000	0.0000	1
2	1.1025	0.9070	2.0500	0.4878	1.8594	0.5378	0.4878	0.9070	2
3	1.1576	0.8638	3.1525	0.3172	2.7232	0.3672	0.9675	2.6347	3
4	1.2155	0.8227	4.3101	0.2320	3.5460	0.2820	1.4391	5.1028	4
5	1.2763	0.7835	5.5256	0.1810	4.3295	0.2310	1.9025	8.2369	5
6	1.3401	0.7462	6.8019	0.1470	5.0757	0.1970	2.3579	11.9680	6
7	1.4071	0.7107	8.1420	0.1228	5.7864	0.1728	2.8052	16.2321	7
8	1.4775	0.6768	9.5491	0.1047	6.4632	0.1547	3.2445	20.9700	8
9	1.5513	0.6446	11.0266	0.0907	7.1078	0.1407	3.6758	26.1268	9
10	1.6289	0.6139	12.5779	0.0795	7.7217	0.1295	4.0991	31.6520	10
11	1.7103	0.5847	14.2068	0.0704	8.3064	0.1204	4.5144	37.4988	11
12	1.7959	0.5568	15.9171	0.0628	8.8633	0.1128	4.9219	43.6241	12
13	1.8856	0.5303	17.7130	0.0565	9.3936	0.1065	5.3215	49.9879	13
14	1.9799	0.5051	19.5986	0.0510	9.8986	0.1010	5.7133	56.5538	14
15	2.0789	0.4810	21.5786	0.0463	10.3797	0.0963	6.0973	63.2880	15
16	2.1829	0.4581	23.6575	0.0423	10.8378	0.0923	6.4736	70.1597	16
17	2.2920	0.4363	25.8404	0.0387	11.2741	0.0887	6.8423	77.1405	17
18	2.4066	0.4155	28.1324	0.0355	11.6896	0.0855	7.2034	84.2043	18
19	2.5270	0.3957	30.5390	0.0327	12.0853	0.0827	7.5569	91.3275	19
20	2.6533	0.3769	33.0660	0.0302	12.4622	0.0802	7.9030	98.4884	20
21	2.7860	0.3589	35.7193	0.0280	12.8212	0.0780	8.2416	105.6673	21
22	2.9253	0.3418	38.5052	0.0260	13.1630	0.0760	8.5730	112.8461	22
23	3.0715	0.3256	41.4305	0.0241	13.4886	0.0741	8.8971	120.0087	23
24	3.2251	0.3101	44.5020	0.0225	13.7986	0.0725	9.2140	127.1402	24
25	3.3864	0.2953	47.7271	0.0210	14.0939	0.0710	9.5238	134.2275	25
26	3.5557	0.2812	51.1135	0.0196	14.3752	0.0696	9.8266	141.2585	26
27	3.7335	0.2678	54.6691	0.0183	14.6430	0.0683	10.1224	148.2226	27
28	3.9201	0.2551	58.4026	0.0171	14.8981	0.0671	10.4114	155.1101	28
29	4.1161	0.2429	62.3227	0.0160	15.1411	0.0660	10.6936	161.9126	29
30	4.3219	0.2314	66.4388	0.0151	15.3725	0.0651	10.9691	168.6226	30
31	4.5380	0.2204	70.7608	0.0141	15.5928	0.0641	11.2381	175.2333	31
32	4.7649	0.2099	75.2988	0.0133	15.8027	0.0633	11.5005	181.7392	32
33	5.0032	0.1999	80.0638	0.0125	16.0025	0.0625	11.7566	188.1351	33
34	5.2533	0.1904	85.0670	0.0118	16.1929	0.0618	12.0063	194.4168	34
35	5.5160	0.1813	90.3203	0.0111	16.3742	0.0611	12.2498	200.5807	35
40	7.0400	0.1420	120.7998	0.0083	17.1591	0.0583	13.3775	229.5452	40
45	8.9850	0.1113	159.7002	0.0063	17.7741	0.0563	14.3644	255.3145	45
50	11.4674	0.0872	209.3480	0.0048	18.2559	0.0548	15.2233	277.9148	50
55	14.6356	0.0683	272.7126	0.0037	18.6335	0.0537	15.9664	297.5104	55
60	18.6792	0.0535	353.5837	0.0028	18.9293	0.0528	16.6062	314.3432	60
65	23.8399	0.0419	456.7980	0.0022	19.1611	0.0522	17.1541	328.6910	65
70	30.4264	0.0329	588.5285	0.0017	19.3427	0.0517	17.6212	340.8409	70
75	38.8327	0.0258	756.6537	0.0013	19.4850	0.0513	18.0176	351.0721	75
80	49.5614	0.0202	971.2288	0.0010	19.5965	0.0510	18.3526	359.6460	80
85	63.2544	0.0158	1245.0871	0.0008	19.6838	0.0508	18.6346	366.8007	85
90	80.7304	0.0124	1594.6073	0.0006	19.7523	0.0506	18.8712	372.7488	90
95	103.0347	0.0097	2040.6935	0.0005	19.8059	0.0505	19.0689	377.6774	95
100	131.5013	0.0076	2610.0252	0.0004	19.8479	0.0504	19.2337	381.7492	100

6.0%

	Single Payment		Equal Payment Series				Gradient Series		
	Compound Amount Factor	Present Worth Factor	Compound Amount Factor	Sinking Fund Factor	Present Worth Factor	Capital Recovery Factor	Gradient Uniform Series	Gradient Present Worth	
N	(F/P,i,N)	(P/F,i,N)	(F/A,i,N)	(A/F,i,N)	(P/A,i,N)	(A/P,i,N)	(A/G,i,N)	(P/G,i,N)	N
1	1.0600	0.9434	1.0000	1.0000	0.9434	1.0600	0.0000	0.0000	1
2	1.1236	0.8900	2.0600	0.4854	1.8334	0.5454	0.4854	0.8900	2
3	1.1910	0.8396	3.1836	0.3141	2.6730	0.3741	0.9612	2.5692	3
4	1.2625	0.7921	4.3746	0.2286	3.4651	0.2886	1.4272	4.9455	4
5	1.3382	0.7473	5.6371	0.1774	4.2124	0.2374	1.8836	7.9345	5
6	1.4185	0.7050	6.9753	0.1434	4.9173	0.2034	2.3304	11.4594	6
7	1.5036	0.6651	8.3938	0.1191	5.5824	0.1791	2.7676	15.4497	7
8	1.5938	0.6274	9.8975	0.1010	6.2098	0.1610	3.1952	19.8416	8
9	1.6895	0.5919	11.4913	0.0870	6.8017	0.1470	3.6133	24.5768	9
10	1.7908	0.5584	13.1808	0.0759	7.3601	0.1359	4.0220	29.6023	10
11	1.8983	0.5268	14.9716	0.0668	7.8869	0.1268	4.4213	34.8702	11
12	2.0122	0.4970	16.8699	0.0593	8.3838	0.1193	4.8113	40.3369	12
13	2.1329	0.4688	18.8821	0.0530	8.8527	0.1130	5.1920	45.9629	13
14	2.2609	0.4423	21.0151	0.0476	9.2950	0.1076	5.5635	51.7128	14
15	2.3966	0.4173	23.2760	0.0430	9.7122	0.1030	5.9260	57.5546	15
16	2.5404	0.3936	25.6725	0.0390	10.1059	0.0990	6.2794	63.4592	16
17	2.6928	0.3714	28.2129	0.0354	10.4773	0.0954	6.6240	69.4011	17
18	2.8543	0.3503	30.9057	0.0324	10.8276	0.0924	6.9597	75.3569	18
19	3.0256	0.3305	33.7600	0.0296	11.1581	0.0896	7.2867	81.3062	19
20	3.2071	0.3118	36.7856	0.0272	11.4699	0.0872	7.6051	87.2304	20
21	3.3996	0.2942	39.9927	0.0250	11.7641	0.0850	7.9151	93.1136	21
22	3.6035	0.2775	43.3923	0.0230	12.0416	0.0830	8.2166	98.9412	22
23	3.8197	0.2618	46.9958	0.0213	12.3034	0.0813	8.5099	104.7007	23
24	4.0489	0.2470	50.8156	0.0197	12.5504	0.0797	8.7951	110.3812	24
25	4.2919	0.2330	54.8645	0.0182	12.7834	0.0782	9.0722	115.9732	25
26	4.5494	0.2198	59.1564	0.0169	13.0032	0.0769	9.3414	121.4684	26
27	4.8223	0.2074	63.7058	0.0157	13.2105	0.0757	9.6029	126.8600	27
28	5.1117	0.1956	68.5281	0.0146	13.4062	0.0746	9.8568	132.1420	28
29	5.4184	0.1846	73.6398	0.0136	13.5907	0.0736	10.1032	137.3096	29
30	5.7435	0.1741	79.0582	0.0126	13.7648	0.0726	10.3422	142.3588	30
31	6.0881	0.1643	84.8017	0.0118	13.9291	0.0718	10.5740	147.2864	31
32	6.4534	0.1550	90.8898	0.0110	14.0840	0.0710	10.7988	152.0901	32
33	6.8406	0.1462	97.3432	0.0103	14.2302	0.0703	11.0166	156.7681	33
34	7.2510	0.1379	104.1838	0.0096	14.3681	0.0696	11.2276	161.3192	34
35	7.6861	0.1301	111.4348	0.0090	14.4982	0.0690	11.4319	165.7427	35
40	10.2857	0.0972	154.7620	0.0065	15.0463	0.0665	12.3590	185.9568	40
45	13.7646	0.0727	212.7435	0.0047	15.4558	0.0647	13.1413	203.1096	45
50	18.4202	0.0543	290.3359	0.0034	15.7619	0.0634	13.7964	217.4574	50
55	24.6503	0.0406	394.1720	0.0025	15.9905	0.0625	14.3411	229.3222	55
60	32.9877	0.0303	533.1282	0.0019	16.1614	0.0619	14.7909	239.0428	60
65	44.1450	0.0227	719.0829	0.0014	16.2891	0.0614	15.1601	246.9450	65
70	59.0759	0.0169	967.9322	0.0010	16.3845	0.0610	15.4613	253.3271	70
75	79.0569	0.0126	1300.9487	0.0008	16.4558	0.0608	15.7058	258.4527	75
80	105.7960	0.0095	1746.5999	0.0006	16.5091	0.0606	15.9033	262.5493	80
85	141.5789	0.0071	2342.9817	0.0004	16.5489	0.0604	16.0620	265.8096	85
90	189.4645	0.0053	3141.0752	0.0003	16.5787	0.0603	16.1891	268.3946	90
95	253.5463	0.0039	4209.1042	0.0002	16.6009	0.0602	16.2905	270.4375	95
100	339.3021	0.0029	5638.3681	0.0002	16.6175	0.0602	16.3711	272.0471	100

7.0%

	Single Payment		Equal Payment Series				Gradient Series		
	Compound Amount Factor	Present Worth Factor	Compound Amount Factor	Sinking Fund Factor	Present Worth Factor	Capital Recovery Factor	Gradient Uniform Series	Gradient Present Worth	
N	(F/P,i,N)	(P/F,i,N)	(F/A,i,N)	(A/F,i,N)	(P/A,i,N)	(A/P,i,N)	(A/G,i,N)	(P/G,i,N)	N
1	1.0700	0.9346	1.0000	1.0000	0.9346	1.0700	0.0000	0.0000	1
2	1.1449	0.8734	2.0700	0.4831	1.8080	0.5531	0.4831	0.8734	2
3	1.2250	0.8163	3.2149	0.3111	2.6243	0.3811	0.9549	2.5060	3
4	1.3108	0.7629	4.4399	0.2252	3.3872	0.2952	1.4155	4.7947	4
5	1.4026	0.7130	5.7507	0.1739	4.1002	0.2439	1.8650	7.6467	5
6	1.5007	0.6663	7.1533	0.1398	4.7665	0.2098	2.3032	10.9784	6
7	1.6058	0.6227	8.6540	0.1156	5.3893	0.1856	2.7304	14.7149	7
8	1.7182	0.5820	10.2598	0.0975	5.9713	0.1675	3.1465	18.7889	8
9	1.8385	0.5439	11.9780	0.0835	6.5152	0.1535	3.5517	23.1404	9
10	1.9672	0.5083	13.8164	0.0724	7.0236	0.1424	3.9461	27.7156	10
11	2.1049	0.4751	15.7836	0.0634	7.4987	0.1334	4.3296	32.4665	11
12	2.2522	0.4440	17.8885	0.0559	7.9427	0.1259	4.7025	37.3506	12
13	2.4098	0.4150	20.1406	0.0497	8.3577	0.1197	5.0648	42.3302	13
14	2.5785	0.3878	22.5505	0.0443	8.7455	0.1143	5.4167	47.3718	14
15	2.7590	0.3624	25.1290	0.0398	9.1079	0.1098	5.7583	52.4461	15
16	2.9522	0.3387	27.8881	0.0359	9.4466	0.1059	6.0897	57.5271	16
17	3.1588	0.3166	30.8402	0.0324	9.7632	0.1024	6.4110	62.5923	17
18	3.3799	0.2959	33.9990	0.0294	10.0591	0.0994	6.7225	67.6219	18
19	3.6165	0.2765	37.3790	0.0268	10.3356	0.0968	7.0242	72.5991	19
20	3.8697	0.2584	40.9955	0.0244	10.5940	0.0944	7.3163	77.5091	20
21	4.1406	0.2415	44.8652	0.0223	10.8355	0.0923	7.5990	82.3393	21
22	4.4304	0.2257	49.0057	0.0204	11.0612	0.0904	7.8725	87.0793	22
23	4.7405	0.2109	53.4361	0.0187	11.2722	0.0887	8.1369	91.7201	23
24	5.0724	0.1971	58.1767	0.0172	11.4693	0.0872	8.3923	96.2545	24
25	5.4274	0.1842	63.2490	0.0158	11.6536	0.0858	8.6391	100.6765	25
26	5.8074	0.1722	68.6765	0.0146	11.8258	0.0846	8.8773	104.9814	26
27	6.2139	0.1609	74.4838	0.0134	11.9867	0.0834	9.1072	109.1656	27
28	6.6488	0.1504	80.6977	0.0124	12.1371	0.9824	9.3289	113.2264	28
29	7.1143	0.1406	87.3465	0.0114	12.2777	0.0814	9.5427	117.1622	29
30	7.6123	0.1314	94.4608	0.0106	12.4090	0.0806	9.7487	120.9718	30
31	8.1451	0.1228	102.0730	0.0098	12.5318	0.0798	9.9471	124.6550	31
32	8.7153	0.1147	110.2182	0.0091	12.6466	0.0791	10.1381	128.2120	32
33	9.3253	0.1072	118.9334	0.0084	12.7538	0.0784	10.3219	131.6435	33
34	9.9781	0.1002	128.2588	0.0078	12.8540	0.0778	10.4987	134.9507	34
35	10.6766	0.0937	138.2369	0.0072	12.9477	0.0772	10.6687	138.1353	35
40	14.9745	0.0668	199.6351	0.0050	13.3317	0.0750	11.4233	152.2928	40
45	21.0025	0.0476	285.7493	0.0035	13.6055	0.0735	12.0360	163.7559	45
50	29.4570	0.0339	406.5289	0.0025	13.8007	0.0725	12.5287	172.9051	50
55	41.3150	0.0242	575.9286	0.0017	13.9399	0.0717	12.9215	180.1243	55
60	57.9464	0.0173	813.5204	0.0012	14.0392	0.0712	13.2321	185.7677	60
65	81.2729	0.0123	1146.7552	0.0009	14.1099	0.0709	13.4760	190.1452	65
70	113.9894	0.0088	1614.1342	0.0006	14.1604	0.0706	13.6662	193.5185	70
75	159.8760	0.0063	2269.6574	0.0004	14.1964	0.0704	13.8136	196.1035	75
80	224.2344	0.0045	3189.0627	0.0003	14.2220	0.0703	13.9273	198.0748	80
85	314.5003	0.0032	4478.5761	0.0002	14.2403	0.0702	14.0146	199.5717	85
90	441.1030	0.0023	6287.1854	0.0002	14.2533	0.0702	14.0812	200.7042	90
95	618.6697	0.0016	8823.8535	0.0001	14.2626	0.0701	14.1319	201.5581	95
100	867.7163	0.0012	12381.6618	0.0001	14.2693	0.0701	14.1703	202.2001	100

8.0%

	Single Payment		Equal Payment Series				Gradient Series		
	Compound Amount Factor	Present Worth Factor	Compound Amount Factor	Sinking Fund Factor	Present Worth Factor	Capital Recovery Factor	Gradient Uniform Series	Gradient Present Worth	
N	(F/P,i,N)	(P/F,i,N)	(F/A,i,N)	(A/F,i,N)	(P/A,i,N)	(A/P,i,N)	(A/G,i,N)	(P/G,i,N)	N
1	1.0800	0.9259	1.0000	1.0000	0.9259	1.0800	0.0000	0.0000	1
2	1.1664	0.8573	2.0800	0.4808	1.7833	0.5608	0.4808	0.8573	2
3	1.2597	0.7938	3.2464	0.3080	2.5771	0.3880	0.9487	2.4450	3
4	1.3605	0.7350	4.5061	0.2219	3.3121	0.3019	1.4040	4.6501	4
5	1.4693	0.6806	5.8666	0.1705	3.9927	0.2505	1.8465	7.3724	5
6	1.5869	0.6302	7.3359	0.1363	4.6229	0.2163	2.2763	10.5233	6
7	1.7138	0.5835	8.9228	0.1121	5.2064	0.1921	2.6937	14.0242	7
8	1.8509	0.5403	10.6366	0.0940	5.7466	0.1740	3.0985	17.8061	8
9	1.9990	0.5002	12.4876	0.0801	6.2469	0.1601	3.4910	21.8081	9
10	2.1589	0.4632	14.4866	0.0690	6.7101	0.1490	3.8713	25.9768	10
11	2.3316	0.4289	16.6455	0.0601	7.1390	0.1401	4.2395	30.2657	11
12	2.5182	0.3971	18.9771	0.0527	7.5361	0.1327	4.5957	34.6339	12
13	2.7196	0.3677	21.4953	0.0465	7.9038	0.1265	4.9402	39.0463	13
14	2.9372	0.3405	24.2149	0.0413	8.2442	0.1213	5.2731	43.4723	14
15	3.1722	0.3152	27.1521	0.0368	8.5595	0.1168	5.5945	47.8857	15
16	3.4259	0.2919	30.3243	0.0330	8.8514	0.1130	5.9046	52.2640	16
17	3.7000	0.2703	33.7502	0.0296	9.1216	0.1096	6.2037	56.5883	17
18	3.9960	0.2502	37.4502	0.0267	9.3719	0.1067	6.4920	60.8426	18
19	4.3157	0.2317	41.4463	0.0241	9.6036	0.1041	6.7697	65.0134	19
20	4.6610	0.2145	45.7620	0.0219	9.8181	0.1019	7.0369	69.0898	20
21	5.0338	0.1987	50.4229	0.0198	10.0168	0.0998	7.2940	73.0629	21
22	5.4365	0.1839	55.4568	0.0180	10.2007	0.0980	7.5412	76.9257	22
23	5.8715	0.1703	60.8933	0.0164	10.3711	0.0964	7.7786	80.6726	23
24	6.3412	0.1577	66.7648	0.0150	10.5288	0.0950	8.0066	84.2997	24
25	6.8485	0.1460	73.1059	0.0137	10.6748	0.0937	8.2254	87.8041	25
26	7.3964	0.1352	79.9544	0.0125	10.8100	0.0925	8.4352	91.1842	26
27	7.9881	0.1252	87.3508	0.0114	10.9352	0.0914	8.6363	94.4390	27
78	8.6271	0.1159	95.3388	0.0105	11.0511	0.0905	8.8289	97.5687	28
29	9.3173	0.1073	103.9659	0.0096	11.1584	0.0896	9.0133	100.5738	29
30	10.0627	0.0994	113.2832	0.0088	11.2578	0.0888	9.1897	103.4558	30
31	10.8677	0.0920	123.3459	0.0081	11.3498	0.0881	9.3584	106.2163	31
32	11.7371	0.0852	134.2135	0.0075	11.4350	0.0875	9.5197	108.8575	32
33	12.6760	0.0789	145.9506	0.0069	11.5139	0.0869	9.6737	111.3819	33
34	13.6901	0.0730	158.6267	0.0063	11.5869	0.0863	9.8208	113.7924	34
35	14.7853	0.0676	172.3168	0.0058	11.6546	0.0858	9.9611	116.0920	35
40	21.7245	0.0460	259.0565	0.0039	11.9246	0.0839	10.5699	126.0422	40
45	31.9204	0.0313	386.5056	0.0026	12.1084	0.0826	11.0447	133.7331	45
50	46.9016	0.0213	573.7702	0.0017	12.2335	0.0817	11.4107	139.5928	50
55	68.9139	0.0145	848.9232	0.0012	12.3186	0.0812	11.6902	144.0065	55
60	101.2571	0.0099	1253.2133	0.0008	12.3766	0.0808	11.9015	147.3000	60
65	148.7798	0.0067	1847.2481	0.0005	12.4160	0.0805	12.0602	149.7387	65
70	218.6064	0.0046	2720.0801	0.0004	12.4428	0.0804	12.1783	151.5326	70
75	321.2045	0.0031	4002.5566	0.0002	12.4611	0.0802	12.2658	152.8448	75
80	471.9548	0.0021	5886.9354	0.0002	12.4735	0.0802	12.3301	153.8001	80
85	693.4565	0.0014	8655.7061	0.0001	12.4820	0.0801	12.3772	154.4925	85
90	1018.9151	0.0010	12723.9386	0.0001	12.4877	0.0801	12.4116	154.9925	90
95	1497.1205	0.0007	18701.5069	0.0001	12.4917	0.0801	12.4365	155.3524	95
100	2199.7613	0.0005	27484.5157	0.0000	12.4943	0.0800	12.4545	155.6107	100

9.0%

	Single Payment		Equal Payment Series				Gradient Series		
	Compound Amount Factor	Present Worth Factor	Compound Amount Factor	Sinking Fund Factor	Present Worth Factor	Capital Recovery Factor	Gradient Uniform Series	Gradient Present Worth	
N	(F/P,i,N)	(P/F,i,N)	(F/A,i,N)	(A/F,i,N)	(P/A,i,N)	(A/P,i,N)	(A/G,i,N)	(P/G,i,N)	N
1	1.0900	0.9174	1.0000	1.0000	0.9174	1.0900	0.0000	0.0000	1
2	1.1881	0.8417	2.0900	0.4785	1.7591	0.5685	0.4785	0.8417	2
3	1.2950	0.7722	3.2781	0.3051	2.5313	0.3951	0.9426	2.3860	3
4	1.4116	0.7084	4.5731	0.2187	3.2397	0.3087	1.3925	4.5113	4
5	1.5386	0.6499	5.9847	0.1671	3.8897	0.2571	1.8282	7.1110	5
6	1.6771	0.5963	7.5233	0.1329	4.4859	0.2229	2.2498	10.0924	6
7	1.8280	0.5470	9.2004	0.1087	5.0330	0.1987	2.6574	13.3746	7
8	1.9926	0.5019	11.0285	0.0907	5.5348	0.1807	3.0512	16.8877	8
9	2.1719	0.4604	13.0210	0.0768	5.9952	0.1668	3.4312	20.5711	9
10	2.3674	0.4224	15.1929	0.0658	6.4177	0.1558	3.7978	24.3728	10
11	2.5804	0.3875	17.5603	0.0569	6.8052	0.1469	4.1510	28.2481	11
12	2.8127	0.3555	20.1407	0.0497	7.1607	0.1397	4.4910	32.1590	12
13	3.0658	0.3262	22.9534	0.0436	7.4869	0.1336	4.8182	36.0731	13
14	3.3417	0.2992	26.0192	0.0384	7.7862	0.1284	5.1326	39.9633	14
15	3.6425	0.2745	29.3609	0.0341	8.0607	0.1241	5.4346	43.8069	15
16	3.9703	0.2519	33.0034	0.0303	8.3126	0.1203	5.7245	47.5849	16
17	4.3276	0.2311	36.9737	0.0270	8.5436	0.1170	6.0024	51.2821	17
18	4.7171	0.2120	41.3013	0.0242	8.7556	0.1142	6.2687	54.8860	18
19	5.1417	0.1945	46.0185	0.0217	8.9501	0.1117	6.5236	58.3868	19
20	5.6044	0.1784	51.1601	0.0195	9.1285	0.1095	6.7674	61.7770	20
21	6.1088	0.1637	56.7645	0.0176	9.2922	0.1076	7.0006	65.0509	21
22	6.6586	0.1502	62.8733	0.0159	9.4424	0.1059	7.2232	68.2048	22
23	7.2579	0.1378	69.5319	0.0144	9.5802	0.1044	7.4357	71.2359	23
24	7.9111	0.1264	76.7898	0.0130	9.7066	0.1030	7.6384	74.1433	24
25	8.6231	0.1160	84.7009	0.0118	9.8226	0.1018	7.8316	76.9265	25
26	9.3992	0.1064	93.3240	0.0107	9.9290	0.1007	8.0156	79.5863	26
27	10.2451	0.0976	102.7231	0.0097	10.0266	0.0997	8.1906	82.1241	27
28	11.1671	0.0895	112.9682	0.0089	10.1161	0.0989	8.3571	84.5419	28
29	12.1722	0.0822	124.1354	0.0081	10.1983	0.0981	8.5154	86.8422	29
30	13.2677	0.0754	136.3075	0.0073	10.2737	0.0973	8.6657	89.0280	30
31	14.4618	0.0691	149.5752	0.0067	10.3428	0.0967	8.8083	91.1024	31
32	15.7633	0.0634	164.0370	0.0061	10.4062	0.0961	8.9436	93.0690	32
33	17.1820	0.0582	179.8003	0.0056	10.4644	0.0956	9.0718	94.9314	33
34	18.7284	0.0534	196.9823	0.0051	10.5178	0.0951	9.1933	96.6935	34
35	20.4140	0.0490	215.7108	0.0046	10.5668	0.0946	9.3083	98.3590	35
40	31.4094	0.0318	337.8824	0.0030	10.7574	0.0930	9.7957	105.3762	40
45	48.3273	0.0207	525.8587	0.0019	10.8812	0.0919	10.1603	110.5561	45
50	74.3575	0.0134	815.0836	0.0012	10.9617	0.0912	10.4295	114.3251	50
55	114.4083	0.0087	1260.0918	0.0008	11.0140	0.0908	10.6261	117.0362	55
60	176.0313	0.0057	1944.7921	0.0005	11.0480	0.0905	10.7683	118.9683	60
65	270.8460	0.0037	2998.2885	0.0003	11.0701	0.0903	10.8702	120.3344	65
70	416.7301	0.0024	4619.2232	0.0002	11.0844	0.0902	10.9427	121.2942	70
75	641.1909	0.0016	7113.2321	0.0001	11.0938	0.0901	10.9940	121.9646	75
80	986.5517	0.0010	10950.5741	0.0001	11.0998	0.0901	11.0299	122.4306	80
85	1517.9320	0.0007	16854.8003	0.0001	11.1038	0.0901	11.0551	122.7533	85
90	2335.5266	0.0004	25939.1842	0.0000	11.1064	0.0900	11.0726	122.9758	90
95	3593.4971	0.0003	39916.6350	0.0000	11.1080	0.0900	11.0847	123.1287	95
100	5529.0408	0.0002	61422.6755	0.0000	11.1091	0.0900	11.0930	123.2335	100

10.0%

	Single Payment		Equal Payment Series				Gradient Series		
N	Compound Amount Factor (F/P,i,N)	Present Worth Factor (P/F,i,N)	Compound Amount Factor (F/A,i,N)	Sinking Fund Factor (A/F,i,N)	Present Worth Factor (P/A,i,N)	Capital Recovery Factor (A/P,i,N)	Gradient Uniform Series (A/G,i,N)	Gradient Present Worth (P/G,i,N)	N
1	1.1000	0.9091	1.0000	1.0000	0.9091	1.1000	0.0000	0.0000	1
2	1.2100	0.8264	2.1000	0.4762	1.7355	0.5762	0.4762	0.8264	2
3	1.3310	0.7513	3.3100	0.3021	2.4869	0.4021	0.9366	2.3291	3
4	1.4641	0.6830	4.6410	0.2155	3.1699	0.3155	1.3812	4.3781	4
5	1.6105	0.6209	6.1051	0.1638	3.7908	0.2638	1.8101	6.8618	5
6	1.7716	0.5645	7.7156	0.1296	4.3553	0.2296	2.2236	9.6842	6
7	1.9487	0.5132	9.4872	0.1054	4.8684	0.2054	2.6216	12.7631	7
8	2.1436	0.4665	11.4359	0.0874	5.3349	0.1874	3.0045	16.0287	8
9	2.3579	0.4241	13.5795	0.0736	5.7590~	0.1736	3.3724	19.4215	9
10	2.5937	0.3855	15.9374	0.0627	6.1446	0.1627	3.7255	22.8913	10
11	2.8531	0.3505	18.5312	0.0540	6.4951	0.1540	4.0641	26.3963	11
12	3.1384	0.3186	21.3843	0.0468	6.8137	0.1468	4.3884	29.9012	12
13	3.4523	0.2897	24.5227	0.0408	7.1034	0.1408	4.6988	33.3772	13
14	3.7975	0.2633	27.9750	0.0357	7.3667	0.1357	4.9955	36.8005	14
15	4.1772	0.2394	31.7725	0.0315	7.6061	0.1315	5.2789	40.1520	15
16	4.5950	0.2176	35.9497	0.0278	7.8237	0.1278	5.5493	43.4164	16
17	5.0545	0.1978	40.5447	0.0247	8.0216	0.1247	5.8071	46.5819	17
18	5.5599	0.1799	45.5992	0.0219	8.2014	0.1219	6.0526	49.6395	18
19	6.1159	0.1635	51.1591	0.0195	8.3649	0.1195	6.2861	52.5827	19
20	6.7275	0.1486	57.2750	0.0175	8.5136	0.1175	6.5081	55.4069	20
21	7.4002	0.1351	64.0025	0.0156	8.6487	0.1156	6.7189	58.1095	21
22	8.1403	0.1228	71.4027	0.0140	8.7715	0.1140	6.9189	60.6893	22
23	8.9543	0.1117	79.5430	0.0126	8.8832	0.1126	7.1085	63.1462	23
24	9.8497	0.1015	88.4973	0.0113	8.9847	0.1113	7.2881	65.4813	24
25	10.8347	0.0923	98.3471	0.0102	9.0770	0.1102	7.4580	67.6964	25
26	11.9182	0.0839	109.1818	0.0092	9.1609	0.1092	7.6186	69.7940	26
27	13.1100	0.0763	121.0999	0.0083	9.2372	0.1083	7.7704	71.7773	27
28	14.4210	0.0693	134.2099	0.0075	9.3066	0.1075	7.9137	73.6495	28
29	15.8631	0.0630	148.6309	0.0067	9.3696	0.1067	8.0489	75.4146	29
30	17.4494	0.0573	164.4940	0.0061	9.4269	0.1061	8.1762	77.0766	30
31	19.1943	0.0521	181.9434	0.0055	9.4790	0.1055	8.2962	78.6395	31
32	21.1138	0.0474	201.1378	0.0050	9.5264	0.1050	8.4091	80.1078	32
33	23.2252	0.0431	222.2515	0.0045	9.5694	0.1045	8.5152	81.4856	33
34	25.5477	0.0391	245.4767	0.0041	9.6086	0.1041	8.6149	82.7773	34
35	28.1024	0.0356	271.0244	0.0037	9.6442	0.1037	8.7086	83.9872	35
40	45.2593	0.0221	442.5926	0.0023	9.7791	0.1023	9.0962	88.9525	40
45	72.8905	0.0137	718.9048	0.0014	9.8628	0.1014	9.3740	92.4544	45
50	117.3909	0.0085	1163.9085	0.0009	9.9148	0.1009	9.5704	94.8889	50
55	189.0591	0.0053	1880.5914	0.0005	9.9471	0.1005	9.7075	96.5619	55
60	304.4816	0.0033	3034.8164	0.0003	9.9672	0.1003	9.8023	97.7010	60
65	490.3707	0.0020	4893.7073	0.0002	9.9796	0.1002	9.8672	98.4705	65
70	789.7470	0.0013	7887.4696	0.0001	9.9873	0.1001	9.9113	98.9870	70
75	1271.8954	0.0008	12708.9537	0.0001	9.9921	0.1001	9.9410	99.3317	75
80	2048.4002	0.0005	20474.0021	0.0000	9.9951	0.1000	9.9609	99.5606	80
85	3298.9690	0.0003	32979.6903	0.0000	9.9970	0.1000	9.9742	99.7120	85
90	5313.0226	0.0002	53120.2261	0.0000	9.9981	0.1000	9.9831	99.8118	90
95	8556.6760	0.0001	85556.7605	0.0000	9.9988	0.1000	9.9889	99.8773	95
100	13780.6123	0.0001	137796.1234	0.0000	9.9993	0.1000	9.9927	99.9202	100

11.0%

	Single Payment		Equal Payment Series				Gradient Series		
	Compound Amount Factor	Present Worth Factor	Compound Amount Factor	Sinking Fund Factor	Present Worth Factor	Capital Recovery Factor	Gradient Uniform Series	Gradient Present Worth	
N	(F/P,i,N)	(P/F,i,N)	(F/A,i,N)	(A/F,i,N)	(P/A,i,N)	(A/P,i,N)	(A/G,i,N)	(P/G,i,N)	N
1	1.1100	0.9009	1.0000	1.0000	0.9009	1.1100	0.0000	0.0000	1
2	1.2321	0.8116	2.1100	0.4739	1.7125	0.5839	0.4739	0.8116	2
3	1.3676	0.7312	3.3421	0.2992	2.4437	0.4092	0.9306	2.2740	3
4	1.5181	0.6587	4.7097	0.2123	3.1024	0.3223	1.3700	4.2502	4
5	1.6851	0.5935	6.2278	0.1606	3.6959	0.2706	1.7923	6.6240	5
6	1.8704	0.5346	7.9129	0.1264	4.2305	0.2364	2.1976	9.2972	6
7	2.0762	0.4817	9.7833	0.1022	4.7122	0.2122	2.5863	12.1872	7
8	2.3045	0.4339	11.8594	0.0843	5.1461	0.1943	2.9585	15.2246	8
9	2.5580	0.3909	14.1640	0.0706	5.5370	0.1806	3.3144	18.3520	9
10	2.8394	0.3522	16.7220	0.0598	5.8892	0.1698	3.6544	21.5217	10
11	3.1518	0.3173	19.5614	0.0511	6.2065	0.1611	3.9788	24.6945	11
12	3.4985	0.2858	22.7132	0.0440	6.4924	0.1540	4.2879	27.8388	12
13	3.8833	0.2575	26.2116	0.0382	6.7499	0.1482	4.5822	30.9290	13
14	4.3104	0.2320	30.0949	0.0332	6.9819	0.1432	4.8619	33.9449	14
15	4.7846	0.2090	34.4054	0.0291	7.1909	0.1391	5.1275	36.8709	15
16	5.3109	0.1883	39.1899	0.0255	7.3792	0.1355	5.3794	39.6953	16
17	5.8951	0.1696	44.5008	0.0225	7.5488	0.1325	5.6180	42.4095	17
18	6.5436	0.1528	50.3959	0.0198	7.7016	0.1298	5.8439	45.0074	18
19	7.2633	0.1377	56.9395	0.0176	7.8393	0.1276	6.0574	47.4856	19
20	8.0623	0.1240	64.2028	0.0156	7.9633	0.1256	6.2590	49.8423	20
21	8.9492	0.1117	72.2651	0.0138	8.0751	0.1238	6.4491	52.0771	21
22	9.9336	0.1007	81.2143	0.0123	8.1757	0.1223	6.6283	54.1912	22
23	11.0263	0.0907	91.1479	0.0110	8.2664	0.1210	6.7969	56.1864	23
24	12.2392	0.0817	102.1742	0.0098	8.3481	0.1198	6.9555	58.0656	24
25	13.5855	0.0736	114.4133	0.0087	8.4217	0.1187	7.1045	59.8322	25
26	15.0799	0.0663	127.9988	0.0078	8.4881	0.1178	7.2443	61.4900	26
27	16.7386	0.0597	143.0786	0.0070	8.5478	0.1170	7.3754	63.0433	27
28	18.5799	0.0538	159.8173	0.0063	8.6016	0.1163	7.4982	64.4965	28
29	20.6237	0.0485	178.3972	0.0056	8.6501	0.1156	7.6131	65.8542	29
30	22.8923	0.0437	199.0209	0.0050	8.6938	0.1150	7.7206	67.1210	30
31	25.4104	0.0394	221.9132	0.0045	8.7331	0.1145	7.8210	68.3016	31
32	28.2056	0.0355	247.3236	0.0040	8.7686	0.1140	7.9147	69.4007	32
33	31.3082	0.0319	275.5292	0.0036	8.8005	0.1136	8.0021	70.4228	33
34	34.7521	0.0288	306.8374	0.0033	8.8293	0.1133	8.0836	71.3724	34
35	38.5749	0.0259	341.5896	0.0029	8.8552	0.1129	8.1594	72.2538	35
40	65.0009	0.0154	581.8261	0.0017	8.9511	0.1117	8.4659	75.7789	40
45	109.5302	0.0091	986.6386	0.0010	9.0079	0.1110	8.6763	78.1551	45
50	184.5648	0.0054	1668.7712	0.0006	9.0417	0.1106	8.8185	79.7341	50
55	311.0025	0.0032	2818.2042	0.0004	9.0617	0.1104	8.9135	80.7712	55
60	524.0572	0.0019	4755.0658	0.0002	9.0736	0.1102	8.9762	81.4461	60

12.0%

	Single Payment		Equal Payment Series				Gradient Series		
	Compound Amount Factor	Present Worth Factor	Compound Amount Factor	Sinking Fund Factor	Present Worth Factor	Capital Recovery Factor	Gradient Uniform Series	Gradient Present Worth	
N	(F/P,i,N)	(P/F,i,N)	(F/A,i,N)	(A/F,i,N)	(P/A,i,N)	(A/P,i,N)	(A/G,i,N)	(P/G,i,N)	N
1	1.1200	0.8929	1.0000	1.0000	0.8929	1.1200	0.0000	0.0000	1
2	1.2544	0.7972	2.1200	0.4717	1.6901	0.5917	0.4717	0.7972	2
3	1.4049	0.7118	3.3744	0.2963	2.4018	0.4163	0.9246	2.2208	3
4	1.5735	0.6355	4.7793	0.2092	3.0373	0.3292	1.3589	4.1273	4
5	1.7623	0.5674	6.3528	0.1574	3.6048	0.2774	1.7746	6.3970	5
6	1.9738	0.5066	8.1152	0.1232	4.1114	0.2432	2.1720	8.9302	6
7	2.2107	0.4523	10.0890	0.0991	4.5638	0.2191	2.5515	11.6443	7
8	2.4760	0.4039	12.2997	0.0813	4.9676	0.2013	2.9131	14.4714	8
9	2.7731	0.3606	14.7757	0.0677	5.3282	0.1877	3.2574	17.3563	9
10	3.1058	0.3220	17.5487	0.0570	5.6502	0.1770	3.5847	20.2541	10
11	3.4785	0.2875	20.6546	0.0484	5.9377	0.1684	3.8953	23.1288	11
12	3.8960	0.2567	24.1331	0.0414	6.1944	0.1614	4.1897	25.9523	12
13	4.3635	0.2292	28.0291	0.0357	6.4235	0.1557	4.4683	28.7024	13
14	4.8871	0.2046	32.3926	0.0309	6.6282	0.1509	4.7317	31.3624	14
15	5.4736	0.1827	37.2797	0.0268	6.8109	0.1468	4.9803	33.9202	15
16	6.1304	0.1631	42.7533	0.0234	6.9740	0.1434	5.2147	36.3670	16
17	6.8660	0.1456	48.8837	0.0205	7.1196	0.1405	5.4353	38.6973	17
18	7.6900	0.1300	55.7497	0.0179	7.2497	0.1379	5.6427	40.9080	81
19	8.6128	0.1161	63.4397	0.0158	7.3658	0.1358	5.8375	42.9979	19
20	9.6463	0.1037	72.0524	0.0139	7.4694	0.1339	6.0202	44.9676	20
21	10.8038	0.0926	81.6987	0.0122	7.5620	0.1322	6.1913	46.8188	21
22	12.1003	0.0826	92.5026	0.0108	7.6446	0.1308	6.3514	48.5543	22
23	13.5523	0.0738	104.6029	0.0096	7.7184	0.1296	6.5010	50.1776	23
24	15.1786	0.0659	118.1552	0.0085	7.7843	0.1285	6.6406	51.6929	24
25	17.0001	0.0588	133.3339	0.0075	7.8431	0.1275	6.7708	53.1046	25
26	19.0401	0.0525	150.3339	0.0067	7.8957	0.1267	6.8921	54.4177	26
27	21.3249	0.0469	169.3740	0.0059	7.9426	0.1259	7.0049	55.6369	27
28	23.8839	0.0419	190.6989	0.0052	7.9844	0.1252	7.1098	56.7674	28
29	26.7499	0.0374	214.5828	0.0047	8.0218	0.1247	7.2071	57.8141	29
30	29.9599	0.0334	241.3327	0.0041	8.0552	0.1241	7.2974	58.7821	30
31	33.5551	0.0298	271.2926	0.0037	8.0850	0.1237	7.3811	59.6761	31
32	37.5817	0.0266	304.8477	0.0033	8.1116	0.1233	7.4586	60.5010	32
33	42.0915	0.0238	342.4294	0.0029	8.1354	0.1229	7.5302	61.2612	33
34	47.1425	0.0212	384.5210	0.0026	8.1566	0.1226	7.5965	61.9612	34
35	52.7996	0.0189	431.6635	0.0023	8.1755	0.1223	7.6577	62.6052	35
40	93.0510	0.0107	767.0914	0.0013	8.2438	0.1213	7.8988	65.1159	40
45	163.9876	0.0061	1358.2300	0.0007	8.2825	0.1207	8.0572	66.7342	45
50	289.0022	0.0035	2400.0182	0.0004	8.3045	0.1204	8.1597	67.7624	50
55	509.3206	0.0020	4236.0050	0.0002	8.3170	0.1202	8.2251	68.4082	55
60	897.5969	0.0011	7471.6411	0.0001	8.3240	0.1201	8.2664	68.8100	60

13.0%

	Single Payment		Equal Payment Series				Gradient Series		
	Compound Amount Factor	Present Worth Factor	Compound Amount Factor	Sinking Fund Factor	Present Worth Factor	Capital Recovery Factor	Gradient Uniform Series	Gradient Present Worth	
N	(F/P,i,N)	(P/F,i,N)	(F/A,i,N)	(A/F,i,N)	(P/A,i,N)	(A/P,i,N)	(A/G,i,N)	(P/G,i,N)	N
1	1.1300	0.8850	1.0000	1.0000	0.8850	1.1300	0.0000	0.0000	1
2	1.2769	0.7831	2.1300	0.4695	1.6681	0.5995	0.4695	0.7831	2
3	1.4429	0.6931	3.4069	0.2935	2.3612	0.4235	0.9187	2.1692	3
4	1.6305	0.6133	4.8498	0.2062	2.9745	0.3362	1.3479	4.0092	4
5	1.8424	0.5428	6.4803	0.1543	3.5172	0.2843	1.7571	6.1802	5
6	2.0820	0.4803	8.3227	0.1202	3.9975	0.2502	2.1468	8.5818	6
7	2.3526	0.4251	10.4047	0.0961	4.4226	0.2261	2.5171	11.1322	7
8	2.6584	0.3762	12.7573	0.0784	4.7988	0.2084	2.8685	13.7653	8
9	3.0040	0.3329	15.4157	0.0649	5.1317	0.1949	3.2014	16.4284	9
10	3.3946	0.2946	18.4197	0.0543	5.4262	0.1843	3.5162	19.0797	10
11	3.8359	0.2607	21.8143	0.0458	5.6869	0.1758	3.8134	21.6867	11
12	4.3345	0.2307	25.6502	0.0390	5.9176	0.1690	4.0936	24.2244	12
13	4.8980	0.2042	29.9847	0.0334	6.1218	0.1634	4.3573	26.6744	13
14	5.5348	0.1807	34.8827	0.0287	6.3025	0.1587	4.6050	29.0232	14
15	6.2543	0.1599	40.4175	0.0247	6.4624	0.1547	4.8375	31.2617	15
16	7.0673	0.1415	46.6717	0.0214	6.6039	0.1514	5.0552	33.3841	16
17	7.9861	0.1252	53.7391	0.0186	6.7291	0.1486	5.2589	35.3876	17
18	9.0243	0.1108	61.7251	0.0162	6.8399	0.1462	5.4491	37.2714	18
19	10.1974	0.0981	70.7494	0.0141	6.9380	0.1441	5.6265	39.0366	19
20	11.5231	0.0868	80.9468	0.0124	7.0248	0.1424	5.7917	40.6854	20
21	13.0211	0.0768	92.4699	0.0108	7.1016	0.1408	5.9454	42.2214	21
22	14.7138	0.0680	105.4910	0.0095	7.1695	0.1395	6.0881	43.6486	22
23	16.6266	0.0601	120.2048	0.0083	7.2297	0.1383	6.2205	44.9718	23
24	18.7881	0.0532	136.8315	0.0073	7.2829	0.1373	6.3431	46.1960	24
25	21.2305	0.0471	155.6196	0.0064	7.3300	0.1364	6.4566	47.3264	25
26	23.9905	0.0417	176.8501	0.0057	7.3717	0.1357	6.5614	48.3685	26
27	27.1093	0.0369	200.8406	0.0050	7.4086	0.1350	6.6582	49.3276	27
28	30.6335	0.0326	227.9499	0.0044	7.4412	0.1344	6.7474	50.2090	28
29	34.6158	0.0289	258.5834	0.0039	7.4701	0.1339	6.8296	51.0179	29
30	39.1159	0.0256	293.1992	0.0034	7.4957	0.1334	6.9052	51.7592	30
31	44.2010	0.0226	332.3151	0.0030	7.5183	0.1330	6.9747	52.4380	31
32	49.9471	0.0200	376.5161	0.0027	7.5383	0.1327	7.0385	53.0586	32
33	56.4402	0.0177	426.4632	0.0023	7.5560	0.1323	7.0971	53.6256	33
34	63.7774	0.0157	482.9034	0.0021	7.5717	0.1321	7.1507	54.1430	34
35	72.0685	0.0139	546.6808	0.0018	7.5856	0.1318	7.1998	54.6148	35
40	132.7816	0.0075	1013.7042	0.0010	7.6344	0.1310	7.3888	56.4087	40
45	244.6414	0.0041	1874.1646	0.0005	7.6609	0.1305	7.5076	57.5148	45
50	450.7359	0.0022	3459.5071	0.0003	7.6752	0.1303	7.5811	58.1870	50
55	830.4517	0.0012	6380.3979	0.0002	7.6830	0.1302	7.6260	58.5909	55
60	1530.0535	0.0007	11761.9498	0.0001	7.6873	0.1301	7.6531	58.8313	60

14.0%

	Single Payment		Equal Payment Series				Gradient Series		
N	Compound Amount Factor (F/P,i,N)	Present Worth Factor (P/F,i,N)	Compound Amount Factor (F/A,i,N)	Sinking Fund Factor (A/F,i,N)	Present Worth Factor (P/A,i,N)	Capital Recovery Factor (A/P,i,N)	Gradient Uniform Series (A/G,i,N)	Gradient Present Worth (P/G,i,N)	N
1	1.1400	0.8772	1.0000	1.0000	0.8772	1.1400	0.0000	0.0000	1
2	1.2996	0.7695	2.1400	0.4673	1.6467	0.6073	0.4673	0.7695	2
3	1.4815	0.6750	3.4396	0.2907	2.3216	0.4307	0.9129	2.1194	3
4	1.6890	0.5921	4.9211	0.2032	2.9137	0.3432	1.3370	3.8957	4
5	1.9254	0.5194	6.6101	0.1513	3.4331	0.2913	1.7399	5.9731	5
6	2.1950	0.4556	8.5355	0.1172	3.8887	0.2572	2.1218	8.2511	6
7	2.5023	0.3996	10.7305	0.0932	4.2883	0.2332	2.4832	10.6489	7
8	2.8526	0.3506	13.2328	0.0756	4.6389	0.2156	2.8246	13.1028	8
9	3.2519	0.3075	16.0853	0.0622	4.9464	0.2022	3.1463	15.5629	9
10	3.7072	0.2697	19.3373	0.0517	5.2161	0.1917	3.4490	17.9906	10
11	4.2262	0.2366	23.0445	0.0434	5.4527	0.1834	3.7333	20.3567	11
12	4.8179	0.2076	27.2707	0.0367	5.6603	0.1767	3.9998	22.6399	12
13	5.4924	0.1821	32.0887	0.0312	5.8424	0.1712	4.2491	24.8247	13
14	6.2613	0.1597	37.5811	0.0266	6.0021	0.1666	4.4819	26.9009	14
15	7.1379	0.1401	43.8424	0.0228	6.1422	0.1628	4.6990	28.8623	15
16	8.1372	0.1229	50.9804	0.0196	6.2651	0.1596	4.9011	30.7057	16
17	9.2765	0.1078	59.1176	0.0169	6.3729	0.1569	5.0888	32.4305	17
18	10.5752	0.0946	68.3941	0.0146	6.4674	0.1546	5.2630	34.0380	18
19	12.0557	0.0829	78.9692	0.0127	6.5504	0.1527	5.4243	35.5311	19
20	13.7435	0.0728	91.0249	0.0110	6.6231	0.1510	5.5734	36.9135	20
21	15.6676	0.0638	104.7684	0.0095	6.6870	0.1495	5.7111	38.1901	21
22	17.8610	0.0560	120.4360	0.0083	6.7429	0.1483	5.8381	39.3658	22
23	20.3616	0.0491	138.2970	0.0072	6.7921	0.1472	5.9549	40.4463	23
24	23.2122	0.0431	158.6586	0.0063	6.8351	0.1463	6.0624	41.4371	24
25	26.4619	0.0378	181.8708	0.0055	6.8729	0.1455	6.1610	42.3441	25
26	30.1666	0.0331	208.3327	0.0048	6.9061	0.1448	6.2514	43.1728	26
27	34.3899	0.0291	238.4993	0.0042	6.9352	0.1442	6.3342	43.9289	27
28	39.2045	0.0255	272.8892	0.0037	6.9607	0.1437	6.4100	44.6176	28
29	44.6931	0.0224	312.0937	0.0032	6.9830	0.1432	6.4791	45.2441	29
30	50.9502	0.0196	356.7868	0.0028	7.0027	0.1428	6.5423	45.8132	30
31	58.0832	0.0172	407.7370	0.0025	7.0199	0.1425	6.5998	46.3297	31
32	66.2148	0.0151	465.8202	0.0021	7.0350	0.1421	6.6522	46.7979	32
33	75.4849	0.0132	532.0350	0.0019	7.0482	0.1419	6.6998	47.2218	33
34	86.0528	0.0116	607.5199	0.0016	7.0599	0.1416	6.7431	47.6053	34
35	98.1002	0.0102	693.5727	0.0014	7.0700	0.1414	6.7824	47.9519	35
40	188.8835	0.0053	1342.0251	0.0007	7.1050	0.1407	6.9300	49.2376	40
45	363.6791	0.0027	2590.5648	0.0004	7.1232	0.1404	7.0188	49.9963	45
50	700.2330	0.0014	4994.5213	0.0002	7.1327	0.1402	7.0714	50.4375	50

15.0%

	Single Payment		Equal Payment Series				Gradient Series		
N	Compound Amount Factor (F/P,i,N)	Present Worth Factor (P/F,i,N)	Compound Amount Factor (F/A,i,N)	Sinking Fund Factor (A/F,i,N)	Present Worth Factor (P/A,i,N)	Capital Recovery Factor (A/P,i,N)	Gradient Uniform Series (A/G,i,N)	Gradient Present Worth (P/G,i,N)	N
1	1.1500	0.8696	1.0000	1.0000	0.8696	1.1500	0.0000	0.0000	1
2	1.3225	0.7561	2.1500	0.4651	1.6257	0.6151	0.4651	0.7561	2
3	1.5209	0.6575	3.4725	0.2880	2.2832	0.4380	0.9071	2.0712	3
4	1.7490	0.5718	4.9934	0.2003	2.8550	0.3503	1.3263	3.7864	4
5	2.0114	0.4972	6.7424	0.1483	3.3522	0.2983	1.7228	5.7751	5
6	2.3131	0.4323	8.7537	0.1142	3.7845	0.2642	2.0972	7.9368	6
7	2.6600	0.3759	11.0668	0.0904	4.1604	0.2404	2.4498	10.1924	7
8	3.0590	0.3269	13.7268	0.0729	4.4873	0.2229	2.7813	12.4807	8
9	3.5179	0.2843	16.7858	0.0596	4.7716	0.2096	3.0922	14.7548	9
10	4.0456	0.2472	20.3037	0.0493	5.0188	0.1993	3.3832	16.9795	10
11	4.6524	0.2149	24.3493	0.0411	5.2337	0.1911	3.6549	19.1289	11
12	5.3503	0.1869	29.0017	0.0345	5.4206	0.1845	3.9082	21.1849	12
13	6.1528	0.1625	34.3519	0.0291	5.5831	0.1791	4.1438	23.1352	13
14	7.0757	0.1413	40.5047	0.0247	5.7245	0.1747	4.3624	24.9725	14
15	8.1371	0.1229	47.5804	0.0210	5.8474	0.1710	4.5650	26.6930	15
16	9.3576	0.1069	55.7175	0.0179	5.9542	0.1679	4.7522	28.2960	16
17	10.7613	0.0929	65.0751	0.0154	6.0472	0.1654	4.9251	29.7828	17
18	12.3755	0.0808	75.8364	0.0132	6.1280	0.1632	5.0843	31.1565	18
19	14.2318	0.0703	88.2118	0.0113	6.1982	0.1613	5.2307	32.4213	19
20	16.3665	0.0611	102.4436	0.0098	6.2593	0.1598	5.3651	33.5822	20
21	18.8215	0.0531	118.8101	0.0084	6.3125	0.1584	5.4883	34.6448	21
22	21.6447	0.0462	137.6316	0.0073	6.3587	0.1573	5.6010	35.6150	22
23	24.8915	0.0402	159.2764	0.0063	6.3988	0.1563	5.7040	36.4988	23
24	28.6252	0.0349	184.1678	0.0054	6.4338	0.1554	5.7979	37.3023	24
25	32.9190	0.0304	212.7930	0.0047	6.4641	0.1547	5.8834	38.0314	25
26	37.8568	0.0264	245.7120	0.0041	6.4906	0.1541	5.9612	38.6918	26
27	43.5353	0.0230	283.5688	0.0035	6.5135	0.1535	6.0319	39.2890	27
28	50.0656	0.0200	327.1041	0.0031	6.5335	0.1531	6.0960	39.8283	28
29	57.5755	0.0174	377.1697	0.0027	6.5509	0.1527	6.1541	40.3146	29
30	66.2118	0.0151	434.7451	0.0023	6.5660	0.1523	6.2066	40.7526	30
31	76.1435	0.0131	500.9569	0.0020	6.5791	0.1520	6.2541	41.1466	31
32	87.5651	0.0114	577.1005	0.0017	6.5905	0.1517	6.2970	41.5006	32
33	100.6998	0.0099	664.6655	0.0015	6.6005	0.1515	6.3357	41.8184	33
34	115.8048	0.0086	765.3654	0.0013	6.6091	0.1513	6.3705	42.1033	34
35	133.1755	0.0075	881.1702	0.0011	6.6166	0.1511	6.4019	42.3586	35
40	267.8635	0.0037	1779.0903	0.0006	6.6418	0.1506	6.5168	43.2830	40
45	538.7693	0.0019	3585.1285	0.0003	6.6543	0.1503	6.5830	43.8051	45
50	1083.6574	0.0009	7217.7163	0.0001	6.6605	0.1501	6.6205	44.0958	50

16.0%

	Single Payment		Equal Payment Series				Gradient Series		
	Compound Amount Factor	Present Worth Factor	Compound Amount Factor	Sinking Fund Factor	Present Worth Factor	Capital Recovery Factor	Gradient Uniform Series	Gradient Present Worth	
N	(F/P,i,N)	(P/F,i,N)	(F/A,i,N)	(A/F,i,N)	(P/A,i,N)	(A/P,i,N)	(A/G,i,N)	(P/G,i,N)	N
1	1.1600	0.8621	1.0000	1.0000	0.8621	1.1600	0.0000	0.0000	1
2	1.3456	0.7432	2.1600	0.4630	1.6052	0.6230	0.4630	0.7432	2
3	1.5609	0.6407	3.5056	0.2853	2.2459	0.4453	0.9014	2.0245	3
4	1.8106	0.5523	5.0665	0.1974	2.7982	0.3574	1.3156	3.6814	4
5	2.1003	0.4761	6.8771	0.1454	3.2743	0.3054	1.7060	5.5858	5
6	2.4364	0.4104	8.9775	0.1114	3.6847	0.2714	2.0729	7.6380	6
7	2.8262	0.3538	11.4139	0.0876	4.0386	0.2476	2.4169	9.7610	7
8	3.2784	0.3050	14.2401	0.0702	4.3436	0.2302	2.7388	11.8962	8
9	3.8030	0.2630	17.5185	0.0571	4.6065	0.2171	3.0391	13.9998	9
10	4.4114	0.2267	21.3215	0.0469	4.8332	0.2069	3.3187	16.0399	10
11	5.1173	0.1954	25.7329	0.0389	5.0286	0.1989	3.5783	17.9941	11
12	5.9360	0.1685	30.8502	0.0324	5.1971	0.1924	3.8189	19.8472	12
13	6.8858	0.1452	36.7862	0.0272	5.3423	0.1872	4.0413	21.5899	13
14	7.9875	0.1252	43.6720	0.0229	5.4675	0.1829	4.2464	23.2175	14
15	9.2655	0.1079	51.6595	0.0194	5.5755	0.1794	4.4352	24.7284	15
16	10.7480	0.0930	60.9650	0.0164	5.6685	0.1764	4.6086	26.1241	16
17	12.4677	0.0802	71.6730	0.0140	5.7487	0.1740	4.7676	27.4074	17
18	14.4625	0.0691	84.1407	0.0119	5.8178	0.1719	4.9130	28.5828	18
19	16.7765	0.0596	98.6032	0.0101	5.8775	0.1701	5.0457	29.6557	19
20	19.4608	0.0514	115.3797	0.0087	5.9288	0.1687	5.1666	30.6321	20
21	22.5745	0.0443	134.8405	0.0074	5.9731	0.1674	5.2766	31.5180	21
22	26.1864	0.0382	157.4150	0.0064	6.0113	0.1664	5.3765	32.3200	22
23	30.3762	0.0329	183.6014	0.0054	6.0442	0.1654	5.4671	33.0442	23
24	35.2364	0.0284	213.9776	0.0047	6.0726	0.1647	5.5490	33.6970	24
25	40.8742	0.0245	249.2140	0.0040	6.0971	0.1640	5.6230	34.2841	25
26	47.4141	0.0211	290.0883	0.0034	6.1182	0.1634	5.6898	34.8114	26
27	55.0004	0.0182	337.5024	0.0030	6.1364	0.1630	5.7500	35.2841	27
28	63.8004	0.0157	392.5028	0.0025	6.1520	0.1625	5.8041	35.7073	28
29	74.0085	0.0135	456.3032	0.0022	6.1656	0.1622	5.8528	36.0856	29
30	85.8499	0.0116	530.3117	0.0019	6.1772	0.1619	5.8964	36.4234	30
31	99.5859	0.0100	616.1616	0.0016	6.1872	0.1616	5.9356	36.7247	31
32	115.5196	0.0087	715.7475	0.0014	6.1959	0.1614	5.9706	36.9930	32
33	134.0027	0.0075	831.2671	0.0012	6.2034	0.1612	6.0019	27.2318	33
34	155.4432	0.0064	965.2698	0.0010	6.2098	0.1610	6.0299	37.4441	34
35	180.3141	0.0055	1120.7130	0.0009	6.2153	0.1609	6.0548	37.6327	35
40	378.7212	0.0026	2360.7572	0.0004	6.2335	0.1604	6.1441	38.2992	40
45	795.4438	0.0013	4965.2739	0.0002	6.2421	0.1602	6.1934	38.6598	45
50	1670.7038	0.0006	10435.6488	0.0001	6.2463	0.1601	6.2201	38.8521	50

18.0%

	Single Payment		Equal Payment Series				Gradient Series		
N	Compound Amount Factor (F/P,i,N)	Present Worth Factor (P/F,i,N)	Compound Amount Factor (F/A,i,N)	Sinking Fund Factor (A/F,i,N)	Present Worth Factor (P/A,i,N)	Capital Recovery Factor (A/P,i,N)	Gradient Uniform Series (A/G,i,N)	Gradient Present Worth (P/G,i,N)	N
1	1.1800	0.8475	1.0000	1.0000	0.8475	1.1800	0.0000	0.0000	1
2	1.3924	0.7182	2.1800	0.4587	1.5656	0.6387	0.4587	0.7182	2
3	1.6430	0.6086	3.5724	0.2799	2.1743	0.4599	0.8902	1.9354	3
4	1.9388	0.5158	5.2154	0.1917	2.6901	0.3717	1.2947	3.4828	4
5	2.2878	0.4371	7.1542	0.1398	3.1272	0.3198	1.6728	5.2312	5
6	2.6996	0.3704	9.4420	0.1059	3.4976	0.2859	2.0252	7.0834	6
7	3.1855	0.3139	12.1415	0.0824	3.8115	0.2624	2.3526	8.9670	7
8	3.7589	0.2660	15.3270	0.0652	4.0776	0.2452	2.6558	10.8292	8
9	4.4355	0.2255	19.0859	0.0524	4.3030	0.2324	2.9358	12.6329	9
10	5.2338	0.1911	23.5213	0.0425	4.4941	0.2225	3.1936	14.3525	10
11	6.1759	0.1619	28.7551	0.0348	4.6560	0.2148	3.4303	15.9716	11
12	7.2876	0.1372	34.9311	0.0286	4.7932	0.2086	3.6470	17.4811	12
13	8.5994	0.1163	42.2187	0.0237	4.9095	0.2037	3.8449	18.8765	13
14	10.1472	0.0985	50.8180	0.0197	5.0081	0.1997	4.0250	20.1576	14
15	11.9737	0.0835	60.9653	0.0164	5.0916	0.1964	4.1887	21.3269	15
16	14.1290	0.0708	72.9390	0.0137	5.1624	0.1937	4.3369	22.3885	16
17	16.6722	0.0600	87.0680	0.0115	5.2223	0.1915	4.4708	23.3482	17
18	19.6733	0.0508	103.7403	0.0096	5.2732	0.1896	4.5916	24.2123	18
19	23.2144	0.0431	123.4135	0.0081	5.3162	0.1881	4.7003	24.9877	19
20	27.3930	0.0365	146.6280	0.0068	5.3527	0.1868	4.7978	25.6813	20
21	32.3238	0.0309	174.0210	0.0057	5.3837	0.1857	4.8851	26.3000	21
22	38.1421	0.0262	206.3448	0.0048	5.4099	0.1848	4.9632	26.8506	22
23	45.0076	0.0222	244.4868	0.0041	5.4321	0.1841	5.0329	27.3394	23
24	53.1090	0.0188	289.4945	0.0035	5.4509	0.1835	5.0950	27.7725	24
25	62.6686	0.0160	342.6035	0.0029	5.4669	0.1829	5.1502	28.1555	25
26	73.9490	0.0135	405.2721	0.0025	5.4804	0.1825	5.1991	28.4935	26
27	87.2598	0.0115	479.2211	0.0021	5.4919	0.1821	5.2425	28.7915	27
28	102.9666	0.0097	566.4809	0.0018	5.5016	0.1818	5.2810	29.0537	28
29	121.5005	0.0082	669.4475	0.0015	5.5098	0.1815	5.3149	29.2842	29
30	143.3706	0.0070	790.9480	0.0013	5.5168	0.1813	5.3448	29.4864	30
31	169.1774	0.0059	934.3186	0.0011	5.5227	0.1811	5.3712	29.6638	31
32	199.6293	0.0050	1103.4960	0.0009	5.5277	0.1809	5.3945	29.8191	32
33	235.5625	0.0042	1303.1253	0.0008	5.5320	0.1808	5.4149	29.9549	33
34	277.9638	0.0036	1538.6878	0.0006	5.5356	0.1806	5.4328	30.0736	34
35	327.9973	0.0030	1816.6516	0.0006	5.5386	0.1806	5.4485	30.1773	35
40	750.3783	0.0013	4163.2130	0.0002	5.5482	0.1802	5.5022	30.5269	40
45	1716.6839	0.0006	9531.5771	0.0001	5.5523	0.1801	5.5293	30.7006	45
50	3927.3569	0.0003	21813.0937	0.0000	5.5541	0.1800	5.5428	30.7856	50

20.0%

	Single Payment		Equal Payment Series				Gradient Series		
	Compound Amount Factor	Present Worth Factor	Compound Amount Factor	Sinking Fund Factor	Present Worth Factor	Capital Recovery Factor	Gradient Uniform Series	Gradient Present Worth	
N	(F/P,i,N)	(P/F,i,N)	(F/A,i,N)	(A/F,i,N)	(P/A,i,N)	(A/P,i,N)	(A/G,i,N)	(P/G,i,N)	N
1	1.2000	0.8333	1.0000	1.0000	0.8333	1.2000	0.0000	0.0000	1
2	1.4400	0.6944	2.2000	0.4545	1.5278	0.6545	0.4545	0.6944	2
3	1.7280	0.5787	3.6400	0.2747	2.1065	0.4747	0.8791	1.8519	3
4	2.0736	0.4823	5.3680	0.1863	2.5887	0.3863	1.2742	3.2986	4
5	2.4883	0.4019	7.4416	0.1344	2.9906	0.3344	1.6405	4.9061	5
6	2.9860	0.3349	9.9299	0.1007	3.3255	0.3007	1.9788	6.5806	6
7	3.5832	0.2791	12.9159	0.0774	3.6046	0.2774	2.2902	8.2551	7
8	4.2998	0.2326	16.4991	0.0606	3.8372	0.2606	2.5756	9.8831	8
9	5.1598	0.1938	20.7989	0.0481	4.0310	0.2481	2.8364	11.4335	9
10	6.1917	0.1615	25.9587	0.0385	4.1925	0.2385	3.0739	12.8871	10
11	7.4301	0.1346	32.1504	0.0311	4.3271	0.2311	3.2893	14.2330	11
12	8.9161	0.1122	39.5805	0.0253	4.4392	0.2253	3.4841	15.4667	12
13	10.6993	0.0935	48.4966	0.0206	4.5327	0.2206	3.6597	16.5883	13
14	12.8392	0.0779	59.1959	0.0169	4.6106	0.2169	3.8175	17.6008	14
15	15.4070	0.0649	72.0351	0.0139	4.6755	0.2139	3.9588	18.5095	15
16	18.4884	0.0541	87.4421	0.0114	4.7296	0.2114	4.0851	19.3208	16
17	22.1861	0.0451	105.9306	0.0094	4.7746	0.2094	4.1976	20.0419	17
18	26.6233	0.0376	128.1167	0.0078	4.8122	0.2078	4.2975	20.6805	18
19	31.9480	0.0313	154.7400	0.0065	4.8435	0.2065	4.3861	21.2439	19
20	38.3376	0.0261	186.6880	0.0054	4.8696	0.2054	4.4643	21.7395	20
21	46.0051	0.0217	225.0256	0.0044	4.8913	0.2044	4.5334	22.1742	21
22	55.2061	0.0181	271.0307	0.0037	4.9094	0.2037	4.5941	22.5546	22
23	66.2474	0.0151	326.2369	0.0031	4.9245	0.2031	4.6475	22.8867	23
24	79.4968	0.0126	392.4842	0.0025	4.9371	0.2025	4.6943	23.1760	24
25	95.3962	0.0105	471.9811	0.0021	4.9476	0.2021	4.7352	23.4276	25
26	114.4755	0.0087	567.3773	0.0018	4.9563	0.2018	4.7709	23.6460	26
27	137.3706	0.0073	681.8528	0.0015	4.9636	0.2015	4.8020	23.8353	27
28	164.8447	0.0061	819.2233	0.0012	4.9697	0.2012	4.8291	23.9991	28
29	197.8136	0.0051	984.0680	0.0010	4.9747	0.2010	4.8527	24.1406	29
30	237.3763	0.0042	1181.8816	0.0008	4.9789	0.2008	4.8731	24.2628	30
31	284.8516	0.0035	1419.2579	0.0007	4.9824	0.2007	4.8908	24.3681	31
32	341.8219	0.0029	1704.1095	0.0006	4.9854	0.2006	4.9061	24.4588	32
33	410.1863	0.0024	2045.9314	0.0005	4.9878	0.2005	4.9194	24.5368	33
34	492.2235	0.0020	2456.1176	0.0004	4.9898	0.2004	4.9308	24.6038	34
35	590.6682	0.0017	2948.3411	0.0003	4.9915	0.2003	4.9406	24.6614	35
40	1469.7716	0.0007	7343.8578	0.0001	4.9966	0.2001	4.9728	24.8469	40
45	3657.2620	0.0003	18281.3099	0.0001	4.9986	0.2001	4.9877	24.9316	45

25.0%

	Single Payment		Equal Payment Series				Gradient Series		
	Compound Amount Factor	Present Worth Factor	Compound Amount Factor	Sinking Fund Factor	Present Worth Factor	Capital Recovery Factor	Gradient Uniform Series	Gradient Present Worth	
N	(F/P,i,N)	(P/F,i,N)	(F/A,i,N)	(A/F,i,N)	(P/A,i,N)	(A/P,i,N)	(A/G,i,N)	(P/G,i,N)	N
1	1.2500	0.8000	1.0000	1.0000	0.8000	1.2500	0.0000	0.0000	1
2	1.5625	0.6400	2.2500	0.4444	1.4400	0.6944	0.4444	0.6400	2
3	1.9531	0.5120	3.8125	0.2623	1.9520	0.5123	0.8525	1.6640	3
4	2.4414	0.4096	5.7656	0.1734	2.3616	0.4234	1.2249	2.8928	4
5	3.0518	0.3277	8.2070	0.1218	2.6893	0.3718	1.5631	4.2035	5
6	3.8147	0.2621	11.2588	0.0888	2.9514	0.3388	1.8683	5.5142	6
7	4.7684	0.2097	15.0735	0.0663	3.1611	0.3163	2.1424	6.7725	7
8	5.9605	0.1678	19.8419	0.0504	3.3289	0.3004	2.3872	7.9469	8
9	7.4506	0.1342	25.8023	0.0388	3.4631	0.2888	2.6048	9.0207	9
10	9.3132	0.1074	33.2529	0.0301	3.5705	0.2801	2.7971	9.9870	10
11	11.6415	0.0859	42.5661	0.0235	3.6564	0.2735	2.9663	10.8460	11
12	14.5519	0.0687	54.2077	0.0184	3.7251	0.2684	3.1145	11.6020	12
13	18.1899	0.0550	68.7596	0.0145	3.7801	0.2645	3.2437	12.2617	13
14	22.7374	0.0440	86.9495	0.0115	3.8241	0.2615	3.3559	12.8334	14
15	28.4217	0.0352	109.6868	0.0091	3.8593	0.2591	3.4530	13.3260	15
16	35.5271	0.0281	138.1085	0.0072	3.8874	0.2572	3.5366	13.7482	16
17	44.4089	0.0225	173.6357	0.0058	3.9099	0.2558	3.6084	14.1085	17
18	55.5112	0.0180	218.0446	0.0046	3.9279	0.2546	3.6698	14.4147	18
19	69.3889	0.0144	273.5558	0.0037	3.9424	0.2537	3.7222	14.6741	19
20	86.7362	0.0115	342.9447	0.0029	3.9539	0.2529	3.7667	14.8932	20
21	108.4202	0.0092	429.6809	0.0023	3.9631	0.2523	3.8045	15.0777	21
22	135.5253	0.0074	538.1011	0.0019	3.9705	0.2519	3.8365	15.2326	22
23	169.4066	0.0059	673.6264	0.0015	3.9764	0.2515	3.8634	15.3625	23
24	211.7582	0.0047	843.0329	0.0012	3.9811	0.2512	3.8861	15.4711	24
25	264.6978	0.0038	1054.7912	0.0009	3.9849	0.2509	3.9052	15.5618	25
26	330.8722	0.0030	1319.4890	0.0008	3.9879	0.2508	3.9212	15.6373	26
27	413.5903	0.0024	1650.3612	0.0006	3.9903	0.2506	3.9346	15.7002	27
28	516.9879	0.0019	2063.9515	0.0005	3.9923	0.2505	3.9457	15.7524	28
29	646.2349	0.0015	2580.9394	0.0004	3.9938	0.2504	3.9551	15.7957	29
30	807.7936	0.0012	3227.1743	0.0003	3.9950	0.2503	3.9628	15.8316	30
31	1009.7420	0.0010	4034.9678	0.0002	3.9960	0.2502	3.9693	15.8614	31
32	1262.1774	0.0008	5044.7098	0.0002	3.9968	0.2502	3.9746	15.8859	32
33	1577.7218	0.0006	6306.8872	0.0002	3.9975	0.2502	3.9791	15.9062	33
34	1972.1523	0.0005	7884.6091	0.0001	3.9980	0.2501	3.9828	15.9229	34
35	2465.1903	0.0004	9856.7613	0.0001	3.9984	0.2501	3.9858	15.9367	35
40	7523.1638	0.0001	30088.6554	0.0000	3.9995	0.2500	3.9947	15.9766	40

30.0%

	Single Payment		Equal Payment Series				Gradient Series		
N	Compound Amount Factor (F/P,i,N)	Present Worth Factor (P/F,i,N)	Compound Amount Factor (F/A,i,N)	Sinking Fund Factor (A/F,i,N)	Present Worth Factor (P/A,i,N)	Capital Recovery Factor (A/P,i,N)	Gradient Uniform Series (A/G,i,N)	Gradient Present Worth (P/G,i,N)	N
1	1.3000	0.7692	1.0000	1.0000	0.7692	1.3000	0.0000	0.0000	1
2	1.6900	0.5917	2.3000	0.4348	1.3609	0.7348	0.4348	0.5917	2
3	2.1970	0.4552	3.9900	0.2506	1.8161	0.5506	0.8271	1.5020	3
4	2.8561	0.3501	6.1870	0.1616	2.1662	0.4616	1.1783	2.5524	4
5	3.7129	0.2693	9.0431	0.1106	2.4356	0.4106	1.4903	3.6297	5
6	4.8268	0.2072	12.7560	0.0784	2.6427	0.3784	1.7654	4.6656	6
7	6.2749	0.1594	17.5828	0.0569	2.8021	0.3569	2.0063	5.6218	7
8	8.1573	0.1226	23.8577	0.0419	2.9247	0.3419	2.2156	6.4800	8
9	10.6045	0.0943	32.0150	0.0312	3.0190	0.3312	2.3963	7.2343	9
10	13.7858	0.0725	42.6195	0.0235	3.0915	0.3235	2.5512	7.8872	10
11	17.9216	0.0558	56.4053	0.0177	3.1473	0.3177	2.6833	8.4452	11
12	23.2981	0.0429	74.3270	0.0135	3.1903	0.3135	2.7952	8.9173	12
13	30.2875	0.0330	97.6250	0.0102	3.2233	0.3102	2.8895	9.3135	13
14	39.3738	0.0254	127.9125	0.0078	3.2487	0.3078	2.9685	9.6437	14
15	51.1859	0.0195	167.2863	0.0060	3.2682	0.3060	3.0344	9.9172	15
16	66.5417	0.0150	218.4722	0.0046	3.2832	0.3046	3.0892	10.1426	16
17	86.5042	0.0116	285.0139	0.0035	3.2948	0.3035	3.1345	10.3276	17
18	112.4554	0.0089	371.5180	0.0027	3.3037	0.3027	3.1718	10.4788	18
19	146.1920	0.0068	483.9734	0.0021	3.3105	0.3021	3.2025	10.6019	19
20	190.0496	0.0053	630.1655	0.0016	3.3158	0.3016	3.2275	10.7019	20
21	247.0645	0.0040	820.2151	0.0012	3.3198	0.3012	3.2480	10.7828	21
22	321.1839	0.0031	1067.2796	0.0009	3.3230	0.3009	3.2646	10.8482	22
23	417.5391	0.0024	1388.4635	0.0007	3.3254	0.3007	3.2781	10.9009	23
24	542.8008	0.0018	1806.0026	0.0006	3.3272	0.3006	3.2890	10.9433	24
25	705.6410	0.0014	2348.8033	0.0004	3.3286	0.3004	3.2979	10.9773	25
26	917.3333	0.0011	3054.4443	0.0003	3.3297	0.3003	3.3050	11.0045	26
27	1192.5333	0.0008	3971.7776	0.0003	3.3305	0.3003	3.3107	11.0263	27
28	1550.2933	0.0006	5164.3109	0.0002	3.3312	0.3002	3.3153	11.0437	28
29	2015.3813	0.0005	6714.6042	0.0001	3.3317	0.3001	3.3189	11.0576	29
30	2619.9956	0.0004	8729.9855	0.0001	3.3321	0.3001	3.3219	11.0687	30
31	3405.9943	0.0003	11349.9811	0.0001	3.3324	0.3001	3.3242	11.0775	31
32	4427.7926	0.0002	14755.9755	0.0001	3.3326	0.3001	3.3261	11.0845	32
33	5756.1304	0.0002	19183.7681	0.0001	3.3328	0.3001	3.3276	11.0901	33
34	7482.9696	0.0001	24939.8985	0.0000	3.3329	0.3000	3.3288	11.0945	34
35	9727.8604	0.0001	32422.8681	0.0000	3.3330	0.3000	3.3297	11.0980	35

35.0%

	Single Payment		Equal Payment Series				Gradient Series		
N	Compound Amount Factor (F/P,i,N)	Present Worth Factor (P/F,i,N)	Compound Amount Factor (F/A,i,N)	Sinking Fund Factor (A/F,i,N)	Present Worth Factor (P/A,i,N)	Capital Recovery Factor (A/P,i,N)	Gradient Uniform Series (A/G,i,N)	Gradient Present Worth (P/G,i,N)	N
1	1.3500	0.7407	1.0000	1.0000	0.7407	1.3500	0.0000	0.0000	1
2	1.8225	0.5487	2.3500	0.4255	1.2894	0.7755	0.4255	0.5487	2
3	2.4604	0.4064	4.1725	0.2397	1.6959	0.5897	0.8029	1.3616	3
4	3.3215	0.3011	6.6329	0.1508	1.9969	0.5008	1.1341	2.2648	4
5	4.4840	0.2230	9.9544	0.1005	2.2200	0.4505	1.4220	3.1568	5
6	6.0534	0.1652	14.4384	0.0693	2.3852	0.4193	1.6698	3.9828	6
7	8.1722	0.1224	20.4919	0.0488	2.5075	0.3988	1.8811	4.7170	7
8	11.0324	0.0906	28.6640	0.0349	2.5982	0.3849	2.0597	5.3515	8
9	14.8937	0.0671	39.6964	0.0252	2.6653	0.3752	2.2094	5.8886	9
10	20.1066	0.0497	54.5902	0.0183	2.7150	0.3683	2.3338	6.3363	10
11	27.1439	0.0368	74.6967	0.0134	2.7519	0.3634	2.4364	6.7047	11
12	36.6442	0.0273	101.8406	0.0098	2.7792	0.3598	2.5205	7.0049	12
13	49.4697	0.0202	138.4848	0.0072	2.7994	0.3572	2.5889	7.2474	13
14	66.7841	0.0150	187.9544	0.0053	2.8144	0.3553	2.6443	7.4421	14
15	90.1585	0.0111	254.7385	0.0039	2.8255	0.3539	2.6889	7.5974	15
16	121.7139	0.0082	344.8970	0.0029	2.8337	0.3529	2.7246	7.7206	16
17	164.3138	0.0061	466.6109	0.0021	2.8398	0.3521	2.7530	7.8180	17
18	221.8236	0.0045	630.9247	0.0016	2.8443	0.3516	2.7756	7.8946	18
19	299.4619	0.0033	852.7483	0.0012	2.8476	0.3512	2.7935	2.9547	19
20	404.2736	0.0025	1152.2103	0.0009	2.8501	0.3509	2.8075	8.0017	20
21	545.7693	0.0018	1556.4838	0.0006	2.8519	0.3506	2.8186	8.0384	21
22	736.7886	0.0014	2102.2532	0.0005	2.8533	0.3505	2.8272	8.0669	22
23	994.6646	0.0010	2839.0418	0.0004	2.8543	0.3504	2.8340	8.0890	23
24	1342.7973	0.0007	3833.7064	0.0003	2.8550	0.3503	2.8393	8.1061	24
25	1812.7763	0.0006	5176.5037	0.0002	2.8556	0.3502	2.8433	8.1194	25
26	2447.2480	0.0004	6989.2800	0.0001	2.8560	0.3501	2.8465	8.1296	26
27	3303.7848	0.0003	9436.5280	0.0001	2.8563	0.3501	2.8490	8.1374	27
28	4460.1095	0.0002	12740.3128	0.0001	2.8565	0.3501	2.8509	8.1435	28
29	6021.1478	0.0002	17200.4222	0.0001	2.8567	0.3501	2.8523	8.1481	29
30	8128.5495	0.0001	23221.5700	0.0000	2.8568	0.3500	2.8535	8.1517	30

40.0%

	Single Payment		Equal Payment Series				Gradient Series		
N	Compound Amount Factor (F/P,i,N)	Present Worth Factor (P/F,i,N)	Compound Amount Factor (F/A,i,N)	Sinking Fund Factor (A/F,i,N)	Present Worth Factor (P/A,i,N)	Capital Recovery Factor (A/P,i,N)	Gradient Uniform Series (A/G,i,N)	Gradient Present Worth (P/G,i,N)	**N**
1	1.4000	0.7143	1.0000	1.0000	0.7143	1.4000	0.0000	0.0000	1
2	1.9600	0.5102	2.4000	0.4167	1.2245	0.8167	0.4167	0.5102	2
3	2.7440	0.3644	4.3600	0.2294	1.5889	0.6294	0.7798	1.2391	3
4	3.8416	0.2603	7.1040	0.1408	1.8492	0.5408	1.0923	2.0200	4
5	5.3782	0.1859	10.9456	0.0914	2.0352	0.4914	1.3580	2.7637	5
6	7.5295	0.1328	16.3238	0.0613	2.1680	0.4613	1.5811	3.4278	6
7	10.5414	0.0949	23.8534	0.0419	2.2628	0.4419	1.7664	3.9970	7
8	14.7579	0.0678	34.3947	0.0291	2.3306	0.4291	1.9185	4.4713	8
9	20.6610	0.0484	49.1526	0.0203	2.3790	0.4203	2.0422	4.8585	9
10	28.9255	0.0346	69.8137	0.0143	2.4136	0.4143	2.1419	5.1696	10
11	40.4957	0.0247	98.7391	0.0101	2.4383	0.4101	2.2215	5.4166	11
12	56.6939	0.0176	139.2348	0.0072	2.4559	0.4072	2.2845	5.6106	12
13	79.3715	0.0126	195.9287	0.0051	2.4685	0.4051	2.3341	5.7618	13
14	111.1201	0.0090	275.3002	0.0036	2.4775	0.4036	2.3729	5.8788	14
15	155.5681	0.0064	386.4202	0.0026	2.4839	0.4026	2.4030	5.9688	15
16	217.7953	0.0046	541.9883	0.0018	2.4885	0.4018	2.4262	6.0376	16
17	304.9135	0.0033	759.7837	0.0013	2.4918	0.4013	2.4441	6.0901	17
18	426.8789	0.0023	1064.6971	0.0009	2.4941	0.4009	2.4577	6.1299	18
19	597.6304	0.0017	1491.5760	0.0007	2.4958	0.4007	2.4682	6.1601	19
20	836.6826	0.0012	2089.2064	0.0005	2.4970	0.4005	2.4761	6.1828	20
21	1171.3556	0.0009	2925.8889	0.0003	2.4979	0.4003	2.4821	6.1998	21
22	1639.8978	0.0006	4097.2445	0.0002	2.4985	0.4002	2.4866	6.2127	22
23	2295.8569	0.0004	5737.1423	0.0002	2.4989	0.4002	2.4900	6.2222	23
24	3214.1997	0.0003	8032.9993	0.0001	2.4992	0.4001	2.4925	6.2294	24
25	4499.8796	0.0002	11247.1990	0.0001	2.4994	0.4001	2.4944	6.2347	25
26	6299.8314	0.0002	15747.0785	0.0001	2.4996	0.4001	2.4959	6.2387	26
27	8819.7640	0.0001	22046.9099	0.0000	2.4997	0.4000	2.4969	6.2416	27
28	12347.6696	0.0001	30866.6739	0.0000	2.4998	0.4000	2.4977	6.2438	28
29	17286.7374	0.0001	43214.3435	0.0000	2.4999	0.4000	2.4983	6.2454	29
30	24201.4324	0.0000	60501.0809	0.0000	2.4999	0.4000	2.4988	6.2466	30

50.0%

	Single Payment		Equal Payment Series				Gradient Series		
	Compound Amount Factor	Present Worth Factor	Compound Amount Factor	Sinking Fund Factor	Present Worth Factor	Capital Recovery Factor	Gradient Uniform Series	Gradient Present Worth	
N	(F/P,i,N)	(P/F,i,N)	(F/A,i,N)	(A/F,i,N)	(P/A,i,N)	(A/P,i,N)	(A/G,i,N)	(P/G,i,N)	N
1	1.5000	0.6667	1.0000	1.0000	0.6667	1.5000	0.0000	0.0000	1
2	2.2500	0.4444	2.5000	0.4000	1.1111	0.9000	0.4000	0.4444	2
3	3.3750	0.2963	4.7500	0.2105	1.4074	0.7105	0.7368	1.0370	3
4	5.0625	0.1975	8.1250	0.1231	1.6049	0.6231	1.0154	1.6296	4
5	7.5938	0.1317	13.1875	0.0758	1.7366	0.5758	1.2417	2.1564	5
6	11.3906	0.0878	20.7813	0.0481	1.8244	0.5481	1.4226	2.5953	6
7	17.0859	0.0585	32.1719	0.0311	1.8829	0.5311	1.5648	2.9465	7
8	25.6289	0.0390	49.2578	0.0203	1.9220	0.5203	1.6752	3.2196	8
9	38.4434	0.0260	74.8867	0.0134	1.9480	0.5134	1.7596	3.4277	9
10	57.6650	0.0173	113.3301	0.0088	1.9653	0.5088	1.8235	3.5838	10
11	86.4976	0.0116	170.9951	0.0058	1.9769	0.5058	1.8713	3.6994	11
12	129.7463	0.0077	257.4927	0.0039	1.9846	0.5039	1.9068	3.7842	12
13	194.6195	0.0051	387.2390	0.0026	1.9897	0.5026	1.9329	3.8459	13
14	291.9293	0.0034	581.8585	0.0017	1.9931	0.5017	1.9519	3.8904	14
15	437.8939	0.0023	873.7878	0.0011	1.9954	0.5011	1.9657	3.9224	15
16	656.8408	0.0015	1311.6817	0.0008	1.9970	0.5008	1.9756	3.9452	16
17	985.2613	0.0010	1968.5225	0.0005	1.9980	0.5005	1.9827	3.9614	17
18	1477.8919	0.0007	2953.7838	0.0003	1.9986	0.5003	1.9878	3.9729	18
19	2216.8378	0.0005	4431.6756	0.0002	1.9991	0.5002	1.9914	3.9811	19
20	3325.2567	0.0003	6648.5135	0.0002	1.9994	0.5002	1.9940	3.9868	20

INDEX

Q

Summary of Useful Excel's Financial Functions (Part A)

Description		Excel Function	Example	Solution
Single-Payment	Find: F Given: P	$=FV(i\%, N, 0, -P)$	Find the future worth of $500 in 5 years at 8%.	$=FV(8\%, 5, 0, -500)$ $=\$734.66$
Cash Flows	Find: P Given: F	$=PV(i\%, N, 0, F)$	Find the present worth of $1,300 due in 10 years at a 16% interest rate.	$=PV(16\%, 10, 0, 1300)$ $=(\$294.69)$
Equal-Payment-Series	Find: F Given: A	$=FV(i\%, N, A)$	Find the future worth of a payment series of $200 per year for 12 years at 6%.	$=FV(6\%, 12, -200)$ $=\$3,373.99$
	Find: P Given: A	$=PV(i\%, N, A)$	Find the present worth of a payment series of $900 per year for 5 years at 8% interest rate.	$=PV(8\%, 5, 900)$ $=(\$3,593,44)$
	Find: A Given: P	$=PMT(i\%, N, -P)$	What equal-annual-payment series is required to repay $25,000 in 5 years at 9% interest rate?	$=PMT(9\%, 5, -25000)$ $=\$6,427.31$
	Find: A Given: F	$=PMT(i\%, N, 0, F)$	What is the required annual savings to accumulate $50,000 in 3 years at 7% interest rate?	$=PMT(7\%, 3, 0, 50000)$ $=(\$15,552.58)$
Measures of Investment Worth	Find: NPW Given: Cash flow series	$=NPV(i\%, \text{series})$	Consider a project with the following cash flow series at 12% ($n = 0, -\$200$; $n = 1, \$150, n = 2, \$300$, $n = 3, 250$)?	$=NPV(12\%, B3:B5)+B2$ $=\$351.03$
	Find: IRR Given: Cash flow series	$=IRR(\text{values, guess})$		$=IRR(B2:B5, 10\%)$ $=89\%$
	Find: AW Given: Cash flow series	$=PMT (i\%, N,$ $-NPW)$		$=PMT(12\%, 3,$ $-351.03)$ $=\$146.15$

Table within IRR example:

	A	B
1	Period	Cash Flow
2	0	-200
3	1	150
4	2	300
5	3	250